Structural Design in Wood

Second Edition

Join Us on the Internet

WWW: http://www.thomson.com
EMAIL: findit@kiosk.thomson.com

thomson.com is the on-line portal for the products, services and resources available from International Thomson Publishing (ITP).

This Internet kiosk gives users immediate access to more than 34 ITP publishers and over 20,000 products. Through *thomson.com* Internet users can search catalogs, examine subject-specific resource centers and subscribe to electronic discussion lists. You can purchase ITP products from your local bookseller, or directly through *thomson.com.*

Visit Chapman & Hall's Internet Resource Center for information on our new publications, links to useful sites on the World Wide Web and an opportunity to join our e-mail mailing list. Point your browser to: **http://www.chaphall.com/chaphall.html** or **http://www.chaphall.com/chaphall/civeng.html** for Civil Engineering

---

A service of 

Structural Design in Wood

JUDITH J. STALNAKER
Associate Professor,
University of Colorado, Denver

ERNEST C. HARRIS
Professor Emeritus,
University of Colorado, Denver

CHAPMAN & HALL

INTERNATIONAL THOMSON PUBLISHING

New York • Albany • Bonn • Boston • Cincinnati • Detroit • London • Madrid • Melbourne
Mexico City • Pacific Grove • Paris • San Francisco • Singapore • Tokyo • Toronto • Washington

Cover design: Curtis Tow Graphics
Illustration on top left corner of cover courtesy of
American Forest & Paper Association, Washington, D.C.

Printed in the United States of America

Chapman & Hall
115 Fifth Avenue
New York, NY 10003

Chapman & Hall
2-6 Boundary Row
London SE1 8HN
England

Thomas Nelson Australia
102 Dodds Street
South Melbourne, 3205
Victoria, Australia

Chapman & Hall GmbH
Postfach 100 263
D-69442 Weinheim
Germany

International Thomson Editores
Campos Eliseos 385, Piso 7
Col. Polanco
11560 Mexico D.F
Mexico

International Thomson Publishing–Japan
Hirakawacho-cho Kyowa Building, 3F
1-2-1 Hirakawacho-cho
Chiyoda-ku, 102 Tokyo
Japan

International Thomson Publishing Asia
221 Henderson Road #05-10
Henderson Building
Singapore 0315

1 2 3 4 5 6 7 8 9 10 XXX 01 00 99 98 97

Library of Congress Cataloging-in-Publication Data

Stalnaker, Judith J.
Structural design in wood /Judith J. Stalnaker, Ernest C. Harris.
-- 2nd ed.
p. cm.
Includes bibliographical references and index.
ISBN 0-412-10631-0 (alk. paper)
1. Building, Wooden. 2. Structural design. I. Harris, Ernest C.
II. Title.
IN PROCESS
624.1' 84--dc20 96-15280
CIP

British Library Cataloguing in Publication Data available

To order this or any other Chapman & Hall book, please contact **International Thomson Publishing, 7625 Empire Drive, Florence, KY 41042.** Phone: (606) 525-6600 or 1-800-842-3636. Fax: (606) 525-7778. e-mail: order@chaphall.com.

For a complete listing of Chapman & Hall titles, send your request to **Chapman & Hall, Dept. BC, 115 Fifth Avenue, New York, NY 10003.**

Contents

Preface

The prime purpose of this book is to serve as a classroom text for the engineering or architecture student. It will, however, also be useful to designers who are already familiar with design in other materials (steel, concrete, masonry) but need to strengthen, refresh, or update their capability to do structural design in wood. Design principles for various structural materials are similar, but there are significant differences. This book shows what they are.

The book has features that the authors believe set it apart from other books on wood structural design. One of these is an abundance of solved examples. Another is its treatment of loads. This book will show how actual member loads are computed. The authors have found that students, more often than not, have difficulty recognizing how load is transferred from one member to another—for example, how to proceed from a specified intensity of floor live load and type and thickness of floor material to knowing the actual load per unit length reaching the beam in question. Worked-out examples and student homework problems will illustrate the process.

Another significant feature that we believe sets this book apart is its inclusion of "structural planning." Most textbooks show only the selection of member proportions or number of connectors in a joint to satisfy a given, completely defined situation. This book, on the other hand, shows the thinking process needed to determine whether or not the member is required in the first place. Following this, the spacing and continuity of the member are decided, its loads are determined, and finally its shape and size are selected.

We believe that illustrating structural planning as well as detailed member and connection design is of considerable value in helping the student make the transition from the often simplistic classroom exercises to problems of the real world. Problems for solution by the student follow the same idea. The first problems in each subject are the usual textbook-type problems, but in most chapters these are followed by problems requiring the student to make structural planning decisions as well. The student may be required, given a load source, to find the magnitude of the applied loads and decide upon a grade of wood. Given a floor plan, the student may be required to determine a layout of structural members. The authors have used most of the problems in their classes, so the problems have been tested.

The book presents many of the design examples in the form of "computation sheets," solutions in the form of actual design office computations. This is intended to reinforce the instruction given in Chapter 1 regarding neatness and orderliness in design computations.

The book refers frequently to three codes—the *National Design Specification for Wood Construction,* the *Standard for Load and Resistance Factor Design for Engineered Wood Construction,* and the *Uniform Building Code.* Wherever possible, however, the basic principles behind code requirements are explained, and where the authors are aware of code shortcomings, that too is pointed out. We refer frequently to code requirements in the belief that theory with no exposure to real life is not good education. What is needed is balance between the theoretical and the practical (the latter is *not* a bad word).

The design method used in the majority of this book is allowable stress design (ASD). In addition, both theory and examples using the newer design method, load and resistance factor design (LRFD), are included. Problems for student solution are specified in a manner that allows for solution by either method. If the instructor so desires, the book can be used for a course dealing exclusively with the LRFD method. In that case, it would be wise for the instructor to require students to use the entire *Standard for LRFD for Engineered Wood Construction.*

In general, the book assumes that the user will be familiar with structural analysis. In fact, the authors intend that the book will reinforce the principles of structural analysis using wood as a vehicle.

The authors use the book in a three semester-hour beginning course and find that sufficient material remains that a second course could be taught from the same book. The prerequisite for the first course should be mechanics of materials as a minimum, including the subjects of shear and moment diagrams, flexural deflections, and axial forces in truss members. A more rigourous course in structural analysis, though desirable as a prerequisite, is not essential.

Chapter 5 (Connections—Nails, Screws, and Bolts) is intentionally in an unconventional position. The authors realize that most books cover connections after member design has been presented. They realize also that (1) because of its position in the course, connection design is something most students learn about almost as an afterthought (as though the subject is of lesser importance than the design of members); and (2) structural failures are far more frequent in the connections than in the members themselves. Further, design of members can be done more effectively if the designer considers how the member will be connected, before making a final selection of member size. With these thoughts in mind, the authors present the subject of connections *before* going on to the design of structural members.

Those who prefer to present the subject in the conventional sequence can merely go on to Chapter 6, delaying the study of Chapter 5 until after Chapter 8 (glulam design). It should work equally well with either sequence.

In the authors' beginning three-hour course in timber structure design, Chapters 1 through 7, 9, 10, 11, most of 8, and parts of 12, 13, and 14 are included.

For students of limited means, it is possible to teach structural design in wood using this text alone. However, the course can be much more meaningful if the student can refer easily to the *National Design Specification,* the *Uniform Building Code,* or the *Standard for LRFD for Engineered Wood Construction.* In our own classes, we require the student to obtain the *National Design Specification* as well as this textbook. (The NDS is handled by our university bookstore, or may be obtained from the American Forest & Paper Association, 1111 19th St., N.W., Suite 800, Washington, DC 20036.

JUDITH J. STALNAKER
Ernest C. Harris
Denver, Colorado

Structural Design in Wood

Second Edition

1

Introduction

From ancient times, wood and stone have been important construction materials. Stone has diminished in importance, but wood is still our most versatile building material. It is reasonable to claim that wood and wood products are among our more important construction materials, ranking along with structural steel and reinforced concrete for building construction. By weight, more wood is used in construction each year than cement or steel (1).

1-1. EVOLUTION OF TIMBER DESIGN

Over the centuries, man learned to use wood effectively, developing rules of thumb to aid in planning and building wood structures. These rules were learned gradually, based only on experience—both successful and unsuccessful. Using these rules, skilled artisans were able to construct in wood, producing durable, long-lasting structures of utility, often of great beauty, and almost always of adequate strength and serviceability. Even today a large volume of wood construction depends, more or less, on such rules of thumb for its design.

More recently, however, engineers and architects have learned to design wood structures in ways that are based on engineering principles. Thus today's designers, using more rigorous design procedures, are able to ensure that a particular design will achieve the desired level of structural safety and stiffness, as well as economy. As an example showing the economy possible with engineered design, the same amount of wood that was needed in the past to build a 320-ft^2 log cabin can be used today to construct a 3500-ft^2 home, with those parts of the trees not suited to lumber production being adequate to manufacture a 30-year supply of paper and tissue products for an average family (2).

Much progress has been made in moving from rule-of-thumb design to present-day engineered designs. Yet our knowledge is not static, and among the design methods shown in this book are some that undoubtedly will someday be replaced by newer and better methods.

As new design methods evolve, greater reliability (with regard to safety) and greater economy (in terms of total volume of wood used) are the usual result. Generally, and this is unfortunate, the improved design methods are more difficult to apply than the earlier methods they supplant, but this is the price to be paid for greater reliability and economy. Occasionally an improved design concept proves so complex that it has to be drastically simplified for the designer to use easily. When this is done, the basic principles involved may become obscured. The "handbook engineer" is not bothered by this, but the more competent *thinking* designer may be prevented from doing the best possible design job.

That the structural designer will play an increasingly important role in all types of wood construction is inevitable. Whether the demand for wood construction and other wood products remains constant or increases, the supply of readily available and usable trees to produce wood of good quality will almost certainly decrease. By one forecast, the demand for various forms of wood and wood products used for buildings is expected to rise by 1–4% per year (3). To meet this demand will require better use of our resources. "Engineered" structures will become more prevalent and those that use wood wastefully will become less common. Structural products manufactured from wood will replace much of the lumber we use today. We see this trend today in the introduction of wood products such as flakeboard, waferboard, plastic/wood laminates, members made from thinly laminated wood, and shop-prefabricated structural components. Engineering know-how will be involved in every step we take toward making better use of wood in construction.

1-2. MATERIAL PROPERTIES

To design effectively, a structural designer must be familiar with the properties and behavior of the material to be used. Unfortunately, the properties and behavior of wood are unlike those for other building materials and much more complex. With the possible exception of soils, wood has more idiosyncrasies than any other material the structural designer uses. Thus, it is even more important in the case of wood than for other construction materials that the designer have a good knowledge of the peculiarities of the material.

Later chapters will show that a wood designer's problems stem partly from the fact that wood is a natural material—one over which, as yet, we have little control. Through forest management, we are beginning to control certain characteristics of the lumber a forest will produce (3). However, we still cannot force a tree to produce a particular quality of material, nor can we compel different trees in a forest to produce exactly the same quality of material. Similar trees, grown in different localities, may produce wood of dissimilar properties. Consequently, we find that the engineering properties of wood are extremely variable. Changes in the environment in which the wood is used further affect its properties and behavior. For these reasons, wood properties and behavior are dealt with at great length in Chapters 2, 3, and 4.

1-3. TYPES OF CONSTRUCTION

Considering only buildings, there are several types of wood construction in use today. These include:

- Light-frame construction
- Post-and-beam construction
- Glulam construction
- Heavy-timber construction
- Pole building construction
- Wood shell and folded-plate construction
- Prefabricated-panel construction
- Log building construction

Commonly, combinations of these types with each other or with structural steel are used.

Light-Frame Construction

Light-frame construction is the type most commonly used for housing. It is used mostly for low structures, but finds occasional use for buildings up to four stories high. Frequently, light-frame construction receives little or no engineering design. Its requirements are spelled out in some detail by various building codes. Generally, knowing the code requirements and working from just a floor plan and perhaps one or two cross-section drawings, skilled carpenters often construct small light-frame buildings without engineering design assistance.

The resulting structure will normally be satisfactory as to safety and durability, but often it may be lacking in economy. With both labor costs and material costs increasing sharply, research has been directed toward designing more efficient structures. It is likely that light-frame construction of the future will require increased engineering design effort to bring about greater economy without sacrificing essential strength, stiffness, and durability.

Today, builders frequently require engineering design for certain components, such as lintels and headers. Loading conditions for these members vary from building to building, so rule-of-thumb methods may be either inadequate or wasteful. In fact, present building codes require that certain types of light-frame structures be engineered.

Light-frame construction (shown by Fig. 1-1) almost always uses lumber of 2-in. nominal thickness (actually $1\frac{1}{2}$ in. thick), such as the familiar 2 × 4s or 2 × 8s. Walls use vertical members called *studs,* which are either 2 × 4s or 2 × 6s and are spaced at 12 in., 16 in., or 24 in. on centers. Floor joists and roof joists or rafters are also 2-in. nominal material placed with their wider di-

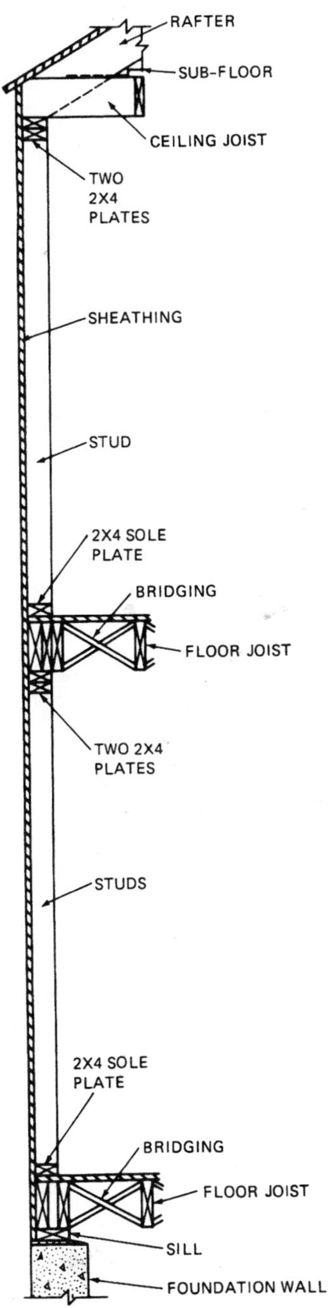

Fig. 1-1. Cross section through light-frame wall.

mension vertical. Where internal geometry of the building permits it, more recent light-frame structures usually have roofs supported by factory-built trusses made of 2 × 4s or 2 × 6s. Trusses and other prefabricated members may also replace solid sawn lumber for floor joists, headers, lintels, etc.

The 12-, 16-, and 24-in. spacings most commonly used in the United States are logical and economical, since most plywood sheets, wallboards, and insulation boards have a 4-ft width. Thus 4 ft, or 48 in., is the usual "module" for building construction in wood.

Post-and-Beam Construction

Post-and-beam construction, also known as timber-frame construction, had its beginnings around 200 B.C. The main, load-supporting members in this type of wood construction are larger than those of light-frame construction, and both vertical members (posts) and horizontal beams are farther apart than the studs and joists of light-frame construction. Posts are usually 8 to 12 ft apart, and horizontal members attached to the posts are spaced vertically in the range from 4 ft to 8 ft (4). More closely spaced floor joists and wall material are connected to and supported by the beams.

If needed for stability, diagonal "knee" braces are used, as shown by Fig. 1-2. Often, in older buildings, the structural members were exposed on the outside of the walls, and the spaces between posts, beams, and knee braces were filled with brick or a kind of plaster. This led to the familiar Tudor style of architecture. Except for the pleasing appearance, however, the filled-in material was troublesome. As the wood expanded or shrunk with moisture changes, wall cracks opened. Also, the wall filling had little insulating value, and its great weight necessitated heavier timber framing to carry the load.

In later forms of post-and-beam construction, the beams support wall material connected to the outside face of the beams. In old barns, for example, vertical wood planks were nailed to the beams to form the outside wall.

When light-frame construction was developed, post-and-beam construction declined in popularity, mostly because light-frame buildings were much cheaper and could be built much faster. Many will argue, however, that a light-frame building is not nearly as durable as a well-built post-and-beam structure.

Post-and-beam construction is staging a comeback in popularity, and this can be attributed partly to the development of new insulating

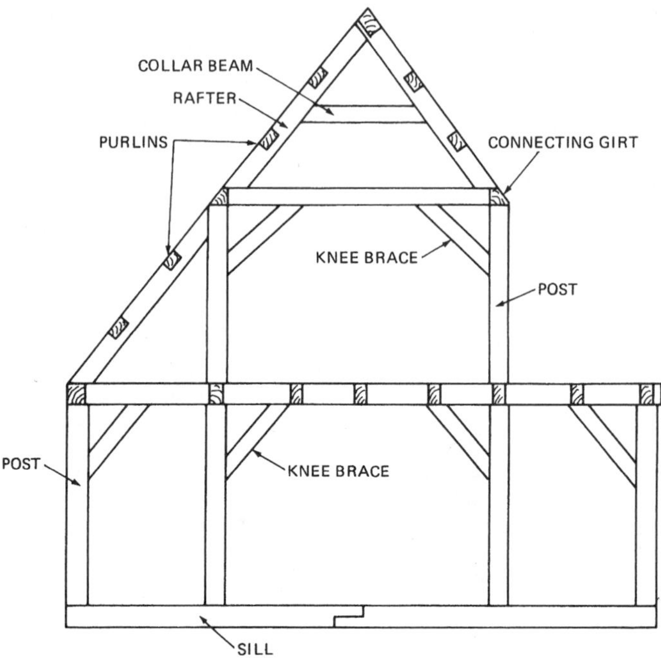

Fig. 1-2. Post-and-beam construction.

wall panels and to prefabrication techniques that help combat the high cost of an otherwise labor-intensive type of construction. Using prefabricated insulated panels for the exterior walls permits placing all main structural members inside the building, protecting those members from the deteriorating effects of weather and condensation.

Purists insist that connections in this type of construction be made using only wood parts. Mortise-and-tenon joints of several varieties were used in the older buildings, and the joints were held together by wood pegs, inserted in holes through the members being joined. Reference 4 shows many types of such connections. Making these connections, however, is time-consuming and requires a very high level of carpentry skill. The authors feel that substituting modern weldments (see Chapter 9) with bolts or other modern timber connectors might produce just as good a structure, permitting construction to proceed more quickly and at less total expense. But, if the appearance of modern connections is offensive, then mortise-and-tenon joints with wood pegs would still be the answer.

Glulam Construction

Because large, good-quality timbers are not readily available today, glued laminated timbers (glulams) are commonly used instead. A glulam is a stack of planks (usually $1^1/_2$ in. thick) glued together under carefully controlled conditions to make a single member of larger cross section. (See Fig. 1-3.) This substitution is good for several reasons: (1) fire resistance of glulams is at least as good as that of heavy sawn timbers; (2) glulams are less apt than sawn timbers to split after they are installed; (3) glulams can be stronger and stiffer than sawn timbers of the same size; and (4) they do not warp badly as sawn timbers frequently do.

Heavy-Timber Construction

Heavy-timber construction is defined by building codes. The *Uniform Building Code* (UBC) (5) defines it as construction in which timber columns are not less than 8 in. in least dimension and beams (except for roof beams) are not less than 6 in. wide by 10 in. deep. Of course, UBC defines heavy-timber construction in

Fig. 1-3. Light-frame construction used in conjunction with glulams. (Courtesy Overly Construction Co. Photograph by authors.)

much greater detail than is given here. Heavy-timber construction requires engineering design. Structures meeting the requirements for heavy-timber construction are much more fire resistant than other types of wood construction. (See Chapter 16.)

In earlier building codes, heavy-timber construction was referred to as "mill building construction." Typically, it was used for manufacturing buildings and storage warehouses. Architects often find these old buildings suitable for conversion to new uses, developing attractive offices, shops, restaurants, theaters, and the like in which the old timber construction is used to aesthetic advantage. Many design engineers and architects are involved in evaluating the old construction and modifying structures to adapt them to modern use.

Heavy-timber construction is not limited to buildings, but may be found in mine structures, docks, bridges, and other structures.

Pole Building Construction

Pole-type construction goes back to prehistoric times in both Europe and North America. Examples of its use in North America are seen in ancient Indian dwelling sites in the southwestern United States. The distinguishing feature of a pole-type building is that its supporting vertical members (round poles) are embedded in the ground, having been either driven into the ground or placed into excavated holes. Horizontal beams and other framing are attached to the poles, as shown later in this book by Fig. 15-5.

Obviously, a big problem with buildings of this type is decay of the embedded poles. However, with the use of chemically treated poles, the problems of decay have been alleviated but not eliminated. As a result, beginning in the 1950s, pole buildings have enjoyed a rebirth in popularity in the United States. For certain applications, pole buildings are suitable and economically feasible.

Pole buildings are used principally in agriculture, but in some localities they are used extensively for housing. They are particularly well suited for housing where the floor must be kept well above ground level to avoid flooding. Along the U.S. Gulf Coast, for example, pole-supported buildings are often better able to resist

Fig. 1-4. The Tacoma Dome erection sequence. Triangular sections of primary glulam beams and secondary purlins are assembled on the ground and then lifted into place, one ring at a time. The dome was erected without dense interior scaffolding. (Courtesy Western Wood Structures, Inc.)

the force of hurricane-driven waves than structures whose supporting walls extend to ground level. The poles, of course, must have adequate bracing or stiffness to resist the wave forces.

The pole acts as a column, transferring vertical load to the soil. If the end area and perimeter of the pole are too small, the hole may be dug to larger diameter and a concrete footing cast in the bottom to reduce the bearing pressure on the soil. Hole depths usually range from 5 ft to 8 ft, and footing diameters are commonly from 16 in. to 24 in. Pole buildings are normally one-story structures, with poles spaced up to 20 ft in each direction. Heavy timber or light framing is fastened to the poles to form walls and roof, and floor if any. With trusses for the roof structure, spans of up to 60 ft are practical, and a few longer spans have been constructed. Although pole buildings date back to ancient times, today's pole buildings are usually designed by engineers rather than by old rules of thumb (6).

Pole construction can also be a partial solution to problems of freezing and thawing beneath the structure, as in permafrost areas. Being a poor conductor, a wood pole carries little heat from the building above into the frozen ground below, so it helps prevent thawing. If the structure is mounted on a platform at the top of the poles, the open cold-air space below also helps prevent thawing of the soil. Poles in this type of environment must be placed into excavated holes, not driven. Having their larger diameter at the bottom, poles provide good resistance to being lifted from the ground by alternate cycles of freezing and thawing.

Wood Shell and Folded-Plate Construction

Wood shells and folded plates may have both economic and aesthetic advantages for roof construction. Roof shells can be of various shapes, the most common being the hyperbolic parabo-

Fig. 1-5. The 530-ft-diameter Tacoma Dome completely framed, with 2-inch timber decking being installed. (Courtesy Western Wood Structures, Inc.)

loid. Other shapes are limited only by the ingenuity and skill of the design engineer and by the capability of the materials used to follow complex surfaces. Finding them aesthetically pleasing, many architects incorporate shell roofs into their designs. The important thing to remember is that *either shells or folded plates require engineering design.* Rules of thumb will not suffice. To design either, an engineer with proper know-how and experience should be engaged (7).

Materials for shell roofs include both boards and plywood sheets. In either case, the materials are used in two or more (usually three) layers. The layers can be connected to each other by special nails, or glue, or both.

Prefabricated-Panel Construction

There are many varieties of prefabricated panel. The object of prefabrication is to reduce cost. Because of carefully controlled factory conditions, prefabricated panels can have nearly uniform quality. They also can speed field erection, thus saving construction cost. Where stressed-skin action is provided, they may also save in total quantity of material used, thereby reducing the total dead load that the structure must be designed to support.

Stressed-skin panels are those in which a plywood skin is glued to supporting members, so that the plywood sheet serves a double purpose: (1) It acts as a beam spanning from one supporting member to another; and (2) being rigidly connected to those supporting members, it acts together with them, the combined member having strength and stiffness exceeding the sums of the strengths or stiffnesses of the separate parts. Stressed-skin panels may have the plywood on one side only or may have both top and bottom plywood sheets attached to the supporting members. Panels are usually manufactured flat, for uses such as wall, roof, or floor panels or interior partitions. But they can also be manufactured with one-way curvature.

Stressed-skin panels always require engineering design unless they are used strictly for non-load-carrying purposes.

Log Building Construction

Log buildings, in several forms, appear to be making a comeback, especially for residential construction. Strangely, many of them are engineered structures. Their popularity is enhanced, of course, by their exterior attractiveness. In many cases they are economically feasible, even though they obviously use more wood than a light-frame building of equal floor area. There are two reasons for this: (1) Less total labor is involved in preparing and installing the wood components; and (2) wood is so effective an insulating material that extra wall insulation is normally not required.

1-4. HYBRID CONSTRUCTION

The authors use the word *hybrid* to describe any construction type that is mixed, and there are numerous examples of this. For example, in construction that is essentially light frame, we may see steel floor trusses substituted for the usual sawn-lumber floor joists, or floor trusses having wood top and bottom chords but metal web members. Or we may see light-frame construction in which headers and lintels will be wood trusses, or even steel beam sections. Other frequent combinations are: (1) a complete structural steel frame, with wood members (either joists or trusses) spanning from steel beam to steel beam to support the plywood or wood composite floor, or (2) wood joists or rafters supported by masonry bearing walls.

So, even though several recognized construction types were listed and defined above, there is absolutely no need that a structure conform strictly to any one of those types. Whenever economy and serviceability are best served by a combination of types, then designers will probably use it.

1-5. TIMBER BRIDGES

Bridges built either wholly or partly of timber include pedestrian, highway, and railway bridges and special types such as pipe bridges. They include trestles, simple girder, arch, or truss bridges, and even suspension bridges. Although wood is an excellent material for many types of bridge, its value for bridges is often not clearly recognized, perhaps because of a fear of either impermanence or weakness. However, properly constructed and maintained wood bridges can be durable. The useful life of modern glulam bridges is estimated to be as high as 50 years, whereas for steel or concrete bridges (because of corrosion by deicing materials) it may be much lower. Timber bridges are not necessarily weak; they can carry very large loads, such as heavily loaded (135-ton) logging trucks (8). Examples showing durability and capability to carry heavy loads and impact are seen in the large number of existing railroad trestles and in the many covered bridges that still carry rural road traffic. The latter owe their long life, of course, to the fact that the main wood structural members and wood roadway deck are protected from the ravages of weather. According to reference 7, bridges of wood have seen useful life of over 500 years!

Timber (usually glued laminated) is a popular material for pedestrian bridges. In these, the entire superstructure—main girders and deck—are wood. The selection of wood over other materials is often made on the basis of aesthetics, but economy, durability, and ease of maintenance also play important roles in determining the choice.

In railway trestles, treated wood piles are frequently used for the foundation, and treated timbers are used for all other structural parts, including the cross ties. Highway bridges spanning up to about 30 ft can be built using sawn timbers for the main longitudinal members (girders). When glued laminated (glulam) girders are used, the practical span limit is higher, spans as long as 150 ft having been constructed. Wood is used frequently for the towers, stiffening trusses, and floor systems of suspension bridges.

Arch bridges have become practical with the development of glulams. This type of timber structure has been used for pedestrian bridges, highway bridges, and pipe bridges. The decision to use wood is often based on appearance of the bridge and the blending of its appearance with the surroundings. Both two-hinged and three-hinged arches are practical using glulams, but fixed-end arches generally are not. For longer spans, wood trusses are practical. These are more commonly used in localities where timber is plentiful but steel is expensive and concrete construction difficult.

Bridge decks are built also using composite wood and concrete. In this type of deck the concrete forms the wearing (top) surface. Wood members below form the tension part of the composite flexural section and the concrete above resists the compression. The wood members also serve as supporting formwork to carry the weight of the wet concrete as the wearing surface is placed.

1-6. NOTES TO STUDENTS

Design Codes

For some of you, this will be the first design course. By *design* we mean the process of choosing the type and quality of material to use, choosing location and spacing of the members, and selecting shape and dimensions for each member so that it will (1) carry the load safely, (2) have the desired stiffness and durability, and yet (3) be as economical as possible. Principles that you learned in previous courses (mechanics of materials, for example) still apply, but now you will have to supplement those principles with design rules agreed upon by committees of experienced engineers and researchers to ensure that timber construction will be safe. These rules make up a standard of practice, sometimes referred to (loosely) as a "design code."

The standard of practice probably having most meaning for this course is the *National Design Specification for Wood Construction* (NDS) (9). The NDS may be one of the required texts for your course; if not, at least it is an available reference. It would be an excellent idea, especially if this is your first design course, to read through the NDS now, not studying its requirements in great detail but just becoming familiar with the types of instruction it gives and where to find things in it.

The NDS sets out the requirements for allowable stress design of wood structures. There are two methods for the design of wood structures: (1) allowable stress design (ASD) and (2) load and resistance factor design (LRFD). The greater portion of *Structural Design in Wood* is devoted to ASD, with LRFD being covered in the last section of many chapters. The standard of practice for LRFD design is the *Standard for LRFD for Engineered Wood Construction* (10).

A student's first exposure to using a design standard is often a confusing experience. Try to realize now, at the outset of this course, that what these two standards do is merely to *define in some detail (for a particular material) how to apply the design principles to which you were introduced in previous courses.* Realizing this, you will appreciate the true purpose of each requirement and will find its application more meaningful.

Accuracy

Your instructor has probably covered the next point before, but it will bear repeating. Just because your calculator shows answers to seven or eight places does not indicate such a high level of accuracy for our purpose in design. The answers your calculator shows are no more accurate than the values you enter into it. Loads and allowable stresses, for example, are rarely known to more than three significant figures, so you should not expect the answers to be significant to more than three figures. In the process of solving a design problem (if it is more convenient than rounding off at each step), you might copy figures as the calculator shows them, that is, to more than three figures. But to retain more than three figures in the *answer* to the design problem would be pure foolishness. By all means, round off the answer to the number of significant figures that is appropriate for the data with which you started.

Occasionally, however, a student learns the above so well that rounding off goes too far, and in this case the answer may suffer. Do not arbitrarily round off to one- or two-figure accuracy early in the problem and then expect to retain three-figure accuracy in the final answer.

Order and Neatness

A designer's computations are usually kept as part of the complete, permanent record of a design project. This is done for the designer's protection, of course. But, equally important, the computations may find additional future use. The structure to which they apply may need to be revised or extended; or it may need to be adapted to a new use in which the loading differs from that of the original design. Consequently, it may be necessary for you (or someone else) to read and understand your computations clearly and easily at some distant future date. If your original computations are neat and orderly, and fully labeled to show exactly what you had in mind, they will be easy to interpret.

Engineers have long been noted for the neatness and orderliness of their work, whether it be longhand computations or working drawings. However, being neat and orderly does not come naturally to most of us; we have to *work* at it. So why wait until professional practice begins to develop this necessary habit of producing neat and orderly work? Wouldn't it make more sense to develop it during student years? To demonstrate what the authors consider proper form, several of the examples in later chapters are shown as typical designer's computation sheets.

A word of caution, however: Neatly recopying a previously made sloppy and disorderly computation is wasting time. Furthermore, the copy may suffer from inaccuracies of the sloppy original. Rather than making a preliminary version of your work, try to *think first, then write.* Orderly thinking is even more important than orderly writing.

REFERENCES

1. Smith, W. F., *Principles of Materials Science and Engineering,* McGraw-Hill, New York, 1986.
2. "Tree Improvement," Georgia Pacific, Portland, OR, 1981.
3. Stone, R. N., and J. F. Saeman, "World Demand and Supply of Timber Products to the Year 2000," *Forest Products Journal,* 27 (10), Oct. 1977.
4. Benson, T., *Building the Timber Frame House,* Scribner's, New York, 1980.
5. *Uniform Building Code,* International Conference of Building Officials, Whittier, CA, 1994.
6. Patterson, D., *Pole Building Design,* American Wood Preservers Institute, McLean, VA, 1969.
7. *Wood Structures, a Design Guide and Commentary,* American Society of Civil Engineers, New York, 1975.
8. Gibas, D., "The Timber Bridge, an Old and New Concept," *Forest Products Journal,* 33 (11/12), Nov./Dec. 1983.
9. *National Design Specification for Wood Construction,* rev. 1991 Ed., American Forest & Paper Association, Washington, DC, 1993.
10. *Standard for Load and Resistance Factor Design for Engineered Wood Construction,* American Society of Civil Engineers, NY, 1996.

2

Wood Structure and Properties

2-1. WOOD AS A STRUCTURAL MATERIAL

For structural applications, wood is most commonly found as either sawn timbers, lumber, or glued laminated members (glulams). In the interest of economy and to permit using wood more efficiently, increasing amounts of wood today find their way into manufactured structural materials or members such as (1) plywood, hardboard, chipboard, flakeboard, waferboard, and plastic/wood laminates (these are known collectively as wood composites); and (2) manufactured members such as plywood-lumber beams or wood trusses. Many of these will be discussed later in this book.

2-2. PROBLEMS IN USE OF WOOD FOR STRUCTURES

For most structural applications, fortunately, wood's advantages far outweigh its disadvantages. However, structural designers must learn to cope with (1) wood's variability and (2) its response to environmental conditions.

Variability is probably the more serious of these. Wood properties vary from species to species, from one position to another in the tree, from one tree to another grown in the same locality, and between trees grown in one locality and those grown in another. Generally, humans have little control over the quality of wood a tree produces, although strides are being made in that direction by means of selective tree farming. It is hoped that by such means straighter and faster-growing trees may be developed, with more nearly uniform properties than are found in trees from natural forests (1).

The moisture content of wood installed in a structure may change with time, eventually reaching an equilibrium moisture content that depends on the average relative humidity of the surroundings. However, as the relative humidity within a building may not be constant, the equilibrium moisture content may vary with time. With any change of moisture content, wood will either shrink or swell—and warp. Collectively these size and shape changes are known as *dimensional instability.* Dimension and shape changes due to moisture change can be reduced or avoided by proper seasoning (drying) and by proper attention to details of design.

Duration of loading causes strength changes: The longer a load remains on a wood member, the weaker the wood member becomes. Luckily, the structural designer can easily consider this problem and compensate for it in the design procedure. Duration of loading has negligible effect on modulus of elasticity but, because of creep, deflections are time-dependent.

Finally, under some conditions wood's durability will be limited. Weathering, decay, insects, or fire can obviously limit the useful life of a wood structure. Yet this problem, too, can be overcome by selecting the proper kind of wood and by proper design, treatment, construction, and maintenance.

2-3. ADVANTAGES OF WOOD AS A STRUCTURAL MATERIAL

Wood's principal advantages over the other common structural materials are (1) its economy, (2) its appearance, and (3) its ease of working and reworking. Other advantages are better durability (for some applications), a high strength/weight ratio, and excellent thermal insulating properties.

A wood structure is frequently less costly to construct than a similar structure in either structural steel, reinforced concrete, precast concrete, or masonry. Of course, a completely valid cost comparison would have to include such costs as operation, maintenance, insurance, and the like, over the entire life of the structure. However, it should be obvious from observing structures all around us that, even when all these factors are considered, wood is frequently the most economical choice for buildings up to three or four stories high.

A wood structure can be either aesthetically pleasing or unattractive, depending entirely on the care that goes into its design. Early U.S. colonial structures provide excellent examples of attractive design and of construction details that ensure good durability despite the fact that wood can decay. Modern glued laminated timber makes possible very attractive buildings. We see numerous examples of this in commercial and institutional buildings, in high-quality residences, and even in interior treatment in larger buildings whose main structural members are of other materials.

Wood can easily be cut, shaped, and finished in the field. Prefabrication at another location is generally not needed, although prefabrication may be employed as a time-saving or money-saving measure. Existing wood structures can be revised or added to more easily than similar structures of any other material.

If proper care is taken in details of design, in maintenance, and in selecting suitable preservative and fire retardant treatments, wood will be durable. If untreated wood is allowed to remain moist in service, however, it will decay. The secret to ensuring durability is fourfold: (1) Use proper species and grades, (2) avoid undesirable environmental conditions, (3) use proper design details, and (4) use whatever treatment is necessary and economically feasible.

When wood is used for all components of a structure, the total weight is often less than for a structure of other materials. This is particularly true in comparison to reinforced concrete. The strength/weight ratio of wood is advantageous wherever dead load is an appreciable part of the total load.

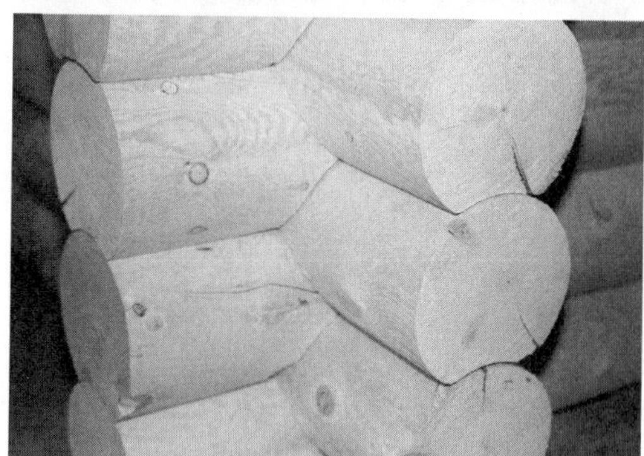

Fig. 2-1. Proper detailing of joints in log building leads to long life. Upper photograph shows old-time construction. Lower photograph (Courtesy Golden Log Homes by CMB, Arvada, Colorado) shows turned logs in a modern log building.

2-4. CLASSIFICATION OF WOOD

Wood is classified as either softwood or hardwood, according to the species of tree from which it is cut. These are not necessarily good names, but they have been used for so long that they are accepted as part of the language by people who work with wood. Softwoods actually include some varieties that are quite hard, while hardwoods include species whose wood is very soft. Softwoods come from the plant group known as gymnosperms. Softwood species include the needle-leaved trees. Most of these are cone bearing (conifers), and most are evergreen, which means that they do not drop their leaves for the winter season. Softwoods also include the scale-leaved evergreens, most of which do not bear cones.

Trees producing hardwoods are of the plant group known as angiosperms. All hardwood trees have broad leaves. Most nontropical hardwood trees are deciduous; that is, they shed their leaves for the winter.

Without getting into biologic structural differences (except for one that is explained later) the above definitions are about as precise as we can get. Perhaps a better way for our purpose is merely to list common North American species of each class. Table 2-1 shows typical softwood species and Table 2-2 typical hardwoods.

The Latin names in the second column of these tables are scientific names. The first part of the scientific name is the genus and the sec-

Table 2-2. Typical North American Hardwoods.

Common Name	Scientific Name	Lumber Name
Northern red oak	*Quercus rubra*	Red Oak
Southern red oak	*Q. falcata*	Red Oak
Black oak	*Q. velutina*	Red Oak
Pin oak	*Q. palustris*	Red Oak
Scarlet oak	*Q. coccinea*	Red Oak
White oak	*Q. alba*	White Oak
Chestnut oak	*Q. prinus*	White Oak
Sugar maple	*Acer saccharum*	Hard maple
Black maple	*A. nigrum*	Hard maple
Red maple	*A. rubrum*	Soft maple
Yellow birch	*Betula alleghaniensis*	Birch
River birch	*B. nigra*	Birch
White ash	*Fraxinus americana*	White ash
Green ash	*F. pennsylvanica*	White ash

Note: The list above is by no means complete.

Table 2-1. Important North American Softwoods.

Common Name	Scientific Name	Lumber Name
Ponderosa pine	*Pinus ponderosa*	Ponderosa pine
Lodgepole pine	*Pinus contorta*	Lodgepole pine
Longleaf pine	*Pinus palustris*	Longleaf yellow pine
Slash pine	*Pinus elliottii*	Longleaf yellow pine
Virginia pine	*Pinus virginiana*	Southern yellow pine
Pitch pine	*Pinus rigida*	Southern yellow pine
Shortleaf pine	*Pinus echinata*	Southern yellow pine
Slash pine	*Pinus elliotti*	Southern yellow pine
Longleaf pine	*Pinus palustris*	Southern yellow pine
Loblolly pine	*Pinus taeda*	Southern yellow pine
Red (or Norway) pine	*Pinus resinosa*	Norway (red) pine
Black spruce	*Picea mariana*	Eastern spruce
Red spruce	*Picea rubens*	Eastern spruce
White spruce	*Picea glauca*	Eastern spruce
Sitka spruce	*Picea sitchensis*	Sitka spruce
Englemann spruce	*Picea engelmannii*	Engelmann spruce
Blue spruce	*Picea pungens*	Engelmann spruce
Douglas fir[a]	*Pseudotsuga menziesii*	Douglas fir
Pacific silver fir	*Abies amabilis*	White fir
White fir	*Abies concolor*	White fir
Eastern hemlock	*Tsuga canadensis*	Eastern hemlock
Western hemlock	*Tsuga heterophylla*	West coast hemlock
Western red cedar[b]	*Thuja plicata*	Western red cedar
Redwood	*Sequoia sempervirens*	Redwood
Bald cypress	*Taxodium distichum*	Cypress

[a]Not a true fir. (Some botanists call it the Douglas tree.)
[b]Not a true cedar. (True cedars—*Cedrus* species—are found only in Europe and Asia.)
Note: The list above is by no means complete.

ond part the species. Ordinarily, scientific names find little use by structural designers. However, scientific names are the only reliable ones for defining precisely a particular species. Names in the first column are called common names. These are the names used by most people when identifying trees, but they are usually less accurate than the scientific names. For example, a species may have different common names in different regions; or, equally confusing, the same common name may be applied to more than one species.

The third column of each table gives still another version—names applied by the lumber industry. These are the names we will use in this book. Under lumber industry names we find several different species listed together under a single name (species group). The justification for this is that the species included under a single lumber industry name grow in the same locality, have similar appearances, and have similar properties. The important thing to notice is that while only the scientific nomenclature is sufficiently accurate for the botanist, lumber industry names are quite suitable for the structural designer's use.

The lists in Tables 2-1 and 2-2 are far from complete. For example, the lumber name red oak, for which Table 2-2 shows five different species, actually includes 14 different North American species. Similarly, the table shows only two species under white oak, but 13 species are actually included under that name. Other important hardwoods (lumber names), not shown in Table 2-2, are hickory, beech, pecan, poplar, cottonwood, aspen, basswood, elm, walnut, and cherry.

To emphasize that hardwoods are not necessarily hard, consider the familiar tropical wood known as balsa. Balsa is actually a hardwood!

2-5. WOOD STRUCTURE

Features of wood structure include those of the wood-forming substance and those of the "whole wood." Features of the wood substance are all microscopic and are of interest to industries that use wood substance (rather than whole wood) as a raw material—the paper industry, for example. On the other hand, except for individual wood cells, most whole-wood features are visible to the naked eye or by aid of merely a hand lens. These are the features that affect wood's suitability for construction, and are the ones that will be explained here.

Wood Cell

Wood's basic structural element is the cell. The hollow, tubular wood cells are many times longer than they are wide. For softwoods, the cell length is about 100 times the width. Cells are approximately rectangular in cross section and have unsymmetrical tapered ends that overlap, in staggered position, with the cells above and below. The cell walls consist principally of cellulose and lignin. Cellulose accounts for 45–50% of the weight of completely dry wood, and lignin for about 20–30%. Another ingredient, hemicellulose, amounts to 20–25% (2). Minor components, which vary from species to species, include natural resins, oils, tannin, and alkaloids. Often, it is these minor components that make a species suitable or unsuitable for some particular application.

Cellulose gives the wood its strength: it is the load-carrying material. Lignin is the "glue" that cements cellulose fibers together, filling the spaces between fibers and stiffening the fibers. The cellulose–lignin combination is actually heavier than water; it has a specific gravity of 1.54. However, wood cells are hollow, so that the specific gravity of whole wood ranges from 0.31 to 0.75 for most woods used in construction. If the cells were not hollow, wood would sink rather than float. A few tropical species do actually sink. Ebony, for example, weighs 70–80 lb/ft^3 (versus 62.4 for water).

If cellulose forms the cell walls in all wood species and is the constituent that gives the wood structural strength, why, then, do wood properties (e.g., strength) vary so much from one species to another? To understand why, we must consider further the structure of the wood itself, in particular the arrangement and shape of its cells. Figure 2-2 is a drawing showing typical cross sections through a piece of wood. Notice that most of the hollow, tubular cells are in a longitudinal position; that is, their long dimension is vertical, parallel to the trunk (stem) of the tree.

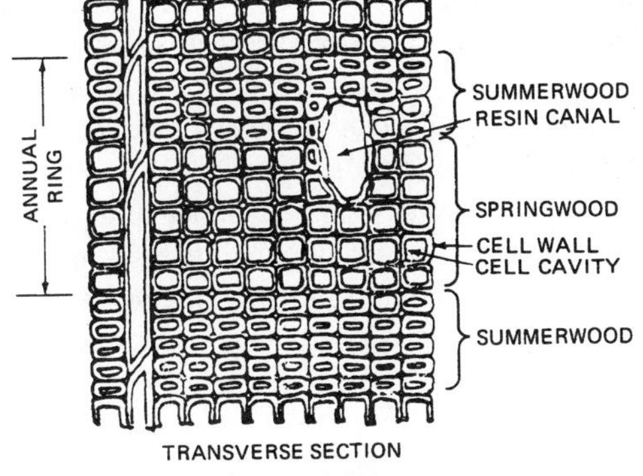

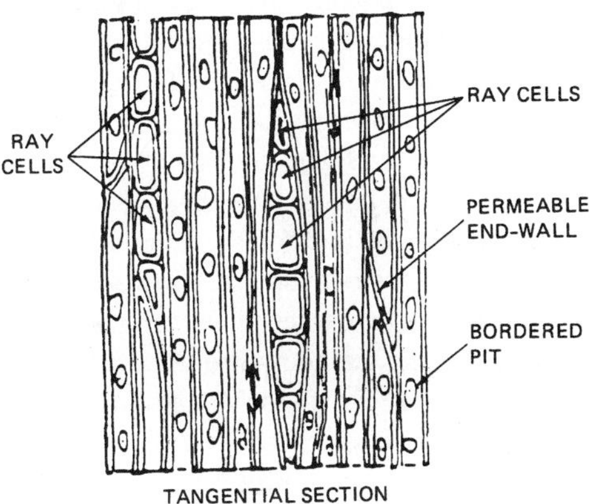

Fig. 2-2. Softwood cross sections. (Courtesy American Society of Civil Engineers, R. J. Hoyle, "Introduction," *Wood Structures, a Design Guide and Commentary, 1975.*)

A few cells, in localized bundles, run radially, that is, parallel to a radial line from the center of the tree trunk to the outside.

Because the wood cell is a hollow tube, it is very efficient for resisting a compressive force parallel to its length, as shown by Fig. 2-3a. For this reason, wood under longitudinal compression has a rather high ratio of strength to weight. However, compression forces normal to the length of the cells easily crush the cells, bending their walls as shown in Fig. 2-3b. Thus, the strength of wood is poor under compression perpendicular to the cell length (called compression perpendicular to the grain).

Longitudinal shear strength of wood is limited by the strength of the lignin that binds adjacent cells together, or by the shear strength of the cell wall, whichever is lower. Longitudinal shear strength of wood is lower than its compression strength in any direction and also lower

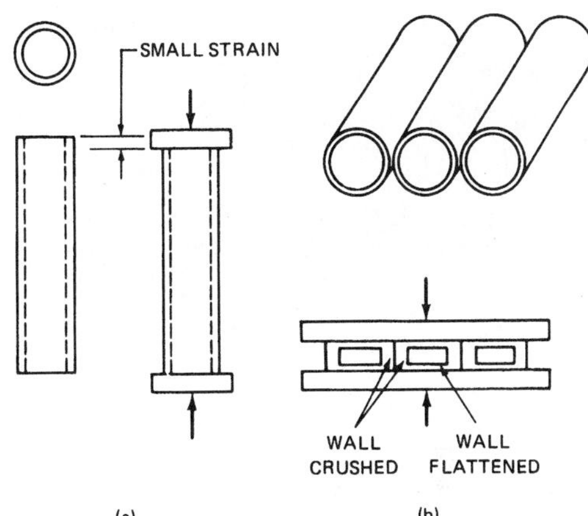

Fig. 2-3. Strength of tubular cells.

than its longitudinal tensile strength. The only strength that is lower than longitudinal shear strength is the transverse tensile strength.

Interestingly, the hollow-cell arrangement has a good effect on both longitudinal compressive strength and bending strength. Given a specified weight of wood in the form of either a beam or a column, its strength is higher than it would be if all void spaces were eliminated and the wood substance moved together to form a solid member of lesser cross-sectional area.

Cell walls are permeable, and the end walls of wood cells are particularly so. In living parts of the tree, fluids necessary to growth can move vertically from one longitudinal cell to another. By capillary attraction and surface tension, fluids are raised by way of these cells all the way from the roots of the tree to the uppermost leaves.

Cambium

Figure 2-4 shows the cross section of an entire tree trunk. New wood cells, formed by cell division, are produced by the cambium layer. As new cells are added on the inner side of the cambium, the diameter of the tree increases. At the same time, the cambium adds new cells to the inner bark, just outside of the cambium layer. Immediately after a cambium cell divides, the newly created cell begins to grow in both length and diameter. During this growth the cell wall is thin and pliable, but after growth ceases a secondary layer is added to the cell wall, thickening it. At this stage the cell wall is primarily cellu-

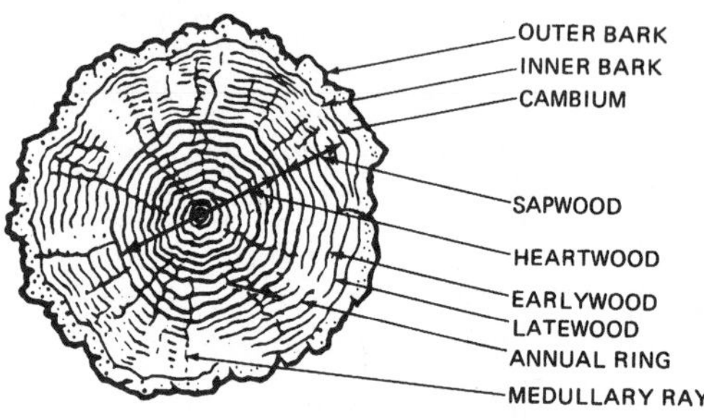

Fig. 2-4. Tree cross section.

lose. After wall thickening ceases, lignin is deposited, cementing the fibers together in a process called *lignification.* The cell now has structural strength that enables it to help carry the weight which later is added to the tree.

Earlywood and Latewood

During springtime, growth is rapid and the newly created cells have relatively large cell openings (cavities) and thin walls. Portions of new wood created during this season are known as earlywood (or springwood). New cells created later in the year, when less moisture is available, have smaller cavities and thicker walls. This portion of the wood is called latewood (or summerwood). Latewood contains more cellulose per unit of cross-sectional area than earlywood; consequently, latewood is stronger than earlywood.

Annual Rings

Each year a band of earlywood and then a band of latewood are added to the tree. Together, these two bands comprise an annual ring. Annual rings are easily visible on all North American hardwoods and softwoods. Generally, the latewood portion of the annual ring is darker than the earlywood. In some tropical woods, grown where differences between seasons are not marked, annual rings may be either obscure or absent.

Medullary Rays

Medullary rays are ribbonlike bundles of cells arranged in a radial direction in the tree. That is, they are perpendicular to the annual rings, running from the center (pith) toward the bark. Medullary rays are prominent in some species but nearly invisible in others. The rays serve a useful structural function: They brace the longitudinal cells so that their buckling strength is higher. Tensile strength perpendicular to the annual rings may be benefited a little, but tangentially (in a direction parallel to the rings and perpendicular to the rays) tensile strength may actually be reduced by the rays.

Heartwood and Sapwood

Wood in the annual rings nearest the outside of the tree—the newer portion—is called sapwood, since it still transports the tree's life fluids. Portions nearer the center no longer carry these fluids or store food, and they are called heartwood. As the tree grows, new annual rings of sapwood are added to the outside, and the innermost (oldest) sapwood rings convert to heartwood. Usually, heartwood is darker than sapwood, the darker color being due to by-products formed as the cells' function for food storage terminates. However, in some species it is hard to distinguish heartwood from sapwood by color alone.

The dark products that color the heartwood are often toxic to insects and fungi; therefore, heartwood of most species resists decay better than sapwood. An exception is sweet gum, whose heartwood decays readily. An exception to the usual color difference is white cedar, whose heartwood is light in color but is very resistant to decay. Unless juvenile wood is involved, the strength properties of lumber sawn from sapwood or from heartwood (each dried to the same moisture content) are the same, since each contains the same amount of cellulose per unit cross-sectional area. However, because of its ability to transport fluids, sapwood may behave differently from heartwood in such properties as permeability and ability to absorb stain or hold paint.

Pores

A structural difference between hardwoods and softwoods is that hardwoods have vessels or pores, while softwoods do not. These are structures whose only function is to carry water.

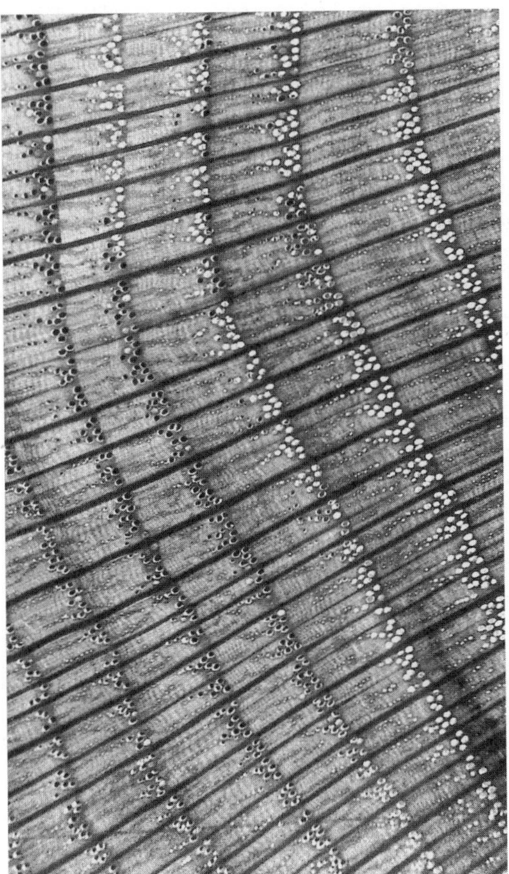

Fig. 2-5. Cross section of strongly ring-porous oak. Note distinct medullary rays. (Photograph by authors.)

Pores are large-diameter structures having thin walls made up of individual cells. In many hardwoods the pores are large enough to be seen easily by the naked eye. Hardwood is classified as either diffuse-porous or ring-porous, according to whether the diameters of the pores are similar throughout the annual ring or larger in the earlywood. Ring-porous wood has the appearance of having pores in the earlywood only. The importance of pores is mainly in the appearance of the wood. Figure 2-5 shows a cross section of ring-porous wood.

2-6. JUVENILE WOOD

A tree grows faster in its early years of life than in later years, and wood produced during those early years has a larger-than-normal percentage of earlywood. This portion of the tree is called juvenile wood. The specific gravity and strength of juvenile wood are less and the longitudinal shrinkage more than for wood grown by the mature tree. Just how much of the tree's interior is juvenile wood depends on the species and other factors, but usually ranges between 5 and 20 annual rings. Genetically improved trees grown on intensively managed tree farms grow fairly rapidly; they can be harvested earlier and, therefore, have a higher percentage of juvenile wood than trees from either natural or regrowth forests (3). Today most trees grown on tree farms are used by the paper industry, so the structural designer is not affected. In the future, however, more of such trees will undoubtedly be used for construction lumber. When this happens, the code-writing committees that establish allowable stresses for the designer's use will have to account for the increased percentage of juvenile wood in the lumber.

2-7. WOOD AXES

Because of its internal structure, wood is orthotropic, which means that its properties (physical and mechanical) differ in the three main, mutually perpendicular directions—longitudinal, radial, and tangential. Figure 2-6 shows the axes in these directions and the three reference planes associated with the axes.

Strength and modulus of elasticity vary in the three directions, and there are six values of Poisson's ratio! Shrinkage (or swelling) occurring as wood's moisture content changes also differs in the three directions; this is what may cause wood to warp as it either dries out or takes on additional moisture.

The nonisotropic nature is probably the most serious characteristic with which the designer of wood structures must cope.

2-8. FACTORS AFFECTING STRENGTH

Major factors affecting strength and stiffness properties of wood are:

1. moisture content
2. specific gravity (indicator of cellulose amount)
3. duration of loading
4. species

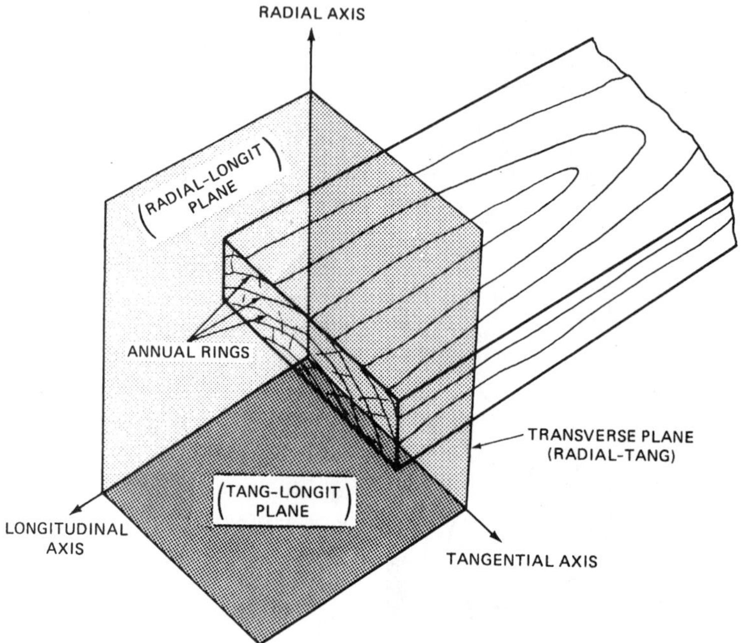

Fig. 2-6. Orthotropic axes and reference planes in wood.

Strength and stiffness properties of construction-sized lumber are affected by all of the above factors, and also by:

5. size and shape of the wood member
6. nature, size, and location of defects (strength-reducing characteristics)

2-9. MOISTURE CONTENT

Moisture content (MC) is expressed as a percentage of the weight of the completely dry wood. To determine MC by the American Society for Testing and Materials (ASTM) standard test (see reference 4), a sample is first weighed, then oven-dried until its weight is constant, cooled, and weighed again. The weight lost during drying is the total weight of water present in the original sample. Expressed as a percent,

$$MC = 100\,(\text{orig. wt.} - \text{dry wt.})/\text{dry wt.} \tag{2-1}$$

Moisture content of seasoned (dried) wood for most construction purposes is equal to or less than 19%; it usually averages about 15%. Green wood, however, can have MC from as low as 40% to as high as 240%! That is, the weight of water in the green wood may be as high as 2.4 times the weight of dry wood (5).

Moisture present in wood may be adsorbed by the cell walls, or after the walls are saturated may occupy the void space within the cell cavities. The first type is called *bound water* and the second *free water.* Adsorbed water softens the cellulose/lignin material of the cell wall. Therefore, it weakens the wood, reducing all strength and stiffness properties except perhaps impact strength. Free water within the cell cavity, on the other hand, has no effect whatever, except to increase the weight of the wood member and possibly to decrease its impact strength when the cell cavities are nearly full.

For practical purposes, we may say that all strength properties decrease as MC increases, but only up to the moisture percent known as the *fiber saturation point* (FSP). This is the moisture content at which the cell wall material is completely saturated but the cell cavities are essentially empty. Any further moisture taken up by the wood is not adsorbed by the cell walls, but merely occupies the cell cavities. The fiber saturation point varies among species, as illustrated by Table 2-3.

Table 2-3. Fiber Saturation Point of Different Woods at Room Temperature.

Species	FSP (%)	Species	FSP (%)
Ash, white	24.0	Pine, loblolly	21.0
Basswood	32.0	Pine, longleaf	25.5
Birch, yellow	27.0	Pine, red	24.0
Cedar, Alaska	28.5	Pine, slash	29.0
Cedar, western red	22.0	Pine, shortleaf	30.0
Douglas fir	26.0	Poplar, yellow	31.5
Fir, red	30.0	Redwood	22.5
Hemlock, western	28.0	Spruce, Sitka	28.5
Larch, western	28.0	Spruce, red	27.0
Oak, white	32.5	Tamarack	24.0
Oak, swamp	31.0	Teak	22.0

Condensed from reference 8, courtesy Van Nostrand Reinhold Company.

Wood is hygroscopic, which means that its moisture content adjusts to reach equilibrium with the temperature and relative humidity of the atmosphere in which it is used. If wood dried at the mill to the usual MC is used in a very dry location, such as the interior of a heated building, it may give up moisture to the atmosphere, shrinking and perhaps warping as it does so. On the other hand, if the wood is used in a very moist location, such as in a laundry or over a swimming pool, it will absorb moisture from the atmosphere, expanding and possibly warping. Such taking on or giving up of moisture is a slow process, so that if atmospheric conditions fluctuate, the MC adjusts to a condition near the average.

The moisture content that remains constant under a given atmospheric condition is called the *equilibrium moisture content,* sometimes abbreviated EMC. Table 2-4 shows average EMC values for various combinations of temperature and relative humidity.

Moisture affects the size of a piece of wood, and, perhaps its shape. With increasing MC

Table 2-4. Moisture Content of Wood in Equilibrium with Stated Dry-Bulb Temperature and Relative Humidity.[2]

Temperature (dry-bulb) (°F)	Relative Humidity (%)																			
	5	10	15	20	25	30	35	40	45	50	55	60	65	70	75	80	85	90	95	98
30	1.4	2.6	3.7	4.6	5.5	6.3	7.1	7.9	8.7	9.5	10.4	11.3	12.4	13.5	14.9	16.5	18.5	21.0	24.3	26.9
40	1.4	2.6	3.7	4.6	5.5	6.3	7.1	7.9	8.7	9.5	10.4	11.3	12.3	13.5	14.9	16.5	18.5	21.0	24.3	26.9
50	1.4	2.6	3.6	4.6	5.5	6.3	7.1	7.9	8.7	9.5	10.3	11.2	12.3	13.4	14.8	16.4	18.4	20.9	24.3	26.9
60	1.3	2.5	3.6	4.6	5.4	6.2	7.0	7.8	8.6	9.4	10.2	11.1	12.1	13.3	14.6	16.2	18.2	20.7	24.1	26.8
70	1.3	2.5	3.5	4.5	5.4	6.2	6.9	7.7	8.5	9.2	10.1	11.0	12.0	13.1	14.4	16.0	17.9	20.5	23.9	26.6
80	1.3	2.4	3.5	4.4	5.3	6.1	6.8	7.6	8.3	9.1	9.9	10.8	11.7	12.9	14.2	15.7	17.7	20.2	23.6	26.3
90	1.2	2.3	3.4	4.3	5.1	5.9	6.7	7.4	8.1	8.9	9.7	10.5	11.5	12.6	13.9	15.4	17.3	19.8	23.3	26.0
100	1.2	2.3	3.3	4.2	5.0	5.8	6.5	7.2	7.9	8.7	9.5	10.3	11.2	12.3	13.6	15.1	17.0	19.5	22.9	25.6
110	1.1	2.2	3.2	4.0	4.9	5.6	6.3	7.0	7.7	8.4	9.2	10.0	11.0	12.0	13.2	14.7	16.6	19.1	22.4	25.2
120	1.1	2.1	3.0	3.9	4.7	5.4	6.1	6.8	7.5	8.2	8.9	9.7	10.6	11.7	12.9	14.4	16.2	18.6	22.0	24.7
130	1.0	2.0	2.9	3.7	4.5	5.2	5.9	6.6	7.2	7.9	8.7	9.4	10.3	11.3	12.5	14.0	15.8	18.2	21.5	24.2
140	0.9	1.9	2.8	3.6	4.3	5.0	5.7	6.3	7.0	7.7	8.4	9.1	10.0	11.0	12.1	13.6	15.3	17.7	21.0	23.7
150	0.9	1.8	2.6	3.4	4.1	4.8	5.5	6.1	6.7	7.4	8.1	8.8	9.7	10.6	11.8	13.1	14.9	17.2	20.4	23.1
160	0.8	1.6	2.4	3.2	3.9	4.6	5.2	5.8	6.4	7.1	7.8	8.5	9.3	10.3	11.4	12.7	14.4	16.7	19.9	22.5
170	0.7	1.5	2.3	3.0	3.7	4.3	4.9	5.6	6.2	6.8	7.4	8.2	9.0	9.9	11.0	12.3	14.0	16.2	19.3	21.9
180	0.7	1.4	2.1	2.8	3.5	4.1	4.7	5.3	5.9	6.5	7.1	7.8	8.6	9.5	10.5	11.8	13.5	15.7	18.7	21.3
190	0.6	1.3	1.9	2.6	3.2	3.8	4.4	5.0	5.5	6.1	6.8	7.5	8.2	9.1	10.1	11.4	13.0	15.1	18.1	20.7
200	0.5	1.1	1.7	2.4	3.0	3.5	4.1	4.6	5.2	5.8	6.4	7.1	7.8	8.7	9.7	10.9	12.5	14.6	17.5	20.0
210	0.5	1.0	1.6	2.1	2.7	3.2	3.8	4.3	4.9	5.4	6.0	6.7	7.4	8.3	9.2	10.4	12.0	14.0	16.9	19.3
220	0.4	0.9	1.4	1.9	2.4	2.9	3.4	3.9	4.5	5.0	5.6	6.3	7.0	7.8	8.8	9.9	*	*	*	*
230	0.3	0.8	1.2	1.6	2.1	2.6	3.1	3.6	4.2	4.7	5.3	6.0	6.7	*	*	*	*	*	*	*
240	0.3	0.6	0.9	1.3	1.7	2.1	2.6	3.1	3.5	4.1	4.6	*	*	*	*	*	*	*	*	*
250	0.2	0.4	0.7	1.0	1.3	1.7	2.1	2.5	2.9	*	*	*	*	*	*	*	*	*	*	*
260	0.2	0.3	0.5	0.7	0.9	1.1	1.4	*	*	*	*	*	*	*	*	*	*	*	*	*
270	0.1	0.1	0.2	0.3	0.4	0.4	*	*	*	*	*	*	*	*	*	*	*	*	*	*

[2]Asterisks indicate conditions not possible at atmospheric pressure.
Source: Courtesy USDA–Forest Service, Forest Products Laboratory.

(again, only up to the fiber saturation point) the cell walls swell and all dimensions of the piece become larger. With reduction of MC, all dimensions become less; that is, the wood shrinks. Swelling and shrinkage are very small in the longitudinal direction, varying from 0.1% to 0.2% as green wood is completely dried (6). However, they are considerably more in the radial direction and highest of all tangentially. Table 2-5 compares, for a few species commonly used for construction, the percentages of shrinkage in radial and tangential directions, as the green wood is completely dried. It is the *difference* of shrinkage and swelling in these two directions that causes wood to warp as its MC changes.

Amount of shrinkage to expect with a given change of MC may be estimated using Tables 2-3 and 2-5. The position of the annual rings is usually not known in advance. Thus, in estimating the dimension change, the designer should assume the rings to be nearly parallel to the dimension in question; that is, the larger dimension change (tangential) should be anticipated. Note that the percentages in Table 2-5 are based on dimensions of the green wood (MC = FSP). (An approximate method often used is based on the fact that the average FSP for various species is about 30%, and the average shrinkage for various species as they dry from EMC to oven-dry is about 6%. Thus, the lateral dimensions of wood change, on the average, by about 1% for each 5% change of MC.) However, it should be recognized that the shrinkage in different pieces of wood is quite variable—from 50% to 150% of that estimated using Tables 2-3 and 2-5.

Table 2-5. Typical Radial and Tangential Shrinkage in Drying from FSP (Green) to Oven-Dry.

Species	Radial[a]	Tangential[a]
Douglas fir		
Coast	4.8	7.6
Interior, west	4.8	7.5
Fir		
California red	4.5	7.9
Hemlock		
Eastern	3.0	6.8
Western	4.2	7.8
Larch		
Western	4.5	9.1
Pine		
Eastern white	2.1	6.1
Ponderosa	3.9	6.2
Longleaf southern	5.1	7.5
Spruce		
Red	3.8	7.8

[a]Percent, based on dimensions when green.
Note: For a more nearly complete list, see reference 6.

Example 2-1

A coast region Douglas fir plank measures 10.25 in. wide when its MC is 10%. What will its width be when MC is raised to 35%?

By Table 2-3, the fiber saturation point for D. fir is 26%, so raising the water content above 26% will have no effect. The change from 10% to 26% will cause the wood to swell. The final dimension will be the same as the dimension, D, when green. From Table 2-5, tangential shrinkage is 7.6% as MC reduces from FSP to dry.

$$10.25 = D[1 - 0.076(16/26)]$$
$$D = 10.75 \text{ in.}$$

Example 2-2

What is the width of the same plank when its MC is raised from 10% to 19%?

The width at 19% would be the width at FSP minus the shrinkage that would occur as the MC was lowered from FSP to 19%.

$$D = 10.75\,[1 - 0.076(26 - 19)/26]$$
$$= 10.53 \text{ in.}$$

Note that, compared to the 10.25 in. width when the plank was installed, it would now be over 1/4 in. wider. To install the planks without any space between might cause trouble if the moisture content did actually rise to 19%.

2-10. SPECIFIC GRAVITY

All strength and stiffness properties depend on how much cellulose is present, so we should expect heavier woods to be both stronger and stiffer than lighter woods. Extensive tests on many species have shown this to be the case. Results of those tests are in the form of equations relating various strength properties to specific gravity (7). These equations are all of the form

$$F = K(SG)^n \qquad (2\text{-}2)$$

Table 2-6. Typical Relationships of Strength of Clear, Straight-Grained Wood to Specific Gravity (Approximate).

Strength Property	Strength Property in Terms of Specific Gravity	
	For Green Wood	For Wood with 12% Moisture
Static bending:		
Bending stress at proportional limit (psi)	$10{,}200SG^{1.25}$	$16{,}700SG^{1.25}$
Modulus of rupture (psi)	$17{,}600SG^{1.25}$	$25{,}700SG^{1.25}$
Modulus of elasticity (million psi)	$2.36SG$	$2.80SG$
Compression parallel to grain:		
Proportional limit (psi)	$5250SG$	$8750SG$
Ultimate (crushing) strength (psi)	$6730SG$	$12{,}200SG$
Modulus of elasticity (million psi)	$2.91SG$	$3.38SG$
Compression perpendicular to grain:		
Proportional limit (psi)	$3000SG^{2.25}$	$4630SG^{2.25}$

Table 2-6 shows typical equations for strength of clear, straight-grained wood.

Equations showing these relationships are interesting, but have limited value in structural design. The reason is that so many other factors (besides specific gravity) affect wood strength, and it is the combination of *all* these factors that controls the actual strength of any member.

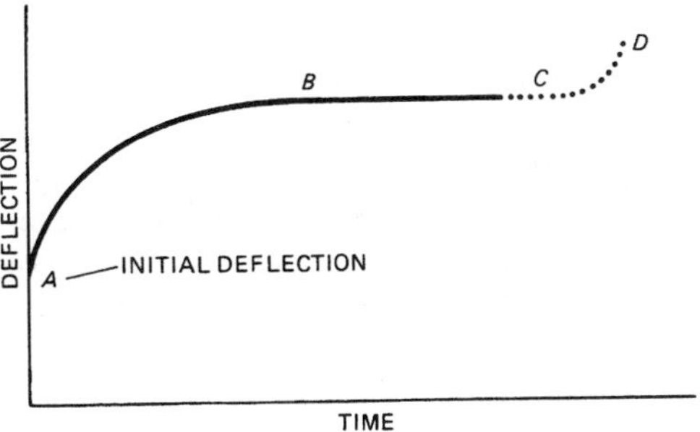

Fig. 2-7. Creep in deflection of a wood beam.

2-11. TIME-DEPENDENT BEHAVIOR OF WOOD

Two time-dependent characteristics of wood are (1) creep under sustained load and (2) a relationship between ultimate strength and load duration. These phenomena are related, but much yet has to be learned about each of them.

Creep

When a wood structural member is loaded, it deflects immediately. If the load remains on the member, deflection increases with time. This increase above and beyond the initial deflection is called creep deflection. Creep can occur in any type of member (beam, column, or tension member) but is of greatest concern for beams. Under the allowable stress magnitudes normally used in structural design, the rate of creep deflection becomes less as time passes, so that after a few years no further increase is seen. Figure 2-7 illustrates the variation of deflection with time, the solid portion of the curve becoming practically horizontal after a few years. If the load were kept on the member for a long enough time, the curve would eventually continue as shown by the dotted line in Fig. 2-7. Similar curves could be plotted for any type of loading, the exact shape depending on the type of loading and level of stress applied.

The complete curve of Fig. 2-7 may be used to define three stages of creep—primary, secondary, and tertiary. During primary creep (segment *AB* of the curve) deflection continues to increase after the load has been applied. Secondary creep is represented by segment *BC*, the essentially horizontal part of the curve. If the load is maintained long enough, tertiary creep occurs as deflection begins to increase again as shown by segment *CD*. During this stage, the rate of creep deflection increases and failure follows (8). For members with very low stress, the overall life of the structure is not long enough for us to observe tertiary creep. For members of very high stress level, however, tertiary creep may be reached

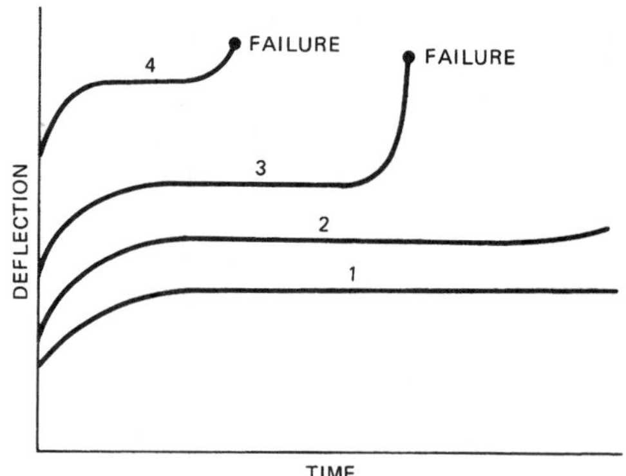

Fig. 2-8. Creep at various stress intensities.

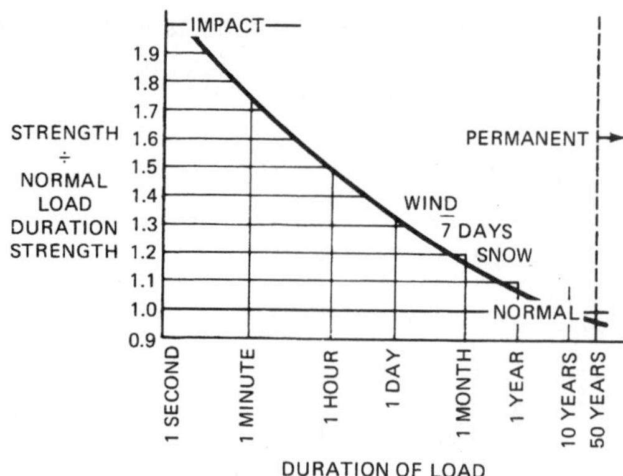

Fig. 2-9. Effect of load duration on strength. (Courtesy Board of Regents, University of Colorado.)

quickly. In this case, collapse will follow unless the increasing deflection alerts the user to either unload the member or reinforce it.

Figure 2-8 shows a set of creep curves for four test beams, each with a different level of applied stress. In the beam represented by curve 1, the applied stress was fairly low, and because the test was not continued long enough, the beam never entered tertiary creep. Curve 2 is for a similar beam, but stressed to a higher level. It just started to enter tertiary creep at the end of the test. Curve 3, for a beam with still higher stress level, shows the beam entering tertiary creep and then failing. Curve 4 is for a beam having a very high initial flexural stress, just slightly below the ultimate stress determined from a conventional short-term flexural test. This beam had a negligible period of secondary creep, then quickly entered tertiary creep and failed.

The beam tested for curve 1 of Fig. 2-8 did not enter tertiary creep. Would tertiary creep eventually occur if the test were continued? Actually, we don't know, for the reason that such a test would take many years—perhaps centuries. It *may* be that tertiary creep will occur eventually, or it *may* be that under such a low stress level tertiary creep will never occur. What is important is this: We know that if the stress level is kept low enough, failure will not occur *during the expected life of the structure.*

Load-Duration Effect

Since the length of time that a load is present on a wood member affects the ultimate or breaking strength of that member, the designer of wood structures must consider that effect. Otherwise, to ensure a desired level of safety, the designer would have to be overly conservative, designing as though all loads were to be in place forever. To consider the load-duration effect, the structural designer uses a method based on the relationship shown by Fig. 2-9. This graph is the result of extensive research by the Forest Products Laboratory (9). Chapter 4 shows how the designer takes into account this load-duration effect.

2-12. STRENGTH-REDUCING CHARACTERISTICS

Because wood structural members are cut from trees rather than being formed from a human-made material, they will have some strength-reducing defects, such as the following: (1) knots, (2) cross grain, (3) checks, (4) shakes, (5) compression wood, (6) wane, and (7) decay.

Knots

Knots are formed by the change of wood structure that occurs where limbs grow from the main stem of the tree. The limb, extending approximately radially in the main trunk, has its own annual rings and rays, and this local arrangement of cells interrupts the normal pattern for the main portion of the tree. The knot has a weakening effect that is illustrated by Fig. 2-10. The grain direction changes severely as the wood fibers pass around the knot (cross grain). In this region, the applied load causes tensile

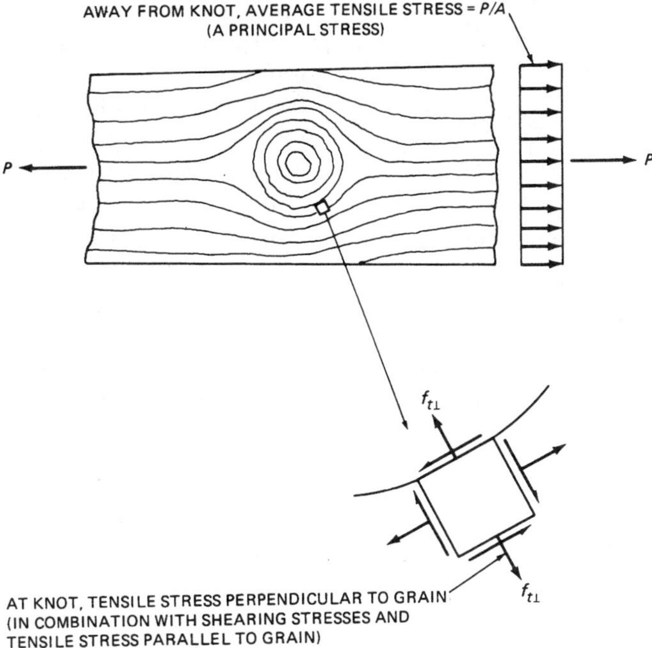

Fig. 2-10. Effect of knot on stresses.

stress components normal to the grain of the knot. Since tensile strength normal to the grain is very low, the knot weakens the member significantly.

Tensile strength is affected most by the knot, but compressive strength is reduced also. Bending strength is reduced too, the amount of reduction depending on where the knot is located on the beam cross section. In drying, wood shrinks more radially than longitudinally, so knots frequently become loose and may even fall out. Then, obviously, they are as damaging to bending, compressive, or shear strength as they are to tensile strength.

The manner in which the tree is cut to produce lumber influences the degree of effect a knot will have on strength. This will be illustrated in Chapter 3.

Cross Grain

When the longitudinal axis of the cells is not parallel to the edge of a piece of wood, the piece is said to have cross grain. Cross grain involves both spiral grain and diagonal grain. Cross grain occurs at knots and other locations. Nearly all lumber has this strength-reducing characteristic to some degree. Figure 2-11 illustrates several cases of cross grain, differing from each other according to the orientation of the annual rings with the faces of the piece of lumber.

Many trees grow in a spiral fashion, so that the direction and degree of cross grain varies along the length of a piece of lumber. This special case of cross grain is called spiral grain. Spiral grain makes wood very difficult to work. Further, as its moisture content changes, the wood warps—all plane surfaces become twisted, warped surfaces. Spiral grain is common and is a hereditary trait in some trees.

Diagonal grain is caused by the way in which a piece of lumber is cut from the log, rather than being due to an inherent defect of the tree itself. However, there will always be variations in grain direction in the tree (cells are not everywhere perfectly parallel), so it is not possible in the mill to cut lumber so as to completely eliminate diagonal grain.

Figure 2-12 shows how cross grain reduces the strength of a piece of wood. Almost everyone has tried to bend a stick with cross grain, only to have it break in the manner shown by Fig. 2-12a. This break occurs because the flexural stress (which is parallel to the surface) has a

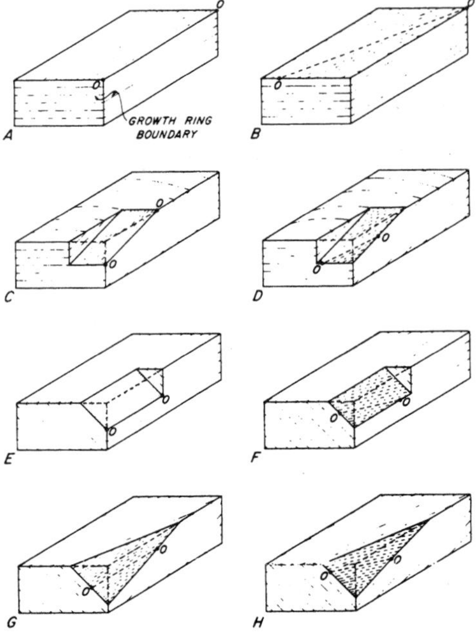

Fig. 2-11. Types of cross grain. (Courtesy of USDA–Forest Service, Forest Products Laboratory.)

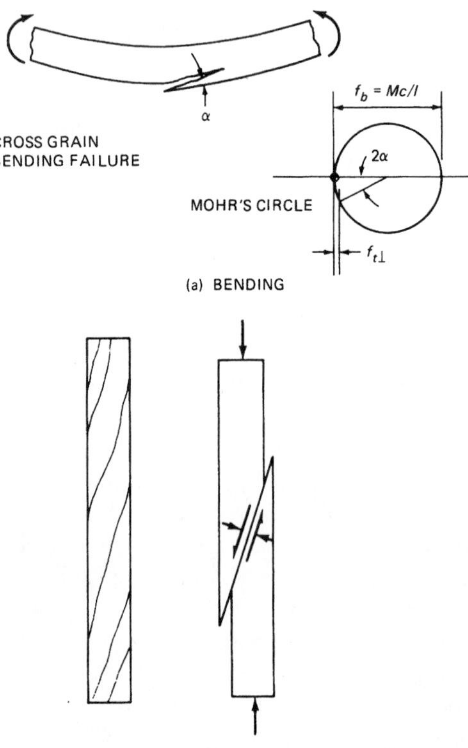

Fig. 2-12. Effect of cross grain on strength.

component perpendicular to the grain, and tensile strength in that direction is very low. The Mohr's circle in Fig. 2-12a shows how appreciable perpendicular-to-grain tension can occur with small cross-grain angles.

Similarly, Fig. 2-12b shows how a compression member with cross grain is weakened. Here, it is either the low strength in compression perpendicular to the grain or the low longitudinal shear strength that precipitates failure.

Strength at an angle to the grain may be estimated by a simple equation called *Hankinson's formula.* This equation, shown in later chapters, is used in designing members and connections where the forces to be transferred are neither parallel to nor perpendicular to the grain.

Shakes and Checks

Sometimes the generic term *splits* is used to cover both shakes and checks. Shakes and checks are longitudinal planes of separation (cracks) in the wood (see Fig. 2-13). Shakes originate in the tree. They may be in the longitudinal/radial plane, in which case they are called *heart shakes* or *rift cracks.* When they are in the longitudinal/tangential plane they are called *ring shakes,* the plane of the break being parallel to the annual rings. Heart shakes are thought to be caused by tensile stresses in the tangential direction, the stresses resulting from growth pressures within the tree. Ring shakes are thought to be caused by shear due to wind, perhaps combined with weakness due to cell damage by frost. Figure 2-14 shows types of shake.

Checks occur as the wood seasons after the tree is felled. They may occur in the log itself or in pieces of lumber cut from the log. Checks are longitudinal cracks, usually in the longitudinal/radial plane but occasionally in the longitudinal/tangential plane. They occur because of stresses resulting from differential shrinkage in the tangential and radial directions during drying, or from uneven drying in different portions of the lumber. Too-rapid drying can cause severe checking. Thus, checking can be reduced by careful attention to drying procedures during lumber production.

Figure 2-15 shows the position and names of types of check. Surface checks may sometimes be removed by planing: however, if drying con-

Fig. 2-13. Checks and shakes in a cottonwood log. (Photograph by authors.)

tinues, they may reappear. The major structural effect of checks is to reduce longitudinal shear strength, often the deciding factor in designing timber beams.

Compression Wood

Compression wood is a form of "reaction wood," in which an unsymmetrical growth pattern has

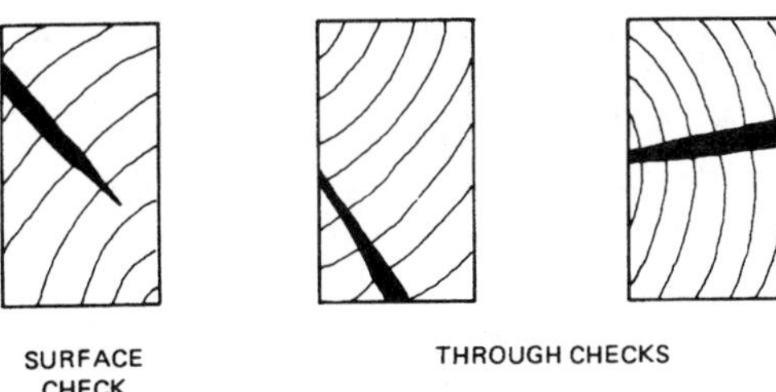

Fig. 2-15. Types of check.

been caused by long-term bending stresses in the live tree. For example, a tree that leans badly during its growth will have to resist a high bending moment caused by the tree's own weight. Compressive stresses due to this bending occur on the lower side, and tensile stresses due to bending occur on the other side. The compressive stresses, being combined with the direct vertical compressive stress, P/A, are the highest stresses for the entire cross section. These high stresses cause the tree to grow thicker annual rings on that side, as shown by Fig. 2-16.

Cell walls in compression wood are thicker, and specific gravity is higher than for normal wood. Also, in the internal microstructure of a compression wood fiber, the angle of inclination

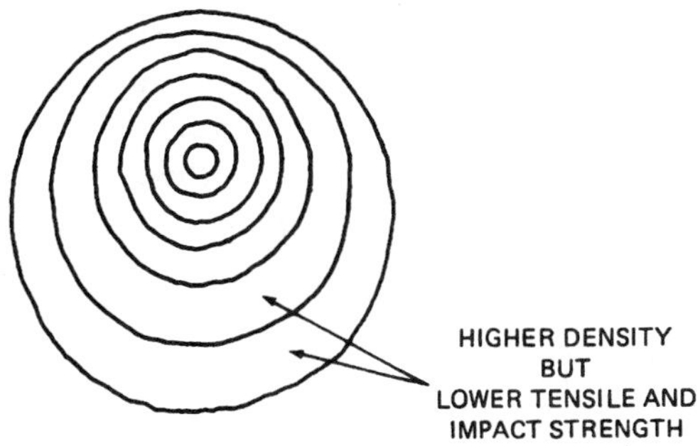

Fig. 2-16. Compression wood.

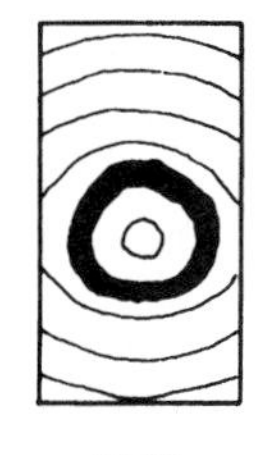

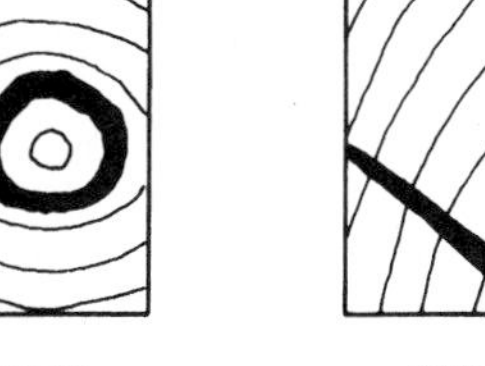

Fig. 2-14. Types of shake.

of its component fibrils is altered in a way that affects physical properties (10, 11). Compression wood is undesirable mainly because, with moisture change, its *longitudinal* shrinkage and swelling are as much as ten times greater than for normal wood.

Wane

Wane is the lumberman's term for the absence of wood. Often wane consists of bark remaining on the surface of the finished lumber. Obviously, the strength is reduced by wane, so wane is taken into account in the allowable design stresses assigned to a given grade of lumber. Inclusions of bark, called bark pockets, occasionally occur within the tree. These result from tree injury that is covered by new wood as the tree grows. The injury giving rise to bark pockets is usually something minor, such as holes made by bird pecks or boring insects.

Decay

Decayed spots represent portions of the wood that should be considered to have no strength, since the wood fibers are interrupted by the decay. Further, decay can easily spread to adjacent areas. Wood with decay should not be used in new construction, but a designer may be called upon to evaluate the strength of an existing structure in which decay is already present. If the decay can be removed or its growth arrested, the structure can conceivably continue to be used.

All of the above defects affect the structural value of the wood, and they are considered in "grading," the process by which wood is evaluated and classified to indicate suitable allowable design stresses. This will be shown in Chapters 3 and 4. Other defects, not mentioned above, affect only appearance and durability.

2-13. THERMAL PROPERTIES OF WOOD

Effect of Temperature on Mechanical Properties

Temperature affects both dimensional stability and strength of wood. Wood expands as its temperature increases, as do other construction materials. Its coefficients of expansion vary with direction, being largest radially and tangentially, and least longitudinally. Expressed as length change per unit of original length per degree Fahrenheit, wood's radial and tangential coefficients of expansion range from $0.000025 \times SG$ (specific gravity) for dense hardwoods to $0.000045 \times SG$ for softwoods such as Douglas fir or redwood (see reference 7). Considering that the specific gravity of these woods is from 0.42 for redwood to 0.68 for common dense hardwoods, the radial and tangential coefficients of expansion are seen to be about 0.000017 to 0.000019 in./in./°F, or about three times as large as for structural steel or concrete.

Wood's longitudinal expansion coefficient is much less. It is independent of *SG* but varies with species from about 0.0000017 to 0.0000025 (12), which is only about 26–42% as much as for other common construction materials.

Exposure to temperatures of 150°F or more for extended periods will permanently weaken wood (12). For temperatures between zero and 150°F, the 70° strength of dry wood changes by $^1/_3 - {}^1/_2\%$ per degree, being weaker at high temperatures and stronger at low. If the exposure to abnormal temperatures in this range is not prolonged, wood's strength properties can be expected to return to approximately their original values when conditions return to normal.

Thermal Conductivity

Wood's thermal conductivity is low compared to that of other common structural materials. In other words, wood can do a better job of insulating against heat loss. Often, therefore, less insulation is needed with wood construction than with other types. The total thermal transmission of a building wall, of course, depends on the properties and thickness of each of the various layers of material used. Engineering information on computing the total thermal transmission (or resistance) of a wall may be found in sources such as the *ASHRAE Handbook* (13).

Thermal conductivity depends on several factors, including direction of heat movement with respect to the grain, specific gravity, moisture

Table 2-7. Thermal Conductivity and Relative Insulating Efficiency of Building Materials.

Material	Conductivity (BTU/hr/ft^2/in. thickness/°F)	Insulating Efficiency Compared to Wood (%)
Average softwood	0.80	100
Clay brick	4.8	16.5
Glass	5.5	14.7
Limestone	6.5	12.3
Sandstone	12.0	6.6
Concrete	12.6	6.3
Steel	312.0	0.25
Aluminum	1416.0	0.06

content, and type and location of defects. Thermal conductivity is about equal in radial and tangential directions, but is about 2.5 times as great along the grain. Table 2-7 shows average values of thermal conductivity for wood and other commonly used construction materials and compares their insulating efficiencies. Notice that wood's thermal efficiency far surpasses that of all other listed materials that might be used for structural members.

2-14. TESTS AND PROPERTIES OF INTEREST TO THE STRUCTURAL DESIGNER

Strength and stiffness are the wood properties of greatest interest to the structural designer. The designer is interested in other properties, of course, such as creep, durability, hardness, dimensional stability, and thermal insulating qualities.

The strength and stiffness properties of most interest in structural design are:

1. compressive strength parallel to the grain
2. modulus of elasticity parallel to the grain
3. tensile strength parallel to the grain
4. compressive strength perpendicular to the grain
5. modulus of rupture (bending strength)
6. longitudinal shear strength (horizontal shear)
7. shear modulus

For visually graded timbers (5 in. × 5 in. nominal and larger), the above properties are determined by testing small clear specimens according to methods spelled out in ASTM D143(4) and ASTM D2555 (14) by the American Society for Testing and Materials. To obtain allowable properties, the strength properties of the small clear specimens are then adjusted to account for defects according to the rules in ASTM D245 (15).

Prior to the 1991 NDS, specified allowables for *dimension lumber* (members with smaller dimension less than 5 in.) were obtained by the small clear specimen approach. Since 1991 allowable properties of American softwood dimension lumber have been determined by tests of full-size specimens.

Standard tests of small clear specimens are described below. These tests are still used for determining properties of timbers and hardwood species of dimension lumber.

Test for Compression Parallel to the Grain

In this ASTM test a clear (defect-free) specimen measuring a full 2 × 2 × 8 in. is used, with the longitudinal axis of the wood parallel to the 8-in. dimension. Longitudinal load is applied, increasing until the compressive stress is well beyond a defined elastic limit or until the specimen fails. From the loads applied and longitudinal strains measured during the test, a stress-strain curve can be plotted, as shown by Fig. 2-17. This curve shows several points of interest to the designer.

Ultimate strength is the highest value of stress reached, shown by the ordinate at point *A*.

Modulus of elasticity, also called Young's mod-

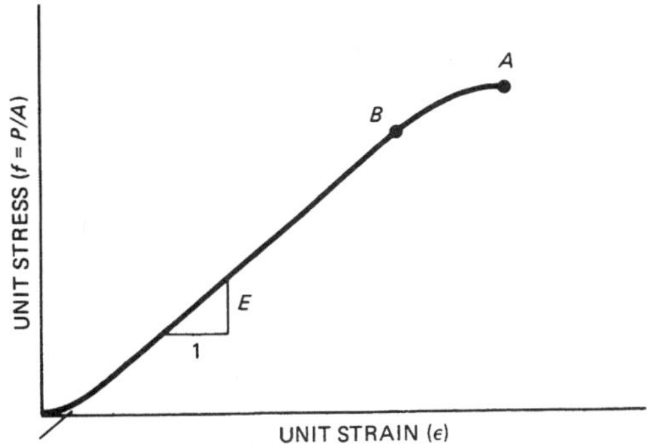

Fig. 2-17. Stress-strain curve, compression parallel to grain.

ulus, is the slope, *E,* of the initial straight-line part of the stress-strain curve. In the U.S. system, it is expressed in pounds per square inch (psi). (Modulus of elasticity can also be found from a bending test, but its value will differ slightly from that found from the compression test. The reason is that the measured beam deflections are affected by shear strains as well as by flexural strains.)

Proportional limit is the value of stress at the upper end of the straight-line portion of the curve, that is, the stress level above which stress is no longer proportional to strain. The proportional limit is sometimes loosely referred to as the elastic limit, but the two are not precisely the same. Point B of the curve in Fig. 2-17 is at the proportional limit.

Notice that at the beginning of the test, the curve of Fig. 2-17 shows a slight reverse curvature. This is due to local crushing of unintentional high spots as the ends of the specimen become seated on the platen of the testing machine. To correct for this, the straight-line portion of the curve may be extrapolated downward to locate an adjusted zero mark. Various types of compression failure seen in this type of test are shown by Fig. 2-18.

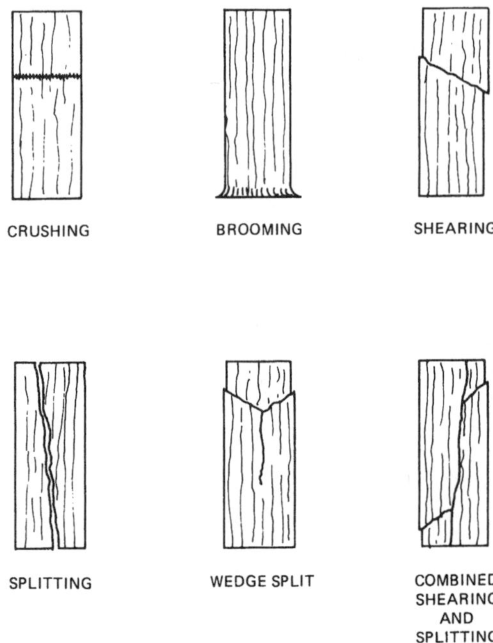

Fig. 2-18. Types of failure under compression parallel to grain.

Test for Tension Parallel to the Grain

The ASTM test for tensile strength parallel to the grain uses a 1-in.-square clear specimen 18 in. long, carefully tapered to measure only 3/8 in. by 3/16 in. at its narrowest portion. A stress-strain curve for this test shows elastic action almost to the breaking point. For defect-free wood, tensile strength parallel to the grain (longitudinally) is appreciably higher than compressive strength in the same direction. However, designers can seldom utilize this high tensile strength, since wood in sizes practical for construction always has strength-reducing defects, such as cross-grain and knots. Tensile strengths in tangential and radial directions are very low compared to the longitudinal strength, about 3% and 5%, respectively.

Test for Compression Perpendicular to the Grain

The ASTM test for compression perpendicular to the grain illustrates clearly how much weaker wood is in compression perpendicular to the grain than in compression parallel to the grain. Figure 2-19 shows schematically how this test is performed, and also a typical stress-strain curve.

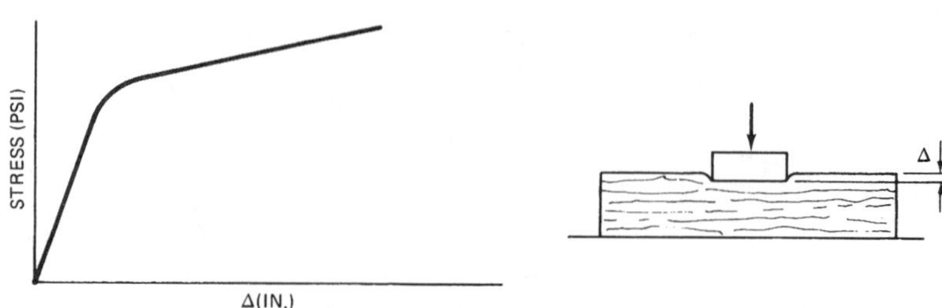

Fig. 2-19. Compression perpendicular to grain.

Notice that there is no true ultimate strength value. As the wood nearest to the surface of the specimen is crushed, it merely becomes consolidated and is harder to crush further. As the load increases, additional wood cells farther from the surface also collapse and the strength increases again. The test could proceed until the load capacity of the testing machine was reached. The wood specimen would still be there, and could accept more load, but of course it would be severely flattened. For this type of loading (in design, we call it *bearing*), the failure criterion is not reaching an ultimate load, but rather reaching a limit beyond which the degree of distortion is considered unacceptable.

Test for Flexure

The ASTM static bending test defined by reference 4 uses a full 2-in.-square clear specimen 30 in. long, tested with a midspan concentrated load and a span of 28 in. The load-deflection curve for this test shows a definite ultimate load. Values that can be computed from this curve are given below.

Modulus of rupture: The modulus of rupture (*MR*) is the value of Mc/I one would compute using the bending moment caused by the ultimate load. Modulus of rupture is *not* the value of extreme-fiber bending stress at failure. Bending stresses under elastic conditions vary as a straight line, as shown by Fig. 2-20a, whereas at ultimate load they vary as shown by Fig. 2-20b. Nevertheless, the *MR* is a useful quantity for comparing bending strengths of different specimens, and it is used to establish allowable stress values for the designer's use.

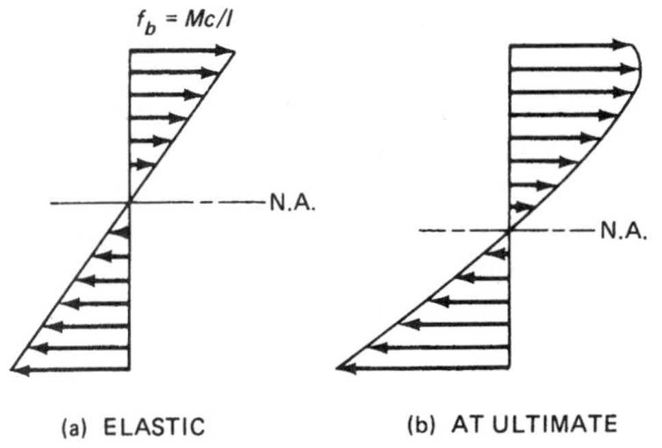

Fig. 2-20. Bending stress variation.

Modulus of elasticity: By equating the theoretical expression for midspan deflection to an observed deflection on the straight-line portion of the load-deflection curve, the modulus of elasticity, E, can be computed. This value of E is the more reliable one to use for computing deflection of other beams, although it may not agree exactly with the values obtained from tension or compression tests.

Test for Shear Strength Parallel to the Grain

Shear strength parallel to the grain is less than in any other direction. In wood beams, the grain is normally parallel to the span, and shear failure occurs along a plane parallel to the span, as in Fig. 2-21. (Wood beams will not fail by shearing at right angles to the span.) Since beams are usually horizontal, the measured ultimate shear stress is often called *horizontal shear strength.* In the ASTM test for shear strength there is no gradual yielding; failure is sudden and only the ultimate load is observed.

ASTM D143 defines other tests also, including tension perpendicular to the grain, impact bending, and hardness. With each test, moisture content of the wood is also determined and recorded.

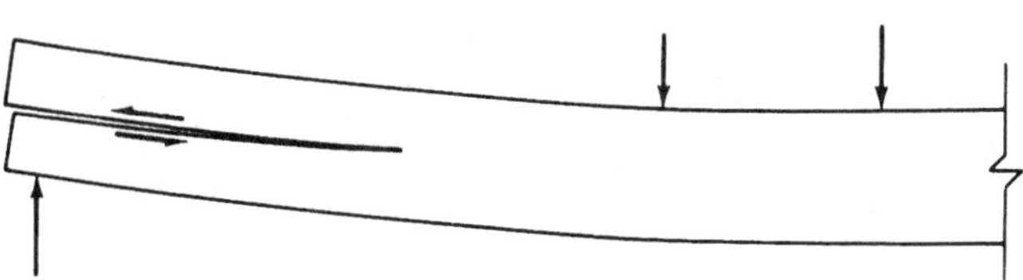

Fig. 2-21. Horizontal shear failure.

REFERENCES

1. "Tree Improvement," Georgia Pacific, Portland, OR, 1981.
2. Thomas, R. J., "Wood Anatomy and Ultrastructure," in *Wood: Its Structure and Properties,* by F. F. Wangaard, Pennsylvania State University, University Park, PA, 1981.
3. Pearson, R. G., and R. C. Gilmore, "Effect of Fast Growth Rate on the Mechanical Properties of Loblolly Pine," *Forest Products Journal,* 30(5), May 1980, pp. 47–54.
4. *Standard Methods of Testing Small Clear Specimens of Timber,* American Society for Testing and Materials, ASTM D143, West Conshohocken, PA, 1994.
5. Tarkow, H., "Wood and Moisture," in *Wood: Its Structure and Properties,* by F. F. Wangaard, Pennsylvania State University, University Park, PA, 1981.
6. *Wood Handbook,* U.S. Forest Products Laboratory, 1974.
7. *Wood Structures, A Design Guide and Commentary,* American Society of Civil Engineers, New York, 1975.
8. Bodig, J., and B. A. Jayne, *Mechanics of Wood and Wood Composites,* Van Nostrand Reinhold, New York, 1982.
9. Wood, L. W., "Behavior of Wood Under Continued Loading," *Engineering News Record,* Dec. 11, 1947.
10. Stamm, A. J., *Wood and Cellulose Science,* Ronald Press, New York, 1964.
11. Mark, R. F., "Molecular and Cell Wall Structure of Wood," in *Wood: Its Structure and Properties,* by F. F. Wangaard, Pennsylvania State University, University Park, PA, 1981.
12. *Timber Construction Manual,* 4th ed., American Institute of Timber Construction, Wiley, NY, 1994.
13. *ASHRAE Handbook of Fundamentals,* American Society of Heating, Refrigerating and Air-Conditioning Engineers, Inc., New York, 1992.
14. *Establishing Clear Wood Strength Values, ASTM D2555-88,* American Society for Testing and Materials, West Conshohocken, PA, 1988.
15. *Establishing Structural Grades and Related Allowable Properties for Visually Graded Lumber, ASTM D245-93,* American Society for Testing and Materials, West Conshohocken, PA, 1993.

PROBLEMS

2-1. A wood sample weighs 67.2 grams (g) when removed from a compression test specimen. After drying in an oven until its weight is constant, it weighs only 56.4 g. What was the moisture content (percent) at the time of the compression test? Was the MC above or below the fiber saturation point?

2-2. The average specific gravity of Douglas fir south may be found from Appendix Table C-1. What is the average unit weight for that species (pounds per cubic foot, or pcf)?

2-3. Figure 2-22 shows two possible ways of nailing of board-and-batten siding for exterior use. Which method would be less troublesome in service, and why?

2-4. Green western larch 2 × 10s are installed as floor joists in a heated building where the average temperature will be 65°F and the relative humidity 25%. The green 2 × 10s measured $9^1/_2$ in. deep when installed. What will be their probable depth when the EMC is reached?

2-5. Planks of 2 × 12 coast D. fir will be used flatwise to form a cover over a water tank. The planks will dried to about 15% MC and will measure $11^1/_4$ in. wide when installed. In use they will be exposed for long periods to 90% relative humidity and 80°F temperature. Why should they not be placed in side-by-side contact with each other? How much space should be allowed between the planks?

2-6. Figure 2-23 shows two arrangements for bolting a wood 2 × 10 to a steel plate. The plank is installed green (unseasoned) and will dry

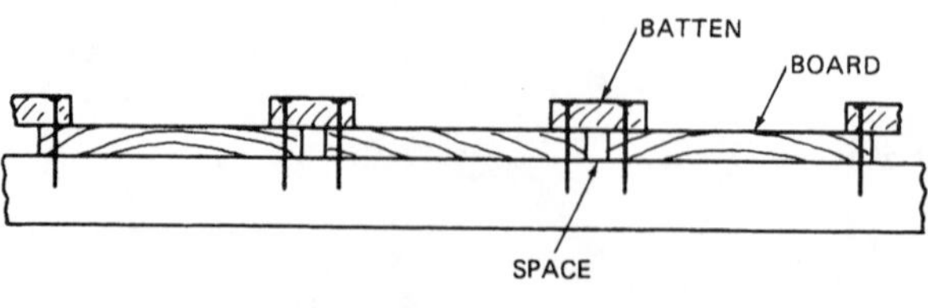

(a)

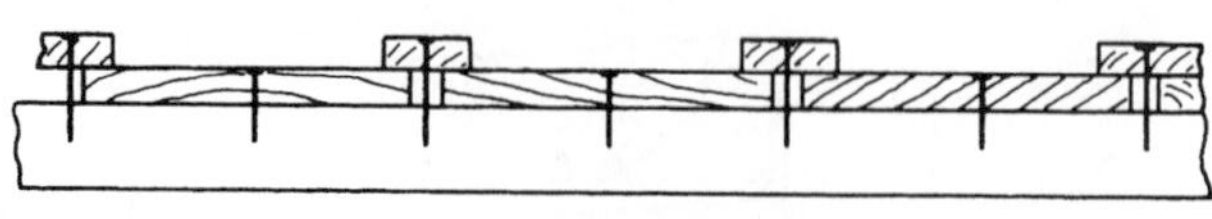

(b)

Fig. 2-22. Two nailing methods for Problem 2-3.

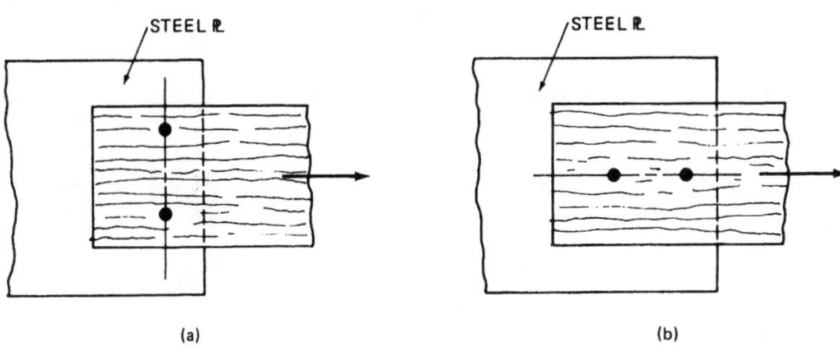

Fig. 2-23. Connections for Problem 2-6.

out after it is connected. Which arrangement is more likely to result in the member splitting, and why?

2-7. Figure 1-1 showed the cross section of a two-story light-frame wall. Assume that the sill is a 2 × 6, floor joists are 2 × 12s, and the length of the studs in each story is 7 ft 10 in. (See Appendix Table B-1 for actual dimensions.)

(a) Recognizing that shrinkage in thickness of the plywood floor sheathing is negligible, how much will the distance from the top of the concrete foundation wall to the underside of the second-floor ceiling joists decrease, assuming that

(1) all wood is ponderosa pine;

(2) MC was 19% when the wood was installed;

(3) average relative humidity in the building and surroundings will be 30%;

(4) average temperature of the wood will be 70°F; and

(5) annual rings are in the worst position for vertical shrinkage.

(b) What sort of trouble could this cause?

2-8. A 2 × 2 clear specimen is tested as a beam on a 28-in. simple span. The beam fails under a midspan vertical concentrated load of 900 lb. Measurements of actual dimensions showed the specimen to be 2.02 in. wide and 1.96 in. deep (vertically). What is the modulus of rupture?

2-9. Assuming that the bending test of Problem 2-8 took five minutes to perform, what would be the probable value of modulus of rupture for a similar specimen under a loading of ten years' duration?

2-10. When the load at the center of the beam of Problem 2-8 was 400 lb, the midpoint of the beam had deflected 0.12 in. Considering that the effect of shear deformation is negligible, what is the modulus of elasticity, *E?*

2-11. For a seasoned wood whose specific gravity is 0.47, what would be the approximate small-clear values for: (a) ultimate compressive strength parallel to the grain; (b) compressive strength (proportional limit) perpendicular to the grain; and (c) modulus of elasticity in compression parallel to the grain?

3

Production and Grading of Sawn Lumber

3-1. LUMBER PRODUCTION

Lumber production has grown from the old axe and two-person saw process to a sophisticated operation involving specialized machinery, such as multiple band saws or laser-beam cutters with digital control, and even optimization techniques to increase the yield of useful material from the log. To the engineer or wood scientist who wishes to specialize in it, wood production presents many interesting technical challenges. The structural designer, however, needs to know little about lumber production, except as it might affect either selection of material or the performance of that material in the completed structure.

Where trees are grown on tree farms, lumber quality and yield are influenced by the care given to the trees—thinning and pruning, for example. Lumber production begins with logging, the felling and trimming of the trees. Following logging the steps in lumber production may include (but not necessarily in this order) debarking, air seasoning, rough sawing and finish sawing (by circular saws, multiple band saws, or laser), as shown in Fig. 3-1, surfacing (by planing or sanding), chemical treatment (preservative or fire retardant), and kiln drying. For the structural designer, the important production features are those dealing with standard sizes, type of surfacing, types of "cuts," and moisture condition.

3-2. STANDARD SIZES OF LUMBER

Softwood lumber is normally available in even-foot lengths, from 6 ft to 24 ft or 26 ft. In some instances, the structural designer can save money for a client by avoiding designs that call for lengths that are just slightly longer than these standard lengths. For example, if a floor system required a total of 48 joists, each 18 ft 3 in. long, the builder would have to purchase 48 pieces each 20 ft long. A total of 84 lin ft of joist material would be wasted, unless a use could be found for the 1 ft 9 in. pieces cut off. Being made aware of the possible saving, the client or the architect might consider the wisdom of reducing the width of the structure by three inches to permit using joists 18 ft long. Hardwoods are also available in 2-ft length increments, and occasionally it is possible to obtain odd-foot lengths.

Quoted lengths of lumber are actual dimensions. However, widths and thicknesses of lumber for structural use are expressed in *nominal* dimensions. These are whole-inch dimensions corresponding approximately to the size of the unseasoned, rough-cut pieces from which surfaced lumber is produced. For example, a piece of wood measuring almost 2 in. thick by 4 in. wide in the rough condition may be surfaced (smoothed) to produce a piece of dressed lumber measuring 1.5 in. by 3.5 in., but this smaller piece is still called a 2×4.

The thickness of wood removed to smooth the surfaces of rough lumber varies with the rough dimension and with the moisture condition of the wood. Larger rough material will probably not be as straight as smaller-sized rough material, and surface defects may be larger. For either reason, a greater wood thickness may have to be removed from large sizes of rough lumber than from smaller sizes. The American Softwood

Fig. 3-1. Band saws reducing log to slabs. (Courtesy Western Wood Products Association.)

Lumber Standard (1) defines the final dimensions required for the finished lumber.

Table 3-1 shows the minimum standard sizes for dressed structural lumber in the United States. Notice that there are two sets of standard dimensions—one for lumber that is surfaced after it is dry and one for lumber that is surfaced while green. The reason for this is that all lumber eventually reaches an equilibrium moisture content (EMC) for the average temperature and relative humidity where it is installed. So lumber surfaced green is made thicker (and wider) than similar lumber that is surfaced when it is dry (19% or less MC). The object of having these two standards is to cause pieces of each variety to be of the same size when they reach a common EMC, the surfaced-green lumber shrinking as it dries and the surfaced-dry mate-

Table 3-1. Minimum Standard Sizes for Dressed Lumber for Structural Use.

	Thickness (in.)			Width (in.)		
		Minimum			Minimum	
	Nominal	Surfaced Dry	Surfaced Green	Nominal	Surfaced Dry	Surfaced Green
Dimension lumber	2[a]	$1\frac{1}{2}$	$1\frac{9}{16}$	2[a]	$1\frac{1}{2}$	$1\frac{9}{16}$
	3	$2\frac{1}{2}$	$2\frac{9}{16}$	3	$2\frac{1}{2}$	$2\frac{9}{16}$
	4	$3\frac{1}{2}$	$3\frac{9}{16}$	4	$3\frac{1}{2}$	$3\frac{9}{16}$
				6	$5\frac{1}{2}$	$5\frac{5}{8}$
				8	$7\frac{1}{4}$	$7\frac{1}{2}$
				10	$9\frac{1}{4}$	$9\frac{1}{2}$
				12	$11\frac{1}{4}$	$11\frac{1}{2}$
				14	$13\frac{1}{4}$	$13\frac{1}{2}$
				16	$15\frac{1}{4}$	$15\frac{1}{2}$
Timbers	5 or more	—	$\frac{1}{2}$ less than nom.	5 or more	—	$\frac{1}{2}$ less than nom.

[a]In 2-in. nominal thickness, widths greater than 14-in. nominal are usually not available.

rial either swelling or shrinking, depending on whether the EMC is more than or less than 19%.

With few exceptions, *nominal* dimensions of lumber normally available in the United States for structural purposes vary by 2-in. increments. Thus (except for nominal 3-in. and 5-in. material that is sometimes available) nominal thicknesses and widths, in inches, are even numbers. Table B-1 in the Appendix of this book shows the standard nominal sizes of sawn lumber and the actual dimensions expected at 19% MC.

Older Size Standards

The dimensions shown in Table 3-1 and Appendix Table B-1 were accepted as the standard for the United States in about 1970. A designer doing remodeling, repair, or preservation work on older buildings should be aware that prior to 1970 the standard dimensions were different. For lumber manufactured before 1970, thickness and width (for lumber of up to 12 in. nominal width) were generally 1/8 in. to 1/4 in. more than shown by Tables 3-1 and B-1. For example, a 2 × 4 measured 1 5/8 in. by 3 5/8 in., as opposed to today's 1 1/2 in. by 3 1/2 in.

However, in buildings dating back to the early part of this century or before, lumber dimensions may not conform even to that pre-1970 standard. For example, in working on restoration and remodeling of an historic building, the authors encountered existing first-floor joists that were actually a full 8 in. deep by 2 1/16 − 2 1/4 in. wide. Second-floor joists were only slightly different. Each were probably referred to as 2 × 8s. When adding modern 2 × 8s to reinforce the floor, it was necessary to install shims under the ends of the new joists to make their tops flush with the tops of the original joists. Designers working with old structures should always *measure* to determine the actual size of existing wood members.

The problem raised by combining lumber of older and present size standards is compounded by the effects of shrinkage. Structural members in place in old structures probably have a very low moisture content, whereas new lumber being added probably has higher MC. Thus, some shrinkage of the newer lumber may be expected, and the amount of shrinkage may add to the difference between old and new dimensions.

3-3. FINISH DESIGNATIONS

Most lumber used in building design is surfaced on both faces and on both edges and is referred to as *S4S* (surfaced on four sides). If a designer specifies a nominal size of lumber but does not define which faces are to be finished, the material supplied will most likely be S4S. If some other combination of surfaced and rough faces is desired, however, the designer may specify it. Other combinations are defined by symbols *S* for sides and *E* for edges. Examples of these combinations include:

S2S1E (surfaced on two sides and one edge)
S2S (surfaced on two sides only)
S1S2E (surfaced on one side and two edges)

If lumber other than S4S is desired, the designer will have to specify it. If the item is not stocked, a special mill order may be required, and the builder may have to wait for the material to be produced. A probable exception is rough material (not surfaced), which is frequently available in some of the larger standard sizes.

3-4. CUTTING PATTERNS

Figure 3-2 shows three patterns in which lumber may be cut from the log. The difference between patterns is only the position of the annual rings relative to the wide face of the piece. The importance of this difference is that shrinkage, warping, abrasion resistance, ease of painting, appearance, and other properties all differ ac-

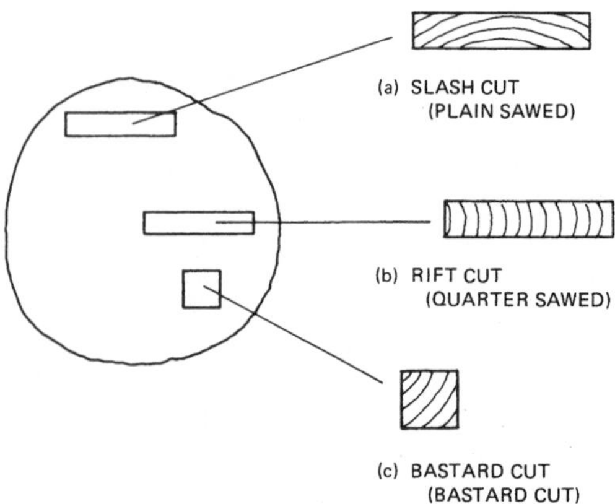

Fig. 3-2. Types of cut (hardwood names in parentheses).

cording to the position of the annual rings on the cross section.

Figure 3-2a shows lumber cut so that the wide face is nearly parallel to the annual rings. In softwoods, this is called *slash cut* lumber, and in hardwoods *plain sawed.* Figure 3-2b shows lumber in which the rings are more nearly perpendicular to the wide face. This is called *rift cut* for softwoods and *quarter sawed* for hardwoods. Of course, there will be many pieces intermediate between these two. Pieces with the rings crossing at about 45 degrees, as shown by Fig. 3-2c, are known as *bastard cut.* Rift-cut lumber is also known as *vertical grain* or *edge grain,* and slash cut as *flat grain.*

In most respects, rift-cut lumber is superior to slash cut. Figure 3-3 shows one reason. As the wood dries, it shrinks, and shrinkage is larger in a transverse direction parallel to the annual rings than in any other direction. Consequently the length along the trace of a ring shortens, and the piece bends (warps) as shown. The slash-cut piece, with its wide face nearly parallel to the rings, distorts appreciably. The rift-cut piece, on the other hand, shows such distortion only on its narrow edges; over the wider dimension it remains flat, although that dimension reduces a little because of shrinkage in a radial direction. The bastard-cut piece deforms to become diamond-shaped, rather than rectangular.

Slash-cut lumber used as floor material is very apt to wear by having whole annual rings pull away from the rest of the piece, leaving the floor with an irregular, often hazardous, surface. Rift-cut material, subject to the same severe use, merely develops small furrows where the exposed edges of the softer springwood erode more easily than the harder summerwood.

If rift-cut lumber is generally superior to slash cut, why not cut the log so that the majority of pieces are rift cut? The answer is that sawing to produce a higher percentage of rift-cut lumber would require turning the log more frequently and would be more expensive. Two common methods used in cutting logs are shown by Fig. 3-4. A third commonly used method is a combination of the two shown. For any of these methods, the majority of the pieces will be either slash cut or bastard cut.

One disadvantage of rift-cut lumber is that it can be severely weakened by knots. As shown by Fig. 3-5, a knot in slash-cut lumber might affect only a small part of the member width, whereas in rift-cut lumber it could affect the entire width.

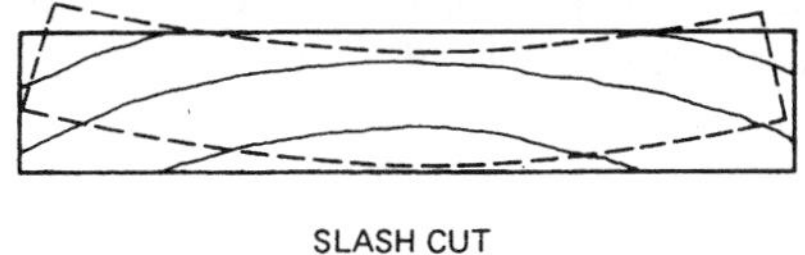

SLASH CUT

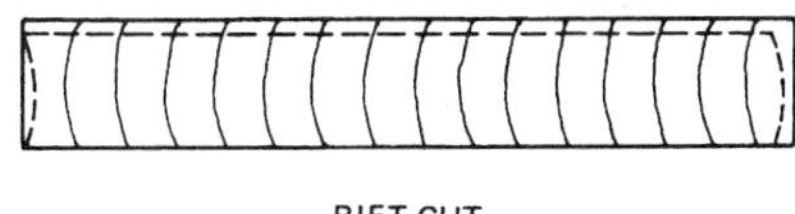

RIFT CUT

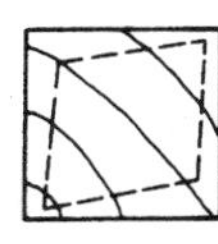

BASTARD CUT

Fig. 3-3. Shape change with loss of MC (exaggerated).

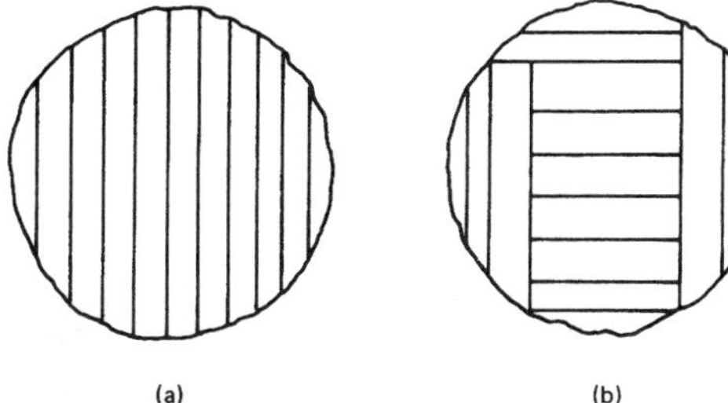

(a) (b)

Fig. 3-4. Two commonly used sawing patterns.

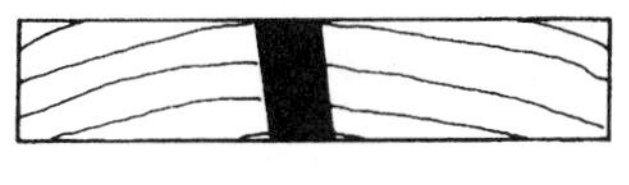

SLASH CUT

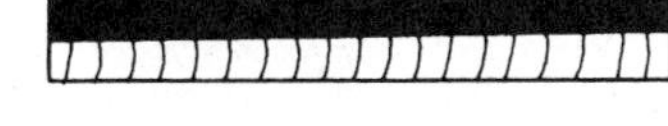

RIFT CUT

Fig. 3-5. Effect of knots on slash-cut and rift-cut lumber.

If the type of cut is not specified, the lumber received will be a mixture of all three cuts, but usually with a majority of slash cut. However, if the better properties of rift-cut lumber are required (as for floors subject to abrasion), the designer can specify that rift-cut lumber be provided, but may expect to pay a higher price than if rift cut were not specified.

3-5. DRYING

Lumber may be either air dried or kiln dried, the object being to reduce the MC (moisture content) to a level close to the equilibrium MC for the conditions under which the lumber will be used. For drying by either method, the lumber is placed in piles, with the layers of wood separated by narrow, intermittent, transverse pieces of wood called *stickers*. The stickers separate the moist pieces of lumber so that air can circulate through the pile, allowing for easy evaporation of adsorbed moisture.

Air drying is an economical method when carried out in well-designed drying yards or sheds, using properly arranged piles, and with favorable drying weather. Air drying may be slow natural seasoning, or it can be accelerated by using fans to circulate the air through the piles of lumber and by using a small amount of heat to raise the air temperature and reduce its relative humidity.

With kiln drying, higher temperatures are used and the rate of air circulation is much faster, and the wood may be dried to any desired MC. Another advantage of kiln drying is that the temperatures used are sufficient to kill any decay-causing organisms present in the wood. Still another advantage is that pitch in resinous woods is set by the heat. However, kiln drying is not without its hazards. The temperature in the kiln must be carefully controlled to avoid damaging the lumber; drying too rapidly can result in shrinkage-caused defects, such as checks, end splits, honeycombing, collapse, warping, and loosening of knots (2).

To What MC Should Lumber Be Dried?

Table 3-2 shows recommended levels of moisture for sawn lumber for various uses and various areas of the United States. Obviously, such a set of "lumped" recommendations cannot provide the optimum MC for each application. The designer should be alert to local conditions or uses that might require using some other MC level in wood for a particular project. Normally, as it leaves the mill, seasoned or dry lumber has a maximum of 19% MC. The average is about 15%.

When Is Lumber Surfaced?

Lumber may be surfaced while the wood is green (undried) or after drying (seasoning). It is sometimes important to know which type of material (surfaced dry or surfaced green) is being

Table 3-2. Recommended Moisture Content at Time of Installation.

	Recommended Percent MC					
	Dry Southwestern States[a]		Damp Southern Coastal States[b]		Remainder of Continental U.S.	
Lumber Use	Avg.	Range[c]	Avg.	Range[c]	Avg.	Range[c]
Interior finish woodwork and softwood flooring	6	4–9	11	8–13	8	5–10
Hardwood flooring	6	5–8	10	9–12	7	6–9
Siding, framing,[d] exterior trim, and sheathing	9	7–12	12	9–14	12	9–14

[a]Eastern California, Nevada, southeastern Oregon, southwestern Idaho, Utah, and western Arizona.
[b]Coastal California (south of San Francisco), coastal N. and S. Carolina, southern Georgia and Alabama, Florida, Louisiana, southern Arkansas, southeastern Texas (Laredo, San Antonio, Dallas).
[c]Range is more important than average for applications where shrinkage effects are critical.
[d]Framing lumber ordinarily not available manufactured to these recommended percentages.
Source: Courtesy American Society of Civil Engineers, R. J. Hoyle, "Introduction," *Wood Structures, A Design Guide and Commentary*, 1975.

used, since the allowable design stresses for the two types may differ (3). One can identify lumber of either type by the grading stamp applied to each piece.

Lumber that is surfaced after it is dried bears a stamp such as "S-DRY," "KD" (kiln dried), or "MC-15," which indicates seasoning to somewhere between 12% and 15% before surfacing. The stamp "S-GREEN" indicates that the wood was surfaced before seasoning.

Though seasoning or kiln drying of lumber up to 4-in. nominal thickness is practical, it is difficult and time-consuming to do so with larger material. Consequently, pieces of larger thickness (such as the Beam-and-Stringer and the Post-and-Timber size groups explained later) are usually furnished unseasoned. This means that such pieces are quite prone to splitting as they dry to their EMC, and it is a good reason for turning to glued laminated members (Chapter 8) rather than using large sawn timbers.

3-6. LUMBER GRADING

Grading is the process of classifying lumber according to *quality for a particular use.* This section will introduce the reader to the principles of lumber grading and to its value to the structural designer. (This section is *not* intended, however, to teach the reader to do lumber grading.) Grading is accomplished through rules that define both desirable characteristics and limits to undesirable characteristics of the lumber, and provide methods of quantifying the undesirable characteristics. Many of the rules governing lumber grading are developed by persons with structural-design expertise, but the actual task of implementing those rules is best left to professionals for whom lumber grading is the principal, full-time job.

The purpose of grading rules is to maintain a standard among lumber mills manufacturing from the same or similar species, so that the quality of the lumber for some particular use can be defined and controlled, regardless of the overall character of the logs from which the lumber is produced. Thus, lumber of a particular grade from different mills will have essentially the same strength and stiffness properties. The structural designer can specify lumber of a particular species group and grade, and can be reasonably assured that the lumber received will be suitable for the intended structural use, regardless of which mill produced it. The grade stamp placed on each piece of lumber will define the quality of material to the user.

History

Lumber grading has been done on some scale for many centuries. Carpentry skill was known in Biblical times. Then, even without formal rules for defining quality, the effect of certain defects was well recognized, and good craftsmen selected lumber carefully to obtain the desired qualities. The first known formalized grading system for selecting lumber was devised in 1754 by a Scandinavian named Swan Alverdson. In the United States, the first grading rules were used in Maine in 1833. As the focus of the lumber industry shifted to the central and western United States, grading rules were refined to adapt them to the new species encountered. By the end of the nineteenth century, grading rules were in use even for west coast species.

Grading rules were developed by various groups, representing manufacturers, distributors, and users of lumber. Naturally, the standards developed by one group often diverged from those of another. In 1924, through the American Lumber Standards Committee, voluntary product standards came into being (see reference 1). Today, even though various regional groups write their own grading rules, those rules fall within the framework of the voluntary standard. As a result, uniform size standards for lumber under 5 in. thick are observed by all U.S. lumber producers. Also, uniform grade names and grading provisions assure designers and users of reliable performance for lumber from most U.S. commercial species.

3-7. TYPES OF GRADING

Grading rules and methods differ according to the use to which the lumber will be put. Properties desired in wood for making furniture are not necessarily the same as those sought in wood for structural members or for interior wall paneling. Thus, different grading rules have been developed for various lumber uses. Even among those for a specific use, grading rules may differ according the size of the piece being graded and the particular grading agency. So grading is related to both size and use. This will be shown later by a partial outline of grading rules of the Western Wood Products Association (WWPA).

In the grading of structural lumber, various strength properties and stiffness are evaluated, the assigned grade name permitting the structural designer to specify a material having a desired level of strength. Two methods used for such grading are visual grading and mechanical grading. Today, visual grading is probably the more common. However, mechanical grading is finding increased use and may soon be the more common method for grading of structural lumber up to 2 in. thick. In visual grading of lumber for structural uses, the grader examines each piece to determine the type, location, and size of various defects that might affect its structural strength. Then according to rules that quantify the effect of each defect, the piece is assigned to a "grade." The grade name or symbol is stamped on the piece, giving the user a means of determining the probable strength and elastic modulus of the material. (Chapter 4 will show how the allowable stresses for design are related to the lumber grade.)

Mechanical grading is used to determine lumber properties by nondestructive testing. This grading method is based on the principle that all strength properties bear some relationship to the modulus of elasticity, E. Figure 3-6 shows, schematically, how a grading machine works. Mechanical grading is now done for only thinner material, of 1 in. to 4 in. nominal thickness. In mechanical grading, each piece is first subject to visual grading. Then it passes through the grading machine, which bends it to a predetermined curvature and measures the force required to deflect it. Deflection and force required are then used to determine the flexural modulus of elasticity, E. Allowable stress levels are determined from the E-value.

Depending on the rules under which it is graded, dimension lumber can be either Machine Stress Rated (MSR) lumber or Machine Evaluated Lumber (MEL). The latter applies to southern pine lumber only.

For either type—visual or machine—grading cannot be considered an exact science. In one case, it depends on visual capability and personal judgment. In the other, it depends on the accuracy of the relationships that we assume to exist between flexural modulus of elasticity and strength properties. Nevertheless, the grading process is sufficiently well defined that pieces graded should not fall more than 5% below grade (4).

3-8. DEFINITIONS

Grading rules involve many terms that the structural designer and other users of graded lumber must understand. Knowing the meaning of these terms may also be helpful to understanding subsequent chapters of this book.

Thickness—the smaller lateral dimension of the piece of lumber, shown in Fig. 3-7. (To call that dimension the "width" might *seem* logical when the member in question is a beam, but that is inconsistent with the terminology for grading. When speaking of the piece of lumber, call the smaller dimension the "thickness.")

Width—the larger lateral dimension, shown in Fig. 3-7. (When the piece is used as a beam, with width vertical, this dimension can also be referred to as "depth.")

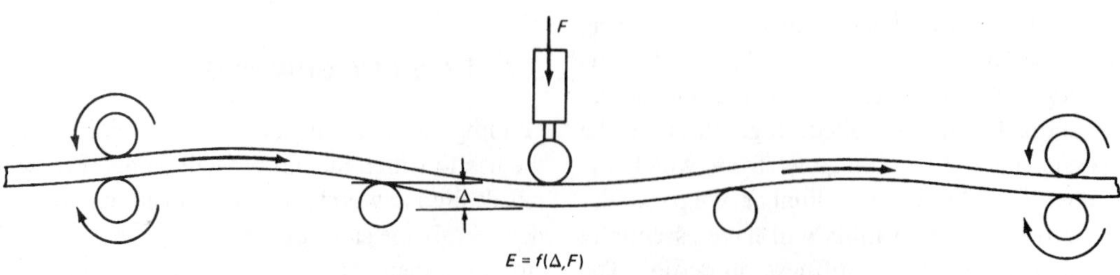

Fig. 3-6. Exaggerated schematic of lumber-grading machine.

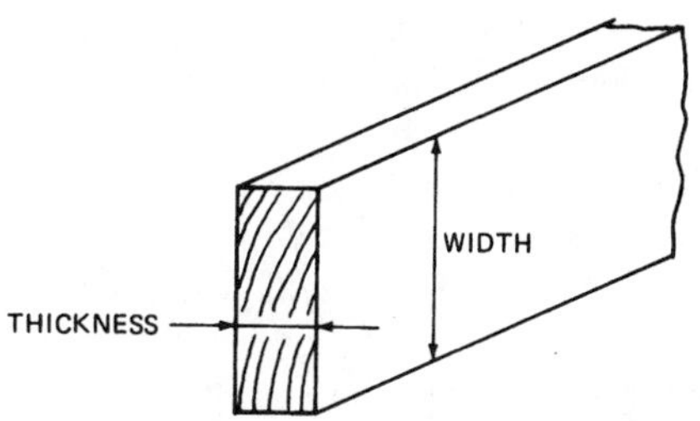

Fig. 3-7. Thickness and width of lumber.

Rough lumber—lumber that has been sawed but has not had its four longitudinal surfaces finished. Saw marks show on each surface.

Dressed lumber—lumber whose surfaces have been finished (by planing or other means) so that the saw marks on the surfaces are removed. Dressed lumber may be S1S, S2S, S1E, or S4S, according to which faces have been finished (surfaced).

Worked lumber—lumber that has been dressed and has also been matched, shiplapped, or patterned. Matched lumber has a groove on one longitudinal edge and a corresponding tongue on the other. The two pieces join together, side by side, as the projecting tongue of one piece is fitted into the groove of the other. It is sometimes called *tongued and grooved* or *dressed and matched.* Shiplapped lumber has a rectangular notch called a rabbet along each edge, the notch being on one face at one edge and on the other face at the other edge. Adjacent pieces overlap slightly as they are installed side by side. Patterned lumber fastens together side by side with something fancier than a plain tongue and groove or even plainer rectangular rabbets.

Dimension Lumber—Surfaced lumber of 2 in. through 4 in. nominal thickness.

Beams and Stringers—abbreviated B&S, pieces 5 in. or more in thickness, with width *more* than 2 in. greater than the thickness. (A 6 × 10 could qualify as B&S, but a 6 × 8 could not.) B&S lumber is graded assuming that its principal use will be in bending, not axial load, although it may actually be used for either purpose.

Posts and Timbers—abbreviated P&T, pieces measuring 5 in. × 5 in. or more, with the width *not* more than 2 in. greater than the thickness. (A 6 × 6 or a 6 × 8 would qualify as P&T, but a 6 × 10 would not.) P&T lumber is graded assuming that its principal use will be in axial load rather than bending, although it may actually be used for either purpose.

3-9. MODERN GRADING RULES

Associations that have produced grading rules for structural lumber include the following:

- Northeastern Lumber Manufacturers Association (NELMA)
- Northern Softwood Lumber Bureau (NSLB)
- Redwood Inspection Service (RIS)
- Southern Pine Inspection Bureau (SPIB)
- West Coast Lumber Inspection Bureau (WCLIB)
- Western Wood Products Association (WWPA)
- National Lumber Grades Authority (for Canada) (NLGA)

Through the voluntary Product Standard (1) mentioned earlier, the grading rules of the agencies listed above are sufficiently similar that a structural designer can be reasonably assured of obtaining the quality lumber desired.

As an example, consider the grading rules of one of the above agencies, the Western Wood Products Association (given in reference 4). The condensed list below shows only those lumber classifications, size/use groups, and grades that might be of interest to the structural designer. Following this list is a rather detailed illustration of the grading rules for just one of these size/use groups. It should be pointed out that some of the grades listed may be hard to obtain. For example, select structural lumber is difficult to obtain in large quantities.

Size/Use Group	Grades
Light Framing and Studs	
Light Framing	Construction
	Standard
	Utility
	Economy
Studs	Stud
	Economy Stud
Structural Light Framing	Select Structural
	No. 1
	No. 2
	No. 3
	Economy

Size/Use Group	Grades
Special Dimension	
Machine-Stress Rated	
Foundation Lumber	
Decking	Selected Commercial
Structural Laminations (see Chapter 8, on glulams) (1″ and 2″ nominal thickness)	
	Laminating One (L1) Laminating One-C (L1-C) Laminating Two (L2) Laminating Two-Dense (L2-D) Laminating Three (L3)
Joists and Planks	
Structural Joists and Planks	Select Structural No. 1 No. 2 No. 3 Economy
Timbers	
Beams and Stringers	Select Structural No. 1 No. 2 No. 3
Posts and Timbers	Select Structural No. 1 No. 2 No. 3

If we were to expand the above list to include factory lumber, boards, paneling, and other items that normally do not interest the structural designer, it would be several times as long.

3-10. EXAMPLE OF VISUAL GRADING OF BEAMS AND STRINGERS

For Douglas fir-larch, seven different grades of Beams and Stringers (B&S) are defined by the WWPA rules—Dense Select Structural, Select Structural, Dense No. 1, No. 1, Dense No. 2, No. 2, and No. 3, the first being the one of highest quality and highest allowable stresses. The NDS specifies allowable stresses for only the first six of these grades.

In WWPA rules for each grade defined for Beams and Stringers, there is first a general statement showing the purpose of the grade. This is followed by a list of characteristics and limiting provisions. For example, the list of characteristics and limits for the Select Structural grade is as follows (4):

Grain—medium.
Stained sapwood.
Firm heart stain—10% of width, or equivalent.
Splits equal in length to half the width of the piece or equivalent of end checks.
Seasoning checks—single or opposite each other with a sum total equal to one-fourth the thickness.
Pinholes limited.
Heavy torn grain.
Pitch streaks.
Slope of grain—1:15 in middle third of length, balance of piece may be 1:12.
Occasional skips—1/16 in. deep, 2 ft in length.
Medium pitch pockets.
Wane—one-eighth of any face, or equivalent slightly more for a short distance.
Shake—one-sixth the thickness on end.
Knots, sound, tight, and well-spaced, are limited as follows:

Width or Thickness (in.)	Narrow Face and Edge of Wide Face. Knots in Middle Third of Length (in.)	Knots at Ends and Along Centerline of Wide Faces (in.)
5	1 1/4	
6	1 1/2	
8	1 7/8	2
10	2	2 5/8
12	2 1/8	3 1/8
14	2 3/8	3 3/8
.	.	.
.	.	.
.	.	.

Knots sizes on narrow faces and at the edge of wide faces may increase proportionally from the size permitted in the middle one-third of the length to twice that size at the ends of the piece, except that the size of no knot shall exceed the size permitted at the center of the wide face.

The above excerpt from the grading rules may appear lengthy, but actually it is somewhat condensed. Obviously then, lumber grading is a job for an expert—one who is thoroughly familiar with the subject. It is a task the structural designer is usually happy to leave to that expert. Nevertheless, knowing something about the basis of grading is helpful to the designer. That knowledge may be useful in deciding borderline cases and may help to avoid construction and design errors.

3-11. GRADING STAMPS

Each piece of graded lumber receives a distinctive stamp telling the assigned grade, species (or species group), moisture condition at time of surfacing, identity of the mill that produced the lumber, and perhaps other information. By the grade stamp, the builder can be assured of getting the specified quality of lumber, and the owner or the building inspector can be assured that the material actually being furnished conforms to the construction specifications.

Figure 3-8 shows one example of a grade stamp for visually graded lumber. In this example stamp, the numeral 12 at the upper left identifies the mill that produced the lumber. (Sometimes the mill is identified by name rather than number.) The logo in the circle at the lower left certifies that the grading was done under WWPA rules and supervision. The species group shows in the square at the lower right. In the bottom center, the label "KD" tells that the wood was kiln dried (seasoned) when it was surfaced. The "1 & BTR" designation shows that the piece bearing this stamp is of Grade No. 1 & Better. It is not necessary for all parts of the grade stamp to be in the same position or form shown by this illustration. Also, the stamp may contain a value for allowable fiber stress. If it does so, the value shown is the "single-member" allowable stress for bending (explained in Chapter 4) about the major axis.

Figure 3-9 shows an example grade stamp for machine stress-rated lumber. This stamp, too, tells which mill produced the lumber and graded

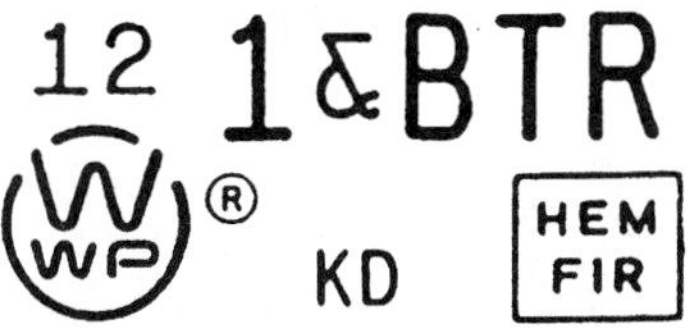

Fig. 3-8. Example of stamp for visually graded lumber. (Courtesy of Western Wood Products Association.)

Fig. 3-9. Example of stamp for machine stress-rated lumber. (Courtesy of Western Wood Products Association.)

it, under what agency rules and supervision it was graded, its moisture condition when it was surfaced, the species or species group, and finally, an allowable bending stress and modulus of elasticity. In this example, E is 1.5 million psi, and the allowable stress for major-axis bending is 1650 psi.

3-12. CAUTION TO DESIGNER AND BUILDER

The grade stamp on a piece of lumber applies only to the entire piece that was graded. A piece cut off from the original graded piece can be of either the same grade, a higher grade, or a lower grade than the stamp implies. Why? The reason is that in grading, the *location* of knots and other defects is considered.

For example, consider the Beam-and-Stringer rules that were outlined above. Assume that you have ordered a 20-ft length of Select Structural 6 × 12. For some reason, you now decide to have that length cut to produce two pieces 10 ft long. Do you now have two 10-ft pieces of Select Structural? Possibly so. But if the original member had a 1:12 grain slope in the end one-thirds of its 20-ft length, wood having that grain slope now appears in the central one-third of the new 10-ft lengths, a location where the limit allowed is 1:15. That limit for the central one-third of length is now exceeded; consequently, the grade of the shorter pieces is *lower* than the grade of the original 20-ft piece. (Further, it may be that only one of the two pieces has the necessary stamp to verify to the building inspector that it is graded lumber. The other 10-ft piece could be rejected as ungraded.)

As another example, assume that you have excess 2 × 12s on the job site, but you need 2 × 10s badly. Can you have the carpenters rip (saw) off a 2-in. strip from the full length of each 2 × 12 to produce 2 × 10s of equal grade? For some of the pieces you probably could not, since knots near the interior of the original width would now become edge knots. In bending members, the most damaging knots are edge knots, that is, knots near the top or bottom extreme fibers, where flexural stresses are the highest. The new 2 × 10s produced by this method would have to be regraded to be certain that they still met the requirements of the grade shown by the grade stamp.

The authors know of a case in which a large number of 2 × 12s were ripped to produce 2 × 10s. The ripped members were installed before the inspector could regrade them. The members having been installed, their upper surfaces were not visible and it could not be confirmed that no edge knots were present. Needless to say, the solution was very expensive for the builder.

As a designer, could you cause trouble for yourself because of the way Beams and Stringers are graded differently in the middle third of their length than in the end portions? Possibly, if you ordered long pieces of sawn lumber with the idea of using them for multiple-span beams. Continuous beams usually have the highest bending moments at their interior supports. In such cases it would be well to check the grading rules to be certain that you have the required quality at all locations along the length of the piece.

3-13. BOARD MEASURE

The quantity of wood in a timber of given length and cross section could be measured in cubic inches or cubic feet. However, the lumber industry in the United States has preferred to use the board foot. A board foot, abbreviated "fbm" for foot-board-measure, is the amount of wood present in a piece one foot long and having an end area of 12 square inches. For example, a 1 × 12 one foot long contains one board foot. So does a 2-ft length of 2 × 3 or a 3-ft length of 2 × 2.

The number of board feet of lumber in a given project is a convenient quantity to use in cost estimating, since lumber prices are often quoted in terms of dollars per Mfbm—cost per 1000 board feet.

Nominal sizes are used in determining the number of board feet. Thus, the fbm figure indicates not the amount of wood present in the finished product, but rather the amount of wood consumed in producing the finished product. Board measure is used routinely for sawn lumber. For glulams, the board measure is that of the pieces from which the glulam is made.

Example 3-1

What is the probable cost of a truckload of lumber that includes the items listed below? Assume that the average price for these items will be $500 per Mfbm. (Today's unit price can be estimated from reports in publications such as the *Engineering News Record.* These prices, however, are usually for carload lots of any one size, so there will be considerable price markup to obtain retail prices for smaller loads.)

170	2 × 4s, 8′ long
38	2 × 4s, 22′ long
85	2 × 10s, 20′ long
16	4 × 4s, 12′ long
85	2 × 12s, 20′ long
1	6 × 14, 22′ long

Consider each item separately. Compute the fraction of a board foot per foot of length, and multiply this by the total length per piece and by the number of pieces. For example, a 2 × 4 has 8 in.^{2} of nominal end area. Each foot of length, therefore, contains $^8/_{12}$ of a board foot. A piece 8 ft long contains 8 times $^8/_{12}$, and 170 such pieces contain 170 (8)($^8/_{12}$) board feet. Tabulating this computation for each item:

$$170\,(2 \times {}^4/_{12})\,(8) = 907 \text{ fbm}$$

$$38\,(2 \times {}^4/_{12})\,(22) = 557 \text{ fbm}$$

$$85\,(2 \times {}^{10}/_{12})\,(20) = 2833 \text{ fbm}$$

$$16\,(4 \times {}^4/_{12})\,(12) = 256 \text{ fbm}$$

$$85\,(2 \times {}^{12}/_{12})\,(20) = 3400 \text{ fbm}$$

$$1\,(6 \times {}^{14}/_{12})\,(22) = \underline{154 \text{ fbm}}$$

$$\text{Total} = 8107 \text{ fbm} = 8.107 \text{ Mfbm}$$

$$\text{Probable cost} = 8.107\,(\$500) = \$4053.50$$

REFERENCES

1. *American Softwood Lumber Standard, Product Standard PS 20-70,* National Institute of Standards and Technology, U.S. Department of Commerce, Washington, DC, 1986.
2. *The Encyclopedia of Wood, Wood as an Engineering Material,* Drake Publishers, New York, 1977.
3. *National Design Specification for Wood Construction,* rev. 1991 Ed., American Forest & Paper Association, Washington, DC, 1993.
4. *Western Lumber Grading Rules 95,* Western Wood Products Association, Portland, OR, 1995.

PROBLEM

3-1. How many fbm are used to produce each of the following?

(a) one 10×16, 24 ft long
(b) one 8×12, 16 ft long
(c) seventy 3×4s, 18 ft long

4

Loads and Design Values

This chapter is in two parts. Part I deals with loads—the forces to which timber structures may be subjected. It concerns the sources of these loads and shows how to compute their magnitudes.

Part II covers the methods the code writers use to establish allowable stresses and shows how the structural designer adjusts these allowables to design by the service load method, that is, by ASD (allowable stress design). A newer method—load and resistance factor design, or LRFD—is also introduced.

Part I. Loads

4-1. GENERAL

Loads and stresses are presented here using the current U.S. system, although at some future date the international system (SI) may be used routinely in the United States. Lumber sizes are given in inch units, with lengths in feet and inches. Appendix Table B-1 shows standard lumber sizes and cross-section properties.

Service loads are the maximum loads one may reasonably expect to occur. Under the service load design method, they are the loads a structure must be designed to resist. Except for loads due to the dead weight of the structure, service loads are usually spelled out by a local building code and tend to be somewhat higher than the *normally expected* loads. For example, the design service load for a classroom is often specified as high as 50 pounds per square foot (psf). However, in a normal classroom situation, with desks occupying about 10 ft^2 each, the weight of the movable desk and the person occupying it actually averages much lower than 50 psf, perhaps as low as 25 psf. The full specified service load would not occur normally, but it might be approached under abnormal conditions. For example, the load intensity near the classroom exit during a fire drill could become even higher than the specified design service load.

Thus the specified service load is conservative for normally expected situations, but does not necessarily cover the extreme case. The margin of safety provided in the allowable stress used for design is usually adequate to prevent catastrophic collapse for overload conditions such as the one illustrated above. The designer should not be too complacent, however; serious failures, often well publicized, do occasionally occur.

Under service load design (also known as either working-stress or allowable-stress design) safety is provided by using an allowable stress that is low enough to protect against (1) variation in material properties, (2) errors in design theory, and (3) uncertainties as to the exact load. The *National Design Specification* (1) is based on the working-stress design method.

The ratio of the load that would cause failure to the load for which the structure is designed is called the factor of safety. Because stress redistribution and inelastic action occur as failure is approached, the factor of safety is *not* the same as the ratio of ultimate stress to allowable stress.

Load Types

Loads are classified according to their source and duration. Types to be considered in timber design are dead load, live load, snow load, ponding load, wind load, seismic load, and occasionally impact. For timber bridges, water pressure and ice loads may also be important.

Design dead loads are computed by the structural designer. Values of other loads, however, are usually specified by various codes. Of these, the *Uniform Building Code* (UBC) (2) is probably the one most widely used in the United States. Local building codes are often patterned after the UBC, differing only with respect to unusual local conditions or more conservative local preferences.

4-2. DEAD LOADS

Of all the loads a designer must consider, dead load is the one that can be determined most precisely, and it generally remains constant for the life of the structure. Dead load is the force due to gravity acting on the mass of the structure itself. Ordinarily, it is not a specified load, but one that must be determined by the designer. For the design of any particular member, it includes the self weight of that member, plus all dead-load forces brought to it by other members or components that it supports. In computing the amount of dead load, a designer should remember to include the weights of items such as ceilings, floor covering, ventilating equipment, light fixtures, fixed partitions, wall panels, and windows. The weights of some of these may have to be estimated, but others may be computed precisely. Also, for office buildings and other buildings in which interior partitions may be added or relocated, the location and amount of dead load due to partition weight is uncertain. To provide for this, UBC requires the designer to consider a uniform dead load of 20 psf in addition to other known floor dead loads. Tables of building material weights are helpful. One such table is Table A-1 in Appendix A.

Our concern is designing wood members, so it will be necessary to estimate the weight per linear foot for various sizes of wood member. One rather precise way of finding unit weight is given by an equation shown by the Supplement to the *National Design Specification,* as follows:

$$\text{Unit wt (lb/ft}^3) = 62.4\,(1 + MC/100)\,[G/(1 + G\,(0.009)\,(MC))] \qquad (4\text{-}1)$$

In the above, MC is the moisture content of the wood in percent, and G is the specific gravity of that species. Specific gravity (of oven-dry wood) for various species can be found in Appendix Table C-1. Wood in the structure will ordinarily contain 11–19% moisture, so the computed weight per cubic foot needs to be adjusted upward for moisture content.

Equation 4-1 corrects for volume change due to change in moisture content. However, the precision implied by that method is seldom justified for the design of structural members, since self-weight of a member is usually only a small percentage of the total load the member carries. With experience, a designer can quickly estimate the member's weight per unit length. In several of the examples in this book, you will see the unit weight of Douglas fir taken as somewhere between 35 and 40 lb per cu ft. This figure comes from the experience of the designer who knows that this is a reasonable estimate.

Another table that may be helpful in estimating self-weight of wood members is Appendix Table B-1. It lists the weights per lineal foot for standard lumber sizes and for various unit weights of wood.

A simpler (and always conservative) method of computing the unit weight of wood is to ignore the volume change, in which case the equation becomes

$$\text{Unit weight (lb/ft}^3) = 62.4\,(G)\,(1 + MC/100) \qquad (4\text{-}2)$$

At the start of design for a particular member, the member's final size is not known, so self-weight can only be estimated. If this estimate is sufficiently wrong as to affect the design (i.e., correcting it will change the choice of member size), the design procedure should be repeated using a corrected estimate. In any case, a faulty estimate of member weight should be corrected before going on to the design of other members whose loads may depend on this estimate.

Example 4-1

What is a reasonable unit weight to use in designing a member of redwood (close grain)?

From Appendix Table C-1, average specific gravity = 0.44 (oven dry).

Average unit weight, oven dry $= 62.4 \times 0.44$
$= 27.5$ pcf.

Unit weight, corrected for 19% moisture
$= 1.19 \times 27.5 = 32.7$ pcf.

Table C-1 does not mention for what *grade* of lumber the value of specific gravity is given. Stronger grades may weigh more than weaker ones. So the designer must still use judgment in rounding off the 32.7 figure.

Example 4-2

Figure 4-1 shows the plan and a cross section of a building floor. Compute the dead load carried by beam A.

This is a second-floor member, so it must support the weights of ceiling, light fixtures, and ventilation ducts for the story below. Assume that D. fir-larch will be used for the joists and beams. Finish floor will be 7/8 in. red oak. (This same building will be used several times in this chapter for computing various types of load.)

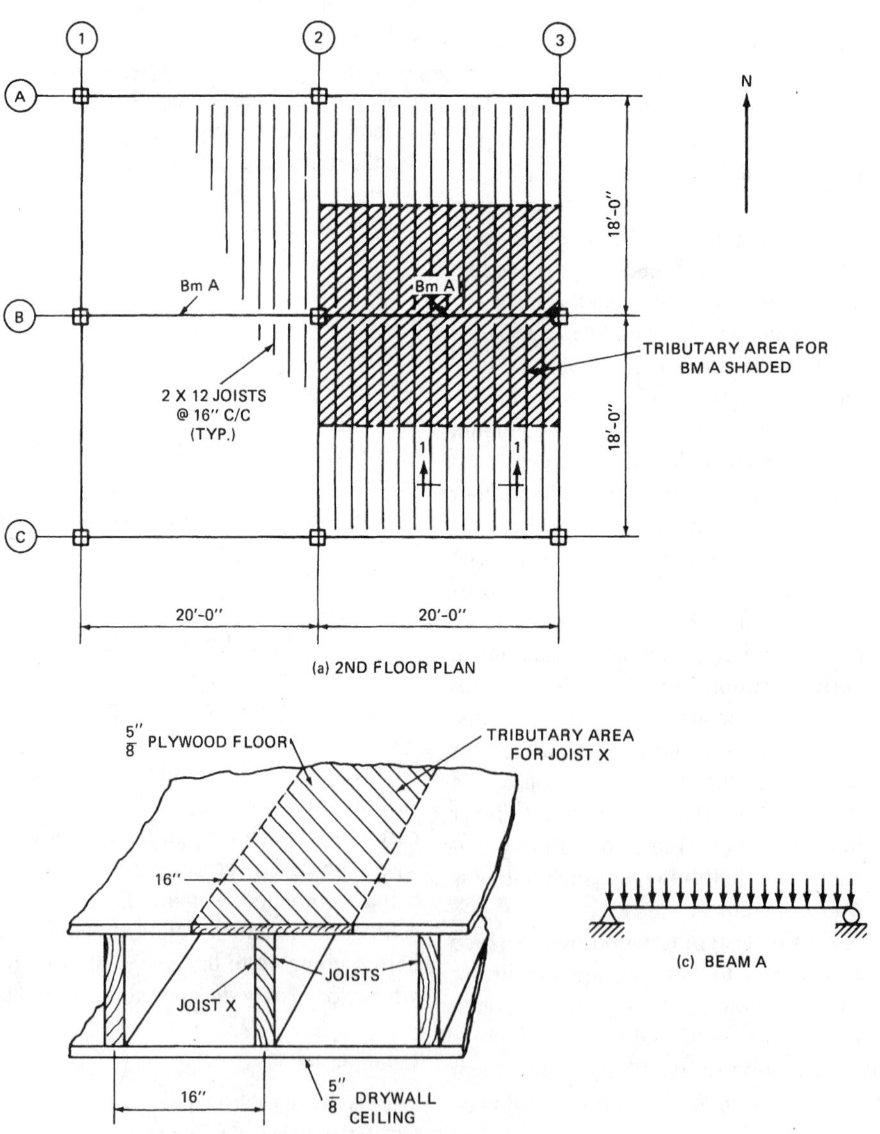

Fig. 4-1. Floor for Examples 4-2, 4-3, and 4-6.

The floor plywood acts as a continuous beam, spanning several 16-in. distances between floor joists and carrying the floor live and dead loads, including the weight of the plywood itself. The joists support this continuous beam, and reactions for the continuous beam of floor plywood become loads on the joists. Next, each joist acts as a beam, carrying all load it receives from the floor plywood, plus the weight of the joist itself.

The joists in turn are supported by the beams, and the joist reactions become loads for the beams, such as beam A. Beam A carries the loads (joist reactions) transferred to it by all joists it supports, plus the weight of beam A itself.

Each joist supports load from one-half the floor plywood span on one side of the joist plus that from one-half the plywood span on the other side (shaded in Fig. 4-1b). This is equivalent to saying that each joist supports the load from a *tributary area* extending from the center of the span on one side of the joist to the center of the space on the other side. Per foot of joist length, this tributary area is

$$1(16/12) = 1.33 \text{ ft}^2$$

The floor dead load per square foot consists of the weight of the plywood plus the weight of the finish flooring material. Assuming the plywood to weigh 40 pcf, the weight of 5/8-in.-thick plywood is approximately

40(0.625/12)	= 2.1 psf
7/8-in. finish floor (SG of red oak = 0.67)	
0.67 (62.4) (0.875/12) (1.19)	= 3.6
Total	= 5.7 psf

The total dead load per square foot of tributary area includes the weight from below—ceiling, light fixtures, and ventilation ducts.

5/8-in. gypsum board (Table A-1)	= 2.5 psf
Lighting and ducts, estimated	= 4
Total approx.	= 6.5 psf

The weight of the joist itself must be estimated. Using the approximate method, D. fir-larch with 19% moisture weighs about 0.50 (62.4) (1.19) = 37 pcf, and the joist weight is

$$37\,(1.5)\,(11.25)/144 = 4.34 \text{ lb/ft}$$

The total dead load per linear foot of joist is

$$4.34 + 1.33\,(5.7 + 6.5) = 20.6 \text{ lb/ft}$$

(Most designers would round this to 21 lb/ft. Had the more accurate method been used, a slightly lower weight would have resulted.)

Since the joists span 18 feet, each joist has a reaction of 9 × 20.6, or 185 lb. Beam A supports joists from each side, so it receives a concentrated dead load of 2 × 185, or 370 lb at each joist location.

Beam A is not yet designed, so we must estimate its weight. Let's assume that the beam will be a 4 × 14. If so, and if it, too, is D. fir-larch, it will weigh about 12 lb/ft. The loads on Beam A are its own weight plus a series of 370-lb concentrated loads as shown by Fig. 4-1c. When solving bending moment due to such a series of closely spaced loads, it is sufficiently accurate to replace them by a uniform load. Thus, the series of 370-lb loads could be replaced by a uniform load of (12/16) × 370, or 278 lb/ft. Adding the beam weight, the total design dead load for beam A is 278 + 12 = 290 lb/ft.

The solution above went into considerable detail so that beginning designers would not lose sight of how load is transferred from one member to the next: from floor to joists, and from joists to main beams. However, the end purpose is to determine a uniform design dead load for beam A, so a simpler approach, demonstrated next, can be used.

Example 4-3

Repeat the above example, but by a simpler and more direct method.

The total tributary area bringing floor dead load to beam A extends from the center of the joist span on one side of beam A to the center of the joist span on the other side. The entire tributary area measures 20 ft by 18 ft. Each square foot of floor area has the following average dead load:

Floor plywood	= 2.1 psf
Finish floor	= 3.6
Ceiling below (including ducts and lighting)	= 6.5
Floor joists (12/16) × 4.34	= 3.3
Total	= 15.5 psf

Total dead load on beam A = (18 × 20 × 15.5) + beam weight. Estimate the beam weight to be 12 lb/ft. Total dead load per foot of beam is

$$w = (18 \times 20 \times 15.5/20) + 12 = 291 \text{ lb/ft}$$

except for round-off differences, the same as found by the lengthy procedure.

4-3. VERTICAL LIVE LOADS

For buildings, vertical live loads include the weight of anything supported by, but not a permanent part of, the structure. Thus, carpeting, furnishings, movable equipment, personnel, and stored materials would all be called *live load.* For bridges, live load is the weight of the traffic the bridge carries—vehicles for a highway or railway bridge, or people for pedestrian bridges.

The weight of snow meets the above definition, but in the United States snow load is usually considered separately from other live loads. This is a reasonable practice, and a convenient one to use in designing timber structures, since allowable stresses for wood vary according to the load duration.

Roof Loads

Roof Live Loads. Even in areas where snow is not expected, some live load must be considered in designing members that support roofs. This live load is based on the realization that the weights of construction supplies and personnel must be carried by the roof members. Appendix Table A-3 shows minimum roof live loads specified by the *Uniform Building Code* (UBC). Note that the improbability of full live load occurring simultaneously over large areas is considered by specifying smaller load intensities for larger tributary areas. ("Method 2" shown by the three columns at the right refers to a reduction method that will be shown here for floor live loads, but which applies to roof live loads also.) Where snow load leads to stronger members or connections, consider snow load instead of roof live load. Use whichever is worse, not both.

The designer should consider *unbalanced* live load wherever it causes a more severe condition than full loading for either members or connections. Examples would be alternate-span loadings for continuous members or loading of one slope only for a pitched roof.

Snow Loads. Design snow load intensity for various localities is usually specified by local building codes for those areas—the *Denver Building Code,* for example. Occasionally, though, terrain and climatic conditions are such that a code-specified snow load is unrealistically low. One such case reaching the authors' attention was for a structure in the Colorado mountains. The specified snow load was 60 psf, but weather records for that particular spot showed 135 psf to be a more reasonable design load. The designer should be alert to such unusual conditions.

Snow load is specified in pounds per square foot *of horizontal projection* rather than pounds per square foot of roof. Also, steeply sloping roofs can hold a lesser depth of snow than can flat roofs or those with a mild slope. Where the specified snow load exceeds 20 psf, UBC allows the following reduction of snow load intensity for each degree of roof slope over 20°.

$$R_s = (S/40) - 1/2 \qquad (4\text{-}3)$$

In this equation, R_s is the snow load reduction in pounds per square foot per degree of slope above 20°, and S is the specified design snow load in pounds per square foot of horizontal projection for a flat roof.

Snow load is normally considered to be uniform over the entire span of members supporting a roof. However, *unbalanced* snow load due to drifting must also be considered. Failures have occurred because of considering only balanced loading. The UBC requires designers to consider the "potential unbalanced accumulation of snow at valleys, parapets, roof structures, and offsets in roofs of uneven configuration." (The main body of UBC does not suggest how the designer might do this, but a UBC appendix gives details on recommended drift depths for various types of structure irregularity.)

Other Roof Loads. In addition to snow, live loads that may affect roof support members include rain, ice, building materials or equipment stored on the roof during construction or repair, and the weight of personnel working on the roof. The last two loads *may* be adequately covered by the minimum specified roof live loads (Table A-3), but the first two are not.

Loads from rain are important in the case of flat, or nearly flat, roofs whose supporting members have fairly long spans and are fairly flexible. *Ponding* on these roofs may cause complete collapse. In ponding, the weight of rainwater causes the member to deflect slightly, forming a small depression in the roof surface. Water filling the depression causes further deflection, allowing more water to accumulate, and so on. Under certain combinations of load and stiffness, the process stops, a point of stability being reached. However, with roof systems of inadequate stiffness, the effect is that of a diverging series. The structure becomes unstable, filling and deflecting continuing until the system collapses. Ponding failure can occur in roof systems of any structural material, particularly in regions where roofs are not designed to carry snow loads. Fortunately, proper attention by the designer can prevent ponding failures. (See Chapter 6.)

Rain (or snow melt water) may also cause significant load when roof drains become blocked, especially if the roofs have parapet walls around their perimeter. If the drains cannot remove water from the roof, the water may accumulate, applying even greater load than the snow load for which the structure was designed. In one case known to the authors, several wood trusses supporting a nearly flat roof had failed. Investigators determined that a softdrink can found on the roof had blocked the drain.

Floor Loads

Building floor live loads are specified by building codes and vary according to building use. Table A-2 (Appendix) shows the UBC minimum floor live loads. A distributed-load intensity is shown for all building types; for some a single concentrated load (located to cause worse effect) also must be considered. As pointed out earlier, the specified design load (service load) is usually conservative. However, the designer should be alert to unusual circumstances in which the actual load might be larger than the specified design load.

Full specified live load is not normally present. It is even less frequent that *large* floor areas receive full design live load simultaneously on each unit of area. Building codes consider this improbability by allowing the designer to use a reduced value of live load in designing members that receive live load from large areas. For example, UBC allows reducing the unit floor live load by 0.08% for each square foot of area in excess of 150 supported by the member, but with limits, as follows:

1. The total reduction may not exceed 40% for members receiving load from one level only, nor 60% for other members.
2. The total reduction may not exceed R (percent), where

$$R = 23.1(1 + D/L) \qquad (4\text{-}4)$$

   (In this equation, D and L are the dead and live loads per unit area, respectively.)
3. Reductions are not allowed for places of public assembly or for areas having design live loads larger than 100 psf.

(Roof live loads other snow may be reduced in the same manner as floor live loads, except the percentage reduction and maximum reduction are given by Method 2 in Table A-3).

Example 4-4

Figure 4-2a shows a building cross section. The governing code specifies snow load of 35 psf of horizontal projection and allows the UBC reduction for roofs sloping more than 20°. In designing the roof beams, what load per lineal foot should be used? (Consider both snow and dead loads.)

The roof slope is 26.6°, so design snow load may be reduced by

$$R_s = (35/40) - 1/2 = 0.375 \text{ psf}$$

for each degree over 20°. Thus, for design, we may use snow load equal to 35 − 0.375 (26.6 − 20) = 32.5 psf *of horizontal projection.*

Figure 4-2b shows a short section of roof decking and a mass of snow resting on it. The thickness (normal to the sketch) is one foot. The volume of snow shown weighs 32.5 lb, and the sloping length of roof on which it acts is 1.0/cos 26.6 = 1.12 ft.

Using the simpler method of Eq. 4-2 and assuming

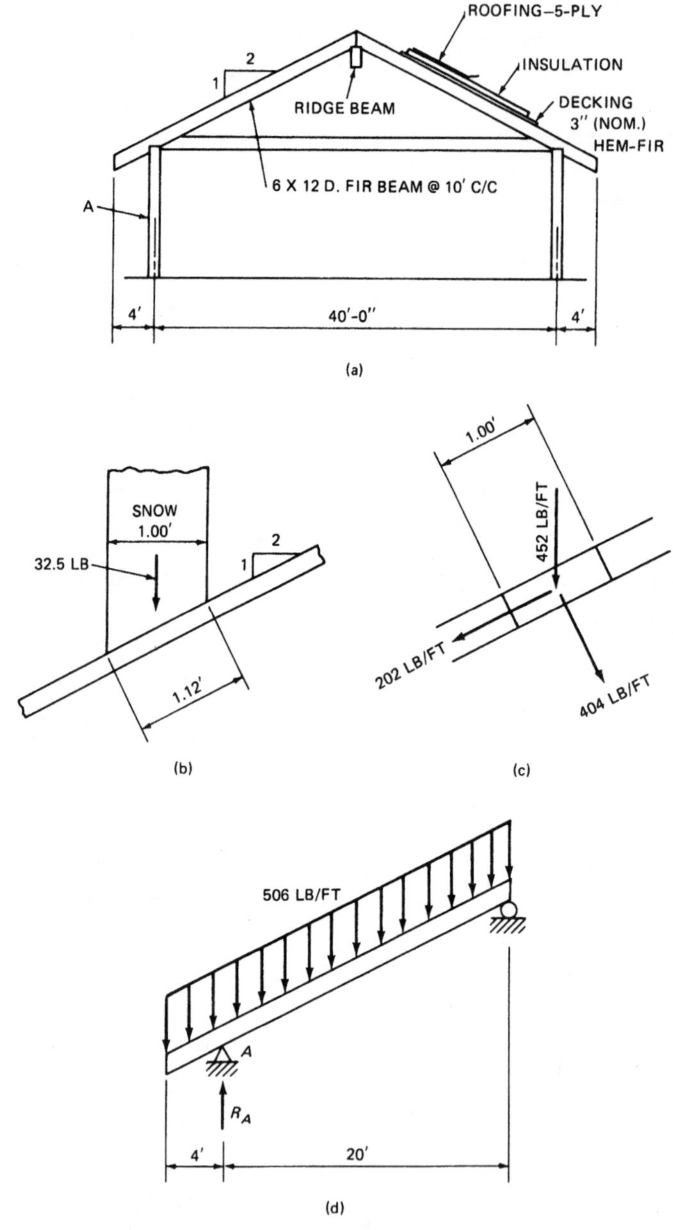

Fig. 4-2. Roof for Examples 4-4 and 4-5.

17% moisture, the unit weight for hem-fir is 0.43 (62.4) (1.17) = 31 pcf

Roof dead loads *per square foot of sloping roof surface* are

Roofing (Table A-1)	= 6.0 psf
Decking (2.5/12) × 31	= 6.5
Insulation (est.)	= 2.0
Total	= 14.5 psf of roof

Snow loads are per square foot of horizontal projection, but dead loads are per square foot of roof area. One or the other of these loads must now be resolved to permit adding them together. In this case, let's express both in pounds per square foot of sloping roof area.

The 32.5-lb snow load acts vertically on a sloping roof area measuring 1.12 ft by 1.0 ft. The snow load per square foot of sloping roof is 32.5/1.12 = 29.0 lb, and the total vertical load per square foot of sloping roof area is 29.0 + 14.5 = 43.5 lb.

The supporting beams (rafters) are at 10-ft centers, so each one-foot length of beam receives 435 lb of load from the roof plus the self-weight per foot of beam. Assuming the beam to weigh 17 lb/ft, the total vertical load on a beam is 452 lb per lineal foot. For computing the beam bending moment, it is sometimes easier if we resolve this vertical load into two components, one transverse to the beam and one parallel, as shown by Fig. 4-2c. The transverse load (404 lb/ft of beam) can be used to determine the beam shear and bending moment. Designers should not forget that the other component, 202 lb/ft, causes axial force in the beam.

Example 4-5

What is the vertical load applied to the column (vertical member) at point *A* of Fig. 4-2a, assuming that the column spacing is 10 ft?

We have already determined the vertical load per square foot of roof and, for snow, per square foot of horizontal projection. Each rafter is supported at one end by the ridge beam and at the other end by a column. From Example 4-4, the total vertical load on the rafter is 452 lb per sloping foot, which amounts to 1.12(452) = 506 lb per horizontal foot. Figure 4-2d shows the rafter and its vertical load. The reaction at point *A* is the load in the column:

$$R_A = 24(506)(12/20) = 7290 \text{ lb}$$

For the design of the column, the self-weight of the column and the weight of any wall material supported by the column would have to be added to this reaction.

Example 4-6

Consider again the floor plan of Fig. 4-1. The specified floor live load is 50 psf, as given by Appendix Table A-2 for offices. What is the design live load per foot of length for beam A?

The tributary area for beam A is 20 ft by 18 ft = 360 ft^2 (one area 20 ft by 9 ft north of beam A and a similar area to the south). The tributary area exceeds 150 ft^2, and the specified live load does not exceed 100 psf. Therefore, the design live load may be reduced. A reduction of 0.08(360 − 150), or 16.8% is allowed based on the area supported by the member. From Example 4-3, we know that the dead load is 15.5 psf, ignoring the weight of beam A. Equation 4-4 gives a limit of

$$R = 23.1(1 + 15.5/50) = 30.3\%$$

The reduction based on tributary area, 16.8%, exceeds neither *R* nor the 40% ceiling for members receiving their load from one level, so the 16.8% reduction controls. The design live load for beam A is

$$50 \text{ psf}(1 - 0.168)(18 \text{ ft}) = 749 \text{ lb/ft}$$

and the total load for design is 749 (live) + 291 (dead), or 1040 lb/ft.

Example 4-7

For the same structure, assume that two such floors (but no roof) deliver live load to column B-2 (at the intersection of grid lines B and 2). What is the design live load for column B-2?

The tributary area per floor for column B-2 is 360 ft^2. For the two supported floors it is 720 ft^2. The allowable live-load reduction for designing the column is 0.08(720 − 150), or 45.6%. The limits are 60% (since the column receives load from more than one level), and *R* = 30.3% by Eq. 4-4. The latter value controls, and the design live load for the interior column is

$$50 \text{ psf}(1 - 0.303)(720 \text{ ft}^2) = 25{,}090 \text{ lb}$$

Example 4-8

What is the total live load for design of column B-2 if it also supports a nearly flat roof and the specified snow load is 30 psf?

No reduction for area is allowed for snow load (although it is allowed for other types of roof live load), so the total design live load per interior column is

$$25{,}090 + (360 \times 30) = 35{,}890 \text{ lb}$$

4-4. WIND LOADS

Wind forces are normal to the surface against which the wind acts. Thus, building walls receive horizontal forces, while horizontal roofs are subject to vertical forces from wind. On sloping roofs the wind forces are normal to the roof and have both horizontal and vertical components. In each case, the force may be either toward the surface (pressure) or away from it (negative pressure, or suction). The importance of suction is shown by failures where roofs are lifted from buildings rather than being blown down into the buildings. The chance of such failure is increased where wall openings allow the wind to increase air pressure inside the building, the internal force from this pressure adding to the outward-directed force from outside suction.

Wind forces for structural design are usually specified by local building codes. Some codes specify the intensity of horizontal wind pressure to be applied to a vertical surface. These codes also include rules for determining pressure or suction intensities on sloping roof surfaces or on the underside of open, overhanging surfaces. Most codes, however, have changed to the method shown by the *Uniform Building Code* (2). UBC's wind force requirements are based on methods first shown by the American National Standards Institute (ANSI) (3). The main points of the UBC method are summarized below.

Figure 4-3 is UBC's map giving basic wind speeds for the United States. The map shows *minimum* wind speeds for design; where local records show higher 50-year wind speeds at a standard distance above the ground, the higher speeds should be used.

Exposure affects the wind pressure a structure will receive. Defined exposure C occurs where the terrain is flat and generally open for one-half mile or more from the site in any full quadrant. Exposure B occurs where at least 20% of the terrain is covered by buildings, forest, or surface irregularities for a distance of one mile or more from the site. Exposure D occurs where the basic wind speed is 80 mph or more, and the terrain is flat and unobstructed, facing large bodies of water one mile or more in width relative to any quadrant of the building. This exposure extends inland for one-fourth mile or ten times the building height, whichever is greater.

The defined exposure and height above the ground are used with Appendix Table A-4 to determine a combined height, exposure and gust factor, C_e. Another UBC table, shown in the Appendix as Table A-5, gives pressure coefficients, C_q, for various parts of the structure. These C_q coefficients allow for pressure computation by either of two methods defined by UBC—the normal-force method or the projected-area

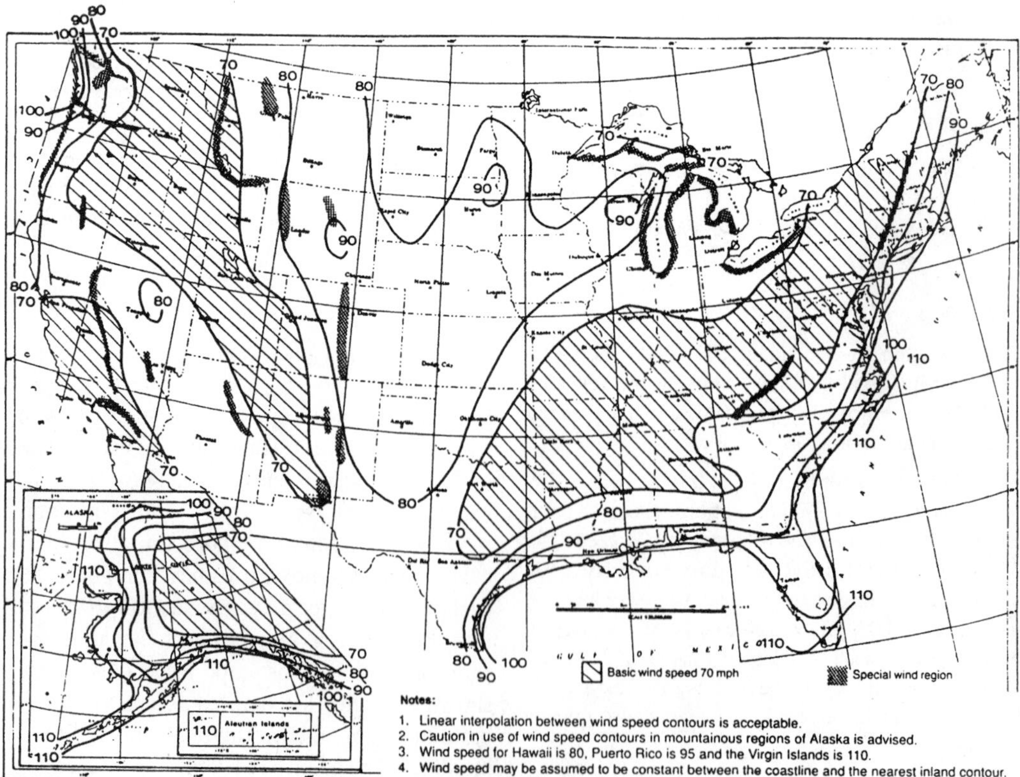

Fig. 4-3. Minimum basic wind speeds in miles per hour (× 1.61 for km/h). (Reproduced from the 1994 edition of the *Uniform Building Code*™, copyright© 1994, with the permission of the publisher, the International Conference of Building Officials.)

method. Next a wind stagnation pressure, q_s, is found from Appendix Table A-6.

Finally, an importance factor, I, is selected. For wind loads, $I = 1.0$ for most structures. However, for essential facilities that (following a windstorm) must be safe and usable for emergency purposes to preserve the health and safety of the public, and for structures supporting or housing hazardous materials, the importance factor for computing wind forces is $I = 1.15$. The above factors are then entered into the following equation to determine the design pressure:

$$p = (C_e)(C_q)(q_s) I \qquad (4\text{-}5)$$

Some codes permit (or even require) design wind loads for certain types of structure to be determined experimentally by wind-tunnel tests.

For structural design, all directions of horizontal wind must be considered. For buildings whose plan is symmetrical about two perpendicular axes, the worst wind direction is parallel to one or the other of these axes. For buildings of square plan, however, a 45° diagonal wind may be a worse condition. For buildings of irregular shape, diagonal winds may control. Also, since wind may cause such buildings to twist, the designer may have to consider the effects of torsion.

4-5. STABILITY UNDER WIND LOADS

The entire structure must be connected to its foundation so that it will be stable under wind load; it must neither slide nor overturn. Overturning moment of the wind forces on the entire structure must not exceed two-thirds of the resisting moment from dead load alone (i.e., the factor of safety against overturning must be at least 1.5). Also, each element (part) of a structure must be strong enough to withstand the wind force and must be adequately connected to the rest of the structure.

The UBC allows using (with some restrictions) either of two methods for determining the applied wind forces. Using these applied loads, the designer can then analyze for stability or for wind-induced stresses.

Method 1, known as the Normal-Force Method, is a theoretically correct method with no simplifying approximations. It may be used for any structure and is the only method allowed for stress analysis of gabled rigid frames. In Method 1, wind forces normal to the surfaces are applied simultaneously to all external surfaces. Figure 4-4a shows the wind pressures and forces for a typical building. These consist of positive pressures on the windward side and negative pressures (suctions) on the leeward side. The lee-side suction for both wall and roof is computed using C_e at the mean height of the roof. Stability against overturning is computed by taking moments about point B for all of the wind forces and for the weight of the building.

Method 2, the Projected-Area Method, is allowed for any building under 200 ft high, but *not* for buildings with gabled rigid frames. Method 2 approximates the true wind forces by using horizontal pressures on the vertical projected area of the entire structure and, simultaneously, vertical pressures on the entire horizontal projected area of the building. Notice that the suc-

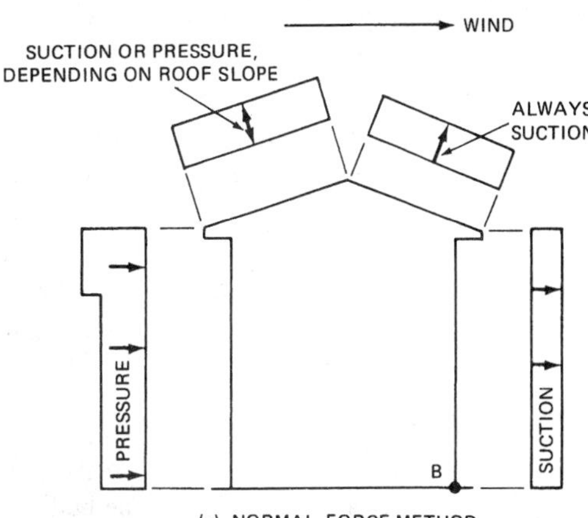

(a) NORMAL-FORCE METHOD

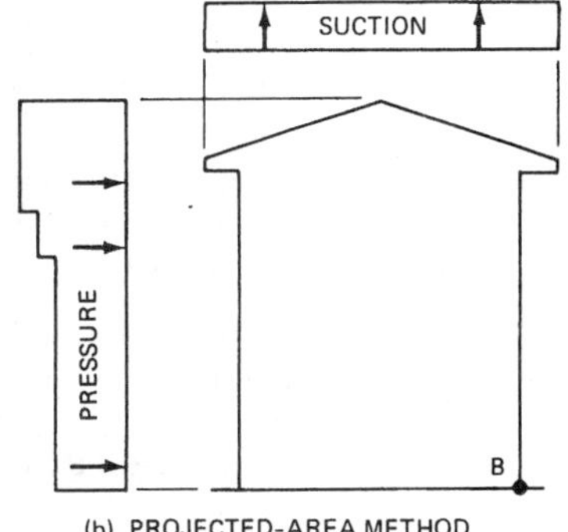

(b) PROJECTED-AREA METHOD

Fig. 4-4. Wind loading methods.

tion and positive pressure are replaced by positive pressure applied to the windward wall only. Appendix Table A-5 provides different pressure coefficients for use with the two methods.

4-6. LOAD COMBINATIONS

What load combinations should the designer consider? For building design by the allowable stress method (ASD), UBC requires the following, in addition to dead load alone:

1. Dead load and floor live load, plus either roof live load or snow.
2. Dead load and floor live load, plus either wind or earthquake.
3. Dead load and floor live load, plus wind, plus one-half snow load.
4. Dead load and floor live load, plus snow, plus one-half wind load.
5. Dead load and floor live load, plus snow, plus earthquake. (With approval by local building officials, snow loads in this combination may be ignored if not over 30 psf and may be reduced if over 30 psf.)

Dead load is present in every combination above, and designers often overestimate dead load. This is a safe practice only for combinations in which the dead-load effect is of the same sign as for other loads. If the effect of a load combination is reduced by dead load, then a conservatively *low* amount of dead load must be used.

Example 4-9

Compute wind forces and check stability of the condominium building shown in Fig. 4-5. The building has two stories, is 60 ft long, and is built without roof overhangs at its ends. The roof is supported by light-frame trusses. Designers estimate the weights of various parts of the building to be as shown in parentheses in Fig. 4-5. Use UBC wind requirements, assuming that the location is an open area where the basic wind speed is 80 miles per hour. Use the Normal-Force Method (UBC's Method 1).

Appendix Table A-4 shows (for Exposure C) the combined height, exposure and gust factors as:

1.06 for heights 0 to 15 ft above the ground

1.13 for heights 15 to 20 ft

1.19 for heights 20 to 25 ft

For the leeward wall and for the leeward roof, C_e is computed using the mean height of *roof,* or 24.29 ft.

From Appendix Table A-6, the stagnation pressure q_s is 16.4 psf. The importance factor is 1.0. Table A-5 gives pressure coefficients, C_q, for various parts of the structure. The above values are used in Eq. 4-5, as tabulated below:

Structure Part	Factor C_e	Coefficient C_q	Pressure (psf) = $(C_e)(C_q)(q_s)(I)$
Windward wall			
Lower 15 ft	1.06	0.8 inward	13.91 pressure
15 to 20 ft	1.13	0.8 inward	14.83 pressure
Above 20 ft	1.19	0.8 inward	15.61 pressure
Leeward wall	1.19	0.5 outward	9.76 suction
(Roof slope is 5 in. vertical to 12 in. horizontal)			
Windward roof	1.19	0.9 outward or 0.3 inward	17.56 suction or 5.85 pressure
Leeward roof	1.19	0.7 outward	13.66 suction

Which of the two alternatives, above, should be used for the windward roof? Regardless of which is used, the resultant force on the windward roof, if extended downward, is seen to pass to the *left* of point *B*. Therefore, to determine stability against overturning, we would use the *suction* alternate, since it increases the overturning (clockwise) moment about *B*. To determine the factor of safety against sliding, however, we would use the *pressure* alternative since it increases the total force acting toward the right.

Amounts of all the wind forces, their distances to point *B*, and their moments about point *B* are computed and entered into the tabulation below.

Force		Horiz (lb)	Vert (lb)	Arm to *B* (ft)	*M* about *B* (ft-k) (Clockwise pos)
$F(1)$	$13.91 \times 15 \times 60$	= 12,519	—	7.50	93.9
$F(2)$	$14.83 \times 5 \times 60$	= 4,449	—	17.50	77.9
$F(3)$	$15.61 \times 1.92 \times 60$	= 1,798	—	20.96	37.7
$F(4)$	$17.56 \times 15 \times 60$	= —	15,804	19.50	308.2
$F(5)$	$17.56 \times 6.25 \times 60$	= −6,585	—	24.29	−160.0
$F(6)$	$13.66 \times 15 \times 60$	= —	12,294	4.50	55.3
$F(7)$	$13.66 \times 6.25 \times 60$	= 5,122	—	24.29	124.4
$F(8)$	$9.76 \times 21.92 \times 60$	= 12,836	—	10.96	140.7
				Total overturning moment =	678.1 ft-k

If base shear is desired, force $F(5)$ should be computed using the pressure value rather than suction. $F(5)$ becomes $5.85 \times 6.25 \times 60 = 2194$ lb acting toward the right. The total base shear (sum of all entries in the "Horiz" column) is then 38,918 lb, or 38.9 kips.

The overturning moment must now be compared to the moment due to dead load alone, a moment that tends to *prevent* overturning. The latter moment

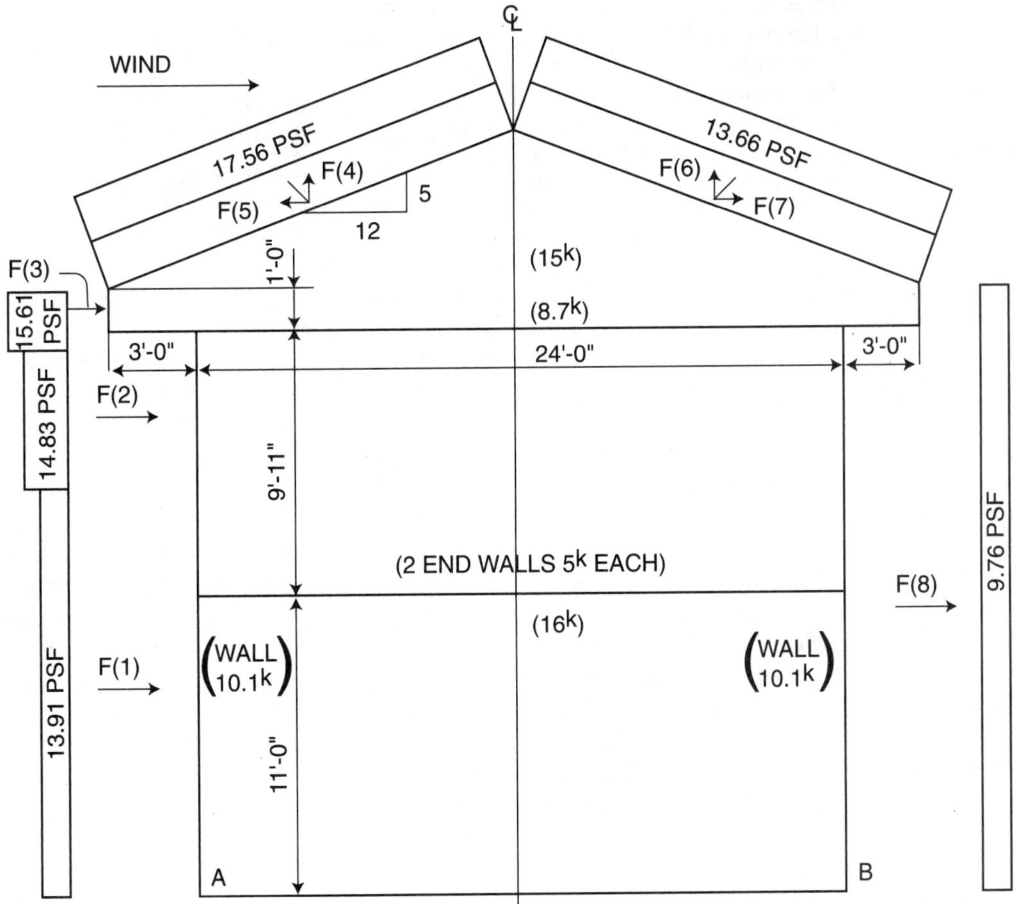

Fig. 4-5. Building for Examples 4-9, 4-10, and 4-11.

should exceed the overturning moment by whatever factor of safety is desired (UBC requires at least 1.5). If the dead load moment does not provide adequate protection against overturning, then special provisions are needed to anchor the structure.

The resultant of all dead loads shown in Fig. 4-5 is 69.9 kips. By symmetry, it acts at the center of the 24-ft building width, so the moment of the dead load about point *B* is

$$12 \times 69.9 = 838.8 \text{ ft-k}$$

The factor of safety against overturning is 838.8/678.1 = 1.24. This is less than the required 1.5, so additional anchorage must be provided.

Example 4-10

What vertical anchor force is required at each corner of the building of Fig. 4-5 to provide stability against overturning?

To provide a factor of safety of 1.5 against overturning requires a "righting" moment of $1.5 \times 678.1 = 1017.2$ ft-k. Dead load provides only 838.8 ft-k of this amount, leaving 178.4 ft-k to be furnished by anchor bolts connecting the ends of the shear walls (points *A* and *B*) to the foundation. Assume the lever arm from the anchor bolts at *A* to the center of the wall at *B* to be 23.7 ft. (This is the approximate distance between centers of opposite walls.) For wind in the direction shown by Fig. 4-5, the foundation and the bolts will provide a resisting couple—a tensile force in the two bolts at *A* and an upward compressive reaction at points *B*. Each force of this couple will be

$$F = 178.4/23.7 = 7.53 \text{ kips}$$

The force in each bolt at corners *A* will be one-half this amount, or 3.76 kips. When the wind is in the opposite direction, anchor bolts at corners *B* will become active.

The anchor bolt forces will also have to be determined for wind in a direction normal to the building ends. In this case, however, those forces are easily seen to be less than for wind as shown by Fig. 4-5.

A safety factor of 1.5 is required also for sliding. Thus, additional anchor bolts at intervals along the end walls would have to resist 1.5 times the base shear of 38.9 kips for the entire building, or one-half this amount for each end wall. Similar bolts to prevent sliding will be required along the walls normal to Fig. 4-5. Chapter 12 discusses anchorage further in the sections covering shear walls.

4-7. SEISMIC LOADS

Seismic loads are forces due to the inertia of a structure as it resists being moved by an earthquake. Although earthquakes can occur anywhere, some localities are more prone than others to damaging ones. In the United States, seismic design is often done following the UBC requirements. The UBC indicates the probable severity of earthquakes by a map dividing the United States into seismic-risk zones, as shown by Fig. 4-6. Zone 0 is one in which no damage is expected. The other zones are assigned a seismic-zone factor, Z, which increases with the severity of expected quake acceleration. The values of seismic zone factor are:

Zone 1 $Z = 0.075$
Zone 2A $Z = 0.15$
Zone 2B $Z = 0.20$
Zone 3 $Z = 0.30$
Zone 4 $Z = 0.40$

Seismic loads on structures include both horizontal and vertical forces. Normally, however, the horizontal forces are of greater concern. Thus, the designer needs to determine the probable horizontal forces caused by the earthquake on various parts of the structure.

UBC allows these lateral forces to be determined by a defined dynamic lateral-force procedure. For certain structures it also allows determining them by a simpler static-force procedure. The static lateral-force procedure may be used for regular structures up to 240 ft high and for irregular structures up to 65 ft high. Fortunately, these limits allow using the static procedure for wood buildings. So, the one described here will be the static lateral-force procedure.

Seismic forces can be in any horizontal direc-

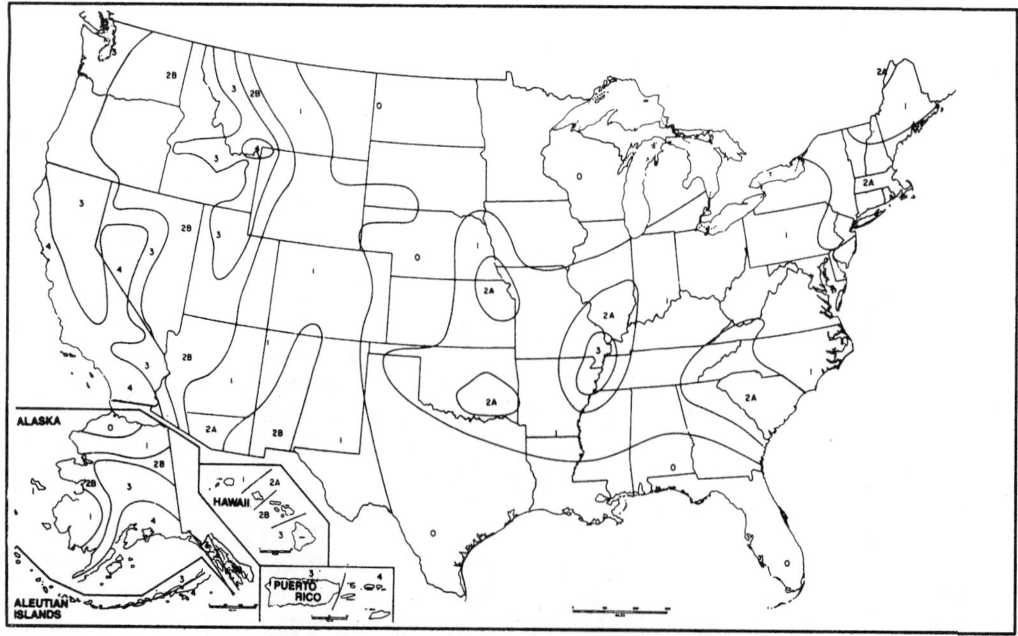

Fig. 4-6. Seismic zone map of the United States (Reproduced from the 1994 edition of the *Uniform Building Code*™, copyright© 1994, with permission of the publisher, the International Conference of Building Officials.)

tion. This means that the resultant forces must be considered parallel to either axis (one at a time) of the structure, or in a diagonal direction. The total seismic force causes a "base shear," V, which is to be determined from the following equation:

$$V = ZICW / R_W \tag{4-6}$$

In Eq. 4-6

Z is the seismic-zone factor, W is the weight of the structure and of certain live loads that must be considered, and the other factors are defined below.

I (importance factor) is 1.25 or 1.5 for essential facilities or facilities holding hazardous materials, and 1.00 for all other structures.

R_W is a factor that varies according to the structural system used. For wood buildings, UBC gives values of R_w as follows:

For *bearing wall* building systems:

8 for light-framed walls having shear panels and not over three stories high

6 for other light-framed walls

For building *frame* systems:

9 for light-framed walls with shear panels and not over three stories high

7 for other light-framed walls

8 for heavy timber braced frames

C, a numerical coefficient depending on both the natural period, T, of the structure and S, a site coefficient for the soil, is calculated by the following:

$$C = 1.25\, S / T^{2/3} \tag{4-7}$$

However, C need not exceed 2.75, and C/R_w shall be not less than 0.075.

Values of S are given by Appendix Table A-7.

For computing the approximate period, T, two methods are given. The one applicable for all buildings is

$$T = C_t\,(h_n)^{3/4} \tag{4-8}$$

in which h_n is the height (ft) of the level that is uppermost in the main portion of the structure, and C_t depends on the type of structure: For wood structures it is 0.020.

Having values for all factors in Eq. 4-6, the total base shear for design can now be calculated. This base shear is, of course, equal to the sum of the inertia forces from portions of the total weight above. A conservative answer for overturning moment is obtained if all of the inertial loads are assumed to act at the top of the structure. However, a more realistic answer is obtained by distributing the load according to the vertical location of the masses.

UBC requires that a portion, F_t, be applied at the top of the structure—in addition to the horizontal force occurring there because of the portion of W located at the top. The added amount is

$$F_t = 0.07\, T\, V \tag{4-9}$$

in which T has the same value as was used to compute C in Eq. 4-7. F_t need not exceed 0.25V. Also, F_t may be taken as zero if T is 0.7 seconds or less.

The balance of the of the base shear is $(V - F_t)$, and this amount is distributed to various heights of the structure according to Eq. 4-10 as follows:

$$F_x = \frac{(V - F_t)\, w_x h_x}{\sum_{i=1}^{n} w_i h_i} \tag{4-10}$$

In this equation, h_x is the height of level x above the building base, w_x is the portion of the total load, w, assigned to level x, and the denominator is the sum of products $w_x h_x$ for each level of the building.

At each level where there is significant mass, a horizontal force F_x corresponding to that level is applied. The F_x corresponding to the level of the top is added to force F_t. From this point on, the overturning moment is calculated by statics.

Both stability and strength under seismic loads must be considered. To check stability, the designer must consider the likelihood of overturning and that of sliding horizontally. The UBC specifies a 1.5 factor of safety against overturning by wind loads, but does not specify a factor of safety for overturning by seismic forces. No factor of safety against sliding is specified, but the authors would use not less than 1.5. Stability and strength must be provided

CHECK E-W SEISMIC FORCES

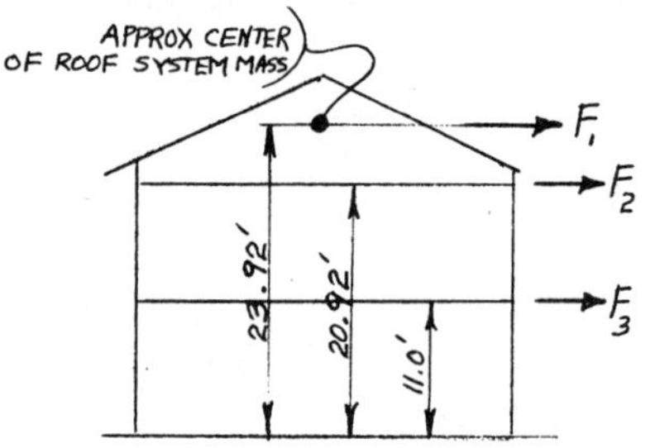

For Zone 3, $Z = 0.30$

$I = 1.0$

$T = C_t h_n^{3/4} = 0.02(23.92)^{3/4} = 0.216$ sec.

$S = 1.0$

$C = 1.25S/T^{2/3} = 1.25(1)/0.216^{2/3} = 3.47$

$3.47 > 2.75$, so use $C = 2.75$

$R_w = 8$

$V = ZICW/R_w = 0.3(1)(2.75)W/8 = 0.103W$

Total $W = 69.9$ k

Base Shear $V = 0.103(69.9) = 7.20$ k

Since $T < 0.7$ sec., $F_t = 0$

Weight of walls (30.2 k) must be distributed to foundation and levels 2 and 3.

To Fdn: $\dfrac{11/2}{20.92} \times 30.2 = 7.94$ k

To Level 3: $\dfrac{(11/2 + 9.92/2)}{20.92} \times 30.2 = 15.10$ k

To Level 2: $\dfrac{9.92/2}{20.92} \times 30.2 = 7.16$ k

30.2

$F_1 = \dfrac{7.20(15)(23.92)}{1033} = 2.50$ k

$F_2 = \dfrac{7.20(7.16 + 8.7)(20.92)}{1033} = 2.31$ k

$F_3 = \dfrac{7.20(15.10 + 16)(11.0)}{1033} = 2.38$ k

$15(23.92) = 359$

$(8.7 + 7.16)(20.92) = 332$

$(16 + 15.10)(11.0) = 342$

$\Sigma w_i h_i = 1033$

OTM $= 2.50(23.92) + 2.31(20.92) + 2.38(11.0) = 134.3$ ft-k (about B)

M (about B) by DL $= 838.8$ ft-k

Factor of Safety $= 838.8/134.3 = 6$ OK

Fig. 4-7. Designer's solution to Example 4-11.

not merely at the base, but at all levels in the structure above its base.

The above is a very condensed summary of the UBC requirements for the entire building. Other UBC sections and tables deal with seismic forces on individual elements and on nonstructural parts of the building.

Example 4-11

Consider again the building of Fig. 4-5, to be built in seismic-risk Zone 3. Preliminary design shows that the dead-load distribution will be approximately as shown by the numbers in parentheses in Fig. 4-5, and that the building will have both horizontal-plane and vertical-plane plywood diaphragms. Site coefficient, S, is taken to be 1.0. Will the building satisfy the UBC requirement for stability against overturning under earthquake conditions?

Figure 4-7 shows the solution as a designer's computation sheet. From this solution we conclude that overturning at the base due to seismic action is no problem, whereas for wind load it was a problem, and additional anchorage had to be provided. (This could be seen early in the solution, but the entire procedure is illustrated here.) Do not assume that this is always the case, as seismic loading often controls the design. Figure 4-7 shows analysis for east–west quake motion only. North–south motion should also be considered, but in this case it will prove even less serious than the east–west motion.

In the solution, note that force $F(1)$ is located at the center of mass for the complete roof system (trusses, sheathing, and roofing), estimated to be 3 ft above the center of the ceiling rafters. Many designers, interpreting the UBC requirements, would place $F(1)$ and $F(2)$ at the height of $F(2)$, the center of the ceiling rafters or the bottom-chord level of the roof trusses. If $F(1)$ is "lumped" with $F(2)$, then, ideally, an upward force at the left and a downward force at the right should be included to compensate for the moment of force $F(1)$ about the level of force $F(2)$.

Part II. Design Values

Structural design in wood today is usually by the allowable-stress method, in which computed "actual" stresses are compared to adjusted design values (allowable stresses). This part of Chapter 4 tells how design code writers determine acceptable design values for various species and grades of wood, and tells how the designer adjusts those values for anticipated conditions of use.

4-8. DESIGN VALUES

A designer of wood structures needs to know (for the species and grade of lumber to be used) the modulus of elasticity parallel to the grain and allowable stress values as follows:

- Flexure
- Compression parallel to the grain
- Compression perpendicular to the grain
- Horizontal shear (i.e., parallel to the grain)
- Tension parallel to the grain

Values of these allowable stresses and elastic modulus are influenced by many factors. For manufactured materials such as steel, quality control is good and little variation occurs among samples meeting the same specification. With wood, this is not so. Both the variation and the *quality* of the wood (measured by the type, size, and location of defects and by specific gravity) affect the strength.

Therefore, to design in wood with reliable safety it is necessary either to use an unreasonably high factor of safety or to choose reasonable allowable stresses based on statistical analysis of the variables involved. The latter is the method used by writers of the *National Design Specification* (see reference 1).

4-9. BASE DESIGN VALUES

For decades, allowable stresses for structural design were based on tests of "small clear" wood specimens, measuring 2 in. × 2 in. in cross section, and defect-free (4,5,6). The small clear test values were then modified to consider conditions of use and the effects of strength-reducing characteristics (defects) found in actual full-sized wood structural members. For the most part, that system worked fairly well, but the need for a better method was apparent. Consequently, in 1977, the U.S. Department of Agriculture (through its Forest Products Laboratory) initiated a massive In-Grade Testing Program.

The program involved testing 70,000 pieces of full-size structural lumber of various dimension grades obtained from U.S. and Canadian mills. After twelve years of testing and research, a series of design values for softwood lumber became available. In-Grade testing removed some of the uncertainty as to the effect of defects and size, and gave designers better values on which to base their structural designs. As a result of the testing program, some design values for visually graded softwood dimension lumber were increased. Lumber grades, however, were not changed.

The test procedures of the In-Grade testing program are now formalized by ASTM Standard D1990-91 (7).

Base design values shown for softwood dimension lumber by the NDS Supplement (8) are based on the results of this In-Grade Testing Program. Design values for other sizes, decking, and hardwood lumber, however, are still based on the former system.

The tabulated base design values depend on the species (or species group), the size category, and the grade of lumber. Values shown are:

F_b (allowable stress for bending, parallel to grain)
F_c (allowable compressive stress parallel to grain)
$F_{c\perp}$ (allowable compressive stress perpendicular to grain)
F_t (allowable tensile stress parallel to grain)
F_v (allowable shear stress parallel to grain)
E (modulus of elasticity parallel to grain)

All of the tabulated values assume dry use and normal-duration load. Normal-duration load is that which will exist to full intensity for not over ten years, either in a single time period or as an accumulation of shorter time periods.

The base design values are given also by other specifications, such as the *Uniform Building Code* (2).

4-10. ADJUSTMENT FACTORS

Because strength of wood members depends on conditions such as moisture content, temperature, and member size, the base (tabular) design values must be adjusted to reflect this. The means for accounting for these conditions is a series of adjustment factors. It is the responsibility of the *structural designer* to determine which adjustment factors apply and to use those factors to determine the adjusted allowables.

Fourteen different types of adjustment factor are listed by the NDS Specification (1). Fortunately, it is seldom necessary to consider more than a few of these simultaneously. These factors are discussed below. In each case, the tabulated design value is *multiplied* by the adjustment factor, and in most cases the adjustment factors must be superimposed (i.e., the tabulated base design value is multiplied by *all* applicable factors).

Load Duration Factor

Figure 2-9 shows how the strength of wood varies with duration of load. If the load duration is known precisely, this curve can be used to determine how to adjust the base design value to obtain an adjusted allowable stress. For ordinary purposes, however, the following adjustment factors, C_D, are used:

For permanent load (load exceeding ten years duration), 0.9
For normal-duration (ten years cumulative duration), as for usual building occupancy loads, 1.0 (i.e., no adjustment)
For 2-month duration (as for snow load), 1.15
For 7-day load (as for construction loads), 1.25
For 10-minute duration (as for wind or earthquake), 1.6 (The 1.6 is what the NDS specifies. The *Uniform Building Code* (ref. 2) specifies 1.33 for earthquake loads, 1.33 for connection design for wind loads, and 1.6 for member design for wind loads.)
For impact loads, 2.0

The load duration factor applies to all allowable stresses except $F_{c\perp}$ and modulus of elasticity, E.

Load duration factors are *not* cumulative with each other. Rather, when a *combination* of load types occurs, the load type having the shortest duration determines the factor to be used for the entire combined loading. Factors are *not* prorated according to the magnitude of various load types. For example, under NDS, if a load combination includes dead load, live load, and wind, the factor for wind (1.6) is used to determine the allowable stress under the *combination* of loads. In this case, however, it is necessary to consider the other possible combinations: dead load plus live load alone, using a factor of 1.0 (ten-year load duration), and dead load alone, using a factor of 0.9. Examples that follow show how to decide which is the critical load combination and how to determine the allowable stress for that combination of loads.

Wet Service Factor

A wet service factor, C_M, applies to all allowable stresses except parallel-to-grain bearing and to modulus of elasticity, E. For sawn lumber, the factor must be applied whenever the moisture content of the wood in use will exceed 19%. For glued laminated members, the factor must be applied whenever the moisture content in use will exceed 16%. Values of these factors are shown at the beginning of each section of the NDS Supplement (8) and of Appendix B-3, B-4, B-5 and B-6. In addition, the wet service factor applies to connection design, where it is a function of moisture conditions both at the time of fabrication and while in service.

Size Factor

Values of the size factor, C_F, for sawn dimension lumber are given by the NDS Supplement (8). They are shown also along with Tables in Appendix B. The size factors for dimension lumber depend on cross-section dimensions and lumber grade, and apply only to F_b, F_c, and F_t.

For larger members (those with least dimension 5-in. nominal or more), those classed either as Post and Timber or as Beam and Stringer, the size factor applies only to the bending value, F_b, and is given as

$$C_F = (12/d)^{1/9} \tag{4-11}$$

in which d is the depth in inches, measured parallel to the load direction. This size factor applies only for members of actual depth exceeding 12 in. For 12-inch and shallower members, C_F is taken as 1.0.

For lumber that is not visually graded, but rather machine-stress-rated (MSR) or machine-evaluated (MEL), there is no size factor.

Volume Factor

For glued laminated members, a volume factor, C_V, (rather than a size factor) is applied to adjust the base design value for bending, F_b. Computation of this factor will be shown in Chapter 8. However, only the smaller of C_V and beam stability factor, C_L, is used.

Beam Stability Factor

The beam stability factor, C_L, is a reducing factor that considers the potential for lateral buckling of laterally unsupported wood beams. Its computation and use will be shown in Chapter 6.

Column Stability Factor

The column stability factor, C_P, is a reducing factor that considers the slenderness of the compression member and its potential for buckling. Its computation and use will be shown in Chapter 7.

Temperature Factor

The temperature factor, C_t, applies to members exposed to prolonged temperatures above 100°F. It adjusts all allowable stresses and modulus of elasticity.

Curvature Factor

Applicable for glued laminated members only, this factor, C_c, and its use will be shown in Chapter 8.

Repetitive Member Factor

Repetitive members are defined as "members in bending, such as joists, trusses, rafters, studs, planks, decking, or similar members that are in contact or not more than 24 inches on centers, are not less than 3 in number, and are joined by floor, roof, or other load-distributing elements adequate to support the design load." (1) This means that repetitive members are those that are close enough alongside each other that, if one member is weaker than normal, the adjacent parallel members can help out. Load that cannot be carried by the weaker member is transferred laterally to the stronger and stiffer members on each side. The design value for a repetitive member is higher than for a *single member,* one that must carry all the load applied to it without depending on assistance from adjacent members. Note that even though studs are listed in the definition above, the factor applies only to their design value for *bending,* F_b.

The repetitive member factor, C_r, applies only to dimension lumber, and its value is 1.15. It is *not* applied to the base design values shown for decking, since those values have already been adjusted for repetitive-member action.

Flat Use Factor

The tabulated base design values are for loads applied perpendicular to the narrow face. Often, however, dimension lumber is used flatwise, that is, with the load applied normal to the wide face of the lumber—bending it in what the authors call "the weaker direction." When dimension lumber is loaded in this manner, a flat use factor, C_{fu}, is applied to the bending value only. The flat use factor is *not* applied to bending values for decking, since flat use has already been considered in the values tabulated for decking.

Form Factor

Section 6-2 discusses the form factor, C_f, which applies to bending members of circular or diamond-shaped cross section.

Shear Stress Factor

This factor, C_H, is used to adjust the allowable horizontal shear stress for sawn lumber or timber members. It is discussed in Chapter 6.

Buckling Stiffness Factor

Applicable only to the modulus of elasticity values for certain trusses, C_T is discussed in Chapter 10.

Bearing Area Factor

Applicable only to the allowable compressive stress perpendicular to the grain, C_b is discussed in Section 6-6.

4-11. TABLES FOR BASE DESIGN VALUES

Appendix Tables B-3 and B-4 show base design values *for dimension lumber* of a few species. Table B-5 shows the values for larger sections of a few species in Post & Timber grades and Beam & Stringer grades. To determine a base design value, the designer must:

1. Select the species (or species group) to be used.
2. Select the applicable table, according to size of the member. (It may be necessary to estimate the size, then design, then correct the computations if the resulting size selected is not the same as the estimated size.)
3. Observe footnotes, if any.
4. Select a grade.
5. Read the base design values needed.

Next the designer must adjust the base design values by multiplying by all applicable adjustment factors. (This will include first an adjustment for member size, which must be estimated. If the estimated size proves incorrect, correct the estimate to continue the design.)

An adjusted allowable stress is denoted by a *prime;* for example, F_b is the symbol for a base allowable bending stress, and F'_b is the symbol for the adjusted allowable—the product of F_b and all adjustment factors that apply.

For a complete listing of base design values for all important U.S. and Canadian lumber species, see the NDS Supplement (8) or the *Uniform Building Code* (2).

Design Values for Mechanically Graded Lumber

Appendix Table B-6 gives design values for machine stress rated (MSR) lumber, and for machine evaluated lumber (MEL). For the first type, the commercial grade designation in the left column actually indicates the allowable bending design value and the modulus of elasticity.

Table B-6 is easy to use, because species is not a consideration and the number of lumber sizes that can be mechanically graded is limited. Table B-6 has its own adjustment factors that are explained in the footnotes to the table.

The next seven examples illustrate the use of the tables in Appendix B and the adjustments required to convert the base (tabular) design values to allowable stresses for design.

Example 4-12

Determine the adjusted allowable bending stress for a 2 × 10 roof rafter of No. 1 Douglas fir-larch. The rafters are spaced 16 inches on centers and have sheets of plywood roof sheathing nailed to them. The roof structure is over a swimming pool, so the moisture content of the wood will certainly exceed 19%. Bending moments for one rafter are:

From dead load alone, 1000 ft-lb
From snow load alone, 1800 ft-lb

Refer to Appendix Table B-3. Find the part of the table for Douglas fir-larch. In that section of the table, read to the right from the grade (No. 1). The base design value shown for bending, F_b, is 1000 psi.

For No. 1 grade in a 10-in. width and 2-in. nominal thickness, Table B-3 shows a size adjustment factor, C_F, of 1.1 for use in determining the allowable bending stress. The rafters meet the requirement for repetitive members, so the basic design value can be adjusted by factor C_r, shown by the table as 1.15.

Since the moisture content of the wood in use will exceed 19%, we must also consider the wet service factor, shown by Table B-3 as 0.85. However, $F_b(C_F) = 1000\ (1.1) = 1100$ psi < 1150 psi, so by the footnote to the table of wet service factors, $C_M = 1.0$.

Up to this point, the adjusted allowable bending stress is

$$1000\ (1.1)\ (1.15)\ (1.0) = 1265 \text{ psi}$$

Now the load-duration factor must be included. There are two load cases:

Dead load alone, $M = 1000$ ft-lb
Dead load plus snow load, $M = 2800$ ft-lb

For the first case, dead load only, the duration factor is 0.9 and the adjusted allowable bending stress is $F'_b = 1265\ (0.9) = 1138$ psi.

For the second case, dead load plus snow load, the duration factor is 1.15 and the adjusted allowable bending stress is

$$F'_b = 1265\ (1.15) = 1455 \text{ psi}$$

Which adjusted allowable stress value should be used? Actually both of them. Unless we use a convenient shortcut (demonstrated later) we must compute the actual stress for each load case and compare it to the adjusted allowable stress for that combination, as is now illustrated.

For dead load alone, the computed actual bending stress is

$$f_b = M/S = 12\ (1000)/21.39 = 561 \text{ psi}$$

This is less than the adjusted allowable of 1138 psi for dead load alone, so the rafter is satisfactory for the first load case.

For dead load plus snow load, the computed actual bending stress is

$$f_b = 12\ (2800)/21.39 = 1571 \text{ psi}$$

This stress exceeds the 1455 psi allowable, so the rafter is *not* satisfactory for combined dead load and snow load.

Since the 2 × 10 did not satisfy *both* load cases, a stronger section must be found. This could be done by either using a larger section (2 × 12, for example) or by selecting a better grade of lumber.

Example 4-13

Repeat the above example, but use the convenient shortcut mentioned above to determine the controlling load combination.

Merely dividing the total bending moment for each load combination by the load duration factor that applies to that combination, we have

For dead load alone: 1000/0.9 = 1111
For dead plus snow: 2800/1.15 = 2435

The quotient for dead load plus snow is the larger, therefore that combination controls the design. The adjusted allowable stress (using $C_D = 1.15$) is determined next and compared to the computed actual bending stress for that combination only.

This shortcut works only if the loading patterns are alike for all cases being compared. In this example, both load cases were for uniform load over the complete span, and the short cut was valid. Had one of the loads been uniform over the entire length and the other either concentrated or applied over only a portion of the length, the method would not be valid.

Example 4-14

No. 3 hem-fir 2 × 6s are be used in the flat position as floor planks to carry a 60-psf total dead load (including the weight of the planks themselves) and a live load of 70 psf. Load will bear directly on the planks, without any means of lateral distribution from one plank to another. Use conditions will be wet, and the planks will probably have moisture content higher than 19%. Which load condition will control the design, and what will be the adjusted allowable flexural stress?

Appendix Table B-3 shows an allowable bending stress of $F_b = 500$ psi. It also shows a flat-use factor of 1.15 for this size of plank as well as a size factor of 1.3. Again, the wet service factor shown is 0.85 and again, the footnote controls, so that the wet-service factor $C_M = 1.0$.

The adjusted allowable bending stress will be

$$F'_b = 500\,(1.15)\,(1.3)\,(1.0)\,C_D = 748C_D$$

For dead load alone: $C_D = 0.9$ and $F'_b = 748\,(0.9) = 673$ psi
For dead plus live loads: $C_D = 1.0$ and $F'_b = 748\,(1.0) = 748$ psi

Alternatively, if we use the shortcut to find out first which load combination controls,

For dead load alone: 60/0.9 = 66.7
For dead plus live: (60 + 70)/1.0 = 130 *Controls*

Adjusted allowable stress for dead plus live = 748 psi (as above).

Example 4-15

The 3 × 8 tension chord of a truss is made from Select Structural southern pine. Use conditions will be dry. The lumber is treated with fire-retardant chemicals. Forces in the chord are:

From dead load: 8 kips
From snow load: 9 kips
From wind load: 5 kips

If the code-specified load combinations are: (1) dead load alone; (2) dead plus snow load; and (3) dead plus wind, which load combination controls, and what is the adjusted allowable tensile stress?

Totals and quotients for these combinations are

For dead alone: 8/0.9 = 8.89
For dead plus snow: (8 + 9)/1.15 = 14.78 *Controls*
For dead plus wind: (8 + 5)/1.6 = 8.12

The base design value from Appendix Table B-4 is 1300 psi. The size factor has already been included in the table of base design tension values for southern pine. So the only additional factors to be considered in the example are the load duration factor (1.15), and a factor for fire-retardant treatment.

The latter factor is not specified by the NDS, which requires that the company doing the fire-retardant treatment specify the applicable factor. For this example, assume that the lumber treatment company states that the lumber will have a 20% loss of tensile strength due to the their treatment. Thus, the adjustment factor for fire-retardant treatment would be 0.80.

The final adjusted allowable tensile stress is

$$F'_t = 1300\,(1.15)\,(0.8) = 1196 \text{ psi}$$

What if the truss itself were used as a repetitive member? Ordinarily, it would be only the top chord (having attached plywood sheathing) that could qualify as a repetitive member. The tension chords, being unattached to each other (insofar as contribution to carrying vertical loads applied to the truss) would not qualify as repetitive; therefore, the 1.15 repetitive member factor is *not* included in the above computations.

Example 4-16

A dense No. 1 Douglas fir-larch 8 × 10 timber, graded under Post and Timber (P&T) rules is used under dry conditions as a short compression member in a storage warehouse. It is subject to the following axial loads:

From dead load,	25 kips
From floor live load,	40 kips
From snow load,	15 kips
From wind load,	29 kips

Assuming that the member is short enough that the column stability factor, C_p, is 1.0 (i.e., there is no likelihood of buckling), what is the allowable compressive stress?

Since the building is used for storage, it is reasonable to assume that the live load duration will be longer than ten years. Therefore, live load will be treated as permanent load. Considering all the load combinations specified by the UBC, the combinations and quotients are as follows:

Permanent $(D + L)$: $(25 + 40)/0.9 = 72.2$
$D + L + S$: $(25 + 40 + 15)/1.15 = 69.6$
$D + L + W$: $(25 + 40 + 29)/1.6 = 58.8$
$D + L + S/2 + W$: $(25 + 40 + 7.5 + 29)/1.6 = 63.4$
$D + L + S + W/2$: $(25 + 40 + 15 + 14.5)/1.6 = 59.0$

The first quotient is the largest, so the first load combination controls, and C_D of 0.9 will be used for the combination. Table B-5 shows the base design value of F_c to be 1200 psi. The size factor for P & T members applies only to the bending value, F_b, not to F_c. So the only adjustment required is for load duration. The adjusted design value (allowable stress parallel to the grain) is

$$F'_c = 0.9\,(1200) = 1080 \text{ psi}$$

If the laterally unsupported length of the member is such that the column stability factor is less than 1.0, the 1080-value above would also be multiplied by C_p to consider the possibility of failure by buckling.

Example 4-17

It is not always possible to apply the above shortcut method directly to the *amounts* of the loads. Rather, bending moments, shears, or axial forces may have to be determined first, so we can identify the controlling load combination. This is the case for the timber beam of Fig.4.8 for use in a wet-process manufacturing plant. Loads shown include the weight of a permanently installed machine. During operation, the machine occasionally imposes an impact load. Which load combination controls for bending, and what is the allowable stress for that combination? For end shear? For end bearing (compression perpendicular to the grain)?

Figure 4-9 shows the solution in the form of a designer's computation sheet. The designer assumed that Beam & Stringer, rather than dimension lumber, will be used. Notice that detailed explanations are not given, but sufficient information is shown that others can check the designer's work. Equally important, the designer can easily determine at any future date the exact basis for the conclusions reached. For operations so simple that they can easily be carried out with a hand calculator (simple-beam reactions, for example) only the answers are shown.

Example 4-18

Machine-stress-rated southern pine 2 × 10s are used as floor joists over a wet manufacturing process. Joists are 12 in. c/c and the attached floor is fairly rigid. The MSR grade is 1800f-1.6E. Dead load totals 80 psf of floor area and live load is 30 psf. Which combination controls, and what are the values of E and allowable stress for bending and for shear?

Appendix Table B-6 shows base design values as follows: $F_b = 1800$ psi and $E = 1{,}600{,}000$ psi. (Both of these are implied also by the grade designation 1800f-1.6E.) Table B-6 does *not* show a design value for shear stress, but footnote 2 of Table 6 tells to use the F_v value shown for No. 2 visually graded dimension lumber of the same species. From Table B-4, we find $F_v = 90$ psi.

Quotient 80/0.9 is less than (80 + 30)/1.0, therefore $(D + L)$ controls and C_D is 1.0.

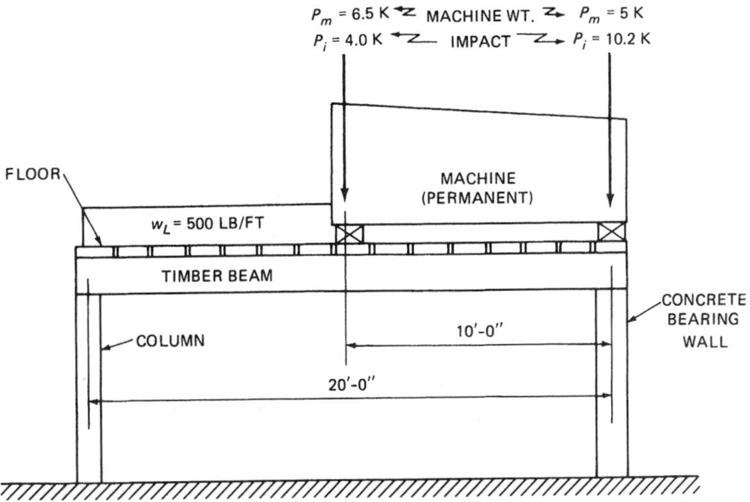

Fig. 4-8. Beam for Example 4-17.

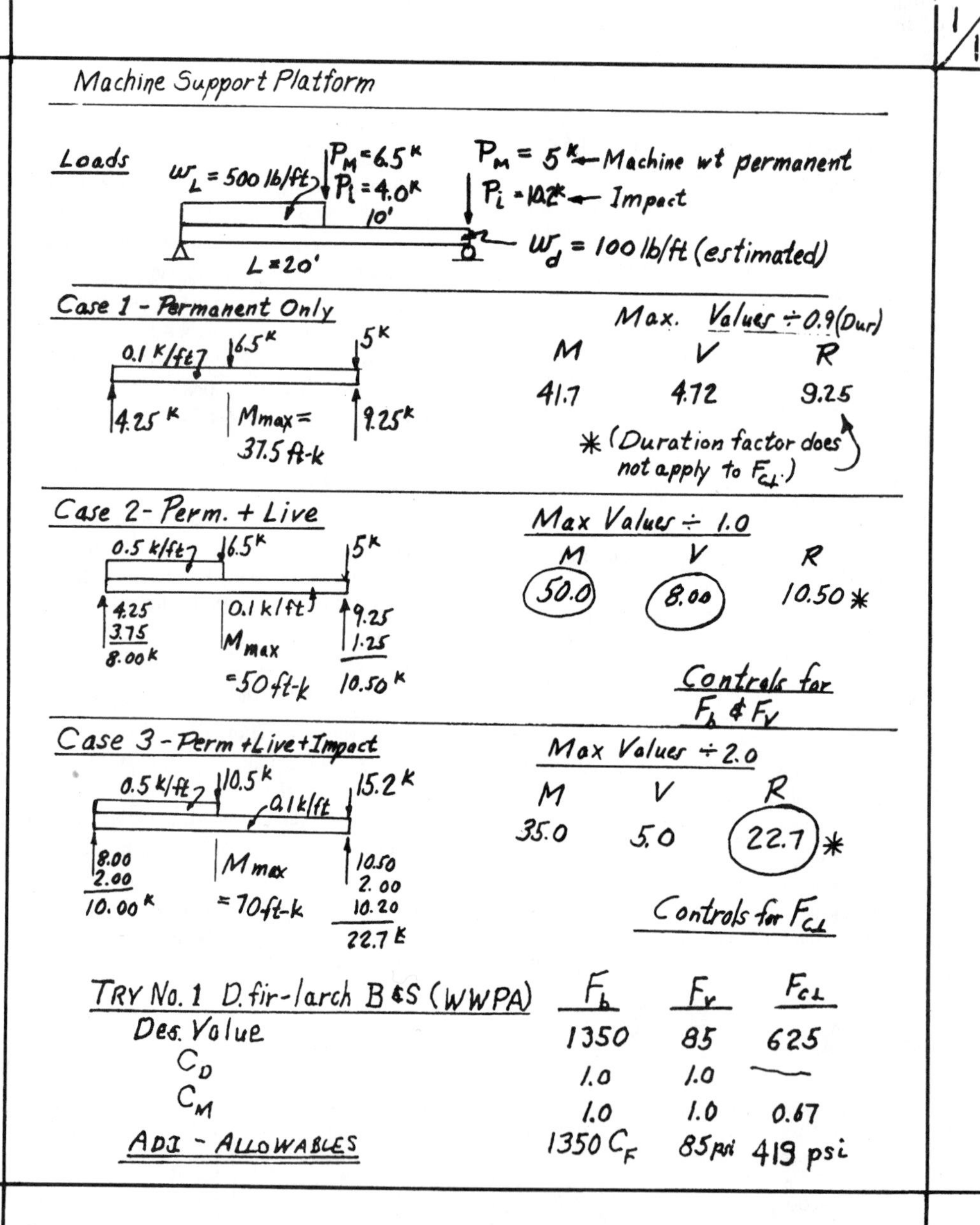

Fig. 4-9. Solution to Example 4-17.

Factors applicable to *bending* are considered first.

> The joists are repetitive members, so C_r is 1.15. No size factor is needed, since the mechanical grading system obtains values by testing the section size in question (i.e., the size effect is already included). The only adjustment factors needed are the repetitive member factor, (C_r), wet service factor, (C_M), and duration factor, (C_D).

The moisture content will likely exceed 19%. According to Table B-6 (since F_b is not less than 1150), factor C_M for bending is 0.85.

The adjusted design value (allowable stress for bending) is

$$F'_b = 1800\,(1.15)\,(0.85) = 1760 \text{ psi}$$

For *shear* stress, the base design value is 90 psi. This must be adjusted for wet service and may be adjusted by shear-stress factor C_H that depends of the presence of splits and size of shakes present on the piece of dimension lumber. Since the piece actually to be used is not known in advance, the only reasonable solution is to assume the worst. In Table B-6, this is seen to be the case with the longest split or shake, for which the factor is 1.00. For cases with shorter defects, the factor C_H would be an *increasing* factor.

Table B-6 shows the wet-service factor for shear as 0.97. The adjusted design value for shear is

$$F'_v = 90\,(0.97) = 87 \text{ psi}$$

The *modulus of elasticity* value is adjusted only by the wet-service factor of 0.9. The adjusted *E*-value is

$$E' = 1{,}600{,}000\,(0.9) = 1{,}440{,}000 \text{ psi}$$

4-12. LOAD AND RESISTANCE FACTOR DESIGN

The procedures shown in sections above are for *allowable stress design* (ASD), in which safety is ensured by limiting the expected stress in the wood to allowable values that are less than the anticipated ultimate strength of the wood. The adjusted allowable stress is equal to the wood's expected ultimate strength divided by a factor of safety.

Wood structural members can also be designed by another method, *load and resistance factor design,* known also as LRFD. In the LRFD approach, safety is ensured by *increasing* the expected loads and then selecting the member so that the stresses due to the increased loads do not exceed the ultimate strength of the wood. The specification that governs design by this method is the *Standard for LRFD for Engineered Wood Construction* (9).

Reinforced concrete prior to the early 1950s was designed almost exclusively by ASD, or *allowable stress design.* During the 1960s the American Concrete Institute Code introduced *strength design* (the same as LRFD) as an alternate acceptable method. Today, the ACI code (10) presents strength design as the preferred design method. The AISC design codes for structural steel are following a similar evolution, and today either ASD (11) or LRFD (12) may be used for designing steel structures. With timber structure design, the move toward design by LRFD has begun.

The LRFD method is based on the requirement that the minimum resistance provided in a structural member must either equal or exceed the force (or moment) due to factored loads on the member. Expressed in equation form,

$$R_u \leq \lambda \phi R' \qquad (4\text{-}12)$$

In Eq. 4-12, λ is the time effect factor for duration of loading. Variability of material and reliability of analysis methods are accounted for by the resistance factor (strength reduction factor), ϕ. R' indicates the adjusted resistance (theoretical strength of a member, such as adjusted bending resistance, M', or adjusted shear resistance, V'. R_u is the force or moment due to factored loads, and is replaced with M_u for bending or V_u for shear.

Neither the loading to be applied nor the resistance (theoretical strength times the strength reduction factor and time effect factor) can be known precisely for all cases. The load may be defined, for example, as the weight of a certain depth of snow, such as 25 psf. The probability is great that the actual snow load will never exceed that amount, but there is always the *possibility* that it may. That is, we cannot say, with 100% assurance, that the snow load will never exceed the quoted amount. Yet, because of the extreme probability that it will not exceed that amount, we regard the quoted snow load as a reliable figure to use in designing ordinary structures. For structures of short life or limited value, or for structures whose failure would not endanger life, we might be willing to use a lesser design load. In other words, we might be willing to accept a greater probability that the design load might be exceeded. To do so would be to use one of the elements of probabilistic design.

Both sides of Eq. 4-12 are subject to variation, so the probability of failure depends on the values used on each side of the equation. This is illustrated graphically by the two distribution curves shown in Fig. 4-10. The curve at the left is for loading, showing a high probability that the ac-

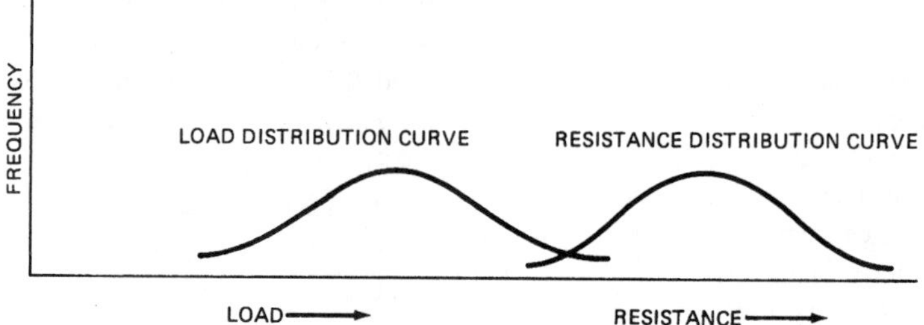

Fig. 4-10. Overlap of load and resistance distribution curves.

tual loading will be at or very near to the design loading. The frequency of much larger or much smaller loadings is less, as shown by the small ordinates at each end of the curve. Plotted on the same graph is a curve for the resistance of the structural member to those loads. The object is to have the resistance larger than the load. The resistance of the structural member cannot be predicted precisely, but it is highly probable that the predicted resistance will be fairly correct. There is a lesser probability that the actual strength will be much less than predicted, and this is shown by the "tail" of low ordinate near the left end of the distribution curve for resistance.

Failure will occur when resistance is less than load. The curves overlap, and the degree of their overlap indicates the probability of failure. Expressed qualitatively only, for greater reliability, the intersection of the two curves should be at a point of very low ordinate.

Reference Strengths

Some of the example problems in the chapters that follow will illustrate design by the LRFD method. To select a member size by LRFD requires knowing *reference strengths* (reference resistances) for various species, grades, and sizes of lumber. Tables showing these will be similar in nature (and volume) to the tables of base design values (for ASD) in the NDS Supplement (8), but the values shown for LRFD will be much higher, representing reliable, lower-bound values of *ultimate* stress.

Reference strengths are determined according to an ASTM standard (13). The ASTM standard provides two procedures for determining reference strengths. The reliability-based procedure uses a set of data from tests and employs reliability (statistical) computations. The other procedure, format conversion, does not require a set of data. Instead, a reference resistance value can be found by multiplying a code-recognized allowable stress by the factor $2.16/\phi$ where ϕ is the resistance factor for the stress property being evaluated. A table of reference strengths is shown here as Appendix Table B-10. Note that reference strengths are expressed in ksi units rather than psi units.

Time Effect Factor

The time effect factors, λ, of LRFD serve the same purpose as the load duration factors of ASD; however, having different bases, the factors have different numerical values. The value of the time effect factor depends on which load combination controls. Table 4-1 lists the various

Table 4-1. Load Combinations and Time Effect Factors.

Load Combination	Time Effect Factor (λ)
$1.4D$	0.6
$1.2D + 1.6L + 0.5(L_r \text{ or } S \text{ or } R)$	0.7 when L is from storage 0.8 when L is from occupancy 1.25 when L is from impact[a]
$1.2D + 1.6(L_r \text{ or } S \text{ or } R) + (0.5L \text{ or } 0.8W)$	0.8
$1.2D + 1.3W + 0.5L + 0.5(L_r \text{ or } S \text{ or } R)$	1.0
$1.2D + 1.0E + 0.5L + 0.2S$	1.0
$0.9D - (1.3W \text{ or } 1.0E)$	1.0

[a] For connections, $\lambda = 1.0$ when L is from impact.
Source: Courtesy American Society of Civil Engineers, *Standard for Load and Resistance Factor Design for Engineered Wood Construction*, 1996.

load combinations that must be considered in LRFD and the corresponding time effect factors.

REFERENCES

1. *National Design Specification for Wood Construction,* rev. 1991 Ed., American Forest & Paper Association, Washington, DC, 1993.
2. *Uniform Building Code,* International Conference of Building Officials, Whittier, CA, 1994.
3. *Minimum Design Loads for Buildings and Other Structures,* ANSI/ASCE 7-93, American Society of Civil Engineers, New York, 1993.
4. *Standard Methods of Testing Small Clear Specimens of Timber,* D143-94, American Society for Testing and Materials, West Conshohocken, PA, 1994.
5. *Standard Methods for Establishing Clear Wood Strength Values,* D2555-88, American Society for Testing and Materials, West Conshohocken, PA, 1988.
6. *Standard Methods for Establishing Structural Grades and Related Allowable Properties for Visually Graded Lumber,* D245-93, American Society for Testing and Materials, West Conshohocken, PA, 1993.
7. *Establishing Allowable Properties for Visually Graded Dimension Lumber from In-Grade Tests of Full-Size Specimens,* D1990-91, American Society for Testing and Materials, West Conshohocken, PA, 1991.
8. *Design Values for Wood Construction,* a Supplement to the National Design Specification, American Forest & Paper Association, Washington, DC, 1993.
9. *Standard for Load and Resistance Factor Design for Engineered Wood Construction,* American Society of Civil Engineers, 1996.
10. *Building Code Requirements for Structural Concrete,* ACI 318-95, American Concrete Institute, Farmington Hills, MI, 1995.
11. *Specification for Structural Steel Buildings* (Allowable Stress Design and Plastic Design), American Institute of Steel Construction, Inc., Chicago, IL, 1989.
12. *Load and Resistance Factor Design Specification for Structural Steel Buildings,* American Institute of Steel Construction, Inc., Chicago, IL, 1993.
13. *Standard Specification for Computing the Reference Resistance of Wood-Based Materials and Structural Connections for Load and Resistance Factor Design,* ASTM D5457-93, American Society for Testing and Materials, West Conshohocken, PA, 1993.

PROBLEMS

4-1. Calculate the approximate dead load for a one- or two-story wood structure (your home, for example). If the species is unknown, assume one.

4-2. Compute design wind loads for the building of Problem 4-1. If you can't measure actual dimensions, estimate them as closely as possible.

4-3. For your seismic-risk zone, calculate the total seismic base shear and overturning moment for the building of Problems 4-1 and 2.

4-4. Does dead load provide adequate safety against overturning for the building of Problems 4-1, 4-2, and 4-3?

4-5. For the structure of Fig. 4-11, what is the total bending moment in one inclined rafter due to snow load, where the specified snow load on a flat roof is 25 psf? Assuming dead load to be 20 psf of roof plus the weight of the rafter, what is the total bending moment in one rafter?

4-6. Compute the bending moment per roof beam (rafter) for the building of Fig. 4-12. Snow load specified is 25 psf of horizontal projection, the maximum wind speed is 70 mph, and the terrain is open. Give answers for all rea-

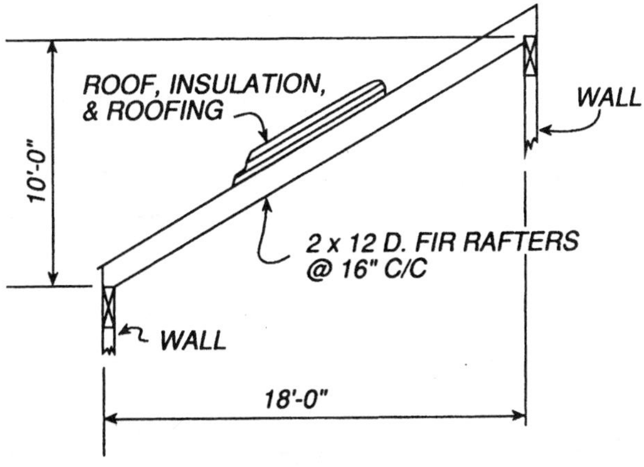

Fig. 4-11. Structure for Problem 4-5.

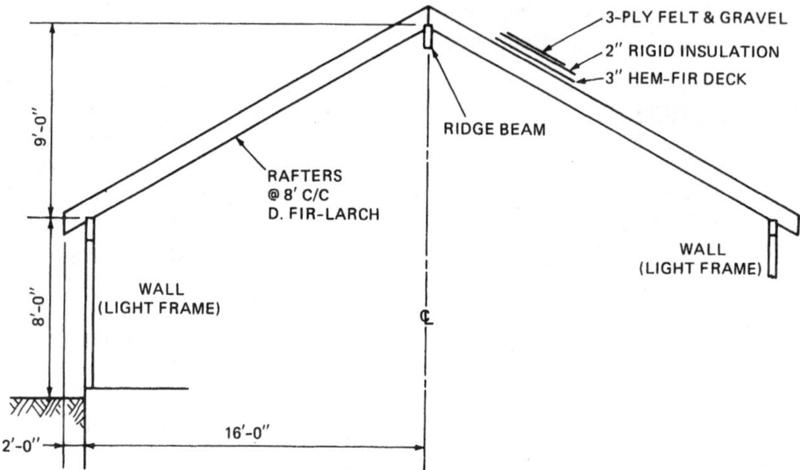

Fig. 4-12. Structure for Problem 4-6.

sonable combinations. Assume that the rafter weighs 17 lb/ft.

4-7. A three-story building with 16- by 20-ft bays is designed for a 70-psf floor live load. Floor joists are 2 × 12 D. fir-larch at 16 in. c/c with 5/8-in. plywood sheathing above. Finish floor weighs about 3 psf. Use the UBC recommendation for weight of partitions. The ceiling below is 1/2-in. drywall. Mechanical and electrical systems add about 10 psf. Beams spanning the 20-ft direction between columns weigh an average of 35 lb per linear ft. The roof system weighs the same as the floor system. Considering allowable reductions, what is the design load ($D + L$) per foot for interior beams supporting the 2 × 12 floor joists?

4-8. For the same building, and for the UBC minimum live load (not snow) on the flat roof, what is the design load ($D + L$) for a bottom-level interior column? Assume each column tier to be a 10 × 10 D. fir, 10 ft long.

4-9. Same as Problem 4-8, but with roof snow load of 30 psf instead of roof live load. What is the bottom-tier column design load ($D + L + S$)? Assume a flat roof.

4-10. If the code specifies a 75 psf floor live load for your classroom and if the floor is supported by beams at 12-ft centers (spanning the short direction), what is the design load ($D + L$) per foot for those beams? Make a reasonable assumption for the self-weight of the beams.

4-11. Estimated dead loads were shown on Fig. 4-5 for use in checking stability and in calculating required anchor-bolt forces in Examples 4-9 and 4-10. Ordinarily the designer tries to be safe by using a high estimate of dead loads. Is it a safe practice here? What would be the required anchor-bolt force if the true dead loads were 10% less than estimated?

4-12. Check the stability versus overturning for the office building of Fig. 4-13. Maximum wind speed is 70 mph and exposure is type B. The building length is 50 ft. Estimated dead loads (walls included) at various levels are shown. Use the Normal-Force Method.

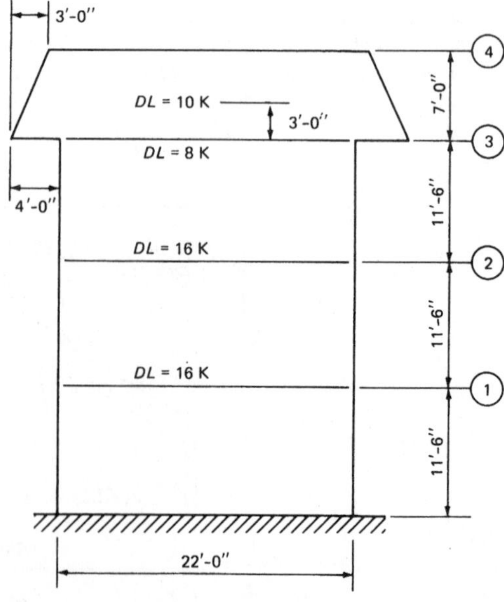

Fig. 4-13. Building for Problems 4-12, 4-13, and 4-14.

4-13. Repeat Problem 4-12 using the Projected-Area Method.

4-14. Check the stability of the building of Problems 4-12 and 4-13 when subject to earthquake. Use seismic-risk zone 2. Assume a box system.

For Problems 4-15 to 4-27, if your instructor requests that the LRFD method be used, find adjusted resistance values $\lambda\phi R'$, rather than adjusted allowable stresses. For either ASD or LRFD, unless it is noted otherwise, the loads specified are to be assumed as of ten-year (normal) duration.

4-15. Find the adjusted design values (allowable stresses) in bending and in shear for visually graded No. 1 southern pine 2 × 4s used as light framing. Spacing is 16 in. center-to-center. Moisture content in service is expected to be about 20%. The 2 × 4s are connected by rigid diaphragm material that can transfer load from one member to the other. The load is of short duration such as snow load.

4-16. Find the adjusted design values (allowable stresses) in compression parallel to the grain and in bending for No. 1 D. fir-larch 6 × 6s, used singly, as columns. Moisture content in service is expected to be about 20%. Wind loads are expected to control the design.

4-17. D. fir-larch 2 × 8s are to be used repetitively as floor joists. They are machine stress rated as 1800f-1.6E. Assuming the load duration to be permanent, what are the adjusted allowable bending and shear stresses, and modulus of elasticity, E'?

4-18. Visually graded southern pine commercial decking (repetitive) will have a 19% maximum service MC. The nominal thickness is 2 in. What is the adjusted allowable bending stress?

4-19. No. 2 D. fir-larch 2 × 6 joists are used to support a light roof. They are placed 12 in. apart and have a plywood deck nailed to their top surface. The loading on the roof is mainly snow load. Moisture in the wood is not expected to exceed 13%. What is the adjusted allowable stress in bending? In shear? In compression perpendicular to the grain?

4-20. If the joists of Problem 4-19 can be relied upon to have *no* splits or shakes in service, what are the adjusted allowable stresses in bending, shear, and compression perpendicular to the grain?

4-21. No. 2 hem-fir 2 × 8s are used flatwise as floor planks in a very moist environment. What are the adjusted allowable stress in bending (F'_b) and adjusted modulus of elasticity (E')?

4-22. Select Structural 4 × 14 redwood beams are used spaced 6 ft apart to support the roof over a laundry room. What are the adjusted allowable stresses for bending, for horizontal shear, and for compression perpendicular to the grain if only dead load is present? What are these allowables when the load includes both snow and dead load?

4-23. If the bending moments for the beams of Problem 4-22 are: dead load, 6,000 ft-lb; snow, 9,000 ft-lb; and wind, 10,000 ft-lb, what is the controlling load combination for the allowable bending stress? Consider all applicable UBC combinations.

4-24. A building floor consists of 3/4-in. plywood connected to No. 3 southern pine 2 × 12 planks, used flat at 12-in. centers. Assume the plywood to weigh 40 pcf. Live load is 40 psf. The planks are treated with fire-retardant chemicals; the treater claims this can be expected to reduce the bending strength by not over 10%. What are the adjusted allowable bending and shear stresses for the planks?

4-25. No. 2 D. fir-larch 2 × 8s are used as compression chords of flat-topped roof trusses at 24-in. centers (repetitive) to support 25 psf dead load (not including weight of the 2 × 8s) and either 20 psf live load or 35 psf snow load under moist-use conditions. Which loading condition controls? What are the adjusted allowable stresses in bending, compression parallel to the grain, and shear? Consider all NDS load combinations.

4-26. What are the allowable stresses in compression parallel to the grain, in bending, and modulus of elasticity, E', for 2 × 6 wall studs of 1500f-1.4E D. fir-larch spaced 16 in. c/c? Assume that the load combination that includes wind (but not snow) will control.

4-27. An existing structure is being investigated for possible increased loading. Beams are 8 × 18, No. 1 D. fir-larch, spanning 16 ft. Some of the beams have end splits 6 in. long on their wide face. Use conditions were, and will continue to be, very dry. Uniform loads are 400 lb/ft live load (reduction already considered) and 200 lb/ft dead load. A midspan concentration is 1 kip for dead load and 1 kip for snow. Which load combination controls? What are the adjusted allowables for: bending, horizontal shear, and compression perpendicular to the grain?

5

Connections—Nails, Screws, and Bolts

5-1. CONNECTION DESIGN

It is important that structural members—beams, columns, and tension members—be designed with due regard to safety and economy. But, no chain being stronger than its weakest link, it is equally important that the connections joining these members to each other be carefully designed. A connection must be able to transfer load from member to member without the connection material itself failing and also without damaging the wood of the members it joins. Design of structural members, addressed in later chapters, may be easier if the designer knows how the member will be connected and knows how the member in question may be weakened by the connection.

There is a wide variety of mechanical connectors (fasteners), ranging from the old wooden pegs to modern custom-made welded assemblies for joining together large glued laminated members. This chapter will cover the more common types—nails, staples, lag screws, wood screws, and bolts. Chapter 9 will present additional information on bolts, and also will cover modern types of timber connectors and specially designed weldments.

5-2. GENERAL PRINCIPLES

Certain principles apply to connection design in general, regardless of whether the fasteners are nails, spikes, screws, lag screws, or bolts. The capacities of each of these fasteners is affected by specific gravity and moisture content of the wood; by dimensions of the connector and the wood members; by the type, number and arrangement of the fasteners; by the position of the fastener relative to the grain; and (in most cases) by the direction of the fastener force relative to the grain.

Some of the above conditions are accounted for in connection design through the use of adjustment factors. Certain adjustment factors that are applicable to both members and connections have already been described in Chapter 4. For design of connections that are laterally loaded, all applicable adjustment factors are multiplied times the base lateral design value (base allowable lateral load), Z, resulting in an adjusted lateral design value, Z'. The base and adjusted withdrawal design values are referred to as W and W,′ respectively.

Base Lateral Design Value

For the dowel-type fasteners discussed in this chapter, the base value, Z, is found by examining several different failure modes, called *yield modes*. These modes are generally referred to by Roman numerals as follows:

I. Mode I is a bearing failure of the wood fibers as the fastener (nail, bolt, etc.) shank presses against the wood. The bearing failure can occur in the wood of either the "main" member (the thicker member in a bolted connection or the member holding the point for a nailed, screwed, or lag screwed connection) or the "side" member.

II. Mode II is a pivoting of the fastener at the shear plane of a single-shear connection with limited localized crushing of wood fibers near the faces of the wood members. This mode does not occur in single-shear nail or screw connections with minimum required fastener penetration.

III. Mode III is a failure of the fastener by yielding in bending at one plastic hinge point per shear plane, accompanied by bearing failure of wood fibers in contact with the fastener. It can occur in either the main or side member.

IV. Mode IV is a failure of the fastener by yielding in bending at two plastic hinge points per shear plane, with limited localized crushing of wood fibers near the shear planes.

The fastener's capacity under each of these yield modes is computed; the smallest of them is the base lateral design value, Z. Containing many variables such as dowel bearing strength of the wood members, yield strength of the fastener in bending, and thickness of the wood members, the yield mode equations are very complex. Luckily, tables of Z values are available for many connections, and therefore the designer will often not have to compute the numerical values of each of the yield modes. The yield mode equations can be found in the *National Design Specification for Wood Construction* (1) or the *Uniform Building Code* (2).

Specific Gravity

Wood with high specific gravity is stronger than wood with low specific gravity. This affects the allowable load for a fastener. For the types of timber connectors covered in this chapter, the effect of specific gravity is accounted for by arranging the tables of lateral and withdrawal design values (see Appendix C) according to wood species.

Moisture Content

Moisture conditions of the wood also affect fastener strength. Appendix Table C-2 gives the adjustment factor, C_M, for this effect. Notice that for many types of fastener, it is not the actual moisture content at any given time, but rather the *change* of moisture content between the time of installation and service that affect fastener strength markedly. This is because any MC change is accompanied by dimensional changes (shrinkage or swelling) that may tend to withdraw or loosen the fasteners, or in some cases, split the wood.

For example, timber structures often are assembled using green lumber. This green lumber (with a high MC) then seasons after the connection is made. If the arrangement of the fasteners does not allow the wood to shrink freely, a significant force develops in each fastener as it acts to prevent the wood from shrinking. This force is present in the wood, also, and may cause the wood to split.

Geometry (Spacing and Distances)

In connection design, important dimensions are center-to-center spacing of fasteners, edge distance, and end distance. Center-to-center distance between fasteners (spacing) and distance from the center of the fastener to the end of the wood member or to its edge (end distance or edge distance) will reduce the connection strength if the distances are too small. This occurs as the wood between fasteners fails in either longitudinal shear or tension perpendicular to the grain. For laterally loaded bolts and lag screws, an adjustment factor called the *geometry factor,* C_Δ, is used when end distances and spacings are less than specified minimums. Because of their smaller diameters and lower loads, wood screws, nails, and spikes are not affected much by fastener arrangement, and no geometry factor is used.

Group Action (Number in a Row)

The number of fasteners in a row parallel to the direction of applied load can affect the strength of a connection. When a row contains several connectors, it is inaccurate to say that the fasteners in that row share the load equally. For laterally loaded bolts and lag screws only, the group action factor, C_g, is applied to account for this. (Factor C_g is applied also for some of the fastener types introduced in Chapter 9.)

Penetration Depth

If the tip of a nail (or similar pointed connector) has too small a penetration depth into the second member pierced, the shear capacity of the connection is reduced. Therefore, for laterally loaded nails, spikes, wood screws, and lag screws, the penetration depth factor, C_d, is used when penetration into the wood is less than a specified minimum penetration.

End Grain (Shank Position Relative to Grain)

When a fastener is driven or screwed into end grain (with the shank parallel to the wood fibers), the capacity is not as great as when the fastener is driven into side grain. For this reason, an adjustment factor called the *end grain factor*, C_{eg}, is used for certain connectors inserted into end grain. For laterally loaded nails, spikes, screws, or lag screws, C_{eg} is 0.67. For lag screws loaded in withdrawal, the adjustment C_{eg} is 0.75.

Direction of Fastener Force Relative to the Grain

Angle of load to grain (Fig. 5-1) affects the strength of fasteners. As load is transferred from one member to another by shear in a screw or bolt, the side of the metal shank presses against the wood. Where this bearing pressure is in the direction of the grain, the resistance will be fairly high; but where it is perpendicular to the grain, resistance will be much lower. This is because the compressive strengths of the wood differ in the two directions. For both bolts and lag screws, shear (lateral) capacity of the connection will be different for parallel-to-grain and perpendicular-to-grain loads.

Often, the fastener causes pressure at some other angle, neither parallel to the grain nor perpendicular. In this case, an allowable dowel bearing value for the fastener is approximated by use of the Hankinson equation. (Nails, spikes, staples, and wood screws are excluded from this requirement, since, for them, joint slip, rather than ultimate strength, is considered the more important criterion.) The Hankinson formula for dowel bearing strength at an angle to grain is

$$F_{e\Theta} = \frac{F_{e\|}F_{e\perp}}{F_{e\|}\sin^2\Theta + F_{e\perp}\cos^2\Theta} \qquad (5\text{-}1)$$

in which $F_{e\|}$ is the dowel bearing strength parallel to grain, $F_{e\perp}$ is dowel bearing strength perpendicular to grain, and θ is the acute angle between the load direction and the grain (longitudinal) direction.

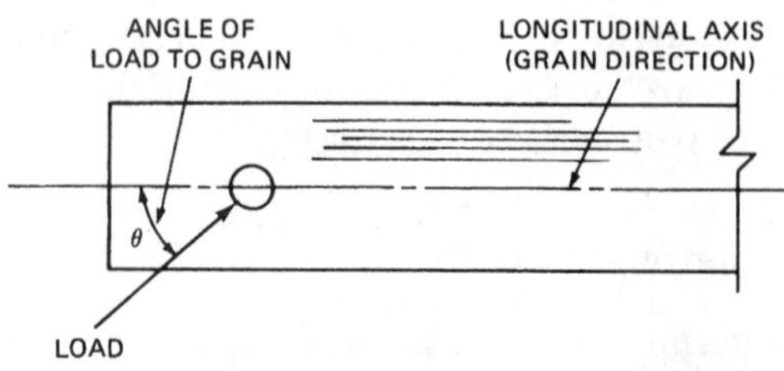

Fig. 5-1. Angle of load to grain.

Net Section

If a hole is drilled in the member (for a bolt or lag screw, for example), the connection reduces the cross-sectional area of the member. In this case, the *net* area of the member may need to be considered when computing the member strength. The cross-sectional area is not considered to be reduced when the fastener merely pushes aside the wood fibers, as does a driven nail.

Load Duration

The load duration adjustment, C_D, discussed in Chapter 4, applies also to connections. Because of the critical nature of connections, the *Uniform Building Code* gives some values of C_D for connection design that are smaller than those given by the NDS. The UBC specifies a C_D of 1.33 for design of connections for wind or earthquake loads rather than the 1.6 given by the NDS. The authors prefer to use the more conservative 1.33 value.

Others

For nails only, other adjustments are the diaphragm factor, C_{di}, and the toe-nail factor, C_{tn}. These will be covered in Section 5-3.

The strength of any connection is improved when the outer pieces joined are steel plates. The added strength results from the stiffness of the plates, which reduces joint slip, bearing deformation, and tilting of the nail, screw, or bolt in the wood as the connection is loaded. Steel side plates have a much higher dowel bearing strength than wood side members, so a larger Z will result if side plates are steel.

Nails, staples, wood screws, or bolts may be used in combination with glue. However, the glued connection is very stiff, whereas nails, for example, must deform appreciably before they resist significant shear. Thus, in a combination joint, the glue tends to resist the total load, the mechanical fasteners not receiving load until the glued connection is broken.

5-3. NAILS AND SPIKES

The variety of available nails and spikes is almost unbelievable—over 10,000 different varieties are produced in the United States during one year (3). An excellent summary of the available types and sizes is given in reference 4. For our purpose, it is sufficient to recognize the three major types (common wire nails, box nails, and common wire spikes) and various sizes shown by Appendix Tables C-3 and C-4. Nail types are depicted in Fig. 5-2, along with some other fasteners covered in this chapter. Common nails have a larger diameter for a given length than either box or threaded nails. Threaded nails have been hardened by heat treating and have greater strength than common nails or box nails of equal diameter.

For lengths up to 9 in., the size of a nail or spike is specified by *pennyweight,* an antiquated term indicating the purchase price, in English pennies, per 100 nails. Today the term is still used, but it may soon be replaced by naming nail sizes by diameter and length.

Nails are usually driven without first drilling a hole, but for either nails or spikes, a pilot hole may be drilled. This will not reduce the allowable load for the fastener, provided the hole diameter is not over 0.9 fastener diameters for species with specific gravity greater than 0.60, or 0.75 fastener diameters for other species.

Allowable loads for nails or spikes depend on several things, some of which are:

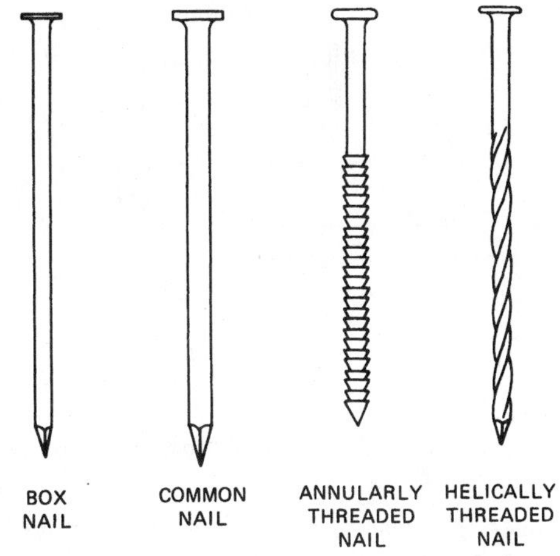

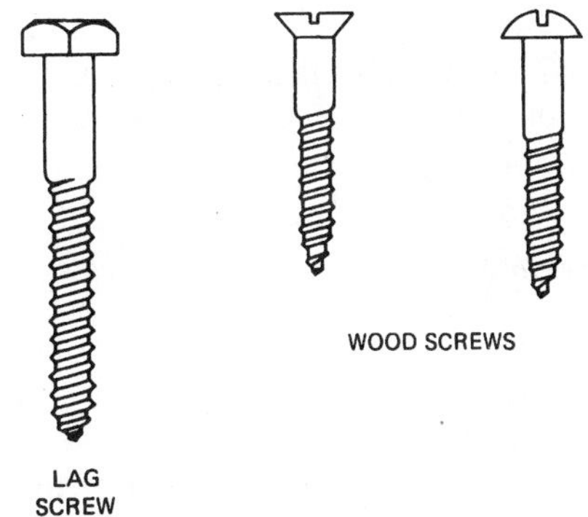

Fig. 5-2. Nail and screw types.

1. Diameter, surface shape, and surface treatment of the shank.
2. Specific gravity and moisture content of the wood.
3. Depth to which the point penetrates into the wood of the member that contains the nail point.
4. Direction of the nail shank relative to the grain.
5. Direction of the load relative to the shank length.

Since they are determined by limits of deformation rather than strength, allowable lateral (shear) loads for nails and spikes are the same regardless of whether the lateral load is parallel to or perpendicular to the grain.

Nails and spikes can transfer either lateral load or load parallel to their length (withdrawal loading), or a combination of the two. They are stronger when driven into side grain than when driven into end grain (Fig. 5-3a). They are

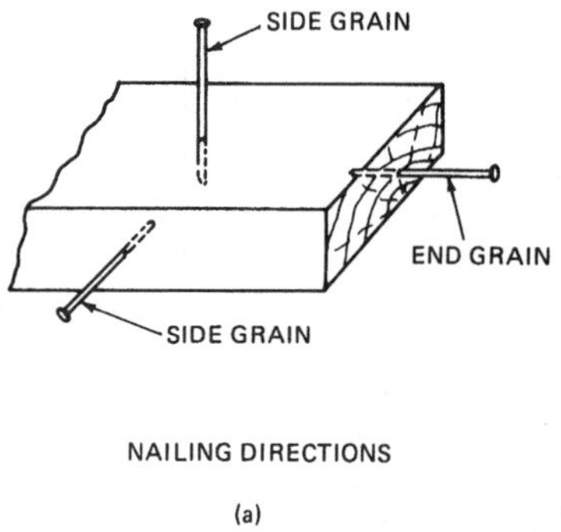

SIDE GRAIN
END GRAIN
SIDE GRAIN
NAILING DIRECTIONS
(a)

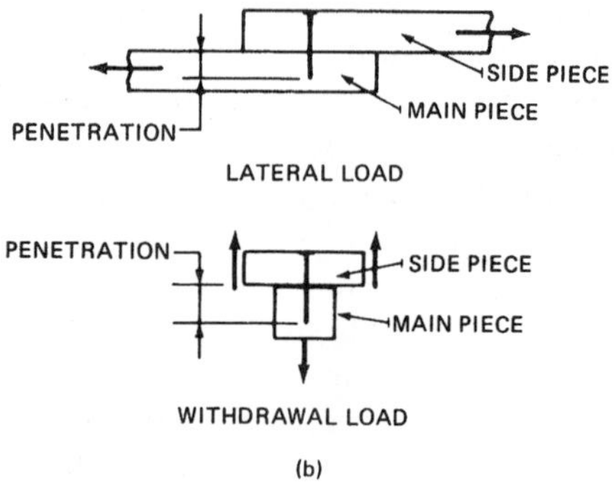

SIDE PIECE
MAIN PIECE
PENETRATION
LATERAL LOAD
PENETRATION
SIDE PIECE
MAIN PIECE
WITHDRAWAL LOAD
(b)

Fig. 5-3. Nailing and load directions. (Courtesy Board of Regents, University of Colorado.)

stronger in shear (lateral load) than in withdrawal (Fig. 5-3b). Withdrawal strength from side grain is much less than for shear, and the designer should avoid this condition if at all possible. Withdrawal strength from end grain is so small and so unreliable that the NDS prohibits using nailed connections that depend on withdrawal from end grain.

Lateral Resistance

The capacity of a nail or spike under lateral (shear) loads is a function of the following yield modes: Mode I in the side member, Mode III in the main member, Mode III in the side member, and Mode IV. The least of the four computed yield mode values is the allowable design lateral load, Z, *in any direction* for common or box nails or spikes driven into side grain of seasoned wood in a single-shear connection. Tables C-3 and C-4 give some values of Z, the work of finding the smallest value from the four yield mode equations already having been done. Because of the complexity of the yield mode equations, it is obvious that designers prefer to use tables such as these. Both the yield mode equations and the tables are limited to nails or spikes having *standard penetration* into the wood member that contains the point of the nail. Penetration is the length of nail in the piece of wood that contains the nail point (Fig. 5-3b). Standard penetration is twelve times the nail or spike diameter, $p = 12D$. For lesser penetrations, the penetration depth factor, as explained below, is used.

The base *lateral* design value must be adjusted as follows, and the adjustment factors are cumulative:

1. For load duration, using the factor, C_D.
2. For less-than-standard penetration. Use penetration depth factor, $C_d = p/(12D)$. In no case, however, should the nail or spike be used to carry load when its penetration is less than $6D$.
3. For moisture content, according to C_M listed in Appendix Table C-2.
4. For double shear (Fig. 5-4), by multiplying by two the Z value found for the weaker of the two shear planes.
5. For laterally loaded toe nails, using the factor $C_{tn} = 0.83$. (For withdrawal loads, use $C_{tn} = 0.67$.)
6. For diaphragm construction, using the diaphragm factor, C_{di} (see Chapter 12).
7. If driven into end grain rather than side grain, a factor $C_{eg} = 0.67$ applies.
8. For temperature, using C_t.

Designers of nailed double-shear connections with wood side members should keep in mind the following: (1) The thickness of the main

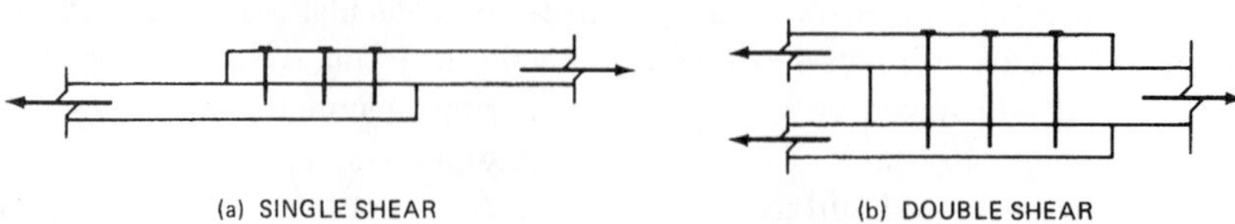

(a) SINGLE SHEAR (b) DOUBLE SHEAR

Fig. 5-4. Single- and double-shear connections.

(center) member must be greater than six nail diameters (6*D*); (2) the penetration depth factor, C_d, is based on penetration into the third member penetrated. As for a single-shear connection, penetration depth cannot be less than 6*D*.

Withdrawal Resistance of Nails in Side Grain

Withdrawal resistance is less than lateral resistance. It depends on the nail or spike diameter, the length of penetration, and the specific gravity of the wood. The withdrawal allowable load *per inch* of penetration into side grain of the member holding the nail point may be computed by

$$W = 1380DG^{2.5} \qquad (5\text{-}2)$$

in which D is the nail diameter (inches) and G is the specific gravity. The allowable value given by Eq. 5-2 is only for common nails, spikes, and box nails, and only for those driven into side grain, either of wood that is seasoned and will remain seasoned, or of wood that is unseasoned and will remain that way in service. This equation is based on one-fifth of the ultimate load from tests. Average values of specific gravity are found in Appendix Table C-1. Nails that will be loaded in withdrawal should be spaced so as to prevent splitting. Woods with lower specific gravity generally do not split as easily as heavier ones, so in the lighter woods, the nail spacing can be less, and the apparent advantage of denser woods, suggested by Eq. 5-2, may be offset.

Nails or spikes are frequently "cement coated." The coating is not cement, but usually a resin that increases the withdrawal resistance. In harder woods, the coating may be damaged by friction during driving. Unfortunately, the effectiveness of the cement coating diminishes with time, so that after only one or two months, withdrawal resistance may be no better than for an uncoated nail. Nails or spikes having grooved or threaded shanks also have greater withdrawal resistance than plain nails, and this added resistance tends to be of long-lasting value (see reference 3).

The structural designer should preferably avoid using nails under withdrawal loading. Furthermore, the NDS prohibits using nails driven into end grain for withdrawal loading. Appendix Table C-2 shows how severely the allowable withdrawal loads must be reduced if moisture content changes occur during service—to only 25% of the normal design allowable.

The following adjustment factors should be used for nails or spikes in withdrawal from side grain: $C_{tn} = 0.67$, C_D, C_M (except for toe-nailed joints), and C_t.

Toe Nailing

Figure 5-5 shows what is meant by *toe nailing.* This method of nailing is commonly used to connect vertical studs to the horizontal plates or sills in light-frame construction. Tests show that the highest strength is obtained if one uses the largest nail that can be driven without splitting the vertical member, having a distance from the end of the member to where the point first enters the wood equal to about one-third the nail length, driving the nail at about 30° with the vertical member, and "burying" the full length of the nail but avoiding excessive damage to the wood. Toe nailing is also more effective when the nails are driven from both sides of the vertical member, that is, cross-slant driving. The allowable design load for laterally loaded toe nails is 0.83 times the lateral loading design value for nails driven into side grain. The allowable design load for toe nails in withdrawal is 0.67 times the withdrawal design load for nails driven into side grain. The same adjustments apply as for regular nailing, with the exception of the adjustment, C_M, for change of moisture content, which is not used for toe nails.

Example 5-1

A 3-in. (nominal) D. fir-larch tension member is spliced using 16 gauge steel side plates, as shown in

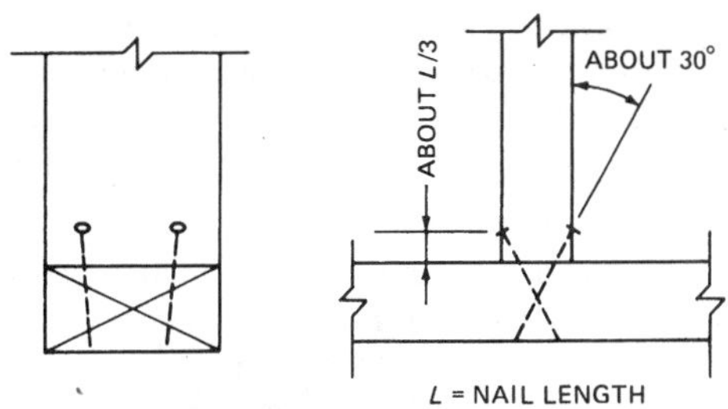

Fig. 5-5 Toe nailing.

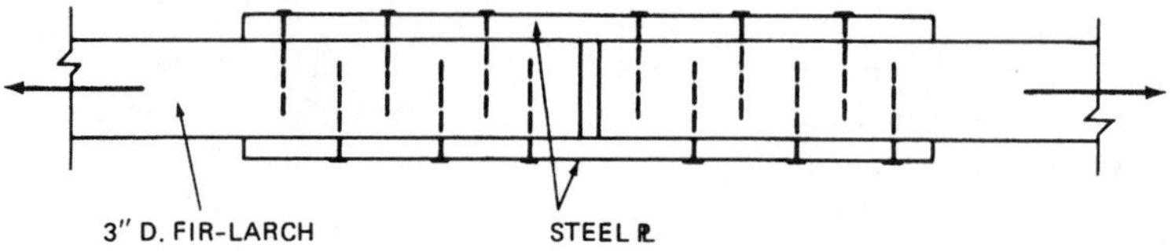

3" D. FIR-LARCH STEEL ℞

Fig. 5-6. Connection for Example 5-1.

Fig. 5-6. The wood was dry when the splice was made, but will be subject to wetting and drying in use. The member load is 400 lb (D = 90 lb, L = 120 lb, and W = 190 lb). How many 8*d* box nails are required through each end of each splice plate?

Find the controlling load combination:

$$90/0.9 = 100$$

$$(90 + 120)/1.0 = 210$$

$$(90 + 120 + 190)/1.33 = 301 \textit{ Controls}$$

Use $C_D = 1.33$

For a single-shear connection,

Lateral design value Z = 70 lb/nail (Appendix Table C-4)
Standard pene. = $12D$ = 12 (0.113) = 1.356 in.
Length of 8*d* box nail = 2.5 in.
Thickness of 16 ga. plate = 0.06 in.
Actual pene. = 2.5 − 0.06 = 2.44 in. > 12D, so $C_d = 1.0$
$C_M = 0.75$
$Z' = C_M C_D Z = (0.75)(1.33)(70) = 69.8$ lb/nail
400/69.8 = 5.7

This, of course, is rounded up to 6 per side of the splice. There are two splice plates, so three nails per end per plate will be needed, for a total of twelve.

Example 5-2

Figure 5-7 shows a single-shear connection of a 2-in. nominal southern pine member to a 4-in. nominal Douglas fir-larch member. How many 16*d* common wire nails are needed to transfer the controlling 400-lb reaction, which is composed of 90 lb dead load and 310 lb snow load? The wood is dry when installed and will remain dry in use.

Find the controlling load combination: 90/0.9 < (90 + 310)/1.15, so snow loading controls and $C_D = 1.15$.

Of the two species, Douglas fir has the smaller specific gravity, so the solution is as if both members were of Douglas fir. For a 1.5-in. side member,

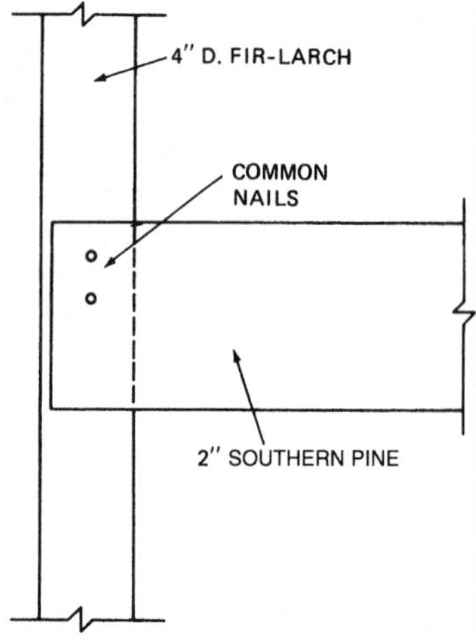

Fig. 5-7. Connection for Example 5-2.

Z = 141 lb/nail.
Standard pene. = $12D$ = 12 (0.162) = 1.94 in.
Length of 16*d* common nail = 3.5 in.
Actual pene. = 3.5 − 1.5 = 2 in. > 1.94, so $C_d = 1.0$
$C_M = 1.0$
$Z' = C_D Z = (1.15)(141) = 162$ lb/nail
400/162 = 2.5 = 3 nails

Example 5-3

Find the number of nails necessary to make a butt splice in which the main member is 2-in. nominal thickness and the side members are 1-in. nominal. The main member carries a 10-year duration load of 1000 lb. Timber is Douglas fir-larch.

(a) Use 8*d* common nails and single shear, and (b) Use 12*d* common nails and double shear.

(a) For a 3/4-in.-thick side member and 8*d* common nails,

Z = 90 lb
Standard pene. = 12D = 12 (0.131) = 1.57 in.
Nail length = 2.5 in., so penetration into main (center) member is 1.5 in. (Nails actually penetrate 1/4-in. into the third member (last member penetrated), but this is discounted as contributing no strength, because 1/4 in. is less than 6D)

$C_d = p/12D = 1.5/1.57 = 0.955$
C_M (dry conditions assumed) $= 1.0$; $C_D = 1.0$
$Z' = (90)(0.955) = 86.0$ lb/nail
$1000/86.0 = 12$ nails per side of the splice

(b) For a 3/4-in.-thick side member, 12*d* common nails, and double shear,

$Z = 2 \times 105 = 210$ lb
Standard pene. $= 12D = 12(0.148) = 1.78$ in.
Nail length $= 3.25$ in., so penetration into third member penetrated $= 0.75$ in.
$6D = 0.89$ in. > 0.75 in.

The minimum penetration requirement has not been met. A double-shear connection cannot be made with 12*d* common nails and these wood members. (In fact, no common or box nail of length at least 3 in. (the sum of the thicknesses of the wood members) can meet the minimum penetration requirement.)

Example 5-4

A single-lap joint joining a pair of Douglas fir 2 × 6s is to be connected with 16*d* common nails. If the joint must transfer 800 lb, normal duration, how many nails are required?

The solution below illustrates how to compute the base lateral design value, *Z*, by choosing the smallest value from the yield mode equations. (The yield mode equations can be found in the NDS or in the *Uniform Building Code*.) A much easier method of finding $Z = 141$ lb is to use Appendix Table C-3.

Data for entry into the yield mode equations:

$F_{em} = F_{es} = 4650$

$R_e = F_{em}/F_{es} = 1$

$F_{yb} = 90{,}000$ psi (Footnote, NDS Table 12.3 B)

$D = 0.162$; $K_D = 2.2$

Pene. $= p = 1.5$ in.

k_1 (by long equation) $= 1.110$

k_2 (by long equation) $= 1.110$

The four yield mode equations are now solved to determine the controlling (least) allowable lateral load Z.

1. $Z = D\, t_s\, F_{es}/K_D = 0.162\,(1.5)\,(4650)/2.2$
   $= 514$ lb
2. $Z = k_1\, D\, p\, F_{em}/[K_D\,(1 + 2R_e)]$
   $= 1.110\,(0.162)\,(1.5)\,(4650)/(2.2 \times 3)$
   $= 190$ lb
3. $Z = k_2\, D\, t_s\, F_{em}/[K_D\,(2 + R_e)]$
   $= 1.110\,(0.162)\,(1.5)\,(4650)/(2.2 \times 3)$
   $= 190$ lb
4. $Z = (D^2/K_D)\,(2F_{em}F_{yb}/(3(1 + R_e))^{1/2}$
   $= (0.162^2/2.2)\ [2 \times 4650 \times 90{,}000/(3 \times 2)]^{1/2}$
   $= 141$ lb *Controls*

This 141 lb is the same as the Table value. Notice that even though 1/2 in. of the nail is "sticking out," this does not affect the answer (i.e., whether penetration is 1.5 in. or 2 in., the base *Z* value is the same). This is because the fourth equation (which is not a function of penetration) controls.

Completing the rest of the solution,

$C_d = 1.5/12D = 1.5/(12 \times 0.162) = 0.772$

$Z' = C_D\, C_d Z = (1.0)\,(0.772)\,(141) = 109$ lb

N required $= 800/109 = 7.3$. Use 8 nails.

Example 5-5

A double-lap connection similar to Fig. 5-4b is made using twenty-four 40*d* common nails through all three members and with their points projecting. All three members are 2-in. nominal Douglas fir-larch that was seasoned at the time of installation and will have a moisture content under 19% in use. What is the allowable permanent load? Would this allowable load change if the nails were clinched (bent over with the shank flat against the side piece)?

For single shear, $Z = 205$ lb/nail
Since double shear, $Z = 2 \times 205 = 410$ lb/nail
Nail length $= 5$ in.
Total thickness of the 3 members is 4.5 in., so penetration into the last member penetrated $= 1.5$ in. (The nails project 1/2 in.)
Standard penetration $= 12D = 12\,(0.225) = 2.70$ in.
$6D = 1.35$ in. < 1.5 in. *OK*
$C_d = p/12D = 1.5/2.70 = 0.556$
$C_D = 0.9$; $C_M = 1.0$
$Z' = C_d C_D Z = (0.556)\,(0.9)\,(410) = 205$ lb/nail
Allowable load $= (205$ lb/nail$)\,(24$ nails$) = 4920$ lb

No increase in allowable load is allowed if nails are crimped.

Example 5-6

What is the allowable load per nail for the withdrawal connection of Fig. 5-3b, if the load is a ten-year duration load consisting of 80% occupancy live load and the remainder dead load? The side piece is 2 in. nominal and the piece containing the point 6 in. nominal. The nails are 12*d* box nails, driven into side grain of No. 2 hem-fir, seasoned when installed, but subject to long periods of exposure to wet use.

Specific gravity of hem-fir = 0.43 and $D = 0.128$ in.
$W = 1380\,(0.128)\,(0.43)^{2.5} = 21$ lb per inch of penetration
$C_D = 1.0$; $C_M = 0.25$
$W' = 0.25\,(21) = 5.25$ lb per inch of penetration
Nail length = 3.25 in.
Penetration = 3.25 − 1.5 = 1.75 in.
Allowable withdrawal load per nail
= (5.25) (1.75) = 9.2 lb

Obviously, this connection is practically worthless for load-carrying members.

Example 5-7

Toe nails are used to connect 2 × 4 hem-fir studs to the sill (bottom) plate in a stud wall. With studs at 16 in. c/c, are two 16*d* common toe nails per stud sufficient to resist a shear load (due to wind) of 112 lb/ft of wall?

For a 1.5-in.-thick side member,
$Z = 122$ lb/nail
Standard pene. = $12D = 12\,(0.162) = 1.94$ in.
Length of a 16*d* nail = $L = 3.5$ in.
If properly driven, penetration into member containing the point is $\frac{2}{3}L = 2.33$ in., which is greater than 1.94 in., so $C_d = 1.0$.
$C_M = 1.0$ (dry conditions assumed); $C_D = 1.33$
$C_{tn} = 0.83$
$Z' = (122)\,(1.33)\,(0.83) = 135$ lb/nail

Load per stud = (112 lb/ft) (1.33 ft) = 149 lb
Nails required = 149/135 = 1.1 = 2 nails (16*d*) per stud

5-4. STAPLES

Wire staples are seeing increasing structural applications, such as attaching plywood diaphragms to framing members (Chapter 12). Staples are frequently used for factory-prefabricated items, such as wood pallets. They are also frequently used in combination with gluing. The reason for the growing popularity of staples is that they can be power-driven, saving hours over the time required for installing hand-driven fasteners.

The NDS does not give design values for staples, but information can be obtained from manufacturers of power staple driving equipment. The American Plywood Association has published information on the strength of staples in joints connecting plywood to lumber, as in a diaphragm (5).

Since the diameter of the staple leg is quite small, staples can be placed quite close together without splitting the wood. When plywood diaphragms are attached to framing members, the crown of the staple should be placed parallel to the long dimension of the framing member so that both legs go into the member.

Staples can have good withdrawal resistance as well as lateral load resistance. As for nails, withdrawal resistance can be improved by using coated staples.

5-5. LAG SCREWS

Lag screws, also known as *lag bolts,* are made with the unthreaded portion of their shanks in the same diameters as machine bolts. Dimensional information is shown by Appendix Table C-5. (This table also shows dimensions for small-sized lag screws that are not recognized in NDS tables of allowable loads.) Lag screws are especially useful for connections where bolts are not possible, either because the material is too thick or the far side of the connected parts is not accessible. Their behavior is much like that of bolts in single shear.

Lag screws are installed in predrilled holes. For the piece receiving the threaded part, the pilot hole should be a little less than the shank diameter: NDS specifies different hole diameters for woods of different specific gravity—65–85% of shank diameter for G greater than 0.60, 60–75% for G between 0.50 and 0.60, and 40–70% for G less than or equal to 0.50. The predrilled hole in parts receiving the shank should be the same diameter as the shank. A washer under the head of the lag screw will protect the wood from damage as the screw is tightened. Installation is by turning the head with a wrench, the threads cutting their way into the wood as the screw is tightened.

Lateral (Shear) Capacity

Lateral-load capacity of a lag screw varies with species (specific gravity) and with angle of load to grain. Lag screws are not used for double-shear connections, so all of the following applies to single-shear connections alone. Shear capacity is a function of yield modes I_s, III_s, and *IV*. Because the yield mode values are functions of angle of load to the grain, three values of Z will result—one for load parallel to grain ($Z_{\|}$), one for load perpendicular to grain in the side member ($Z_{s\perp}$), and one for load perpendicular to grain in the main member ($Z_{m\perp}$). Appendix Table C-6 shows design values (allowable loads) parallel to the grain and perpendicular to the grain for lag screws in joints with a wood side piece (the part nearest the head of the lag screw), and Appendix Table C-7 shows similar values for joints with a steel side piece. When the load is neither parallel nor perpendicular to the grain, and the lag screw is installed in side grain of the main member, Hankinson's formula (Eq. 5-1) is used to determine the allowable load for the lag screw.

The base lateral design value is adjusted using the following factors:

1. Load duration, C_D
2. Wet service, C_M
3. Group action, C_g (see below)
4. Geometry, C_Δ
5. Penetration depth, C_d
6. End grain. If screwed into end grain rather than side grain, $C_{eg} = 0.67$ for lateral loading. (It is different for withdrawal loading.)
7. Temperature, C_t

Group Action (Number in a Row)

Group action factor, C_g, applies to both lag screws and bolts. If there is a large number of lag screws or bolts in a row (see Fig. 5-8 for definition of row), each additional fastener is of diminishing effectiveness, and each does not carry an equal share of the load. In structural steel, because of yielding, it is an acceptable design assumption for connections of reasonable length to say that the fasteners share the load equally. In wood this is not the case.

In 1975, a paper by C. O. Cramer (6) presented a method by which actual loads on each connector in a row parallel to the load might be computed. His method considered the deformation of each fastener (see Fig. 5-9) and of the wood members between fasteners. Using an equation relating the elongation of the side pieces and main piece, he derived a means for calculating the fastener loads. This method led to reduction factors that allow the designer to make the simplifying assumption that all fasteners in the row share load equally (even though we know that the fasteners at the ends of the row take considerably more than their equal share of the load transferred). Appendix Tables C-8 and C-9 show this modification for bolts and for lag

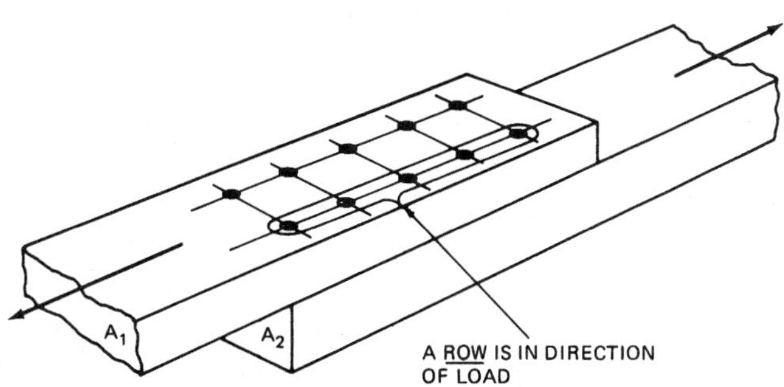

Fig. 5-8. Two-row connection.

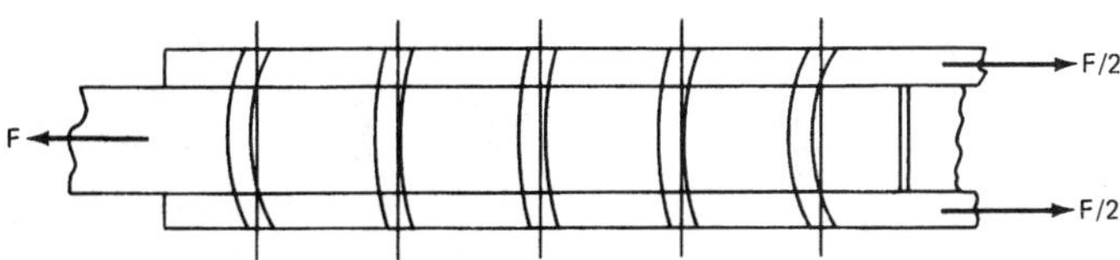

Fig. 5-9. Deformation in a row of bolts.

screws. Footnotes to the tables show how the ratio of areas is computed.

Note that the effectiveness of a fastener can range from 100% to less than 50%. Obviously, a penalty is paid for using a large number of fasteners in a row. One alternative is to use more rows but fewer lag screws per row. For connecting tension members this alternative has a disadvantage also—the net area of the member is reduced as additional rows are added. This effect can be reduced, however, by staggering the lag screws of one row relative to those of the next, as shown by Fig. 5-10. If the stagger is at least four fastener diameters, the net section reduction by the second row does not have to be added to that by the first row.

Geometry (Spacing and End Distance)

End distance and spacing must be sufficient that tearout does not control the connection strength. If these distances are less, then the allowable load must be reduced by application of the geometry factor, C_Δ. Or, worded conversely, if the anticipated load on a fastener is less than the maximum adjusted allowable load, both spacing and end distance may be reduced. (These same principles and rules apply to both lag screws and bolts.)

Figure 5-11 summarizes the NDS rules for bolts and lag screws loaded either parallel or perpendicular to the grain. The NDS avoids specifying spacing requirements for lag screws with loads at some other angle to the grain. (A guide might be found, however, in the practice shown by Chapter 9 for shear plates and for split-ring connectors.) End distances in Fig. 5-11 are for ends cut at 90° to the length. What should the end distance be when the end is cut diagonally? The NDS does not say for lag

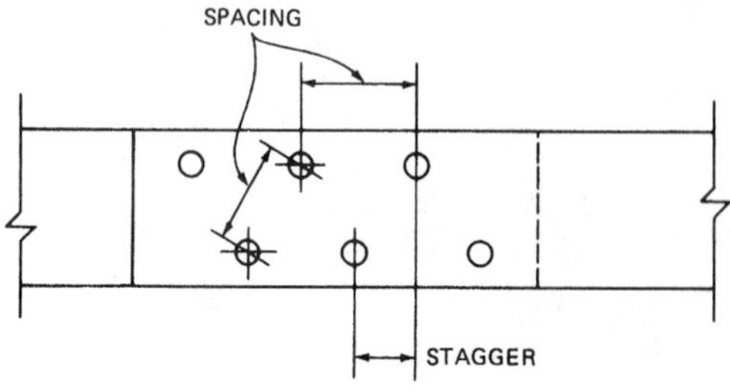

Fig. 5-10. Spacing and stagger.

screws (or bolts); however, it does present a method of measuring such end distances for timber connectors. The authors feel that this is a conservative approach, suitable for bolts or lag screws also.

Penetration Depth

Standard penetration of a lag screw (not including the length of the tapered tip) is eight times the shank diameter, $p = 8D$. For lesser penetration the penetration depth adjustment factor, $C_d = p/(8D)$, is used for laterally loaded connections. In no case should penetration depth be less than $4D$ in a laterally loaded lag screw connection.

Withdrawal Capacity

For capacity against withdrawal, lag screws are much more effective than nails. For shorter lengths of penetration of the threaded part into the wood, it is the wood that fails, allowing the screw to withdraw. For longer penetration of the threaded part, the wood does not fail, but the screw itself fails in tension. Thus, the tensile strength of the screw is a maximum beyond which the withdrawal strength cannot go, no matter how much the effective penetration length (see dimension *T–E* in Appendix Table C-5).

For embedment lengths in which the wood fails, withdrawal resistance is affected by specific gravity of the wood. This is reflected by the following equation for withdrawal design value,

$$W = 1800D^{0.75}G^{1.5} \tag{5-3}$$

which gives allowable withdrawal load, *W*, per inch of penetration of the effective threaded length. (The unthreaded shank does not add to withdrawal strength.) In the equation, *D* is the unthreaded shank diameter, and *G* is the wood specific gravity. In addition to determining an allowable load based on strength of the wood, the designer must check to make sure that the withdrawal load does not exceed an allowable based on tensile strength of the threaded shank.

Since the yield strength of material used for lag screws is usually 45,000 psi, the allowable withdrawal load based on tensile strength is (45,000) (thread root area/factor of safety). The

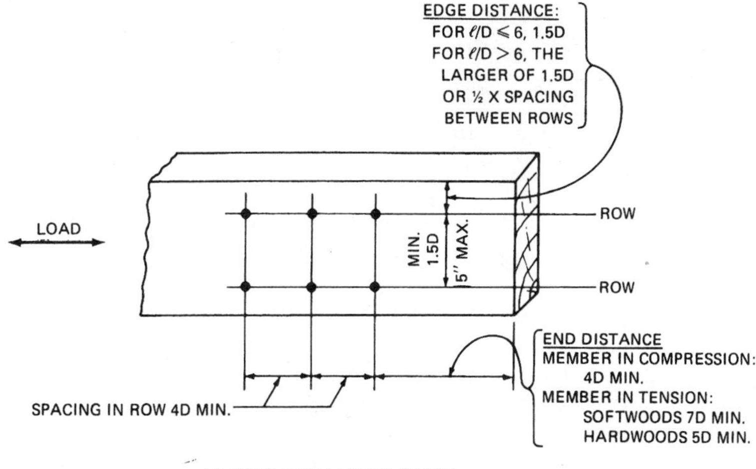

(a) LOAD PARALLEL TO GRAIN

EDGE DISTANCE: 4D MIN. FOR EDGE TOWARD WHICH FASTENER IS PUSHING
CONTROLLED BY OTHER MEMBER BUT 5" MAX.
EDGE DISTANCE 1.5D MIN.
ROW SPACING FOR ℓ/D = 2, 2.5D MIN. FOR ℓ/D ≥ 6, 5D MIN. (INTERPOLATE) 5" MAX.
ROW ROW
LOAD

(b) LOAD PERPENDICULAR TO GRAIN (OF HORIZONTAL PIECE)

Fig. 5-11. Spacing, end distance, and edge distances for lag screws and bolts. (D = shank diameter, l = length in main member.)

NDS does not specify values for this allowable, but the designer can easily calculate it for any desired safety factor. Older editions of the NDS pointed out that the approximate tensile value of the lag screw would be developed when the penetration of the threaded portion into the piece containing the point was 7–11 diameters, depending on the wood specific gravity.

The adjustment factors that are used for withdrawal loading of lag screws are C_D, C_M, C_t, and $C_{eg} = 0.75$.

Example 5-8

A Douglas fir 2 × 6 is connected to a 4 × 6 of the same species in a single-lap connection as shown by Fig. 5-12. The permanent axial load to be transferred is 2.08 kips. Choose a size, length, and arrangement of lag screws to make the connection.

The maximum length of lag screw that can be used for this connection is 5 in., the total thickness of the two members. Many combinations of diameter, length, and number of screws will be satisfactory. In Appendix Table C-5, we see that either 4- or 5-in. lengths could be used.

Try a 4-in. length. Next, assume a diameter. For 3/8-in.-diameter lag screws, Appendix Table C-6 shows a lateral design value of $Z = 400$ lb per screw. This is for a single-shear connection, with a 1.5-in. wood side piece, parallel-to-grain loading, and ten-year duration of load. Penetration of the lag screw is $T - E = 2.28$ in. (Table C-5). Because this is less

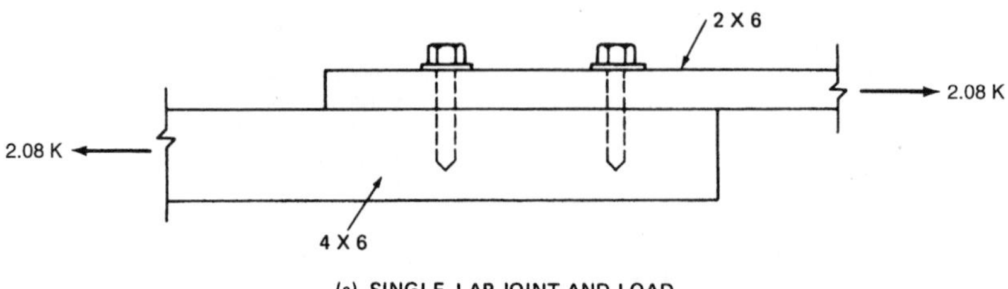

(a) SINGLE-LAP JOINT AND LOAD

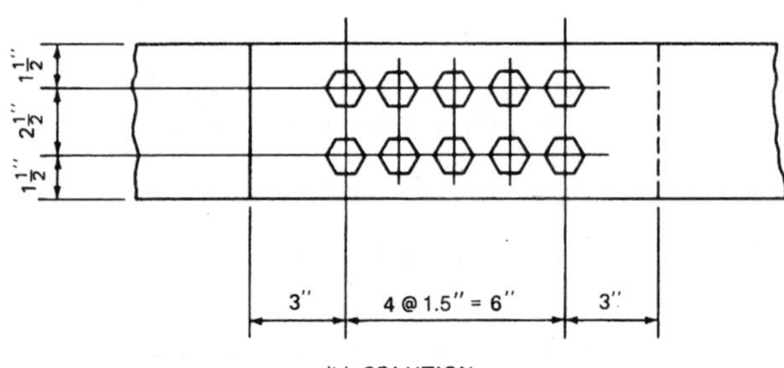

(b) SOLUTION

Fig. 5-12. Connection for Example 5-8.

than $8D$ = 3 in., C_d = 2.28/3 = 0.76. Adjusted for load duration and penetration depth, the allowable is Z' = (400) (0.9) (0.76) = 274 lb per lag screw. If 3/8-in. by 4-in. lag screws are used, the approximate number required will be

$$2080/274 = 7.6, \text{ say } 8$$

The arrangement of the screws must now be considered. The two possibilities are to use (1) a single row of eight screws or (2) two rows of four screws each. Use Appendix Table C-8 to determine the group action factor, C_g, for number in a row. Area A_m is defined as the area of the main piece, which is the member receiving the point of the lag screw. Area A_s is the area of the side piece.

$$A_m = (3.5)(5.5) = 19.25 \text{ in.}^2$$

$$A_s = (1.5)(5.5) = 8.25 \text{ in.}^2$$

Extrapolating (using A_s/A_m = 0.43) on Table C-8, C_g = 0.61 for eight in a row and C_g = 0.87 for four in a row. The adjusted allowable loads per lag screw are

(0.61) (274) = 167 lb if there are eight per row,

(0.87) (274) = 238 lb if used in two rows of four each

Assuming that we still prefer a single row, the approximate number required now becomes 2080/167 = 13. But for 13 in a row, the adjustment factor is less than 0.61, which makes the required number of lag screws larger than 13. Obviously, there is too great a penalty to be paid for insisting on a single row in this case.

Try two rows of five each. The extrapolated C_g is 0.795 and the allowable per lag screw is (0.795) (274) = 218 lb, and the total number required is 2080/218 = 9.5. Two rows of five 3/8-in. by 4-in. lag screws each will be satisfactory, providing all spacing and edge distance requirements can be met. (If these requirements can be met, then the geometry factor is 1.0.)

Use Fig. 5-11 to determine required spacing and edge distances. The spacing between rows must be at least 1.5 diameters, which is 0.56 in. center-to-center. Try 2.5 in. for a practical spacing between rows. Ratio L/D for the 3/8-in. by 4-in. screws exceeds 6, so the required edge distance is not less than 1.5 diameters (0.56 in.) nor less than one-half the actual spacing between rows (1.25 in.). The latter controls. The sum of two edge minimum distances plus the 2.5-in. spacing between rows is only 5 in.; but the member is 5.5 in. wide, so the edge distances may be increased. A practical solution would be to use 1.5-in. edge distances with the two rows 2.5 in. on centers.

Spacing in each row, parallel to the load direction and the grain, must be at least four diameters, or 1.5 in.

Douglas fir is a softwood, and the member is in tension, so the required end distance is seven times the lag screw diameter, or 2.63 in., say 3 in. Figure 5-12b shows the final answer to the example.

Example 5-9

Figure 5-13 shows a connection of a 2 × 8 bracing member to a beam. Load in the brace is 2.1 kips (D = 0.5 k and W = 1.6 k). The wood will be dry both at the time of fabrication and in service. Design a lag screw connection.

Find the controlling load combination:

0.5/0.9 = 0.56

2.1/1.33 = 1.58 *Controls.* Use $C_D = 1.33$.

The unthreaded shank length should preferably be at least 1.5 in. (the thickness of the side piece) to keep the threaded portion out of the side piece. Try ½-in. × 5-in. lag screws. Dowel bearing strengths (from the NDS) for Douglas fir-larch are 5600 psi and 3150 psi for load parallel to and perpendicular to the grain, respectively. The load in the 4 × 16 beam (the "main" member since it receives the point of the lag screw) is at 50° to the grain. So, an intermediate dowel bearing strength for the 4 × 16 is computed by the Hankinson formula, as follows:

$$F_{em} = (5600)(3150)/(5600 \sin^2 50° + 3150 \cos^2 50°)$$

$$= 3845 \text{ psi}$$

This value is then used in the yield mode equations. The controlling design value from the yield mode equations (computations not shown here) is Z = 532 lb.

The penetration of the threaded portion into the main member is 2.69 in. Therefore the penetration depth factor, C_d, equals 2.69/8D = 2.69/(8 × 0.5) = 0.673. The wet service factor C_M = 1.0 from Table C-2. Adjusted for penetration depth, wind, and wet service the allowable is

(0.673) (1.33) (1.0) (532) = 476 lb per screw.

Approximate N req. = 2100/476 = 4.41 screws

Try five screws, arranged as shown on Fig. 5-13. One row has three lag screws. Extrapolating on Table C-9, we find the group action factor, C_g, to be 0.94. The adjusted allowable load is now 0.94 × 476 = 447 lb/screw.

N required = 2100/447 = 4.7

Use five, ½-in. × 5-in. lag screws

Spacing, edge distance, and end distance requirements are the same as for bolts. Following the rules shown by Fig. 5-11, the dimensions on Fig. 5-13 should be

b and c = not less than 4 diameters, or 2 in.
d = not less than four diameters, or 2 in. (as for a load perpendicular to the edge of the horizontal member)

In addition, dimension 2c should not exceed 5 in. unless vertical shrinkage of the 4 × 16 is checked and found to be negligible. Dimension a is calculated as shown by Fig. 5-13b. A reasonable dimension (instead of the 2.76 in.) would be a = 3 in. Actual dimensions would probably be much greater than these minimums.

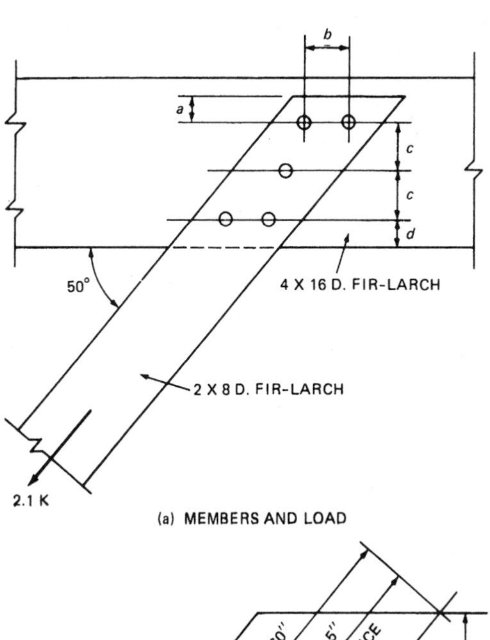

Fig. 5-13. Connection for Example 5-9.

Example 5-10

A horizontal 2 × 6 (laid flatwise) is connected crosswise to the bottom edge of D. fir-larch 2 × 10 ceiling joists at 16 in. c/c using lag screws. The 2 × 6 is to support a vertical load of 400 lb, placed anywhere along the length of the 2 × 6. Assuming one screw at each location where the 2 × 6 passes beneath a joist, what size and length of lag screw should be used? Load duration is permanent.

Since the load can be placed anywhere, the maximum withdrawal load for one screw is 400 lb plus the weight of a 16-in. length of the 2 × 6, or about 403 lb.

Many different diameters could be used, but the larger ones are impractical since the lead hole for installing them would remove too much wood from the tension side of the joist cross section. Try a ¼-in.-diameter lag screw. For a specific gravity of 0.50,

$W = 1800\,(0.25)^{0.75}\,(0.50)^{1.5} = 225$ lb per inch of thread penetration.

Required penetration of the threaded part into the piece containing the point (the joist), for $C_D = 0.9$, is $403/(0.9 \times 225) = 1.99$ in. Try ¼-in. × 4 in. lag screws with $T - E = 2.34$ in. (Appendix Table C-5). The table shows that the length of the unthreaded portion of these lag screws is equal to the thickness of the 2 × 6 (1.5 in.), so the full 2.34 in. will be in the joists.

The lag screw will be installed in side grain, so C_{eg} does not apply. Penetration depth factor does not apply for withdrawal connections. Tensile allowable for the 0.173-in. shank root diameter (Appendix Table C-5), allowing a safety factor of 2, is $45{,}000\,(0.173^2)\,(\pi/4)/2 = 529$ lb $> W'$. Therefore tensile strength of the steel does not control.

$W' = (0.9 \times 225$ lb per in.$)\,(2.34$ in.$) = 474$ lb $>$ 403 lb

Use one ¼-in. × 4-in. lag screw per joist crossing.

5-6. WOOD SCREWS

Wood screws are used mainly for connecting finish materials, cabinetry, and the like. They are often used to hold together pieces being glued. They have occasional structural uses, however, and can carry either transverse (lateral) or withdrawal loads. Screw sizes are expressed in gauge numbers that indicate the size of wire from which they are manufactured. Screw heads may be either round or countersunk, as shown in Fig. 5-2.

Allowable *withdrawal* load, *W*, in pounds per inch of embedment of the threaded part into the member containing the point is given by the following equation:

$$W = 2850DG^2 \qquad (5\text{-}4)$$

The allowable applies to screws installed in side grain. (Withdrawal loads from end grain are not allowed.) For determining embedment length of the threaded part, it is helpful to know that the thread length is approximately two-thirds the total wood screw length. For withdrawal, the only applicable adjustment factors are C_D, C_M, and C_t.

Allowable *lateral* loads (lateral design values), *Z*, are shown by Appendix Table C-14. These values have been determined based on the same three yield modes as for lag screws. The *Z* values are for screws having a standard penetration of seven diameters into side grain of the piece containing the point. Penetrations of less than $4D$ should not be used. For penetrations between $7D$ and $4D$, the allowable lateral load is reduced using factor $C_d = p/(7D)$. The allowable lateral load for wood screws in end grain (rather than side grain) is reduced using factor $C_{eg} = 0.67$. In addition to these two adjustment factors, the factors C_D, C_M, and C_t apply for laterally loaded wood screws.

To have structural value, wood screws must be installed in predrilled holes. NDS requires:

For screws loaded in withdrawal—A lead hole diameter of about 90% of the root diameter in wood with specific gravity *G* greater than 0.60, and about 70% for *G* between 0.50 and 0.60. For *G* less than or equal to 0.50, no lead hole is required.

For wood screws loaded laterally—The lead hole diameter in species with $G > 0.60$ must be the same as the shank diameter for the part receiving only the shank and the same as the root diameter for the part receiving the threaded part. For other species, the required dimensions are 7/8 as great.

Field construction supervision will have to be especially good to ensure that these required diameters are actually provided. The designer using wood screws for structural purposes should question whether he or she can really exercise control over what diameters will be used.

Screws should be installed by rotating using a screwdriver; they should not be hammer-driven. A lubricant is used to make installation easier. Soap is often used, but this may be undesirable in that it holds moisture and can promote decay. Wax would be better and equally effective.

5-7. BOLTED CONNECTIONS

Bolt Types

Three types of bolts are commonly used in wood structures, but only one of them is suitable for major structural purposes. The word *bolt,* used without a modifier, indicates the type shown by Fig. 5-14a. This type, which may have either a square head or a hexagonal head, is suitable for structural purposes. If made of low-carbon steel, such bolts are sometimes called *common bolts.* Bolts made of high-strength steel may also be used.

Figures 5-14b and c show two other types of bolt that are frequently used in wood construction, although they are of minimal structural value. The carriage bolt is useful where the head must be smooth or where one side of the connection cannot easily be reached to hold the bolt head while the nut is tightened. A short square section of the bolt shank cuts into the edge of the hole so that the bolt cannot rotate freely. The disadvantage to the carriage bolt is that no washer can be used under the head, so the head is drawn into the wood as the nut is tightened. Overtightening can easily damage the wood member and allow the bolt to rotate. Used carefully and without overtightening, the carriage bolt is useful for nonstructural and minor structural applications.

The stove bolt, shown in Fig. 5-14c, is threaded for the full length of shank. The threaded shank does not provide good resistance in bearing against the side of the hole. Stove bolts should not be used for structural purposes where one of the members connected tends to slide relative to the other, that is, in shear connections.

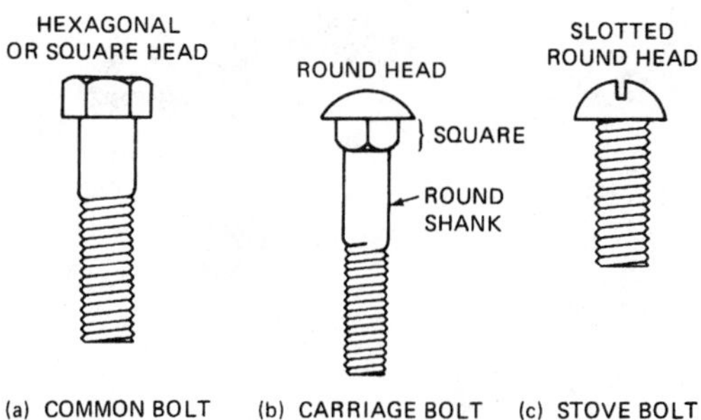

Fig. 5-14. Bolt types.

Everything that follows in this section will apply only to common bolts, the type shown by Fig. 5-14a.

Types of Bolted Connection

Bolts may be used alone to transfer load, or they may be used merely to hold the parts in firm contact while other devices (keys, timber connectors, or plugs) transfer the load. Bolts may be used in shear-type connections or the load may have a tension component.

In shear connections, load transfer is by bearing (compression) of the bolt shank against the wood at the side of the hole and by shear in the bolt shank. Friction between the connected parts may be sufficient to transfer some load immediately after the bolts are tightened. But this capability cannot be counted on, since shrinkage may reduce the thickness of the members being joined. When this happens, the joint is loosened and compression between the parts is reduced, so that friction cannot be developed. Bolted connections that are subject to rotation (combined vertical load and moment, for example) are covered in Chapter 9.

Shear (Lateral) Capacity

Design of bolted connections to resist shear forces is very similar to design of the other connectors covered in this chapter. First, the base design value, Z, is found from Appendix Tables C-10 or C-11 (for single-shear connections) or from Tables C-12 or C-13 (for double-shear connections). Then the base value is multiplied by all applicable adjustment factors.

The adjustment factors that apply to laterally loaded bolted connections are the following:

1. Load duration, C_D.
2. Wet service, C_M, using Table C-2. (See Figs. 5-15 and 5-16.)
3. Group action, C_g, for number of bolts in a row. This is the same as for lag screws (see Tables C-8 and C-9).
4. Geometry, C_Δ, if end distances and spacing are less than the required minimums.
5. Temperature, C_t.

Fig. 5-15. Poor detail of bolted connections caused 145 of the 150 such connections in this building to split the wood. (Photograph by authors.)

Fig. 5-16. Note how wood shrinkage caused the two-bolt connection to split the wood. (Photograph by authors.)

The base design value for connections of more than three members is found by considering the joint to be a series of single-shear connections. The smallest design value for any single-shear plane, multiplied by the number of shear planes, gives the base design value for the connection. This is equivalent to saying that the design value at each of the shear planes should be taken to equal the smallest value for any shear plane.

Hole Diameter

To realize the full allowable design value of the bolt, it is important that the hole be prepared properly. Sharp tools must be used, and the rate of drilling (feed) must be properly controlled to ensure a smooth-sided hole.

For proper bearing of the bolt shank against the wood of a shear-type connection, the hole diameter should be only slightly larger than that of the bolt. The NDS specifies that bolt hole diameters be 1/32–1/16 in. larger than the bolt diameter. However, others recommend that hole diameters vary with the bolt size and with expected service moisture conditions. Generally, for bolts used in wood that will be at 6% MC or less, the hole diameter is equal to the bolt diameter plus 1/16 in. for bolts up to 3/4 in. diameter and plus 3/32 in. for larger bolts. For bolts in wood at higher service moisture content, the difference should be less. At 12% MC, holes 1/32 in. larger than the bolt diameter are recommended for bolts up to 3/4 in., and 1/16 in. for bolts larger than 3/4 in. For wet service, only 1/32 in. should be added for all sizes of bolt.

Washers

Washers are necessary under both the head and the nut. They serve two purposes. The force caused by tightening is spread over an area larger than the area of contact between the head or nut with the wood, thus preventing the wood from being crushed. Washers also permit the nut to be rotated easily as it is tightened. Where the head or nut bears against a steel plate of sufficient thickness to spread the load, a washer is not needed.

If the bolt is loaded in tension (or for a threaded steel rod used as a tension member) it may be necessary to use a washer that is larger than standard size and thickness. The necessary diameter can be calculated easily, as follows:

$$\text{Req. net bearing area} = \text{tensile load}/F'_{c\perp} \qquad (5\text{-}5)$$

The net area provided is the area enclosed by the outer perimeter of the washer minus the area of the hole in the wood or in the washer, whichever is larger.

Example 5-11

A horizontal 2 × 12 joist is connected to a 4 × 4 vertical post by three 1/2-in. bolts in a single vertical row. Both joist and post are D. fir-larch that is seasoned when installed and will be exposed to the weather in use. Based on the connection strength only, what is the allowable vertical reaction for a load combination that includes snow?

Load will be perpendicular to the grain in the side member (the joist), but parallel to the grain in the main member (the post). By Appendix Table C-10, the smaller of the design values for each of these conditions is $Z_{s\perp} = 370$ lb.

There are three bolts in a row, and Appendix Table C-8 must be used to determine a group action (C_g) adjustment factor. Area $A_m = 3.5\ (3.5) = 12.25$ in.^{2} The side member (joist) has bolt loads perpendicular to the grain, so area A_s is the product of member thickness times the "overall width of the fastener group." There being only one row of fasteners, the overall width is taken as the minimum parallel to grain spacing of the fasteners. Minimum parallel to grain spacing is four diameters, or 4(1/2) = 2.0 in. Thus, area $A_s = 1.5\ (2.0) = 3.0$ in.^{2} The ratio $A_s/A_m = 0.245$, and by extrapolating on Table C-8, the group action factor (for three in a row) is

$$C_g = 0.91 - (0.96 - 0.91)\ (0.255/0.5) = 0.88$$

The adjusted allowable load per bolt is

$$Z' = C_D C_M C_g Z = (1.15)\ (0.75)\ (0.88)\ (370) = 281 \text{ lb}$$

Allowable reaction (based on connection only) is

$$3\ (281) = 843 \text{ lb}$$

(This must be compared to the allowable reaction based on shear in the joist. See Chapter 6.)

Example 5-12

A hem-fir 2 × 6 (No. 1) tension member carries load of D = 1.52 k and W = 4.56 k. Design a tension splice using ½-in. bolts and two ¼-in. steel splice plates. The wood is seasoned when installed, but is exposed to the weather.

Find the controlling load combination:

$$(1.52 + 4.56)/1.33 > 1.52/0.9, \text{ so } C_D = 1.33$$

The base value parallel to the grain is Z = 900 lb

Estimating that $C_g = 0.75$, the approximate $Z' = C_D C_M C_g Z$ = (1.33) (0.75) (0.75) (900) = 673 lb.

Approx. N req. = 6080/673 = 9 bolts

Try 10 bolts (2 rows of 5 each)

Group action adjustment, C_g, for 5 in a row:

A_m = 1.5 (5.5) = 8.25 in.^{2}
Assume that side plates are of 5-in. width, so
A_s = 2 (5) (¼) = 2.5 in.^{2}
A_m/A_s = 8.25/2.5 = 3.3

Extrapolating from Table (using A_m of approximate 8 in.^{2}), the group action factor is

$$C_g = 0.77 - (0.06/6)(12 - 3.3) = 0.68$$

$$Z' = (1.33)(0.75)(0.68)(900) = 610 \text{ lb}$$

Req. N = 6080/610 = 9.97 bolts

OK Use 10 bolts (2 rows of 5 on *each* side of the joint in the 2 × 6)

Check net area in steel plates:

A net = 2 (¼) (5 − 2 × ⅝) = 1.88 in.^{2}
Tensile stress = 6080/1.88 = 3230 psi
Allowable for A36 steel = 29,000 psi *OK*

Check net area in wood member

A net = 1.5 (5.5 − 2 × 9/16) = 6.56 in.^{2}
Tensile stress = 6080/6.56 = 927 psi
Allowable (see Chapter 7) = $F_t' = C_D C_M C_F F_t$ = (1.33) (1.0) (1.3) (600) = 1040 psi *OK*

(For consistency, a load duration factor of 1.33 was used here, although, since member strength rather than bolt strength is being checked, many designers would use C_D = 1.6.)

Example 5-13

Same as Ex. 5-12, but using unseasoned wood that will dry to less than 19% MC in service.

The allowable load per bolt will be affected by seasoning of the wood after installation (see Appendix Table C-2). If a single plate (on each face of the splice) is used to contain two rows of bolts, as in Example 5-12, the allowable load per bolt has a reduction factor of C_M = 0.4. The reason is that high tensile stresses perpendicular to the grain are set up as the steel plates restrain the wood from shrinking.

On the other hand, if two narrow plates are used on each face of the splice (one narrow plate per row of bolts) to replace each wider plate, the factor is C_M = 1.0. Obviously, to use the separate (split) plates is the better solution, insofar as number of bolts is concerned. In this case, (assuming C_g = 0.75), the approximate $Z' = C_D C_M C_g Z$ = (1.33) (1.0) (0.75) (900) = 898 lb. Approx. N req. = 6080/898 = 6.8 bolts

Try 8 bolts (2 rows of 4 each):

Assuming two side plates each ¼ in. × 2 in. on each face of the 2 × 6 (four plates total),

A_m/A_s = 8.25/2.0 = 4.125

C_g = 0.77 − (0.6/6) (12 − 4.125) = 0.69

N req. = 6080/(1.33 × 1.0 × 0.69 × 900) = 7.4 bolts

OK Use 8 bolts per end (2 rows of 4; i.e., 8 bolts on each side of the joint in the 2 × 6)

Example 5-14

Figure 5-17 shows a two-piece member spliced to a three-piece member. All wood members are D. fir-larch. The loads shown are ten-year (normal) duration. How many ⅝-in. bolts are required, assuming that they can be placed in two rows?

For shear plane No. 1 or No. 4 (main and side members each of thickness 2.5 in.), the single-shear design value is 1007 lb/bolt (from yield mode equations).

For shear plane No. 2 or No. 3 (main member thickness = 2.5 in. and side member thickness = 1.5 in.), the single-shear design value is 850 lb/bolt. The latter, being the smaller, controls.

Since there are four shear planes, the unadjusted design value for the connection is (850) (4) = 3400 lb per bolt. The approximate number of bolts required is (12 kips)/(3.4 kips per bolt) = 3.5 bolts. *Try 2 rows of 2 bolts each.*

The smaller adjustment factor, C_g, will result from considering the side member to be 1.5-in. thick and the main member to be 2.5-in. thick.

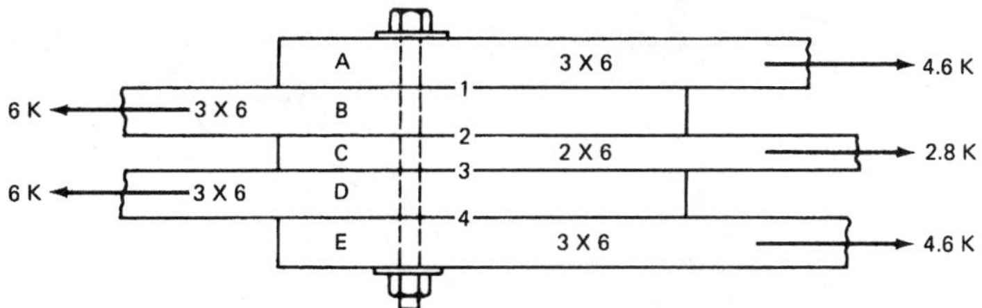

Fig. 5-17. Connection for Example 5-14.

$A_s = 1.5\ (5.5) = 8.25\ \text{in.}^2$

$A_m = 2.5\ (5.5) = 13.75\ \text{in.}^2$

$A_s/A_m = 0.6$ and $C_g = 0.99$

N req. $= 12{,}000/(0.99 \times 850 \times 4) = 3.6$ bolts

OK Use 4 bolts (2 rows of 2 bolts each)

Example 5-15

A double-shear connection as shown in Fig. 5-9 joins a 3 × 6 member to two 2 × 6 side pieces with a single row of five 5/8-in. bolts. All wood is seasoned D. fir-larch, but frequently will be wet in service. What is the allowable permanent load for the connection? What is the controlling net area for tension in the wood members?

For one 5/8-in.-diameter bolt, Appendix Table C-12 shows a double shear design value of $Z = 1760$ lb parallel to the grain. Use Table C-8 to determine C_g:

$A_s = (2)\ (1.5)\ (5.5) = 16.5\ \text{in.}^2$

$A_m = (2.5)\ (5.5) = 13.75\ \text{in.}^2$

$A_m/A_s = 13.75/16.5 = 0.833$

Using a value of 13.75 in the second column of the table, and then interpolating, C_g is found to equal 0.92.

For load duration, $C_D = 0.9$

For wet service (Table C-2), $C_M = 0.67$

$Z' = C_g C_D C_M Z = (0.92)\ (0.9)\ (0.67)\ (1760)$
$= 976$ lb/bolt

Allow. load $= (5)\ (976) = 4880$ lb

For the net area of main member, assume a 21/32-in. hole diameter (bolt diameter plus 1/32 in. for wet service):

A (net) $= 2.5(5.5 - 21/32) = 12.11\ \text{in.}^2$

If the tensile stress in the wood based on this net area is greater than the allowable tensile stress, then the allowable load for the connection will have to be reduced below 4880 lb (see Chapter 7).

Example 5-16

Figure 5-18 shows an inclined tension member connected to a beam. The inclined member is a hem-fir 2 × 8, and the beam is a D. fir-larch 4 × 16. Tensile force F is 2.1 kips ($D = 0.42$ k and $L = 1.68$ k). Assuming two rows of 3/4-in.-diameter bolts, how many bolts per row are needed? In the space available, can they be positioned so as to satisfy all spacing, edge distance, and end distance requirements?

This problem is complicated in two ways: Different species are used, and the load is parallel to the grain of one member but at an angle to the grain of the other. Since the specific gravity of hem-fir is smaller than that of D. fir-larch, the problem must be solved assuming both members to be hem-fir.

The thinner member in a bolted connection is called the "side" member. The 2 × 8 is the side member and has load parallel to the grain, so $F_{es} = 4800$ (from the NDS).

The 4 × 16 has load at 60° to the grain. Using $F_{e\perp} = 2050$, the Hankinson equation gives

$$F_{em} = (4800)\ (2050)/(4800 \sin^2 60° + 2050 \cos^2 60°) = 2393 \text{ psi}$$

Using this value in the lengthy computations of the yield mode equations results in a design value of 326 lb/bolt.

Approximate number req. = 2100 lb/326 lb per bolt = 6.4
Try 8 bolts (4 per row)

The group-action adjustment for four bolts per row is computed as follows: Assume spacing between rows = 3.5 in. Also assume that the load is perpendicular to the grain in the angle-to-grain member (a conserva-

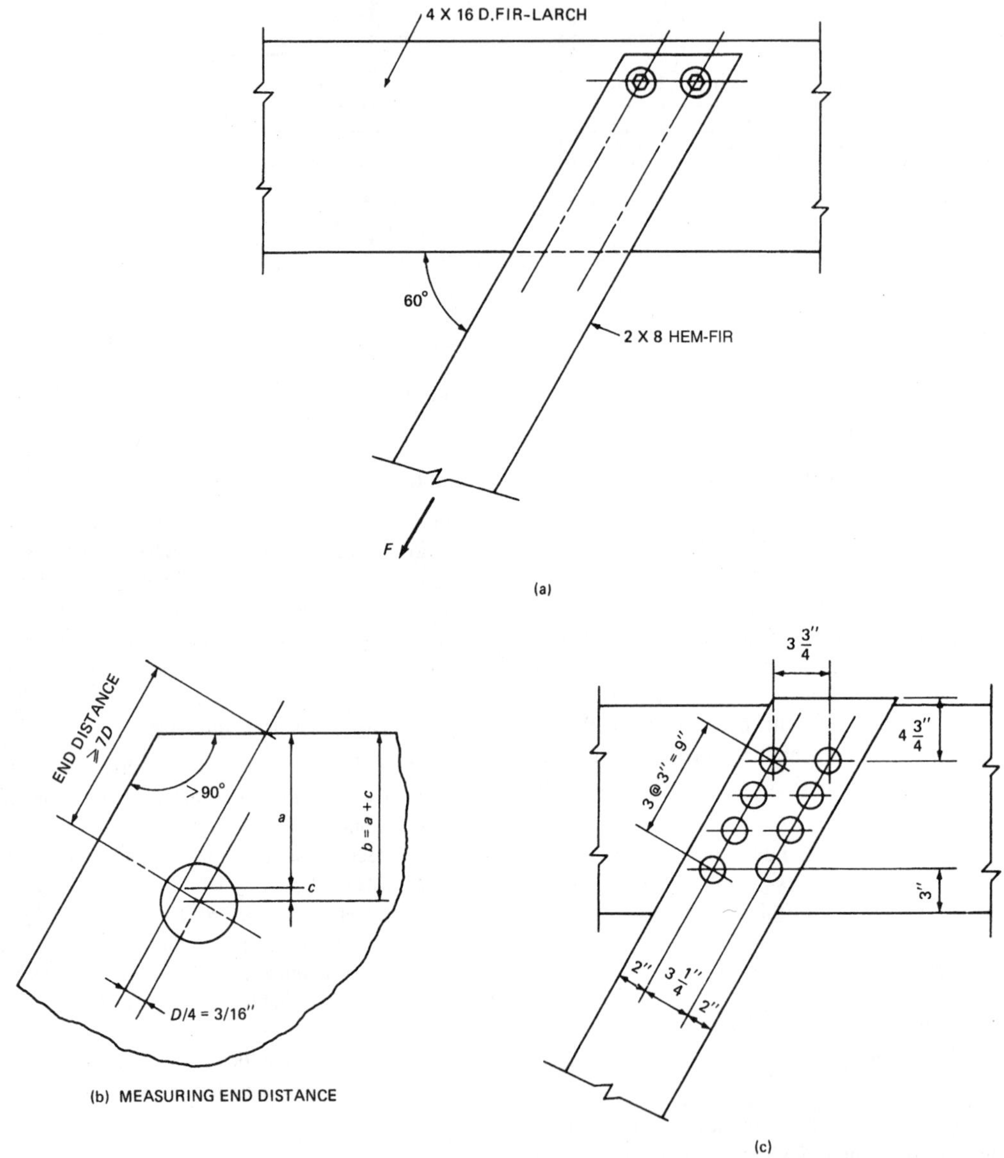

Fig. 5-18. Connection for Example 5-16.

tive assumption for calculating A_m). The "overall width of the fastener group" is 3.5 in.

$$A_m = 3.5\ (3.25) = 11.375\ \text{in.}^2$$

$$A_s = 1.5\ (7.25) = 10.875\ \text{in.}^2$$

$$A_s/A_m = 0.956$$

Factor for four per row (interpolated from Table C-8) is $C_g = 0.95$.

$$N\ \text{req.} = 2100/(0.95 \times 326) = 6.8$$

Use 8 bolts (2 rows of 4)

5-8. CONNECTION DESIGN BY LRFD

Engineers who have designed connections by the allowable stress design (ASD) method already have many of the skills necessary for connection design by the load and resistance factor design (LRFD) method. All of the many adjustment factors such as penetration depth factor, end grain factor, and group action factor are equally applicable to LRFD. The one exception is the load duration factor that is used in ASD. It is not used in LRFD, but the time effect factor,

λ, is used instead. In LRFD, the symbol for lateral resistance, Z', refers to the resistance of the entire connection, rather than to the resistance of an individual connector. Also, the symbol for withdrawal resistance, Z_w', refers to the total withdrawal resistance rather than to the resistance per inch of penetration. Connections are designed such that

$$Z_u \leq \lambda \phi_z Z' = \lambda(0.65)Z' \qquad (5\text{-}6)$$

where Z_u is the factored load, λ is the time effect factor (see Section 4-12), and Z' is the adjusted connection resistance. Design in LRFD is usually done in kip, rather than pound, units.

For nails and wood screws, lead hole requirements are the same by each method. For bolts, hole diameters are specified to be $D + {}^1\!/_{32}$ in. for $D < 0.5$ in. and $D + {}^1\!/_{16}$ in. for $D \geq 0.5$ in. Lag screws have nearly identical hole requirements by each method.

Unlike ASD, LRFD specifies spacing requirements for nails and wood screws. The minimum spacing between fasteners in a row is $10D$ for wood side plates and $7D$ for steel side plates. Minimum spacing between rows is $5D$. (A "row" is parallel to the grain.) For bolts and lag screws, the spacing, end distance, and edge distance requirements given by Fig. 5-11 are still applicable.

Withdrawal Resistance

The reference withdrawal resistance in side grain of each connector is listed below, where D = shank diameter (in.), G = wood specific gravity, p = penetration (in.), and n_f is number of fasteners.

$$\text{For nails, } Z_w = 4.59DG^{2.5}pn_f \qquad (5\text{-}7)$$

$$\text{For wood screws, } Z_w = 9.47DG^{2}pn_f \qquad (5\text{-}8)$$

$$\text{For lag screws, } Z_w = 5.98D^{0.75}G^{1.5}pn_f \qquad (5\text{-}9)$$

For each type of connector, the value from the equation is adjusted using the factors outlined earlier in this Chapter for the respective connector type. The LRFD standard requires a minimum penetration for fasteners loaded in withdrawal as follows: *For lag screws*—The lesser of one in. or one-half the threaded length. *For wood screws*—The lesser of one in. or one-half the nominal screw length.

Lateral Resistance

The reference lateral resistance for each of the connectors is determined in the same manner as the design value is in ASD. That is, several yield mode equations are evaluated, and the smallest value is chosen as the reference strength. Except for a constant (of 3.3), the yield mode equations for each connector are identical by the two design methods, therefore the LRFD lateral resistance values will be 3.3 times the ASD design values.

Example 5-17

Using the LRFD method, choose size and number of lag screws for the following connection: A 2 × 10 hem-fir beam connected to the side of a 4 × 4 hem-fir vertical post. The end reaction of the beam is $D = 40$ lb, $L = 165$ lb (storage), and $S = 310$ lb. The member is green when installed, and the connection will be exposed to the weather.

Find controlling load combination (see Section 4-12):

$$1.4D = 1.4(0.04) = 0.056 \text{ k}$$

$$1.2D + 1.6L + 0.5S = 1.2(0.04) + 1.6(0.165) + 0.5(0.31) = 0.467 \text{ k}$$

$$1.2D + 1.6S + 0.5L = 1.2(0.04) + 1.6(0.31) + 0.5(0.165) = 0.626 \text{ k}$$

Dividing each of these by the corresponding time effect factor, the three results are 0.056/0.6 = 0.093; 0.467/0.7 = 0.667; and 0.626/0.8 = 0.783. The latter controls, so $Z_u = 0.626$ kips and time effect factor, $\lambda = 0.8$. From Appendix Table C-5, a lag screw of length 4 in. has an unthreaded shank length of $S = 1.5$ in., so that all of the threaded portion will be in the post. Many different combinations of length and diameter are possible.

Try ${}^3\!/_8$-in. diameter by 4 in. lag screws

$$\text{Standard pene.} = 8D = 8(^3\!/_8) = 3 \text{ in.}$$

$$\text{Actual pene.} = T - E = 2.28 \text{ in.}$$

$$C_d = p/(8D) = 2.28/3 = 0.76$$

$$C_M = 0.75 \text{ (Appendix Table C-2)}$$

Next, find the reference lateral resistance, Z. Load is perpendicular to the grain in the beam and parallel to the grain in the post. Because the perpendicular-to-grain resistance is the smaller resistance, that is the value that is used. If a LRFD table for fastener design strength is not available, using 3.3 times the value on the corresponding ASD table (in this case, Appendix Table C-6) will be acceptable. Using Table C-6, $Z = 3.3(230) = 759$ lb $= 0.759$ kips per lag screw.

$$\text{Approximate } n_f \text{ req.} = 0.626/(\lambda\phi_z Z C_d C_M)$$
$$= 0.626/[(0.8)(0.65)(0.759)(0.76)(0.75)]$$
$$= 2.8 \text{ lag screws}$$

Determine the group action factor, C_g, for three in a row. Area of the main member (member receiving the point) is (3.5) (3.5) = 12.25 in.^{2} and area of the side member is (1.5) (9.25) = 13.88 in.^{2}, so $C_g = 0.98$.

$$Z' = ZC_dC_MC_gn_f$$
$$= (0.759)(0.76)(0.75)(0.98)(3) = 1.27 \text{ k}$$

$$\lambda\phi_z Z' = (0.8)(0.65)(1.27)$$
$$= 0.662 \text{ k} > 0.626 \text{ k} \qquad OK$$

Example 5-18

A 2 × 12 header is to be nailed to a 2 × 8 joist by nailing through the header into end grain of the joist, using three 30*d* common nails in a vertical row. The wood is D. fir-larch. What is the maximum factored reaction for the joist, assuming load case of $1.2D + 1.6L$ controls? Use the LRFD method.

For this load combination (assuming live load is from occupancy), the time effect factor is 0.8.

Nail length = 4.5 in. and nail diameter = 0.207 in.

$$\text{Standard pene.} = 12D = 12(0.207) = 2.48 \text{ in.}$$

$$\text{Actual pene.} = 4.5 - 1.5 = 3 \text{ in.} > 2.48, \text{ so } C_d = 1.0$$

$$\text{End grain factor} = C_{eg} = 0.67$$

$$C_M = 1.0 \text{ (dry conditions assumed)}$$

Using for the LRFD resistance value 3.3 times the ASD design value given on Appendix Table C-3, the lateral resistance is (3.3) (186) = 614 lb = 0.614 kips per nail.

$$Z' = ZC_{eg}n_f = (0.614)(0.67)(3) = 1.23 \text{ k}$$

$$Z_u \le \lambda\phi_z Z' = (0.8)(0.65)(1.23) = 0.640 \text{ k}$$

Maximum factored reaction is 0.64 kips.

Example 5-19

Using LRFD, determine if a 1/4-in. × 3-in. lag screw, through a 1/4-in. steel plate and into side grain of a D. fir-larch member will carry a withdrawal load of D = 50 lb and L = 320 lb (occupancy).

Find the controlling load combination:

$$1.4(0.05) = 0.07 \text{ k}$$

$$1.2(0.05) + 1.6(0.32) = 0.572 \text{ k}$$

Considering the values of λ that correspond to each combination (λ = 0.6 and 0.8), it is obvious that the latter load combination controls, so $Z_u = 0.572$ k and the time effect factor is 0.8.

Penetration = $T - E$ = 1.84 in. (See Appendix Table C-5.)

$$C_M = 1.0 \text{ (assumed dry)}$$

The reference withdrawal resistance (by Eq. 5-9) is

$$Z_w = Z_w' = 5.98D^{0.75}G^{1.5}pn_f = 5.98(0.25)^{0.75}(0.50)^{1.5}(1.84)(1) = 1.38 \text{ k}$$

$$\lambda\phi_z Z' = (0.8)(0.65)(1.38) = 0.718 \text{ k} > Z_u = 0.572 \text{ k } OK$$

Example 5-20

A 4 × 10 of No. 1 D. fir-larch is to be spliced using 1/2-in. diameter bolts and two 1/4-in. thick steel side plates. The tensile loads on the member are D = 1.9 k, S = 2.8 k, and W = 1.7 k. The wood will be partially seasoned at the time of installation, but will be dry in use. Find the number of bolts required using the LRFD method.

Find the controlling load combination:

$$1.4D = 1.4(1.9) = 2.66 \text{ k}$$

$$1.2D + 1.6S + 0.8W = 1.2(1.9) + 1.6(2.8) + 0.8(1.7) = 8.12 \text{ k}$$

$$1.2D + 1.3W + 0.5S = 1.2(1.9) + 1.3(1.7) + 0.5(2.8) = 5.89 \text{ k}$$

Considering the effect of λ by dividing each of these by the λ which corresponds to that load combination (λ = 0.6, 0.8, and 1.0, respectively), the largest quotient will be from the second load combination. Therefore, factored loading of $Z_u = 8.12$ kips controls and time effect factor is 0.8.

Using 3.3 times the entry in Appendix Table C-13, the reference resistance for a 3.5-in. main member thickness is (3.3) (1510) = 4980 lb = 4.98 k.

Assuming a single row of fasteners, $C_M = 1.0$ (see footnote to Table C-2)

Approx. n_f req. $= 8.12/(\lambda\phi_z Z)$
$= 8.12/[(0.8)(0.65)(4.98)] = 3.1 = 4$ bolts

Find group action factor for four in a row (using 6-in.-wide plates):

$A_m = (3.5)(9.5) = 33.25 \text{ in.}^2$

$A_s = (2)(0.25)(6) = 3 \text{ in.}^2$

$A_m/A_s = 33.25/3 = 11.1$

From Table C-9, $C_g = 0.95$

$Z' = ZC_g n_f = (4.98)(0.95)(4) = 18.9$ k

$\lambda\phi_Z Z' = (0.8)(0.65)(18.9) = 9.83 \text{ k} > 8.12 \text{ k}$
OK

Use four ½-in.-diameter bolts each side of splice.

Example 5-21

Solve Ex. 5-1 again, this time using the LRFD method. Recall that this example involves a 3-in. (nominal) tension member spliced using 16 gauge steel side plates and 8*d* nails.

The load combinations are:

$1.4D = 1.4(90) = 126 \text{ lb} = 0.126 \text{ k}$

$1.2D + 1.6L = 1.2(90) + 1.6(120) = 300 \text{ lb}$
$= 0.3 \text{ k}$

$1.2D + 1.3W + 0.5L = 1.2(90) + 1.3(190) + 0.5(120) = 415 \text{ lb} = 0.415 \text{ k}$

The largest quotient (load divided by corresponding time effect factor) is given by the last combination, so factored load $Z_u = 0.415$ kips and time effect factor, $\lambda = 1.0$. The reference lateral resistance is 3.3 times the ASD design value; i.e., $Z = 3.3(70 \text{ lb}) = 231 \text{ lb} = 0.231$ k/nail. The penetration depth factor, C_d, and wet service factor, C_M, have the same values in both ASD and LRFD. From Example 5-1, those values are 1.0 and 0.75 respectively.

Try 4 nails per side of the splice.
The design resistance is

$$\lambda\phi Z' = \lambda\phi(C_M C_d Z n_f)$$

$$= (1.0)(0.65)(0.75)(1.0)(0.231)(4)$$

$$= 0.450 \text{ kips} > Z_u = 0.415 \text{ k } OK$$

One reason that the ASD solution gives a more conservative answer for this problem is the authors' decision to use a load duration factor of 1.33 rather than 1.6 in ASD solutions involving wind load.

Example 5-22

Solve Example 5-6 again, this time using the LRFD method. Recall that the problem deals with a nailed connection in withdrawal. Find the greatest factored load per nail that may be applied.

It is obvious that the load combination of $1.2D + 1.6L$ will control, as it has a larger coefficient on L than the other load combination equations. Table 4-1 shows that the corresponding time effect factor, $\lambda = 0.8$.

As in Example 5-6, $C_M = 0.25$ and penetration equals 1.75 in.

From Eq. 5-7,

$$Z_w = 4.59DG^{2.5}pn_f$$

$$= 4.59(0.128)(0.43)^{2.5}(1.75)(1)$$
$$= 0.125 \text{ kips per nail}$$

Design resistance is

$$\lambda\phi Z_w C_M = (0.8)(0.65)(0.125)(0.25)$$
$$= 0.016 \text{ kips per nail}$$

The largest factored load per nail that may be applied is $Z_u = 0.016$ kips.

REFERENCES

1. *National Design Specification for Wood Construction,* rev. 1991 Ed., American Forest & Paper Association, Washington, DC, 1993.
2. *Uniform Building Code,* International Conference of Building Officials, Whittier, CA, 1994.
3. *Wood Structures, A Design Guide and Commentary,* American Society of Civil Engineers, New York, 1975.
4. Hornbostel, C., *Construction Materials—Types, Uses, and Applications,* Wiley, New York, 1978.
5. *Plywood Diaphragms,* Research Report 138, American Plywood Association, Tacoma, WA, 1983.
6. Cramer, C. O., "Load Distribution in Multiple-Bolt Tension Joints," *Journal of the Structural Division, American Society of Civil Engineers,* 94(ST5). Proc. paper 5939, New York, 1975.

PROBLEMS

5-1. Two 2-in. nominal hem-fir planks are to be nailed together, with the nails driven into side grain. The planks are seasoned but will be wet in use. Loads are $D = 100$ lb and $S = 350$ lb.

What is the allowable lateral load per 16*d* common nail? How many nails are required?

5-2. Two 2 × 10 hem-fir planks are to be spliced in a double-shear connection using two 2 × 10 side pieces of the same material. The wood is partially seasoned at the time of installation and will be seasoned in use. Using the smallest common nail that will fully penetrate all three pieces, what is the allowable lateral load per nail if $D + S$ load controls?

5-3. Common nails (16*d*) are loaded in withdrawal from side grain. The side piece is 2 in. nominal and the piece containing the point is 4 in. nominal. The wood is D. fir-larch, seasoned at the time the connection is made, but subject to wetting and drying. What is the allowable seven-day-duration load per nail?

5-4. Figure 5-19a shows the end of a 2 × 8 joist supported by a 2 × 12 header. The species is D. fir-larch. What is the allowable reaction $(D + L)$ for the joist, based on the strength of three 20*d* common nails in end grain as shown?

5-5. Figure 5-19b shows an alternate detail in which the end of the joist bears on a 2 × 4 that is nailed into side grain of the 2 × 12 header. If the 2 × 4 is nailed to the header using four 16*d* common nails, what is the allowable reaction $(D + L)$ for the joist?

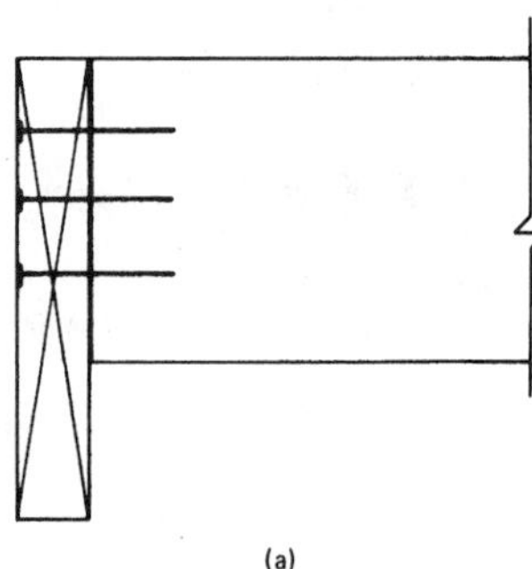
(a)

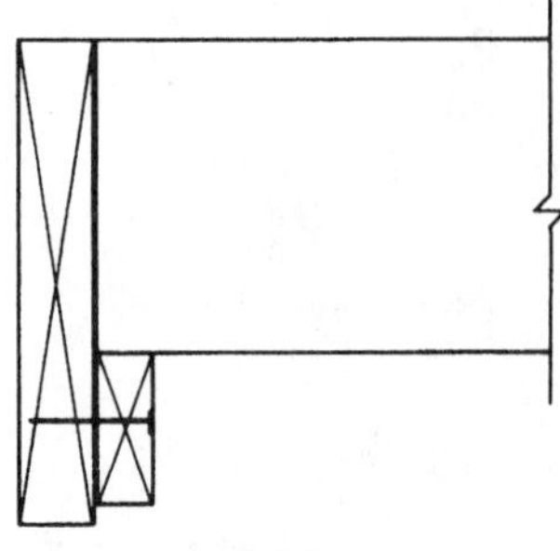
(b)

Fig. 5-19. Details for Problems 5-4, 5-5, and 5-6.

5-6. Assume that the joist of Problem 5-4 is connected to the header with only toe nailing, using four 16*d* common nails. What is the allowable joist reaction? Of the three details considered in Problems 5-4, 5-5, and 5-6, which is best, and why?

5-7. Wind pressure of 20 psf acts on a light-frame wall. The wall studs are 2 × 6 hem-fir, spaced 16 in. c/c and spanning 7 ft 6 in. between top and bottom plates, which also are 2 × 6 hem-fir. The ends of each stud are connected by toe nailing. To resist the shear due to wind load, what number of 16*d* box nails are required, if placed in a cross-slant pattern (like Fig. 5-5)?

5-8. A pair of D. fir-larch 2 × 4s is used together as a single vertical member at the end of a high wall. (This member is called a "chord.") A nailed splice is made such that one 2 × 4 is continuous. Common nails (16*d*) are used to transfer the tension load of 1960 lb (all due to wind) across the splice. Conditions of wood at installation and in use are dry. How many 16*d* nails are required each end of the splice?

5-9. A 1/8- by 3-in. perforated steel plate is nailed to the top and bottom edges of a No. 2 grade 4 × 12 D. fir-larch, to give the wood member better bending strength. The plates are nailed to the wood using two rows of 10*d* common nails. What is the maximum satisfactory nail spacing per row where the total shear (due to $D + L$) is $V = 1300$ lb? Use VQ/I (in which I = moment of inertia of the transformed gross section) to compute the longitudinal shear force per linear inch between the steel plate and the wood beam.

5-10. A joist is nailed to a D. fir-larch 4 × 4 post as shown in Fig. 5-7. The wood was dry when the connection was made, but will be subject to wetting and drying in use. The end reaction of the joist is $D = 120$ lb plus $L = 390$ lb. How many 20*d* common nails are required?

5-11. A nailed connection resists a withdrawal load (from side grain) of $D = 45$ lb plus $S = 165$ lb. The side piece is 2 in. nominal and the piece containing the point is 4 in. nominal. The wood members are hem-fir, seasoned and exposed to dry use. How many 16*d* box nails are required?

5-12. How many 14-gauge staples should be used to connect the short edge of an 8-ft by 4-ft plywood sheet to the roof rafters if the shear force along that edge due to wind is 400 lb/ft?

The penetration of the staples into the D. fir-larch rafters is 1.5 in. Assume the adjusted allowable lateral load is 100 lb for a 14-gauge staple with 2-in. penetration into the main piece, and only 75 lb for 1-in. penetration, with intermediate values interpolated.

5-13. Find the number of 3/8-in. by 5-in. lag screws (if used in a single row) to resist a lateral force of $D = 250$ lb plus $S = 1400$ lb between a 2-in. nominal and a 4-in. nominal piece of hem-fir. Both pieces are loaded parallel to the grain. Conditions of installation and use are both wet.

5-14. Same as Problem 5-13, except D. fir-larch.

5-15. Two 4 × 12s of Select Structural D. fir-larch are to be spliced using lag screws and two 1/2-in.-thick steel side plates, with the lag screws in single shear. The tensile loads on the member are 1.4 kips from dead load, 2 kips from snow, and 1.6 kips from wind. Use two rows in the top plate and two rows (offset) in the bottom plate. How many 3/8-in.-diameter by 3-in.-long lag screws are required? Use all UBC load cases.

5-16. A 2 × 12 D. fir-larch beam is to be connected to the side of a same-species 4 × 4 vertical post using lag screws in a single row. The end reaction of the beam is 200 lb due to live load, 350 lb due to snow, and 45 lb due to dead load. The member will be seasoned when installed, but will have MC above the fiber saturation point in use. How many 5/8-in.-diameter by 5-in. lag screws are required?

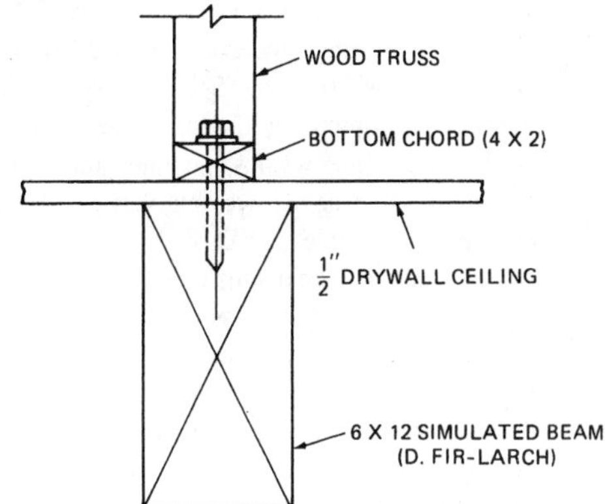

Fig. 5-20. Connection for Problem 5-17.

5-17. Architects revising the interior of an existing building want the appearance of a heavier type of construction. One feature of this treatment is to add wood beams that really are not supporting members. Actually, the only way to hold these beams is to hang them from the bottom chord of existing wood trusses above, as shown by Fig. 5-20. If 1/4-in.-diameter by 4-in. lag screws are used, what would be the maximum spacing of the lag screws? (Assume that the net section of the truss chord is adequate.)

5-18. Figure 5-21 shows a wood sign supported by two timber poles. The total weight of the sign is

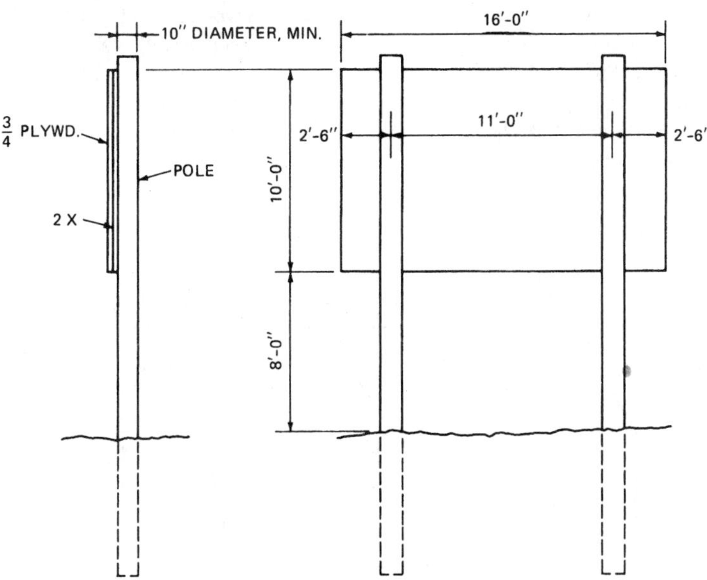

Fig. 5-21. Sign and support for Problem 5-18.

1400 lb. The sign will be in an open area (UBC Exposure C) where the recommended design wind speed is 70 mph. (See Section 4-4 for method of computing pressure due to wind loads.) The builder wishes to connect the sign to the poles using lag screws, one vertical row in each pole. The poles are D. fir-larch, seasoned when installed, but exposed to weather. Assuming that the lag screws pass through 3/4-in. plywood and a 2-in. nominal member before entering the poles, how many 5/8-in.-diameter by 4-in. lag screws are required on each pole?

5-19. A hook carrying a withdrawal load is fastened to the underside of a hem-fir wood member with four 12 gauge (0.216-in.-diameter) wood screws, 2.5 in. long. Find the safe load on the hook if all load is dead load. The base plate of the hook is 1/6 in. thick. The connection will be exposed to weather.

5-20. Find the number of 18 gauge (0.294-in.-diameter) wood screws 3 in. long to resist a lateral force of $D = 60$ lb plus $L = 170$ lb between a 1-in. nominal and 4-in. nominal piece of hem-fir. Both pieces are loaded parallel to the grain. Conditions of installation and use are dry. Use Appendix Table C-14.

5-21. A bolted tension lap joint is made joining two 2 × 8 D. fir-larch planks. The planks are seasoned but will be wet in use. Loads are $D = 1.4$ k and $S = 4.4$ k. How many 1-in.-diameter bolts are required if they are placed in two rows? What is the required spacing between the rows? What is the required spacing between bolts in a row?

5-22. Design a tension lap joint between a 2 × 6 and a 3 × 6 Select Structural hem-fir member. Use 5/8-in.-diameter bolts in 2 rows. Load is $W = 5.68$ k and $D = 1.22$ k. The connection is made from seasoned wood, but will be exposed to weather in use. Check the tensile stress on the net section. Make a layout sketch showing all necessary dimensions.

5-23. A 4 × 12 of Select Structural D. fir-larch is to be spliced using 1/2-in-diameter bolts and two 1/4-in.-thick by 4-in.-wide steel side plates. The tensile loads on the member are 3 kips from dead load, 4 kips from snow, and 3 kips from wind, and $C_M = 1.0$. How many bolts are required if a single row is used? Use all UBC load cases. (See Sec. 4-6.)

5-24. Design a tension splice of a D. fir-larch 3 × 12 main member using 3/4-in.-diameter bolts and wood side members that are of the same cross section as the main member. Load is $D = 6.5$ k and $S = 20$ k. Conditions of installation and use are dry. Use either two or three rows of bolts. Sketch a detail of the joint labeling all distances. Check the tensile stress on the net section knowing that the adjusted allowable, F_t', is 1150 psi.

5-25. A vertical No. 1 D. fir-larch 4 × 10 and an identical diagonal member (30° angle between them) are connected using two 7/8-in. bolts in a row parallel to the diagonal. If the wood is installed and used dry, what is the allowable tension load, F, in the diagonal member, based on the strength of the bolted connection? $C_D = 1.0$.

5-26. Compute the spacing, edge distance, and end distances required for the bolts of Problem 5-25. Show the bolt arrangement and these dimensions on a scale sketch of the connection.

5-27. A wood 2 × 8 D. fir-larch ledger is bolted to a concrete wall as shown by Fig. 5-22. Plywood floor sheathing nailed to the top of the ledger applies vertical load ($D + L$) of 130 lb/ft of wall and horizontal load (parallel to the wall, from diaphragm action) of 200 lb/ft due to

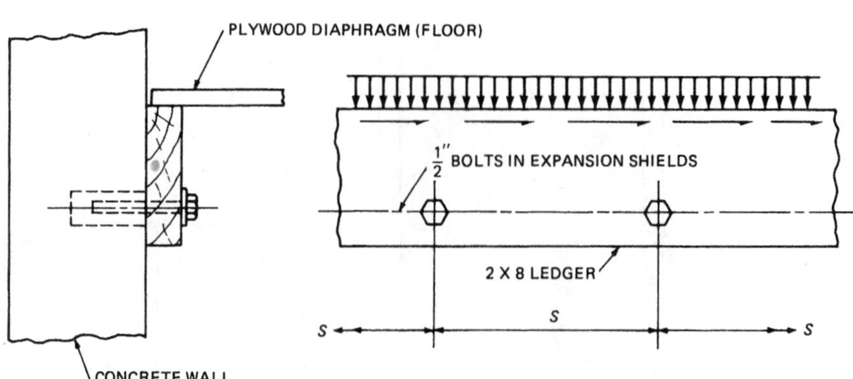

Fig. 5-22. Ledger bolts for Problem 5-27.

wind. What spacing, *s*, is required for the ½-in. bolts joining the ledger to the wall? (The bolt value in the wood is based on a main-member thickness equal to twice the ledger thickness.) The wood is installed and used dry.

5-28. Two 3 × 12s of Select Structural D. fir-larch are to be spliced in a double-shear connection using 2 × 12 side members and ½-in. bolts. The ten-year duration load (dead plus occupancy live) in the members is 13 kips tension. Design the splice using two rows of bolts. Compute the net area of the members, and check the tensile stress on the net section.

5-29. Design a bolted connection that joins the brace in Fig. 5-13 to the beam, assuming that the load is $D = 1.0$ k plus $W = 3.5$ k. (Do not necessarily use five bolts, unless that is what you consider the best choice.) Make a layout sketch showing all necessary dimensions.

6

Selecting Sawn-Timber Beams

6-1. INTRODUCTION

Structural members stressed primarily in bending are called *beams.* Horizontal beams of dimension lumber, such as 2 × 8s, used repetitively in bending are usually referred to as *joists,* whereas large beams that support other beams are called *girders.* Beam sections are generally chosen on the basis of bending and then checked for other possible failure modes. Beam failure could be a result of bending, shear, lateral buckling, bearing, deflection under service loads, or deflection from creep or ponding. Beam design consists of selecting a species, grade, and cross-sectional dimensions to prevent any of these types of failure from occurring.

Some of these failures might be catastrophic, while others might affect only aesthetics or serviceability. This chapter will discuss each of these failure modes in the sequence in which they would normally be considered.

Beam sections are specified by nominal dimensions, but cross-sectional properties such as area and section modulus are always based on actual dimensions. Unless bending moment due to self-weight is trivial, one must include beam weight as a load on the beam. In this case, the specific gravities given in Appendix Table C-1 will be helpful.

6-2. DESIGN FOR FLEXURE—LATERALLY SUPPORTED BEAMS

If not braced to prevent their compression surface from moving sidewise, wood beams may buckle laterally due to instability. Fortunately, most wood beams and joists are braced at close enough intervals along their compression face that lateral buckling is prevented. If this is the case, then the beam stability factor, C_L, is 1.0.

Lateral support may be provided by blocking (short pieces of dimension lumber cut to fit between joists), or cross bracing called *bridging,* as shown in Fig. 6-1. The diagonal bridging members are sometimes light-gauge metal, but are more often cut from wood 1 × 3s or 1 × 4s for material economy. Bridging not only keeps the joists upright (prevents rotation), but also helps distribute loads to adjacent joists. Lateral support can be provided by well-nailed plywood sheathing or by nailing one-inch nominal boards (subfloor) diagonally over the joists. Commercial metal hangers, besides providing a bearing seat, can also prevent the ends of beams from rotating.

The NDS (1) lists minimum requirements for lateral support to permit using C_L equal to 1.0. These requirements, developed from past experience, are:

1. If the ratio of depth to thickness (based on nominal dimensions) of the beam is not more than 2 to 1, no lateral support is required.
2. If the ratio is 3 or 4 to 1, the ends must be held in position to prevent rotation.
3. If the ratio is 5 to 1, one edge must be held in line for its entire length.
4. If the ratio is 6 to 1, full-depth bridging or blocking must be installed at not over 8-ft centers. There are two alternatives to this: (a) Both edges must be held in line, or (b) the ends must be held in position to prevent rotation, and also the full length of the compression edge of the member must be laterally supported by subflooring or sheathing.

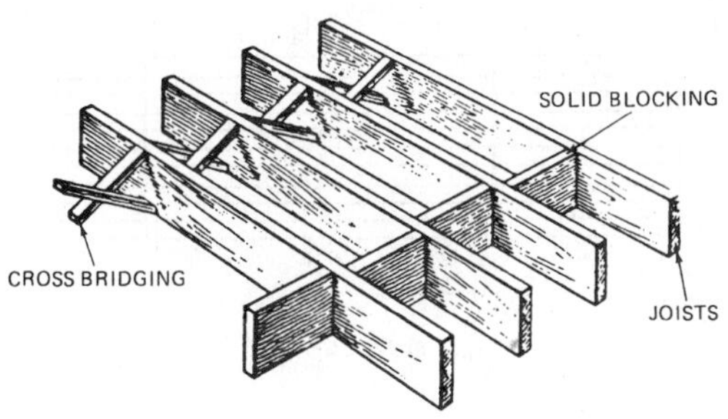

Fig. 6-1. Means of lateral support of beams.

Design Procedure

When designing sawn-timber beams by the allowable (working) stress design method, the computed extreme fiber stress ($f_b = M/S = Mc/I$) must not exceed the adjusted allowable bending stress:

$$f_b \leq F'_b \tag{6-1}$$

In this equation and other equations to follow, the primed symbol for allowable stress (F'_b in this case) indicates the particular allowable stress including all applicable adjustments. Some of these adjustments (discussed in Chapter 4) are size factor (C_F), load duration factor (C_D), and wet service (moisture) factor (C_M). In bending-moment calculations for simple beams, the span is to be taken as the distance center-to-center of *required* bearing areas.

Should there be a notch in a bending member, the section modulus at that location should be reduced accordingly for calculating flexural stress. The notch depth must not exceed one-sixth the depth of the member. Notches should be avoided, especially on the tension side of the member (except at the ends), and must not be located in the middle one-third of the span where bending moment is the greatest.

The steps in the design procedure are as follows:

1. Determine which load combination controls and find C_D.
2. Estimate beam weight per foot.
3. Solve for actual maximum bending moment.
4. Estimate adjusted allowable bending stress using *estimates* of C_F and C_L.
5. Solve for section modulus required.
6. Choose a trial section.
7. Verify weight per foot, moment, adjustment factors, and section modulus required.
8. Recycle as needed.
9. After selecting a section for bending, check for shear, deflection, and bearing (if required).

Form Factor

For circular and diamond-shaped beams, there is an adjustment to the allowable bending stress called the form factor, C_f. Early experiments showed that a circular log can support the same load as a square beam of the same cross-sectional area, even though the circular log has a section modulus only 1/1.18 times as large. A square beam loaded diagonally to its face can support the same load as if it were loaded perpendicular to a face. Consequently, a circular beam has $C_f = 1.18$ and a diagonally loaded square beam has $C_f = 1.414$. The adjusted allowable bending stress is multiplied by this factor. Use the section properties of the actual circle or diamond cross section.

Decking

Tongued and grooved (T&G) decking, used for floors and roofs, is another type of wood bending member. It is popular today in conjunction with glulam beams and arches or heavy-timber beams because the underside of the decking forms an attractive exposed ceiling for the room below at no additional cost. Tongues and grooves distinguish decking from dimension lumber used as floor planks. A projection (tongue) on the edge of one board fits into a depression (groove) in the adjacent board as seen in Fig. 6-2. The tongues and grooves constrain each board to deflect when adjacent boards deflect; that is, tongued-and-grooved members are repetitive.

Common sizes of sawn-lumber decking are 2 × 6, 3 × 6, or 4 × 6 nominal. Requirements for manufacture and design of decking are given in AITC Standard 112 (2). The 2-in. nominal has a single tongue and groove as in Fig. 6-2a,

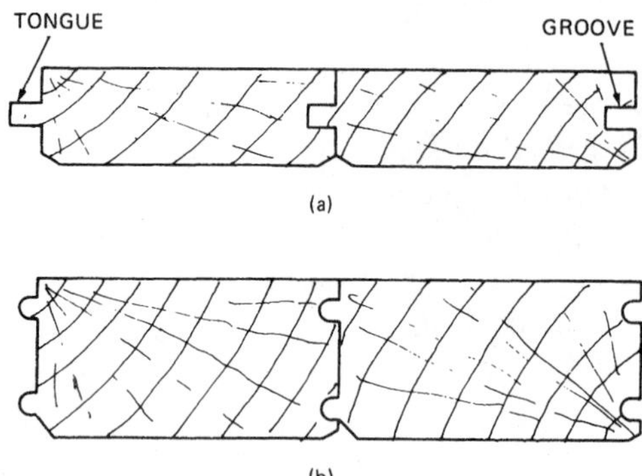

Fig. 6-2. Sawn lumber decking.

whereas the heavy-timber decking has double tongues and grooves as in Fig. 6-2b.

Appendix Table B-3 gives allowable bending stresses for 4-in. nominal decking. Except for redwood, this allowable bending stress should be multiplied by a size factor of 1.10 for 2-in. nominal decking and by a factor of 1.04 for 3-in. decking. Bending stress values shown are already for flat use, so no adjustment for flat use is needed for decking. Decking is designed based on bending and deflection only. Shear stress is low and is usually disregarded. Because decking is bent in the plane of least flexural rigidity, it has no tendency to buckle laterally, so C_L is always 1.0.

Decking may be laid in several different arrangements distinguished by location of end joints. The lay-up arrangements (see Fig. 6-3) and basis for design are:

1. Simple span: If all pieces bear on only two supports and end joints occur atop every joist for every piece of decking, then the arrangement is merely a simple span. The design for flexure and deflection would proceed exactly as for a simple span beam; that is, max $M = w\ell^2/8$ and max $\Delta = 5w\ell^4/(384EI)$.

2. Two-span continuous: If all end joints occur in line at alternate supports, as shown by Fig. 6-3b, the arrangement is two-span continuous. It is designed as a two-span continuous beam with both spans fully loaded; that is, max $M = w\ell^2/8$ and max $\Delta = w\ell^4/(185EI)$.

3. Combination simple and two-span continuous: In this lay-up, shown by Fig. 6-3c, every other piece in the *end* span is a simple span. All other pieces are continuous over two spans. The

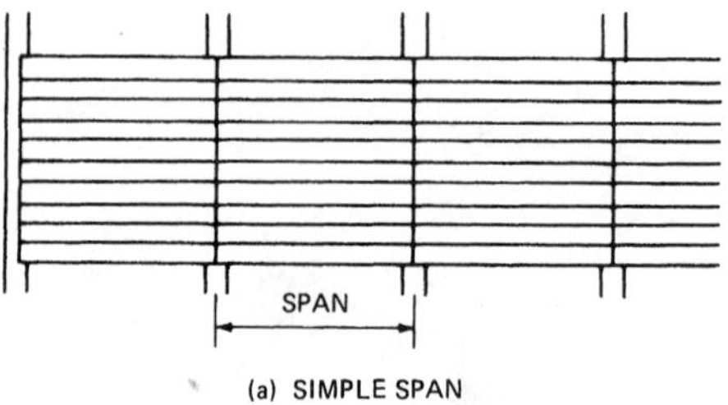

(a) SIMPLE SPAN

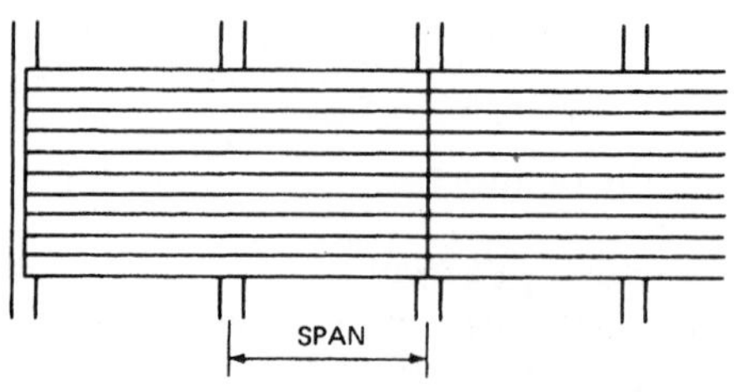

(b) TWO-SPAN CONTINUOUS

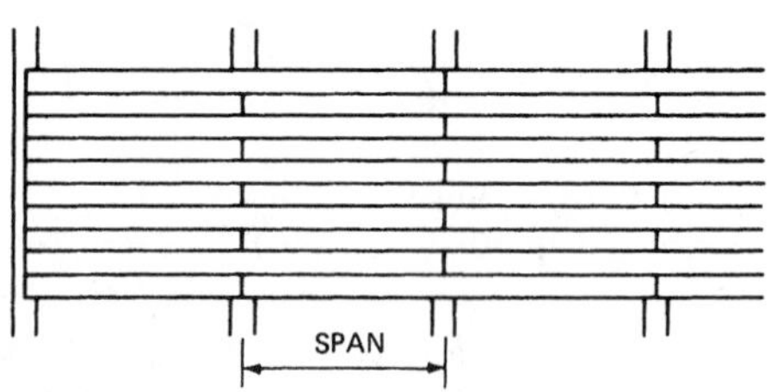

(c) COMBINATION SIMPLE AND TWO-SPAN CONTINUOUS

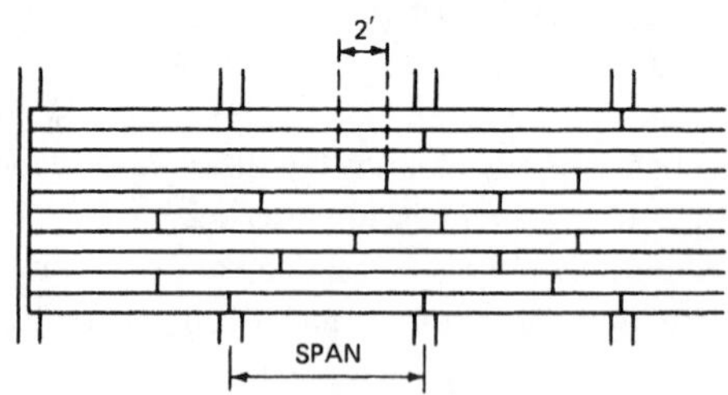

(d) CONTROLLED RANDOM LAY-UP

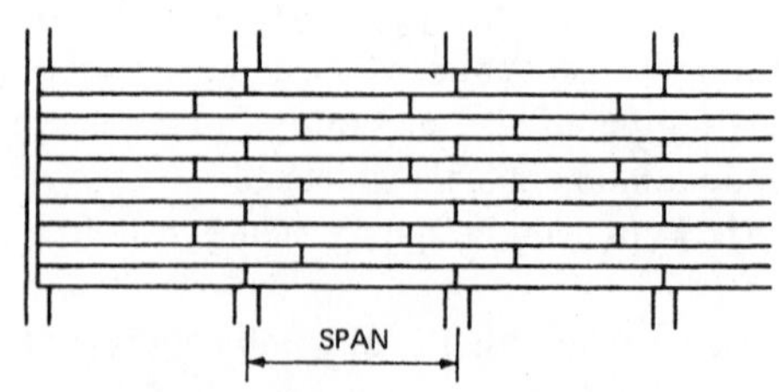

(e) CANTILEVERED PIECES INTERMIXED

Fig. 6-3. Layup patterns of decking. (Courtesy of the American Institute of Timber Construction (AITC).)

net effect is that all end joints occur over supports, but for adjacent pieces they are in alternate courses. Design for bending is based on a two-span continuous beam with both spans loaded. Design for deflection is based on the average of the maximum deflections of a simple span and a two-span continuous beam; that is, max $M = w\ell^2/8$ and max $\Delta = w\ell^4/(109EI)$.

4. Controlled random lay-up, shown by Fig. 6-3d (see reference 3 for more details): Use a maximum moment of $w\ell^2/10$ for flexural design, but use only two-thirds (tables in reference 3 use 69%) of the actual moment of inertia in bending and deflection calculations for 2-in. nominal and 80% of the moment of inertia for 3- and 4-in. nominal. These reductions are based on previous experience with random joints. Maximum deflection is considered to be as for a uniformly loaded three-span beam (max $\Delta = w\ell^4/(145EI)$. With the reduction in moment of inertia, this amounts to $w\ell^4/(100EI)$ for 2-in. nominal and $w\ell^4/(116EI)$ for thicker decking.

5. Cantilevered pieces intermixed, as in Fig. 6-3e: Flexural design is the same as for controlled random lay-up, including the reduction in moment of inertia. The deflection design formula is $\Delta = w\ell^4/(105EI)$ with no reduction of moment of inertia.

Example 6-1

A simple laterally supported beam (D. fir-larch) has a span of 16 ft and supports a uniform load of 480 lb/ft. The load is of ten-year (normal) duration. Assuming No. 1, B&S, the tabulated allowable bending stress, F_b, is 1350 psi. Choose a section based on flexure only.

Estimate $C_F = 0.95$ (an estimate between 0.95 and 0.98 is a good first trial); $C_D = 1.0$

$$F'_b = (0.95)\,(1350) = 1282 \text{ psi}$$

Estimate that the cross-sectional area will be about one square foot, so the beam weight (assuming 19% MC) is approximately

$$w = 0.50 \times 62.4 \times 1.19 = 37 \text{ lb/ft}$$

$$M = w\ell^2/8 = 517(16)^2/8 = 16{,}500 \text{ ft-lb}$$

$$S \text{ req} = M/F'_b = (16{,}500)\,(12)/1282 = 154 \text{ in.}^3$$

A 6 × 14 beam has a section modulus of 167 in.^{3} (see listing of section moduli in Appendix Table B-1) and weighs

$$(37)\,(74.25)/144 = 19 \text{ lb/ft}$$

For a 6 × 14, $C_F = (12/13.5)^{1/9} = 0.987$. Estimates for C_F and beam weight were both conservative, so the 6 × 14 is a safe choice. If the computations are repeated using these new values,

$$M = 499(16)^2/8 = 16{,}000 \text{ ft-lb}$$

$$S \text{ req} = 12\,(16{,}000)/(0.987 \times 1350) = 144 \text{ in.}^3$$

There is no smaller beam with a section modulus of at least 144, so the 6 × 14 is the final selection. Other sections, wider and shallower than the 6 × 14, could be found that would have the necessary bending strength. However, the chosen section has the smallest area and, therefore, probably the smallest cost.

Example 6-2

In an apartment building, the living room floor beams (joists) extend 5 ft past the exterior wall, forming a balcony that is 5 ft wide and 16 ft long. The loads on the balcony are:

5 psf dead load, including beam weight

60 psf live load due to people, furniture, and stacked firewood

38 psf snow load

Using machine-stress-rated lumber of grade 1200f-1.2E and assuming these joists will be spaced at 16 in. c/c, what size joist should be used considering flexure only? Assume that the negative bending moment at the exterior wall controls the design and that there is full lateral support.

Permanent loads total 5 psf (just the dead load).

Ten-year duration loads = 5 + 60 = 65 psf

Two-month duration loads = 5 + 60 + 38 = 103 psf

Dividing each of these combinations by the load-duration factor for the type of shortest duration, the largest quotient will be for the loading case that controls.

5 / 0.9 = 5.6

65 / 1 = 65

103 / 1.15 = 89.6 *Controls,* so $C_D = 1.15$

The tributary width to each joist is 16 in. (i.e., each joist carries load equal to that occurring on an 8-in. width on each side of the joist).

$w = (103 \text{ psf}) (16 / 12 \text{ ft}) = 137 \text{ lb/ft}$

$M = w\ell^2 / 2 = (137) (5)^2 / 2 = 1710 \text{ ft-lb}$

$F_b = 1200 \text{ psi}$

$C_D = 1.15$

$C_r = 1.15$ (repetitive member)

$F_b' = 1200(1.15) (1.15) = 1587 \text{ psi}$

$S \text{ req} = 1710 (12) /1587 = 12.9 \text{ in.}^3$

Use 2 × 8s (S = 13.1 in.^{3}) at 16 in. center-to-center. Note that C_F does not apply to MSR lumber.

Example 6-3

Exposed floor beams for a ski lodge will be made from 16-in.-diameter logs whose unadjusted F_b = 1250 psi. If the beams span 30 ft, what is the allowable permanent uniform load per foot?

Beams of circular cross section have no tendency to buckle, so C_L = 1.0. Cross-sectional area = 201 in.^{2}, so the equivalent square has a side of 14.2 in. Using the equivalent square, the size factor is

$C_D = 0.9$
$C_F = (12/14.2)^{1/9} = 0.98$
$F_b' = 0.9(0.98)(1250) = 1102 \text{ psi}$
$S = 14.2^3/6 = 477 \text{ in.}^3$
Allowable $M = F_b' S = 1102(477)/12 = 43{,}800$ ft-lb $= w\ell^2/8$
Allowable w = 389 lb/ft (including beam weight)

An alternate to using the equivalent square is to use section properties of the actual circular section.

Example 6-4

A wood house built in the early 1900s is to be converted to use as a private club. The building code requires increased load capacity because of change in use. A two-span continuous beam (each span = 13 ft) supporting the first floor consists of four 2 × 12s nailed together. Because of random joints where pieces of lumber abut, only three of the four 2 × 12s are effective at any one cross section. The lumber has been identified as Douglas fir corresponding to No. 2 grade. Measurements show the 2 × 12s to be actually 1¾ in. by 11⅝ in. Find the allowable uniform load on the continuous beam, assuming full lateral support.

Assuming that the duration of loads till now has exceeded ten year cumulative,

Factor for load duration = 0.9

$C_F = 1.0$ (from Table B-3)

$F_b' = 0.9(1.0) (875) = 788 \text{ psi}$

$S = (3) (1.75) (11.625)^2/6 = 118 \text{ in.}^3$

Maximum moment in a two-span beam is the same as for a single-span beam so

$w(13^2)/8 = F_b' S = 788(118)/12$

Allowable w = 367 lb/ft

Example 6-5

Douglas fir-larch T&G decking for a floor will be laid up in a controlled random pattern (Fig. 6-3d). This Commercial grade 2 × 6 decking supports a uniform live load (normal duration) of 50 psf. Self-weight is the only dead load. Moisture content will not exceed 19%. What is the farthest distance apart the supporting joists may be placed (based on bending stress)?

$E = 1{,}700{,}000$ psi (Table B-3)

$F_b = 1650$ psi (repetitive)

$C_D = 1.0$

$C_F = 1.10$

$F_b' = (1.10) (1650) = 1815 \text{ psi}$

Using a specific weight of 37 pcf and an actual thickness of 1.5 in.,

$w_D = (1.5) (37) / 12 = 5 \text{ psf}$

Considering a one-foot width (approximately two planks),

$w_D = (5 \text{ psf}) (1 \text{ ft}) = 5$ lb/ft per ft width

$w_L = (50 \text{ psf}) (1 \text{ ft}) = 50$ lb/ft per ft width

$I = (12)(1.5)^3/12 = 3.375 \text{ in.}^4$ per ft width

$c = 0.75$ in.

Expressing permissible span, ℓ, in inches:

$f_b = (0.1\, w\ell^2) (c) / (21 / 3)$

$= 0.1 (55 / 12) (\ell^2) (0.75) / 2.25 = 0.153\ell^2$

Let $f_b = F_b' = 1815$ psi:

$\ell = 109$ in. (based on bending stress)

A maximum permissible span based on deflection will be discussed in Example 6-13.

6-3. DESIGN FOR FLEXURE—LATERALLY UNSUPPORTED BEAMS

Most beams are laterally supported by such things as attached plywood or decking, or by blocking, bridging, or diagonal bracing. If the distance between points of lateral support is too great, as in Fig. 6-4, the beam can buckle (twist and move sidewise). In this case we reduce the allowable bending stress to a value less than used for laterally braced beams.

One must take care to distinguish between the span of a beam and its unbraced (laterally unsupported) length. The span of a beam is the distance between supports—the walls, columns, or girders that the beam bears on. The unbraced length is the distance between *lateral* (sidewise) restraints. Lateral supports prevent movement perpendicular to the plane of the loads.

In the following, the design procedure specified by the NDS is explained first. Then the theoretical basis for the design equations is given for those who wish to understand the fundamentals of beam buckling.

Design Procedure

To analyze a beam with only intermittent points of lateral support, calculate the beam stability factor, C_L, which is an adjustment to allowable bending stress. Finding C_L requires knowing F_b^*, K_{bE}, R_B, and F_{bE}. F_b^* is simply the tabulated allowable bending stress multiplied by all applicable adjustment factors except for C_{fu}, C_V (glulams only), and C_L. K_{bE} equals 0.438 for visually graded lumber and machine-evaluated lumber, but equals 0.609 for glulams and other products for which the coefficient of variation for elastic modulus, E, does not exceed 0.11.

Find the slenderness ratio, R_B, and F_{bE} from the following equations:

$$R_B = \sqrt{\ell_e d / b^2} \quad (6\text{-}2)$$

$$F_{bE} = K_{bE} E' / R_B^2 \quad (6\text{-}3)$$

The slenderness ratio for beams is similar to the ℓ/r ratio for columns. In Eq. 6-2, b = thickness (width normal to plane of loads), d = depth, and ℓ_e = effective unbraced length. Effective unbraced length, ℓ_e, depends on the shape of the bending moment diagram and is defined in Table 6-1. Effective unbraced length is also a function of the actual unbraced length ℓ_u (actual distance between points of adequate lateral support).

If a bending member is used in a flatwise position, slenderness factor will be of no concern since the compression face will have no tendency to buckle. Also, for circular beams such as logs, and for beams whose depth does not exceed their width, conditions of lateral support need not be considered because such members cannot buckle due to bending. They are stable.

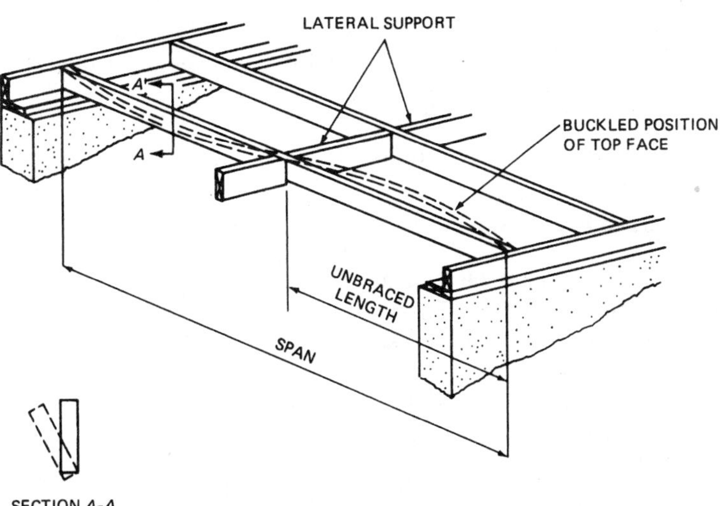

Fig. 6-4. Lateral buckling of a beam.

Table 6-1. Effective Length, ℓ_e, for Bending Members.

Cantilever[1]	when $l_u/d < 7$	when $l_u/d > 7$
Uniformly distributed load	$l_e = 1.33\, l_u$	$l_e = 0.90\, l_u + 3d$
Concentrated load at unsupported end	$l_e = 1.87\, l_u$	$l_e = 1.44\, l_u + 3d$
Single Span Beam[1]	**when $l_u/d < 7$**	**when $l_u/d > 7$**
Uniformly distributed load	$l_e = 2.06\, l_u$	$l_e = 1.63\, l_u + 3d$
Concentrated load at center with no intermediate lateral support	$l_e = 1.80\, l_u$	$l_e = 1.37\, l_u + 3d$
Concentrated load at center with lateral support at center	$l_e = 1.11\, l_u$	
Two equal concentrated loads at 1/3 points with lateral support at 1/3 points	$l_e = 1.68\, l_u$	
Three equal concentrated loads at 1/4 points with lateral support at 1/4 points	$l_e = 1.54\, l_u$	
Four equal concentrated loads at 1/5 points with lateral support at 1/5 points	$l_e = 1.68\, l_u$	
Five equal concentrated loads at 1/6 points with lateral support at 1/6 points	$l_e = 1.73\, l_u$	
Six equal concentrated loads at 1/7 points with lateral support at 1/7 points	$l_e = 1.78\, l_u$	
Seven or more equal concentrated loads, evenly spaced, with lateral support at points of load application	$l_e = 1.84\, l_u$	
Equal end moments	$l_e = 1.84\, l_u$	

[1]For single span or cantilever bending members with loading conditions not specified above.

$l_e = 2.06\, l_u$ when $l_u/d < 7$
$l_e = 1.63\, l_u + 3d$ when $7 \le l_u/d \le 14.3$
$l_e = 1.84\, l_u$ when $l_u/d > 14.3$

Equation 6-3 requires the adjusted value of E. The only adjustment factors that apply to modulus of elasticity are C_M, C_t, and C_T. E is never adjusted for load duration.

When all preliminary calculations have been done, find the beam stability factor from

$$C_L = \frac{1+\left(F_{bE}/F_b^*\right)}{1.9} - \sqrt{\left[\frac{1+\left(F_{bE}/F_b^*\right)}{1.9}\right]^2 - \frac{F_{bE}/F_b^*}{0.95}} \tag{6-4}$$

Theory of Elastic Stability of Beams

For a laterally unsupported beam, if the flexural rigidity in the plane of bending is large in comparison with the lateral flexural rigidity, the beam will buckle laterally at a certain critical load in the elastic range. The critical bending moment is an ultimate load at which sudden instability occurs. To determine the critical load for the case of bending moment that is uniform along the length of a beam, consider the beam of Fig. 6-5a with applied end moments, M_o, about axes parallel to the Z-axis. The beam ends are free to rotate (about the Z-axis), but are fixed against twisting (about the X-axis).

In the deflected position, the compression edge of the beam has moved sidewise, as the cross section in Fig. 6-5b indicates. The displacement in the Z-direction is denoted as w, and $dw/dy = \beta$ is angle of twist (4). From equilibrium of the buckled beam, the bending and twisting couples at any cross section are

$$M_x \simeq M_o \frac{dw}{dx}$$

$$M_y \simeq M_o \frac{dw}{dy} = M_o\beta$$

$$M_z \simeq -M_o$$

But by the theory of elastic stability,

$$M_y = -EI_y \frac{d^2w}{dx^2} = M_o\beta$$

$$M_x = GK\frac{d\beta}{dx} = M_o\frac{dw}{dx}$$

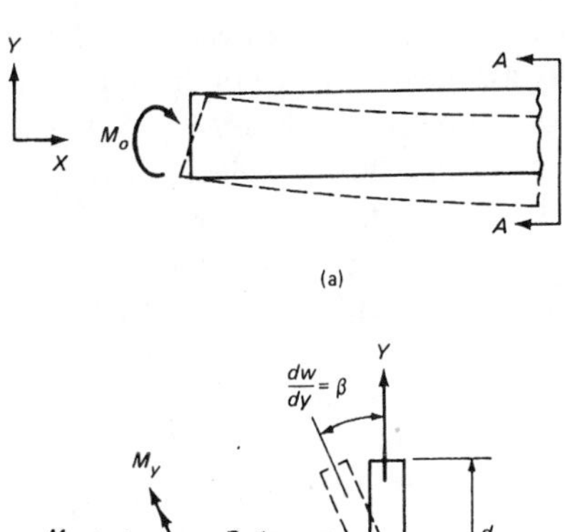

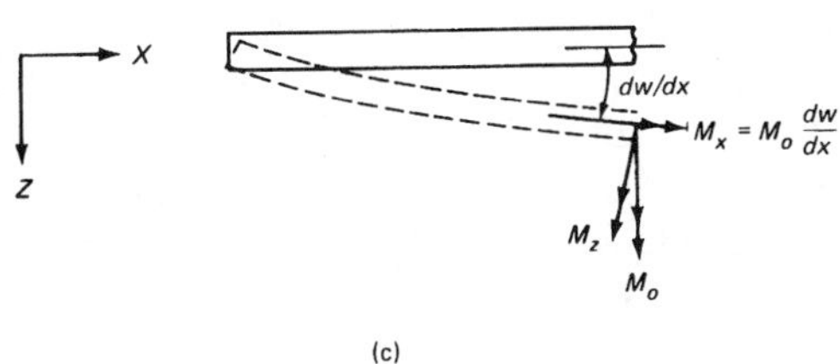

Fig. 6-5. Laterally unsupported beam.

where K is the torsional stiffness factor for a rectangular cross section (K equals about $0.26db^3$ for wood beams of practical size). Differentiating the latter with respect to x,

$$GKd^2\beta/dx^2 = M_o(d^2w/dx^2)$$

Substituting,

$$GKd^2\beta/dx^2 = M_o(-M_o\beta/(EI_y))$$

$$d^2\beta/dx^2 + M_o^2\beta/(EI_yGK) = 0$$

By inspection, the solution to the differential equation will require that β be the sum of a sine function and a cosine function. Solving the equation gives

$$Critical\ M = \pi\sqrt{EI_yGK} / \ell$$

$$Critical\ f_b = M_{cr}d / (2I_x) = \pi d\sqrt{EI_yGK} / 2I_x\ell$$

$$= \frac{\pi d\sqrt{E(db^3/12)(G)(0.26db^3)}}{2(bd^3/12)(\ell)}$$

$$Critical\ f_b = Cb^2 / \ell d$$

where C is constant for a given species and grade of wood.

This result shows that critical flexural stress for lateral buckling is proportional to b^2 (square of beam thickness) and inversely proportional to both beam length and depth. It is interesting to contrast this critical stress for beams with critical buckling stress for a column. For a column, the critical stress is inversely proportional to the *square* of the length, rather than the length.

Equation 6-4 is based (in part) on the lateral buckling theory just presented. The allowable stress, computed using adjustment factor, C_L, is a function of $b^2/\ell d$.

Example 6-6

A simply supported 6 × 18 beam (B&S) with an unadjusted F_b = 1550 psi and E = 1,500,000 psi is loaded with a uniform load of seven days duration. The beam will be in a humid location where moisture content is expected to remain about 20%. The span is 40 ft. The beam is laterally braced at its ends and at midspan only. Based on flexure, what uniform load can be superimposed?

C_M (for both F_b and E) = 1.0

C_D = 1.25 (from Fig. 2-9)

$C_F = (12 / 17.5)^{1/9} = 0.959$

For an assumed unit weight of 37 pcf, the beam weight is about
(37) (5.5) (17.5) /144 = 25 lb /ft

ℓ_u = (20) (12) = 240 in.

ℓ_e = 1.63 (240) + 3 (17.5) = 444 in.

$R_B^2 = (444)(17.5)/5.5^2 = 257$

$F_b^* = (1550)(1.0)(1.25)(0.959) = 1858$ psi

$F_{bE} = K_{bE}E'/R^2_B = 0.438\ (1,500,000) / 257$
$= 2556$

By Eq. 6-4, $C_L = 0.911$
$F_b' = (0.911)(1858) = 1693$ psi

Allowable $M = F'_b S$

$w\,(40)^2 / 8 = 1693\,(280.7) / 12$

Allowable $w = 198$ lb/ft

Subtracting the weight of the beam, the allowable superimposed load is 198 − 25 = 173 lb/ft

Example 6-7

Design a Douglas fir beam, spanning 22 ft, to support two concentrated loads of 0.77 kips each (from a stairway). Use either No. 1 & Btr or Dense No. 1 grade. Additionally the beam supports 100 lb/ft, not including self-weight. The first point load is 7 ft from the left support and the second is 10 ft from the left. Lateral support exists only at the location of the concentrated loads. Consider only flexure. Fig. 6-6 shows the solution in computation sheet form.

6-4. DESIGN OF BEAMS FOR SHEAR

Timber beams are relatively weak in shear parallel to the grain, so their shear strength needs to be calculated. Shear failure causes a long horizontal split at beam mid-depth (Fig. 2-21). To design to prevent shear failure, one compares the computed actual shear stress with the adjusted allowable shear stress. That is, the beam is satisfactory in shear if

$$f_v \le F'_v \qquad (6\text{-}5)$$

where $f_v = VQ/Ib$ (for rectangular members $f_v = 1.5V/A$). The tabulated allowable shear stress F_v is based on the assumption that a full-length split is present, so if end splits are known to be minor and will not grow in length, the allowable shear stress may be multiplied by C_H, the shear stress factor. Values of C_H can be found in Appendix Tables B-3 to B-6. The largest value of C_H is 2.0 (for members having no splits or shakes). If the designer decides to apply the shear stress factor to a redwood member, F_v = 80 psi should be used. Similarly, F_v = 90 psi should be used when applying C_H to a southern pine member. Figure 6-7 shows how to measure the size of splits and checks when applying the shear stress factor.

Some of the NDS provisions for shear are based on the 1933 work of Newlin, Heck, and March (5). Their experiments on checked beams showed horizontal shear stresses to be lower than usually assumed by virtue of the upper and lower halves of the checked beam each acting as independent beams. This "two-beam action" is the basis of NDS section 4.4.2.2 and NDS Appendix E. Because the experiments were for beams with severe (artificially created) splits, two-beam action cannot be expected in glued laminated beams that have relatively small checks. Keenan verified the lack of two-beam action in glulams by a finite element study (6). He also found from laboratory and finite element work that, contrary to popular assumption, shear strength was not increased by the application of compressive stress perpendicular to the grain as caused by the beam reaction.

Shear strength is affected by beam size and shear span (distance from reaction to first concentrated load). Researchers are attempting to develop ways of considering a size effect for shear when designing, just as the size effect for bending adjusts the allowable bending stress. Accumulating data from many sources, Keenan (6) found that shear strength was a function of sheared area, A_s, defined as the product of beam width and the shear span. He proposed, for Douglas fir, an equation for the maximum actual shear stress at failure as

$$\tau_{max}(\text{psi}) = 1382 - 486 \log (A_s/4) \qquad (6\text{-}6)$$

Another researcher, Longworth, has discovered that the relationship between maximum shear stress at failure and sheared area is also affected by beam width (7). He found that failure stress was more properly a function of sheared volume.

Basic Method for Shear Calculations

The most common case where shear stress will need to be computed is for uniform loads on a simple-span beam. The critical section is at a distance d (depth of beam) from the face of the support. Using statics, the shear, V, at this location is $w\ell/2$ − wd. Then the actual shear stress can be computed as $f_v = VQ/Ib$ (or, for a rectangular cross section, $f_v = 1.5V/A$). Actual shear stress can then be compared with the adjusted allowable shear stress, F'_v.

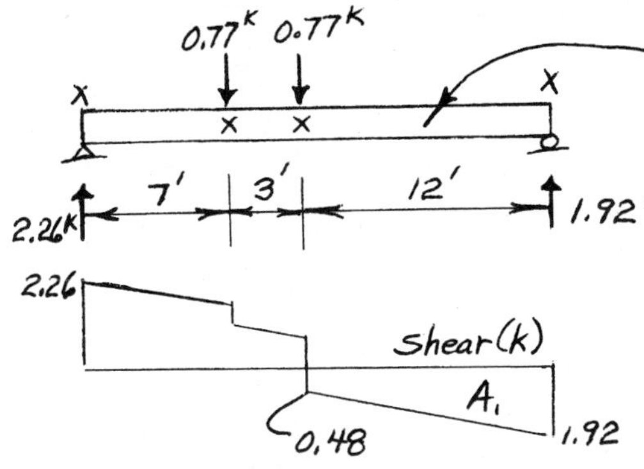

0.77^K 0.77^K

$w = 0.10 + \text{bm wt (est. } 0.02)$
$= 0.12\ K/ft$

Lateral support at ends & conc. loads only (Points x)

2.26^K 7′ 3′ 12′ 1.92^K

2.26

shear(k)

A_1

0.48 1.92

$M_{max} = A_1 = \frac{12}{2}(0.48 + 1.92)$
$= 14.4\ \text{ft-K}$

Try timbers (B&S) Dense No. 1 Douglas fir-larch

$F_b = 1550\ \text{psi}$; Assume $C_L \times C_F = 0.95$

Approx $S_{req} = 12 \times 14{,}400 / (0.95 \times 1550) = 117\ \text{in.}^3$

Try 6×12 ($S = 121\ \text{in.}^3$)

$C_F = 1.0$ since $d < 12''$

$l_u = 144''$; $l_u/d = 144/11.5 = 12.5$

$l_e = 1.63(144) + 3(11.5) = 269''$

$R_B^2 = 269 \times 11.5 / 5.5^2 = 102$

$F_{bE} = 0.438(1{,}700{,}000)/102 = 7300$

$F_b^* = (1.0)(1550) = 1550$; $C_L = 0.987$

$F_b' = 0.987(1550) = 1530\ \text{psi}$; $f_b = \frac{M}{S} = 1428\ \text{psi}$ OK

Try No. 1 & Btr grade 4×16 (Dim. lumber with $S = 135.7\ \text{in.}^3$)

$F_b = 1150\ \text{psi}$

$C_F = 1.0$, so $F_b^* = 1150\ \text{psi}$

$f_b = M/S = 1273\ \text{psi}$ — Obviously NG even without computing C_L

4×16 is largest available dimension lumber section

Use 6×12, Dense No. 1 grade

(Self wt $= 37(5.5 \times 11.5)/144 = 16\ \text{lb/ft} < 20$ assumed OK)

Fig. 6-6. Calculation sheet for Example 6–7.

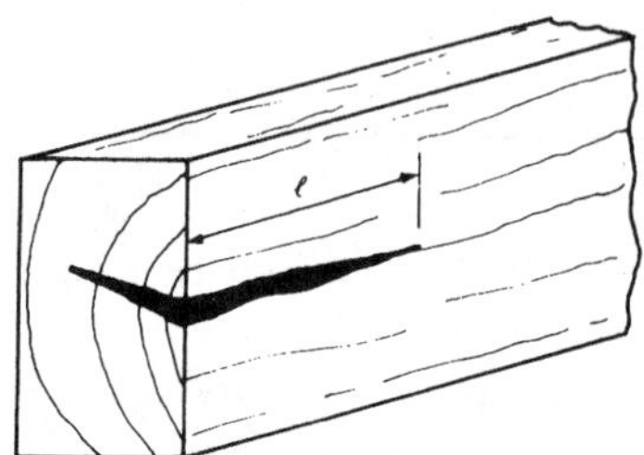

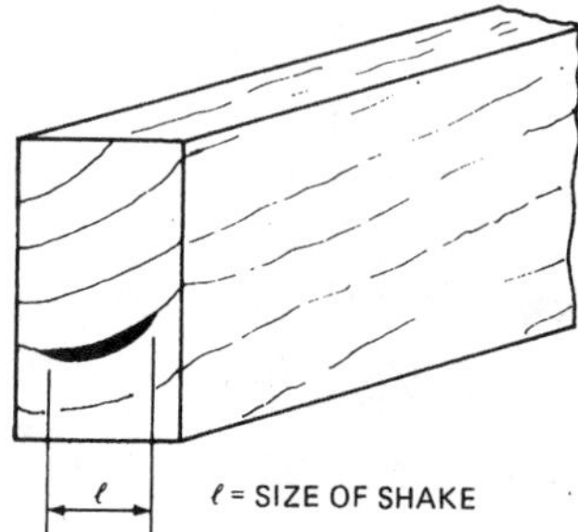

Fig. 6-7. Splits and shakes.

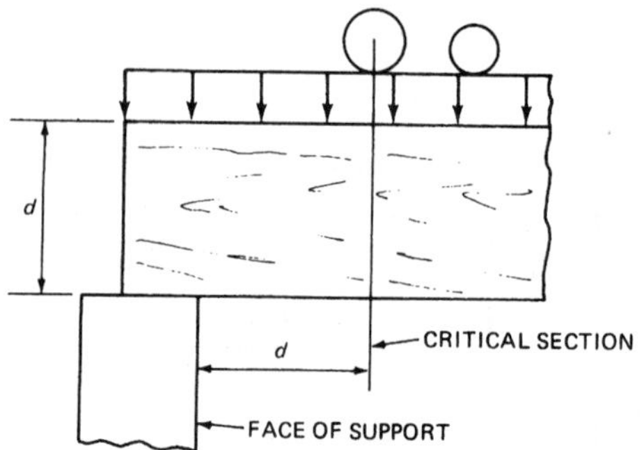

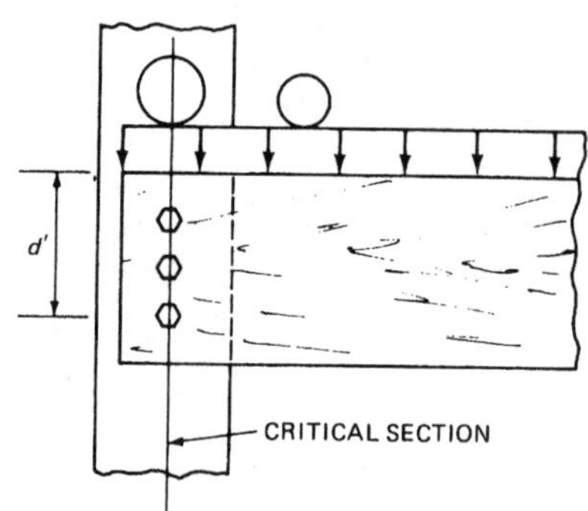

Fig. 6-8. Critical sections for shear.

For beams with concentrated loads, the NDS suggests two methods of checking wood beams for shear. The first way is by methods of statics. There are two cases for solving for shear, V, by this method:

A. If the beam has loads applied to one surface and is supported by bearing on the opposite surface: Compute shear, V, at a critical section distance d (depth of beam) from the face of the support, locating the movable loads (if any) to cause largest possible V at the critical section. This case is illustrated in Fig. 6-8a.

Having computed the maximum shear force, V, at the critical section, calculate the maximum shear stress, f_v, and compare it with the allowable shear stress, F'_v as adjusted (including C_H if it applies). If f_v exceeds F'_v but does not exceed $2F'_v$ (adjusted by all except C_H) then the alternate method *may* be used, if desired, for sawn lumber.

B. If the beam is not supported by bearing pressure on the face opposite the loads, but rather is supported by bolts or other fasteners: Use a critical section *at the connection,* and locate the moving loads to cause the largest possible shear, V, at that point. Figure 6-8b shows this case. The design procedure for this case is described later in the section "Design of Joints in Shear."

Alternate Method for Shear

If case A above shows the beam to have insufficient shear strength, then for sawn lumber (but not glulams) the designer may choose to follow the two-beam provisions of NDS section 4.4.2.2 instead. This section is applicable to case A, above, only; it does not apply to case B. Nor does it apply to notched beams. (There is considerable disagreement among design professionals about the validity of the entire concept of two-beam action.)

For two-beam action on a simple span, shear is calculated differently for uniform loads than for concentrated loads. For uniform loads, end shear is taken to be the same as shear at distance d from the support, that is,

$$V = w\ell_c/2 - wd \qquad (6\text{-}7)$$

in which ℓ_c is the clear span between faces of supports. For ℓ_c/d ratios of about 3 to 35 this value is unnecessarily conservative, so Eq. 6-7 may be multiplied by K_v where

$$K_v = 0.95 + \frac{\sqrt{\ell_c / d}}{250} - 1.32\,(d / \ell_c) + 11.5\,(d / \ell_c)^3 \qquad (6\text{-}8)$$

For a concentrated load, *P*, located at distance *x* from the face of the support, the maximum shear force calculated by statics would be $P(\ell_c - x)/\ell_c$. This is multiplied by an algebraic function to reflect the experiments of Newlin et al (5). The maximum shear force, under two-beam action, due to a single concentrated load then becomes

$$V = \frac{P(\ell_c - x)(x / d)^2}{\ell_c[2 + (x / d)^2]} \qquad (6\text{-}9)$$

For two-beam action when there are moving concentrated loads rather than stationary ones, the loads should be positioned so as to cause the maximum *V* (from Eq. 6-9). This is usually done by trial and error using Eq. 6-9. But if there is a single moving load (or an outstandingly large load in a group of moving loads), placing that load at $3d$ or $\ell/4$ (whichever is closer) from the support will result in maximum or near maximum *V*, eliminating the need for trial and error.

To conclude that the beam is satisfactory in shear when *V* due to all loads is computed by this second method, the maximum shear stress computed for two-beam action is compared to F'_v. (Two-beam action is based on having severe checking, so C_H *cannot* be used here.)

End Notches

Sometimes at a support, the lower face of a beam is notched to bring the top of the beam to the proper elevation (Fig. 6-9). Notching a beam on the tension side is not recommended as the notch induces tensile stress normal to the grain; this, in combination with horizontal shear, tends to cause splitting. If an end notch on the tension face is used, the depth of this end notch must not exceed one-fourth the beam depth. In the shallow portion next to the notch, the shear stress is larger than $1.5V/A$, and is taken as:

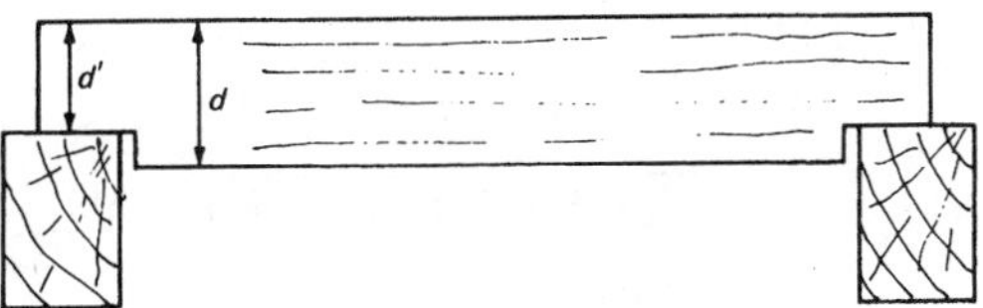

Fig. 6-9. Notched beam.

$$f_v = \left(\frac{3V}{2bd'}\right)\left(\frac{d}{d'}\right) \qquad (6\text{-}10)$$

For notched beams the shear force, *V*, is usually considered to be the maximum shear at the face of the support, not the lesser shear at a distance *d* from the support. The NDS does not allow two-beam action to be considered for beams with end notches.

Design of Joints in Shear

Case B above describes shear in beams at end connections, where the beam is not supported by bearing on the lower face of the beam, but rather by bolts, lag screws, or other fasteners. The critical section is at the connector group, as shown in Fig. 6-8b. The shear stress at the critical section is calculated by Eq. 6-10, except that d' now equals the depth of the member less the distance from the unloaded edge of the member to (1) the center of the nearest bolt or lag screw (for bolts or lag screws) or to (2) the nearest edge of the nearest connector (for split rings and shear plates).

For the special case when the joint is at least five times member depth from the end, the allowable shear stress may be increased 50%; actual stress is calculated by Eq. 6-10, except that the ratio d/d' in the equation is taken as 1.0.

Example 6-8

A hem-fir Select Structural grade 10 × 18 beam with clear span of 16 ft and a 4-in. bearing length at each end carries a dead load of 80 lb/ft plus its own weight (specific weight is about 30 pcf). The live loads are two 8-kip concentrated wheel loads separated by a distance of 5 ft; these loads may be in any location on the beam. The cumulative duration of the live loads is less than two months. Is the beam OK for shear?

Beam weight = (9.5)(17.5)(30)/144

= 35 lb/ft

Total uniform dead load = 80 + 35 = 115 lb/ft

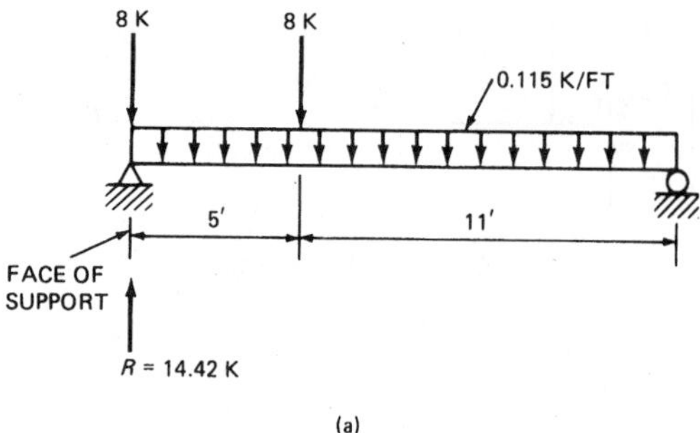

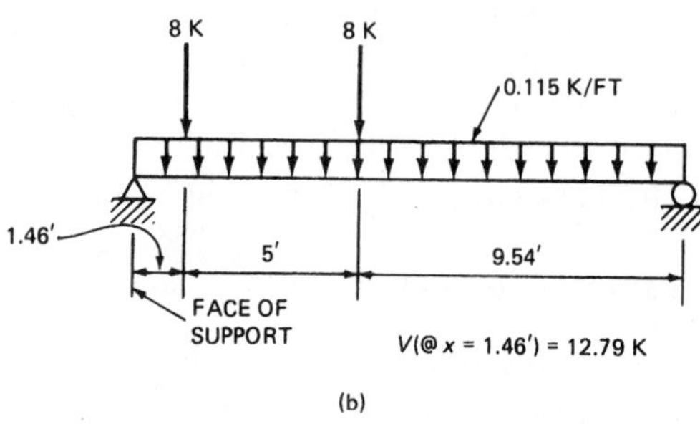

Fig. 6-10. Beam for Example 6–9.

For steel beams one would calculate the maximum shear, V, due to the two moving loads by placing one load at the face of the support as shown in Fig. 6-10a, in which case the maximum shear would be 14.42 k. However, the beam is supported by bearing on its bottom face, so for this wood beam the critical section is at a distance d from the face of the support. For maximum shear V at this location, the loads would be positioned as in Fig. 6-10b, with one load a distance $d =$ 17.5 in. = 1.46 ft from the support face. Now the shear at the critical section calculates as 12.79 k.

$$f_v = 1.5V/A = 1.5\,(12{,}790)/(9.5)\,(17.5)$$
$$= 115 \text{ psi}$$
$$F_v = (1.15)\,(70) = 80 \text{ psi} < 115 \qquad \textit{No good}$$

Example 6-9

Select a No. 1 hem-fir beam (specific wt = 30 pcf) to carry a uniform dead load of 210 lb/ft and uniform snow load of 250 lb/ft on a clear span of 18 ft. Beam ends are supported on a concrete wall. (Lateral support is at ends only.) The solution is shown by the designer's computation sheet in Fig. 6-11.

Example 6-10

Select a D. fir-larch beam to support an occupancy live load of 230 lb/ft and a dead load of 40 lb/ft that includes an estimate of beam self-weight. The beam has sufficient lateral support so that $C_L = 1.0$. The simple span for computing bending is 19 ft. (If the bearing length is not known, it will be conservative to use for shear the same span length as used in bending.) Notch the beam ends, if necessary, so that the distance from top of support to top of beam is 10 in. Choose the section based on flexure and check it for shear.

Assuming that the section will be in the Beam & Stringer size/use class, choose No. 1 grade.

$$F_b = 1350 \text{ psi}$$
$$F_v = 85 \text{ psi}$$

Since 270/1.0 > 40/0.9, $D + L$ controls.
Assuming $C_F = 1.0$,

$$F_b' = (1.0)\,(1350) = 1350 \text{ psi}$$
$$M = w\ell^2/8 = 270\,(19)^2/8 = 12{,}180 \text{ ft-lb}$$
$$S\text{ req} = M/F_b' = 12{,}180\,(12)/1350 = 108 \text{ in.}^3$$

After scanning Table B-1, try a 6 × 12 ($S = 121$ in.^{3}; $C_F = 1.0$). The greatest end notch depth allowed in a 6 × 12 is (1/4)(11.5) = 2.88 in. Therefore, the 1.5-in. notch we need to make will be acceptable provided the shear stresses remain sufficiently low.

The maximum shear (at the notch, not at d from the support face) is

$$V = w\ell/2 = (270)(19/2) = 2570 \text{ lb}$$
$$f_v = \left(\frac{3(2570)}{2(5.5)(10)}\right)\left(\frac{11.5}{10}\right)$$
$$= 80.6 \text{ psi} < F_v = 85 \text{ psi } \textit{OK}$$

Use a 6 × 12, No. 1 grade.

Example 6-11

A 12-ft clear simple span beam (6 × 12 hem-fir) is loaded with a uniform ten-yr-duration load of 785 lb/ft including self-weight. The beam has been well seasoned and no further drying is expected in service. For this reason, no checks (splits) are anticipated (none are present now). F_v from Appendix Table B-5 is 70 psi. Check the beam for shear.

At a distance $d = 11.5$ in. from the support face,

$$V = (785)\,(12/2 - 11.5/12) = 3960 \text{ lb}$$
$$f_v = (1.5)\,(3960)/(5.5 \times 11.5) = 94 \text{ psi}$$

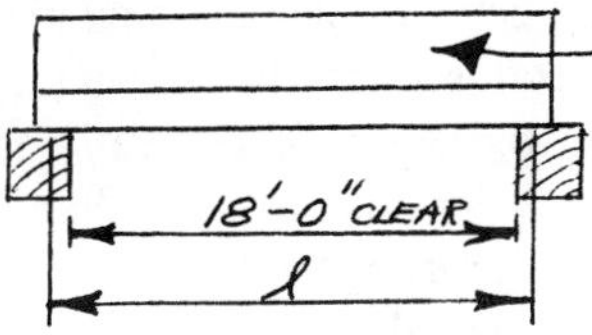

LOADS: DEAD 210 LB/FT
BM WT 25 (EST.)
235 LB/FT
SNOW 250 LB/FT

18'-0" CLEAR

ℓ

$\frac{235}{0.9} = 261;\quad \frac{250+235}{1.15} = 422 > 261$

TRY No. 1 HEM-FIR

D+S CONTROLS & $C_D = 1.15$

ASSUME B&S AND $C_F = C_L = 1.0$

$F_b' = 1.15\,(1050) = 1208$ psi

ASSUME 3" REQ'D BEARING LENGTH AT EACH END

THEN SPAN $= 18 + 3/12 = 18.25$ ft

$M_{D+S} = 485\,(18.25)^2/8 = 20{,}190$ ft-lb

S REQ'D $= 20{,}190\,(12)/1208 = 201$ in.^{3}

TRY 6×16 ($S = 220$; $C_F = 0.972$)

$F_b^* = 0.972\,(1.15)(1050) = 1174$ psi

$\ell_e = 1.63\,(18.25)(12) + 3\,(15.5) = 403$ in.

$R_B^2 = 403\,(15.5)/5.5^2 = 206$

$F_{bE} = 0.438\,(1{,}300{,}000)/206 = 2764$ psi

$C_L = 0.97$

$F_b' = 0.97\,(1174) = 1139$ psi

S REQ'D $= 12\,(20{,}190)/1139 = 213$ in.$^3 < 220$ OK

CHECK SHEAR

$V_{@d} = 485\left(9 - \frac{15.5}{12}\right) = 3740$ lb

$f_v = 1.5\,(3740)/(5.5 \times 15.5) = 66$ psi

$F_v = 1.15\,(70) = 80$ psi OK

VERIFY WT

$30\,(5.5 \times 15.5/144) = 18$ lb/ft < 25 assumed OK

USE 6×16

Fig. 6-11. Computation sheet for Example 6–9.

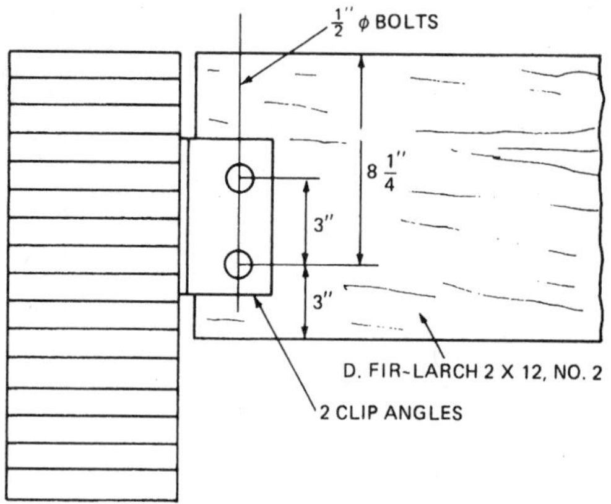

Fig. 6-12. Beam for Example 6–12.

$$C_H = 2.0$$

$$F_v = (70)(2.00) = 140 \text{ psi} > 94 \text{ psi} \qquad \mathit{OK}$$

Note that the NDS permits doubling the allowable shear stress in this case. However, using this increased allowable seems unwise in the authors' opinion. Conditions that may cause a beam to develop checks or splits in service just cannot be predicted with any certainty.

Example 6-12

The Douglas fir beam of Fig. 6-12 is to be supported at each end by two ½-in.-diameter bolts through two ¼-in.-thick steel angles. What is the maximum allowable reaction based on the bolt value and based on shear in the beam? $C_D = 1.0$. Do not apply C_H.

Based on bolt value (double shear):

Allowable load per bolt = 470 lb

Allowable reaction = 2 (470) = 940 lb

Based on shear in the beam:

$$F_v = F'_v = 95 \text{ psi}$$

$$f_v = 1.5\,(R/bd')\,(d/d')$$

$$\text{Allowable } R = (2/3)(95)(1.5 \times 8.25) \times (8.25/11.25) = 575 \text{ lb} \qquad \textit{Controls}$$

Clip-angle connections of this type are not advisable for heavily loaded members or with large numbers of bolts, because vertical shrinkage may cause splitting.

6-5. DEFLECTION

Besides having strength, a beam should also have adequate stiffness so that it will not deflect or sag too much. Excessive deflections can lead to broken windows, cracked plaster, ponding of water, or just bad appearance. Architectural or structural damage is not the only reason for limiting deflection. Also, the human factor needs to be considered. People feel uncomfortable walking across a floor that is saucershaped, no matter how safe and strong the floor is.

Because wood will creep under long-term loading (see Chapter 2), one cannot simply calculate live- and dead-load deflection by mechanics formulas and expect the result to be the maximum over the life of the beam. To account for creep over the long term, the immediate deflection due to long-term loads is magnified.

Assuming that all live loads are short-term loads, the NDS recommends separating the effects of the short-term (live) loads and long-term (dead) loads, calculating the total deflection over the long term as follows:

$$\Delta_{\text{Total}} = \Delta_{\text{Live}} + 1.5\,\Delta_{\text{Dead}}$$
$$= \Delta_L + 1.5\,\Delta_D \qquad (6\text{-}11)$$

This equation assumes that the additional deflection due to creep will equal one-half the initial, immediate dead-load deflection. This equation is applicable to either glulams or seasoned lumber. For unseasoned lumber, the factor 1.5 in the equation is replaced with 2.0. This same method of calculating long-term deflection was in use as early as 1930 (8).

The designer must keep computed deflections within limits specified by authorities such as the *Uniform Building Code* (9) and the AITC (reference 3). The UBC deflection limits are reproduced in Table 6-2. According to this table, the UBC assumes additional deflection due to dead-load creep to be one-half of the immediate dead-load deflection for seasoned lumber—the same as the NDS. Notice also that the UBC does not consider dead-load deflection that takes place immediately to be of any consequence. This is because the immediate deflection due to dead load occurs before the fragile materials (glass, plaster, or gypsum board) are applied. However, designers in many offices include the immediate dead-load deflection, producing a more conservative design.

Table 6-2. UBC Deflection Requirements.

Maximum Allowable Deflection for Structural Members[a]

Type of Member	Member Loaded with Live Load Only (*L.L.*)	Member Loaded with Live Load Plus Dead Load (*L.L.* + *K.D.L.*)
Roof member supporting plaster or floor member	*L*/360	*L*/240

[a]Sufficient slope or camber shall be provided for flat roofs in accordance with UBC Section 1605.6

L.L. = Live load
D.L. = Dead load
K = Factor as determined below
L = Length of member in same units as deflection

Value of *K*

Wood	
Unseasoned	Seasoned[b]
1.0	0.5

[b]Seasoned lumber is lumber having a moisture content of less than 16% at time of installation and used under dry conditions of use such as in covered structures.

Table 6-3 shows deflection limits recommended by the AITC. Notice that the limits are less restrictive for industrial buildings than for commercial buildings. (In Table 6-3, applied load means all types of live load.)

The designer generally calculates beam deflection using simple formulas, a few of which are given in column (a) of Table 6-4. These formulas are for flexural deflection only, that is, deflection due to the stretching and compressing of fibers parallel to the beam's axis. However, shear stresses cause additional deflection called *shear deflection.* Shear deflection occurs as plane cross sections become warped due to nonuniform distribution of the shear stresses. Formulas for shear deflection are given by column (b) of Table 6-4.

Table 6-3. AITC Recommended Limits for Deflection of Beams.

Type of Beam	Type of Load	
	Applied Loads Including Live, Snow, and Wind	Total of Dead Load Plus Applied Loads
	Fraction of Span Length	
Roof beams		
Industrial buildings	1/180	1/120
Commercial or institutional buildings		
With plaster below	1/360	1/240
No plaster below	1/240	1/180
Floor beams[a]	1/360	1/240
Bridge stringers		
Highway bridges	1/300	
Railway bridges	1/400 to 1/300	

[a]These limits for floor beams are intended to minimize plaster cracking and to ensure comfort to persons walking. They may not be adequate to eliminate levels of vibration that disturb building occupants. More rigid limits should be observed in such cases.
(Courtesy of the American Institute of Timber Construction (AITC).)

Table 6-4. Maximum Deflections.

	(a) Bending Deflection	(b) Shear Deflection
Uniform load, simple span	$\frac{5w\ell^4}{384EI}$	$\frac{Kw\ell^2}{8AG}$
Center conc. load, simple span	$\frac{P\ell^3}{48EI}$	$\frac{KP\ell}{4AG}$
End conc. load, cantilever	$\frac{P\ell^3}{3EI}$	$\frac{KP\ell}{AG}$
Uniform load, cantilever	$\frac{w\ell^4}{8EI}$	$\frac{Kw\ell^2}{2AG}$

Note: $K = 1.20$ for a rectangular section and 1.11 for a circular section.

It is common practice to assume that shear deflection is a minor part of the total deflection and to ignore it when designing. However, this may not be a good practice for all wood beams. Shear deflection contributes significantly to total deflection for beams with small span-to-depth ratios, box beams, and I-beams with thin webs. Also, shear deflection increases as the ratio of modulus of elasticity to shear modulus (*E*/*G*) increases (10). For example, in beams of basswood, shear deflection for a deep beam may be more than half the total deflection! It is improbable, however, that a shear deflection this large would occur in a beam of species and span commonly used in construction. These species generally have an E/G ratio between 6 and 20 (11). As a rule of thumb, the designer should include shear deflection for a rectangular beam if the span-to-depth ratio is 15 or less.

Ponding

The accumulation of rainwater or snowmelt on roofs is a serious problem that should not be overlooked when designing roof decking, beams, and girders. Progressive deflection and rapid collapse as water accumulates is called *ponding*. The chance of ponding can be minimized by constructing building roofs so that they slope at least 1/4 in. vertical per foot horizontal after long-term deflection. This does not completely eliminate the chance of ponding, however. Ponding can still occur due to inadequate capacity of drainage pipes in downpours or due to water confinement by parapet walls or ice dams. Beams and purlins designed for light loads generally have low stiffness and are most susceptible to ponding.

All members of a flat or nearly flat roof system must be designed with a great enough stiffness to prevent ponding. One simple rule of thumb is that any roof member should be stiff enough that a 5-psf load causes no more than 1/2 in. deflection (12). Kuenzi and Bohannan (13) developed a method, verified experimentally, for estimating the effects of ponding. The method does not account for dead-load creep and resulting permanent set (which one may wish to consider, especially for unseasoned members). In this method, the before-ponding deflections and bending and shear stresses are multiplied by a magnification factor to estimate the increased deflections and stresses under ponding loads. This magnification factor, M_F (see reference 3) is given by

$$M_F = \frac{1}{1 - W'\ell^3 / \pi^4 EI} \qquad (6\text{-}12)$$

where M_F = factor for magnifying stresses and deflections under all but ponding loads to determine stresses and deflections under all loads including ponded water

W' = ponding load due to a one-inch depth of water equaling 5 psf times the tributary area, lb/in.

ℓ = span of member, in.

E = modulus of elasticity of member, psi

I = moment of inertia of member, in.^{4}

Example 6-13

Based on bending stress, a maximum permissible span for 2 × 6 Douglas fir T&G decking was determined in Example 6-5. What is the maximum permissible span for this same decking based on a deflection limit of $\ell/180$ due to dead load plus live load? Which controls?

Recall that total load, w, was 55 lb/ft per foot width. For controlled random lay-up,

$$\Delta = w\ell^4/[145E(2I/3)]$$

$$\simeq w\ell^4/100EI$$

$$= (55/12)\ell^4/[100(1{,}700{,}00)(3.375)]$$

$$= 7.99 \times 10^{-9}\ell^4$$

Set equal to deflection limit of $\ell/180$

$$\ell/180 = 7.99 \times 10^{-9}\ell^4$$

$$\ell = 88.6 \text{ in. (based on deflection limit)}$$

The maximum span based on bending was 109 in., so the maximum permissible span of decking is 88.6 in. Actual spacing will be some convenient round number less than 88.6 in.

Example 6-14

A seasoned 2 × 12 Douglas fir floor joist (No. 1 & Btr grade) has a uniform dead load of 100 lb/ft including joist weight, and a live load of 60 lb/ft on a simple span of 16 ft. Find short- and long-term flexural deflection, and check against UBC deflection limitations. (Span/depth ratio = (16) (12)/11.25 = 17, so shear deflection can be assumed to be insignificant.)

$$\Delta_L = \frac{5(60)(16^4)(12^3)}{384(1{,}800{,}000)(178)} = 0.28 \text{ in.}$$

$$\Delta_D = \frac{5(100)(16^4)(12^3)}{384(1{,}800{,}000)(178)} = 0.46 \text{ in.}$$

(or $\Delta_D = \Delta_L\,(100/60)$)

Total initial $\Delta = 0.28 + 0.46 = 0.74$ in.

Total long-term $\Delta = 0.28 + (1.5)\,(0.46) = 0.97$ in.

For live load only,

UBC limit $= \ell/360 = (16)\,(12)/360 = 0.53$ in.

0.28 in. < 0.53 in. *OK*

For live plus one-half dead ($\Delta_L + 0.5\,\Delta_D$),

UBC limit $= \ell/240 = (16)\,(12)/240 = 0.80$ in.

$0.28 + (0.5)\,(0.46) = 0.51$ in.

0.51 in. < 0.80 in. *OK*

Example 6-15

A 6 × 16 unseasoned beam is used on a 14-ft span. Loads are a 2000-lb central concentrated live load and $w_D = 440$ lb/ft including beam weight. The value of E from Table Appendix B-5 for the species = 1,400,000 psi. Assuming $G = E/13$, find the total (shear and flexural) long-term deflection.

Initial flexural deflection:

$$\Delta_L = \frac{2000(14^3)(12^3)}{48(1{,}400{,}000)(1.03)(1707)}$$

$$= 0.08 \text{ in.}$$

$$\Delta_D = \frac{5(440)(14^4)(12^3)}{384(1{,}400{,}000)(1.03)(1707)}$$

$$= 0.15 \text{ in.}$$

Notice that the average value of modulus of elasticity has been multiplied by 1.03 to obtain a pure bending modulus. When bending and shear deflections are calculated separately, it would be more nearly correct to use a pure bending modulus even though the difference is minor. Published values of E in Appendix B Tables are 97% of the true value so that, for average span/depth ratios, the designer will not have to calculate shear deflection separately.

Initial shear deflection (Table 6-4):

$$\Delta_L = \frac{1.2(2000)(14 \times 12)}{4(5.5 \times 15.5)(1{,}400{,}000 \times 1.03 / 13)}$$

$$= 0.01 \text{ in.}$$

$$\Delta_D = \frac{1.2(440 / 12)(14 \times 12)^2}{8(5.5 \times 15.5)(1{,}400{,}000 \times 1.03 / 13)}$$

$$= 0.02 \text{ in.}$$

Total long-term deflection which includes effect of creep (Eq. 6-11) is

$$(0.08 + 0.01) + (2)\,(0.15 + 0.02) = 0.43 \text{ in.}$$

Note that shear deflection causes 0.05 in., or almost 12% of the total long-term deflection for a beam of these proportions. If shear deflection were ignored, the deflection computed using $E = 1{,}400{,}000$ psi would be 0.40 in., only 7% less than the 0.43 in. value.

Example 6-16

No. 1 hem-fir 6 × 18 beams ($F_b = 1050$ psi, $F_v = 70$ psi, $E = 1{,}300{,}000$) are used at 8-ft centers to support an essentially flat roof system. Check the beams for the effect of ponding. The beams have been designed for dead load plus snow load. Span = 22.5 ft; dead load = 19 psf (which includes an allowance for weight of the beam); snow load = 30 psf; allowable total-load short-term deflection = $\ell/180$.

$D + S$ controls, since $19/0.9 < 49/1.15$

$C_F = (12/17.5)^{1/9} = 0.96$

$F'_b = (1.15)\,(0.96)\,(1050) = 1160$ psi

$F'_v = (1.15)\,(70) = 80$ psi

Under dead and snow loads (but not ponded water),

$w = (19 + 30)(8) = 392$ lb/ft

$M = (392)\,(22.5)^2/8 = 24{,}800$ ft-lb

$f_b = M/S = (24{,}800)\,(12)/280.7 = 1060$ psi

For shear at a distance d from support,

$$V = w\ell/2 - wd = (392)(11.25) - (392)\,(17.5/12) = 3840 \text{ lb}$$

$$f_v = 1.5V/A = (1.5)\,(3840)/96.25 = 59.8 \text{ psi}$$

$$\Delta(\text{total}) = \frac{5(392)(22.5)^4(12)^3}{384(1{,}300{,}000)(2456)} = 0.71 \text{ in.}$$

Calculating ponding magnification factor (Eq. 6-12),

$$W' = (5 \text{ psf})(22.5 \text{ ft})(8 \text{ ft}) = 900 \text{ lb}$$

$$M_F = \frac{1}{1 - \dfrac{(900)(22.5)^3(12)^3}{\pi^4(1{,}300{,}000)(2456)}} = 1.06$$

Magnify stresses and deflections to find effect of dead, snow, and ponding (due to melted snow) loads combined.

$$f_b = (1060)\,(1.06) = 1120 < 1160 \qquad OK$$

$$f_v = (59.8)\,(1.06) = 63.4 < 80 \qquad OK$$

$$\Delta \text{ (total)} = (0.71)\,(1.06) = 0.75 \text{ in.}$$

$$\ell/180 = (22.5)\,(12)/180 = 1.5 \text{ in.} \qquad OK$$

6-6. DESIGN FOR BEARING PERPENDICULAR TO GRAIN

The ends of a beam may bear directly on a supporting member, which could be either another wood member, a masonry wall, or a steel hanger, girder, or bearing plate. A beam may also be stressed in bearing when a member above transfers its load to the beam. A bearing (i.e., compression perpendicular to grain) failure is not a collapse; it is merely a crushing of fibers and generally not a serious failure. To prevent this crushing, the reaction, R, divided by the bearing area must not exceed the adjusted $F_{c\perp}$. In equation form,

$$f_{c\perp} = R/A_b \leq F'_{c\perp} \qquad (6\text{-}13)$$

The bearing area equals the width of beam, b, times the bearing length, as seen in Fig. 6-13. Therefore, solving for bearing length in the above equation yields

$$\text{req } \ell_b = R/F'_{c\perp}b \qquad (6\text{-}14)$$

The allowable stress in compression perpendicular to the grain is not to be adjusted for load duration.

Having unloaded area adjacent to an area loaded perpendicular to the grain causes bearing strength to be higher than it would be if the entire block of wood were loaded. The code provides for this increase by raising the allowable stress for bearing areas that are both less than 6 in. long and not nearer than 3 in. to the end of a member. This is accomplished by allowing the tabular design value in compression perpendicular to grain to be multiplied by the bearing area factor, C_b, where

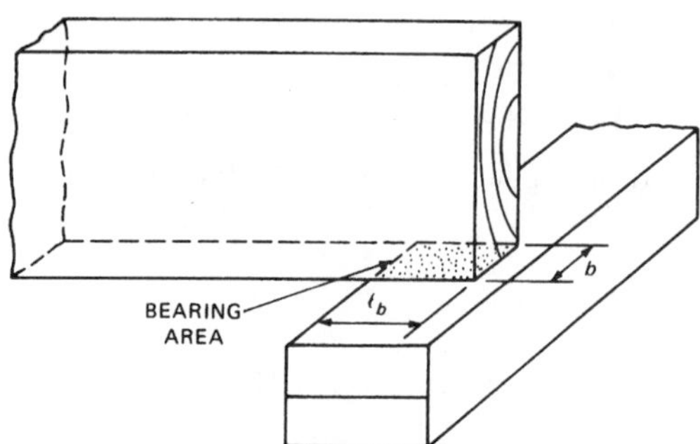

Fig. 6-13. Beam bearing.

$$C_b = (\ell_b + 0.375)/\ell_b \qquad (6\text{-}15)$$

In this equation, ℓ_b is the length of bearing in inches measured along the grain of the wood. This is equivalent to saying that the effective bearing length is 3/8 in. more than the actual bearing length. No allowance is made for the fact that pressure over the supporting area will probably not be uniform.

The design code does not specify a minimum bearing length, but common sense would indicate a minimum of 3 or 4 in. The bearing length provided should be large enough so that even when construction tolerances (location of support; beam length) are considered, the beam will not slip off the support.

Base design values in Appendix B for compression perpendicular to the grain are based on a deformation of 0.04 in. If deformation is critical, one may design sawn-lumber beams according to NDS Section 4.2.6, which limits the allowable stress (based on a deformation of 0.02 in.) to

$$F_{c\perp 0.02} = 5.60 + 0.73\,F_{c\perp} \qquad (6\text{-}16)$$

In this equation, $F_{c\perp}$ is the value from the Appendix B Tables.

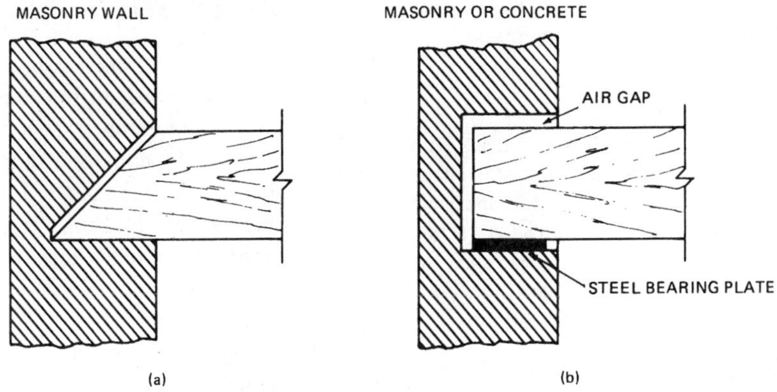

Fig. 6-14. Two types of beam bearing.

Figure 6-14 shows two ways of framing timber beams into masonry walls. The method in Fig. 6-14a is known as a fire-cut, because floor joists framed in this manner can collapse in a fire without pulling the wall down along with them. The bearing plate in Fig. 6-14b may be used for larger beams to distribute the reaction over a larger area of masonry. The small space shown above the beam prevents wall load on the beam, prevents masonry from prying loose under beam deflection, and provides aeration to help prevent decay.

Bearing stresses may need to be considered for members other than beams. For example, the bearing capacity (perpendicular to grain) of a bottom plate (usually a 2 × 4) having 2 × 4 studs nailed to it can be checked using the equations of this section. The increase in allowable stress obtained from Eq. 6-15 would apply. The bearing capacity of the stud itself, parallel to grain, would also have to be considered (see Chapter 7).

Bearing at an Angle to Grain

Sometimes bearing stresses are applied at an angle (neither parallel nor perpendicular) to the grain. In this case, one needs to use the Hankinson formula, which accounts for the strength of wood being greater for end-grain bearing than for compression perpendicular to the grain. If θ is the angle between the load direction and the direction of the grain, then the allowable stress normal to the inclined surface is

$$F_\theta' = \frac{\left(F_g'\right)\left(F_{c\perp}'\right)}{F_g' \sin^2\theta + F_{c\perp}' \cos^2\theta} \qquad (6\text{-}17)$$

In this expression, F'_g is the adjusted allowable bearing stress parallel to the grain (Appendix Table B-9), and $F'_{c\perp}$ is the adjusted (but not for load duration) allowable compressive stress perpendicular to grain.

Example 6-17

No. 1 D. fir-larch 4 × 16 beams, with a 12-ft simple clear span, have been designed to support a storage platform for sacks of cement. These heavy sacks and the beam's self-weight cause a uniform load of 1100 lb/ft. The wood is initially dry and remains dry in service. The beam ends will rest on steel connection brackets that have been embedded in masonry pilasters. To prevent a bearing failure of the Douglas fir, what length of connection bracket (bearing length) will be required?

Table B-3 shows $F_{c\perp} = 625$ psi. Because of dry conditions, there is no moisture adjustment of this value. Load duration adjustment is not used for $F_{c\perp}$. Nor is C_b applicable here (since the bearing area will be at the end of the member, not 3 in. away from the end).

The required bearing length is estimated to be 4 in. Assuming that the entire 12-ft, 8-in. length of beam can be loaded, the maximum reaction is

$$R = 1100\,(12.67)/2 = 6970 \text{ lb}$$

From Eq. 6-14,

$$\text{req } \ell_b = (6970)/[(625)(3.5)] = 3.19 \text{ in.}$$

In this case, it is an unnecessary refinement to recycle. Use a bearing length of 3¼ in. or more.

Example 6-18

A roof rafter (4 × 12 D. fir-larch No. 2) bears on a steel connection bracket atop a concrete wall as shown in Fig. 6-15a. As seen in Fig. 6-15b, the acute angle between the direction of the grain and direction of the load is 50°. Snow loading has controlled the rafter design. Find the allowable compressive stress on the horizontal bearing surface of the rafter.

From Table B-9, F_g = 2020 psi, and from Table B-3, $F_{c\perp}$ = 625 psi. Using the Hankinson formula,

$$F'_\theta = \frac{(1.15 \times 2020)(625)}{(1.15)(2020)\sin^2 50° + (625)\cos^2 50°}$$

$$= 895 \text{ psi}$$

Note that the end-grain allowable is modified for load duration, but the perpendicular-to-grain allowable is not.

Example 6-19

A temporary wood bridge is made necessary by a crane breakdown and is to carry wheeled bins of partially completed parts. Design the timber beams. No. 3 southern pine is available at a local lumber yard. The design is shown in computation sheet form by Fig. 6-16. A 3 × 8 is selected for bending. This section is then checked for shear and for end bearing and is found satisfactory.

6-7. FLOOR SYSTEM DESIGN

In the preceding sections design of bending members has been broken into small parts in order to concentrate on one aspect of the design at a time. Design is actually a more integrated process, usually involving the design of systems rather than member selection. True design involves choosing from many different layouts of the structural members before actually deciding upon sizes of members.

In the design of a complete floor system the planks, decking, or plywood to which the load is directly applied are designed first. Next the joists supporting the first-designed members are chosen. Then the designer selects the beams that support the joists. Each member selected is, in some way, dependent on other members of the system. For example, the spacing chosen for joists will depend on how far the decking can span. These principles are best developed in an example.

Example 6-20

A floor system is to be designed for the storage building in Fig. 6-17. Floor live load = 100 psf and dead load = 10 psf plus weight of wood members. Assume both dead and live loads are of permanent duration since this is a storage building. Masonry walls are 12 in. thick. Dry conditions of use.

Alternates for the floor deck are: (1) planks, (2) T&G decking, and (3) plywood. Some alternates for the beam/joist system are:

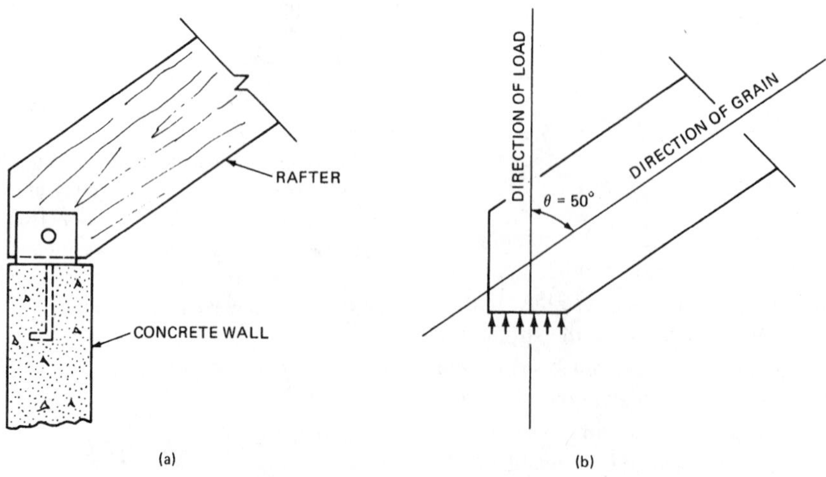

Fig. 6-15. Bearing at an angle to the grain for Example 6–18.

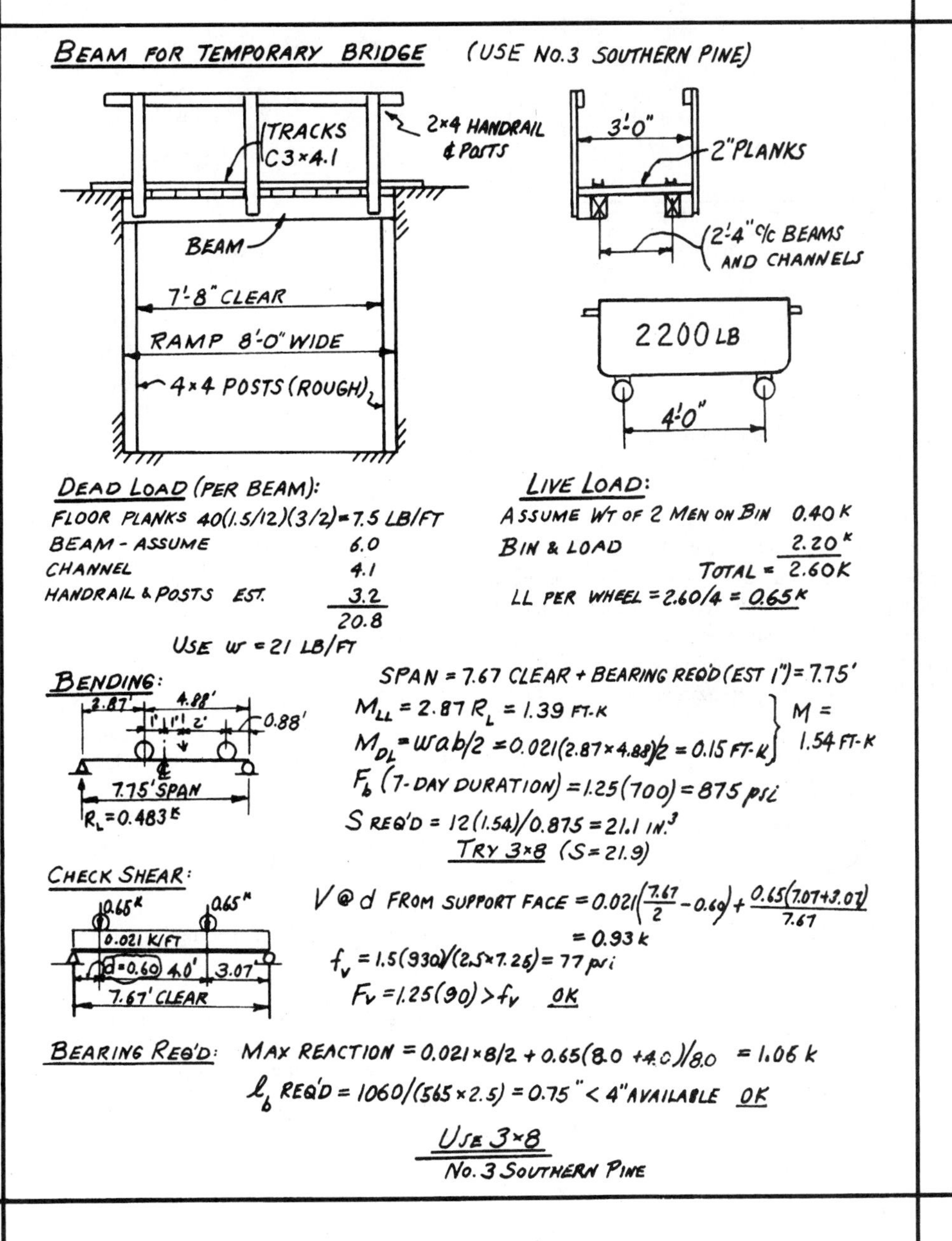

Fig. 6-16. Computation sheet for Example 6–19.

1. Several large beams running N–S and spanning 18 ft clear, with joists E–W.
2. Beams running N–S but with supports at walls and a center column (reducing beam span to one-half); joists E–W.
3. Beams down the E–W centerline, supported on a row of posts, with joists N–S having one end on the beam and one end on the masonry wall.

Using steel beams in combination with wood joists is another option the designer may wish to consider.

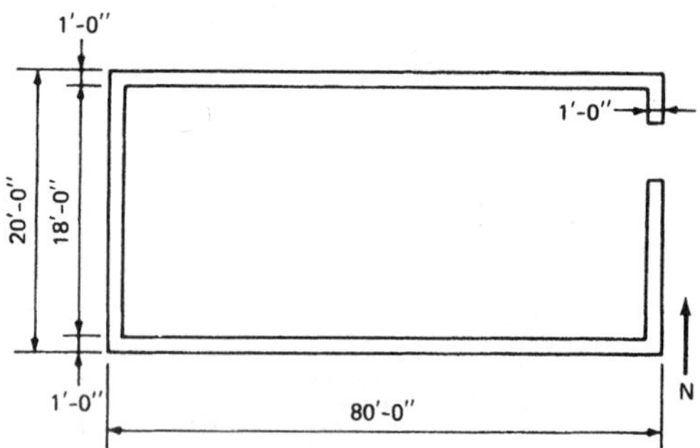

Fig. 6-17. Plan view of storage building for Example 6-20.

Assuming we wish no obstructing columns, let us design components for the first of the beam/joist systems above.

Floor design: No. 2 D. fir-larch planks, 2-in. nominal, have been chosen for the deck material. If the designer calculates for a 12-in. width of floor, the actual width of plank will be immaterial (except as it may affect the allowable stresses). Based on bending strength, determine the allowable span of the planks.

Using 37 pcf as the approximate weight of D. fir-larch,

$$w = 100 + 10 + 37(1.5/12) = 115 \text{ psf}$$

$$S \text{ per ft} = 12(1.5)^2/6 = 4.5 \text{ in.}^3/\text{ft}$$

$$F_b = 875 \text{ psi}$$

If 2 × 8s are used,

$$F'_b = C_D C_F C_{fu} F_b$$
$$= 0.9(1.2)(1.15)(875) = 1087 \text{ psi}$$

$$M \text{ allow. per ft} = 1087(4.5)/12 = 408 \text{ ft-lb per ft}$$

For four or more spans (Appendix Table A-8),

Maximum $M = 0.107\,w\ell^2$, so, based on bending,

$$\text{Allow } \ell = \sqrt{408/(115 \times 0.107)} = 5.76 \text{ ft}$$

Because adjustment factors C_F and C_{fu} vary with plank width, this allowable span would change if a different width of plank were used.

Based on a deflection limit of $\ell/240$, determine the allowable span of the planks. Per foot width,

$$I = 12(1.5)^3/12 = 3.375 \text{ in.}^4/\text{ft}$$

For four or more spans (Appendix Table A-8),

$$\Delta = w\ell^4/154EI$$

But, since all load is long-term,

$$\Delta = 1.5\,w\ell^4/154EI$$

$$\frac{12\ell}{240} = \frac{1.5(115)(\ell^4)(12^3)}{154(1.6)(10^6)(3.375)}$$

$$\text{Allow } \ell = 5.18 \text{ ft}$$

Deflection rather than bending, controls. Planking cannot span more than 5.18 ft; that is, joists may not be spaced farther apart than 5.18 ft.

Check shear in planks. If joists are 2-in. nominal, plank clear span = 5.18 − 1.5/12 = 5.06 ft maximum.

$$\text{Critical } V = 115(5.06/2 - 1.5/12)$$
$$= 277 \text{ lb/ft width}$$

$$f_v = 1.5(277)/(12 \times 1.5) = 23 \text{ psi}$$

$$F_v = 0.9(95) = 86 \text{ psi } OK$$

Joist design: To choose joists, the designer must know the spacing of the beams because this defines the length of the joists. The length (clear) of the building in the longitudinal direction is 78 ft. Try 11-ft beam spacing × 7 spaces = 77 ft. (Joists will cantilever slightly over beams at end walls.)

Assuming beams will be 9½ in. wide, the clear span of the joists = 11.0 ft − 9.5/12 = 10.21 ft. Assume that the required bearing length of the joists is 2 in., then

Joist span = 10.21 + 2/12 = 10.38 ft

Try joist spacing = 16 in. c/c

Try No. 2 D. fir-larch.

Estimated joist wt = 4 lb/ft

$$w = (16/12)(115) + 4 = 157 \text{ lb/ft}$$

$$M = 157(10.38)^2/8 = 2114 \text{ ft-lb}$$

Assuming $C_F = 1.0$ for now,

$$F'_b \text{ (repetitive)} = F_b C_r C_D = 875(1.15)(0.9)$$
$$= 906 \text{ psi}$$

Req $S = 28.0$ in.^{3}

Possible section: 2 × 12 ($S = 31.6$ in.^{3} and $C_F = 1$)

Try joist spacing = 36 in. c/c

Estimated joist wt = 11 lb/ft

$w = (36/12)(115) + 11 = 356$ lb/ft

$M = 4795$ ft-lb

Assuming $C_F = 1.0$,

F'_b (nonrepetitive) = 875(0.9) = 788 psi

Req $S = 73.0$ in.^{3}

Possible section: 3 × 14 ($S = 73.2$ in.^{3} and $C_F = 0.9$)

Recycling with $C_F = 0.9$, Req $S = 81.1$
3 × 14 NG

Possible section: 4 × 12 ($S = 73.8$ in.^{3} and $C_F = 1.1$)

A 4 × 12 is obviously OK since C_F is greater than the assumed C_F.

There are many other joist spacings (up to a limit of 5.18 ft) that could be tried. Choosing between them will probably be based on cost and/or availability. In this case, the 2 × 12s not only used less volume of lumber but are usually easier to obtain. *Tentatively use 2 × 12 joists (No. 2) at 16 in. c/c.*

Check joist bearing length:

$F_{c\perp} = 625$ psi

For $w = 157$ lb/ft, reaction = 157(11/2) = 864 lb

Req $\ell_b = 864/(1.5 \times 625) = 0.92$ in.

Check joist deflection:

Span = 11 − 9.5/12 (est) + 0.92/12 = 10.29 ft

Δ (long term) = $1.5(5)\, w\ell^4/384EI$

$$= \frac{1.5(5)(157)(10.29)^4(12)^3}{384(1.7 \times 10^6)(178)} = 0.20 \text{ in.}$$

$= \ell / 629$ *OK*

Check joist shear:

Clear span = 11 − 9.5/12 = 10.21 ft

Critical $V = 157(10.21/2 - 11.25/12) = 654$ lb

$f_v = 1.5(654)/[1.5(11.25)] = 58$ psi

$F'_v = 0.9(95) = 86$ psi *OK*

Use 2 × 12 joists at 16 in. c/c.

Beam design: Clear distance between the masonry walls is 18 ft. Use a steel plate to distribute the load to the masonry walls, where the plate is set back 1 in. from the interior face of the wall. Assume bearing length = 3 in.

Beam span = 18 + 2/12 + 3/12 = 18.42 ft

w (from joists) = 11(157)/(16/12) = 1295 lb/ft
Beam wt (est) = 45
w = 1340 lb/ft

$M = 1340(18.42)^2/8 = 56{,}800$ ft-lb

Try No. 1 Douglas fir B & S (WWPA).

Est. $F'_b = 0.9(1350) = 1215$ psi

Req $S = M/F_b = 561$ in.^{3}

Try 12 × 18 ($S = 587$ in.^{3}).

$C_F = (12/17.5)^{1/9} = 0.96$

$F'_b = 0.9(0.96)(1350) = 1170$ psi

Beam wt = 37 pcf(11.5)(17.5)/144 = 52 lb/ft

> est 45

Corrected $w = 1295 + 52 = 1347$ lb/ft

Corrected $M = 57{,}100$ ft-lb

$f_b = M/S = 1170 \text{ psi} = F'_b$ *OK*

Check beam bearing length:

Req $\ell_b = 1347(9)/[625(11.5)]$

= 1.69 in. < 3 in. assumed *OK*

Check beam shear:

Critical $V = 1347(18.17/2 - 17.5/12)$

= 10,270 lb

$f_v = 1.5(10{,}270)/(11.5 \times 17.5) = 76.5$ psi

$F_v = 0.9(85) = 76.5$ psi *OK*

Check beam long-term deflection

$$\Delta = \frac{1.5(5)(1347)(18.42)^4(12)^3}{384(1{,}600{,}000)(5136)}$$

= 0.64 in.

$= \ell / 347$ *OK*

Use 12 × 18 beams.

6-8. BEAM DESIGN BY LRFD

The process of choosing beams for strength and serviceability by the LRFD method employs different steps than that of the ASD method. Each LRFD design will involve the *time effect factor,* λ, which was explained in Section 4-12. Service loads will have to be factored (except for deflection calculations), and the largest ratio of factored load to time effect factor chosen for design. For a trial section the applied factored load effect (e.g., M_u or V_u) will need to be compared with the resisting strength (e.g., $\lambda\phi_b M'$, or $\lambda\phi_v V'$). Resistance factors, ϕ, that are used in design of beams are:

$\phi_b = 0.85$—flexure
$\phi_v = 0.75$—shear or torsion
$\phi_s = 0.85$—stability
$\phi_c = 0.90$—compression $\perp$ to grain

For designers who have designed concrete structures by the strength design method or steel structures by the LRFD method, designing wood structures by LRFD will present little problem. One of the two main differences from steel or concrete is that for wood the time effect factor must be used. In addition, each reference value will need to be adjusted by all applicable adjustment factors, resulting in an adjusted reference value. Reference values are given in kip units (rather than pound units).

When there is not continuous lateral support, steps for beam design (strong-axis bending) are as follows:

1. Compute factored moment, M_u, and find controlling load combination and time effect factor, λ. (See Table 4-1.)
2. Look up reference strength, F_b, and use it to choose a trial section.
3. For the trial section, find values of all adjustment factors except C_L.
4. Find F_b^* (adjusted by all except C_L and C_v).
5. Find ℓ_e (effective unbraced length) and M_e (elastic lateral buckling moment) $= 2.40E'_{y05}I_y/\ell_e$
6. Find $\alpha_b = \phi_s M_e/(\lambda\phi_b S_x F_b^*) = M_e/(\lambda S_x F_b^*)$
7. Find $$C_L = \frac{1+\alpha_b}{2c_b} - \sqrt{\left(\frac{1+\alpha_b}{2c_b}\right)^2 - \frac{\alpha_b}{c_b}} \qquad (6\text{-}18)$$
   where $c_b = 0.95$
8. Find nominal resistance $M' = C_L S_x F_b^*$
9. Verify that required M_u does not exceed $\lambda\phi_b M'$
10. Check shear and deflection.

For members having continuous lateral support of the compression flange, $C_L = 1.0$, and steps 5, 6, and 7 above can be eliminated.

Many of the same adjustment factors that were used in ASD are also applicable in LRFD. The repetitive member factor, C_r, that was used in ASD has a counterpart in LRFD called the *load-sharing factor* (which is denoted by the same symbol as in ASD). As before, the adjustment applies to bending resistance of assemblies consisting of three or more framing members spaced not more than 24 in. on center, and connected by load distributing elements, such as sheathing. The value of the load sharing factor, C_r, is 1.15 for sawn-lumber framing members, 1.05 for glued laminated timber, and varies from 1.04 to 1.15 for prefabricated I-joists (depending on variability of flange material).

For design by the LRFD method, two values of modulus of elasticity are needed. The modulus, E_{05}, that 5% of the pieces in the given grade fall below is used for strength and stability calculations. For deflection calculations, the average or mean modulus, E, is used. For visually graded lumber, ref. 14 gives the following equation showing the relationship between the adjusted values of the two different moduli.

$$E'_{05} = 1.03E'(1 - (1.645)(0.25)) \qquad (6\text{-}19)$$

Example 6-21

By the LRFD method, choose a nominal 8-in.-thick member for a beam of 20-ft span. The beam carries a uniform dead load (including an estimate of self-wt) of 50 lb/ft and occupancy live load of 600 lb/ft. Use No. 1 D. fir-larch and base the choice on flexure alone. The beam has lateral support at the ends and midspan only.

Service moments are:

$M_D = 0.050\ (20)^2/8 = 2.5$ ft-k

$M_L = 0.600\ (20)^2/8 = 30.0$ ft-k

Load combinations:

$1.4D$; $M_u = 1.4(2.5) = 3.5$ ft-k

$1.2D + 1.6L$; $M_u = 1.2(2.5) + 1.6(30) = 51$ ft-k

The latter controls and time effect factor $\lambda = 0.8$ (Table 4-1).

$$\phi_b = 0.85$$

The reference strength, F_b, from Appendix Table B-10 is 3.43 ksi. (*Note:* The symbol used for reference strength is the same as a symbol used in ASD. However it does *not* stand for allowable stress when used in LRFD.)

Assuming adjustment factors will reduce the reference value to, say, 3.1 ksi,

$$S \text{ req} = M_u/(3.1 \times \lambda \times \phi_b)$$
$$= 51(12)/(3.1 \times 0.8 \times 0.85) = 290 \text{ in.}^3$$

Try 8 × 16 with $S = 300$ in.^{3}

$$C_F = (12/15.5)^{1/9} = 0.972$$

$$F_b^* = F_b C_F = 3.43\ (0.972) = 3.33 \text{ ksi}$$

$$\ell_u/d = (10 \times 12)/15.5 = 7.7, \text{ therefore}$$

$$\ell_e = 1.63\ \ell_u + 3d = 1.63(10 \times 12) + 3(15.5) = 242 \text{ in.}$$

$E = 1600$ ksi. Putting this value into Eq. 6-19 gives $E'_{05} = 970$ ksi.

$$M_e = 2.40E'_{05}I_y/\ell_e = 2.40\ (970)\ (545)/242 = 5240 \text{ in-k}$$

$$\alpha_b = M_e/(\lambda S_x F_b^*) = 5240/(0.8 \times 300 \times 3.33) = 6.56$$

By Eq. 6-18, $C_L = 0.991$

$$M' = C_L S_x F_b^* = (0.991)\ (300)\ (3.33)/12 = 82.5 \text{ ft-k}$$

$$\lambda\phi_b M' = (0.8)\ (0.85)\ (82.5) = 56.1 \text{ ft-k} > \text{required } M_u = 51 \text{ ft-k} \quad \textit{OK}$$

Example 6-22

By the LRFD method, find the required bearing length for the 8 × 16 D. fir-larch beam selected in Ex. 6-21.

The load combination of $1.2D + 1.6L$ obviously controls, and the factored uniform load is

$$w_u = 1.2(0.050) + 1.6(0.600) = 1.02 \text{ k/ft}$$

The factored end reaction is $R_u = (1.02)\ (20)/2 = 10.2$ k

Reference strength, $F_{c\perp} = 1.20$ ksi (Table B-10)

(Although the load duration factor is not used for bearing calculations in ASD, the time effect factor, λ, *does* apply to LRFD calculations.)

$$\text{Resisting strength} = \lambda\phi_c P_\perp = \lambda\phi_c F_{c\perp} b\ell_b$$
$$= (0.8)\ (0.90)\ (1.20)\ (7.5)\ (\ell_b) = 6.48\ \ell_b$$

Equating factored end reaction to resisting strength and solving, required bearing length

$$\ell_b = 1.57 \text{ in.}$$

Example 6-23

By the LRFD method, check the 8 × 16 D. fir-larch beam of Ex. 6-21 for shear. Bearing length actually provided at each end is 2.5 in., so the clear span is 19.58 ft.

$$w_u = 1.02 \text{ k/ft from Ex. 6-22.}$$

As was the case with ASD, the critical section for shear is at a distance d from the face of the support. At the critical section,

$$V_u = w\ell_c/2 - wd$$
$$= (1.02)\ (19.58/2) - (1.02)\ (15.5/12) = 8.7 \text{ kips}$$

The tabular value for resisting shear (reference strength) is $F_v = 0.24$ ksi by Table B-10.

$\phi_v = 0.75$; $\lambda = 0.8$; all adjustment factors are 1.0.

Rearranging the mechanics of materials equation expressing that shear stress (for a rectangular section) equals $1.5V/A$,

$$V' = (0.24)\ (A/1.5) = (0.24)\ (7.5 \times 15.5/1.5) = 18.6 \text{ k}$$

Adjusted shear resistance, $\lambda\phi_v V' = (0.8)\ (0.75)\ (18.6) = 11.2 \text{ k} > V_u = 8.7 \text{ k}$ *OK*

Example 6-24

Repeat Example 6-2, but use the LRFD method.

Factored combinations to be considered are:

1. $1.4D = 1.4(5) = 7$ psf
2. $1.2D + 1.6L + 0.5S = 1.2(5) + 1.6(60) + 0.5(38) = 121$ psf

The time effect factors are 0.6 for the first combination and 0.8 for the second. Dividing each factored load by the proper time effect factor, it becomes obvious that the second combination will control the design. (The 0.8 time effect factor was used, as for occupancy live load, since the duration of storage for firewood will be fairly short.)

As in Example 6-2, the tributary width for each joist is 16 in. Thus, each joist carries the load from a 16-in. width of floor.

$w_u = 121(16/12) = 161$ lb/ft $= 0.161$ k/ft
$M_u = 0.161(5^2)/2 = 2.01$ ft-k

The only applicable adjustment factor is the *load sharing factor, C_r*, which is 1.15, just as for design under ASD. The tabular reference bending strength for the grade of wood specified in Example 6-2 is 3.05 ksi and the adjusted reference strength is

$F'_{bx} = 1.15(3.05) = 3.51$ ksi.
$S \text{ req} = M_u/(\lambda\phi_b F'_{bx})$
$= 12(2.01)/(0.8 \times 0.85 \times 3.51) = 10.1 \text{ in.}^3$

Use 2 × 8 joists ($S = 13.1$ in.^{3}) @ 16 in. centers.

These joists are satisfactory in bending at the support end of the cantilever portion. As for ASD, they should also be checked for shear and for deflection.

Example 6-25

Consider again the laterally supported beam of Example 6-10. Repeat its design using LRFD. Recall that this beam has its ends notched to 10 inches deep.

Using the same grade of timber as assumed for Example 6-10, the reference strength for bending is $F_b = 3.43$ ksi and for shear is $F_v = 0.24$ ksi.

Load combinations to be considered under LRFD are:

1. $1.4D = 1.4(40) = 56$ lb/ft; $56/0.6 = 93.3$
2. $1.2D + 1.6L = 1.2(40) + 1.6(230) = 416$ lb/ft; $416/0.8 = 520$ *Controls*

Factored load divided by time effect factor is larger for the second load combination, so that load combination controls. Select the section for bending, then check it for shear.

Assuming that size factor, C_F, is 1.0, the adjusted strength F'_b is equal to the reference bending strength, $F_b = 3.43$ ksi.

To select a section for bending, set

$$\lambda\phi_b M' \geq M_u$$

$$0.8(0.85)M' = 0.416(19^2)/8 \text{ ft-k}$$
$$= 12(18.77) \text{ in.-k}$$

Required adjusted moment resistance is

$$M' = 18.77(12)/(0.80 \times 0.85) = 331 \text{ in-k}$$

Required section modulus is

$$S = 331/F'_b = 331/3.43 = 96.5 \text{ in.}^3$$

Based on bending alone, a 6 × 12 (which *does* have $C_F = 1.0$) would be satisfactory. With the notched ends, however, it is likely that shear will control the design. As was the case in Example 6-10, the deepest notch permitted by the LRFD code is 11.5/4 = 2.88 in. The reduced depth at the ends is to be 10 inches, so the required notch depth is only 1.5 in.

The end reaction due to factored loads is

$$V_u = 0.416(19/2) = 3.95 \text{ kips}$$

For the section containing the notch, the adjusted notched section shear resistance for the 6 × 12 is given by

$$V' = (2/3)F'_v bd_n)(d_n/d)$$

$$V' = (2/3)(0.24)(5.5 \times 10)(10/11.5)$$
$$= 7.65 \text{ kips}$$

$$\lambda\phi_v V' = 0.8(0.75)(7.65) = 4.59 \text{ kips}$$

This exceeds the required value of 3.95 kips, so the 6 × 12 is satisfactory in shear as well as in bending.

REFERENCES

1. *National Design Specification for Wood Construction,* rev. 1991 Ed., American Forest & Paper Association, Washington, DC, 1993.
2. *Standard for Tongue-and-Groove Heavy Timber Roof Decking,* AITC 112-81, American Institute of Timber Construction, Englewood, CO, 1993.
3. *Timber Construction Manual,* 4th ed., American Institute of Timber Construction, Wiley, NY, 1994.
4. Gerstle, K. H., *Basic Structural Design,* McGraw-Hill, New York, 1967.
5. Newlin, J. A., G. E. Heck, and H. W. March, "Wood-Beam Design Method Promises Economies," *Engineering News-Record,* 110, May 1993, pp. 594–596.
6. Keenan, F. J., "Shear Strength of Wood Beams," *Forest Products Journal,* 24(9), Sept. 1974, pp. 63–70.
7. Longworth, J., "Longitudinal Shear Strength of Timber Beams," *Forest Products Journal,* 27(8), Aug. 1977, pp. 19–23.
8. Jacoby, Henry S., and Roland P. Davis, *Timber Design and Construction,* 2nd ed., Wiley, New York, 1930.
9. *Uniform Building Code,* International Conference of Building Officials, Whittier CA, 1994.
10. Biblis, E. J., "Shear Deflection of Wood Beams," *Forest Products Journal,* 15(11), Nov. 1965, pp. 492–498.
11. Hoyle, R. J., *Wood Technology in the Design of Structures,* 4th ed., Mountain Press Publishing Co., Missoula, MT, 1978.
12. Haussler, R. W., "Roof Deflection Caused by Rainwater Pools," *Civil Engineering, ASCE,* 32(10), Oct. 1962, pp. 58–59.
13. Kuenzi, E. W., and B. Bohannan, "Increases in Deflection and Stresses Caused by Ponding of Water on Roofs," *Forest Products Journal,* 14(9), Sept. 1964, pp. 421–424.

14. *Standard for Load and Resistance Factor Design for Engineered Wood Construction,* American Society of Civil Engineers, 1996.

PROBLEMS

Assume ten-year (normal) duration of load and consider beam weight in problem solutions unless indicated otherwise. Unless stated or implied differently in the problem, moisture content of the wood in use is less than 19%.

6-1. A laterally supported beam of No. 1 hem-fir has a controlling bending moment of 8200 ft-lb (which includes an estimate of beam weight) from combined dead and wind loads. For bending strength, what is the shallowest satisfactory S4S section of 6-in. nominal thickness?

6-2. A No. 1 hem-fir beam, laterally supported, spans a clear distance of 20 ft and has superimposed dead load of 100 lb/ft and snow load of 150 lb/ft. Assuming the required bearing length at each end is 2 in., select the most economical section (the one of least cross-sectional area) considering bending alone. Do not neglect self-weight.

6-3. A laterally supported beam of 15-ft span carries at each one-third point concentrated loads that include 6 kips *D,* 4 kips occupancy *L,* and 2 kips *S.* Choose the most economical No. 1 hem-fir section, considering bending only. (*Hint:* Select ignoring beam weight, then verify your choice considering it.)

6-4. What permanent uniform load can be put on a 12-in.-diameter log used as a beam if the span is 12 ft? Consider only bending stresses. Neglect log weight. Unadjusted F_b = 1500 psi.

6-5. A uniform load of 400 lb/ft is applied to a log (specific wt = 40 pcf) with F_b = 1600 psi. If the span is 12 ft, what is the required log diameter (based on flexure)?

6-6. A single 4 × 14 beam of No. 2 D. fir-larch spans 24 ft and has lateral support at its ends only. Considering flexure only, what is the allowable superimposed permanent uniform load per foot?

6-7. A 6 × 18 D. fir-larch timber (specific wt = 33 pcf), Select Structural grade, is used as a floor beam to carry uniform dead plus occupancy live load on a 40-ft span. Lateral support is at ends and center only. Use is dry. What is the allowable load per foot that can be added over and above the beam's own weight? Consider flexure only.

6-8. A single 6 × 14 beam of No. 2 D. fir-larch spans 30 ft and has lateral support at its ends only. Considering flexure and shear only, what is the allowable superimposed permanent uniform load per foot?

6-9. A No. 1 hem-fir beam spanning 24 ft supports a permanent midspan concentrated load of 1600 lb. There are lateral supports at the ends and center. MC in use will be greater than 19%. Based on bending, choose the most economical safe section whose smallest cross-sectional dimension is 6 in. nominal. Neglect beam weight.

6-10. Considering bending only, select the most economical No. 2 D. fir-larch section for the rafter of Fig. 4-12. The only loads are dead load and snow load of 30 psf of horizontal projection. Roof deck provides lateral support to the rafter.

6-11. Roof rafters (repetitive members) are to be placed 24 in. c/c with a slope of 3/12. Snow load is 30 psf on a horizontal plane and dead load is 15 psf of sloping roof surface. Rafters have a horizontal span of 13 ft from ridge beam to exterior wall and are laterally supported. Based on flexure only, what is the smallest acceptable No. 2 hem-fir rafter?

6-12. A box beam (see Fig. 6-18) has No. 1 D. fir-larch 2 × 6 top and bottom flanges glued to 1/2-in. plywood webs. Assume that the top flange is laterally supported over its full length. Calculate the depth, *d,* required to resist an applied bending moment of 42 ft-kips assuming the contribution of the plywood to flexural strength is negligible. (Because one flange has nearly uniform tension and the other nearly uniform compression, the allowable stress will be the smaller of F'_c or F'_t.)

6-13. A 2 × 8 D. fir-larch beam of clear span 18 ft has an applied uniform *D* + *L* load of 70 lb/ft.

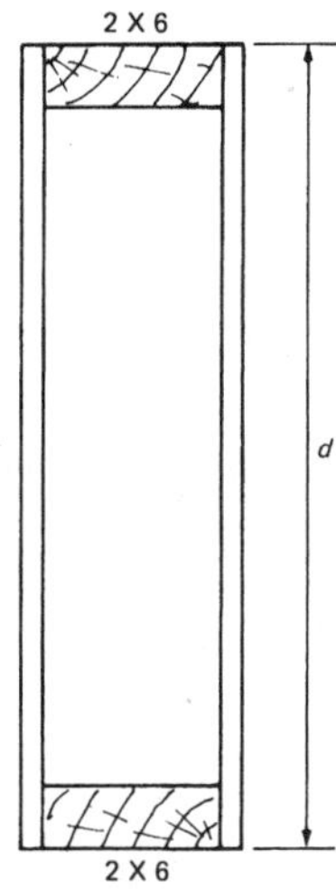

Fig. 6-18. Beam for Problem 6-12.

Assuming a specific weight of 38 pcf, compare the maximum shear stress at the critical section to the allowable shear stress.

6-14. An existing structure is being investigated for possible increased loading. Beams are 8 × 18, No. 1 D. fir-larch, spanning 16 ft (clear span of 15.5 ft). Some beams have end splits 8 in. long on their wide face. Use conditions were, and will continue to be, very dry. Uniform loads are 400 lb/ft occupancy live load (reduction already considered) and 200 lb/ft dead load. A midspan concentrated load is 1 k for dead load and 1 k for snow. Are the beams satisfactory for bending? For shear (using C_H)? (*Caution:* The controlling combination may not be the same for shear as for bending.)

6-15. Hem-fir beams (Select Structural) are used 6 ft apart on 18-ft spans to support the nearly flat roof over a laundry room, where temperature will be about 85°F and relative humidity will be 90%. The roof consists of three-ply felt and gravel on 2-in. nominal wood decking with 3-in. rigid insulation. Snow load is 32 psf. Select, based on flexure and shear, the most economical safe section whose smallest cross-sectional dimension is 4 in. nominal. Consider beam weight.

6-16. For a 2 × 8 No. 2 D. fir-larch joist (repetitive member), simply supported at its ends, prepare a plot of the allowable uniform load (based on flexure and shear only) versus the span length of the joist. $C_L = C_H = 1.0$.

6-17. Prove that the No. 2 D. fir-larch 4 × 16 beam of Fig. 6-19 is or is not satisfactory in shear. Use only the basic method; do not use two-beam action or C_H. All loads are permanent.

6-18. A 6 × 10 No. 1 hem-fir beam has a 10-ft clear span. A single concentrated load of 2800 may move to any position on the span. The beam is part of a temporary bridge that will only be used one month before being dismantled. Disregarding beam weight, is the beam satisfactory for shear?

6-19. A 6 × 14 beam of No. 2 Douglas fir-larch spans 30 ft and has lateral support at its ends only. The beam is part of the roof system of a health clinic and has drywall attached to its underside. The beam carries 20 lb/ft *DL* (from drywall, roofing materials, and joists) in addition to its self weight, as well as 23 lb/ft snow load. Is the beam OK for shear?

6-20. At the ends, a No. 1 hem-fir 4 × 6 beam measures 5 in. vertically because it has been notched on the lower face. What is the maximum allowable shear, *V*, if ten-year (normal) duration of load controls? Do not use C_H.

6-21. Determine the maximum service ($D + L$) load reaction for an 8 × 10 S4S beam that has a 1.75-in.-deep notch cut at each end. MC in use will be greater than 19%. $F_v = 70$ psi. Do not apply the shear stress factor C_H. How would the answer change if a gradual decreasing of the cross section had been used rather than a notch?

6-22. Prove that the beam of Fig. 6-20 is or is not satisfactory in shear.

6-23. A nominal 6 × 12 beam of hem-fir is notched such that the cross section of the end measures 9-in. vertically. Center-to-center of required bearing lengths is 14 ft 2 in. The total (ten-year-duration) load on the beam is 230 lb/ft. Is the beam satisfactory for shear?

6-24. A 2 × 8 No. 1 D. fir-larch (repetitive) joist is simply supported, laterally supported, and uniformly loaded. Calculate the length and load for which bending and shear capacity of the member are reached simultaneously.

6-25. An 8 × 10 hem-fir beam ($F_v = 70$ psi) is simply supported on a clear span of 9 ft 6 in. It is used at 21% maximum moisture content. Neglect self-weight. Do not apply C_H or two-beam-action provisions. Based on shear strength alone, compute the total allowable $D + L$ uniform load (a) using maximum shear at the face of the support and (b) using critical shear at a distance *d* from the face of the support.

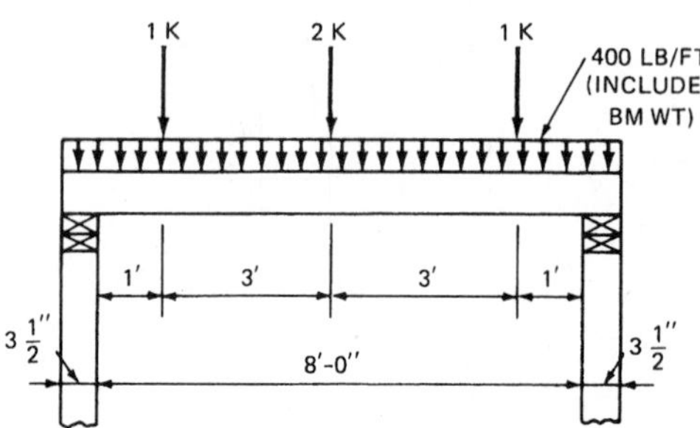

Fig. 6-19. Beam for Problem 6-17.

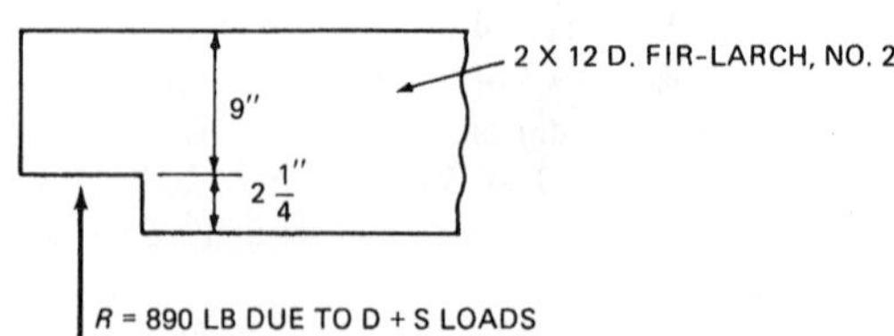

Fig. 6-20. Beam for Problem 6-22.

6-26. An 8 × 10 hem-fir beam (F_v = 70 psi) is simply supported on a clear span of 9 ft 6 in. It is used at 21% maximum moisture content. Neglect self-weight. Do not apply C_H. Based on shear strength alone, compute the maximum value of a single moving concentrated load that may be applied (a) using the usual shear formula with critical section as defined in NDS Section 3.4.3.1 and neglecting two-beam action, and (b) using the recommended procedure of locating the load at three beam depths from the support or at the quarter point, whichever is the closer to the support as in NDS Section 4.4.2.2.

6-27. The composite timber beam shown in Fig. 6-21 is joined by ½-in. by 5-in. lag screws (allowable shear force = 660 lb/screw). For V = 1500 lb:

(a) Calculate the shear stress at the joint.

(b) Find the required lag screw spacing for this stress.

(c) Calculate the shear stress at the neutral axis.

(d) Is the section satisfactory?

6-28. A 2 × 8 No. 2 D. fir-larch beam carries 50 lb/ft snow load plus dead load of 10 lb/ft plus self-weight on an 18-ft span. It is supported at its ends by bolted connections to 4 × 4 D. fir posts. Assume dry conditions. Design the bolted connection, making sure that the beam will not be overstressed in shear. Give bolt size and locations.

6-29. Design a Select Structural beam D. fir-larch (assumed specific wt 37 pcf) of 2-in. nominal thickness to carry a single concentrated load (ten-yr duration) of 2500 lb at midspan. The beam is to be fastened to a 4 × 4 post with ¾-in. bolts. Span length is 10 feet. Neglect deflection. The bolt design value is $Z_{s\perp}$ = 590 lb per bolt in single shear. Minimum edge distance to be used (measured to center of bolt) is 3 in. at the top of the beam and 1½ in. at the bottom of the beam. Use C_H as for an unsplit member. The beam cannot be considered a repetitive member.

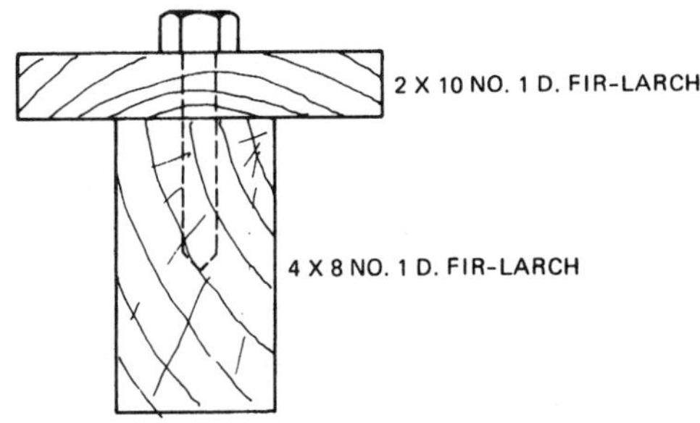

Fig. 6-21. Beam for Problem 6-27.

6-30. Four-inch nominal decking having tabular F_b = 2000 psi and E = 1,800,000 psi is arranged in a two-span continuous layup on a gentle sloped roof. Deflection is limited to $\ell/240$ with roof beams spaced at 12 ft c/c. What is the allowable uniformly distributed two-month duration (snow) load?

6-31. Douglas fir-larch Select Dex grade WCLIB decking, 3-in. nominal thickness, is surfaced and used at 19% maximum MC. Layup is controlled random. Load is D = 20 psf plus self-weight and S = 83 psf on the nearly flat roof. Deflection is limited to $\ell/180$. What is the maximum allowable span for the decking?

6-32. Simple-span 2-in. nominal decking (D. fir-larch WWPA Commercial grade) is to be used on a roof with slope of 5/12. Loads are S = 30 psf of horizontal projection and D = 10 psf of sloping surface. Deflection is limited to $\ell/240$. What is the maximum allowable decking span? Decking will be laid with its longitudinal axis parallel to slope, so the component of load parallel to the roof will cause compression. However, ignore the compression force for this problem.

6-33. A 6 × 12 (D. fir-larch, Select Structural) is used as a floor beam on a 15-ft simple span. Based on deflection alone and using the UBC requirements for long-term deflection, what is the allowable load per linear foot of beam (including the beam weight)? Assume that the load is of ten-year duration and the wood is seasoned.

6-34. What is the maximum uniformly distributed $D + L$ load that can be placed on a 2 × 12 S4S No. 2 hem-fir joist on a 16-ft span if a 2.5-in.-deep notch is cut into each end? Neglect weight of joist. Joists are on 16-in. centers. Do not apply the shear stress factor, C_H. Maximum deflection is limited to $\ell/240$ for the total of dead and live load. Check bending, shear, and deflection.

6-35. Compute by NDS the total long-term deflection for the roof beam in Problem 6-19. Does the section satisfy the AITC limits?

6-36. A 6 × 14 beam of No. 1 WWPA D. fir-larch supports a total uniform load of w kips per foot. (a) Find (as a function of w) the short-term deflection due to bending and due to shear for a simple span of 12 ft. Assume E/G = 13. What percent of the total deflection is shear deflection? (b) Repeat for a span of 24 ft.

6-37. An unseasoned 6 × 10 No. 2 D. fir-larch beam spans 18 ft. Assume $E/G = 13$. The live load is 100 lb/ft. The superimposed dead load is 150 lb/ft. (a) Determine the total long-term deflection, disregarding shear deflection. (b) Repeat including shear deflection. Remember to raise E by 3%.

6-38. A 4 × 16 No. 3 redwood floor beam spans 15 ft. Assume $E/G = 13$; $L = 250$ lb/ft and D = 250 lb/ft plus beam weight.

(a) Determine total deflection over the long term, disregarding shear deflection.
(b) Repeat including shear deflection.
(c) What percent of total deflection is shear deflection?
(d) Under the UBC, is the beam satisfactory for deflection?

6-39. No. 2 D. fir-larch 4 × 12 beams for a mildly sloped roof system are to be checked for the effect of ponding. Dead load (including self-weight) is 18 psf and snow load is 25 psf. Beams are used at 4 ft c/c and span 16 ft. Allowable total-load deflection = $\ell/180$. Are the beams satisfactory?

6-40. A southern pine 6 × 10 beam supports a uniform snow load of 170 lb/ft plus total dead load of 100 lb/ft. The beam is No. 2 SR grade. The clear distance between supports is 18 ft. Determine the required bearing length at each end of the beam. Check the beam for shear and flexure.

6-41. For the roof beam of Problem 6-19, compute the required bearing length.

6-42. A No. 1 redwood 6 × 12 beam supports a 350-lb/ft uniform load of cumulative duration 15 hours. The clear distance between supports is 20 ft. Determine the required bearing length at each end of the beam if the structure is to be located in a very humid climate.

6-43. For the 4 × 12 No. 2 D. fir-larch beam of Fig. 6-22, what bearing length, ℓ_b, is required?

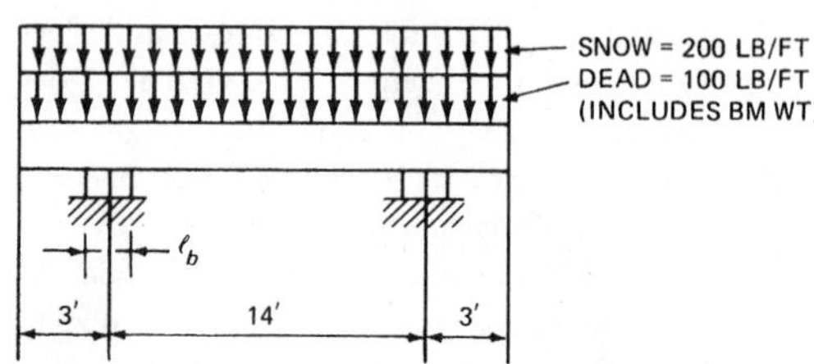

Fig. 6-22. Beam for Problem 6-43.

6-44. A building floor consists of ½-in. plywood resting on (but not rigidly connected to) No. 2 southern pine 2 × 12 planks (surfaced dry, used at 19% maximum MC), used flat at 12 in. c/c. Planks as long as 24 ft can be found easily. The planks, in turn, are supported by beams. Live load is 50 psf and the plywood weighs 1.5 psf. The planks are treated with fire-retardant chemicals. Assume the fire-retardant treatment reduces F_b, F_v, and E by 10%. Limit live-load deflection of planks and beams to $\ell/240$. (a) How far apart may the beams be placed (nearest safe whole foot)? (b) Design an interior beam of No. 1 (WWPA) D. fir-larch, assuming a clear span of 17 ft 6 in., with ends supported on masonry walls.

6-45. Design a complete floor system for the storage building shown in Fig. 6-17. Loads are: D = 10 psf plus weight of wood members, and L = 100 psf. Dry conditions. The system should have a row of columns (you determine how many) on the east-west centerline with beams in an east-west direction and joists north-south. Use T & G decking for the structural floor.

6-46. A 12 × 22 ft platform is to be constructed of No. 2 Douglas fir-larch. Floor is to be planking (2 × 10s laid side by side). Live load is 100 psf. Design the platform as shown in Fig. 6-23. (a) Calculate maximum allowable joist spacing based on capacity of planking. (b) Choose a layout of joists so that all spacings (not necessarily five) will be equal and design the joists, J-1. (c) Design the beams, B-1.

6-47. A No. 1 hem-fir 6 × 14 beam spans a clear horizontal distance of 20 ft and has superimposed dead load of 100 lb/ft and snow load of 150 lb/ft. The beam has a slope of 20%. Moisture content of the wood in use is 25%. If the beam is supported in a manner similar to Fig. 6-15, find the required bearing length.

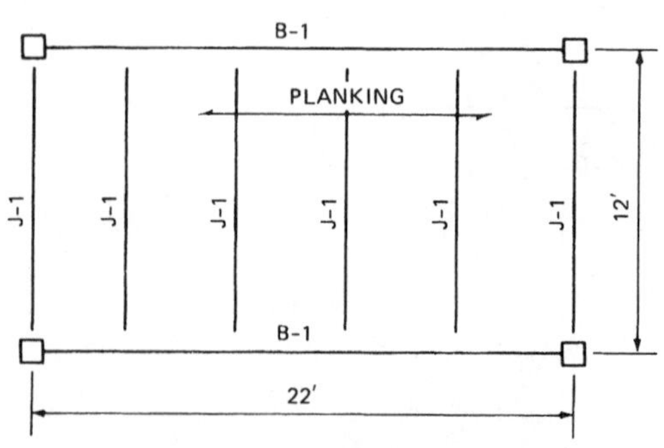

Fig. 6-23. Plan for Problem 6-46.

7

Selecting Sawn-Timber Compression and Tension Members

7-1. WOOD COLUMNS

Columns are generally thought of as the vertical supporting members of buildings. However, there are other structural members that act as columns—the piers of a bridge or the compression chord of a truss, for example. Generally columns are compression members, but they can also have combined compression and bending or can even have tensile axial force under loading that causes uplift. For purposes of design, we define a *column* as a structural member whose primary loads are axial compression.

The short, thick wood column will fail by crushing of its fibers. The longer, more slender column will fail by buckling due to lateral instability. Column design equations must be able to account for each of these failure modes.

Wood columns are usually of square cross section or round. Other shapes can be made but rarely are. Columns may be made of a single, solid piece of wood or formed from several pieces of wood. In the latter category are glulam columns (discussed in Chapter 8), spaced columns, and built-up columns. Spaced columns are assembled from two or more long members that are separated at the ends and midlength by spacer blocks. Built-up columns are formed by nailing or bolting two or more members to form either a solid cross section or other shape, such as a hollow, box-shaped section.

It is interesting to compare the specific strength (ratio of strength to unit weight) of both wood and steel columns. For geometrically similar (but not equal) sections of equal weight, the radius of gyration varies inversely as the square root of specific gravity (1). The strength of longer columns is highly dependent on the square of radius of gyration. Therefore, longer columns of wood (having lesser specific gravity than steel) will have superior specific strength over steel.

Figure 7-1 shows the results of comparing light column sections of wood and steel that are of equal weight and length. For the two sections compared, the advantage lies with steel only for very short lengths where strength is not influenced much by slenderness ratio, ℓ/r.

History of Wood Column Design

The historical development of theory for columns of any material is a subject that alone could fill an entire book. Starting in the 1700s various linear, parabolic, and other formulas were proposed for design. A few examples will show how design practice for wood columns developed.

Henry Dewell, author of a 1917 timber design book (2), felt that the American Railway Engineering Association (AREA) recommendations for working stresses were the best available at that time. This association used a single linear formula for both long and intermediate column design, as follows:

$$F'_c = F_c(1 - \ell/60d) \qquad (7\text{-}1)$$

where F'_c is the allowable stress on the column (considering that it may buckle), F_c is the allowable stress on a small block of wood, and d is the least cross-sectional dimension. A maximum of

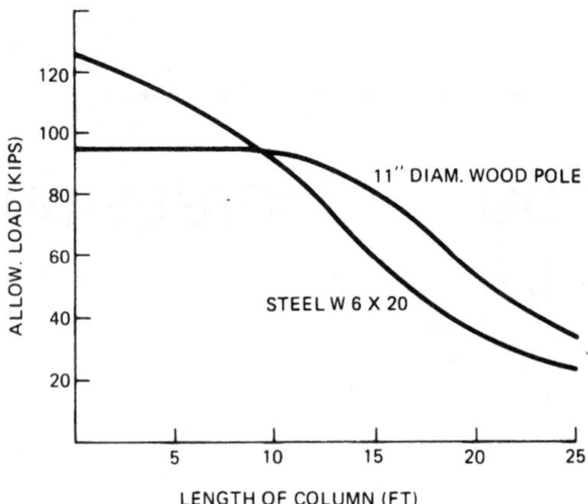

Fig. 7-1. Comparison of design strength of steel and timber columns of equal weight and length (wood F_c = 1000 psi).

60 was allowed for ℓ/d, the ratio of column length to least lateral dimension d. For "short" columns (then considered to have ℓ/d less than 15) the working stress for column design was 75% of F_c. The AREA equations and allowable stresses were for use in designing railway bridges and trestles.

Dewell also proposed using a version of the Gordon-Rankine formula, which he developed by curve-fitting test data accumulated from other sources. The proposed column formula,

$$F_c' = \frac{5000 / 3.5}{1 + \dfrac{\ell^2}{1750d^2}} \qquad (7\text{-}2)$$

contained a factor of safety of 3.5 and was to be used for all columns having ℓ/d ratios between 10 and 60.

Today's *National Design Specification for Wood Construction* (3) recognizes the different failure modes of columns (crushing, crushing and buckling, or buckling alone) by imposing an adjustment factor—the column stability factor, C_p. This factor will be nearly 1.0 for short columns that fail by crushing, but will be smaller for long, slender columns. The column stability factor first appeared in the NDS in the 1991 edition. Earlier editions, from 1977 through 1986, gave three equations for column design—one each for short, intermediate, and long columns. The intermediate formula (for columns not extremely stocky or extremely slender) was derived empirically at the Forest Products Laboratory and presented in a 1930 technical bulletin (4). The intermediate formula appeared in the 1947 NDS, but disappeared from the specification for several years before reappearing in 1977. During the interim, the specification contained only the long-column and short-column equations. It view of the series of changes in design methods, it is fair to assume that further changes will occur.

7-2. COLUMN DESIGN

Sections 3.6 through 3.10 of the *National Design Specification for Wood Construction* (3) deal with column design. NDS Appendices A, G, and H can be especially helpful also.

We know that the deflected shapes of buckled columns will differ depending on end support conditions. In most cases the effective length (distance between inflection points) will differ from the actual unbraced length. Figure 7-2 shows both theoretical and recommended design values for coefficient of effective length, K_e. These coefficients can be multiplied by actual length or length between lateral supports to obtain effective column length ($K_e\ell = \ell_e$). The recommended values recognize that perfectly fixed ends are impossible; therefore, the recommended coefficients (for other than pinned ends) are slightly larger than the theoretical values.

Actual support conditions seldom are equivalent to one of these idealized end conditions (fixed, pinned, or free). Given a column whose ends are laterally supported but not rigidly fixed, designers often assume that actual (unbraced) length equals effective length. This assumption (that K_e equals 1.0) is conservative for columns in which sidesway is prevented.

Research by Neubauer (5) shows that assuming a K_e equal to 1.0 is overly conservative for flat-ended 2 × 4 studs. He first tested columns with the squared ends pressed directly between the upper and lower platens of the testing machine. Then he tested columns that had pointed ends formed from steel channels and angles. For intermediate and long columns with flat ends, the maximum load carried was two to four times as great as for pointed ends. Neubauer con-

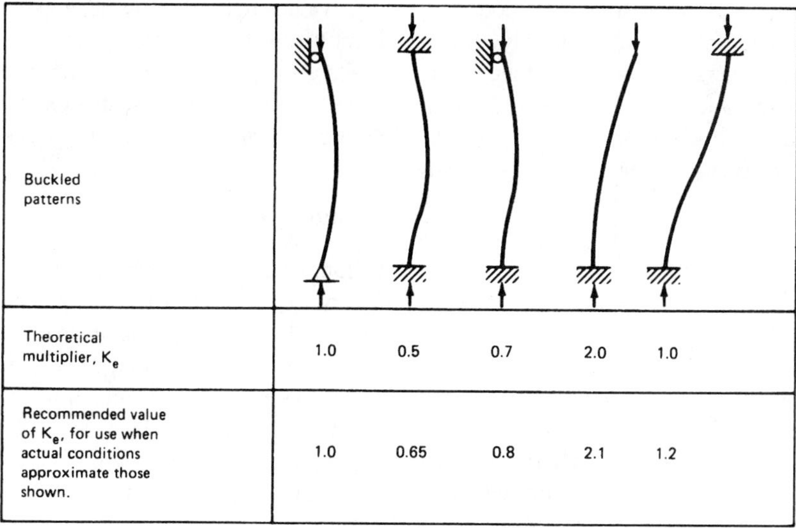

Fig. 7-2. Effective-length coefficients for compression members.

cluded that the increased strength should be recognized by a reduction in effective length.

Slenderness Ratio

Columns of all materials are classified by their *slenderness ratio.* A slenderness ratio of a column is the ratio of effective unbraced length to radius of gyration (ℓ_e/r). This ratio may differ for the two principal axes of the cross section. If so, the larger of the two slenderness ratios is the critical one. Radius of gyration, *r,* and cross-sectional dimension, *d,* are proportional to each other in a rectangular column ($d = r\sqrt{12}$). Because most wood columns are rectangular, the NDS classifies them, for simplicity, by ℓ_e/d ratio rather than ℓ_e/r ratio. A large slenderness ratio indicates a greater instability and tendency to buckle under lower axial load.

If the unbraced length of a column is the same in both directions, the column will buckle about the *w*-axis (Fig. 7-3a). In other words, its middle will move in the *s*-direction. Because of the tendency to buckle about the *w*-axis, this axis will be called the *weak axis,* and the *s*-axis will be called the *strong axis.*

There may be a different effective column length for buckling about each axis. An example to

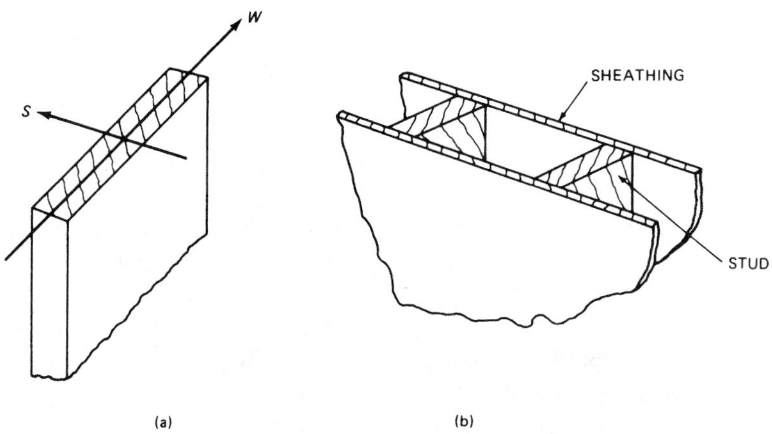

Fig. 7-3. Weak- and strong-axis buckling.

illustrate this is a plywood-sheathed wood shear wall like the one in Fig. 7-3b. The plywood is nailed to the narrow faces of the studs along the entire length of the stud. Therefore, the stud cannot buckle about its weak axis and the effective column length for weak-axis buckling is essentially zero. (Studs are the vertical members in a light-frame wall, spaced at 16-24 in. on center and of small cross section. See Fig. 1-1.) When the studs are so braced, their controlling (larger) slenderness ratio will be the one about their strong axis.

Design Equations for ASD

Wood columns are designed in the allowable stress design (ASD) method by ensuring that the compressive axial stress in the column does not exceed the allowable adjusted compression stress parallel to grain; i.e.,

$$f_c \leq F'_c \tag{7-3}$$

Applicable adjustment factors for compression stress parallel to grain are load duration, wet service, temperature, size, and column stability. To find the column stability factor, C_p, one must first find

$$F_{cE} = K_{cE} E'/(\ell_e/d)^2. \tag{7-4}$$

In this equation K_{cE} is 0.3 for visually graded lumber and machine-evaluated lumber, and 0.418 for products with $COV_E < 0.11$. Also, be cautioned that E is *never* adjusted for load duration. After calculating the value of F_{cE} from Eq. 7-4, the designer then finds C_p from the following equation:

$$C_P = \frac{1+\left(F_{cE}/F_c^*\right)}{2c} - \sqrt{\left[\frac{1+\left(F_{cE}/F_c^*\right)}{2c}\right]^2 - \frac{F_{cE}/F_c^*}{c}} \tag{7-5}$$

In Eq. 7-5, the value of c is 0.8 for sawn lumber, 0.9 for glued laminated timber, and 0.85 for round timber piles. Also, F_c^* in the equation is the tabulated parallel-to-grain compression design value multiplied by all applicable adjustment factors except C_p.

Slenderness is limited by NDS by the requirement that the ℓ_e/d ratio cannot exceed 50, except during construction, when it may be as large as 75.

Design Procedure

Column design is a trial-and-error procedure. Worrying about what to use for a guess or estimate will only waste time. Instead, time is saved by making a quick estimate and then analyzing that trial section to get a much better second estimate. Following are the necessary steps for designing a solid column with axial load only (no applied bending moment):

1. Estimate the allowable stress.
2. Use the estimate to solve for the required area and select a trial section. Do not choose a section with $\ell_e/d > 50$.
3. Decide whether or not the trial section is satisfactory by either
   a. comparing the actual stress f_c for that section to the adjusted allowable stress F'_c, or
   b. comparing the actual area of the trial section to the required area using F'_c computed for the trial section.

   A safe section has $f_c \leq F'_c$, or has actual area greater than or equal to required area.
4. Try other sections and recycle.

After the section has been chosen, three other checks may be necessary:

1. If the member has holes for split rings or bolts, the compressive stress at the net section must be computed to see that it does not exceed the short-column stress, F_c^*.
2. The column may have an end-bearing area that is less than the column cross-sectional area. In this case the designer must check the bearing stress parallel to grain (end-grain bearing stress). This check ensures that the fibers at the end of the column do not crush or broom. To prevent a bearing parallel-to-grain failure,

$$f_g \leq F'_g \tag{7-6}$$

Design values for bearing stress parallel to grain can be found in Appendix Table B-9. Applicable adjustments to the tabular values are load duration and temperature only. If the end-grain bearing stress exceeds 75% of the adjusted allowable value, then a bearing plate of metal or other rigid material must be used. These design values are quite large, so it is unlikely that end-grain bearing will control unless the column end has been reduced in cross section for some reason, or the column bears on a surface of lesser area than the column.

3. If the column or stud bears on a wood bottom plate or sill, the bearing capacity of the plate must be checked. The plate or sill may not be strong enough in compression perpendicular to the grain to support the load. Bearing capacity of certain grades of concrete floors (6) should be checked also. The load on a column may have to be limited to the capacity of the material the column must bear on.

All of the above assumes that the column is straight. If there is significant initial crookedness, the axial load will cause bending moment and the member must be designed as a beam-column. (Refer to Sections 7-6 and 7-7.)

Example 7-1

A column with pinned ends is to be designed from No. 1 Douglas fir-larch. The unsupported length of the column is 13 ft 0 in. The total load, which includes snow load, is 75 kips ($D = 40$ k and $S = 35$ k). Find the best square wood column.

Determine which loading condition controls.

$40/0.9 = 44.4$ for dead (permanent) load

$75/1.15 = 65.2$ for dead plus snow load
Controls

A square section (5 in. × 5 in. or larger) is classified as P & T, and Appendix Table B-5 shows

$F_c = 1000$ psi and $E = 1{,}600{,}000$ psi

If $C_p = 1.0$, then $F'_c = C_D \times F_c = 1.15\ (1000) = 1150$ psi. Guess that the allowable is somewhat less than 1150—say 1050 psi.
Then

A req $= 75{,}000/1050 = 71.43$ in.^{2}

d req $= 8.45$ in.

Try 10 × 10 nominal column with $d = 9.5$ in.

$\ell_e/d = 12 \times 13/9.5 = 16.4$

C_F is not applicable for compression for members 5 × 5 and greater.

$F_{cE} = 0.3(1{,}600{,}000)/16.4^2 = 1785$ psi

$F^*_c = 1.15(1000) = 1150$ psi

Using $c = 0.8$ in Eq. 7-5, $C_p = 0.818$

$F'_c = (0.818)(1150) = 941$ psi

$f_c = 75{,}000/(9.5 \times 9.5) = 831 \text{ psi} < 941$ psi
OK

The 10 × 10 is the smallest satisfactory section. It is not necessary to recycle. The next smaller section is an 8 × 8 with $d = 7.5$ in. It will have a slenderness ratio greater than 16.4 and therefore F'_c less than 941 psi. For the 8 × 8, $f_c = 75{,}000/7.5^2 = 1333$ psi, so it is obvious that the 8 × 8 would be too small.

Example 7-2

A steel column base (as shown in Fig. 7-4) for the column of Ex. 7-1 is to be designed. Find the minimum value of dimension b based on bearing parallel to grain.

From Appendix Table B-9,

$F_g = 1480$ psi

$f_g = P/A = 75{,}000/(9.5 \times b)$

Setting equal and solving for b,

Minimum $b = 5.3$ in.

Example 7-3

A 10 × 16 No. 1 D. fir-larch column is 22 ft long. About the strong axis it is braced only at its ends;

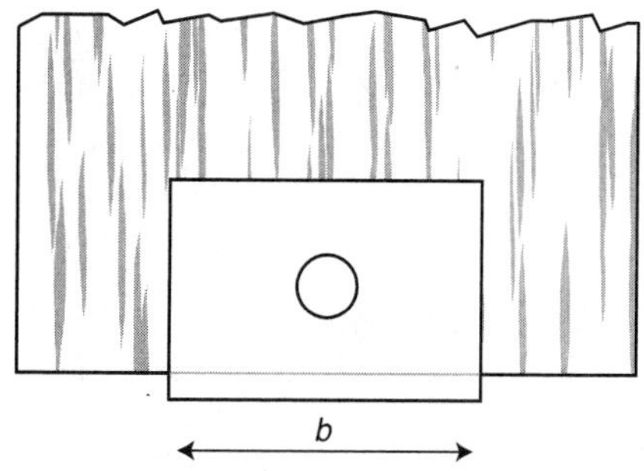

Fig. 7-4. Column base for Example 7-2.

however, about the weak axis it is attached to a bracing member at the 15-ft point. Two 3/4-in. bolts installed in 25/32-in. diameter holes connect the brace to the column, as shown by Fig. 7-5a. The total load is 111 kips with snow load controlling. Is the column satisfactory for the design loads?

A 10 × 16 member is classified as Beam & Stringer since 16 > 10 + 2. By Appendix Table B-5, $F_c = 925$ psi and $E = 1{,}600{,}000$ psi. Load duration factor, C_D, is 1.15. Assuming pinned ends,

About the weak axis, $l_e/d = 12 \times 15/9.5 = 18.9$

About the strong axis, $l_e/d = 12 \times 22/15.5 = 17.0$

The larger ratio controls.

$$F_{cE} = 0.3(1{,}600{,}000)/18.9^2 = 1344 \text{ psi}$$

$$F_c^* = 1.15(925) = 1064 \text{ psi}$$

Using $c = 0.8$, $C_p = 0.765$

$$F_c' = 0.765(1064) = 814 \text{ psi}$$

$$f_c = 111{,}000/(9.5 \times 15.5) = 754 \text{ psi} < 814 \text{ psi}$$
OK

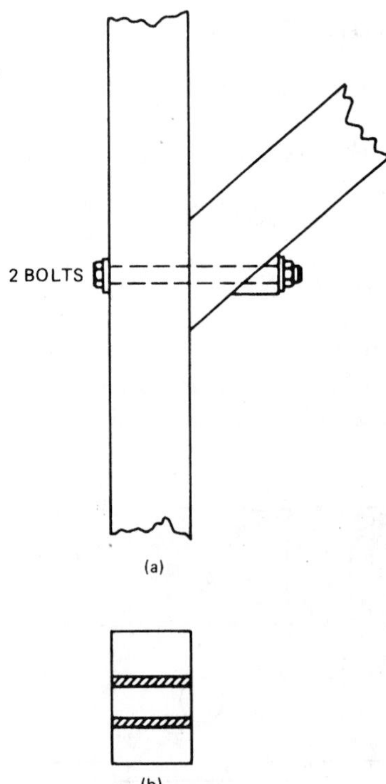

Fig. 7-5. Column for Example 7-3.

Check stress on the net section at the drilled holes (Fig. 7-5b).

$$f_c = P/A_{net}$$
$$= 111{,}000/[9.5 \times 15.5 - 2(25/32)(9.5)]$$
$$= 838 \text{ psi} < F_c^* = 1064 \text{ psi} \qquad OK$$

In this example, the holes are small relative to the size of the cross section, so the short-column allowable, F_c^*, is not exceeded at the section containing the hole. However, if the holes were larger, the strength at the net section might control the design.

7-3. ROUND AND TAPERED COLUMNS

The strength of a round column, or of a round column with bending, has historically been assumed equal to that of a square member having the same area of cross section. This simplification is still allowed today because of ease of calculation (NDS Section 3.7.3). Therefore, to design a round column, one may find the minimum size of square column that would satisfy all requirements and then provide a round shape of equal area.

Alternatively (See NDS Appen. H, Sect. H.3), the designer may replace d in Eq. 7-4 with $r\sqrt{12}$, where r stands for the radius of gyration of the round column. The first method gives an adjusted allowable compressive stress 1-5% greater. Therefore, the equivalent-square method is only slightly in error, but on the unsafe side.

Tapered columns can be found in varieties such as shown in Fig. 7-6. For tapered columns

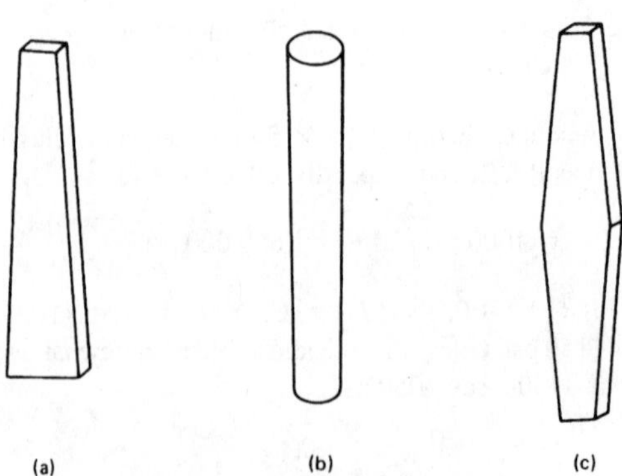

Fig. 7-6. Tapered column types.

like those in the figure, a particular cross section is assumed to be critical. The column is then designed as if the entire column had this cross section. However, one must also check to ensure that the compressive stress at the smallest cross section of the tapered column does not exceed the allowable stress for a short column, F_c^*.

In each plane, the dimension, d, of the critical cross section can be found from

$$d = d_{min} + (d_{max} - d_{min})[a - 0.15\,(1 - d_{min}/d_{max})] \quad (7\text{-}7)$$

For simply supported columns the value of a is taken as 0.50 if the column is tapered toward one end only and as 0.70 if it is tapered toward both ends (as in Fig. 7-6c). Values of a for other support conditions can be found in NDS Section 3.7.2.

Round sections are often tapered. Examples are telephone poles or timber piles. Usually round, tapered columns are designed as equivalent square columns.

Example 7-4

A round column is to be designed for a rustic lodge lobby. The wood to be used has $E = 1{,}600{,}000$ psi and $F_c = 1000$ psi. The column will be 17 ft long. What diameter (to the nearest half inch) is necessary to support a ten-year duration load of 65 kips, consisting of 50 kips L (occupancy) and 15 kips D? Assume pinned ends. Use the equivalent square column method.

Whenever a trial-an-error solution is required (as is the case here), it is convenient to tabulate the solution. (Advantages of the tabulated solution are that it is compact and is easy for the designer or other person to check.) Figure 7-7 shows a designer's computation sheet for this example. An 11-in.-diameter round column (untapered) would be the smallest satisfactory section.

Example 7-5

Find the allowable axial load of a round, tapered column of length 20 ft. The column has a diameter of 12 in. at the base, tapering to 8 in. at the top. The column has $E = 1{,}600{,}000$ and $F_c = 925$ psi. The load will be of more than ten years cumulative duration. Use the equivalent square method.

Min. Diam. = 8 in.; min. d of equiv sq $= \sqrt{\pi D^2/4} = 7.09$ in.

Max. Diam. = 12 in.; max. d of equiv sq = 10.63 in.

Find d at critical section:

$$d = 7.09 + (10.63 - 7.09)[0.50 - 0.15(1 - 7.09/10.63)]$$

$$d = 8.68 \text{ in.}$$

$$F_c^* = C_D\,F_c = 0.9\,(925) = 832.5 \text{ psi}$$

$$\ell/d = 12(20)/8.68 = 27.65$$

$$F_{cE} = 0.3(1{,}600{,}000)/(27.65)^2 = 627.8$$

$$C_P = 0.587$$

$$F_c' = 0.587\,(832.5) = 489 \text{ psi}$$

$$\text{Allow. } P = (A)(F_c') = (8.68^2)(489) = 36{,}800 \text{ lb}$$

Check compressive stress at small end.

$$P/A = 36{,}800/\pi(4)^2 = 732 \text{ psi} < F_c^* = 832 \text{ psi} \quad OK$$

Allow. $P = 36{,}800$ lb

Example 7-6

Rework Ex. 7-5 with the more accurate method, which uses the properties of the actual circular cross section.

D = Diameter of crit. section

$$= 8 + (12 - 8)\,[0.50 - 0.15\,(1 - 8/12)]$$

$$= 9.80 \text{ in.}$$

(*Note:* This corresponds exactly to the location of the critical section for the equivalent square method.)

$$r = \sqrt{I/A} = D/4 = 9.80/4 = 2.45 \text{ in.}$$

Substitute $r\sqrt{12}$ for d:

$$r\sqrt{12} = 8.49 \text{ in.}$$

$$F_{cE} = 0.3\,(1{,}600{,}000)/(20 \times 12/8.49)^2 = 600.7$$

$$F_c^* = 0.9(925) = 832.5$$

$$C_p = 0.570$$

$$F_c' = 0.570(832.5) = 475 \text{ psi}$$

$$\text{Allow. } P = 475 \times 9.80^2\pi/4 = 35{,}800 \text{ lb}$$

This allowable load is about 97% of that found using the equivalent square method.

$P = 65\text{ K} \quad C_D = 1.0$

$E = 1{,}600{,}000\text{ PSI}; \quad F_c = 1000\text{ PSI}; \quad L = 17'$

HINGED ENDS, $\therefore \ell_e = 17(12) = 204''$

$F_c{}^* = 1000\text{ PSI}$

$$F_{cE} = \frac{0.3(1{,}600{,}000)}{(204/d)^2} = 11.53d^2 = 11.53S^2$$

FOR CIRCULAR SECTION,

S (SIDE OF EQUIV. SQUARE) $= \sqrt{\pi D^2/4}$

TRY D=	EQUIV. S	A	F_{cE}	C_p	F_c'	$A_{REQ.}$	
9"	7.98"	63.62	734.2	0.577	577	113	NG
10"	8.86"	78.54	905.1	0.655	655	99.2	NG
10.5"	9.31	86.59	999.4	0.691	691	94.1	NG
11"	9.75"	95.03	1096	0.722	722	90.0	OK

Use 11" DIAMETER

Fig. 7-7. Computation sheet for Example 7-4.

7-4. SPACED COLUMNS

Spaced columns are composed of two or more individual members separated at the ends and middle by blocking as shown in Fig. 7-8. The end spacer blocks that separate the two (or more) members of a spaced column provide increased fixity of the individual pieces at the ends. This increased fixity causes the distance between inflection points in the buckled column to be much less than it would be for the elements acting as independent columns. Therefore, the strength of a spaced column of two members is not merely twice that of a single member but, rather, may be many times that of a single member.

Frequently, compression members of trusses are spaced columns. In a truss the end spacer blocks are often other web or chord members of the truss. (See Chapter 10.)

A longer spacer block at the ends provides for better fixity, thereby affecting the buckled shape. (With the longer blocks, distance between inflection points in the buckled shape will decrease.) Better fixity leads to a higher strength; therefore, the code provides for two end-fixity conditions. If the centroid of connectors in the end block is within $\ell/20$ from the column end, this is known as *condition a.* If the centroid of connectors is between $\ell/20$ and $\ell/10$ from the ends (better fixity), it is *condition b.* A

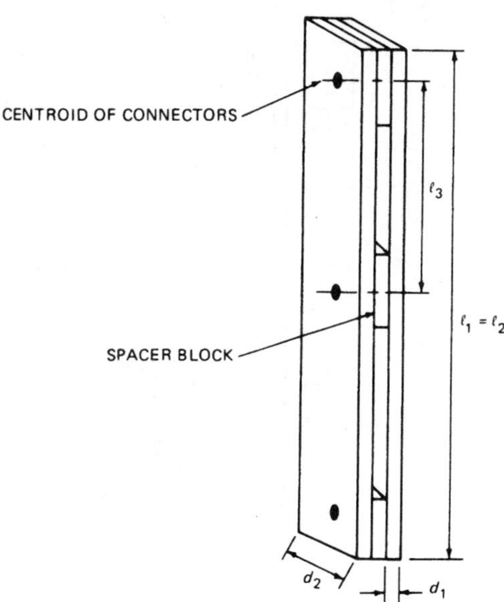

Fig. 7-8. Spaced column.

factor K_x for use in design formulas has a value of 2.5 for condition *a* and 3.0 for condition *b*.

The NDS assumes that the end spacer blocks will be joined to the long individual column members with split-ring or shear plate timber connectors (see Chapter 9), rather than nails or bolts. The British Standard Code of Practice (7) allows for the spacer blocks to be joined with nails, bolts, screws, timber connectors, or glue. The British Code considers gluing to produce the stiffest spaced columns and therefore allows spaced columns joined by gluing to have a lesser effective length than those joined by other methods. The NDS requires that when there is more than one spacer block at midlength the connection be made with a timber connector; a single middle block may be joined with bolts, nails, or timber connectors. Spacer blocks must be as thick as or thicker than the individual pieces composing the column.

To find the load on an end-block split-ring connector, the designer chooses a value of spacer-block constant from Table 7-1. The required connector capacity is the spacer-block constant times the area of an individual column element. To choose size and/or number of split rings, this required capacity is then compared to the split-ring design values in Table C-17. The procedure is illustrated in Example 7-7.

The increased strength of a spaced column is considered to be effective only for buckling in the plane parallel to the small dimension, as in Fig. 7-9a. The column is designed in this plane as a spaced column. For buckling in the other plane (see Fig. 7-9b) the column is designed as a solid column. Therefore, every spaced-column design requires considering two cases—one for each plane of possible buckling—to see which case controls the design. For each of these cases the allowable load of an individual member is determined first. The allowable load for the spaced column is the product of the allowable load for one member times the number of members. Limits allowed for slenderness ratios in spaced columns are (refer to Fig. 7-8):

80 for ℓ_1/d_1
50 for ℓ_2/d_2
40 for ℓ_3/d_1

Design Equations

For checking buckling as a spaced column, most design equations are identical to those for an ordinary column. The exception is the equation for

Table 7-1. Shear Constants for Designing Spacer-Block Connections in Spaced Columns.

Ratio ℓ_1/d_1 of Individual Element of Spaced Column	Shear Constant per End per Plane			
	Species Grouping			
	A	B	C	D
11 or less	0	0	0	0
	(INTERPOLATE for ratios between 11 and 60)			
60 or above (maximum ratio allowed = 80)	468	399	330	261

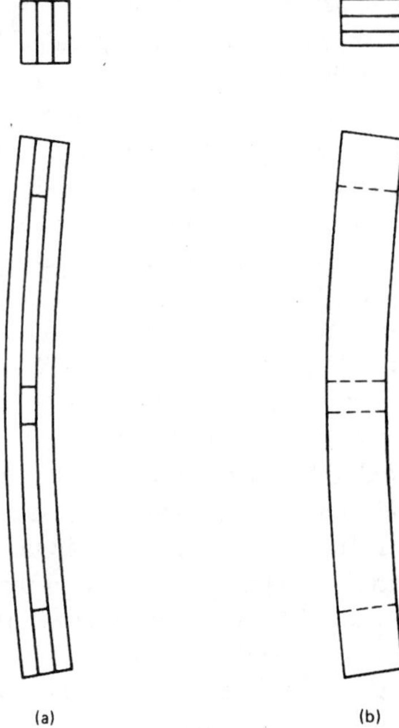

Fig. 7-9. Spaced-column buckling about two different axes.

F_{cE}, which now incorporates the connector fixity constant, K_x. Eq. 7-4 is replaced with

$$F_{cE} = K_x K_{cE} E'/(\ell_e/d)^2 \qquad (7\text{-}8)$$

It is a good idea to check spaced columns to see that the short-column allowable compressive stress, F_c^*, is not exceeded on the net area. Removal of wood for the split-ring connectors could cause the actual stress on the net area to exceed this allowable value.

Example 7-7

Solve for the capacity of the spaced column shown by Fig. 7-10 under a load of seven days cumulative duration. The two members are No. 1 & Btr Douglas fir-larch 4 × 6s. Choose connectors.

Assume actual length equals effective length.

Load duration factor = C_D = 1.25. By Appendix Table B-3, E = 1,800,000 psi; F_c = 1500 psi; For compression, C_F = 1.1. Check end fixity class.

$\ell/20 = 144/20 = 7.2$ in.

6 in. < 7.2 in., so this is condition a and $K_x = 2.5$

Check about a–a axis:

Treat as a spaced column for buckling about this axis:

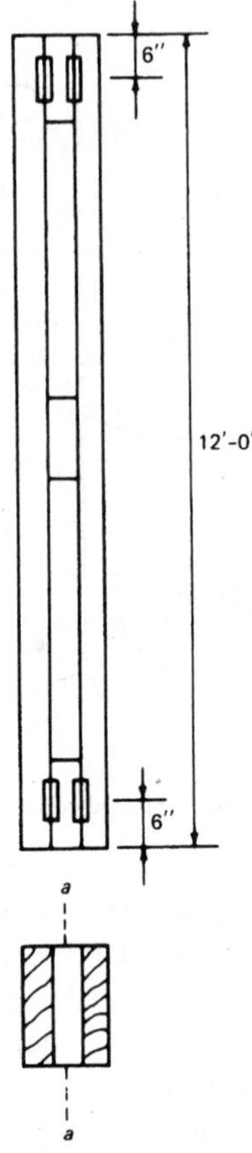

Fig. 7-10. Spaced column for Example 7-7.

$\ell_1/d_1 = 144/3.5 = 41.14$

$F_{cE} = 0.3(2.5)(1{,}800{,}000)/41.14^2 = 797.6$ psi

$F_c^* = 1.25(1500) = 1875$ psi

$C_p = 0.379$

$F_c' = 0.379(1875) = 711$ psi

Check about other axis: Treat as a solid column for buckling in this direction.

$\ell_2/d_2 = 144/5.5 = 26.18$

$F_{cE} = 0.3(1{,}800{,}000)/26.18^2 = 787.9$ psi

$F_c^* = 1875$ psi as before

$C_p = 0.375$

$F_c' = 0.375(1875) = 703$ psi

The second axis checked controls.

Allow. $P = 2(703)(3.5 \times 5.5) = 27{,}100$ lb

Check net area under $P = 27{,}100$ lb

$$A \text{ net} = 3.5 \times 5.5 - A \text{ (bolt shank hole)} - A \text{ (ring groove)}$$

$$= 3.5 \times 5.5 - (13/16)(3) - (0.5)(4.50)$$

(Assuming 4-in. split ring. See Fig. 9-19.)

$$= 14.56 \text{ in.}^2$$

At the net section,

$$f_c = 27{,}100/14.56 = 1861 \text{ psi}$$

$$F_c^* = 1875 \text{ psi} \qquad OK$$

Choose split rings:

$$\ell_1/d_1 = 41.14$$

For timber connectors, D. fir-larch is in group B woods, so by interpolation on Table 7-1,

$$\text{Spacer block constant} = K_s = 245 \text{ psi}$$

$$\text{Connector capacity req.} = (3.5 \times 5.5)(245)$$

$$= 4716 \text{ lb on each plane}$$

One 4-in.-diameter ring has capacity = 5260(1.25) = 6575 lb. Use one 4-in.-diameter split ring at each of the four end-block joints.

(a)

(b)

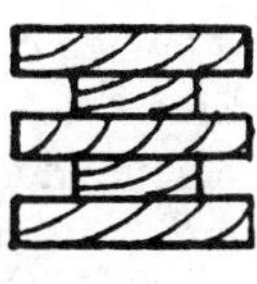

(c)

(d)

Fig. 7-11. Built-up column cross sections.

Table 7-2. Capacities of Built-up Columns.

ℓ/d	Percent of Solid-Column Capacity
6	82
10	77
14	71
18	65
22	74
26	82

7-5. BUILT-UP COLUMNS

Columns made by nailing or bolting two or more pieces of wood together to form a solid column are known as *built-up* or *layered* columns. Sketches of some of the many possible cross sections appear in Fig. 7-11. Because of slip between adjacent surfaces as the fasteners deform, no built-up column will be as strong as a solid (i.e., one-piece) column of the same dimensions and quality of wood. Care must be taken so that the pieces are adequately spiked or bolted together. (Glued laminated columns are made of several pieces of wood, but they are not classed as built-up. Their glued surfaces allow for excellent transfer of shear, so that the glulam column has the same capacity as a solid column of the same quality of material.)

The percent of solid-column strength of a built-up column has been found to be a function of ℓ/d ratio (8,9) as shown in Table 7-2. This table is based on experiments on columns with cover plates as in Fig. 7-11a or b and should be used only for columns with cover plates. Tests by Dewell (2) indicated that built-up columns without cover plates were far inferior to those with cover plates. For columns without cover plates he recommended using a built-up column capacity of 80% of the mean (average) of the strengths computed (1) as a solid column and (2) as a summation of the capacities of the individual pieces composing the column.

Both of the guidelines concerning the strength of built-up columns mentioned in the previous paragraph were developed more than a half century ago. The most recent and comprehensive study of built-up columns was undertaken by Malhotra and Van Dyer (10). Their theoretical and experimental program applied to columns in elastic and inelastic ranges and to columns of all cross-sectional configurations, taking into account the effect of interlayer slip. The design procedures resulting from the study have a more rational basis than the method suggested by Table 7-2, but are fairly complex.

Because of its light weight and economy, two researchers singled out the box column (similar to Fig. 7-11b, except hollow) for study (11). They found that strength varied directly with the number of nails. The columns tested were formed from 1 × 4 or 1 × 6 boards, using six-penny ringed nails. With optimum nailing (about 68 nails on an 8-ft-tall column composed of four 1 × 6s), each column had strength equal to or greater than that of a solid pinned column of the same net cross-sectional area.

The only type of built-up column addressed by the NDS is that of Fig. 7-11d. A built-up column of this type, if nailed according to NDS-specified requirements, is considered to have 60% of solid column strength (if bolted, 75%).

Example 7-8

A one-story building has No. 2 hem-fir 4 × 8 interior columns that are 10 ft 0 in. long. The building owner wants to put an open-air restaurant on the roof, increasing the load on the columns.

(a) What is the allowable ten-year duration load on each existing column using the NDS?
(b) If each column were built up by adding 2-in. nominal material on each face, as shown by Fig. 7-12, what ten-year duration load could then be allowed? Assume that the elements would be well nailed together.

(a) For No. 2 hem-fir, $E = 1{,}300{,}000$ psi, $F_c = 1250$ psi, and C_F for compression = 1.05.

$\ell_e/d = 12 \times 10/3.5 = 34.3$

$F_{cE} = 0.3(1{,}300{,}000)/34.3^2 = 331.5$ psi

$F_c^* = 1.05(1250) = 1313$ psi

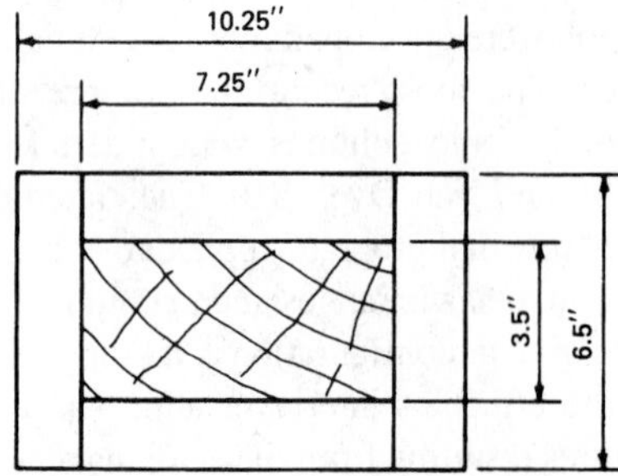

Fig. 7-12. Column cross section for Example 7-8.

$C_p = 0.238$

$F_c' = 0.238(1313) = 312$ psi

Find allowable load for the 4 × 8 column:

$$\text{Allow. } P = (F_c')(A) = (312)(3.5 \times 7.25) = 7920 \text{ lb}$$

(b) The added material will be 2 × 8s (whole or ripped) of the same grade as the original column. For 2 × 8s, C_F is 1.05 as before, so F_c^* remains the same.

For the built-up member, $\ell_e/d = 12 \times 10/6.5 = 18.5$

$F_{cE} = 0.3(1{,}300{,}000)/18.5^2 = 1140$ psi

$F_c^* = 1313$ psi (same)

$C_p = 0.640$

$F_c' = 0.640\,(1313) = 840$ psi

Find load allowed for a *one-piece* column of this size.

$$\text{Allow. } P = (840)(6.5 \times 10.25) = 55{,}970 \text{ lb}$$

For the built-up column, using Table 7-2, this value should be reduced to

$$\text{Allow. } P = 55{,}970(0.66) = 36{,}900 \text{ lb}$$

Or, if the NDS reduction is used,

$$\text{Allow. } P = 55{,}970(0.60) = 33{,}600 \text{ lb}$$

NDS does not address this pattern of built-up column, but the authors believe the pattern of Fig. 7-11b may be superior to that of Fig. 7-11d.

7-6. BEAM-COLUMNS

Many columns are subjected to bending in combination with axial compression loads. In fact, because of nonuniform bearing, misalignment, or member crookedness, the load on many columns probably does not pass through the centroid of the cross section. This causes the column to be stressed both in axial compression and flexure. Some designers (12) even feel that, because perfectly axial load is only a theoretical possibility, all wood columns should be designed for a minimum eccentricity.

Columns loaded in such a way that they must

resist both compression and bending are called beam-columns. The top chord of a truss supporting roof loads between panel points (from joists or from roof sheathing nailed directly to the chord) is one example of a beam-column. Another example is an exterior column carrying vertical gravity loads and also subjected to horizontal wind or seismic loads. A column whose vertical load is applied onto a side bracket is also a beam-column.

P-Δ Effect or Moment Magnification

Bending moments in beam-columns can result from either lateral forces, applied moment loads, or eccentrically applied end loads. Whatever the cause, these initial bending moments cause the member to deflect, and additional bending moment occurs equal to the product of the end load, P, and deflection Δ. Figure 7-13 illustrates the cause of this P–Δ effect, also referred to as *moment magnification.* The result of this action is an increase in bending moment and the flexural stress, f_b, that bending moment causes.

Beam-Column Design

Columns with bending moment as well as axial load are designed using an *interaction equation.*

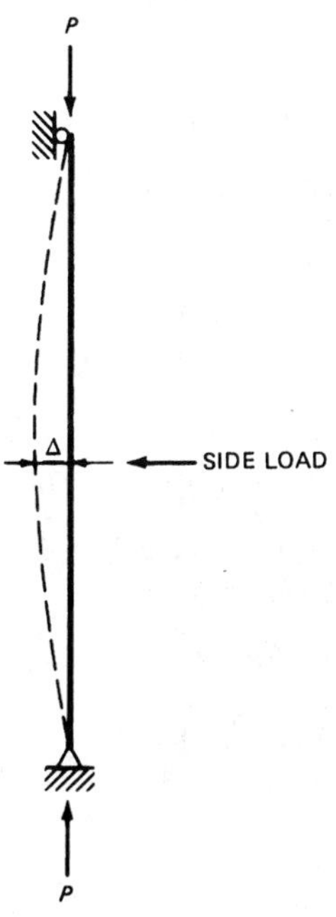

Fig. 7-13. P-Δ effect.

The designer uses trial and error to find a satisfactory section. For axial compression combined with uniaxial bending, the interaction equation is:

$$\left(\frac{f_c}{F_c'}\right)^2 + \frac{f_b}{F_b'[1 - (f_c / F_{cE})]} \leq 1.0 \quad (7\text{-}9)$$

In this equation, F_{cE} of the denominator is computed using the larger cross-section dimension (called d_x) when bending is about the strong axis, but is computed using the smaller cross-sectional dimension, d_y, when bending is about the weak axis. The magnification factor $1/(1 - f_c/F_{cE})$ in Eq. 7-9 accounts for the P–Δ effect.

Special provisions exist for very light truss members (2 × 4 or smaller) that are subjected to combined bending and compression. If such a member is stiffened by plywood nailed to its narrow face, it is allowed an increase in stress. These special provisions (NDS Section 4.4.3) will be covered in Chapter 10.

Sometimes bending about both axes (biaxial bending) is present either alone or in combination with axial compression. In that case the interaction equation is:

$$\left(\frac{f_c}{F_c'}\right)^2 + \frac{f_{bx}}{F_{bx}'[1 - (f_c / F_{cEx})]} + \frac{f_{by}}{F_{by}'[1 - (f_c / F_{cEy}) - (f_{bx} / F_{bE})^2]} \quad (7\text{-}10)$$

The bending stresses in Eqs. 7-9 and 7-10 are those due to bending moment due to transverse (side) loads or to end moments from frame action. When the bending moment is the result of eccentricity of load, a different interaction equation should be used. That equation is

$$\left(\frac{f_c}{F_c'}\right)^2 + \frac{f_{bx} + f_{bex}[1 + 0.234(f_c / F_{cEx})]}{F_{bx}'[1 - (f_c / F_{cEx})]} + \{f_{by} + f_{bey}[1 + 0.234(f_c/F_{cEy}) + (0.234/F_{bE}^2)(f_{bx} + f_{bex})^2]\} \div \{F_{by}'[1 - (f_c/F_{cEy}) - ((f_{bx} + f_{bex})/F_{bE})^2]\} \quad (7\text{-}11)$$

where $f_{be} = Pe/S$ is the bending moment due to eccentricity of load. As this equation shows, the portion of the bending stress that is due to eccentricity of load has a larger magnification than the remainder of the bending stress. In general, however, the increase is only around 4%.

Example 7-9

A roof truss with top-chord slope 4:12 supports snow load and dead loads due to decking, roofing, ceiling, and insulation. The top chord of the truss receives a uniform vertical load applied to the narrow face. Therefore, this chord has to carry both axial and bending loads. The top chord is continuous over three sloping 5-ft 7-in. spans, and the most heavily loaded of the three carries a compression load of 4300 lb and a uniform load of 125 lb per horizontal foot. Will a 2 × 6 of D. fir-larch, No. 2 (placed with its 1.5-in. dimension horizontal) be sufficient for this member?

Assume, conservatively, that coefficient of effective length $k_e = 1$.

For bending, $C_F = 1.3$ and $F_b = 875$ psi

For compression, $C_F = 1.1$ and $F_c = 1300$ psi

$E = 1{,}600{,}000$ psi

The horizontal distance between panel points is 5.58 × (12/12.65) = 5.29 ft.

Bending: Decking nailed to the top (narrow) face prevents lateral buckling, so that the allowable bending stress need not be adjusted for instability using C_L. Also note that if it were known that the trusses were at 24-in. centers or closer, the repetitive member factor, C_r, would have been used.

$$F'_b = C_F C_D F_b = 1.3(1.15)(875) = 1308 \text{ psi}$$

$$M = 0.1w\ell^2 \text{ (Table A-8)}$$

$$M = 0.1(125)(5.29)^2 = 349.8 \text{ ft-lb}$$

$$f_b = 12(349.8)/7.56 = 555 \text{ psi}$$

Axial Compression: To find F'_c the ℓ_e/d ratio we will use will be for column buckling in the strong direction. Because of the nailed decking, the member is not able to buckle in the weak direction.

$$\ell_e/d = 5.58(12)/5.5 = 12.2$$

$$F_{cE} = 0.3(1{,}600{,}000)/12.2^2 = 3225 \text{ psi} = F_{cEx}$$

$$F^*_c = 1.1(1.15)(1300) = 1645 \text{ psi}$$

By Eq. 7-5, $C_p = 0.864$

$$F'_c = 0.864(1645) = 1421 \text{ psi}$$

$$f_c = 4300/(1.5 \times 5.5) = 521 \text{ psi}$$

Substituting in interaction Eq. 7-9:

$$\left(\frac{521}{1421}\right)^2 + \frac{555}{1308[1 - (521/3225)]}$$

$$= 0.64 < 1.0 \qquad OK$$

Based on combined stresses, a 2 × 6 is sufficient for the entire top chord, because the span analyzed was the most heavily loaded portion of the chord. If this truss used bolts or split rings as connectors, it would be necessary to check the stress at the net section. The axial load divided by net area would need to be compared to the allowable short column stress, F^*_c.

Example 7-10

An 8 × 8 No. 2 hem-fir (P&T, WWPA graded) member, 12 ft long, has simply supported ends and carries vertical load of 3.2 k dead and 4.2 k snow, applied at the face of the column. Is it satisfactory?

Appendix Table B-5 shows for this member, $F_c = 375$ psi, $F_b = 525$ psi, and $E = 1{,}100{,}000$ psi. 3.2/09 < 7.4/1.15, so the load combination that included both *D* and *S* controls.
As a column,

$$\ell_e/d = 12 \times 12/7.5 = 19.2$$

$$F_{cE} = 0.3(1{,}100{,}000)/19.2^2 = 895 \text{ psi}$$

$$F^*_c = 1.15(375) = 431 \text{ psi}$$

$$C_p = 0.873$$

$$F'_c = 0.873(431) = 376 \text{ psi}$$

$$f_c = 7400/7.5^2 = 132 \text{ psi}$$

As a beam,

Since $d = b$, $C_L = 1.0$

Since $d < 12$ in., $C_F = 1.0$

$$F'_b = (1.0)(1.0)(1.15)(525) = 604 \text{ psi}$$

The eccentric load causes $M = Pe = 7400(7.5/2)$ $= 27{,}750$ in.-lb

$$f_{be} = Pe/S = 27{,}750/70.3 = 395 \text{ psi}$$

Substituting into Eq. 7-11,

$$\left(\frac{132}{376}\right)^2 + \frac{(395)(1 + 0.234(132/895))}{(604)(1 - (132/895))}$$

$$= 0.123 + 0.654(1.213) = 0.92 < 1.0 \qquad OK$$

Note that in this case the bending stress was magnified by 21.3%.

Example 7-11

An 8 × 12 column of No. 1 (B&S) Douglas fir-larch, graded under WCLIB rules, is used to carry a vertical floor load of 18 kips (ten-year duration) placed 4 in. off center, as shown by Fig. 7-14. In addition, the column is subject to a uniform wind load of 400 lb/ft from the attached exterior wall system. Assuming the wall system does not provide lateral support, is the member satisfactory as a beam-column?

Reactions to the eccentric vertical load and wind load are computed first, and are shown by Fig. 7-14b. Then the shear diagram is used to solve for the maximum bending moment of 13.02 ft-k, shown by Fig. 7-14d. Of this 13.02 ft-k, 3.45 ft-k = (0.428)(8.07) is due to eccentric load and 9.57 ft-k is due to transverse load.

The column must be analyzed (a) for action as a beam-column under combined dead, live, and wind loads; and (b) for action as a beam-column under eccentric vertical load alone.

Case a:

F_b = 1350 psi, F_c = 925 psi, E = 1,600,000, and C_D = 1.6

As a column,

Larger $\ell_e/d = 14(12)/7.5 = 22.4$

$F_{cEy} = 0.3(1{,}600{,}000)/22.4^2 = 957$ psi

$F_c^* = 1.6(925) = 1480$ psi

$C_p = 0.528$

$F_c' = 0.528(1480) = 781$ psi

Actual $f_c = 18{,}000/(7.5 \times 11.5) = 209$ psi

As a beam,

(Note that both wind and eccentric vertical load cause strong-axis bending.)

$\ell_u/d = 14(12)/11.5 = 14.6 > 14.3$

$\ell_e = 1.84(14 \times 12) = 309.1$ in.

$R_B^2 = 309.1(11.5)/7.5^2 = 63.2$

$F_{bE} = 0.438(1{,}600{,}000)/63.2 = 11{,}090$ psi

$C_F = 1.0$

$F_b^* = (1.0)(1.6)(1350) = 2160$ psi

Substituting in Eq. 6-4, $C_L = 0.988$

$F_b' = 0.988(2160) = 2130$ psi

Due to transverse load,

$f_b = M/S = 9570(12)/165.3 = 695$ psi

Due to eccentric load,

$f_{be} = 3450(12)/165.3 = 250$ psi

The final item needed is F_{cE}, computed using the larger cross-sectional dimension, d = 11.5 in.

$F_{cEx} = 0.3(1{,}600{,}000)/(14 \times 12/11.5)^2 = 2249$ psi

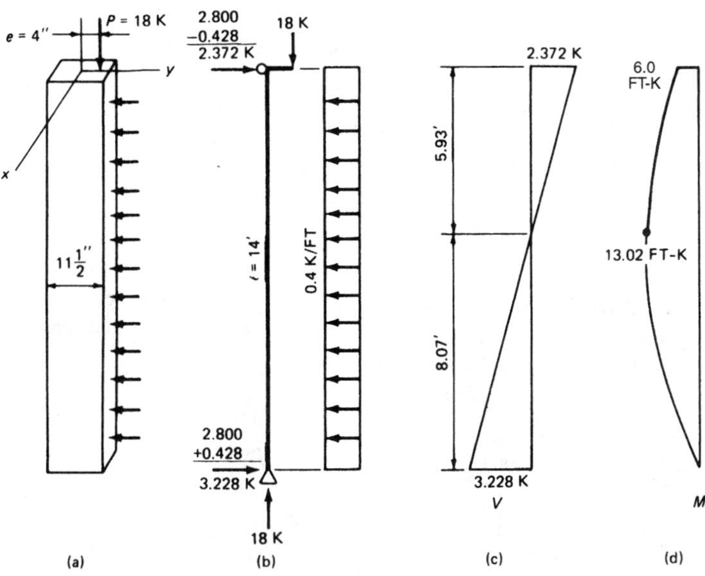

Fig. 7-14. Beam-column for Example 7-11.

Substituting into interaction Eq. 7-11,

$$\left(\frac{209}{781}\right)^2 + \frac{695 + 250[1 + 0.234(209 / 2249)]}{(2130)(1 - (209 / 2249))}$$

$= 0.56 < 1.0$ *OK for Case a*

Case b:
Load duration factor = 1.0 (*Case b* does not include wind).
As a column,

$F_{cE} = 957$ (same), but $F_c^* = 1.0(925) = 925$ psi

So $C_p = 0.703$ and $F_c' = 0.703(925) = 650$ psi

$f_c = 209$ psi (same)

As a beam,

$F_{bE} = 11{,}090$ (same), but $F_b^* = F_b = 1350$ psi

$C_L = 0.993$

$F_b' = 0.993(1350) = 1341$ psi

If vertical load acts alone, maximum moment (at the top end) is

$Pe = (18\text{k})(4/12 \text{ ft}) = 6$ ft-k

Stress f_{be} due to eccentric vertical load is

$f_{be} = 6{,}000(12)/165.3 = 436$ psi

Substituting into Eq. 7-11,

$$\left(\frac{209}{650}\right)^2 + \frac{436(1 + 0.234(209 / 2249))}{(1341)(1 - (209 / 2249))}$$

$= 0.47 < 1.0$ *Ok for Case b*

Example 7-12

A No. 1 D. fir-larch 8 × 12 (B&S), 16 ft long, is subject to an eccentric load at the top ($e = 7$ in.) of 10 k causing uniform strong-axis bending and a side load of 0.9 k at midheight, which causes weak-axis bending. Use $C_D = 1$. The member has lateral support at the ends only. Is the section satisfactory?

$F_b = 1350$ psi; $F_c = 925$ psi; $E = 1{,}600{,}000$ psi.
Axial compression:

Larger $\ell_e/d = 16(12)/7.5 = 25.6$

$F_{cE} = 0.3(1{,}600{,}000)/25.6^2 = 732 \text{ psi} = F_{cEy}$

$F_c^* = 925$ psi

$C_p = 0.605$

$F_c' = (0.605)(925) = 560$ psi

$f_c = 10{,}000/7.5 \times 11.5 = 116$ psi

Strong-axis bending:

$\ell_u/d = 16(12)/11.5 = 16.7$

$\ell_e = 1.84 l_u = 353$ in.

$R_B^2 = 353(11.5)/7.5^2 = 72.2$

$F_{bE} = 0.438(1{,}600{,}000)/72.2 = 9706$ psi

$C_F = 1.0$

$F_b^* = 1350$ psi

$C_L = 0.992$

$F'_{bx} = 0.992(1350) = 1339$ psi

$f_{bex} = Pe/S_x = 10{,}000(7)/165.3 = 423$ psi

Weak-axis bending:

$F'_{by} = 1350$ psi ($C_L = 1.0$)

(Had this been a 2 × or 4 × member, an adjustment for flatwise use would have been applied.)

$M_y = P\ell/4 = (900)(16)/4 = 3600$ ft-lb

$f_{by} = M_y/S_y = 3600(12)/107.8 = 401$ psi

Compute F_{cE}:

For strong axis, $F_{cEx} = 0.3(1{,}600{,}000)/(16 \times 12/11.5)^2 = 1722$ psi

For weak axis, $F_{cEy} = 732$ psi (as above)

Interaction equation:
Eq. 7-11 will be used. Note that the strong-axis bending stress is due to eccentric load and therefore has larger magnification than the weak-axis bending stress.

$$\left(\frac{116}{560}\right)^2 + \frac{423(1 + 0.234(116 / 1722))}{1339(1 - (116 / 1722))}$$

$$+ \frac{401}{1350(1 - (116 / 732))}$$

$= 0.74 < 1.0$ *OK*

7-7. COLUMNS OR BEAM-COLUMNS WITH INITIAL CURVATURE

The preceding discussion of column design assumes the member is initially perfectly straight. Occasionally the designer may be called upon to investigate the capacity of a column whose initial crookedness is not minor enough to disregard. In this case, some results from structural stability theory (13, 14) are of use.

A differential equation based on equilibrium of the initially bent column can be solved for the total deformation of the loaded column. Assuming that the initial shape of a pinned-end column is a half sine wave with midheight deflection of *a*, the total mid-height deflection after load application is

$$\Delta_T = \frac{a}{1 - P / P_E} \qquad (7\text{-}12)$$

In this equation, P is the applied load and $P_E = P_{\text{Euler}} = \pi^2 EI/\ell^2$. Inspection of the denominator of Eq. 7-12 shows that as P approaches P_E (the Euler load), deformation becomes infinitely large. A perfect column will remain straight up to the Euler load; an initially deformed column begins to deform further as soon as any load is applied. The deformation is relatively insignificant at low loads, but increases rapidly as the applied load increases.

If the only load on the crooked column is an axial load, then the magnified bending moment due to initial crookedness is $P\Delta_T$. The designer can then use the interaction equation, Eq. 7-9, with no magnification factor.

For analysis of the initially curved column with eccentrically applied end loads, the designer can use superposition. The eccentrically applied end loads can be replaced by the equivalent set of loads consisting of axial load P and end moments *Pe*. The maximum deflection at midheight of the beam-column is then

$$\Delta_{max} = \frac{a}{1 - P / P_E} + e \sec \frac{\pi}{2}\sqrt{\frac{P}{P_E}} - e \qquad (7\text{-}13)$$

where the first term is due to initial crookedness and the second and third to eccentricity of load. The maximum moment at midheight of the beam-column is

$$M_{max} = P(e + \Delta_{max})$$

$$= P\left(\frac{a}{1 - P / P_E} + e \sec \frac{\pi}{2}\sqrt{\frac{P}{P_E}}\right) \qquad (7\text{-}14)$$

Similar expressions for bending moment can be developed for other load types on initially crooked columns.

Creep in Columns

For columns where dead load is a major portion of the total load, creep may be of concern to the designer. However, a study (15) shows that crookedness, rather than creep, is the primary cause of secondary bending stress in columns. Column curvature and material (modulus of elasticity) variability were found to be much more significant than creep. The study concluded that creep could safely be ignored if the interaction equation (rather than simple column equation) were used in design with an estimate (say ½ in.) of expected initial crookedness.

7-8. TENSION MEMBERS

Wood is required to resist tensile stresses when it is used in trusses or in the tension flange of a plywood-lumber beam. Other tension members are the chords of horizontal and vertical diaphragms (Chapter 12). In the design of a tension member, the adjusted allowable tensile stress, F_t', must not be exceeded. The tensile stress in the member is calculated using the *net area.* This is the cross-sectional area remaining after bolt holes or projected connector holes are subtracted from the gross cross-sectional area of the member. Example 5-15 shows a joint for a member stressed in tension. This example points out that both the capacity of the wood in tension and the capacity of the bolts in shear must be checked.

Adjustments that may apply to allowable tensile stress are load duration, wet use, temperature, and size factor (the latter for dimension lumber only).

Example 7-13

Because of poor soil conditions, a different support system has been chosen for the sign of Problem 5-18 and Fig. 5-21. Lateral bracing (dotted line in Fig. 7-15a) prevents movement parallel to the sign. The sign is continuously bolted to the upper part of the posts. Normal to the sign, diagonal braces will be used as shown in Fig. 7-15b. Design both the vertical and diagonal supports using D. fir-larch of No. 1 grade for any dimension lumber members and Dense No. 1 grade for larger members. The vertical posts are attached to concrete foundations.

Recall that the sign width is 16 ft and the wind speed is 70 mph. For Exposure C the combined height, exposure, and gust factor coefficient, C_e, is 1.06 for heights of 0 to 15 ft, and increases linearly to 1.10 at the 18-ft level (highest point of the sign). For simplicity, conservatively use C_e = 1.10 for the entire depth of the sign. From Eq. 4-4,

$$p = C_e C_q q_s I = (1.10)\,(1.4)\,(12.6)\,(1.0) = 19.4 \text{ psf}$$

F_1 = wind force resisted by one post = 19.4(10) (16/2) = 1550 lb

Sign wt = 1400 lb

Estimated post wt = (24 lb/ft)(18 ft) = 430 lb

F_2 = 1400/2 + 430 = 1130 lb

Using a free body of the vertical post of Fig. 7-15b,

$$\Sigma M_B = 0 = -F_1\,(13) + 7F_{AC}/\sqrt{2}$$

$$F_{AC} = 4070 \text{ lb (tension)}$$

$$F_{AB} = F_2 + F_{AC}/\sqrt{2} = 1130 + 2880 = 4010 \text{ lb}$$

The shear and bending moment diagrams for the post are shown in Figs. 7-15c and 7-15d.

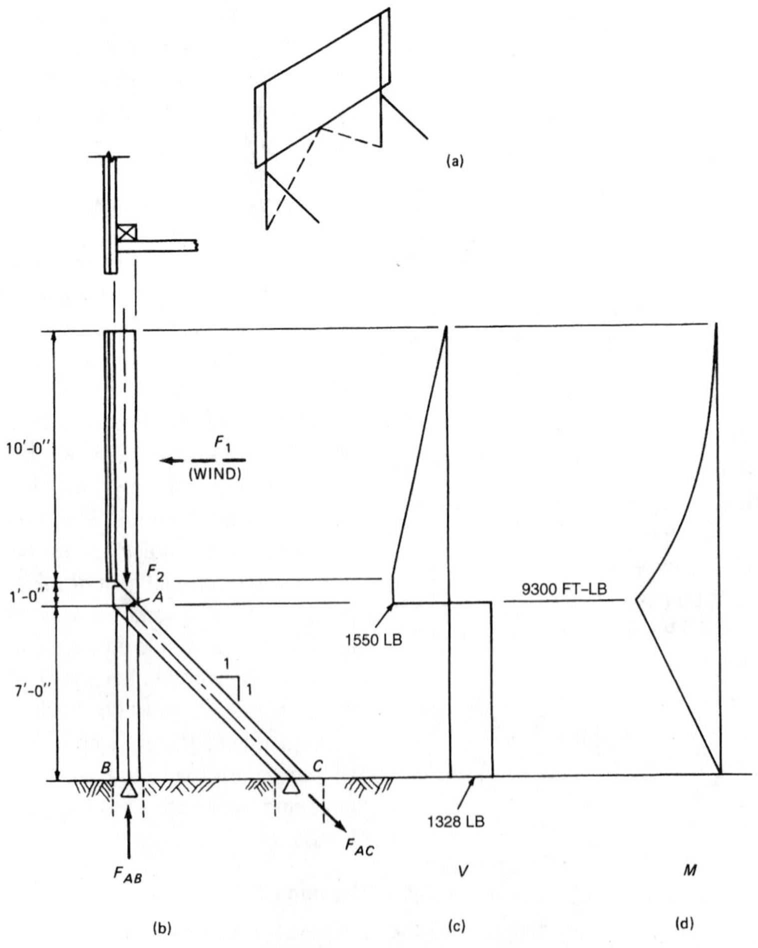

Fig. 7-15. Sign support for Example 7-13.

Design of diagonals AC:

$F_{AC} = 4070$ lb tension

For a 2 × 4, $F'_t = C_F C_D F_t = 1.5\,(1.6)\,(675) = 1620$ psi

Net area req. = 4070/1620 = 2.51 in.^{2}

Assuming 2 rows of 3/4-in.-diam. bolts (hole diam. = 0.81 in.),

Net area provided = (1.5) [(3.5) − 2(0.81)] = 2.82 in.^{2} > 2.51 *OK*

Tentatively use 2 × 4 diagonal

Check diagonal for wind reversal:

$F_{AC} = 4070$ lb comp.

$\ell_e/d = 7\sqrt{2}\,(12)/1.5 = 79.2 > 50$ *NG*

Try 3 × 8:

$\ell_e/d = 7\sqrt{2}\,(12)/2.5 = 47.5$

$E = 1{,}700{,}000$ psi; $F_c = 1450$ psi; $C_F = 1.05$; $C_D = 1.6$

$F_{cE} = 0.3\,(1{,}700{,}000)/(47.5)^2 = 226$ psi

$F^*_c = (1.05)\,(1.6)\,(1450) = 2436$ psi

$C_p = 0.091$

$F'_c = 0.091\,(2436) = 222$ psi

$f_c = 4070/(2.5 \times 7.25) = 225$ psi

Since $f_c > F'_c$, we should increase the size of the member; however, it is so close that many engineers would say OK.

Design posts (beam-columns):

Try 6 × 8 (P&T). First check the bottom 7 ft of post and then the top 11 ft. (Buckling of the lower portion is about the weak axis. For the top part, the post can buckle only about its strong axis.) For the bottom 7 ft, assume that the ends are free to rotate, but not translate, so the coefficient of effective length = 1.0. Lower portion, as a column,

$K_e\ell/d = (1.0)\,(7)\,(12)/5.5 = 15.3$

$E = 1{,}700{,}000$ psi; $F_c = 1200$ psi

$F_{cE} = 0.3\,(1{,}700{,}000)/(15.3)^2 = 2180$ psi

$F^*_c = (1.6)\,(1200) = 1920$ psi

$C_p = 0.733$

$F'_c = 0.733\,(1920) = 1407$ psi

$f_c = 4010/(5.5 \times 7.5) = 97.2$ psi

Lower portion, as a beam,

$\ell_u/d = 7\,(12)/7.5 = 11.2$

$\ell_e = 1.63\,(7 \times 12) + 3\,(7.5) = 159$ in.

$R_B^2 = (159)\,(7.5)/(5.5)^2 = 39.4$

$F_{bE} = 0.438\,(1{,}700{,}000)/39.4 = 18{,}900$ psi

$F^*_b = (1.6)\,(1400) = 2240$ psi

$C_L = 0.993$

$F'_b = 0.993\,(2240) = 2224$ psi

$M = (5 + 1)F_1 = 6(1550) = 9300$ ft-lb

$f_b = 9{,}300\,(12)/51.6 = 2163$ psi

$F_{cEx} = 0.3\,(1{,}700{,}000)/(7 \times 12/7.5)^2 = 4066$ psi

Substituting in interaction Eq. 7-9,

$$\left(\frac{97.2}{1407}\right)^2 + \frac{2163}{2224(1 - 97.2 / 4066)} = 1.0 \le 1.0 \qquad \textit{Bottom part is OK}$$

Check top:

Coeff. of effective length = 2.1 (Fig. 7-2)

$K_e\ell/d = 2.1\,(11 \times 12)/7.5 = 37.0$

$F_{cE} = 0.3\,(1{,}700{,}000)/(37.0)^2 = 373$

$F^*_c = (1.6)\,(1200) = 1920$ psi

$C_p = 0.186$

$F'_c = 0.186\,(1920) = 357$ psi

$P = 1400/2 + (24\text{ lb/ft})\,(11\text{ ft}) = 964$ lb

(To assume P as an end load is conservative. Actually it is distributed along the column length.)

$f_c = 964/(5.5 \times 7.5) = 23.4$ psi

$F'_b = 2224$ psi (same as for bottom portion)

$f_b = 2163$ psi (same as for bottom portion)

(Weak-direction buckling is not possible; the sign is continuously bolted to the upper part of the columns.)

The interaction equation gives

$$\left(\frac{23.4}{357}\right)^2 + \frac{2163}{2224(1 - 23.4 / 4066)} = 0.98 < 1.0 \qquad \textit{Upper part is OK}$$

Use 6 × 8 for post (larger dimension normal to plane of sign).

7-9. COMBINED TENSION AND BENDING

Occasionally wood members are stressed in combined axial tension and bending. Combined tension and bending might occur, for example, when the bottom chord of a roof truss supports the ceiling below directly. For combined tension and bending, the NDS has the following two requirements, the second of which is a lateral-stability criterion:

$$\frac{f_t}{F_t'} + \frac{f_b}{F_b^*} \leq 1.0 \tag{7-15}$$

$$\frac{f_b - f_t}{F_b^{**}} \leq 1.0 \tag{7-16}$$

In the second equation, F_b^{**} is the tabular bending stress multiplied by all applicable adjustment factors except C_v (which would apply only to glued laminated timber).

In these interaction equations for combined tension and bending, net section properties should be used where the combined tension and bending moment occur at a section reduced by bolt holes or connector grooves. If the combined tension and bending occur where these holes or grooves are absent, use gross section properties. In any case, also check for tension alone and bending alone, using the appropriate section properties (net or gross) where the maximum values of tension or bending moment occur.

7-10. COLUMN DESIGN BY LRFD

The steps in load and resistance factor design of columns are similar (but not identical) to those of the allowable stress method. An adjustment factor for column stability, C_p, is used, although it has different form than in ASD. For a trial section, the designer first finds the Euler load

$$P_e = \pi^2 E'_{05} I/(K\ell)^2 \tag{7-17}$$

Using this equation, both P_{ex} and P_{ey} need to be found (unless it is obvious which is more critical), with the smaller controlling the design. The modulus of elasticity referred to in the equation is the adjusted fifth percentile value (i.e., the value that 95% of the pieces exceed). E'_{05} is the modulus of elasticity to be used for strength or stability (buckling) calculations. A different modulus (E'), which accounts for shear deflection, is used for deflection calculations. Reference 16 gives the following equation showing the relationship between E'_{05} and E'.

$$E'_{05} = 1.03E' \, (1 - (1.645)(\mathrm{COV}_E)) \tag{7-18}$$

For visually graded lumber, $\mathrm{COV}_E = 0.25$. For glulams, $\mathrm{COV}_E = 0.11$ and the 1.03 in Eq. 7-18 is changed to 1.05.

Next in the design process, the designer computes the resisting load, P'_0 of a short (zero-length) column as follows:

$$P'_0 = AF^*_c \tag{7-19}$$

where F^*_c is the reference compression strength multiplied by all applicable adjustments. In the next design step, the ratio $\propto_c$ is computed, where

$$\propto_c = \phi_s P_e/(\lambda \phi_c P'_0) \tag{7-20}$$

The resistance factor for stability is $\phi_s = 0.85$ and for compression is $\phi_c = 0.90$. Using the result of Eq. 7-20, the column stability factor, C_p is found from the following equation:

$$C_P = \frac{1+\propto_c}{2c} - \sqrt{\left(\frac{1+\propto_c}{2c}\right)^2 - \frac{\propto_c}{c}} \tag{7-21}$$

In Eq. 7-21, the constant c has the same values as in the ASD equation for column stability adjustment (Eq. 7-5). Finally, the trial section is adequate if the factored load is less than the design resistance, where design resistance is the product of adjusted resistance P', resistance factor, ϕ_c, and the time effect factor, λ. In equation form,

$$P_u < \lambda \phi_c P' = \lambda \phi_c C_p P'_0 \tag{7-22}$$

Combined Bending and Compression for LRFD

Just as in ASD, an interaction equation is used for combined axial compression and bending (either uniaxial or biaxial). That interaction equation follows:

$$\left(\frac{P_u}{\lambda\phi_c P'}\right)^2 + \frac{M_{mx}}{\lambda\phi_b M_x'} + \frac{M_{my}}{\lambda\phi_b M_y'} \leq 1.0 \tag{7-23}$$

Terms in the equation that have not been defined elsewhere (or that need special explanation) are:

P'—adjusted resistance for axial compression acting alone (without moments) for the axis with the larger Kl/r ratio (i.e., the axis with the larger $(Kl)^2/I$ ratio). The column stability factor should be included as one of the adjustments.

M_x', M_y'—adjusted flexural resistances for strong and weak axes, respectively from the procedure in Section 6-8 with $C_b = 1.0$.

M_{mx}—magnified applied factored moment about the strong axis. This moment must include the P-Δ effect that was discussed in Section 7-6. The simplest way to do this (without using a computer program) is to compute the magnified moment by $M_{mx} = B_{bx}M_{bx} + B_{sx}M_{sx}$, where M_{bx} is the factored first-order moment from loads that result in no appreciable sidesway, and M_{sx} is the factored first-order moment from loads that do cause sidesway.

M_{my}—magnified applied factored moment about the weak axis ($M_{my} = B_{by}M_{by} + B_{sy}M_{sy}$)

Finding the multiplier factors B_{bx}, B_{sx}, and so on, requires information too lengthy to present here. The definitions of these factors can be found in the *Standard for LRFD for Engineered Wood Construction* (16). The procedure for designing a beam-column is illustrated in Example 10-4 of the chapter on trusses.

Example 7-14

A lower-story column, 18 ft long, receives load from the roof as well as the floor above. Loads are: 60 k snow, 45 k floor live load, and 20 k dead load. The column will be No. 1 D. fir (P&T). By the LRFD method, choose the best square column, assuming pinned ends.

Load combinations (See Table 4-1):

$$1.4D = 1.4(20) = 28 \text{ k}$$

$$1.2D + 1.6L + 0.5S = 1.2(20) + 1.6(45) + 0.5(60) = 126 \text{ k}$$

$$1.2D + 1.6S + 0.5L = 142.5 \text{ k} \qquad \textit{Controls}$$

The last two combinations have identical time effect factors, so the third combination controls, and time effect factor (Table B-10) $\lambda = 0.8$ The reference strength for compression is $F_c = 2.40$ ksi and $E_{05} = 970$ ksi by Eq. 7-18. Other than column stability factor, C_p, no other adjustments apply. Estimate that $C_p = 0.65$.

$$P_u = \lambda\phi_c P'$$

$$142.5 = 0.8(0.90)(2.40 \times A \times 0.65)$$

$$\text{Approx req. } A = 127 \text{ in.}^2$$

Try a 12 × 12 (A = 132 in.²)

$$K\ell = 12(18) = 216 \text{ in.}$$

$$P_e = \pi^2 E_{05}' I/(K\ell)^2$$

$$= \pi^2(970)(11.5^4/12)/(216)^2 = 299 \text{ kips}$$

The resisting strength of a zero-length column is

$$P_0' = AF_c^* = (11.5 \times 11.5)(2.40) = 317.4 \text{ kips}$$

$$\alpha_c = \phi_s P_e/(\lambda\phi_c P_0')$$

$$= 0.85(299.0)/(0.8 \times 0.90 \times 317.4) = 1.11$$

$$C_p = \frac{1+\alpha_c}{1.6} - \sqrt{\left(\frac{1+\alpha_c}{1.6}\right)^2 - \frac{\alpha_c}{0.8}}$$

$$= 0.726$$

$$P' = C_p P_0' = 0.726\,(317.4) = 230 \text{ kips}$$

$$\text{Design Resistance} = \lambda\phi_c P' = 0.8(0.90)(230) = 166 \text{ k} > \text{Required strength of } 142.5 \text{ k} \qquad \textit{OK}$$

Example 7-15

Solve Example 7-4 again, this time using the LRFD method. Assume the reference resistance $F_c = 2.4$ ksi. (This is $2.16/\phi_c = 2.16/0.90$ times as great as the allowable stress given in Ex. 7-4.) Recall that this problem involves the design of a 17-ft-long round column.

Load combinations are

$$1.4D = 1.4(15) = 21 \text{ k}$$

$$1.2D + 1.6L = 1.2(15) + 1.6(50) = 98 \text{ k}$$

When divided by their respective time effect factors (0.6 and 0.8) from Table 4-1, the latter gives the larger quotient. So the latter load case controls and $\lambda = 0.8$.

For visually graded lumber, the fifth-percentile value of elastic modulus (by Eq. 7-18) is

$$E'_{05} = 1.03E' \,(1 - (1.645)\,(0.25))$$

For our problem, $E' = 1600$ ksi, so $E'_{05} = 970$ ksi
Try 11-in. diameter cross section.

$$K\ell = 12(17) = 204 \text{ in.}$$

$$I = \pi D^4/64 = \pi(11)^4/64 = 719 \text{ in.}^4$$

$$P_e = \pi^2 E'_{05} I/(K\ell)^2$$

$$= \pi^2\,(970)(719)/(204)^2 = 165 \text{ kips}$$

The resisting strength of a zero-length column is

$$P'_0 = AF^*_c = (\pi(11^2)/4)\,(2.4) = 228 \text{ kips}$$

$$\alpha_c = \phi_s P_e/(\lambda\phi_c P'_0)$$

$$= 0.85\,(165)\,/\,(0.8 \times 0.90 \times 228) = 0.854$$

By Eq. 7-21 (with $c = 0.85$ for a pole),

$$C_p = 0.661$$

$$P' = C_p P'_0 = (0.661)\,(228) = 151 \text{ kips}$$

$$\text{Design resistance} = \lambda\phi_c P' = 0.8(0.9)\,(151)$$

$= 109$ k $>$ factored load of 98 k *OK*

Example 7-16

Solve Example 7-9 again, this time using the LRFD method. Assume that percentages of dead load and snow load are 20% and 80% respectively. Recall that this example involves a beam-column serving as the top chord of a roof truss. Also recall that the total service (unfactored) axial force is 4.3 kips compression and total vertical service uniform load is 0.125 kips per horizontal foot. Reference strength F_c = (1.3 ksi)(2.16/0.90) = 3.12 ksi. Reference strength F_b = (0.875 ksi)(2.16/0.85) = 2.22 ksi.

The load case of $1.2D + 1.6S$ will be examined first.

Axial Compression:

The unfactored force consists of $P_D = 0.20(4.3) = 0.86$ k and $P_s = 0.80(4.3) = 3.44$ k.

Total factored compression force

$$= P_u = 1.2(0.86) + 1.6(3.44) = 6.54 \text{ k}$$

Substituting $E' = 1600$ ksi (from Ex. 7-9) into Eq. 7-18, the fifth-percentile value of E is $E'_{05} = 970$ ksi. Weak-axis buckling is prevented by the decking, so strong axis moment of inertia will be used for buckling computations.

$$I_x = (1/12)(1.5)(5.5)^3 = 20.8 \text{ in.}^4$$

$$K\ell = 12(5.58) = 67.0 \text{ in.}$$

$$P_{ex} = \pi^2 E'_{05} I_x/(K\ell)^2$$

$$= \pi^2(970)(20.8)/(67.0)^2 = 44.4 \text{ kips}$$

The resistance of a zero-length column is

$$P'_0 = AF^*_c = (1.5 \times 5.5)(C_F \times F_c)$$

$$= (1.5 \times 5.5)(1.1 \times 3.12) = 28.3 \text{ kips}$$

$$\alpha_c = \phi_s P_e/(\lambda\phi_c P'_0)$$

$$= 0.85(44.4)/(0.8 \times 0.90 \times 28.3) = 1.85$$

By Eq. 7-21, $C_p = 0.854$

Adjusted compression resistance is

$$P' = C_p P'_0 = (0.854)\,(28.3) = 24.2 \text{ kips}$$

Bending:

The uniform dead load (unfactored) $= w_D = 0.20(0.125) = 0.025$ k/ft and $w_S = 0.80(0.125) = 0.100$ k/ft. Total factored uniform load

$$= w_u = 1.2(0.025) + 1.6(0.100) = 0.190 \text{ k/ft}$$

$$M_{ux} = 0.1 w_u \ell^2 = 0.1(0.190)\,(5.29)^2 = 0.532 \text{ ft-k}$$

Next the magnified moment (magnified for second-order effects) must be found. Multiplier B_{bx} is

$$B_{bx} = C_{mx}/(1 - P_u/(\phi_c P_{ex}))$$

$$= 1.0/(1 - 6.54/(0.90 \times 44.4))$$

$$= 1.20$$

(In the above equation, the value of C_{mx} is 1.0, because the member is subjected to transverse load and has pinned ends.)

The factored moment, including magnification is

$$M_{mx} = B_{bx} M_{bx} = B_{bx} M_{ux}$$

$$= (1.20)\,(0.532) = 0.638 \text{ ft-k} = 8.86 \text{ in-k}$$

Since decking provides lateral support to the truss chord, $C_L = 1.0$ and adjusted bending resistance is

$$M'_x = S_x F'_{bx} = (7.56)\,(C_F \times F_{bx})$$

$$= (7.56)\,(1.3 \times 2.22) = 21.8 \text{ in-k}$$

Interaction Equation for combined axial compression and bending:

$$\left(\frac{P_u}{\lambda\phi_c P'}\right)^2 + \frac{M_{mx}}{\lambda\phi_b M_x'}$$

$$\left(\frac{6.54}{(0.8)(0.90)(24.2)}\right)^2 + \frac{8.86}{(0.8)(0.85)(21.8)}$$

$$= 0.74 < 1.0 \qquad OK$$

(If analyzed for the load case of dead load alone, the member would again be found to be satisfactory.)

REFERENCES

1. Desch, H. E., *Timber: Its Structure and Properties,* 5th ed., St. Martins Press, New York, 1973.
2. Dewell, H., *Timber Framing,* Dewey Publishing Co., San Francisco, 1917.
3. *National Design Specification for Wood Construction,* rev. 1991 Ed., American Forest & Paper Association, Washington, DC, 1993.
4. Newlin, J. A., and J. M. Gahagan, "Tests of Large Timber Columns and Presentation of the Forest Products Laboratory Column Formula," U.S. Dept. Agr. Tech. Bulletin 167, 1930.
5. Neubauer, L. W., "Full-Size Stud Tests Confirm Superior Strength of Square-End Wood Columns," *Transactions of the American Society of Agricultural Engineers,* 15(2), 1972, pp. 346-349.
6. *Building Code Requirements for Structural Concrete* ACI-318, American Concrete Institute, Farmington Hills, MI, 1995.
7. Council for Codes of Practice, British Standards Institution, *British Standard Code of Practice: The Structural Use of Timber,* 1971.
8. U.S. Forest Products Laboratory, *Wood Handbook,* U.S. Dept. of Agriculture, Washington, DC, 1940.
9. U.S. Forest Products Laboratory, *Wood Handbook,* U.S. Dept. of Agriculture, Washington, DC, 1974.
10. Malhotra, S. K., and D. B. Van Dyer, "Rational Approach to the Design of Built-up Timber Columns," *Wood Science,* 9(4), 1977, pp. 174-186.
11. Neubauer, L. W., and V. K. Arora, "Proposed Strength Formulas for Wood Box Columns," *Forest Products Journal,* 24(1) 1974, pp. 31–35.
12. Gurfinkel, G. R., *Wood Engineering,* 2nd ed., Kendall/Hunt Publishing Co., Dubuque, IA, 1981.
13. Chajes, Alexander, *Principles of Structural Stability Theory,* Prentice-Hall, Englewood Cliffs, NJ, 1974.
14. Timoshenko, S. P., and J. M. Gere, *Theory of Elastic Stability,* McGraw-Hill, NY, 1961.
15. Itani, R. Y., M. C. Griffith, and R. J. Hoyle, "The Effect of Creep on Long Wood Column Design and Performance," *Journal of Structural Engineering, ASCE,* 112(5), May 1986, pp. 1097–1114.
16. *Standard for LRFD for Engineered Wood Construction,* American Society of Civil Engineers, New York, 1996.

PROBLEMS

In the following problems, assume moisture content in service to be less than 19% unless stated or implied otherwise. When solving these problems, remember that the lumber classification Beams and Stringers (B&S) refers to lumber of thickness 5-in. nominal (or more) with width more than 2 in. greater than thickness and that Posts and Timbers (P&T) refers to square or nearly square sections (nominal width not more than 2 in. greater than thickness). Live loads are normal (not over ten-year) duration unless noted.

7-1. A 6 × 8 hem-fir column, Select Structural grade (P&T) is 12 ft long. It carries a dead load of 20 k. Use is under wet conditions and $K_e = 1.0$. What snow load can it carry in addition to the 20 k dead load?

7-2. An 8 × 10 column, 13 ft long, of No. 2 D. fir-larch (P&T) is laterally braced at midheight in the weak direction. Assuming pinned ends, what axial wind load can it carry in addition to a dead load of 16 k? Assume pinned ends.

7-3. Find the allowable axial load for a 4 × 8 column, No. 2 D. fir-larch, overall length 18 ft, with weak-way bracing (i.e., one-way only) at midheight. Ends are pinned. Controlling load combination is $D + L + W$. Use is wet.

7-4. A 10 × 12 column, 14 ft long, has one end fixed and one end free. It is made of Select Structural D. fir (P&T). What is the largest service $D + L + S$ load that it can carry?

7-5. A nominal 8-in.-wide (least cross-sectional dimension) column, 9 ft long, is to be No. 1 D. fir-larch. The coefficient of effective length, $K_e = 1.0$. If the load is 14 k dead load and 43 k live load, determine the column size. Make your first trial using P&T, then if a larger section is required, use B&S.

7-6. A No. 1 hem-fir nominal 10-in.-wide (least cross-sectional dimension) column is 8 ft 4 in. long. The coefficient of effective length, $K_e = 1.0$. The load on the column is 25 k dead and 57 k live load. Moisture content in use is expected to be 20%. Determine the column size. Use P&T for your first trial.

7-7. Same as Problem 7-6, except K_e is determined as for a column that is fixed at one end and free to both translate and rotate at the other end.

7-8. Design a square column (pinned ends) of No. 1 D. fir-larch, (P&T) 9 ft long. The loads are $D = 25$ k and $L = 56$ k. Column will be in a location of extreme humidity.

7-9. Design a square column (pinned ends) 23 ft long of No. 1 D. fir-larch (P&T) to carry a load of $D = 11$ k and $L = 30$ k.

7-10. Same as Problem 7-9 except moisture content in service will be greater than 19%.

7-11. An 8 × 10 column of No. 1 D. fir-larch (P&T) carries $D = 10$ k and $L = 48$ k. If the column rests on a 3-in. by 7.5-in. steel plate at its bottom end, is it satisfactory for end-grain bearing? If not, what size plate is required?

7-12. If the column of Problem 7-11 is 9 ft long, assuming pinned ends, is it satisfactory for buckling?

7-13. The column of Problem 7-11 will have a hole for a 7/8-in. bolt (7.5 in. long) near the upper end to connect it to a column cap. At the section where the hole is located, is the stress level acceptable? (The actual compressive stress on the net area must not exceed F_c^*.)

7-14. The end of a 6 × 6 southern pine column carrying axial load of 9 k D plus 32 k L rests on a U-shaped steel bracket (post base). What bearing area must the post base provide to the bottom end of the column?

7-15. A 22-ft-long column (pinned ends) has axial load of $D = 13$ k plus $S = 34.5$ k. It is laterally braced in all directions at both ends and the midheight. Choose the most economical square column using No. 1 D. fir-larch (WWPA).

7-16. A 21-ft-long column (pinned ends) has least cross-sectional dimension of 6 in. nominal. The member must carry dead load of 19 k plus occupancy live load of 24 k. The column is braced at the midheight in the weak direction. Choose the column of least area using No. 1 D. fir-larch (WWPA). (*Caution:* Allowable stresses for P&T and for B&S are different.)

7-17. The column in Problem 7-3 has 2.85 in.^{2} of wood removed to install a split ring. Is the column satisfactory for compressive stress on the net area?

7-18. A 10-ft-high stud wall carries an axial load due to $D + S$. Studs are 2 × 4 No. 3 D. fir-larch at 16 in. c/c. They are supported about the weak axis by sheathing nailed to the narrow face. Find the allowable load per foot of wall.

7-19. An 8-in.-diameter round column is to be 12 ft long. The column has $E = 1{,}600{,}000$ psi and $F_c = 950$ psi. Assume pinned ends and a load duration of seven days. Using the equivalent square method, find the allowable concentric load.

7-20. Same as the previous problem, except use the alternate method.

7-21. A round column (pinned ends) is to be designed for a ski lodge. The wood to be used has $E = 1{,}300{,}000$ psi and $F_c = 950$ psi. The column will be 20 ft long. What diameter (to the nearest ½ in.) is necessary to support a load of $S = 61$ k and $D = 20$ k? Use the equivalent square method.

7-22. Same as the previous problem, except use the alternate method.

7-23. Design a round column to carry a load of $D = 30$ k and $W = 95$ k. Unsupported length is 12 ft and $K_e = 1.0$. Neglect any taper. $E = 1{,}600{,}000$ psi and $F_c = 1100$ psi. Use either of the two methods.

7-24. A round, tapered column (pinned ends) 15 ft long has a diameter at the bottom of 10 in., tapering to 7 in. at the top. Conditions of use will be wet. $E = 1{,}400{,}000$ psi and $F_c = 1000$ psi. What service load can it carry if the duration of load is 2 months? $F_g = 1220$ psi.

7-25. Find the allowable ten-year axial load for an 8-ft-long, hinged-end tapered column. ($E = 1{,}600{,}000$ psi and $F_c = 1000$ psi.) The thickness is a constant 5.5 in., but the width is 11.5 in. at the bottom, tapering to 5.5 in. at the top.

7-26. A square, tapered column with pinned ends is 12 ft long. The width of the column at the bottom is 14 in., tapering to 10 in. at the top (actual dimensions). The wood to be used has $E = 1{,}400{,}000$, $F_c = 950$ psi, and $F_g = 1380$ psi. Find the safe load the column can carry if the load combination includes snow. Make sure to check the column in parallel-to-grain bearing at the top where the load is transferred into the small end of the column.

7-27. A simply supported column tapers toward both ends. In one plane, the column is a constant thickness of 7.5 in. In the other plane, the midheight of the column measures 13.5 in., tapering to 9.5 in. at each end. Wood properties are $E = 1{,}600{,}000$ psi, $F_c = 1000$ psi, and $F_g = 1480$ psi. What ten-year-duration service load can the column carry?

7-28. A column has a rectangular cross section everywhere, but it is tapered so that it has a 4-in. by 6-in. cross section at the small end and 6-in. by 8-in. cross section at the large end (actual dimensions). Moisture content in use is expected to be 20%. For $\ell_e = 10$ ft, $E = 1{,}500{,}000$, and $F_c = 1100$ psi, determine the maximum permissible axial $S + D$ load.

7-29. A spaced column consisting of two 3 × 8s is to be 8 ft 6 in. long. Using No. 2 hem-fir and connector condition *b*, compute the allowable service load, *P*, assuming that snow is the controlling load.

7-30. For the column in the previous problem, choose split-ring connectors to carry the load you computed. How far from the end should the connectors be placed? Check to determine if net area is sufficient.

7-31. Solve for the capacity of a spaced column consisting of two D. fir-larch 4 × 8s (No. 1) under load of two months cumulative duration. Use is under wet conditions. The column length is 12 ft and the centers of the connectors are 12 in. from each end. (Assume that the connectors are strong enough, whatever the load you solve.)

7-32. Design the top chord of a truss (length = 6 ft 9 in.) as a spaced column. Compression load is 12 k dead plus 28.6 k live, and $K_e = 1.0$. Use 3-in. nominal material of No. 2 D. fir. Assume connector condition *a*. Choose split-ring connectors. Assume net area is sufficient.

7-33. Design a spaced column of overall length 10 ft 2 in. ($K_e = 1.0$) to carry 19 k *D* plus 31 k *L*. Assume condition *b*. Use No. 1 & Btr hem-fir dimension lumber. Try to keep the ratio of width to thickness of the individual members about 3 to 1. Choose split-ring connectors. Check net area.

7-34. A built-up column, 9 ft long, like Fig. 7-11a, is composed of six 2 × 6s of No. 1 D. fir-larch. Assuming it is properly connected, what ten-year duration load can it safely support? What load could a 9-ft-long 6 × 6 solid column of No. 1 D. fir-larch support?

7-35. A built-up column, like Fig. 7-11d, consisting of three 2 × 6s is nailed together with 30*d* common nails. The column is 8 ft 5 in. long and is No. 1 hem-fir. Using NDS requirements, what $D + S$ load can it support? If your instructor requests, describe the required nailing pattern.

7-36. Design a built-up column, 8 ft long, like Fig. 7-11d, of No. 2 hem-fir. Use NDS requirements. The column will carry snow load of 3.3 k and dead load of 1.2 k. Use either 2 × 4s or 2 × 6s, if possible. Assume nailing is sufficient to develop built-up column strength.

7-37. A No. 1 hem-fir (WWPA) beam-column is 8 ft 2 in. long. The 40-kip, permanent-duration end load is eccentric by 6 in. Find the most economical square (P&T) section.

7-38. A nominal 10 × 12 No. 1 (WWPA) hem-fir column is 8 ft 4 in. long. A 3-in. eccentricity at each end of the 33-kip end load causes bending about the strong axis. There is also a central side load of 1.5 k applied to the smaller face. For ten-year duration of load, is this section with pinned ends satisfactory? Draw the bending moment diagram as part of your solution.

7-39. Repeat the previous problem if the 33-kip load is eccentric at the top end only. Draw the bending moment diagram as part of your solution.

7-40. A 17-ft-long column supports a vertical load of $D = 4.5$ k plus $S = 13.5$ k as well as a uniform horizontal wind load of 175 lb/ft. Chose a square column (P&T) of No. 1 D. fir-larch.

7-41. A No. 1, 8 × 12 (B&S) D. fir-larch column has an effective length of 12 ft. A 4-in. eccentricity at each end of the 20-kip $D + L$ load causes bending about the strong axis. What uniform wind load that also causes strong-axis bending can be superimposed?

7-42. Repeat the previous problem, except the wind load causes weak-axis bending.

7-43. Design a column (pinned ends) 14 ft long to carry a concentric $D + L$ load of 42 k and a moment due to wind of 8.2 ft-k. Try 8-in.-nominal (smallest cross-sectional dimension) B&S of Dense Select Structural D. fir-larch. If this proves inadequate, try 10-in. nominal.

7-44. Design a member 10 ft long to carry a concentric compressive force of 19.2 k and a transverse force of 1.4 k at midheight. Assume ten-year load duration. Wood is Select Structural Douglas fir-larch.

7-45. A 10 × 10 No. 1 (WCLIB) hem-fir beam-column supports an axial load *P* (dead plus live). It also supports a uniform horizontal wind load of 200 lb/ft. If the column is 20 ft long, what is the allowable *P*?

7-46. Prove that the following beam-column is (or is not) satisfactory: An 8 × 10 No. 1 (WWPA) hem-fir member, 16 ft long, braced against weak-axis buckling by attached plywood siding. The loads are uniform horizontal wind load of 150 lb/ft and a concentric axial load $(D + L)$ of 20 k.

7-47. A door header of 14-ft span is to be designed using 2-in.-nominal members of No. 2 D. fir-larch. It is expected that two or even three members side by side will be necessary. Bending loads are $D = 64$ lb/ft and $L = 120$ lb/ft. In addition, there is a wind load of 7 k axial tension. Choose the required members.

7-48. Find the allowable axial tension load (due to wind) in a 2 × 8 member if there are two rows of 3/4-in.-diameter bolts at the net section. Lumber is No. 2 southern pine. Assume holes 1/32 in. larger than the bolt.

7-49. A 2-in.-nominal tension member (No. 3 Douglas fir-larch) carries a wind load of 5200 lb. Using $C_D = 1.6$, choose the proper size member if a single row of 1-in.-diameter bolts are used. Bolt holes will be 1/16 in. larger than the bolt diameter.

7-50. A 10 × 10 No. 1 D. fir-larch column has an effective length of 28 ft. Assuming that we wish to design for a minimum eccentricity of 0.5 in., what is the allowable load for the column?

(a) Use Eq. 7-11, assuming an end load that is eccentric by 0.5 in.

(b) Use $(f_c/F'_c)^2 + f_b/F'_b \leq 1.0$, calculating f_b using the total deflection from Eq. 7-12.

(c) What is the allowable load if $e = 0.0$?

7-51. Same as Problem 7-50, except $\ell_e = 17$ ft.

7-52. A 10 × 10 No. 1 D. fir-larch column has no initial crookedness, but has end loads that are eccentric by 8 in. in a direction causing strong-axis bending only. Effective length is 25 ft.

(a) Using a computer, graph magnification factor $[1 + 0.234\,(f_c/F_{cE})]/[1 - (f_c/F_{cE})]$ vs. P. (This is the magnification factor from Eq. 7-11.)

(b) To the same scale, graph

$$\sec\left(\frac{\pi}{2}\sqrt{\frac{P}{P_E}}\right) \text{ vs. } P.$$

8

Glued Laminated Members

Glued laminated timbers were used as early as the late 1800s. During and since World War II, they have been used extensively. Casein adhesives were the most frequently used until the early 1970s, when they were generally replaced by synthetic resins suitable either for exterior or interior use. Glued laminated members (glulams) have seen increasing use in both buildings and bridges. Structures that would not have been feasible using only sawn-timber members have proved practical and successful using glulams.

8-1. GLULAMS

Structural glued laminated timber is defined by the *National Design Specification* (1) as "an engineered, stress-rated product of a timber laminating plant, comprising assemblies of suitably selected and prepared wood laminations bonded together with adhesives." Obviously, glulams for reliable structural use are neither do-it-yourself items nor items that can be manufactured at the job site.

Requirements for the manufacture of glulams are defined by the national consensus standard ANSI/AITC A190.1–1991 (2). Two associations, the American Institute of Timber Construction and Engineered Wood Systems, have the responsibility for quality assurance and technical research. They help control quality of the finished product by inspecting and approving laminating plants and operations.

Most glulams are made by gluing together the wide faces of 2-in. nominal lumber, as shown by Fig. 8-1. One-inch nominal is also used but mostly for arches. In bending, glulams are most effective when manufactured so that the wide face of the laminations is parallel to the neutral axis, that is, with the applied loads normal to the wide face of the laminations, as in Fig. 8-1a. They may also be used with the loads parallel to the wide face of the laminations (Fig. 8-1b), although the strength, stiffness, and economy advantages of glued laminated timber may be lost when that is done. Glulams may also be used as tension members or as compression members. Laminations for members to be used primarily for axial loading are selected by criteria different from those for members to be used primarily for bending.

The inner laminations of wide members may consist of side-by-side pieces, ordinarily not glued along their edges. When the edges are not glued, the longitudinal joints of one lamination must be staggered relative to those of adjacent laminations, as shown by Fig. 8-1c, and the transverse distance between such joints in adjacent layers must be at least equal to the lamination thickness. If the member will be subject to torsion, however, the edge (longitudinal) joints must be glued, since torsion will cause longitudinal shear stresses along those joints.

The structural designer must know enough about glulams to be able to select grade, size, shape, and perhaps the species of the assembled glulam member, but ordinarily is not required to participate in selecting the lamination grades or the makeup of the glulam cross section. That part of the design is done by the laminator, who will produce members meeting the geometric and strength requirements specified by the structural designer. The laminator's service may also include design and manufacture of the hardware (connection material) needed to install the members.

Members may also be made using very thin laminations, much thinner than the thickness of

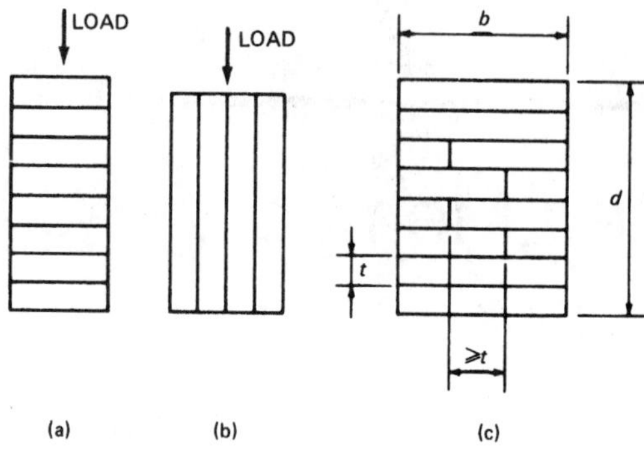

Fig. 8-1. Glulam cross sections.

a nominal 1-in. board. These are called *laminated veneer lumber* (LVL). Members of this type are available in certain proprietary products, although the NDS does not yet give specific requirements for their design. They are not considered to be glued laminated timber.

8-2. ADVANTAGES OF GLULAMS OVER SAWN TIMBERS

Glulams were originally developed to produce curved members such as large arches. Later they were used to solve the problems faced in obtaining sawn timbers of both large size and good quality. Yet as those problems were solved, other advantages were realized. Today we recognize not only that glulams may be superior to sawn timbers, but also that they provide for better utilization of our timber resources. Glulams have the following advantages over sawn timbers:

1. *Greater strength:* Lamination material is carefully selected, so that a glulam member can be of large size yet practically free from defects. More importantly, those defects that are present can be located (on the cross section) where they will do the least damage and can be dispersed along the length of the member. Thus, their effect at any given cross section is minimized and a higher modulus of rupture is obtained than would be possible in an equivalent size of sawn lumber. The highest grade of wood is used in the highest-stressed layers. In a beam, the tension laminations are of highest quality, while those near the neutral axis are of lower quality.

2. *Larger cross sections:* Glulams can be assembled with larger section moduli and moments of inertia than are feasible with sawn timbers. With the disappearance of virgin forests, large sawn timbers are becoming more and more difficult to find.

3. *Longer spans:* These are possible because of the two above factors. For example, arched roof structures with spans of up to 500 ft have been built using glulams.

4. *Freedom from warping and checking:* Large sawn timbers almost always develop severe checks. These are caused by internal stresses that are largely the result of nonuniform drying. Large sawn timbers also are prone to warp, especially in a twisting pattern. By comparison, individual laminations of a glulam are generally only 2 in. in nominal thickness so that they can be fairly uniformly dried. They tend to remain straight and to have only minor checking. In the assembled glulam member no single lamination can warp independently, so the surface of each lamination is restrained to conform in shape with those to which it is glued.

5. *Diversity of shapes:* Glulams can be manufactured to many outlines, making possible architecturally attractive structures. Tapered beams, curved beams, arches, and rigid frames are all possible. Until glulams were developed, these shapes were normally possible only in structural steel or reinforced concrete (see Fig. 8-2).

6. *Varying section properties:* By tapering its depth or by changing the grade of lamination material used, the final member can be given a strength capacity that varies along the length according to the strength required.

7. *Camber:* Under dead load alone, a beam may have objectionable deflection. With sawn timbers, there is nothing that can be done to correct this. With glulams, however, it is a simple matter during assembly to bend the laminations before gluing, so that the resulting beam will have a slight upward bend (camber) to offset the downward deflection expected when the full dead load is applied.

8. *Fire resistance:* The fire rating of large glulam members is equal to that of heavy sawn timbers, equaling or often surpassing the ratings for structures of other materials. The fire resistance of glulam members is far greater than that of conventional light-frame construction, in

Fig. 8-2. Intersecting glulam highway bridges. (Photograph by Mike Douglas.)

which most members are of 2-in. nominal thickness.

9. *Economy:* Glulam members are often more economical than similar members of structural steel or reinforced concrete. Glulams may be more costly than sawn timbers (if timbers of the required size are available at all), but the glulams' other advantages often outweigh a slight increase in first cost. For a particular structure, of course, the answer must be obtained by comparative designs and is influenced by factors such as cost of transportation, erection cost, and the need for fireproofing.

As compared to sawn timbers, glulams do present some problems, but these are largely problems of manufacture only and do not often affect the structural designer. One such problem is that of residual stresses. If the swelling and shrinkage characteristics of adjacent laminations differ, then with changes of moisture content that may occur during use, the glulam can develop internal residual stress. Similarly, the presence of abnormalities such as cross grain and compression wood will cause unequal behavior by adjacent laminations and give rise to residual stress. Solving this problem is a task of the laminator's engineers, but not of the structural designer who selects the size and grade of glulam for a particular application.

Example 8-1

Compare the sizes of a glulam and a sawn-timber beam to carry a normal-duration load of 800 lb/ft on a simple, laterally supported span of 25 ft. Disregard self-weight and consider only flexural strength.

$$\text{Maximum } M = 800\ (25^2/8) = 62{,}500 \text{ ft-lb}$$

For the glulam, assume visually graded western species, combination 22F-V3. (The meaning of this terminology will be explained later.) For this combination, the tabulated unadjusted allowable bending stress is 2200 psi, as the combination name implies. Among the adjustments that must be considered is the volume factor (C_v), which is similar to the size factor, C_F, for a sawn-timber beam.

Ignoring, for now, the volume factor,

$$S \text{ req.} = 12\ (62{,}500)/2200 = 340.9 \text{ in.}^3$$

From Appendix Table B-2, a glulam 5.125 in. wide by 21 in. deep has $S = 377$ in.^{3} However, for this member size and length, the volume factor (see Section 8-6) is

$$C_v = K_L\ (1292/Lbd)^{1/x}$$

in which $x = 10$ (specified for western species), L is 25 ft, b and d are width and depth of the cross section (inches), and K_L is specified as 1.0 for uniformly distributed load.

$$C_v = 1.0\ (1292/(25 \times 5.125 \times 21)^{1/10} = 0.929$$

When designing a glulam beam, the designer uses whichever is smaller—C_v or the beam stability factor C_L. In this example, C_L is 1.0, so the adjusted allowable bending stress is

$$F'_b = 2200\ (0.929) = 2044 \text{ psi.}$$

The actual bending stress for the 5.125 in. by 21 in. glulam is

$$12\ (62{,}500)/377 = 1989 \text{ psi} < 2044 \text{ psi allowed.}$$

OK

For the sawn timber, assume D. fir-larch, B&S, Select Structural grade, which has a tabulated (unadjusted) allowable flexural stress of 1600 psi. Ignoring the size factor, the required S is 12 (62,500)/1600 = 468.8 in.^{3} To provide this would take a 6 × 24 or an 8 × 20 timber. Try the 6 × 24, since it has the smaller area. For this member, the depth is 23.5 in. and the *size* factor is 0.928. The adjusted allowable stress is 0.928 (1600) = 1485 psi. The required S is now 505 in.^{3} (vs. 506 provided). The 6 × 24 is still OK.

In this example, the next-to-top grades of each were assumed. The comparison could be modified, of course, by choosing other grades. That the glulam should be the more costly is suggested by the amount of wood consumed. The glulam is made by assembling 14 pieces of 2 × 6, which are later reduced to 5$^1/_8$ in. wide. The 2 × 6s measure less than their name implies, but wood measuring a full 2 in. by 6 in. was used to manufacture them. Assuming the beam's overall length to be 26 ft, the total amount of wood required initially to produce the glulam is

$$14\,(2 \times 6/12)\,(26) = 364 \text{ fbm}$$

The sawn 6 × 24, on the other hand, used only 312 fbm.

The glulam used more wood, and its cost is further affected by the labor involved in its production. So, selection of the glulam over the sawn timber must be based on factors other than cost. For example, appearance, availability, or freedom from warping could justify a higher initial cost.

8-3. GLULAM PRODUCTION

Glulams are made using various species of wood, including the following:

Softwoods	
Douglas fir-larch	Southern pine
Douglas fir south	Western Woods
Hem-fir	(Canadian softwood sp.)

Hardwoods	
Ash, white	Oak, red
Beech	Oak, white
Birch	Poplar, yellow
Cottonwood, eastern	Sweetgum
Elm	Tupelo
Hickory	

Of the species listed, Douglas fir and southern pine are probably the most commonly used. There is no reason why species other than those listed above cannot be used. All that is needed is to find suitable species and to develop appropriate standards for production. Other species are being experimented with and we may expect to see the above list expanded in the future. One reason that the list will expand is that good material for tension laminations is becoming increasingly difficult to find.

Researchers are seeking ways to increase the strength and stiffness of glulams by improving the material used for the lamination on the tension side of the member. One method under study is that of building up a tension lamination by gluing thin, clear (defect-free) strips of wood along each edge of a wider piece of wood (3). Thus far, it has been discovered that the edge strips cause the resulting lamination to have significantly higher strength than a single piece of wood of the same total width. Apparently the reason is that edge defects weaken a piece of lumber more than interior defects. Clear edge strips prevent having defects near the edges.

Another way of enhancing the strength and stiffness is by using tension laminations of high-strength, fiber-reinforced plastic. Developed at Oregon State University, this method has been used successfully on an Oregon Bridge (4).

Adhesives

Adhesives for glulams must be able to withstand horizontal (longitudinal) shearing stresses due to load, in combination with stresses from causes such as shrinkage, swelling, or bending to produce a curved member. To be effective the glue must be stronger than the wood is in shear and in tension perpendicular to the grain. Performance requirements for glulam adhesives are spelled out by reference 2.

The AITC specifications (5) permit either dry-use adhesives or wet-use adhesives. Casein adhesives are for dry use only. They should not be used where the moisture content of the member will exceed 16%, either repeatedly or for long periods. Although permitted by the codes, casein glues are not often used.

Phenol, resorcinal, or melamine base adhesives are specified for wet use, but may be used for dry situations also. They should *always* be used where the glulam member will have over

16% moisture in service. This would include all cases of exposed (outdoor) members. It would also include members exposed to a moist environment: swimming pools, laundries, and manufacturing buildings where the processes cause high humidity.

Grades and Combinations

Wood for glulam laminations is graded under special rules. Grading may be either visual or by machine. Visual grading is based on the occurrence and spacing of defects such as knots and cross grain. Most glulams today are made from visually graded lumber. The trend will probably be toward greater use of machine-graded laminations, in which the strength is assumed to be a function of measured modulus of elasticity, E. Machine-graded lumber, of course, is also graded visually for edge defects.

The first and perhaps most critical step in glulam manufacture is to select the proper grades of laminations. For a bending member, for example, this involves finding a lamination of suitable tensile strength to serve as the outermost lamination on the tension side of the member. Normally this would be the bottom lamination. Wood just above the bottom layer is not so highly stressed. Thus the second layer need not be of such high quality as the bottom one. Layers near the neutral axis need only provide sufficient shear strength, the flexural stresses there being minimal. Laminations near the top surface, of course, must be strong enough to resist the high flexural compressive stresses.

The laminations are stacked according to the above reasoning, following criteria spelled out by the AITC specifications. Where moment reversals occur, as in continuous beams, it may be desirable to use those combinations that have the same allowable tensile flexural stress on both the top and bottom surfaces of the beam. One example of such a combination is western species 24F-V8.

Figure 8-3 shows the cross section of a typical glulam beam and indicates the grades of laminations used. Such groupings of laminations by strength are specified by the AITC specifications. These groupings are called combinations and are designated by terms such as *Combination 22F-V4*. (Appendix Table B-7 shows some of the combinations available for members subject primarily to bending.) In this system of nomenclature, the numerals preceding the first letter indicate the allowable flexural stress—the *22,* for example, meaning 2200 psi. The letter V indicates visual grading of the individual lamination; a letter E is used to show machine grading. The *4* of the combination number indicates the fourth combination in that series that has the 2200 psi bending strength.

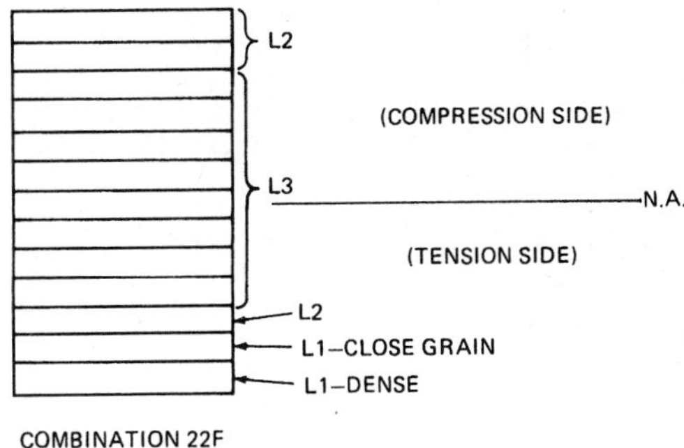

Fig. 8-3. Typical glulam makeup.

Combinations for members subject to primarily axial load are shown by Appendix Table B-8. Glulam columns are designed exactly as solid columns, except that F_c and E are from Table B-8.

When specifying a particular combination listed by the tables, the designer should verify that such material is actually available. Older publications listed allowable bending stresses as high as 3000 psi, but the present NDS gives a limit of 2400 psi for 24F—the strongest combination. If the designer specifies the allowable stresses required, rather than specifying a particular combination, the member desired is often easier to obtain.

Use of Mixed Species

In the interest of better utilization of the stronger woods, glulams may be made using stronger species for the outer laminations and weaker species for the interior ones. The weaker woods will have lower E-values than the stronger ones. Thus, when the laminator's engineers select wood species and grades to use in a glulam section, the effect of elastic properties, as well as strength, must be considered. This can be done

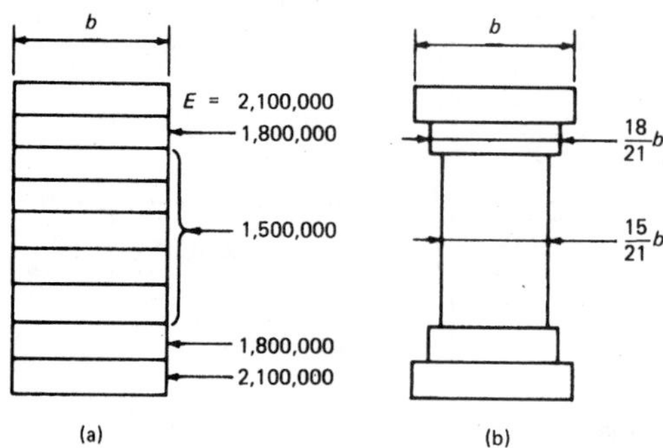

Fig. 8-4. Transformed section.

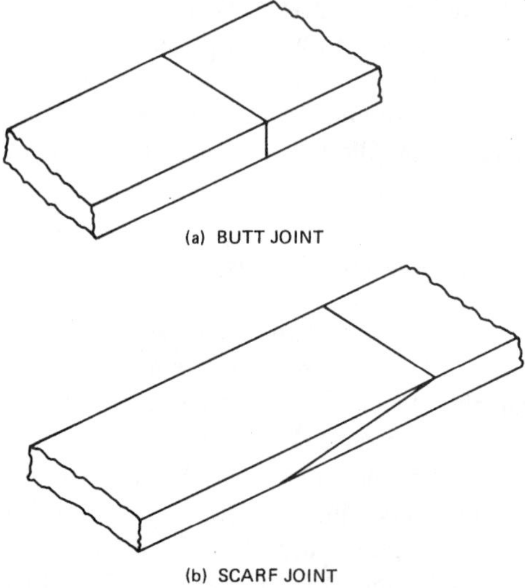

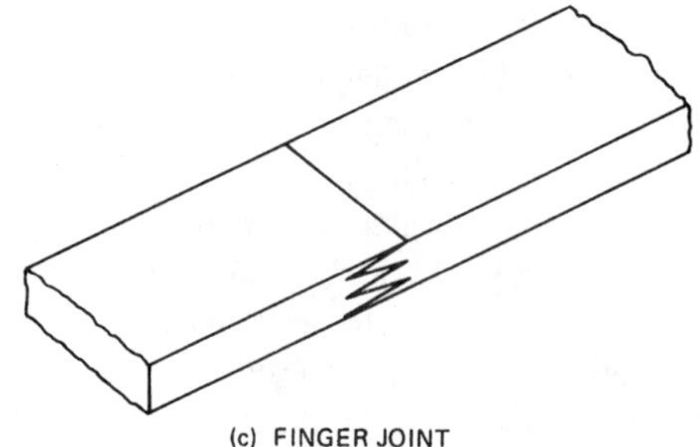

Fig. 8-5. Types of end joint.

by the transformed-section method—a method familiar to designers of reinforced concrete.

Figure 8-4a shows a glulam section made up of woods of different values of modulus of elasticity, E. In Fig. 8-4b, laminations having lower E-values are replaced by lesser areas of wood having the same E as the outer laminations. The section shown by Fig. 8-4b is the transformed section. Stresses in the outer laminations due to bending may be computed using the moment of inertia value of the transformed section in the flexure formula, $f_b = Mc/I_t$. To determine stresses in weaker laminations, the answer given by My/I_t for the transformed section must be multiplied by the ratio of E(inner)/E(outer).

This type of computation is made only by those who design the makeup of the glulam section to conform to a particular combination designation. It is not done by the designer who selects a glulam section for a particular application.

End Joints

Laminations may be joined as needed to obtain sufficient length or to change the grade of lamination along the length of the member. Three types of end joint are used, as shown by Fig. 8-5.

Butt Joint. The butt joint is cheap and easy to make, but it is not a good structural joint. It has no value whatever for transmitting tensile stress. It can transfer up to 80% of the allowable compression for the lamination, but if the butt joint were used for compression laminations of a glulam beam, there would be appreciable deformation before the joint becomes effective. The joint could be made more effective by careful fitting of a steel bearing plate between the abutting ends. This would be a costly procedure, however, and rarely justified. Consequently, butt joints are not permitted in structural glued laminated members.

Scarf Joint. The scarf joint of Fig. 8-5b can come close to being a full-strength splice. The efficiency of scarf joints varies with the slope of the joint. For tension scarf joints in bending members, the following efficiencies may be used: slope of 1 in 12 or flatter, 90%; slope of 1 in 10, 85%.

Current standards from which glued laminated members are produced require qualification of the end joints to meet a strength criterion.

Finger Joint. The most commonly used joint today is the finger joint, shown by Fig. 8-5c. The finger joint can carry an appreciable percentage of the tensile capacity of the spliced lamination. The value of the joint factor depends on the exact geometry of the joint and on the manufacturing process. The long sloping edges of the "fingers" act like miniature scarf joints, but the ends of the fingers cannot be perfectly sharp. Finger joints may approach (but do not equal) the joint efficiency of well-made scarf joints.

Despite the fact that they are not quite as efficient as scarf joints, finger joints waste less lumber and are easier to produce. The finger joint is made by passing the pieces to be spliced though a machine whose rapidly rotating cutters form the ends to the overlapping finger pattern (see Fig. 8-6). The members are coated with glue and pressed together and the glue curing is accelerated. All of this—cutting, gluing, fitting together, and initial curing—can be done as the pieces move through the machine without stopping. Proof-testing of the finger-jointed splice may also be done as the lamination travels through the machine, and this has become common practice (6).

If scarf or finger joints are used, the strength-reducing effects of the knots are most often the controlling factors in the strength of a lamination.

Fig. 8-6. Finger joint. Used to splice laminations for glulams. (Courtesy National Casein Co.)

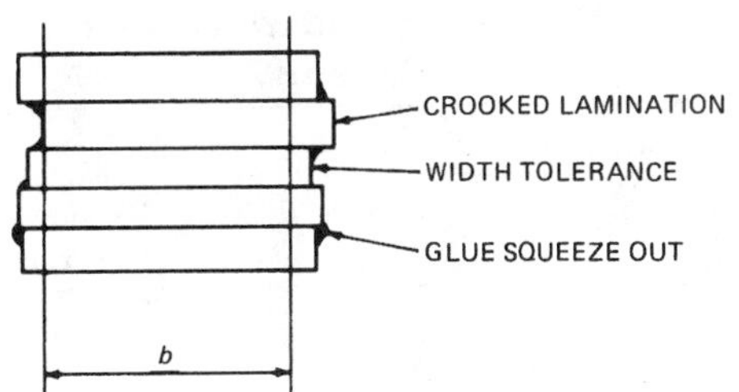

Fig. 8-7. Glulam irregularities (before planing).

Assembly and Gluing

After the laminations have been prepared, they are glued, assembled in forms to provide the proper curvature or camber, and clamped together until the glue is sufficiently cured to allow handling of the member. After gluing and clamping, the edges of the member are irregular, as shown in Fig. 8-7. The irregularities are caused by glue squeeze-out, width tolerances, and lateral crookedness of the laminations. To remove the irregularities, the members are smoothed by planing each side surface. The finished glulam will have smooth, planed sides, and its top and bottom surfaces will be the original smooth, wide faces of the boards used.

8-4. STANDARD SIZES OF GLULAM

Glulam laminations are made from standard widths of sawn lumber. Wider widths of lamination are harder to force into alignment with each other; thus greater irregularity is expected after gluing wide glulams than narrower ones. Consequently, a greater amount must be removed (by planing or by sanding) from the sides of wider glulams to obtain the desired straight, smooth side surfaces.

Table 8-1 shows the standard widths for finished glulams made in the United States. Note that the widths for southern pine glulams differ slightly from those for western species. Also, glulams manufactured prior to about 1970 may be 1/8 inch wider than those of today.

Glulam depths today are multiples of the lamination thickness, that is, multiples of either 1.5 inches or 3/4 inches (multiples of 1.375 inches for southern pine).

A glulam is specified by its *actual* finished di-

Table 8-1. Standard Widths of Glulams.

Standard Actual Widths		
Western Species	Southern Pine	Nominal Width of Laminations
2-1/2	—	3 in.*
3-1/8	3	4
5-1/8	5	6
6-3/4	6-3/4	8
8-3/4	8-1/2	10
10-3/4	10-1/2	12
12-1/4	—	14
14-1/4	—	16

* Made by splitting a sawn 2 × 6.

mensions. Although a 2 × 6 piece of S4S lumber does not measure 2 in. by 6 in., a 6-3/4 × 15 glulam *does* measure what its name implies, 6 3/4 in. by 15 in.

Different degrees of side smoothness may be obtained by specifying an appearance grade. Standard appearance grades are Industrial, Architectural, and Premium Architectural. Textured side surfaces may also be specified, and these may reduce the width somewhat, giving slightly smaller area, section modulus, and moment of inertia. The manufacturer of the glulam should be consulted to determine how much width will be removed to obtain the textured surface.

8-5. LIMITS OF CURVATURE

Glulams are produced without softening the wood laminations to permit easy bending. To avoid damaging the laminations as they are bent, the curvature must be limited. Thicker laminations cannot be bent as sharply as thinner ones. Curvature must also be limited so that high residual stresses are not present in the finished member. Residual stresses may also result from drying. Whatever their cause, residual stresses reduce the bending capacity of the member, since the residual stresses and those due to bending moment from the beam's loads are additive.

AITC (7) *recommends* that the minimum radius of curvature, measured to the inside of the curve, be limited as follows: For 2-in. nominal lamination thickness (1.5-in. actual) of all species, not less than 27 ft 6 in. For 1-in. nominal (3/4-in. actual), not less than 9 ft 4 in. for southern pine, or less than 7 ft 0 in. for western species.

However, NDS *requires* that in no case shall the ratio of lamination thickness to inside-face radius of curvature, t/R_i, exceed 1/100 for hardwoods or southern pine, or 1/125 for other softwoods.

8-6. ALLOWABLE STRESSES AND ADJUSTMENTS

Appendix Table B-7 gives allowable stresses for glulam members stressed primarily in bending. The table shows values for both visually graded and E-rated material, and for both western species and southern pine. The lumping of various species under a single name is justified by the fact that the laminations themselves are stress graded; thus the strength of the assembled glulam timber depends solely on the *combination* of lamination grades used and on the compatibility of mixed species for gluing.

The design values for glulams shown by Table B-7 were developed following procedures now given by ASTM Standard D-3737-91 (8). These procedures involve adjusting the strength of clear (defect-free) lumber to assess the weakening effects of features such as knots, cross grain, member size, and number of laminations. The same design values are accepted by the NDS.

To use Table B-7, one must first find the portion of the table that covers the species and type of grading to be used (visually graded western species, for example). The combinations that may be available are listed under this classification. As explained earlier, the numeral at the beginning of the combination, when multiplied by 100, shows the allowable tensile flexural stress. However, the allowable is to be used only at locations where the tensile flexural stress is on the same side of the member as the tension lamination. An example of such designation would be *20F-V4,* in which the letter *V* indicates visually graded lamination material. For this combination the table shows an allowable flexural stress of 2000 psi, *provided the tensile bending stress is on the same face as the tension lamination.* If the beam is continuous, so that somewhere along its length flexural tensile stresses are on the opposite side (top rather than bottom, for example), the allowable tensile flexural stress at that location is only 1000 psi. Reading to the right in the table, various other allowable stresses are shown.

The table also gives allowable stresses for bending about the *YY*-axis. Notice that the allowables for bending about this axis are lower than those for the *XX*-axis. The reason is that, for bending about the *YY*-axis, even the lower-strength inner laminations are at the extreme fiber, where they are subject to the maximum flexural strains. For bending in this direction, most of the strength advantage of the glulam is lost.

Appendix Table B-7 gives UBC's values for allowable stress for shear. However, in 1995 the NDS raised all F_{vxx} values of 165 psi to 190 psi.

The right-hand portion of Appendix Table B-7 gives allowable stresses due to axial loads. These allowables should be used for members that were graded primarily for bending but also have some applied axial load. For members whose loads are *primarily* axial, rather than bending, glulams graded for that type of use (Table B-8) should be specified.

Reference 9 shows values of allowable stress for various hardwoods. Before specifying hardwood glulams, the designer should make certain that there are hardwood laminating plants in the locality.

Adjustments

As for sawn timber members, the tabulated design values are unadjusted allowable stresses. The adjustment factors that apply to glulams are:

C_D	Load duration factor
C_M	Wet service factor
C_t	Temperature factor
C_V	Volume factor (instead of size factor)
C_L	Beam stability factor
C_{fu}	Flat use factor
C_c	Curvature factor
C_p	Column stability factor
C_b	Bearing area factor

Two of these, C_V and C_c, are for glulams only. Except for the flat use factor and wet use factor, all others have the same values as for sawn timber members.

Volume Factor. For glulam design, a *volume* factor is used rather than the size factor, as for sawn timbers. The volume factor applies to the allowable *bending* stress only, and is given by NDS as

$$C_V = K_L\,(21/L)^{1/x}\,(12/d)^{1/x}\,(5.125/b)^{1/x}, \text{ but not } > 1.0$$

Obviously, this reduces to

$$C_V = K_L\,(1292/Lbd)^{1/x} \qquad (8\text{-}1)$$

In either of the above two equations, L is the length (feet) between points of zero bending moment, d is the member depth (inches), b is the finished width (inches) of the widest piece used to make the glulam, x is 21 for southern pine and 10 for all other species, and K_L is a loading-condition coefficient as shown below.

Values of K_L for beams of single span are as follows:

With one concentrated midspan load	1.09
With uniformly distributed load	1.0
With two equal 1/3-point concentrated loads	0.96

For continuous and cantilever beams, use K_L = 1.0 regardless of loading. The volume and stability factors are *not* applicable in combination. Use whichever of these two factors is the smaller.

Fire Retardant. Treatment of wood with fire retardant reduces its strength by 10–25%. The effect varies with species and treatment method, according to whether the laminations are treated before gluing or the assembled glulam is treated after gluing. The NDS does not spell out the percentage by which the allowable stresses are to be reduced. Rather, the designer should consult the manufacturer doing the fire retardant treatment and drying to obtain a recommended design value reduction.

Temperature. As explained in Chapter 2, the strength of wood is affected by abnormal temperatures. For combined high temperature and high MC, the effect is the greatest. The NDS adjustment factors applying to glulams include C_t, temperature factor, as given in Chapter 2. Chapter 4 of the *Wood Handbook* (10) and the *NDS Commentary* (11) give further helpful information.

Slenderness. Lateral buckling of glulam beams may be considered using Eq. 6–4, as shown for sawn-timber members in Chapter 6. However, since glulams have smaller coefficients of variation for modulus of elasticity, term K_{bE} in that equation may be taken as 0.609, rather than the smaller value given for sawn timbers.

When computing C_L for glulam beams, use E_{yy}, not E_{xx}. (For deflection calculations, use E_{xx}.) Also, use only the smaller of C_L and C_V, not both.

Column buckling is considered using Eqs. 7–4 and 7–5 with term K_{cE} taken as 0.418 and c taken as 0.9.

Curvature Factor. In curved beams, flexural stresses across the depth are not linear. Also, in curved glulams, residual stresses are caused when the laminations are bent prior to their assembly. When bending moment is applied to the completed curved glulam, flexural stresses due to that bending moment are added algebraically to the already-present residual stresses. To account for both of these, the tabulated allowable bending stress is multiplied by

$$C_c = 1 - 2000\,(t/R_i)^2 \qquad (8\text{-}2)$$

in which t is the actual thickness of a curved lamination and R_i is its inside-face radius.

Wet Service Factor. When glulam members are used with their moisture content 16% or greater, each tabulated allowable is multiplied by the wet service factor, C_M. For example, this factor is 0.8 for bending.

Flat Use Factor. When glulams have a width of 12 in. or less measured parallel to the wide face of the laminations, F_{byy}, the allowable bending stress for bending about the *YY*-axis, is multiplied by this factor.

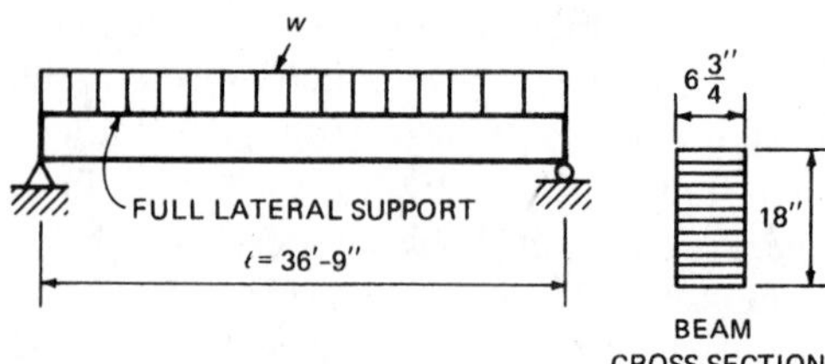

Fig. 8-8. Beam for Example 8-2.

Example 8-2

Determine the adjusted allowable bending and shear stresses for the glulam beam shown by Fig. 8-8. The member is of combination 22F-V4, western species, and the loading is mainly snow load. The member is to be used in a paper mill, where the atmospheric humidity is very high.

Appendix Table B-7 gives 2200 psi and 190 psi for unadjusted allowable stresses in bending and horizontal shear, respectively. These allowables must be adjusted, using all adjustment factors that are applicable. For load duration, C_D is 1.15. The volume factor is

$$C_V = 1.0\,[1292/(36.75 \times 6.75 \times 18)]^{1/10} = 0.883$$

Wet service factors, C_M, are 0.8 for bending and 0.875 for shear.

The adjusted allowable stresses are

$$F'_b = 2200 \times 1.15 \times 0.883 \times 0.8 = 1787 \text{ psi},$$

and

$$F'_V = 190 \times 1.15 \times 0.875 = 191 \text{ psi}$$

Example 8-3

What is the adjusted allowable bending stress for a western species glulam, combination 22F-V4, size 6.75 in. wide × 30 in. deep, used on a 30-ft span, having lateral support at its ends only, and carrying only a central, concentrated, normal-duration load plus the beam weight? The beam will have wet use, defined for glulams as having moisture content exceeding 16%.

For bending, the wet service factor $C_M = 0.80$ and for modulus of elasticity, $C_M = 0.833$. Since the load is of normal duration, $C_D = 1.0$.

C_V and C_L must *each* be determined, although only the smaller of the two will be used. The volume factor is

$$C_V = 1.09[1292/(30 \times 6.75 \times 30)]^{0.1} = 0.934$$

Terms needed to compute the beam stability factor, C_L, are:

$$F_b^* = 2200 \times 0.8 = 1760 \text{ psi}$$

$$K_{bE} = 0.609 \text{ for glulams}$$

Effective unbraced length $\ell_e = 1.37\,\ell_u + 3d$
$= 583.2$ in.

Use $E_{yy} = 1{,}600{,}000$ psi to compute F_{bE}.

$$R_B^2 = \ell_e\,(d)/b^2 = 583.2 \times 30/6.75^2 = 384$$

$$E' = 0.833 \times 1{,}600{,}000 = 1{,}333{,}000 \text{ psi}$$

$$F_{bE} = 0.609 \times 1{,}333{,}000/384 = 2114 \text{ psi}$$

To simplify Eq. 6–4, term $F_{bE}/F_b^* = 1.20$

Substituting in Eq. 6–4, beam-stability factor $C_L = 0.879$.

The stability factor is less than the volume factor, so the stability factor will be used instead of the volume factor. The adjusted allowable bending stress is

$$F_b' = 2200 \times 0.8 \times 0.879 = 1547 \text{ psi}$$

8-7. SUGGESTED DESIGN PROCEDURE

In most respects the design procedure for an untapered glulam member is the same as for sawn-timber members. The points of difference are that (1) a volume factor (rather than a size factor) must be used, (2) a curvature factor may have to be applied, and (3) deflection more often controls than it does for sawn-timber beams. A suggested design procedure is as follows:

1. Assume a combination (first make sure that glulams of that quality will actually be available).
2. Assume a weight per foot for the member, to be confirmed or corrected later in the design procedure.
3. Assume a tentative value for the volume factor, C_V, and solve for a tentative required section modulus.
4. Solve for a required area of cross section to avoid overstress in shear.
5. Solve for the required moment of inertia to keep deflection within the allowable limit.
6. Select a section having the required section modulus, area, and moment of inertia.
7. Verify that all assumptions made are close enough that correcting them will not change the selection (or correct the assumptions in a further trial).
8. If bending deflection controlled the selection, consider now whether including *shear deflection* will change your decision.

If the member is tapered or curved, additional steps will be required.

Example 8-4

Select a glulam beam section for the conditions shown by Fig. 8-9. This beam is one of many supporting a building floor system. The floor live load is 200 psf. Dead load is from 3-in. nominal decking, plus about 2 psf for floor finish. (The decking will give top lateral support to the glulam beams.) The building will be used for long-term storage, exceeding ten-year normal duration.

Assume that western species combination 20F-V4 will be available, for which the tabulated design values are 2000 psi for bending and 190 psi for shear. The load carried by one beam will equal the weight of dead load and live load in the tributary area from the center of one spacing between beams to the center of the adjacent spacing between beams. Since the beams are spaced equally, the tributary area is equal to the beam span times the beam spacing. The load per foot for one beam is, therefore,

Beam weight (assumed): 40 lb/ft

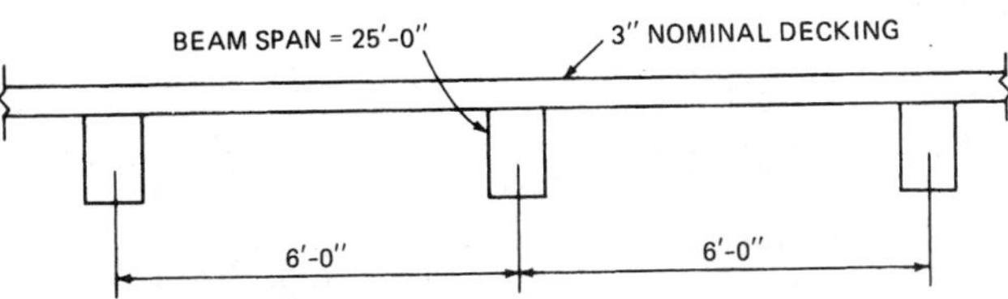

Fig. 8-9. Cross section, floor system for Example 8-4.

Floor weight:

Decking (assuming 35 pcf × 2.5/12)	7.29 psf	
Floor finish	2.0 psf	
Floor wt, total =	9.29 psf	
Multiply by 6-ft spacing		=56 lb/ft
Live load:		
6 (200)		= 1200 lb/ft
Total		= 1296 lb/ft

The bending moment on one beam is

$$M = 1.296\ (25)^2/8 = 101.2 \text{ ft-k}$$

Assume the volume factor for this beam to be 0.95. Thus, if that assumed factor were correct, the adjusted allowable bending stress for permanent loading would be

$$F'_b = 2000 \times 0.9 \times 0.95 = 1710 \text{ psi}$$

and the required section modulus would be

$$S = 12 \times 101.2/1.710 = 710.2 \text{ in.}^3$$

The maximum end reaction is 12.5 (1.296) = 16.2 kips. Ignoring (for now at least) the fact that shear, V, at the critical section is less than the reaction,

$$\text{Req area} = 1.5 \times 16.2/(0.9 \times 0.190) = 142 \text{ in.}^2$$

Assume that the allowable long-term deflection is to be limited to $\ell/240$, or 1.25 in., and that both dead and live loads are long term. E_{xx} is 1,600,000 psi. A factor of 1.5 is used to approximate the effect of creep due to long-term loading.

$$\text{Req } I = 5\ (1.296)\ (25^4)\ (12^3)/(384\ E \times 1.25/1.5) = 8543 \text{ in.}^4$$

Table B-2 (Appendix) shows that a glulam measuring 6.75 in. by 25.5 in. will satisfy all three requirements, having $A = 172.1$, $S = 732$, and $I = 9327$. This section apparently might be used. However, we must first verify that all assumptions made were satisfactory.

The actual weight for this beam (using D. fir and a specific gravity of 0.51, as for D. fir with MC = 5%) will be (0.51) (62.4) (6.75 × 25.5/144) = 38 lb/ft. The 40 lb/ft assumed is OK, and correcting it will not change the beam selection.

The actual volume factor for this beam is

$$C_V = 1.0\ [1292/(25 \times 6.75 \times 25.5)]^{0.1} = 0.887$$

This is less than assumed, so calculations must be made to determine a new required section modulus, S.

$$F'_b = 2000 \times 0.9 \times 0.887 = 1597 \text{ psi}$$

$$\text{New } S \text{ req} = 12 \times 101.2/1.597 = 760.4 \text{ in.}^3$$

The originally selected section has $S = 732$, which is not sufficient, so a larger section must be tried. The next size available, 6.75 in. by 27 in. has $S = 820$, which is satisfactory.

Example 8-5

Repeat Example 8-4, but with a span of 62 ft.

Assuming a first estimate (as for Ex. 8-4) of w = 1.296 k/ft,

$$M = 1.296\ (62)^2/8 = 622.7 \text{ ft-k}$$

$$V < R = 1.296 \times 31 = 40.2 \text{ k}$$

Assuming that volume factor, C_V, is 0.85, the adjusted allowable bending stress will be 2.0 (0.9) (0.85) = 1.53 ksi. Therefore, for the first trial solution,

$$S \text{ req} = 12 \times 622.7/1.53 = 4884 \text{ in.}^3$$

$$A \text{ req} = 1.5 \times 40.2/(0.9 \times 0.190) = 353 \text{ in.}^2$$

The moment of inertia required to limit deflection to $\ell/240 = 3.10$ in. is

$$I \text{ req} = 5\ (1.296)\ (62)^4\ (12)^3/(384 \times 1600 \times 3.10/1.5) = 130{,}300 \text{ in.}^4$$

As a first trial section, consider 10.75 × 54, the smallest 10.75-in.-wide section meeting all three of the above requirements. Next, check the assumptions made above, recompute, and either verify that the first choice is satisfactory, or make another selection.

$$\text{Weight} = 32\ (10.75 \times 54)/144 = 129 \text{ lb/ft (vs. 40 assumed)}$$

$$C_V = 1.0\ [1292/(62 \times 10.75 \times 54)]^{0.1} = 0.717 \text{ (vs. 0.85 assumed)}$$

Corrected values are:

$$w = 1.385 \text{ k/ft}$$

$$M = 665.5 \text{ ft-k}$$

$$F'_b = 2.0\ (0.9)\ (0.717) = 1.291 \text{ ksi}$$

For the second trial,

$$S \text{ req} = 12 \times 665.5/1.291 = 6186 \text{ in.}^3$$

$$A \text{ req} = 353\ (1.385/1.296) = 377 \text{ in.}^2$$

$$I \text{ req} = 130{,}300\ (1.385/1.296) = 139{,}200 \text{ in.}^4$$

The smallest 10.75-in.-wide section meeting all three of the above requirements is 10.75 × 60. The weight and volume factor for that section should now be checked again to verify, in a third trial, that it will be satisfactory.

If this is done, $w = 1.399$ k/ft; $M = 672.2$ ft-k; $C_V = 0.709$, and the required properties are:

$$A \text{ req} = 353\ (1.399/1.296) = 381 \text{ in.}^2$$

$$S \text{ req} = 12\ (672.2)/(2.0 \times 0.9 \times 0.709) = 6321 \text{ in.}^3\text{, and}$$

$$I \text{ req} = 130{,}300\ (1.399/1.296) = 140{,}700 \text{ in.}^4$$

All three requirements are satisfied. If saving headroom (vertical space required) is important, the designer might investigate ways of reducing the 60-inch depth. Possibilities are (1) try a wider beam, and (2) try a smaller spacing between beams.

Example 8-6

Figure 8-10 shows a simple glulam beam with an overhang at one end. The top of the beam is laterally supported by the floor plywood. The dead load shown includes an estimate for the beam weight. Thus the moment values shown are total bending moments. The load duration is normal. Select a glulam size, using western species combination 20F-E3. This combination is not balanced; therefore, it has two different values for allowable tensile stress due to bending.

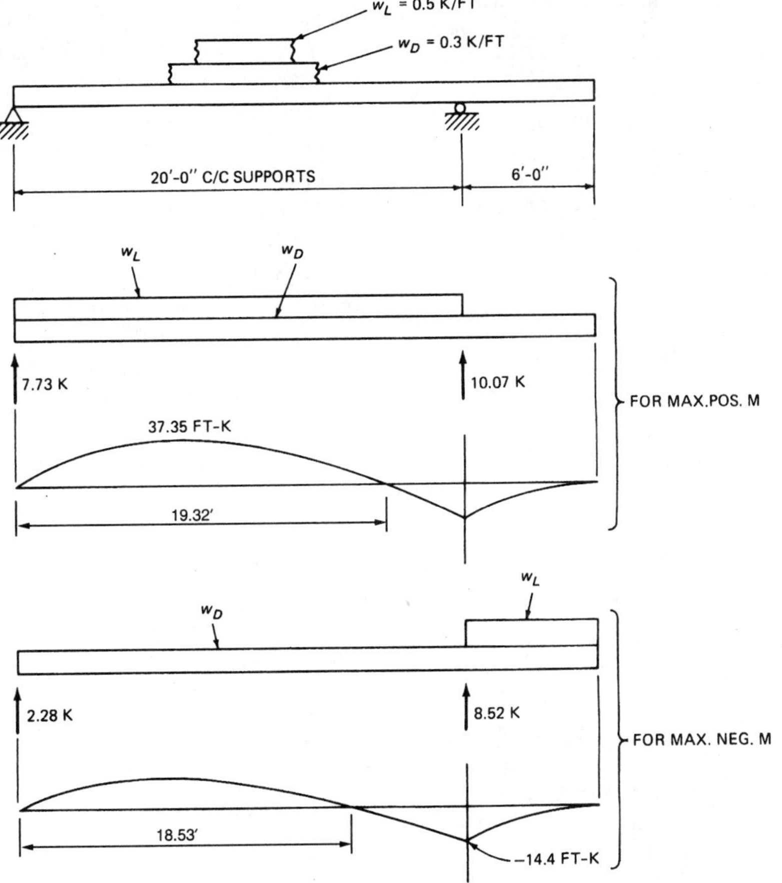

Fig. 8-10. Glulam with overhang for Example 8-6.

Allowable unit stresses shown by Appendix Table B-7 are:

Horizontal shear, 190 psi

For tension due to bending

Where the bottom fibers are in tension, 2000 psi

Where the top fibers are in tension, 1000 psi

The only adjustment factor needed is the volume factor, C_V, which is estimated to be 0.95. If this factor is correct, the required section modulus values are:

For positive bending, 12 (37.35)/(0.95 × 2.0) = 235.9 in.^{3}

For negative bending, 12 (14.4)/(0.95 × 1.0) = 181.9 in.^{3}

Try a 6.75 in. × 15 in. section. Check the value of C_V assumed. In Eq. 8-1, length L is the distance (ft) between points of zero bending moment. These distances are 19.32 ft for positive M and 15.20 ft for negative M. The smaller value calculated for C_V is 0.96. The assumed 0.95-value was close enough; changing it will not lead to a change of section.

The maximum shear (left of the right support) is 8.27 k, and the shear at beam depth away from the *face* of that support (assuming the support width to be 3.5 in.) is

$$V = 8.27 - 0.8\ (1.25 + 1.75/12) = 7.15\text{ k}$$

The computed shear stress is

$$f_v = 1.5 \times 7150/(6.75 \times 15) = 106\text{ psi}$$

which is well below the 190-psi allowable.

Deflection must be checked at two locations. As an approximation, the maximum total deflection in the 20-ft span is less than it would be for the same span without an overhang. That is, deflection is less than

$$(5/384)\ (800)\ (20^4)\ (1728)/(1{,}700{,}000 \times 1898) = 0.893\text{ in.}$$

or for live load alone, less than 5/8 of 0.893, or 0.558 in., which is only 1/430 of the 20-ft span. Satisfactory.

At the end of the overhang, the maximum deflection will occur when the overhang is fully loaded but the 20-ft span has only dead load. Computation by the conjugate beam method (not shown here) indicated a negligible downward deflection (0.06 in.) at that point. Perhaps more significant is the possible upward deflection at the right end when the overhang carries dead load only but the 20-ft span is fully loaded. This deflection is found to be about 0.4 in. upward. Whether or not this much deflection is objectionable will depend on the detail manner in which the beam is to be used. If the overhang is an interior balcony, not connected to anything else, the deflection may not be harmful. If it is an exterior balcony, water might not drain properly from the overhang. If the overhang is connected to other items, those items might be damaged by the deflection of the overhang.

8-8. BIAXIAL BENDING

Beams can be loaded to have simultaneous vertical and horizontal bending moment. If the allowable bending stresses for bending about the X-axis and for bending about the Y-axis were the same, one could merely compute the maximum combined flexural stress (i.e., the sum of M_x/S_x and M_y/S_y), and then compare it to the common allowable. For glulams, however, the allowable stress for bending about the Y-axis (perpendicular to the wide face of the laminations) is much less than for bending about the X-axis. No allowable value for the combined bending stress is given, so it is necessary to use an interaction equation to decide whether or not the member is satisfactory.

The NDS interaction equation is for axial compression in combination with bending about each axis. Omitting the terms related to axial compression (i.e., setting the axial compression equal to zero) that equation for biaxial bending reduces to

$$f_{b1}/F'_{b1} + f_{b2}/\{F'_{b2}[1 - (f_{b1}/F_{bE})^2]\} \leq 1.0 \qquad (8\text{-}3)$$

The term $F_{bE} = K_{bE}E'/R_B^2$, in which $R_B^2 = \ell_e d/b^2$.

Subscript *1* in Eq. 8-3 refers to stresses due to bending about the strong axis (i.e., bending due to loads normal to the narrow face of the member). Subscript *2* refers to stresses caused by bending about the other axis.

Figure 8-11 shows a straight line for that equation. Points below the line represent combinations of bending moment about the two axes that are satisfactory, whereas points above the line represent combinations that are not satisfac-

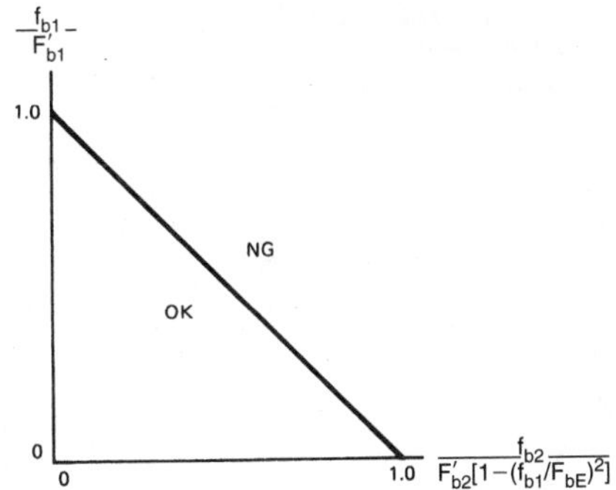

Fig. 8-11. Basis of interaction equation for biaxial bending.

tory. If laterally unsupported lengths or shapes of moment diagram differ for the two axes, adjusted allowable stresses, F'_{b1} and F_{b2}, may be different.

Example 8-7

A 6.75 in. by 24-in. glulam, combination 20F-V9, is used on a 32-ft span with lateral support at the ends only. It has a diagonal concentrated load of 4.9 kips (not over ten-year duration) at midspan. The load is inclined at 23° from the vertical. Self-weight of the beam is about 40 lb/ft. Is the beam satisfactory in bending?

Components of the inclined load are

$$\text{Vertical } P = 4.9 \cos 23° = 4.51 \text{ k}$$

$$\text{Horizontal } P = 4.9 \sin 23° = 1.91 \text{ k}$$

Bending moments are

$$M_x = 4.51 \times 32/4 + 0.040\,(32^2/8) = 41.2 \text{ ft-k}$$

$$M_y = 1.91 \times 32/4 = 15.3 \text{ ft-k}$$

Flexural stresses are

$$f_{b1} = M_x/S_x = 12 \times 41.2/648 = 0.763 \text{ ksi}$$

$$f_{b2} = M_y/S_y = 12 \times 15.3/182 = 1.009 \text{ ksi}$$

The effective length for axis *1* is

$$\ell_e = 1.37\, l_u + 3d = 1.37\,(12 \times 32) + (3 \times 24) = 598 \text{ in.}$$

For axis *1*, slenderness ratio R_B, squared, is

$$[\ell_e d/b^2] = 598 \times 24/6.75^2 = 315$$

$$F_{bE} = K_{bE}E/R_b^2 = 0.609 \times 1{,}400{,}000/315 = 2707 \text{ psi}$$

$$F_b^* = 2000 \text{ psi}$$

$$\text{Ratio } F_{bE}/F_b^* = 1.35$$

Substituting in the equation for beam stability factor gives $C_L = 0.907$. (This factor applies only to strong-axis bending, since lateral buckling cannot occur for bending about the flatwise axis.)

$$C_V = 1.09\,[1292/(32 \times 24 \times 6.75)]^{0.1} = 0.949$$

This is larger than C_L, so $C_L = 0.907$ controls.

$$F'_{b1} = 2000 \times 0.907 = 1814 \text{ psi}$$

$$F'_{b2} = F_{b2} = 1400 \text{ psi}$$

$$f_{b1}/F_{bE} = 0.763/2.707 = 0.282$$

Inserting the above terms into the interaction equation (Eq. 8-3) gives

$$0.763/1.814 + 1.009/[1.4\,(1 - (0.282)^2)] = 1.20 > 1.0 \quad NG$$

Had the beam proved satisfactory for biaxial bending, what other conditions should be investigated? For sure, shear should be checked. Notice from Appendix Table B-7 that the allowable shear stresses for the two directions are unequal. An interaction equation should probably be used, but NDS does not address this situation for shear. It would be conservative, however, to limit the sum of shear stresses computed for the two axes to the smaller of the tabulated allowable shear stresses, multiplied by all applicable adjustment factors.

8-9. CANTILEVER BEAM SYSTEMS

Two examples of cantilever beam systems are shown schematically by Fig. 8-12. Figure 8-12a shows a two-span system, in which the left beam is extended a distance *C* beyond its interior support. From the right end of that cantilevered portion (overhang) another beam is supported, this beam having a span *B*. At the end of the overhang, the connection allows one member to rotate freely relative to the other. It is referred to as a *pinned connection* even though usually an actual pin is seldom used. (A detail for this type of connection may be found in Chapter 9.)

Why arrange beams in this manner? Why not merely use two simple beams, one of span *A* and

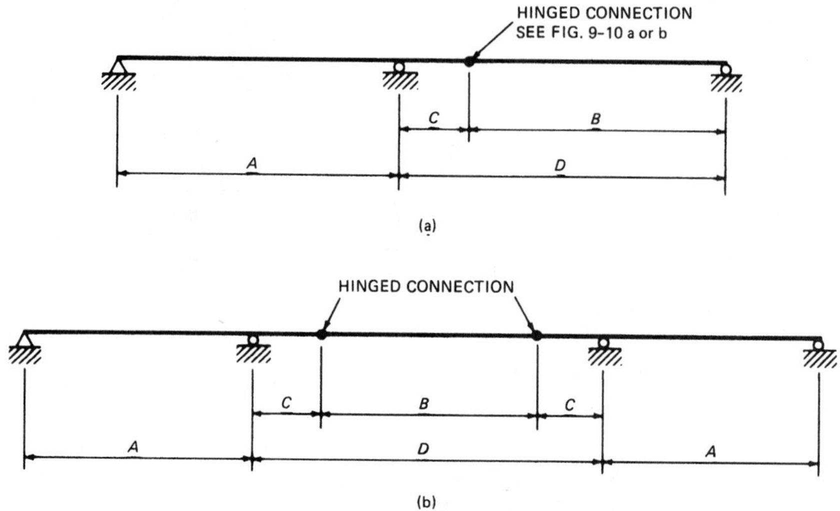

Fig. 8-12. Types of cantilever construction.

the other of span *D?* Reasons for preferring the cantilever system are:

1. Economy, caused by bending moments being less than for simple beams of spans *A* and *D.*
2. With the same size of member, longer feasible spans for the cantilever system than for the simple, one-span beams. Longer spans mean more usable floor area below.
3. Possible ease of construction as compared to a system that uses only single-span beams.

The advantages listed above are even more evident in the three-span system shown by Fig. 8-12b. The central beam, extending from one pin connection to the other, is called a *suspended span.* Another possibility, not shown here, is a three-span system in which a central span extends beyond its supports at each end, the cantilever overhangs each serving as supports for suspended end spans.

If the only load condition to be considered were full uniform load, and if the glulam to be used had the same strength for negative bending as for positive, then it would be easy to optimize the design by choosing the best length, *C,* for the overhang. However, the *Uniform Building Code* (12) and others require that partial live loads be considered. Thus, for each of the systems shown by Fig. 8-12, two load cases would have to be used by the designer:

Case 1—Full load on all spans
Case 2—Dead load on all spans, with live load either in the center span only or in both end spans

Dimension *C* that is most economical for case 1 will be different from the dimension that is best for case 2. Some compromise value will be the optimum.

Choice of the best length of overhang is affected by the glulam makeup. If the glulam has tension laminations on its bottom surface only, the allowable negative bending moment is less than the allowable positive. On the other hand, if the glulam is designed and fabricated with tension laminations on the top in portions of the length where negative bending moment is expected, and tension laminations on the bottom where positive bending moment is expected, the allowable positive and negative bending moments will be alike. The best overhang length, *C,* may be larger in the latter case than in the former. The larger overhang makes the system more effective, so for cantilever construction it may be advisable to specify combinations in which top and bottom allowable stresses are alike. These combinations can be identified from the third and fourth columns of Appendix Table B-7. For all other combinations, the allowable negative bending moment will be one-half the positive.

What are the disadvantages of cantilever beam systems? Possible larger deflection is one disadvantage, and with it come the potential problems of vibration and ponding. These disadvantages do not necessarily cancel the advantages, but they are factors that the designer must consider carefully. Larger deflection can be partially offset by initial camber, *which must be specified by the designer.* For the beam of Fig. 8-12b, for example, deflection would have to be computed at the interior of the end span, the end of the overhang, and the center of the suspended span. At those points, camber should be at least equal to $1\frac{1}{2}$ times the long-term dead load deflection. If the beam system must support a roof (for which ponding is a concern) the UBC requires either that the roof be sloped to drain away all water after long-term deflection of all roof framing members or that the supporting beams be cambered additionally (beyond the camber required for 1.5 times dead load) to ensure that the roof will drain even under full design loads.

Where a cantilever beam system is used to support a floor or a roof, the structural floor or roof material (if attached to the glulams) usually has enough diaphragm stiffness to ensure lateral stability of the glulam over those portions of the length where the top side is in compression. For the portion that has negative bending moment, however, the compression (bottom) side is laterally unsupported. For that portion lateral stability must be checked, using the distance between the inflection point and the pin connection as the laterally unsupported length. If this distance proves to be too long (so that the required reduction of allowable stress is too much) it may be necessary to provide one or more intermediate lateral braces so as to reduce the laterally unsupported length.

Suggested Design Procedure

The first step, of course, is to try to find a satisfactory length of overhang, one that will give economy yet moderate deflections. This may have to be done by trial and error, completing alternate designs using the following procedure, or it may be dictated by other factors—architectural, for example.

1. Estimate the weight of the member. Usually, it will be best to keep the same member size throughout, although the suspended span can be made of a smaller section if appearance is not important.
2. Solve the maximum bending moments of each sign; do this for each span and for each load condition. It will often help to sketch the bending-moment diagrams.
3. Choose a tentative section.
4. Check any assumptions made and recycle as required.
5. Compute deflections and specify the required camber.
6. Check for bearing at the supports. At the support next to the overhang, the reaction may be large enough that the compression stress perpendicular to the grain exceeds the allowable.
7. Design the connection at the "hinge."
8. Check horizontal shear at the cantilever end and at each side of all points of support.

Example 8-8

Choose a D. fir-larch glulam section for the cantilever beam floor system shown by the tentative layout of Fig. 8-13. Use a combination that allows equal positive and negative bending moments. Limit short-term total deflection at midspan to $l/180$. As instructed by the UBC, consider both full loading and partial loading. Dead load from the floor sheathing and joists is 20 psf; specified live load is 45 psf (after considering any allowable reductions).

The solution to this example is shown in designer's computation sheet form by Fig. 8-14. Three load cases are considered—all spans fully loaded, live load on end spans only, and live load on center span only. Deflection of the exterior span is estimated by computing the deflection for a simple span using the same section. (The actual deflection will be slightly less.) For the cantilever and the suspended span, it is solved by AISC handbook equations (13).

Notice (on Fig. 8-14) that the designer selected a section that will be overstressed by a small percentage, actually only 3.6%. Is this an acceptable practice?

It has been common practice for many decades to allow overstress of up to 10% for building structures and a lesser amount for bridges. This custom, however, is diminishing, so that today some engineers allow no leeway at all, preferring to be limited to *exactly* the specified allowables. The specification is

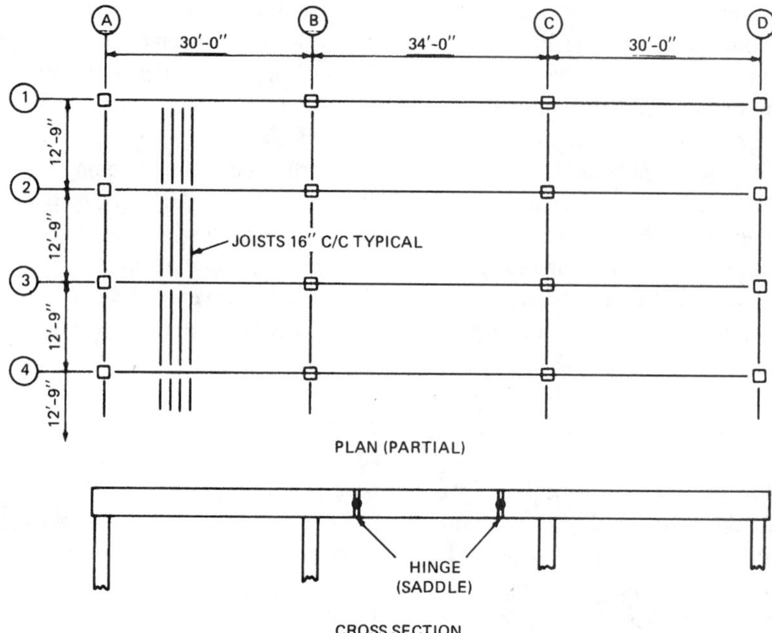

Fig. 8-13. Trial floor system for Example 8-8.

still a compilation of compromises among the writers of the specification, so great precision cannot be attributed to the values quoted. But with the increase of litigation over structural failures, many engineers feel uncomfortable about stretching the specification in any manner. Perhaps the engineer who selected the 6.75 in. by 25.5 in. glulams of this example should consider instead the next larger size.

However, accepting the design shown by Fig. 8-14, it is interesting to compare the cantilever system design with one using three separate simple-span glulam beams, with columns at the same locations as in Fig. 8-14.

For the left and right spans, say the span is 29.5 ft c/c of required bearing lengths.

$$M = 0.87\ (29.5)^2/8 = 94.6 \text{ ft-k}$$

For use in Eq. 8-1, L, the distance between points of zero bending moment, is now the same as the span itself, or 29.5 ft. Completing a design for the left and right simple spans, we find that glulams 6.75 in. by 27 in. are satisfactory. Similarly, for the center simple span (33.5 ft c/c of required bearing lengths) a 6.75 in. by 31.5 in. glulam is required. These sizes for the simple-span system compare to 6.75 in. by 25.5 in. for each span of the cantilever system.

Based on net volumes of the finished glulams, the cantilever system saves—for each beam line—164 fbm over the volume for simple beams having the same loads and spans as the cantilever system. The cantilever system also saves six inches of building height!

Example 8-9

Check lateral stability of the beam selected in Example 8-8. Is lateral bracing required for the bottom of the glulam?

From Fig. 8-14, the designer's solution to Example 8-8, we can compute the lengths over which the beam *bottom* is in compression. Load case III produces the largest negative bending moment. Furthermore for case III the distance from the inflection point to the pin connection is larger than for either case I or case II. So case III controls. However, NDS Section 3.3.3.4 requires that the bottom of the beam be laterally supported at the point of bearing on the interior column.

Thus, the maximum *laterally unsupported* length over which the bottom of the beam is in compression is 13.32 ft. Ratio $\ell_u/d = 12\ (13.32)/25.5 = 6.27$, so the effective length (see Ch. 6) is

$$\ell_e = 2.06\ (12 \times 13.32) = 329 \text{ in.}$$

The above value is for the portion in the outside span. This should be compared with the value for the cantilever portion between the column and the pin. For

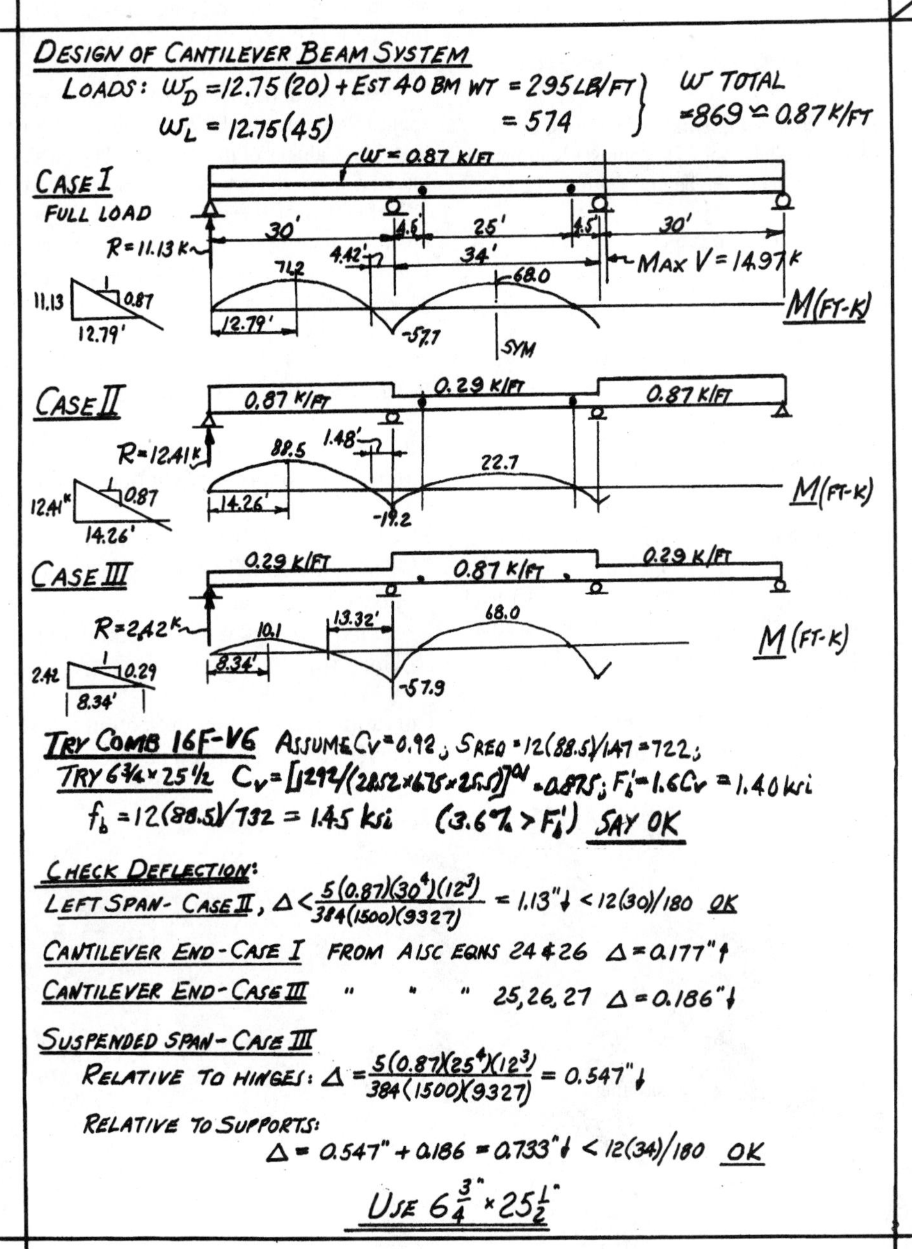

Fig. 8-14. Computation sheet for Example 8-8.

that part, there is both uniform load and a concentrated load. So, using the larger effective length factor—the one for a concentrated load at the cantilever end,

$$\ell_e = 1.87\ (12 \times 4.5) = 101 \text{ in.}$$

The larger value, 329 in., will control.

$$R_B^2 = 329\ (25.5)/6.75^2 = 184$$

$$F_b^* = 1600 \text{ psi};\ F_{bE} = 0.609\ (1{,}400{,}000)/184 = 4634$$

$$C_L \text{ (by Eq. 6–4)} = 0.978$$

The distance between points of zero bending moment is 13.32 ft + 4.5 ft, or 17.82 ft, so the volume factor is

$$C_V = [1292/(17.82 \times 6.75 \times 25.5)]^{0.1} = 0.917$$

The volume factor, being smaller, controls, and it is used, but factor C_L is not. The adjusted allowable stress for this portion of the beam length is

$$F_b' = 1.6 \times 0.917 = 1.47 \text{ ksi}$$

The actual bending stress for that length of beam is

$$f_b = 12\,(57.9)/732 = 0.949 \text{ ksi} < 1.47 \qquad OK$$

Example 8-10

How should the beams of Examples 8-8 and 8-9 be cambered?

Deflections were computed by the designer, as shown by the computation sheet in Fig. 8-14. For the left span, the computed downward deflection due to dead plus live loads was slightly less than 1.13 in. The portion due to dead load alone is slightly less than

$$(295/870)\,(1.13) = 0.38 \text{ in.}$$
(short-term deflection)

Camber at approximately the center of the left span should compensate for long-term dead-load deflection, or 1.5 (0.38) = 0.57 in. A camber of 5/8 in. or more should be specified.

Dead-load deflection (short-term) of the suspended span, relative to the hinges, is (295/870) (0.547) = 0.185 in. Camber of the suspended beam should be at least 1.5 times this amount, or 0.278 in. Say 3/8 in.

Deflection of the cantilever end is very small in this particular instance. For this beam, camber of the cantilever end will be omitted. However, in designs of cantilever beam systems, deflection of this point may sometimes be important.

8-10. CURVED GLULAMS

The design of curved glulams (see Fig. 8-15) differs from that of straight ones in three ways:

1. Curvature limits must be checked.
2. The allowable bending stress must be adjusted for curvature.
3. Radial stresses must be computed and provided for.

Curvature limits given in Section 8-5 should be followed to prevent residual stresses from being

Fig. 8-15. Curved glulam pedestrian bridge. Deck is vertically laminated lumber. (Photograph by authors.)

too high and to prevent damaging the laminations as they are bent to the desired shape during fabrication. The allowable bending stress for the glulam must then be reduced, so that the sum of the flexural stress due to loads and the residual stress due to bending during fabrication does not exceed a reasonable total for the quality of wood used.

When bending moment is applied to a beam that is initially curved in the plane of bending, radial stresses (as well as flexural stresses) occur. These radial stresses may be either tensile or compressive.

For graphic evidence of tensile radial stress in a curved member, see Fig. 8-16, which shows what happens when a model of a curved glulam is bent. When the applied bending moment tends to straighten the bundle, the lamina-

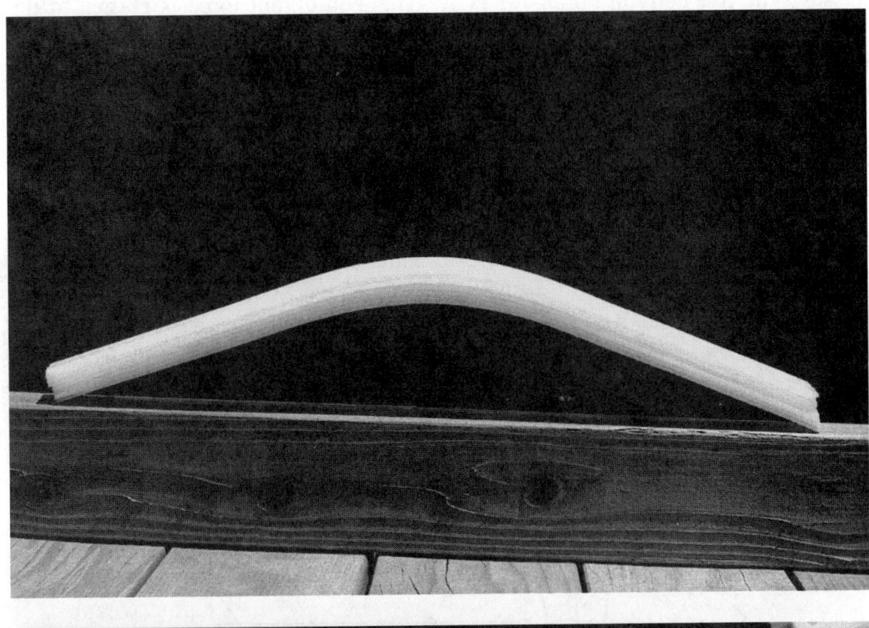

Fig. 8-16. Platic model of curved glulam. Top: unloaded. Bottom: with load tending to straighten the member. (Photograph by authors.)

tions try to move apart. Under the opposite direction of applied bending moment, however, the laminations are pressed more firmly together. This effect can easily be demonstrated by the reader by merely holding together firmly several sheets of paper in an initially curved shape. Bending the bundle one way separates the pages; bending it the other way compresses them together.

How can we compute the magnitude of these radial stresses? The free body in Fig. 8-17a is a short segment of a curved beam having bending moments that tend to straighten the segment. To the right is a diagram of stresses indicated by the flexure formula. (Curved-beam theory need not be used to compute the flexural stresses, since the radius of curvature is always large compared to the glulam depth.) Resultant forces T and C are equal to each other.

$$T = C = M/(2d/3) \qquad (8\text{-}4)$$

An element (shaded) from below the neutral axis of the segment is now removed as a free body as shown in Fig. 8-17b. Since the two forces T are not parallel to each other, their resultant is a radial force, F:

$$F = 2T \sin \alpha \qquad (8\text{-}5)$$

Force F has its largest value when the removed element extends all the way to the neutral axis, as shown.

An equal but opposite force must be applied to hold the removed element in equilibrium. This equilibrant force is normal to the interior surface of the removed element, and is equal to the product of the radial stress and the area of that surface. Thus,

$$f_r = F/(2R_c\alpha b) = 3M/(2R_c bd) \qquad (8\text{-}6)$$

where R_c is the radius to the centerline of the glulam depth.

The above equations are for constant-section curved beams only; they do not apply to tapered curved beams, which are discussed later.

When the bending moment due to load tends to *straighten* the beam (as in Fig. 8-17), the radial stress is tensile. When the bending moment

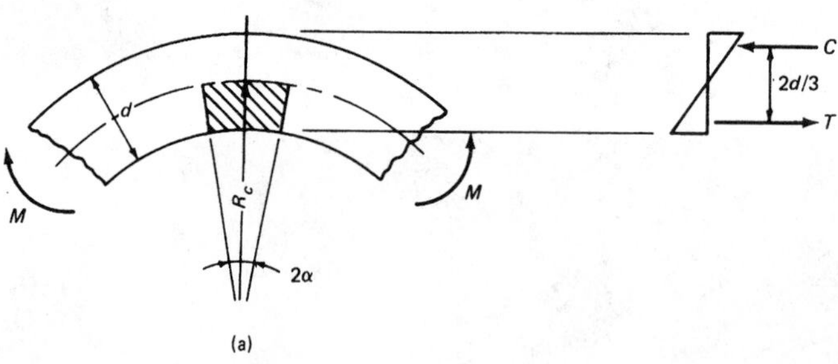

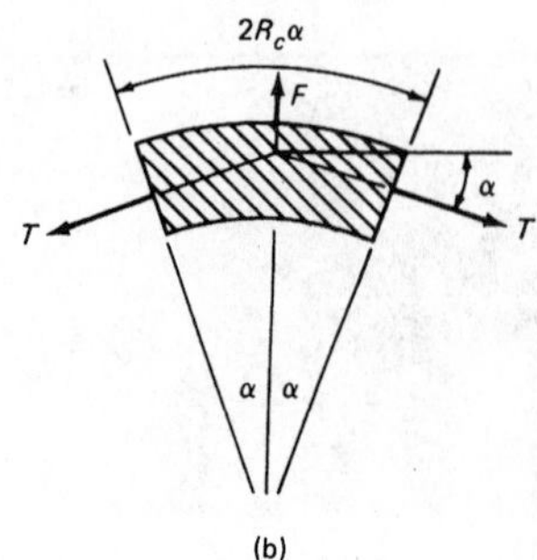

Fig. 8-17. Radial force in curved glulam.

tends to increase the beam's curvature (i.e., opposite to that shown by Fig. 8-17), it causes compressive radial stress.

Radial tensile stresses are perpendicular to the grain of the wood lamination. In this direction the wood is weaker in tension than the glue that joins the laminations; therefore, it is the tensile strength of the wood itself that needs to be considered. The NDS allowable values for this radial *tensile* stress are:

For southern pine under any loading

$$F'_{rt} = (1/3)F'_V$$

For D. fir-larch, D. fir south, hem-fir, western woods, and Canadian softwood species—

Under wind or earthquake loading

$$F'_{rt} = (1/3)F'_V$$

Under other load conditions

$$F'_{rt} = (15 \text{ psi})\,(C_D)\,(C_M)\,(C_t)$$

For the latter group of species, under loading that does *not* include wind or earthquake, when f_{rt} exceeds 15 $C_D C_M C_t$ but does not exceed $(1/3)F'_V$, mechanical reinforcing shall be provided to resist the entire computed radial tensile stress. The mechanical reinforcing could consist of bolts, epoxy-embedded steel dowels such as rebars, or externally applied fittings to resist forces in a radial direction.

Radial compressive stresses result when moment applied to a curved glulam tends to increase its curvature (reduce its radius of curvature). Allowable values of radial compressive stress are the same as the adjusted values for compression perpendicular to the grain.

Suggested Design Procedure for Curved Glulams

Design is necessarily by trial and error. The member depth is a dimension to be determined, and this affects adjustment factors for both volume and curvature. Frequently, when the design is commenced, only the outer radius is known, so any other dimensions needed must be estimated. A suggested procedure is as follows:

1. Make the usual estimate of member weight. Compute tentative bending moments and shears.
2. Ignoring curvature and volume factors for now (or making estimates of them), solve for an approximate required section modulus and choose a trial depth.
3. Use this trial depth (or slightly more) to compute both the minimum radius and the radius at the centerline of the member depth. Verify that these radii meet the specified limits for radius and for ratio t/r_i.
4. Determine the adjusted allowable bending stress, including adjustments for curvature, volume, and other applicable factors.
5. Select a section for flexure and verify that it is satisfactory in shear.
6. Compute the radial stress and compare it to the adjusted allowable.
7. Verify all assumptions made above, and correct if necessary.
8. Recycle as required to step 3 (or to step 1 if the estimated member weight was badly in error).

The designer should be alert to the fact that a curved member may also have axial load. If this happens, axial load must be considered in the design, along with bending and shear.

Example 8-11

A curved (arched) glulam beam of 36-ft simple span carries its own weight plus a single, normal (ten-year) duration, concentrated load of 20 kips at midspan. The radius of curvature of the beam centerline is 44 ft. Moisture content of the wood will exceed 16%. Select a beam size, using combination 22F-V4, western species and 2-in. nominal laminations. Bracing connected to the top of this beam and the adjacent ones provides lateral support for the compression side of the beam.

Simple geometry calculations show that the rise of the beam at midspan is 3.85 ft. The length of arc, from center to center of supports is 37.09 ft, measured along the beam centerline.

Several things are unknown and must be assumed to start the selection: beam wt per ft, C_c, C_V, and inside radius of the bottom (most severely curved) lamination. Estimated values of these are:

Wt per ft, assume 70 lb per ft of beam length. (This averages 70 (37.09/36.0) = 72 lb per ft of span.)

C_C, assume 0.95

C_V, assume 0.95

Bending moments are:

Due to own weight, $M = 0.072\ (36^2)/8$
$= 11.7$ ft-k

By concentrated load, M
$= 20\ (36/4) = 180$ ft-k

Approximate Total $M = 191.7$ ft-k

C_M for bending (from Table B-7) is 0.8 for moisture content exceeding 16%.

The adjusted allowable bending stress, using the above factors, is

$$F'_b = 2.2\ (0.95)\ (0.95)\ (0.8) = 1.59 \text{ ksi}$$

Approximate S req $= 12\ (191.7)/1.59 = 1447$ in.^{3}

Try a glulam measuring 8.75 in. by 33 in.

Corrected values for this size glulam are next determined.

Using specific gravity of 0.50 and assuming a moisture content of 16%, unit wt = (62.4) (0.50) (1.16) = 36 lb/ft^3. Corrected wt per ft of beam length = 36 (8.75 × 33)/144 = 72 lb/ft, and w = 72 (37.09/36) = 74 lb/ft of span due to weight of the beam itself. Preliminary design of the top bracing system shows that it will weigh about 22 lb per ft, so the total dead load is 74 + 22 = 96 lb/ft.

$$\text{Total } M = 0.096\ (36^2)/8 + 20\ (36)/4 = 196 \text{ ft-k}$$

Inside radius of the bottom lamination = 12 (44) − 33/2 = 511.5 in.

$$t/R_i = 1.5/511.5 = 1/341 < 1/125 \qquad OK$$

(Note that R_i also complies with the AITC recommendation that it should exceed 27.5 ft.)

$$C_C = 1 - 2000\ (1.5/511.5)^2 = 0.983$$

$$C_V = 1.09\ [1292/(36 \times 8.75 \times 33)]^{0.1} = 0.885$$

Adjusted allowable bending stress

$$F'_b = 2.2\ (0.983)\ (0.885)\ (0.8) = 1.531 \text{ ksi}$$

S req $= 12\ (196)/1.531 = 1536$ in.^{3} < actual S = 1588 *OK*

Next, the section selected must be checked to see if it is satisfactory for shear and for radial stress.

The end reaction is approximately 18 (0.096) + 20/2 = 11.73 k. (To be ridiculously precise, the reaction could be divided into an axial compression component and a shear component. The effect would be trivial in this case.)

Shear, computed at beam depth from the support (assuming bearing length of 4 in.), is

$$V = 0.096\ (18 - 35/12) + 10 = 11.45 \text{ k}$$

$$f_v = 1.5[11{,}450/(8.75 \times 33)] = 59 \text{ psi}$$

$$< 190\ (0.875) = 166 \text{ psi} \qquad OK$$

Example 8-12

Check the section selected in Example 8-11 for radial stress. The loads tend to straighten the beam, so the radial stress is tensile.

$$f_r = 3M/(2Rbd)$$

in which R is the radius at mid-depth, or 528 in.

$$f_r = 3\ (12 \times 196)\ (1000)/[2 \times 528 \times 8.75 \times 33] = 23.1 \text{ psi}$$

$$F'_v/3 = 0.875\ (190)/3 = 55 \text{ psi, and}$$

$$15C_DC_MC_t = 15\ (1.0)\ (0.8) = 12 \text{ psi}$$

The radial stress exceeds 12 psi but does not exceed 55 psi; therefore we can use the section, but must provide mechanical reinforcement to resist the entire computed radial tensile stress. This can be done using bolts, lag screws, epoxy-embedded metal dowels, or externally applied metal fittings (see Fig. 8-18). Assume that epoxy-embedded rebars will be used.

The load to be resisted by each rebar is merely the total of all radial stresses (at the neutral plane of the beam) on an area equal to the connector spacing (pitch) times the beam width. For this beam, at midspan, the connector force would be

$$23.1 \times 8.75\ \text{p} = 202\ \text{p}$$

Try 3/8-in. rebars, grade 60. From the ACI specifications (14) we find that the allowable tensile stress for grade-60 rebars is 24,000 psi. For a No. 3 grade-60 rebar, the allowable tensile load is (0.11) (24,000) or 2640 lb. Setting $202p$ equal to this allowable force per bar, the maximum allowable pitch is 2640/202 = 13.1 in.

To use No. 3 rebars at 12-in. centers would be a practical solution.

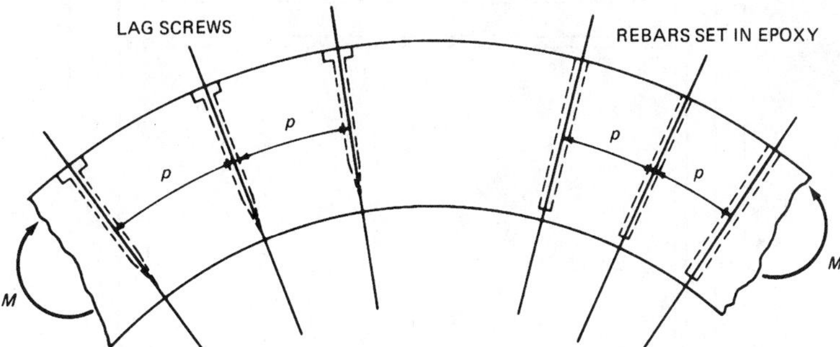

Fig. 8-18. Two types of mechanical means of resisting radial tension.

How long would the connectors have to be? The radial tensile stress is maximum at the beam's neutral axis. Above and below this point, the radial stress reduces parabolically. At 1.5 in. from the bottom, it will be only $23.1[1 - (15/16.5)^2] = 4.0$ psi, which is less than the 12-psi limit allowed if mechanical connectors are not used. Obviously, the rebar (or lag screw) need not penetrate the bottom lamination.

Along the beam, away from the point of maximum moment, the bending moment diminishes, as does the intensity of radial tensile stress. The spacing of mechanical connectors can be increased as the distance from the point of maximum bending moment increases. Better still, at points beyond which the computed radial tensile stress is less than the 12-psi allowable, the mechanical connectors can be eliminated completely.

Whichever system of mechanical connectors is used, holes are made in the beam, and these holes reduce the section modulus. Therefore, *net* section properties should be determined and the suitability for bending recomputed. For example, if the holes for the No. 3 rebars in this beam are 1/2 inch in diameter, and if the holes are assumed to penetrate the entire depth, the net section modulus is computed to be 1497 in.^{3} (down from 1588).

Example 8-13

For a more realistic example, consider the building whose outline shows in Fig. 8-19. There are several ways in which the roof of this building could be supported. Five such schemes are shown by Fig. 8-20. The number of supporting members to be used depends on the strength and stiffness of the deck, and also on architectural requirements. Whether or not to use intermediate columns is also a matter to be decided by the architect or by the owner. Of these five, let's consider scheme II. In this scheme no intermediate columns are used and the roof support members are glulams with a short curved segment to give the roof a two-way pitch. Also, the architect has suggested a radius of 36 ft, measured to the top of the roof. Try western species combination 20F-V3, for which the tabular allowable stresses are 2000 psi for bending (1000 psi where tension due to bending is on the top face) and 165 psi for shear.

Deflection of roof support members should be limited. Assume for this case that the desired limit for live-load deflection is $\ell/240$. Ponding need not be considered, since the roof has appreciable (3:12) pitch.

Roof loads consist of the deck weight, 9 psf for roofing and insulation and 30 psf of horizontal projection for snow. The weight of the beam itself, of course, must be estimated. Assume that it has already been decided that the roof deck will be nominal 3-in. hem-fir selected decking and that the insulation will be 3-in. thick.

How many beams should we use? This will depend, of course, on the strength and stiffness of the roof deck. The adjusted allowable bending stress of the decking is

$$F'_b = 1600\,(1.15)\,(1.04) = 1914 \text{ psi.}$$

Modulus of elasticity is

$$E = 1{,}500{,}000 \text{ psi.}$$

Load per sq ft of roof surface is:

Decking (assume 30 pcf) 30 (2.5/12) =	6.3 psf
Roofing and insulation	9.0 psf
Total w_d =	15.3 psf of roof surface

Converting this to pounds per foot of horizontal projection,

15.3 (12.37/12) =	15.8 psf
Snow	30.0 psf
Total load =	45.8 psf of horizontal projection

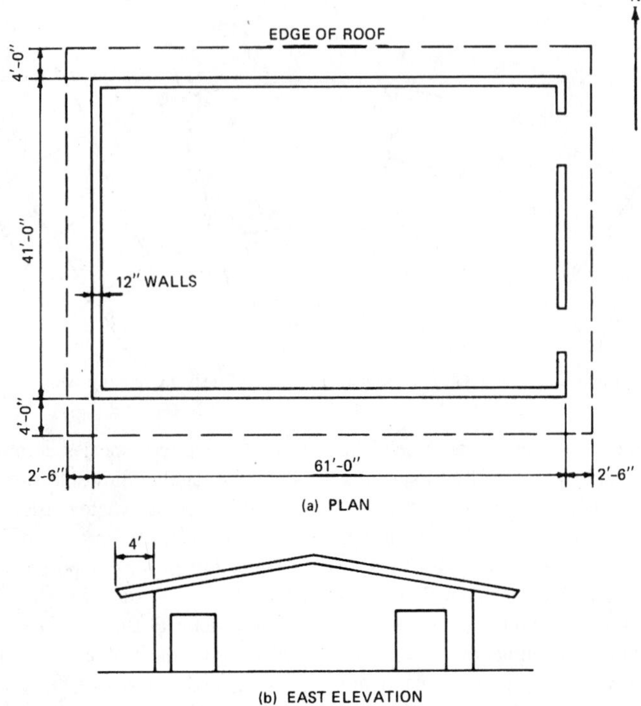

Fig. 8-19. Building plan and elevation for Example 8-13.

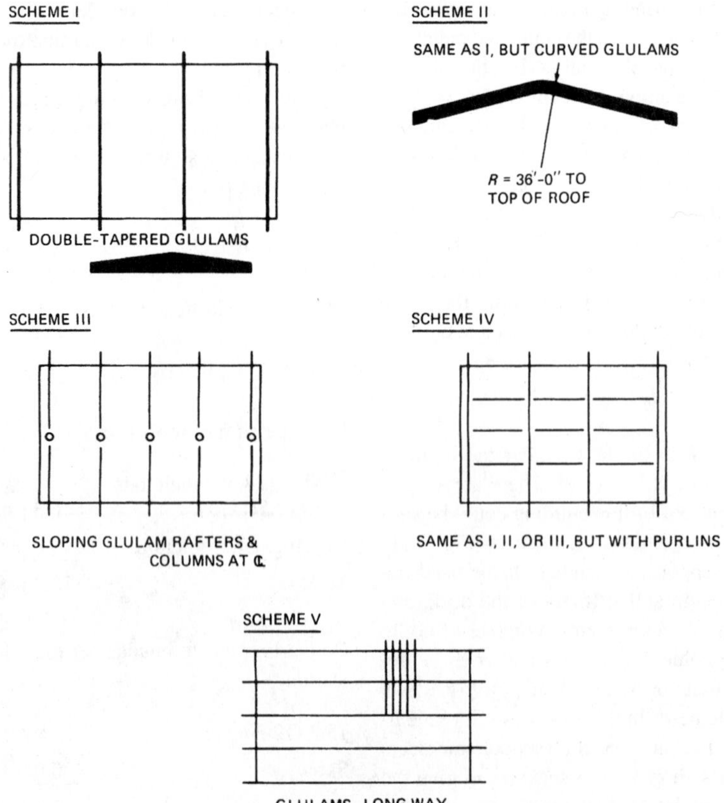

Fig. 8-20. Five framing schemes for building of Fig. 8-19 and Example 8-13. (Courtesy of Board of Regents, University of Colorado.)

Roof *deck* bending will be caused only by the component of this load that is normal to the roof surface. Per square foot of roof deck, this will be

$$45.8 \times 12/12.37 = 44.4 \text{ psf}$$

Assuming that the roof deck will be continuous over two spans

$$\text{Actual } M = 44.4\ell^2/8$$

$$S \text{ furnished by 3-in. deck} = 12 \times 2.5^2/6 = 12.5 \text{ in.}^3/\text{ft}$$

$$\text{Allowable } M = 12.5\,(1914/12) = 1994 \text{ ft-lb/ft}$$

Setting this allowable M equal to $44.4\ell^2/8$, the allowable ℓ based on strength of the decking is 18.95 ft. (That is, the glulam beams might be placed as far apart as 18 ft 11 in.) For two-span continuous decking, the deflection is $w\ell^4/185EI$. Equating this to the assumed total deflection of $\ell/180$,

$$\frac{12\ell}{180} = \frac{44.4\ell^4(12)^3}{185 \times 1{,}500{,}000 \times 12(2.5)^3 / 12}$$

From this, the maximum ℓ for which the deflection limit is not exceeded is found to be 15.56 ft. Deflection controls the solution and the supporting beams must be spaced at not over about 15 ft 6 in. apart. Looking at the floor plan of Fig. 8-19, we see that the end beams will be about 60 ft on centers. This length could be divided into four 15-ft lengths. So try glulams on 15-ft centers.

Now that a tentative spacing for the glulam beams is known, we can turn our attention to the design of the glulams themselves. Each interior glulam will carry load from a 15-ft width of roof; the end ones (considering the effect of the roof overhang at the end of the building) will carry load from about 10.8 ft of roof. The load per foot of horizontal length for each interior glulam will be:

From roof: 45.8 (15)	= 687
Glulam wt, estimated	= 45
Total	= 732 lb/horizontal ft

For dead load alone, this total will be 732 − 30 (15), or 282 lb/ft.

$$732/1.15 > 282/0.9$$

so the load combination that includes snow will control the design.

Assuming that the span will be 40.0 ft (from center to center of the walls), and considering the effect of the roof overhangs at the north and south walls, the midspan bending moment will be 139.0 ft-k. Using an adjusted allowable stress of 2000 × 1.15 = 2300 psi, a first approximation of the required section modulus is

$$\text{Req } S = 12\,(139.0)/2.3 = 725.2 \text{ in.}^3$$

Volume factor and curvature factor were ignored in the first approximation, above, so now they must be estimated to obtain a more accurate value of the required section modulus. A section modulus of slightly more than the required 725.2 would be provided by a 6.75 × 25.5-in. glulam, but when volume and curvature factors are considered, that member will probably prove unsatisfactory. So for the next trial, let's consider the next deeper section, 6.75 × 27 in. For this member, the volume factor is approximately

$$C_V = 1.0\,(1292/[39.0 \times 27 \times 6.75])^{0.1} = 0.843$$

(The length, 39.0 ft, above, is from point of zero moment to point of zero moment.)

The top of the glulam will be laterally supported by the decking, so C_L is 1.0. The assumed weight per foot is close enough. Allowing 1 in. for roofing thickness and 3 in. for insulation, the inside (minimum) radius of the 27-in. glulam will be

$$12\,(36) - 1 - 3 - 2.5 - 27 = 398.5 \text{ in.}$$

For the 27-in. glulam and this minimum radius of curvature, the curvature factor will be

$$C_c = 1 - 2000\,(1.5/398.5)^2 = 0.972$$

Thus, the adjusted allowable stress for the second trial is

$$F'_b = 2000 \times 1.15 \times 0.843 \times 0.972 = 1885 \text{ psi.}$$

The required section modulus is

$$\text{Req } S = 12\,(139.0)/1.885 = 885 \text{ in.}^3$$

The trial beam has $S = 820 \text{ in.}^3 < 885$ *NG*

A third trial, 6.75 × 28.5-in., with $C_V = 0.839$ and $C_c = 0.971$, will prove satisfactory for bending.

Shear check: The maximum shear at distance d from the face of the support is about

$$V = (19.5 - 28.5/12)\,(732) = 12{,}540 \text{ lb}$$

$$f_v = 1.5\,(12{,}540)/(28.5 \times 6.75) = 98 \text{ psi} < 1.15 \times 190 \quad OK$$

Negative bending at the support must be checked, since the flexural tension there is on the top surface, and the tabular allowable bending stress is only 1000 psi.

$$M = 732 \times 4.50^2/2 = 7410 \text{ ft-lb}$$

$$f_b = 12 \times 7410/914 = 97 \text{ psi}$$
$$< 1.15 \times 1000 = 1150 \text{ psi} \qquad (C_V = 1.0)$$
OK

Check radial tension: The centerline radius using the 28.5-in. section will be

$$397 + 28.5/2 = 411 \text{ in.}$$

$$f_r = 3\,(12)\,(139{,}000)/(2 \times 411 \times 6.75 \times 28.5)$$
$$= 31.6 \text{ psi}$$

Since this exceeds 15 (C_D), the member will have to be mechanically reinforced for the entire radial tensile stress, or a different member size selected.

Check deflection: Since the roof is nearly flat, it will be fairly close to estimate the deflection by computing it for a simple beam of 40-ft span. (Because of the roof overhangs, the actual deflection will be slightly less than this amount.) Snow load deflection is

$$5\,(450)\,(40^4)\,(1728)/(384 \times 1{,}600{,}000 \times 13{,}020)$$
$$= 1.24 \text{ in.} < \ell/240 = 2.00 \text{ in.} \qquad OK$$

Long-term dead load deflection will be about

$$1.5\,(1.24)\,(732 - 450)/450 \text{ or } 1.17 \text{ in.}$$

8-11. TAPERED GLULAMS

Glulam beams are frequently tapered to provide a desired roof pitch. More often than not the bottom edge of the beam is horizontal and the top sloped. A design complication is that stress conditions are more complex than in a straight, parallel-edged beam. Consider a double-tapered beam such as shown by Fig. 8-21. Ignoring the local vertical compression that occurs at load points, the resultant stress at either surface, top or bottom, is parallel to that surface. It is a *principal* stress. Were the material isotropic, one might compute the principal stress and design the member merely by comparing the computed principal stress to an allowable stress value.

Wood is *not* isotropic, however, and the allowable stresses depend on direction relative to the wood grain. Consequently, an interaction equation must be used. The interaction equation used in designing tapered glulams is the spheroidal equation

$$\left(\frac{f_x}{F_x}\right)^2 + \left(\frac{f_y}{F_y}\right)^2 + \left(\frac{f_{xy}}{F_{xy}}\right)^2 \le 1.0 \qquad (8\text{-}7)$$

Terms in capital letters in the above equation are allowable stresses and those in lowercase are the computed or "actual" stresses. The definitions are:

f_x = computed horizontal flexural stress at the tapered surface, equal to M/S.
f_y = computed stress perpendicular to the grain and equal to $f_x \tan^2 \theta$, compressive if the taper cut is on the compression surface, and tensile if the taper cut is on the tension side (not recommended).
f_{xy} = actual shear stress = $f_x \tan \theta$.
F_x = tabular bending design value, multiplied by all applicable adjustment factors except C_V, C_L, and C_i (the latter defined below).
F_y = tabular design value for compression perpendicular to the grain, multiplied by all applicable adjustment factors (for a taper cut on the tension side—not recommended—$F_y = F_{rt}$).
F_{xy} = tabular design value for shear parallel to the grain, multiplied by all applicable adjustment factors.

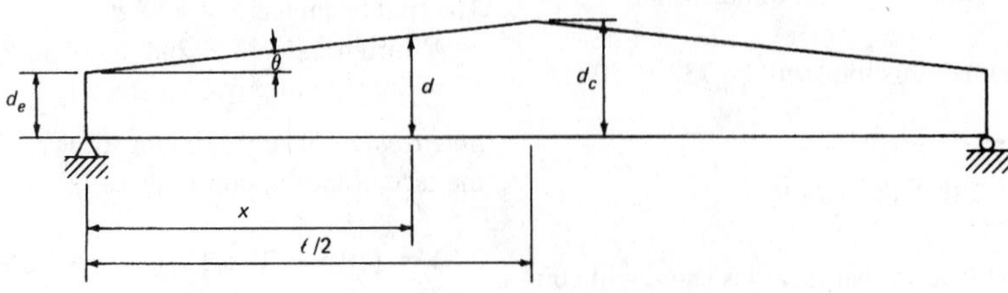

Fig. 8-21. Double-tapered glulam.

By substituting for the above in the interaction equation (8-7) and setting the total equal to unity (optimum design) a sometimes simpler form is obtained, as follows:

$$F'_b = F^*_b\, C_i \tag{8-8}$$

The last factor, C_i, is called the "interaction factor." For our purpose in this textbook, we will use the basic interaction equation, 8-7, and the definitions given above. However, the professional designer of tapered glulams would find additional material in the AITC *Timber Construction Manual* (7) helpful, and would probably use Eq. 8-8.

The definitions shown above for terms of Eq. 8-7 were obtained by elastic analysis of a small element of the tapered beam (15). They were derived assuming the material to be isotropic, but were later shown experimentally to apply to orthotropic materials also, including wood. Technically, the definitions shown above are limited to the specific case of a beam with a constant degree of taper. However, by replacing the tangent of the angle of slope by (dy/dx) in the expressions for f_{xy} and f_y, the definitions could be used for members in which the taper is not constant (16).

It is normal to think of shear stress as being highest at the neutral axis and zero at the extreme fiber, following the rule $f_v = VQ/Ib$. For a tapered beam, however, that is true only at the support of a simply supported beam or at the free end of a cantilever beam. Elsewhere, the location of maximum shear stress f_{xy}, is closer to the tapered face. Except for locations very close to the support, the maximum value of shear stress f_{xy} occurs at the extreme fiber of the tapered face. This effect is included in the definitions given above. For one particular condition—symmetrical double taper and a single midspan concentrated load—the shear stress distribution will be as shown by Fig. 8-22.

Critical Cross Section

To check a given design, all that is required is to compute values for all six factors in the interaction equation and then substitute them in the equation. If the sum of the squared terms does not exceed 1.0, the design is considered satisfactory. But at what cross section should these stresses be checked? Since the first term of the interaction equation usually is much larger than the others, the cross section that controls is usually the one for which the computed flexural stress, f_x, is the largest. Considering the double-tapered beam of Fig. 8-21, the flexural stress at any cross section is

$$f_x = M/S \tag{8-9}$$

Under uniform load, the bending moment at any distance x from the left support is

$$M_x = w\ell x/2 - wx^2/2 \tag{8-10}$$

Using d to designate the depth of the member at any point,

$$d = d_e + x(d_c - d_e)/(\ell/2) \tag{8-11}$$

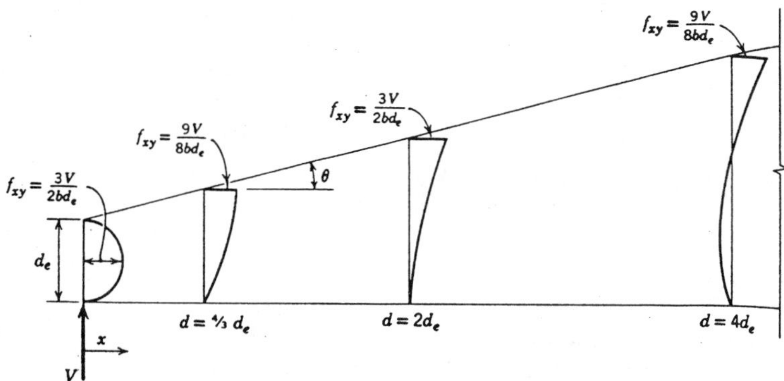

Fig. 8-22. Shear-stress distribution in symmetrical double-tapered glulam with concentrated midspan load. (Courtesy of the American Institute of Timber Construction (AITC).)

The flexural stress M/S at distance x from the support is f_x

$$= \frac{\frac{w\ell}{2}\left(\frac{\ell}{2}\right)\left(\frac{d-d_e}{d_c-d_e}\right) - \frac{w}{2}\left(\frac{\ell^2}{4}\right)\left(\frac{d-d_e}{d_c-d_e}\right)^2}{bd^2/6} \tag{8-12}$$

To find the maximum value of flexural stress at distance x from the support, the above expression is differentiated with respect to depth d and the derivative set equal to zero. After a little algebraic work it shows that, for a symmetrical double-tapered beam under uniform load, the maximum value of flexural stress will occur where the beam depth is

$$d = (d_e/d_c)(2d_c - d_e) \tag{8-13}$$

For a symmetrical double-tapered beam with only a concentrated load at midspan, the maximum stresses will occur where the depth is twice the end depth.

The method shown above would be suitable for other shapes of beam or types of loading. In most cases, rather than deriving an equation as was done above, it is much easier merely to compute stresses and apply the interaction equation at several locations until the location giving the highest total is found. Such a successive-approximation solution is quite satisfactory.

Unusual Loadings or Shapes of Beam

In the years since the above method was derived, much progress has been made in application of the finite-element method. The authors suggest that problems of unusual loadings or shapes of glulam might be solved more reliably using finite-element analysis than by the above definitions. The nonisotropic nature of wood can easily be modeled in standard, available FEM programs. It would still be necessary, of course, to enter the resulting stress values into the interaction equation (Eq. 8-7) to determine whether or not the design is satisfactory.

Tapered-Beam Design Procedure

Normally, the slope of one face relative to the other is known. Usually the upper surface slopes to match the roof, while the lower face of the member is horizontal. Since shear will probably control at the end having the smallest depth, a good method involves finding the depth there first. A suggested sequence of steps is as follows:

1. Based on shear, choose a width and the minimum acceptable depth at the support. Since the depth there is not yet known, the designer must either calculate the shear, V, at the support or estimate the depth so as to be able to compute shear, V, at the critical section.
2. Based on the desired geometry (roof slope or architectural detail) determine a trial value for depth at the deepest point of the span.
3. Check the deflection for this beam. Both flexural and shear deflections should be considered and the total should not exceed the specified limit. (Shear deflection may, of course, prove to be a negligible part of the total.)
4. At the point of maximum flexural stress, if the location of that point is known, compute the stresses and use the interaction formula to determine whether the section is satisfactory. If the location of the maximum-stress section is not known, compute the stresses and use the interaction formula at sufficient, closely spaced locations along the span to be certain that the member is not overstressed at any location.
5. Recycle as necessary to find a member that satisfies all strength and stiffness requirements.
6. Finally, determine the required camber.

Deflections of Tapered Beams

Of the above steps, the only one that may be laborious is computing the deflection. Tapered-beam deflections due to bending may be computed by any method that takes into account the variable moment of inertia—the moment area or conjugate beam methods, for example. Since this is a time-consuming procedure, an approximate method has been developed. In this method, deflection of the tapered beam is approximated as the deflection of an equivalent prismatic member. The depth of the equivalent member is determined by

Table 8-2. Constant for Deflection of Tapered Beams with Uniform Loading.

Taper		C_{dt}
Double	When $0 < (d_c - d_e)/d_e \leq 1.0$	$1 + (d_c - d_e)(0.66/d_e)$
Double	When $1.0 < (d_c - d_e)/d_e \leq 3.0$	$1 + (d_c - d_e)(0.62/d_e)$
Single	When $0 < (d_c - d_e)/d_e \leq 1.1$	$1 + (d_c - d_e)(0.46/d_e)$
Single	When $1.1 < (d_c - d_e)/d_e \leq 2.0$	$1 + (d_c - d_e)(0.43/d_e)$

Source: Ref. 16

$$d = C_{dt} d_e \tag{8-14}$$

in which d_e is the end depth (small end, if single taper), and C_{dt} is a constant from Table 8-2. This approximation includes the effect of shear deflection (7, 16).

Example 8-14

Design a double-tapered glulam of western species combination 20F-V3 to support the roof of the building shown by Fig. 8-19. This type of roof support will be scheme I as shown by Fig. 8-20. Assume that the roof slope required is 1/2 in./ft. Because the roof slope for the double-tapered glulam is less than for the curved glulam designed in Example 8-13, the total load per horizontal foot will be less: 725 lb per ft, rather than 732 (as for Ex. 8-13).

Following the sequence of steps in the suggested procedure, above:

1. End shear, $V = 725\ (20) = 14{,}500$ lb
   Allowable $F_v = 1.15 \times 190 = 218$ psi
   Req $A = 1.5\ (14{,}500)/218 = 100$ in.^{2}
   At the end try $8\frac{3}{4} \times 13.5$ in. (area $= 118.1$ in.^{2})
2. For a roof slope of $\frac{1}{2}$:12, the required center depth is

$$20\ (1/2) + 13.5 = 23.5 \text{ in.}$$

3. Check deflection:
   $(d_c - d_e)/d_e = (23.5 - 13.5)/13.5 = 0.74$
   So, $C_{dt} = 1 + (0.74)\ (0.66) = 1.49$
   Equiv $d = (13.5)\ (1.49) = 20.12$ in.
   Equiv $I = (8.75)\ (20.12^3)/12 = 5939$ in.^{4}

$$\Delta = \frac{5(725)(40^4)(1728)}{384(1{,}600{,}000)(5939)} = 4.39 \text{ in.}$$

This maximum deflection under total load is obviously much too high. To correct it we must increase either the depth or the width of the section, or both. The desired limit for total deflection is about $\ell/180$, or 2.67 in. After several trials, we find that a section measuring $8\frac{3}{4} \times 18.5$ in. at the support, and tapered to 28.5 in. deep at midspan will meet this limit. Deflection computations are as follows:

Try $8\frac{3}{4} \times 18.5$ in. at support, tapered to 28.5 in. deep at midspan.

$(d_c - d_e)/d_e = (28.5 - 18.5)/18.5 = 0.541$
So, $C_{dt} = 1 + (0.541)\ (0.66) = 1.36$
Equiv $d = (18.5)\ (1.36) = 25.16$ in.
Equiv $I = (8.75)\ (25.16^3)/12 = 11{,}610$ in.^{4}
$\Delta = 2.25$ in. < 2.67 in.; *OK*

4. Check interaction equation at point of maximum f_x. Maximum stress occurs where the beam depth (under uniform load) is

$$d = (18.5/28.5)\ (2 \times 28.5 - 18.5) = 25.0 \text{ in.}$$

By similar triangles, this depth occurs at 13 ft from the support, at which location the bending moment is

$$M = 725\ (20)\ (13) - 725\ (13)^2/2 = 127{,}200 \text{ ft-lb}$$
$$S = 8.75\ (25^2)/6 = 911 \text{ in.}^3$$

Terms for the interaction equation are:

$f_x = M/S = 12\ (127{,}200)/911 = 1676$ psi
$f_{xy} = 1676\ (10/240) = 70$ psi
$f_y = 1676\ (10/240)^2 = 2.9$ psi
$F_x = 2000\ (1.15) = 2300$ psi
$F_{xy} = 218\ (1.15) = 251$ psi
$F_y = 560$ psi

Note that F_y uses the allowable for *compression* perpendicular to the grain, since the tapered surface is on the compression side of the beam. Substituting in the interaction equation,

$$(1676/2300)^2 + (70/251)^2 + (2.9/560)^2$$
$$= 0.61 < 1.0 \qquad \textit{OK}$$

5. Determine required camber. The total short-term deflection at midspan is computed above as 2.25

in. Of this total, 2.25 (725 − 450)/725 = 0.85 in. is dead-load deflection. The long-term dead-load deflection will be about 1.5 times this amount, or about 1.28 in. Camber of at least that much should be provided.

8-12. MEMBERS BOTH TAPERED AND CURVED

Glulam beams that are both double tapered and curved can be used for roof members. Usually, the top edge of these beams is formed by two straight lines meeting at the apex, and the bottom edge is curved. The result is a variable cross section, making design and analysis difficult. The complexity is somewhat reduced by a series of design aids (graphs and tables) available in the *Timber Construction Manual* (reference 7). From these aids, the designer can calculate trial values for end depth, centerline depth, and bottom slope, and can find approximate values of stress and deflection for the trial section.

A curved tapered member (such as the Tudor arch of Fig. 8-23) may be made up of two parts: a curved beam of constant section and a haunch that is attached mechanically.

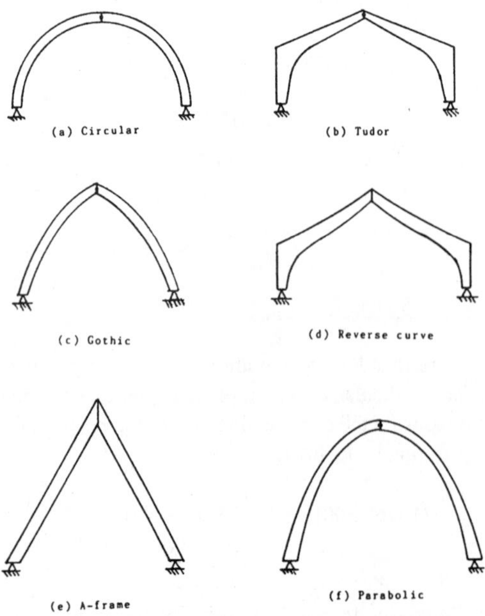

Fig. 8-23. Shapes used for three-hinged arches.

8-13. THREE-HINGED ARCHES

The three-hinged arch is probably the most commonly used glulam member that is both tapered and curved. At first glance, a three-hinged arch looks like a rigid frame. If constructed in steel, a rigid frame would probably be used. The difficulty of making fixed-end connections in timber, however, makes the three-hinged arch a much better choice. Figure 8-23 shows several shapes that may be used. The advantage of one shape over another is mainly a question of aesthetics, although selecting a suitable shape can also affect economy.

The three-hinged arch is a statically determinate structure; therefore its analysis is quite simple. If the shape of the member centerlines is known and the location of all loads known, then the resulting bending moment, shear, and axial load can be computed at any cross section. With these known, the member can be checked at each cross section as a member subject to combined bending and axial load.

At the outset of the design process, however, *the designer does not yet know the shape of the member centerline.* Therefore, the design procedure is necessarily a trial-and-error procedure. An initial assumption must be made as to the shape. A procedure for doing this will be shown by Example 8-15. In many cases the architect will have a fairly good idea as to a shape that will satisfy aesthetic and functional requirements, and this shape may be used by the structural designer for the first trial analysis and design.

Design of a three-hinged arch will be slow and cumbersome, unless the designer follows an orderly procedure. The following is suggested as a sequence of design steps (17):

1. Determine the applied loads. This will include snow and wind loads, as well as roof dead loads. If these loads are brought to the arch member by roof decking, they will be uniformly distributed loads; if roof purlins are used, the loads will reach the arch member as concentrated loads. The weight of the arch member itself must be estimated (and later confirmed or revised).

2. Sketch *to scale* the estimated outline of the arch members, so that the location of the mem-

ber centerlines can be estimated. At this step be certain that the limits on allowable curvature are not exceeded.

3. For each combination of loads to be considered, compute the arch reactions. These will include both horizontal and vertical reaction at each end, and also the force transferred from one segment of the arch to the other at the central "hinge." (The central connection may be actually pinned, or it may be merely a connection that is incapable of transferring bending moment from one segment to the other.)

4. Determine the area required at the base to resist shear. Setting the allowable longitudinal shear stress equal to the computed unit stress gives

$$\text{Req}\, A = bd = 1.5V/F_v \qquad (8\text{-}15)$$

5. For the first trial, use a crown depth $d_c = 1.5b$ and tangent-point depth $d_t = 1.5d$(base).

Members of three-hinged arches are subject to both axial compression and bending moment. Therefore an interaction equation must be used in design. For designing three-hinged arches, the American Institute of Timber Construction (AITC) recommends using Eq. 8-16 (simpler than the NDS equation for beam columns), as follows:

$$f_c/F'_c + f_{bx}/F'_{bx} \leq 1.0 \qquad (8\text{-}16)$$

AITC also recommends that the volume factor, C_V, for the arch segments be solved by assuming that the ratio of member effective length to member depth equals 21. Carrying this assumption into the volume-factor equation presented earlier results in

$$C_V = K_L(5.125/b)^{1/x}\,(12/d)^{2/x} \qquad (8\text{-}17)$$

Since the depth, d, may vary from point to point along the arch member, so also does the volume factor.

Example 8-15

Select glulam sizes for the three-hinged arch shown by Fig. 8-24. The arches are spaced 12 ft c/c. The roof consists of 3-in. decking, with insulation and roofing, giving dead load of 17 psf of roof surface. Specified snow load for the locality is 30 psf of horizontal projection. Design wind speed is 80 mph and the importance factor is 1.0. Use UBC-specified load combinations.

Try western species combination 24F-V8, for which the tabulated allowable stresses are:

Bending, 2400 psi (either top or bottom)

Shear, 190 psi

Compression parallel to the grain, 1650 psi

E_{xx} = 1,800,000 psi; E_{yy} = 1,600,000 psi; E for axial compression = 1,600,000 psi

UBC snow load reduction (Eq. 4–1) for each degree of slope above 20° is

$R_s = 30/40 - 1/2 = 0.25$ psf of horizontal projection

Reduced snow load $= 30 - 0.25\,(26.57 - 20) = 28.4$ psf

Snow load on each arch is 12 (28.4) = 341 lb per horizontal foot.

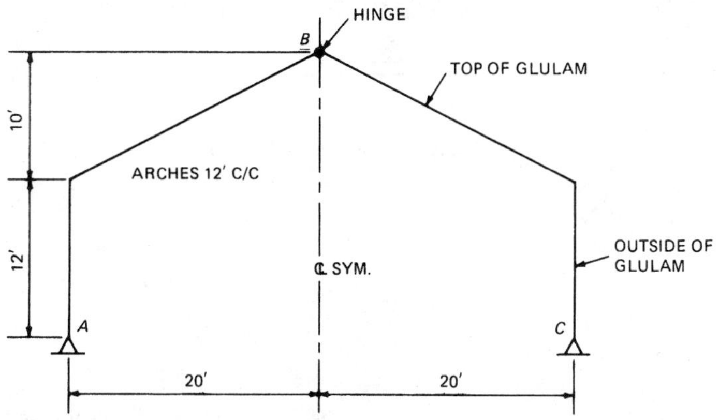

Fig. 8-24. Three-hinged arch for Example 8-15.

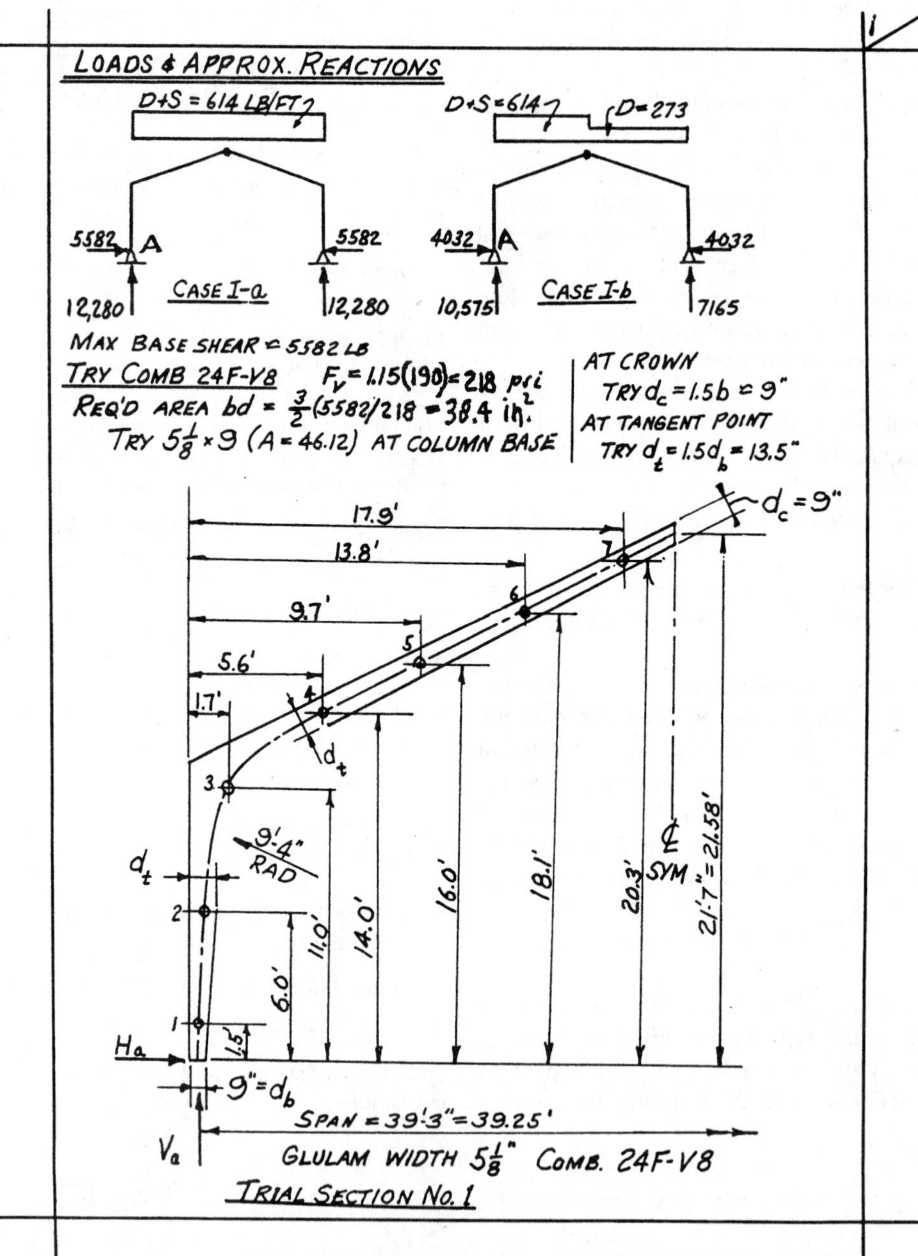

Fig. 8-25a. Designer's calculations for Example 8-15.

For each square foot of horizontal projection, the roof area is 13.42/12 ft.². Estimating the weight of the glulam to average 45 lb per horizontal foot, the total dead load is

$$12\,(17)\,(13.42/12) + 45$$
$$= 273 \text{ lb per horizontal ft}$$

UBC requires considering the following load cases:

Case I—$D + S$
- Case I-a, D + full snow
- Case I-b, D + unbalanced snow

Case II-$D + W$

Case III-$D + W + \frac{1}{2}S$

Case IV-$D + S + \frac{1}{2}W$

The designer feels that either case I-a or case I-b will control the design, and considers those load cases

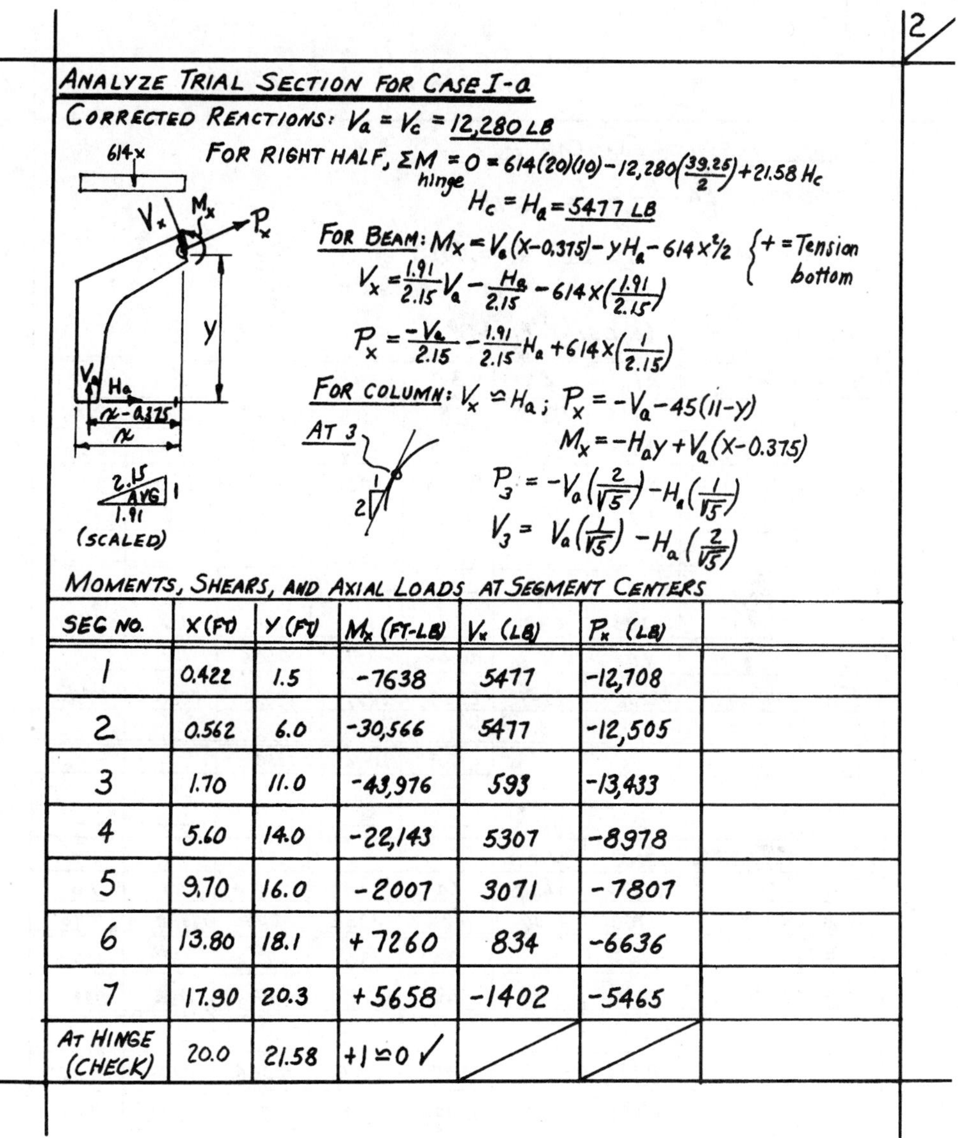

2/

ANALYZE TRIAL SECTION FOR CASE I-a

CORRECTED REACTIONS: $V_a = V_c = 12{,}280$ LB

FOR RIGHT HALF, $\Sigma M_{hinge} = 0 = 614(20)(10) - 12{,}280\left(\frac{39.25}{2}\right) + 21.58\,H_c$

$H_c = H_a = 5477$ LB

FOR BEAM: $M_x = V_a(x - 0.375) - yH_a - 614x^2/2$ {+ = Tension bottom

$$V_x = \frac{1.91}{2.15}V_a - \frac{H_a}{2.15} - 614x\left(\frac{1.91}{2.15}\right)$$

$$P_x = \frac{-V_a}{2.15} - \frac{1.91}{2.15}H_a + 614x\left(\frac{1}{2.15}\right)$$

FOR COLUMN: $V_x \cong H_a$; $P_x = -V_a - 45(11 - y)$

$$M_x = -H_a y + V_a(x - 0.375)$$

AT 3

$$P_3 = -V_a\left(\frac{2}{\sqrt{5}}\right) - H_a\left(\frac{1}{\sqrt{5}}\right)$$

$$V_3 = V_a\left(\frac{1}{\sqrt{5}}\right) - H_a\left(\frac{2}{\sqrt{5}}\right)$$

MOMENTS, SHEARS, AND AXIAL LOADS AT SEGMENT CENTERS

SEC NO.	X (FT)	Y (FT)	M_x (FT-LB)	V_x (LB)	P_x (LB)	
1	0.422	1.5	-7638	5477	-12,708	
2	0.562	6.0	-30,566	5477	-12,505	
3	1.70	11.0	-43,976	593	-13,433	
4	5.60	14.0	-22,143	5307	-8978	
5	9.70	16.0	-2007	3071	-7807	
6	13.80	18.1	+7260	834	-6636	
7	17.90	20.3	+5658	-1402	-5465	
AT HINGE (CHECK)	20.0	21.58	+1 ≅ 0 ✓			

Fig. 8-25b. Designer's calculations, continued.

first. The analysis for case I-a and a first trial design section are shown by Fig. 8-25a. Loads have already been computed above. From these, approximate arch reactions are calculated, based on the shape shown by Fig. 8-24. The approximate horizontal reaction at point *A* is then used to determine a tentative glulam size ($5^1/_8 \times 9$ in.) based on shear at the column base. With this preliminary information, a shape for the first trial section is estimated, as shown in Fig. 8-25a for trial no. 1. The designer divided this trial section into seven segments (for the half arch) for checking stresses and for calculating deflections. The dimensions on the computation sheet are obtained by scaling. Exact dimensions will be needed, of course, for final construction drawings.

Now that a tentative shape for the arch centerline is available, corrected reactions are calculated, as shown in Fig. 8-25b. The vertical reactions are unchanged, but the horizontal reaction for case I-a reduces to 5477 lb. This corrected value of horizontal reaction is then used in computing values of bending moment, shear, and axial load at each segment center. Figure 8-25b shows a table of these values (but not their computation) for case I-a.

To decide whether the trial section is satisfactory, interaction of combined bending and axial-load stresses

CHECK STRESSES – LOAD CASE I-a

INTERACTION EQ: $\frac{f_c}{F_c'} + \frac{f_{bx}}{F_{bx}'} \leqslant 1.0$

LATERAL SUPPORTS: COLUMN – BASE, ②, ③

BEAM – FULL LENGTH

BY AITC: $C_V = (5.125/b)^{0.1}(12/d)^{0.2}$

SEGMENT	1	2	3	4	5	6	7
DEPTH d (in.)	10.13	13.5	28	13.5	12.22	10.94	9.66
AREA (in.²)	51.9	69.2	143.5	69.2	62.6	56.1	49.5
I (in.⁴)	444	1051	9375	1051	779	559	385
S (in.³)	87.7	156	670	156	128	102	79.7
BENDING ℓ_u	72"	72"	—	—	—	—	—
C_L	0.99	0.99	1.0	1.0	1.0	1.0	1.0
C_V	1.0	0.977	0.844	0.977	0.996	1.0	1.0
$F_{bx}' = 2400(1.15)C$	2732	2697	2329	2697	2749	2760	2760
AXIAL COMP.			$F_c^* = 1.15 \times 1650 = 1898$				
ℓ_e	72"	72"					
ℓ_e/d_2	14	14	—	—	—	—	—
$F_{cE} = \frac{0.418(1{,}600{,}000)}{14^2}$	= 3412	3412	—	—	—	—	—
C_P	0.896	0.896	1.0	1.0	1.0	1.0	1.0
$F_c' = 1.15(1650)C_P$	1700	1700	1898	1898	1898	1898	1898
$f_c = P/A$	245	181	94	130	125	118	110
$f_b = 12M/S$	1045	2351	788	1703	188	854	852
INTERACTION							
f_c/F_c'	0.144	0.106	0.050	0.068	0.066	0.062	0.058
f_{bx}/F_{bx}'	0.383	0.872	0.338	0.631	0.068	0.309	0.309
SUM	0.527	0.978	0.388	0.699	0.134	0.371	0.367
OK?	✓	✓	✓	✓	✓	✓	✓

RADIAL COMPESSION AT HAUNCH – PT. 3: $R_c = 9.33 + \frac{1}{2}(28/12) = 10.5'$

$$f_r = \frac{3M}{2R_c bd} = \frac{3(12 \times 43{,}976)}{2(12 \times 10.5)(5.125 \times 28)} = 43.8 \text{ psi} < 650 \quad \text{OK}$$

Fig. 8-25c. Designer's calculations, continued.

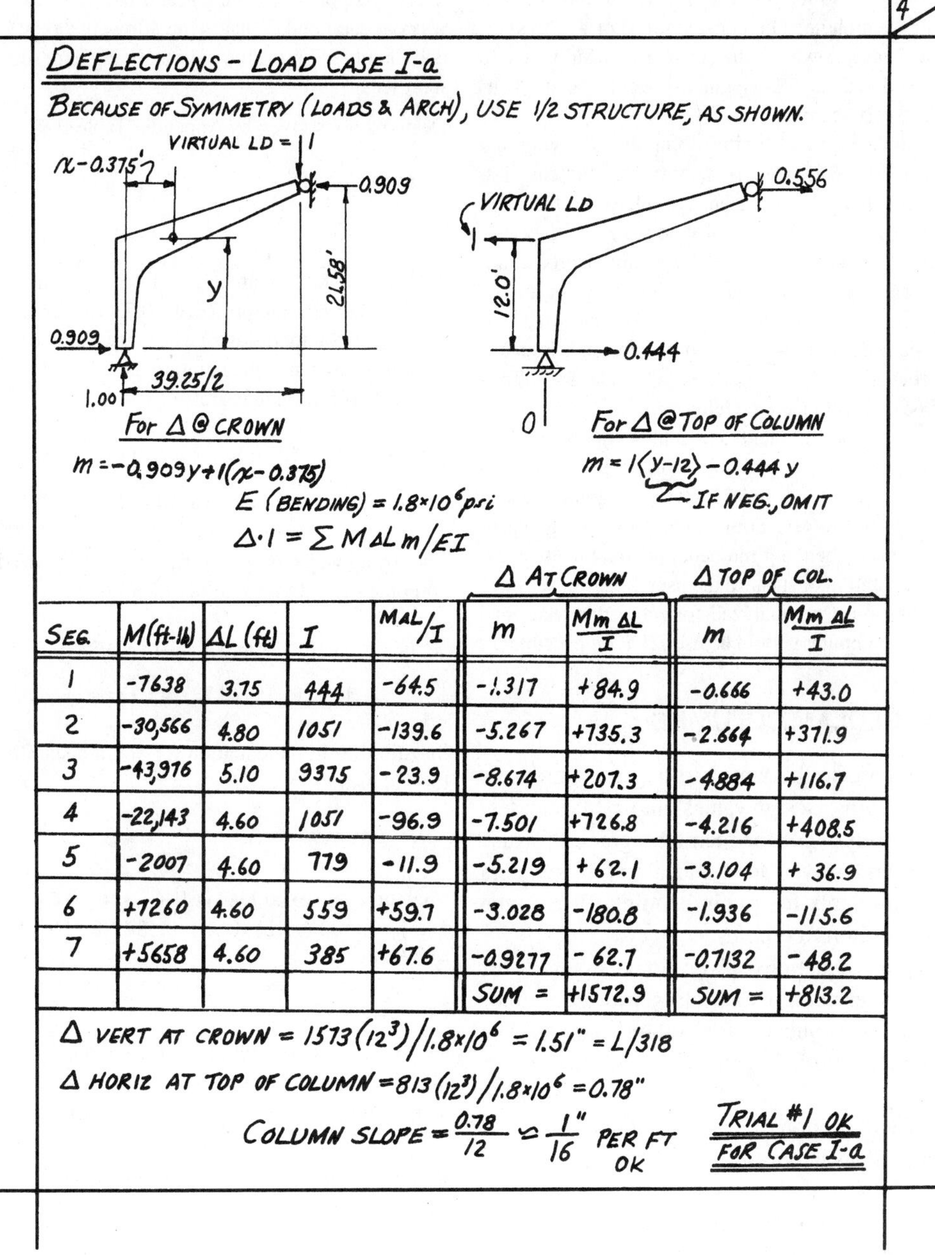

SEG.	M (ft-k)	ΔL (ft)	I	MΔL/I	Δ AT CROWN m	Δ AT CROWN MmΔL/I	Δ TOP OF COL. m	Δ TOP OF COL. MmΔL/I
1	-7638	3.75	444	-64.5	-1.317	+84.9	-0.666	+43.0
2	-30,566	4.80	1051	-139.6	-5.267	+735.3	-2.664	+371.9
3	-43,976	5.10	9375	-23.9	-8.674	+207.3	-4.884	+116.7
4	-22,143	4.60	1051	-96.9	-7.501	+726.8	-4.216	+408.5
5	-2007	4.60	779	-11.9	-5.219	+62.1	-3.104	+36.9
6	+7260	4.60	559	+59.7	-3.028	-180.8	-1.936	-115.6
7	+5658	4.60	385	+67.6	-0.9277	-62.7	-0.7132	-48.2
					SUM =	+1572.9	SUM =	+813.2

Δ VERT AT CROWN = 1573 (12³)/1.8×10⁶ = 1.51" = L/318

Δ HORIZ AT TOP OF COLUMN = 813 (12³)/1.8×10⁶ = 0.78"

COLUMN SLOPE = 0.78/12 ≈ 1"/16 PER FT OK

TRIAL #1 OK FOR CASE I-a

Fig. 8-25d. Designer's calculations, continued.

is checked at each segment center. These calculations, tabulated in Fig. 8-25c, show trial section no. 1 to be satisfactory for combined stresses under load case I-a.

Finally, radial compressive stresses at the haunch are computed, and these too are satisfactory.

Having shown that the trial section is strong enough, the next step is to compute its deflection. Two deflections are of interest: vertical deflection at the hinge, and horizontal deflection at the top of the column. Unless a computer matrix analysis program is used, the easiest way to compute these deflections is by virtual work. A shortcut of this method permits a tabular solution in which actual writing of moment equations and integrating are avoided.

Figure 8-25d shows the deflection computation by this method for case I-a. Because for load case I-a both load and structure are symmetrical, only the left half of the arch need be considered. (For any unsymmetrical case, however, the entire arch would have to be considered: i.e., 14 segments would be used instead of only seven.)

Example 8-15 ends at this point after showing that the first trial section is satisfactory for load case I-a, but the design is by no means complete. All the other load cases (unless it can easily be seen that they will not control) must be checked by a similar procedure. If the trial section is found unsatisfactory for any of these cases, the trial section must be modified and analysis for that particular loading repeated. Because the procedure is lengthy, tables of standard solutions have been developed (see reference 7).

The laborious moment area analysis shown by Fig. 8-25 can easily be replaced by a computer solution. All that is needed is a simple analysis program that can handle plane frames. Some such programs give answers for only bending moment and axial loads, leaving computation of shear to the user. In any case, since the frame must be analyzed for more than one load case, the computer should be used if at all possible.

8-14. GLULAM COLUMNS

Glulams are well adapted to use as columns as well as beams. Design values (allowable stresses) are given by NDS for members graded for use mainly under axial load and also for members intended mainly for use in bending. Either type may be used for columns.

Column equations given in Chapter 7 for sawn-timber columns are also (with two minor differences) applicable for glulam columns. The two minor differences are:

1. In the definition of term K_{cE}, the factor of 0.3 for sawn-timber columns is replaced by 0.418 for glulams.
2. In the equation for column stability factor C_p, factor c is 0.9, rather than 0.8 as for sawn timbers.

These differences are due to the lesser variability (coefficients of variation) for glulams as compared to sawn timbers.

Example 8-16

A glulam of combination 22F-V1, western species is selected for use as a column. The section size is 8.75 × 10.5 in. The member is 25 ft long, with weak-way lateral support at midheight and lateral support both ways at each end. What is the allowable axial load if due mainly to snow, the column being exposed to the weather?

Design values given by Appendix Table B-7 are:

$$F_c = 1100 \text{ psi}$$

$$E = 1{,}300{,}000 \text{ psi} = E_{yy}$$

Adjustment factors are $C_D = 1.15$ for load duration and wet service factors $C_M = 0.73$ for compression and 0.833 for modulus of elasticity, E. The column stability factor must be computed.

The two ℓ_e/d values are:

For the X-axis (strong way) 12 (25)/10.5 = 28.6

For the Y-axis (weak way) 12 (12.5)/8.75 = 17.1

The strong way controls. Terms F_{cE} and F_c^*, needed to compute the column stability factor, are

$$F_{cE} = 0.418\,(1{,}300{,}000 \times 0.833)/28.6^2 = 553 \text{ psi}$$

$$F_c^* = 1100\,(1.15)\,(0.73) = 923 \text{ psi}$$

Substituting in the equation for column stability factor and using $c = 0.9$,

$$C_p = 0.537$$

The allowable axial load is $F_c\,(C_D)\,(C_M)\,(C_p)\,(8.75 \times 10.5) = 1100\,(1.15)\,(0.73)\,(0.537)\,(8.75 \times 10.5) = 45{,}560$ lb

Example 8-17

Consider the same column as for the previous example, but with 30.5 kips of normal-duration axial load and bending moment of 7.5 ft-kips (about the weak axis) due to wind. Is it satisfactory?

First, solve for terms to be substituted in the interaction equation (Eq. 7–9).

Actual stresses are:

$$f_c = 30{,}500/(8.75 \times 10.5) = 332 \text{ psi}$$

$$S_y = 10.5\,(8.75)^2/6 = 134 \text{ in.}^3$$

$$f_{by} = 12\,(7500)/134 = 672 \text{ psi}$$

$$F_{cEy} = 0.418\,(1{,}300{,}000 \times 0.833)/17.1^2 = 1548 \text{ psi}$$

$$C_D = 1.6 \text{ (NDS value)}$$

$$F_c^* = 1100\,(1.6)\,(0.73) = 1285 \text{ psi}$$

Substituting the above values in Eq. 7–5, the column stability factor is

$$C_p = 0.823$$

So $F'_c = 1285 \times 0.823 = 1058$ psi

Since the bending is about the weak axis, lateral buckling due to bending is not possible, and C_L is 1.0. Other factors for weak-axis bending are $C_M = 0.8$ and $C_{fu} = 1.04$. The tabular allowable for weak-axis bending (Y-axis) is 1050 psi. The adjusted allowable stress is

$$F'_{by} = 1050\,(1.6)\,(0.8)\,(1.04) = 1398 \text{ psi}$$

Substituting the above values in interaction Eq. 7–9 gives

$$(332/1058)^2 + 672/[1398\,(1 - 332/1548)]$$
$$= 0.710 < 1.0 \qquad OK$$

8-15. GLULAM DESIGN BY LRFD

Glued laminated beams are designed by the load and resistance factor design method in almost the same manner as structural lumber (solid-sawn) members would be. For example, the design procedure presented in Section 6–8 for LRFD design of solid-sawn members applies equally to design of glulams. The major difference is that for glulams, additional adjustment factors (C_v and C_c) are required that are not considered for solid-sawn members. Design of a glulam beam is illustrated by the following example.

Example 8-18

Repeat the curved glulam design of Example 8-11 for bending only, using the LRFD method. From Example 8-11, preliminary unfactored moment for dead load is $M = 11.7$ ft-kips and for live load, 180 ft-kips.

Load combinations to be considered for the LRFD solution are:

1. $M_u = 1.4M_D = 1.4\,(11.7) = 16.4$ ft-kips
2. $M_u = 1.2M_D + 1.6M_L$
   $= 1.2\,(11.7) + 1.6\,(180) = 302.0$ ft-kips

The second of these combinations controls, and $\lambda = 0.8$.

Tabular reference strength, F_b, is 5.59 ksi.
As for Example 8-11, try an 8.75×33-in. glulam.

$$F'_b = (F_b)\,(C_M)\,(C_C)\,(C_V)$$
$$= 5.59 \times 0.8 \times 0.983 \times 0.885 = 3.89 \text{ ksi}$$

$$M' = C_L S_x F_b' = 1.0 \times 1588 \times 3.89/12$$
$$= 515 \text{ ft-k}$$

The design resistance is

$$\lambda\phi_b M' = (0.8)\,(0.85)\,(515) = 350 \text{ ft-k}$$
$$> M_u = 302 \text{ ft-k}$$

required for the controlling load combination. So the initial selection appears to be satisfactory. However, as for the ASD solution, all assumptions made must be verified and necessary corrections made. Also, using LRFD, the section must be checked for shear and for radial tension.

Example 8-19

Continue Example 8-18 to determine whether the section selected is satisfactory for shear.

For each of the two load combinations to be considered, shear V_u (due to loads) is computed at 33 inches (member depth) from the face of the support. Assuming that the required bearing length for the beam will be about 4 inches at each end, the values of factored shear are computed at $33 + 2 = 35$ inches from the center of the end bearing.

1. $V_u = 1.4\,(0.096)\,(18 - 35/12) = 2.03$ kips (D only)
2. $V_u = 1.2\,(0.096)\,(18 - 35/12) + 1.6\,(10)$
   $= 17.7$ kips *Controls*

Next, compute the flexural shear resistance of the section. The adjusted flexural shear resistance, V', is

$$V' = (2/3)\,F'_v\,bd$$

In this equation term F'_v is the adjusted horizontal shear strength (shear reference strength multiplied by each applicable adjustment factor). Recalling from Example 8-11 that the MC in use will exceed 16 percent, Factor $C_M = 0.875$. All other adjustment factors will be unity.

The reference strength for shear is $F_v = 0.545$ ksi.

$$V' = (2/3)\,(0.545 \times 0.875)\,(8.75 \times 33) = 91.8 \text{ k}$$

The time effect factor for the controlling load case is $\lambda = 0.8$, and factor ϕ_v for shear is 0.75. Thus, the *design* shear resistance for the section selected in Example 8-19 is

$$\lambda\phi_v V' = 0.8\,(0.75)\,(91.8) = 55.1 \text{ kips}$$

Since this exceeds the required value of 17.7 kips, the section is satisfactory in shear.

REFERENCES

1. *National Design Specification for Wood Construction,* rev. 1991 Ed., American Forest and Paper Association, Washington, DC, 1993.
2. *Structural Glued Laminated Timber,* ANSI/AITC A190.1-1991, American Institute of Timber Construction, Englewood, CO, 1991.
3. Johnson, J. W., "Edge Gluing Douglas Fir Lumber to Improve Strength Properties," *Forest Products Journal,* 28(11), Nov. 1978, pp. 37–41.
4. *Emerging Technology,* American Society of Civil Engineers, Aug./Sept., 1994, New York.
5. *Design Standard Specifications for Structural Glued Laminated Timber of Softwood Species,* AITC 117-93, American Institute of Timber Construction, Englewood, CO, 1993.
6. Eby, R. E., "Proof Loading of Finger Joints for Glulam Timber," *Forest Products Journal,* 31(1), Jan. 1981, pp. 37–41.
7. *Timber Construction Manual,* 4th ed., American Institute of Timber Construction, Wiley, NY, 1994.
8. *Establishing Stresses for Structural Glued Laminated Timber (Glulam),* ASTM D3737-91, American Society for Testing and Materials, West Conshohocken, PA, 1991.
9. *Standard Specification for Hardwood Glued Laminated Timber,* AITC 119-94, American Institute of Timber Construction, Englewood, CO, 1994.
10. *Wood Handbook,* U.S. Department of Agriculture, Washington, DC, 1974.
11. *NDS Commentary,* Commentary on the 1991 Edition, American Forest and Paper Association, Washington, DC, 1993.
12. *Uniform Building Code,* International Conference of Building Officials, Whittier, CA, 1994.
13. *Manual of Steel Construction,* 9th ed., American Institute of Steel Construction, Chicago, IL, 1989.
14. *Building Code Requirements for Structural Concrete,* ACI 318-95, American Concrete Institute, Farmington Hills, MI, 1995.
15. *Wood Structures, a Design Guide and Commentary,* ASCE Structural Division, American Society of Civil Engineers, New York, 1975.
16. Maki, A., and E. Kuenzi, "Deflection and Stresses of Tapered Wood Beams," Forest Products Laboratory Research Paper FPL-34, 1965.
17. "Mathematical Solution of 3-Hinged Arch Design," Technical Note 22, American Institute of Timber Construction, Englewood, CO, 1993.

PROBLEMS

Note: Unless otherwise noted, beams have full lateral support, dead load mentioned does not include weight of the beam, and live load is of normal duration.

8-1. A glulam beam is to carry live load of 700 lb/ft and superimposed dead load of 200 lb/ft on a 24-ft simple span. Select a section using combination 24F-V3 western species. Consider flexure and shear. Compare to a sawn timber of the best grade D. fir-larch.

8-2. Glulam beams of 30-ft span are spaced 14 ft apart. Dead load is 8 psf of roof. Live load is 25 psf. Deflection limits are $\ell/360$ for live load alone and $\ell/240$ for total load. Consider long-term effect for dead load. Choose a section with $b = 6.75$ in. using combination 24F-V3 western species.

8-3. Choose a combination 20F-V4 western species glulam to carry vertical permanent load of 1.0 k/ft total (including beam weight) on a 20-ft simple span. Limit long-term total vertical deflection to $\ell/360$: (a) with laminations horizontal; (b) with laminations vertical.

8-4. What is the allowable bending moment (strong-axis bending) for a $6\frac{3}{4} \times 24$-in. glulam of southern pine, combination 22F-V4, subject to permanent uniform load and laterally supported only at the ends of a 30-ft simple span?

8-5. Same as Problem 8-4, except western species.

8-6. Select a 22F-V4 western species glulam to carry a 12-k normal duration concentrated load at the center of a 28-ft simple span, with lateral support at the ends and midspan only.

8-7. Try a section with $b = 10.75$ in. for the biaxially loaded beam of Example 8-7. Choose the most economical depth for that width section.

8-8. Choose a combination 22F-V1 western species glulam section, subject simultaneously to wind bending moments $M_x = 35.0$ ft-k and $M_y = 9.7$ ft-k. Span is 24 ft.

8-9. Same as Problem 8-8, except southern pine.

8-10. Figure 8-26 shows a three-span cantilever beam system. Dead load is 0.4 k/ft, including the beam weight. Live load is 0.8 k/ft, placed in either a full-loading or a partial-loading pattern. Deflection limit under long-term loading is 1.5 in. Choose 20F-V4 western species glulam sections.

8-11. For a similar system to that of Fig. 8-26, but with spans of 20 ft, 25 ft, and 20 ft, choose the best locations for the hinges and select glulams of western species 20F-V2. Load is 1.6 k/ft total, normal duration, present all along the beams. (*Hint:* Since top-fiber allowable tensile bending stress is one-half of that for the bottom fiber, hinges should be located such that the negative bending moment at the

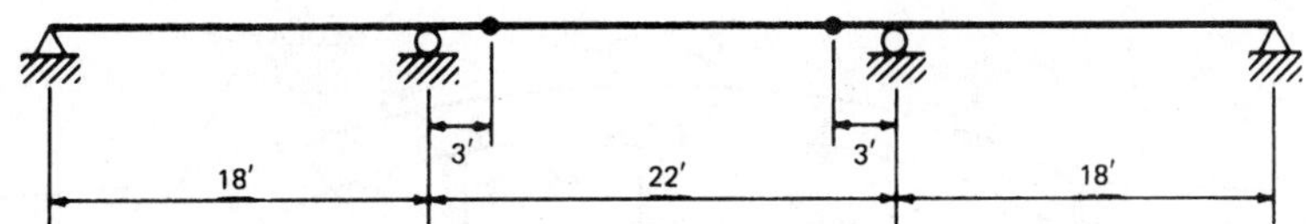

Fig. 8-26. Cantilever system for Problem 8-10.

support is one-half the maximum positive bending moment.

8-12. Rework the curved glulam design of Example 8-11, changing the concentrated load to 5 kips. If mechanical connectors are required, use lag screws and check bending strength of the net section.

8-13. A curved glulam beam with 3/4-in. laminations, 24 in. deep by 8¾ in. wide, combination 20F-V3 western species, 34-ft span, has an inside radius of 15 ft. Normal-duration bending moment of 100 ft-k causes tension on the concave side. Is the beam OK in flexure? Is it OK for radial stress? Show proof.

8-14. Select a glulam section for the curved beam of Fig. 8-27. Design mechanical connectors if they are needed. Loads are: dead, 200 lb/ft of horizontal projection plus beam weight; snow, 600 lb/ft of horizontal projection. Use western species combination 22F-E1. What bearing length is required at the reaction? (*Note:* The bearing stress is not perpendicular to the grain.)

8-15. A western species combination 22F-V7 double-tapered glulam of 60-ft span is 8¾ in. wide, 30 in. deep at the ends, and 55 in. deep at the center. The controlling load includes snow and is 650 lb/ft, including the beam weight. Check the effect of combined stresses at distance $x =$ 15 ft and $x = 20$ ft from the end.

8-16. A double-tapered glulam is to span 50 ft and have a top-surface slope of ½ in. per ft. Beams are on 12-ft centers. Loads are: snow, 40 psf of horizontal projection; dead, 26 psf of roof plus beam weight. Use southern pine combination 22F-E1. Choose width and depths. What end-bearing length is needed?

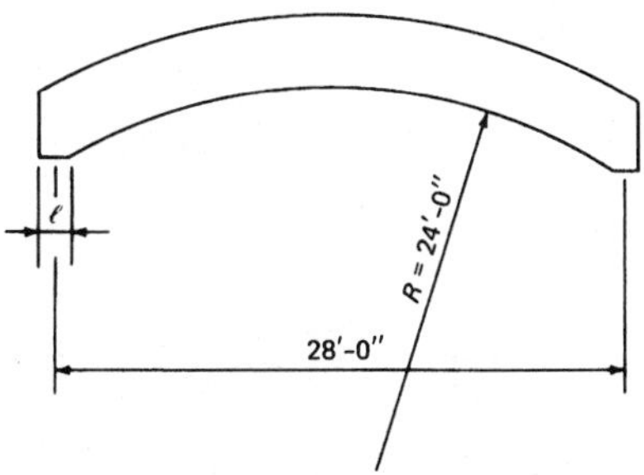

Fig. 8-27. Curved beam for Problem 8-14.

8-17. Prove for a double-tapered glulam with a single concentrated load at midspan that the maximum flexural stresses occur where depth, d, is twice the end depth, d_e.

8-18. Design a wood roof system for a building 70 ft wide by 150 ft long. Use ½-in. plywood on joists (at 16 in. c/c) that are parallel to the building length. Choose the number and spacing of beams. Either one row or two rows of interior columns will be acceptable. The roof is nearly flat. Sufficient slope to permit drainage and prevent ponding can be provided by varying the vertical position of joist hangers. Use glulams wherever possible. Limit long-term deflection to $\ell/200$. Snow load is 40 psf. Insulation and roofing weigh 12 psf. Suspended ceiling weighs 11 psf.

8-19. Design the glulam beams (using western species combination 20F-V3) for scheme V of Fig. 8-20. The roof joists slope from north to south at 1.0 in. per ft. The glulams are horizontal and spaced at 10 ft c/c. Also design the joists (at 4 ft c/c) of No. 2 D. fir-larch.

8-20. For the building shown in cross section by Fig. 8-28, choose a size of 2-in. nominal roof joist of No. 2 D. fir-larch. Then

(a) Choose glulams (western species 20F-V4) for the side-span beams.

(b) Choose glulams for the curved center-span beams.

(c) Design the glulam columns, using western species combination 10.

Snow load is 30 psf. Roof is 1/2-in. plywood on joists at 16-in. centers. Roofing, insulation, and ceiling weigh 17 psf. Exterior walls and diaphragm action by the plywood roof sheathing provide lateral bracing for the top end of the columns.

8-21. For the three-hinged arch of Example 8-15, analyze and check for load case I-b. If the trial section of Example 8-15 is found unsatisfactory, revise the section and recalculate for

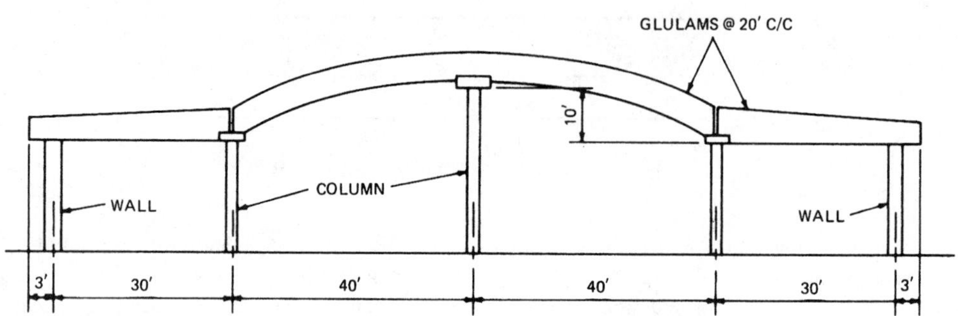

Fig. 8-28. Building for Problem 8-20.

case I-b. (Note that all 14 segments must be considered for this unsymmetrical load case.)

8-22. Same as Problem 8-21, but for load case II.

8-23. Choose a glulam section, Douglas fir combination 3, for a column 19 ft long to support an axial load of $D = 38$ kips and $S = 125$ kips. Choose a section that is as close as possible to being square.

8-24. If the loads of Problem 8-23 are applied at the face of the column (i.e., eccentrically, causing strong-axis bending) what depth column would be required if $b = 12.25$ in.?

8-25. Design a Douglas fir glulam column, combination 3, of 25-ft overall length, with weak-way bracing at midheight. Column ends are pinned. Axial load is $D = 32$ kips and $S = 130$ kips.

9

Bolts, Timber Connectors, and Special Weldments

Bolts for resisting direct shear or tension loads were covered in Chapter 5. This chapter introduces the use of bolts under more complex loading conditions.

A wide variety of metal connection devices makes it possible to use wood members more effectively and in situations where formerly only structural steel members were practical. Examples include beam and joist hangers, special welded steel assemblies, shear plates, split rings, and spiked grids. Many of these will be discussed in this chapter.

Some proprietary types of connector are selected by *catalog engineering,* that is, merely using allowable loads and dimension limits provided by the manufacturer's literature. These will not be covered in detail here. For other types, however, design either is entirely the responsibility of the structural designer or requires the designer's expertise to choose the connector number and arrangement.

The list of available connectors is long, but it will continue to grow, being limited only by the ingenuity of designers and the desire of manufacturers to meet new needs and develop new market opportunities.

9-1. BOLT GROUPS SUBJECT TO MOMENT

In wood construction, it is preferable to avoid applying bending moment to groups of fasteners, but occasionally it cannot be avoided. An example would be a bracket connection as shown by Fig. 9-2. To develop a method of analysis, refer to Fig. 9-3.

First, replace the eccentric load in Fig. 9-3 by an equal, centrally applied load and a moment equal to the original load times its eccentricity ($M = Fe$). The moment, acting alone, rotates the side member about the centroid of the fastener group. As it rotates, the side member pushes each bolt with a force, T. Force T applied to each bolt by the moment is in a direction normal to the radius, r, from that bolt to the center of rotation (Fig. 9-3b). Assuming that the bracket does not deform severely, the distance each bolt is moved by the moment is proportional to radius, r. Forces T are proportional to the distance moved; therefore, forces T are proportional to radii, r. Force T required to move each bolt is also proportional to the stiffness of the material that deforms to permit the bolt to move. This stiffness cannot be defined precisely, but is assumed to be proportional to the allowable shear load, S, for the bolt. This is a reasonable approximation since allowable bolt loads are based on slip rather than ultimate strength.

Putting the above in equation form, the ratio of forces, T, for two bolts is

$$\frac{T_1}{T_2} = \frac{r_1 \times S_1}{r_2 \times S_1} \qquad \text{(a)}$$

or

$$T_2 = T_1(r_2 \times S_2)/(r_1 \times S_1) \qquad \text{(b)}$$

The sum of the moments of all the T-forces about the center of rotation is equal to the applied moment Fe. Thus

$$M = T_1 r_1 + T_2 r_2 + \ldots + T_n r_n \qquad \text{(c)}$$

Fig. 9-1. Hinged connection for glulam arch bridge. (Photograph by Mike Douglas.)

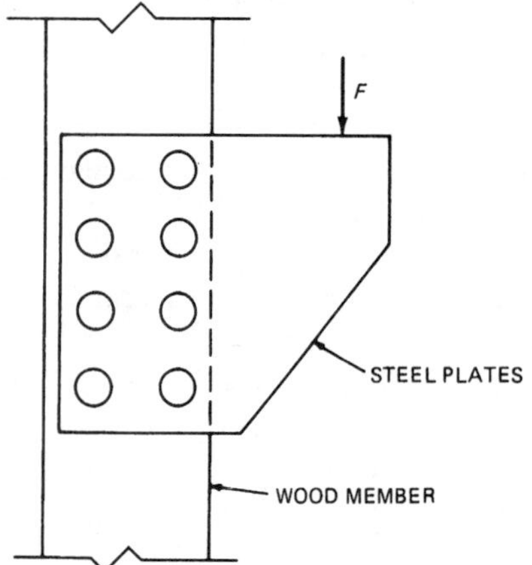

Fig. 9-2. Bolted bracket.

Next, substitute an expression like (b) for all except the first term of (c). Multiply the first term of (c) by r_1S_1/r_1S_1. Then, collecting terms, expression (c) becomes

$$M = \frac{T_1}{r_1S_1}\sum r_i^2 S_i \tag{9-1}$$

The summation term in this equation can be referred to as the polar moment of inertia of the strengths, *S,* about the center of rotation. Calling this summation *J,* Eq. 9-1 becomes

$$T_1 = Mr_1S_1/J \tag{9-2}$$

Had the above been derived by writing expression (b) in terms of any other bolt force *T,* say T_i, then a more general form of the equation would result, as follows:

$$T_i = Mr_iS_i/J \tag{9-3}$$

Each force, *T,* computed by Eq. 9-3 will be in the direction shown by Fig. 9-3b, normal to the distance, *r,* for each bolt. These are the components of bolt force due to $M = Fe$, acting alone. The central load, *F,* also must be accounted for, and its effect is a downward force, $V = F/n$, on each bolt. To find the total force on any bolt, add *V* and *T* for that bolt vectorially.

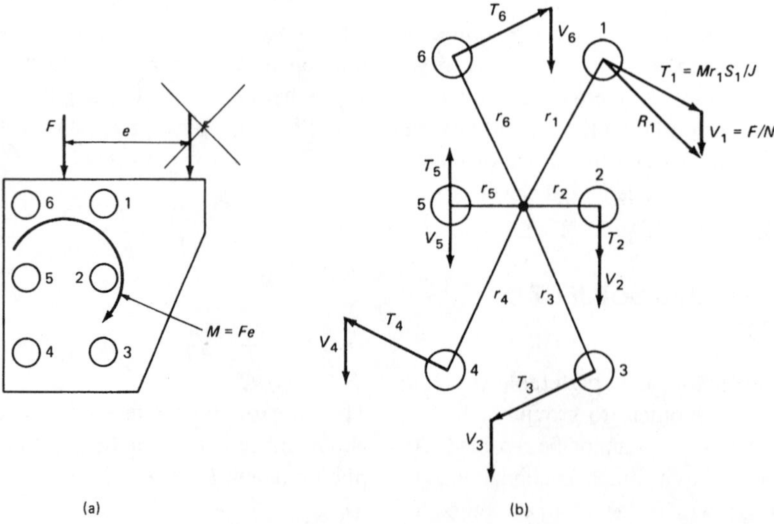

Fig. 9-3. Bolt forces on connection with moment.

Since the strengths, S, are in both the numerator and denominator of Eq. 9-3, it is often simpler to use *relative* rather than actual strengths of the bolts in various directions. This is illustrated in Example 9-1.

The method shown above is not precise. For one thing, it uses a strength, S, based on the direction of force, T, due to moment only; whereas strength, S, in the direction of the *resultant* is not considered. Looking to the future, it may be that an ultimate strength method, similar to that now available for structural steel design, may be developed for joints such as this.

Example 9-1

Figure 9-4 shows a bracket, consisting of two 1/4-in. steel plates, connected by six 5/8-in. bolts to a 6 × 12 D. fir-larch column. Is the connection satisfactory for the 3-k load shown?

The eccentric load of 3 k is first replaced by its equivalent—a 3-k load acting at the centroid of the bolt group, and a clockwise moment, $M = 3(9) = 27$ in.-k.

Considering the effect of moment alone, the direction of forces, T, will be parallel to the grain for bolts nos. 2 and 5. For bolts 1, 3, 4, and 6, force T will be at an angle

$$\theta = \tan^{-1}(4/3) = 53.1°$$

The NDS gives dowel-bearing strengths parallel to the grain and perpendicular to the grain of 5600 psi and 2800 psi, respectively. By Hankinson's equation, dowel-bearing strength at 53.1° to the grain is

$$F_{e\theta} = (5600)(2800)/(5600 \sin^2 53.1° + 2800 \cos^2 53.1°) = 3416 \text{ psi}$$

Substituting in the yield-limit equations of the NDS (not shown here), the allowable force T_1 is 1582 lb and allowable force parallel to grain, T_2, is 1963 lb. Rather than actual strengths, we will use their *relative* values. Letting $S_1 = 1.00$, the relative strengths for bolts 2 and 5 will each be 1963/1582 or 1.241. Calculate the polar moment of inertia, J, of the relative strengths as tabulated.

Bolt	S	r(in.)	Sr^2
1	1.00	5	25.00
2	1.241	3	11.17
3	1.00	5	25.00
4	1.00	5	25.00
5	1.241	3	11.17
6	1.00	5	25.00
			$J = 122.34$ bolts × in.^{2}

Bolt forces due to moment alone: Bolts 1, 3, 4, and 6 are alike:

$$T_1 = Mr_1S_1/J = 27(5)(1.0)/122.34 = 1.103 \text{ k/bolt}$$

$$T_2\ (or\ T_5) = 27(3)(1.241)/122.34 = 0.822 \text{ k/bolt}$$

Due to direct vertical load alone:

Each bolt receives 3.0/6 = 0.500 k

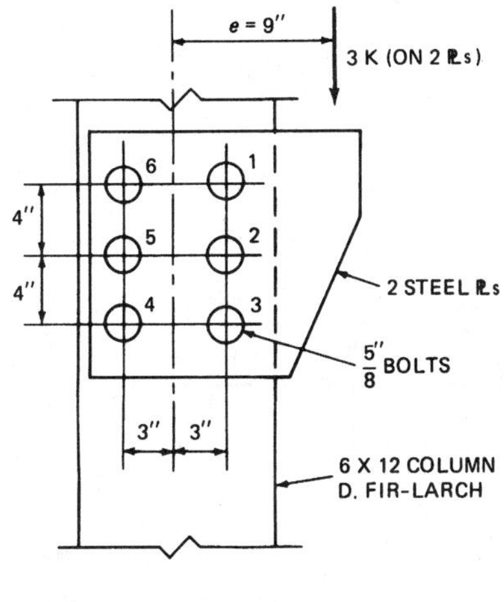

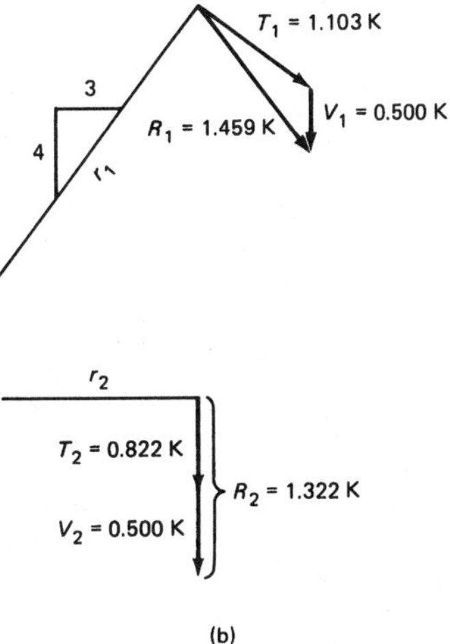

Fig. 9-4. Connection for Example 9-1.

Total forces are found by adding vectorially as shown by Fig. 9-3b for bolt nos. 1 and 2.

Resultant forces are:

R_1 or $R_3 = 1.459$ k (these are the most highly loaded bolts)

$R_2 = 1.322$ k

The allowable force on bolts 1 and 3 can be redetermined now that the direction of the resultant is known (37.2° to the grain), by first finding the dowel-bearing strength at this angle.

$$F_{e\theta} = (5600)(2800)/(5600 \sin^2 37.2° + 2800 \cos^2 37.2°) = 4101 \text{ psi}$$

From the yield limit equations of the NDS,

allowable $R_1 = 1784 \text{ lb} > 1459 \text{ lb}$ *OK*

For bolt 2, the allowable is

$1963 \text{ lb} > 1322 \text{ lb}$ *OK*

As a refinement, one could recycle, using the above allowables to compute new relative strengths, *S*, and then solve for a revised *J* and new resultant bolt loads. However, the additional effort would bring about little change in the answers; furthermore, the method itself is only approximate, so recycling would probably not improve the answer significantly.

9-2. CONNECTIONS WITH MOMENT-BOLTS IN TENSION AND SHEAR

Figure 9-5 shows a bracket consisting of a wood block bolted to the face of a wood column. Again, the actual load is replaced for analysis by an equal load at the shear plane and a moment $M = Fe$. The effects of the vertical load and of the moment are computed separately. Assume that each bolt receives an equal portion of the shear, *V*. The limit for *V* is the allowable bolt load, Z'.

Figure 9-5b shows a cross section at the shear plane of the bolts. The shaded area at the bottom is the area over which the bracket is pressing against the column. The moment causes tension in the bolts above the compressive area. Two different materials are involved: The steel bolts are stretching, and the wood of the block and column are being shortened (horizontally) by the compressive stress. To analyze, the actual cross section is replaced by a transformed section, shown by Fig. 9-5c. In the transformed section, each square inch of steel is replaced by *n* times as much area of imaginary wood, *n* being the ratio of the modulus of elasticity of steel to that of wood perpendicular to the grain. Herein lies the greatest error of the method, since *E* perpendicular to the grain is normally not known. Fortunately, the procedure is not extremely sensitive to the value of *n*.

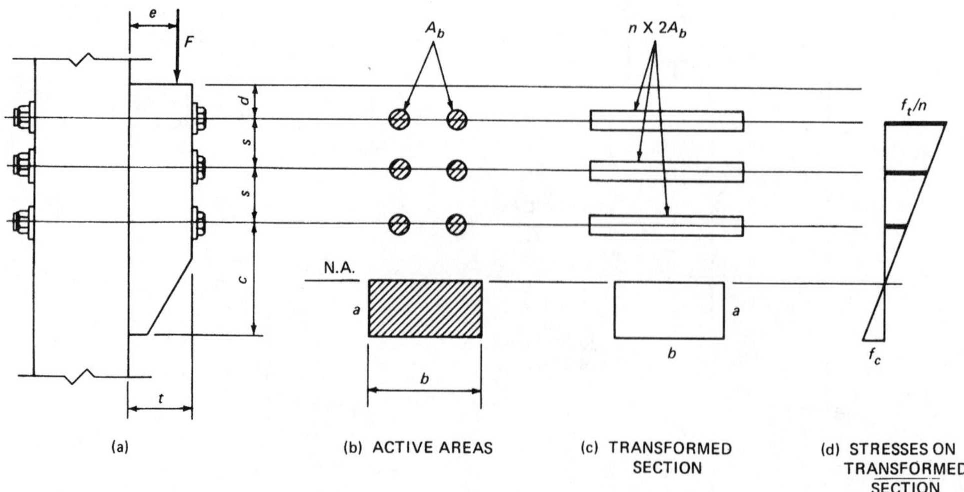

Fig. 9-5. Bracket; bolts in shear and tension (f_t = stress in steel).

To find the centroid, which is the neutral axis for bending, merely equate the moment of the area above the neutral axis to the moment of the area below. Knowing the location of the centroid (dimension a on Fig. 9-5c), compute the moment of inertia of the transformed section, and then use the flexure formula to find stresses at any level, as shown in Fig. 9-5d. Notice that the stress value, My/I, computed at each bolt location must be multiplied by n to convert it from a transformed-section stress to a steel stress. The steel stress, multiplied by the area of the bolt shank, is the tensile force in one bolt at that level.

Knowing both the shear stress and the tensile stress in the bolts, how does one decide whether or not the combined stresses are acceptable? Wood design specifications (such as the NDS) do not answer this. The authors suggest, however, that the interaction equations given by the AISC Specification (1) for bolts in combined tension and shear could be used.

Example 9-2

For the bracket of Fig. 9-5a, assume that: load $F = 7.5$ k; eccentricity $e = 3$ in.; six $^3/_4$-in.-diameter bolts are used; bracket and column are both No. 1 D. fir-larch; dimensions are: $b = 9.5$ in., $c = 10$ in., $d = 2$ in., $s = 4$ in., and $t = 5^1/_2$ in.; and the column is a 10×10. What are the forces on the most highly stressed bolts? Is the connection satisfactory?

The 7.5-k vertical load is assumed to be shared equally by the six bolts, so that each receives a vertical shear of $7.5/6 = 1.25$ k. The allowable shear, Z, from the NDS yield limit equations (not shown here) is 1611 lb., so the bolt group is satisfactory for the vertical load alone, located at the shear plane.

Modulus of elasticity perpendicular to the grain for structural woods varies with species, and values used range from about 4% to about 11% of the modulus parallel to the grain. Using E parallel to the grain of 1,600,000 psi (Appendix Table B-5), the transverse E is somewhere between 64,000 and 176,000 psi. Try an intermediate value, say 120,000 psi. Then ratio $n = 29{,}000{,}000/120{,}000 = 242$. For the transformed section, replace each pair of bolts by $2n$ times the area of the bolt shank, or $2(242)(0.442) = 214$ in.^{2}, represented by the broad areas of Fig. 9-5c. Dimension a is unknown. Distances of the bolt areas above the neutral axis are:

Top bolts (18 in. $- a$)

Next pair down (14 in. $- a$)

Bottom pair (10 in. $- a$)

The total moment of the areas above the neutral axis, about the neutral axis (N.A.) is

$$214[(18 - a) + (14 - a) + (10 - a)]$$
$$= 8988 - 642a$$

The moment of the area below, about the neutral axis, is

$$9.5a(a/2)$$

Equating, $8988 - 642a = 4.75a^2$.

The solution to this quadratic equation is $a = 12.79$ in. This puts the neutral axis above the lowest pair of bolts, so the assumption used in the above equations was incorrect. Repeat the calculation, assuming the neutral axis to lie between the bottom and middle pairs of bolts. Then the equation of moments of areas about the neutral axis becomes

$$214[(18 - a) + (14 - a)] = 4.75a^2$$

and the solution is $a = 13.87$ in., which is consistent with the assumption that a is between 10 and 14 in. The moment of inertia of the transformed section (about the neutral axis) is

$$9.5(13.87^3)/3 = 8450$$
$$214(14 - 13.87)^2 = 4$$
$$214(18 - 13.87)^2 = \underline{3650}$$
$$I_t = 12{,}104 \text{ in.}^4$$

Stress on the top part of the transformed area is Mc/I_t, or

$$7500(3)(18 - 13.87)/12{,}104 = 7.68 \text{ psi}$$

This is a level of stress for the imaginary wood that was substituted for steel at the location of the top bolts. Stress in the steel is n times as large, or

$$f_s = 7.68 \times 242 = 1860 \text{ psi (tension)}$$

The AISC specification (1) gives an allowable tensile stress for A-307 bolts, when combined with shear, as

$$26 \text{ ksi} - 1.8f_v, \text{ but not over 20 ksi}$$

So the allowable tensile stress is

$$26{,}000 - 1.8 \times 1250/0.442 = 20{,}910 \text{ psi}$$

Use 20,000 max.

The computed tensile stress is much less than this allowable, so the bolts are satisfactory for combined shear and tension.

Using the same transformed section, we may estimate the compressive stress perpendicular to the grain at the bottom edge of the bracket as

$$f_{c\perp} = Ma/I_t = 22{,}500\ (13.87/12{,}104) = 26 \text{ psi}$$

which is well below the allowable shown by Table B-5, regardless of grade.

9-3. STITCH BOLTS

Bolts under load tend to split the wood they connect. Splitting may occur because of a wedging action as the bolts are pushed parallel to the grain, or it may result because the bolt-bearing pressures normal to the grain cause tensile stress in that direction. The allowable loads for bolts are established knowing that this tendency exists. Nevertheless, if the designer is concerned about the potential for splitting under the action of the bolts, or of splitting by other causes (drying shrinkage, for example), which might reduce the effectiveness of the bolts, then *stitch* bolts may be helpful.

Figure 9-6 shows a tension splice with metal side plates. Failure would likely involve splitting and formation of wood plugs sheared from the wood members. Vertical stitch bolts would help prevent that failure.

There is no doubt that the stitch bolts improve the ultimate strength of the connection. However, the authors are not aware of any quantitative information that would help determine appropriate increased allowable loads for the main bolts.

9-4. JOIST AND BEAM HANGERS AND FRAMING ANCHORS

These are, by far, the most commonly used of modern timber connectors. Figure 9-7 shows a sampling of the variety produced by one manufacturer.

Figure 9-8 shows a joist hanger and its typical application. In this type of hanger, the joist transfers its reaction to the hanger through bearing, but load transfer to the header is by shear in special nails through the hanger flanges. In other types of hanger, the outstanding legs project up and over to bear on the top surface of the header (the double member in Fig. 9-8), in which case the vertical element of the hanger acts as a tension member carrying the entire reaction. Standard manufactured hangers for 2-in. nominal members are made of sheet metal, often galvanized. Those for larger members are made from folded or welded steel plates. Catalog information normally gives the structural designer all information needed to select the proper type and size of hanger.

Framing anchors are especially useful in light-frame construction. Used as in Fig. 9-9a, they perform the same function as joist hangers. In Fig. 9-9b, they make a connection that otherwise would require toe nailing. Wind may cause severe uplift on roof rafters. Framing anchors

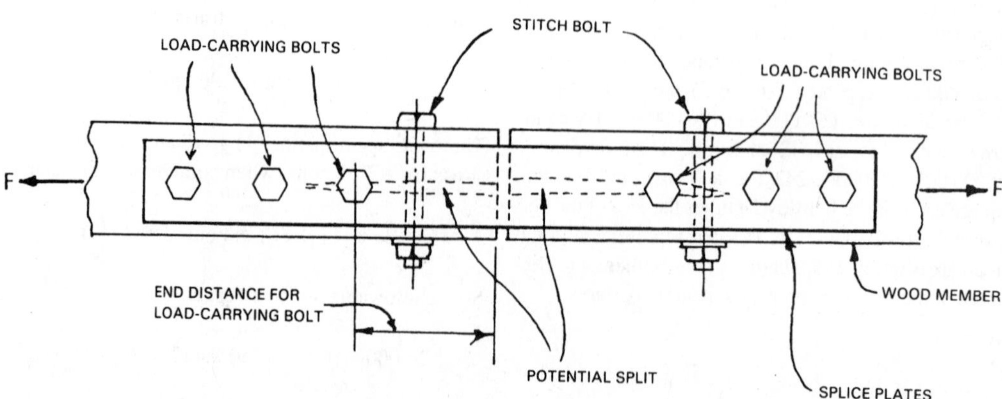

Fig. 9-6. Connection reinforced by stitch bolts (shows two alternate positions for stitch bolts).

Fig. 9-7. A variety of commercial sheet metal joist hangers. (Photograph by authors.)

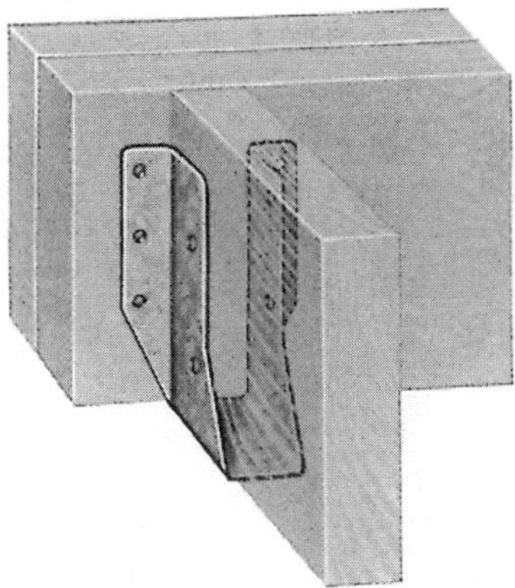

Fig. 9-8. Light-gauge metal joist hanger. (Courtesy of Cleveland Steel Specialty Company, Cleveland, OH.)

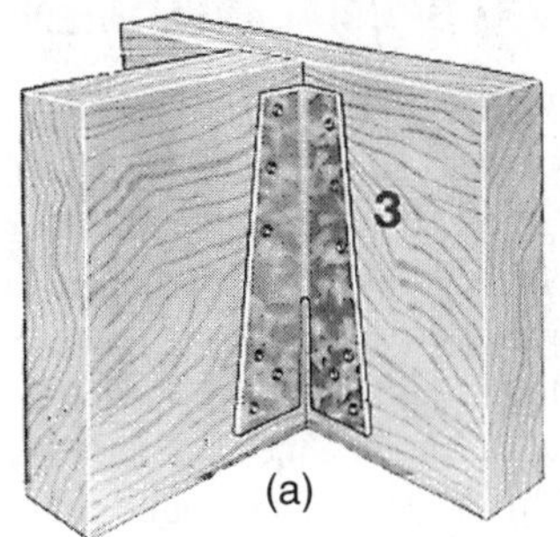

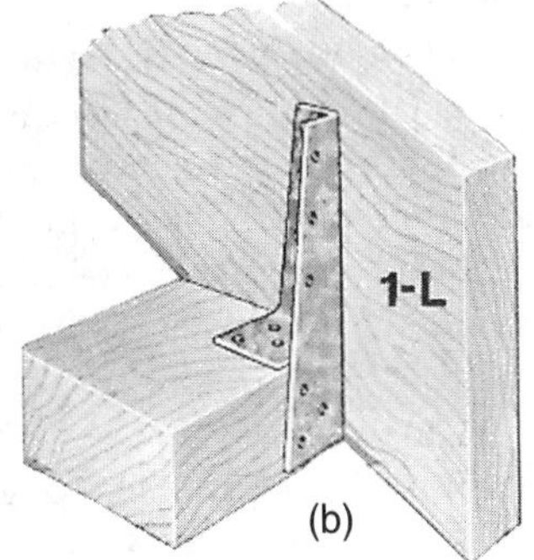

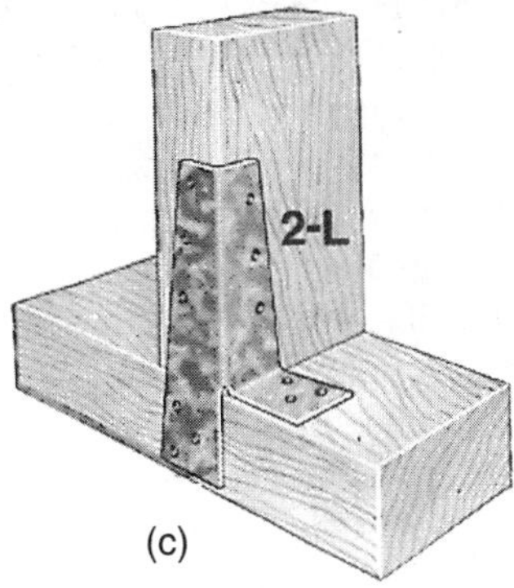

Fig. 9-9. Framing anchors. (Courtesy of Cleveland Steel Specialty Company, Cleveland, OH.)

often can transfer the required downward reaction from the rafter to the structure below, whereas toe nailing might be entirely inadequate. In Fig. 9-9c, framing anchors connect the wall studs to the bottom plate. Here, too, framing anchors replace toe nailing. The roof uplift must be transferred somehow all the way down into the foundation, so that uplift reaction (or a portion of it) must also be transferred from stud to lower plate. Again, framing anchors do this job better than toe nailing.

A structural designer ordinarily would not be involved in designing these connectors unless working for a hanger manufacturer. In that case, the design process would be largely a matter of testing to determine allowable loads for new hangers or anchors. However, designers of wood structures must often determine *whether* to use framing anchors and, if so, the number and type required.

Unless shown otherwise, the usual load-duration modifications apply to the allowable load tables shown by the manufacturers' literature. Caution: The designer must make sure that using a duration factor larger than 1.0 does not overstress the metal parts of the connection. Load-duration factors do *not* apply to design in metals.

9-5. SPECIAL WELDMENTS

Figure 9-10 shows a few of many standard types of weldment. The two types of saddle shown are used in cantilever construction to transfer the reaction of a suspended span to the overhanging member. A tension tie is needed when expected longitudinal load as well as vertical must be carried from one member to the other. Thus, the detail in Fig. 9-10b would not be suitable in seismic zones. Ordinarily, and in the absence of the need to resist calculated longitudinal loads, we would prefer not to restrain to prevent slight longitudinal shrinkage. In that case, the simple saddle of Fig. 9-10b would have enough flexibility to permit some shrinkage. In either case, the vertical load causes horizontal forces in parts of the saddle. In the simple saddle of Fig. 9-10b, for example, forces F are each equal to $R(e/d)$.

Figures 9-10c and d show two of many types of beam hanger. (Beam hangers are made in light-gauge sheet metal as well as structural steel.) Figure 9-10g shows the connection of wood beams to the top of a pipe column. The two arch-base details of Figs. 9-10e and f represent extremes for pinned-end arch construction. The first type would be used if stress conditions in the arch made it absolutely necessary to allow free end rotation. The second type would be used if some small amount of end moment could be tolerated. By no means, however, is it a fixed-end connection. Since the bolts through the arch are spaced as close together as possible ("minimum spacing") and the hole diameter is slightly larger than the bolt diameter, end rotation of the arch occurs with negligible end moment.

Finally, Fig. 9-10h shows a typical anchorage for a wood column to a concrete foundation. In another version of this connection, the U-strap has a hole in its center, and the entire assembly is placed over a single anchor bolt that is embedded in the foundation.

Most of the details in Fig. 9-10 are from the AITC *Timber Construction Manual* (2), which shows a large variety of other standard details. Hardware items of those details are not routinely stocked, but in many cases manufacturers can produce them from standard designs. If the situation is unusual in any way, however, the designer of the wood structure will probably have to design the special connector required. Figure 9-11 shows an example, which was designed for an architect by his structural engineering consultant. Figure 9-12 shows another special weldment.

The design of special weldments involves both timber design and structural steel design. The process can be broken into the following requirements:

1. Protect the wood members. This usually means providing either adequate bearing length or adequate spacing, end distance, and edge distance for bolts to carry the beam reaction. The member may be shorter than expected, and the members to which it connects may not be precisely located. So construction tolerance must be considered, providing added bearing length or (if the reaction is carried by bolts) added end distances.
2. Protect the steel parts. This requires providing the proper bending strength for parts against which the wood members bear, providing the necessary tensile or compressive strength, and meeting the AISC requirements (see reference 1) for preventing local buckling of the steel.
3. Provide bolts or other devices to hold the wood parts in proper position in the weldment or to transfer member forces to the weldment.
4. Detail the weldment and the bolts or other devices so as to allow the wood members to shrink (or swell) without restraint.
5. In all of the above, remember that load-duration modifications do *not* apply to the structural steel parts themselves.

Welded connections sometimes appear complicated. The complexity of analysis is reduced, however, by proper use of free bodies. This is shown by Example 9-4.

Example 9-3

Design a hanger to support an 8¾- × 33-in. sloping glulam purlin, combination 20F-V4, from a horizontal glulam girder, 8¾ × 42 in. Both members are Douglas fir. The 15-k vertical reaction is due to snow plus dead load. The following shows only major elements of the solution.

Bearing stress allowable (not to be modified for load duration) is that for compression perpendicular to the

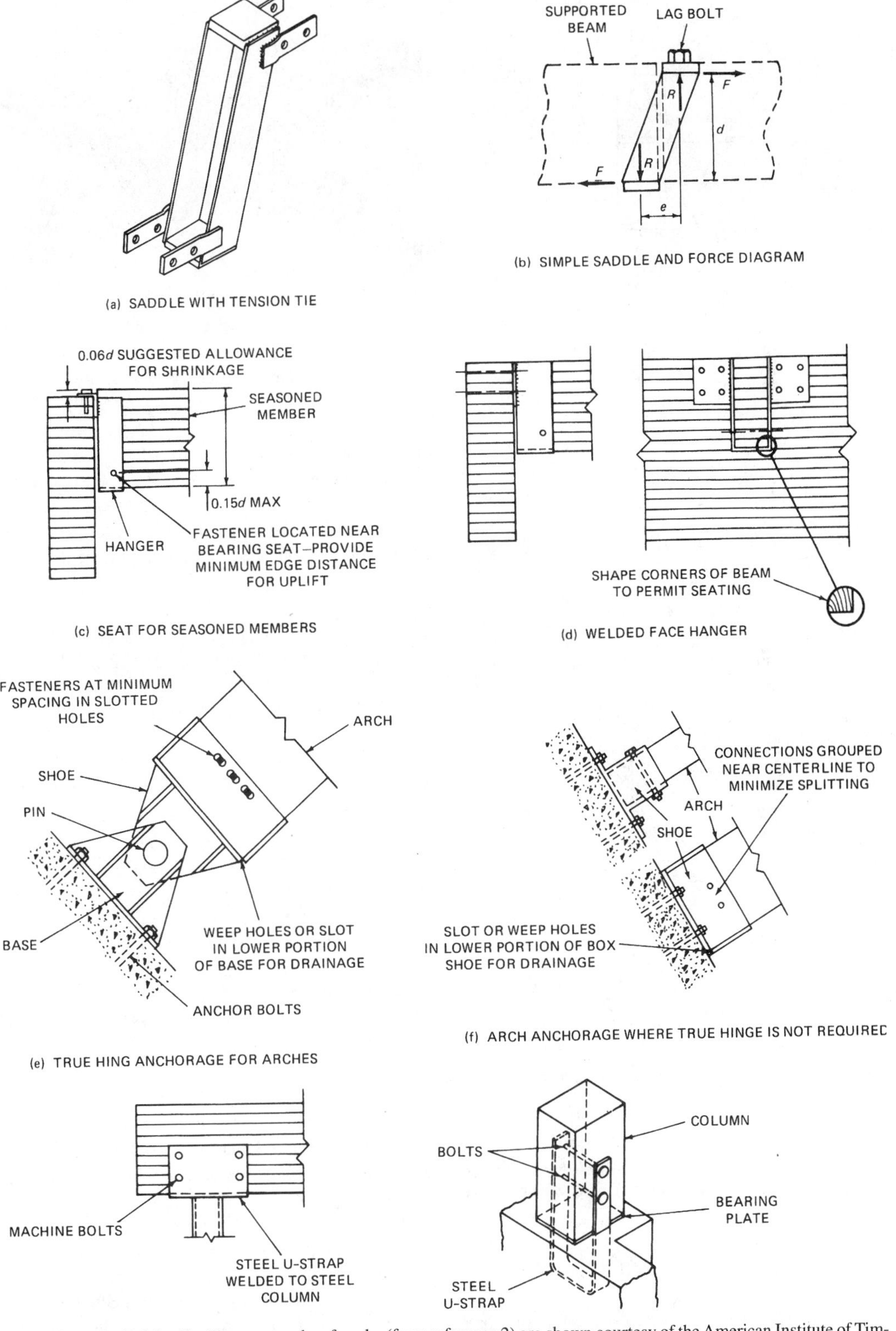

Fig. 9-10. Typical details. Figures a, c, d, e, f, and g (from reference 2) are shown courtesy of the American Institute of Timber Construction (AITC). Figures b and h are from an earlier edition of reference 2. Designers should check the latest edition of reference 2 for other details.

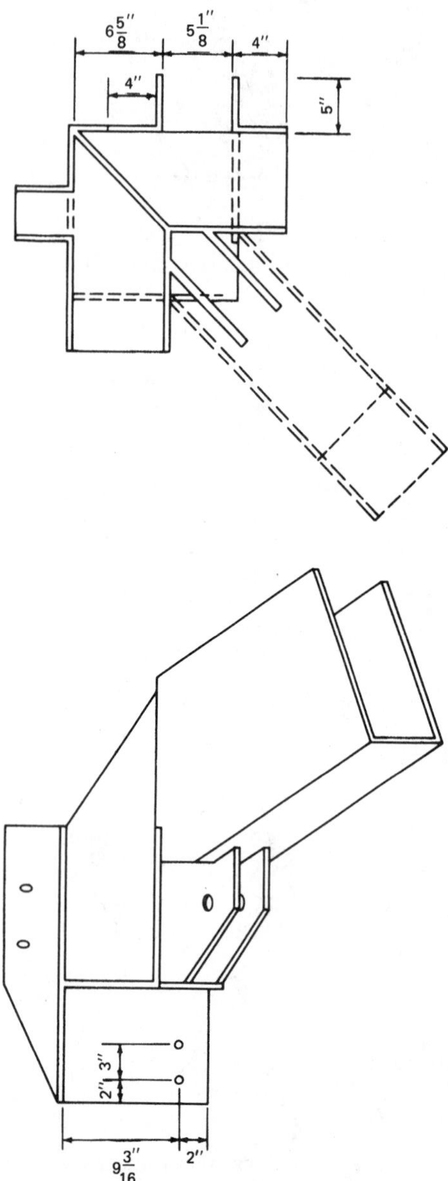

Fig. 9-11. Two views of a complex weldment. (Courtesy of V. Hairabedian, AIA.)

grain. On the compression face (top) this is 560 psi, while on the tension face (bottom) it is 650 psi. The bearing area required is

$$15{,}000/650 = 23.1 \text{ in.}^2$$

Bearing length must be 23.1/8.75 = 2.64 in. or more. Tentatively, try a 5-in. bearing length (seat width).

A possible detail is as shown by Fig. 9-13. Allowing $^1/_8$-in. clearance for inserting the purlin into the

Fig. 9-12. Complex weldment for joining glulam arches in fabric-covered dome structure. (Photograph by authors.)

hanger, dimensions w should be $8^7/_8$ in. Assuming the vertical side members to be $^3/_8$-in. thick, the transverse steel seat the purlin bears on can be considered a simple beam of 9.25-in. span. If the purlin bears uniformly on the seat, the seat bending moment will be approximately

$$M = \left(\frac{15,000}{8.75 \times 5}\right) 9.25^2 / 8$$

$$= 3667 \text{ in.-lb/inch of plate width}$$

Allowing a 27-ksi bending stress (for A36 steel), the required section modulus is

$$S = 3667/27{,}000 = 0.136 \text{ in.}^3 \text{ per inch width}$$

$$bt^2/6 = 0.136; \qquad b = 1.0''$$

$$\text{Required } t = 0.903 \text{ in.}$$

At first glance, this thickness seems excessive. What can be done about it? Four possibilities occur to the authors.

1. Consider the plate deflection when calculating the bending moment. (The plate deflection decreases the pressure near the center, concentrating it more near the edges of the supported glulam. This might deform the edges of the glulam. In any case, stress conditions would be difficult or impossible to compute accurately. Unless physical testing is done to verify the strength, it would be better to avoid this solution.)

2. Consider continuity of the vertical side plates with the bottom plate. (From Fig. 9-13c, however, we see that the vertical members have little flexural stiffness compared to the bottom plate. So continuity will have little effect in reducing the bottom plate bending

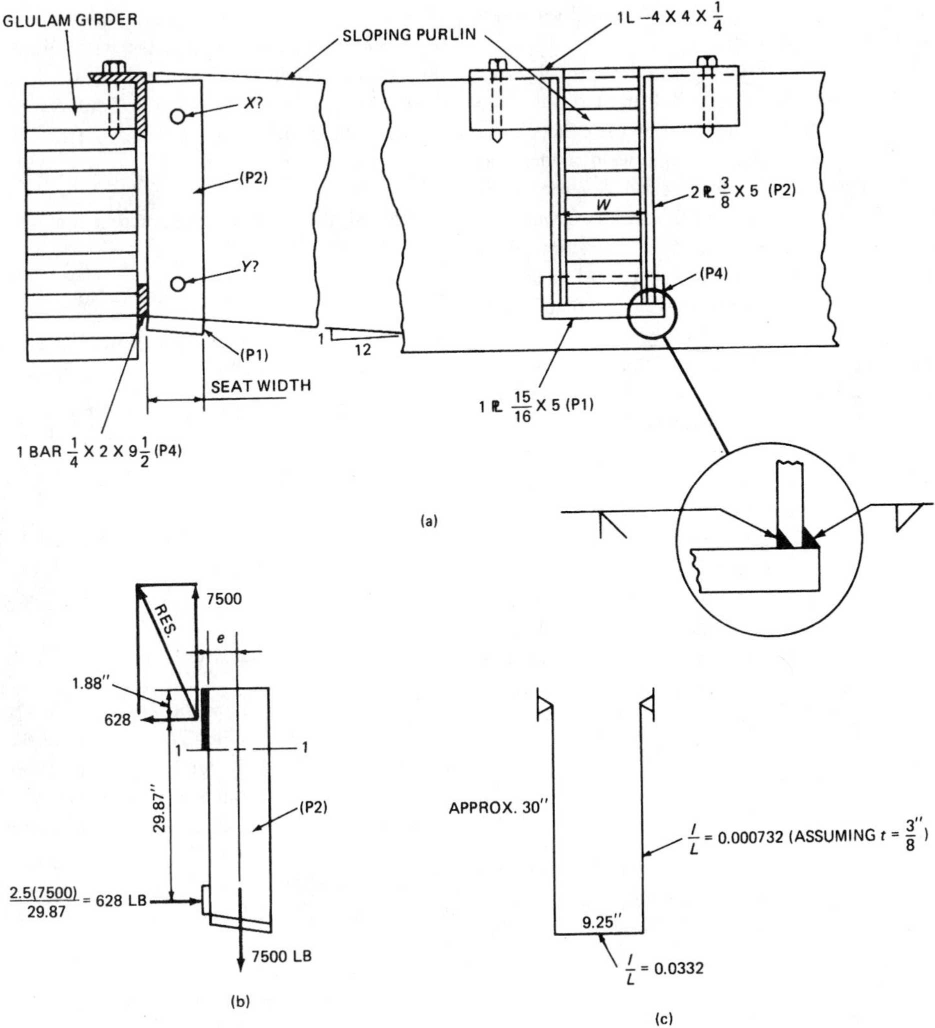

Fig. 9-13. Special hanger for Example 9-3.

moment, and the moment computed above is not overconservative.)

3. Stiffen the bottom plate. (This could be done only if the projecting stiffener were not objectionable.)

4. Realize that this is a large hanger, and *use* a 15/16-in. plate. (That is what will be done here.)

Each vertical side plate is subjected to the forces shown by Fig. 9-13b. Stress in the side plate on horizontal section 1–1 at the lower edge of the upper weld is $P/A + M/S$, where (from forces below the section)

$$M = 628\ (29.87 - 1.88) = 17{,}580 \text{ in.-lb}$$

$$f = 7500/5t + 17{,}580/(t \times 5^2/6)$$

$$= 5719/t \text{ psi}$$

Setting this equal to an allowable tensile stress of 22,000 psi, the required thickness t is 0.26 in.

From reference 1, the limit for ratio b/t (width to thickness) is $95/\sqrt{F_y} = 15.8$ for A36 steel. Using this limit, the minimum allowable t is 5.0/15.8 = 0.316 in. To meet both requirements, use $t = 3/8$ in.

The top angle is sized so that its bearing stress on top of the glulam girder will not exceed the allowable for the girder (560 psi) and thickness is determined so that the bending strength of its horizontal leg is adequate. The angle may be shimmed out from the face of the girder so that the angle fillet does not deform the edge of the girder. Lag screws shown carry no computed load, but merely keep the hanger in position on the girder.

Weld joining each vertical plate to the top angle

carries the resultant of a 7500-lb vertical load and a 628-lb horizontal load.

Weld joining the bottom plate to the vertical plates must be adequate to transfer 7500 lb at each end of the bottom plate. Because the bottom plate acts as a beam, its ends rotate. Thus, a single weld at each end would be subject to a "wedge" action, severely overstressing the weld. To prevent this, use two welds as shown and assume the entire 7500-lb reaction to be resisted by the inner weld. The inner weld must be a groove weld (not a simple fillet weld) so that it does not interfere with the seating of the beam.

Something—bolts, lag screws, or even nails—should be used to prevent the rafter from sliding out of the hanger. Should the holes provided for this be in position x or position y?

Sad experience has shown that position x is bad. Glulam beams restrained there by nails suffered longitudinal splits, looking much like shear failures. The cause was vertical shrinkage, which raised the bottom of the beam off its intended seat. The reaction, then carried by the nails rather than by beam bearing, caused perpendicular-to-grain tension that split the glulam. Position y is the better position.

Example 9-4

Figure 9-14a shows two roof purlins, plus vertical and diagonal bracing members, all intersecting at and supported by a sloping rafter. All members are D. fir glulams. Each purlin transfers a 3-k vertical reaction to the rafter (dead + snow). Wind loads in the vertical and diagonal braces (reversible—one tension, the other compression) are 2.4 k and 4 k, respectively. For equilibrium, either purlin may be required to resist the horizontal component of the diagonal load. Purlins, vertical, and diagonal are all $5^1/_8$ in. wide. Design a welded and bolted connection detail.

Reference 2 does not suggest a detail for this situation, so the designer must plan the complete connection. One solution—not necessarily the best, and by no means the only one—is shown by Fig. 9-14b. The connection is in two separate parts, one on each side of the rafter. A cross plate could be welded between the two parts to join them together, but the resulting assembly might be more difficult to handle than two separate pieces.

The left portion is shown by Fig. 9-14c. Load applied to this part is the 3-k downward purlin force ($D + S$), plus a possible horizontal 3.2 k (horizontal component of the 4-k force in the diagonal) either way due to wind.

First, consider only the vertical reaction to the purlin. Load will be perpendicular to the grain. Using a dowel-bearing strength perpendicular to the grain of 2600 psi for D. fir-larch, 3/4-in. diameter bolts, and 0.3125-in.-thick steel side plates, yield limit equations (not shown here) will give an allowable perpendicular to the grain of $Z = 1864$ lb. For the vertical reaction of the purlin, N req $= 3000/(1.15 \times 1864) = 1.40$ bolts.

Second, for combined vertical reaction and horizontal wind load:

Resultant force $= 4.39$ k, at 43.2° to grain

Using Hankinson's equation, dowel-bearing strength at 43.2°

$= (5600)(2600)/(5600 \sin^2 43.2° + 2600 \cos^2 43.2°)$

$= 3635$ psi

From yield limit equations, allowable (at 43.2° to grain) $Z = 2414$ lb

N req $= 4390/(1.33 \times 2414) = 1.37$ bolts.

Use two 3/4-in.-diameter bolts, as shown.

Axial load in the purlin can be tensile, so the required end distance in the purlin is $7D = 5.25$ in. if the bolts carry their full allowable load. However, the required end distance can be reduced (using the geometry factor) by the ratio of actual bolt load to full allowable parallel to the grain. Yield-limit equations give a parallel-to-grain allowable $Z = 3267$ lb. So the required end distance is $(5.25)(3200)/(1.33 \times 3267) = 3.87$ in. Use 4 in.

The two plates marked P1 are spaced $5^1/_4$ in. apart, allowing $^1/_8$ in. more than the width of the purlin between them. Plates P1 are welded to plate P2. Allowing $^1/_2$-in. clearance between the end of the purlin and plate P2, each line of weld joining the plates must resist a 1.5-k vertical load at an eccentricity of 4.5 in. and a horizontal load of 1.6 k, as shown on the free-body diagram of plate P1 in Fig. 9-14d. The weld resists the applied loads with a vertical force, V, a horizontal force H, and a moment, M. A second load case ($D + S$ only) will not control.

How thick should the plates be? Thick enough that the bolt bearing value in the steel plates will be adequate, and preferably thick enough to meet the AISC width/thickness limit of $b \leq 15.8t$ to avoid local buckling of the edge of the A36 plate. For plates P1, $^3/_8$-in. thickness would be adequate for a 6-in. plate width. For simplicity, make plate P2 the same thickness.

The right portion, Fig. 9-14e, is slightly more complicated. The purlin connection to plates P3 is the same as for the left portion. The diagonal has either 4 k tension or 4 k compression, but no shear. The adjusted

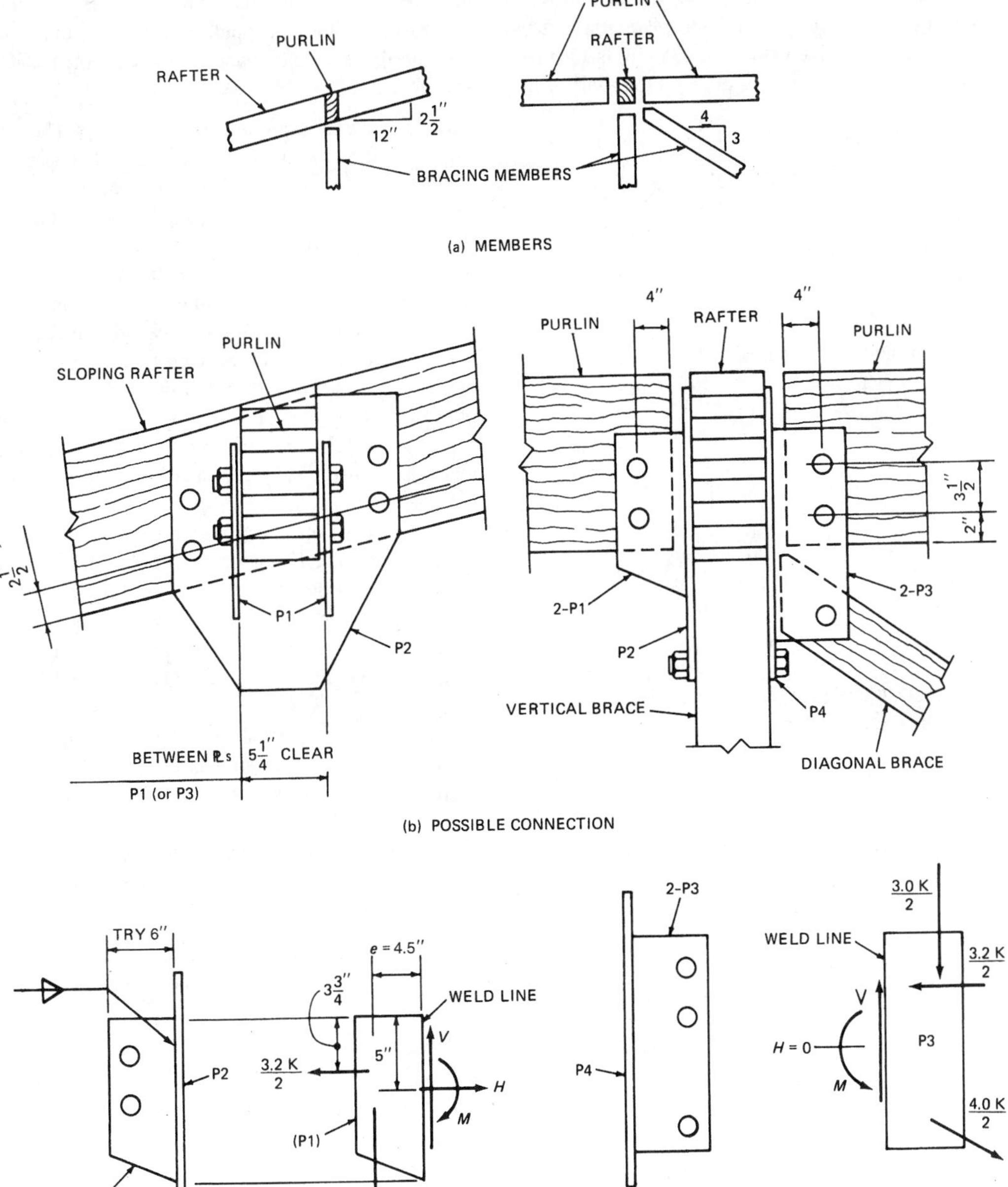

Fig. 9-14. Connection for Example 9-4.

double-shear value of one 3/4-in. bolt in the diagonal is 1.33 (3267) = 4350 lb, so only one 3/4-in. bolt is needed. The end distance required in the diagonal is 7(3/4)(4.0/4.35) = 4.8 in. Plotting the detail to scale, we conclude that the plates P3 must be 6 in. wide. The *b*/*t* criterion suggests that plates P3 be 3/8 in. thick.

Welds joining plates P3 to P4 are loaded eccentrically, by the diagonal as well as the purlin. Figure 9-14f shows one plate P3 as a free body. In designing the weld, two load cases must be recognized: the vertical 1.5 k (*D* + *S*) alone, and the combination of all three factors (*D* + *S* + *W*) shown.

The vertical member transfers 2.4 k wind load (either tension or compression) into plates P2 and P4. One ¾-in. bolt (value with wind 4350 lb) is adequate.

Loading for bolts joining plate P4 to rafter:

Under $D + S$

Purlin reaction = 3.0 k

Under $D + S + W$:

Purlin reaction =	3.0 k down
Vertical component of 4 k tension in diagonal =	2.4 k down
One-half of vertical member force =	<u>1.2 k</u> up
Sum =	4.2 k on one shear plane

$3.0/1.15 = 2.61 \qquad 4.2/1.33 = 3.16$

The loading including wind controls. Angle of load to grain = 78.2° (90 minus roof slope of 11.8°).

Using Hankinson's equation, dowel-bearing strength at 78.2°

$$= (5600)(2600)/(5600 \sin^2 78.2° + 2600 \cos^2 78.2°)$$

$$= 2660 \text{ psi}$$

From yield-limit equations, single-shear allowable at 78.2° to the grain is $Z = 1008$ lb. From Table C-9, a conservative estimate of group action factor is 0.97, so the adjusted allowable, Z', is $0.97(1.33)(1008) = 1300$ lb.

$$N \text{ req} = 4400/(1300) = 3.4 \text{ bolts}$$

Try 4 bolts. These bolts must also pass through plate P2 on the other side of the rafter.

Bolts joining the left portion to the rafter must carry 3.0 k ($D + S$), or 4.2 k ($D + S + W$). The latter controls, and N req = 4200/1300 = 3.2 bolts. For symmetry, use four ¾-in. bolts passing through both plates, P2 and P4. These plates will have to be wide enough to satisfy bolt spacing and edge distance requirements. The lower edge of the rafter is the loaded edge, and the distance of the bolts from that edge must be at least $4D$, or 3 in.

As an alternate to connecting the purlins as shown, they could be supported on horizontal bearing plates welded between plates P1 and plates P3.

Example 9-5

In the glulam chapter, Example 8–8 covered the design of a cantilever beam system. The solution was in the form of a designer's computation sheet, shown by Fig. 8–14. Using loads and sections from that solution, design a saddle connection to support the suspended span.

Another designer's computation sheet, shown here by Fig. 9-15, shows a solution to this part of the problem. There being no longitudinal loads to contend with, the designer assumed first that a simple saddle (as in Fig. 9-10b) might suffice. This would require two 7-in. long lag screws to resist the horizontal forces, F. So the designer continued assuming a saddle in which the required forces F are resisted by end-grain bearing between the glulam and transverse steel bars. It was necessary to estimate the bar width in order to compute force F. For greater accuracy in designing the bearing plates, the designer might consider two-way plate action, using equations such as are developed in reference 3.

In the final step, the designer verified that ⅝- × 4½-in. bars would be satisfactory for the slanting side parts of the saddle. Notice that, since the allowable steel stresses for tension and for weak-axis bending are not alike, the designer used an interaction equation rather than merely adding $P/A + Mc/I$ to find the total stress.

Figure 9-15 does not show design of the necessary welds, although that certainly is a part of the problem. If appearance is of little concern, the detail of Fig. 9-10a might be used instead.

If the top seat of the saddle would interfere with deck material, the designer would show that seat recessed into a dap in the top of the left glulam.

9-6. SHEAR PLATES AND SPLIT RINGS

Screws and bolts have a common fault—they slip until they come into bearing, following which they cause bearing deformation in the wood. Consequently, it is difficult to realize the full load capacity of wood members when bolts or screws are used. An early attempt to correct this inadequacy was the Kübler dowel, shown by Fig. 9-16. This double-tapered dowel of cast iron or oak was embedded, one-half in each of the two wood members being connected. A bolt passing through both wood members and the dowel held the pieces together, but did not transfer shear. The dowel (up to 4 in. wide) presented a large bearing area to the wood member and could transfer large loads from one member to the other without severe bearing deformation. The Kübler dowels probably inspired develop-

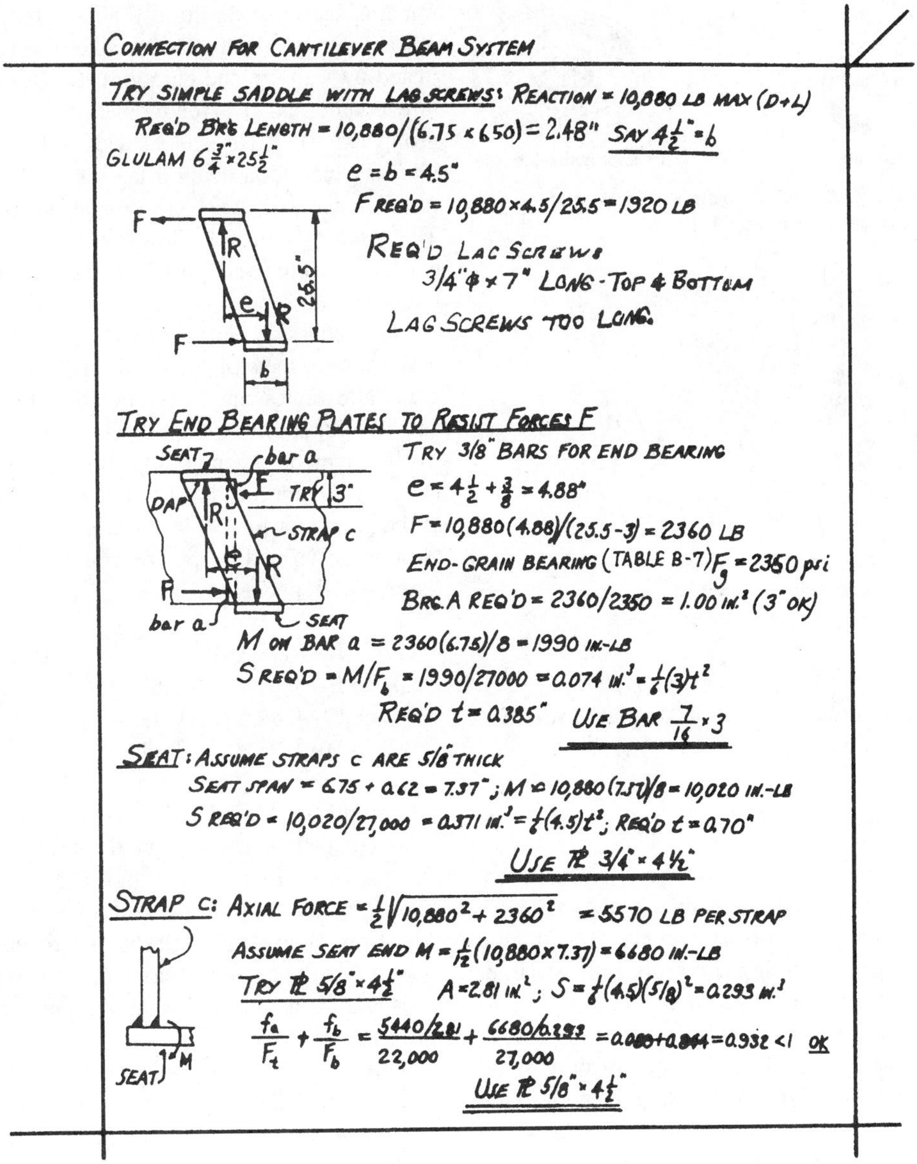

CONNECTION FOR CANTILEVER BEAM SYSTEM

TRY SIMPLE SADDLE WITH LAG SCREWS: REACTION = 10,880 LB MAX (D+L)

REQ'D BRG LENGTH = 10,880/(6.75 × 650) = 2.48" SAY $4\frac{1}{2}$" = b

GLULAM $6\frac{3}{4}$" × $25\frac{1}{2}$"

e = b = 4.5"

F REQ'D = 10,880 × 4.5/25.5 = 1920 LB

REQ'D LAG SCREWS
3/4"φ × 7" LONG - TOP & BOTTOM

LAG SCREWS TOO LONG.

F, R, e, R, 25.5", F, b

TRY END BEARING PLATES TO RESIST FORCES F

SEAT, bar a, F, TRY 3", DAP, R, STRAP c, e, R, F, bar a, SEAT

TRY 3/8" BARS FOR END BEARING

$e = 4\frac{1}{2} + \frac{3}{8} = 4.88"$

F = 10,880(4.88)/(25.5 − 3) = 2360 LB

END-GRAIN BEARING (TABLE B-7) F_g = 2350 psi

BRG. A REQ'D = 2360/2350 = 1.00 IN.² (3" OK)

M ON BAR a = 2360(6.75)/8 = 1990 IN.-LB

S REQ'D = M/F_b = 1990/27000 = 0.074 IN.³ = $\frac{1}{6}(3)t^2$

REQ'D t = 0.385" USE BAR $\frac{7}{16}$ × 3

SEAT: ASSUME STRAPS c ARE 5/8" THICK

SEAT SPAN = 6.75 + 0.62 = 7.37"; M = 10,880(7.37)/8 = 10,020 IN.-LB

S REQ'D = 10,020/27,000 = 0.371 IN.³ = $\frac{1}{6}(4.5)t^2$; REQ'D t = 0.70"

USE PL 3/4" × $4\frac{1}{2}$"

STRAP c: AXIAL FORCE = $\frac{1}{2}\sqrt{10{,}880^2 + 2360^2}$ = 5570 LB PER STRAP

ASSUME SEAT END M = $\frac{1}{12}$(10,880 × 7.37) = 6680 IN.-LB

TRY PL 5/8" × $4\frac{1}{2}$" A = 2.81 IN.²; S = $\frac{1}{6}(4.5)(5/8)^2$ = 0.293 IN.³

$\frac{f_a}{F_t} + \frac{f_b}{F_b} = \frac{5440/2.81}{22{,}000} + \frac{6680/0.293}{27{,}000}$ = 0.088 + 0.844 = 0.932 < 1 OK

SEAT, M

USE PL 5/8" × $4\frac{1}{2}$"

Fig. 9-15. Computation sheet for Example 9-5.

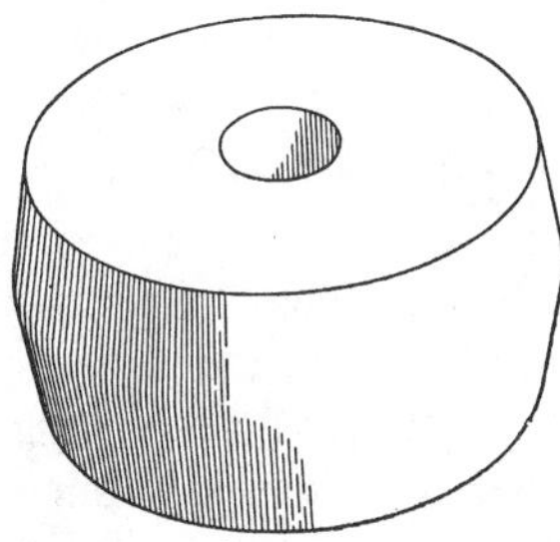

Fig. 9-16. Kübler dowel, a forerunner of shear plates and split rings.

ers of modern split-ring and shear-plate connectors. These too, present fairly large bearing areas and thus avoid causing large bearing deformations in the wood.

Figure 9-17 shows two types of shear plate and Fig. 9-18 shows the split-ring connector. These connectors reached peak popularity during World War II, when material shortages made it necessary to use timber for long-span structures that until then had been practical only in structural steel. Today, the use of split rings has

(a) Pressed steel (b) Cast malleable iron

Fig. 9-17. Shear plates. (Courtesy of Cleveland Steel Specialty Company, Cleveland, OH.)

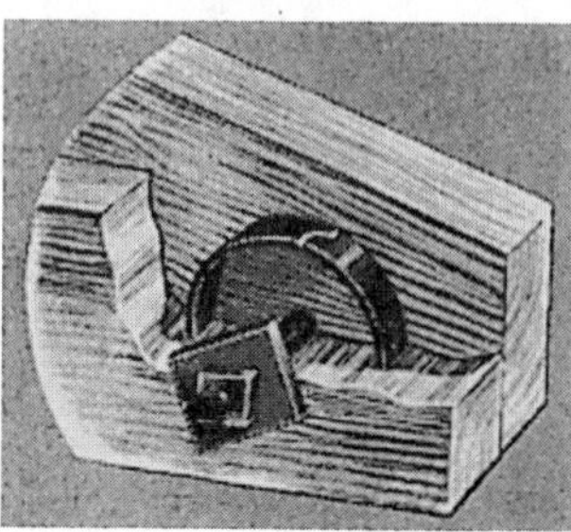

Fig. 9-18. Split-ring connector. (Courtesy of Cleveland Steel Specialty Company, Cleveland, OH.)

declined somewhat, but shear plates still find many applications.

9-7. SHEAR PLATES

Shear plates are of two varieties—pressed steel plates of 2⅝-in. outside diameter, and malleable iron ones 4 in. in outside diameter. The smaller plates are used with ¾-in. bolts and the larger ones with either ¾- or ⅞-in. bolts.

A shear plate is installed in a precut groove and dap so that its flat surface is flush with the face of the wood member and its flange is in the groove. Two nails through the small holes in the shear plate prevent it from dropping out during handling. Load is transferred from the member into the shear plate by wood bearing pressures against the sides of the flange. It is then transferred to the bolt by bearing of the shear plate against the bolt shank. The bolt then carries the load to the other member—again by way of a shear plate if the other member is wood.

Shear plates are useful for connecting wood members to steel members or for joining together wood members that have to be disconnected and reinstalled.

Appendix Table C-17 shows allowable loads for shear plates installed in seasoned wood that remains seasoned during use and is subject to normal-duration loads. These values must be adjusted by C_D (for load duration), C_M (for moisture conditions), C_t (for temperature), C_g (for group action), C_Δ (for end distance and spacing), C_d (for penetration depth if lag screws are used rather than bolts), and C_{st} (for metal side plates). The resulting adjusted allowable load must not exceed the allowable based on the strength of the steel bolt (see footnote to Table C-17).

The metal side plate factor, C_{st}, is permitted only for 4-in. shear plates and for the parallel-to-grain allowables only, as follows: 1.18, 1.11, and 1.05 for species groups A, B, and C, respectively. No metal side plate adjustment is allowed for group D, nor for the allowable perpendicular to the grain, nor for 2⅝-in. shear plates. Species Groups are defined by specific gravity, G, of different wood species as follows:

Group A—$G \geq 0.62$
Group B—$0.49 \leq G \leq 0.61$
Group C—$0.42 \leq G \leq 0.48$
Group D—$0.31 \leq G \leq 0.41$

Interpolation on Table C-17 is permitted for intermediate thicknesses, but the lower thickness of wood listed in each block is the minimum thickness allowed, and the larger allowable load shown in each block (or by the footnote) is the absolute maximum. Table C-18 gives end distance and spacing requirements as well as geometry factors.

Table C-17 gives allowables both perpendicular to and parallel to the grain. For allowable loads at other angles, first complete all adjustments, *then* use Hankinson's formula, as follows:

$$N' = P'Q'/(P' \sin^2 \theta + Q' \cos^2 \theta) \qquad (9\text{-}4)$$

9-8. SPLIT RINGS

Split rings have tapered thickness, so special tools are needed to cut the grooves to receive them. The bolt merely holds the parts together firmly, but does not carry load from one member to the other. Each member must be grooved, so the split ring is not used to connect wood members to metal ones.

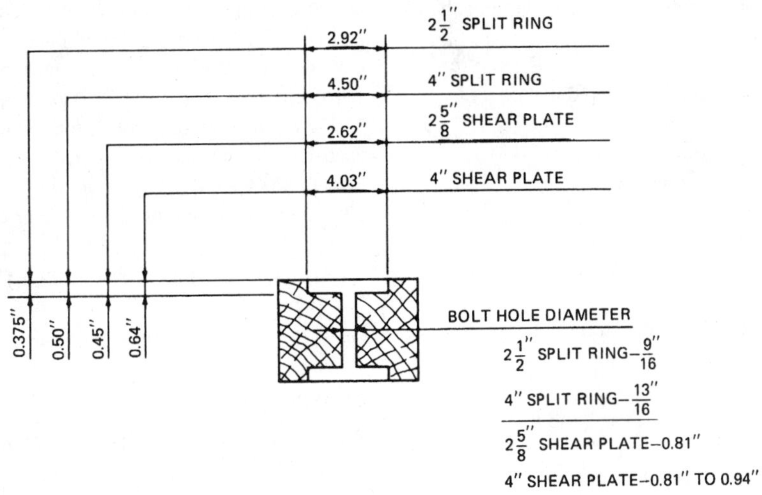

Fig. 9-19. Net area of wood member with split rings (or shear plates) in two faces.

Figure 9-19 shows dimensions that are helpful in determining the net area remaining in a wood member after installation of a split ring (or shear plate) and its accompanying bolt. Appendix Table C-19 shows allowable loads for split rings installed in seasoned wood that remains seasoned in service and is subject to normal-duration loading. The table gives values for load both parallel to and perpendicular to the grain. For allowable loads at other angles, use Hankinson's equation (Eq. 9-4). (All adjustment factors apply equally to values parallel to and perpendicular to the grain, so Hankinson's equation may be solved first, if desired.) Adjustment factors to consider are load duration, wet service, temperature, group action, and geometry. If lag screws are used rather than bolts to hold the assembly together, then the penetration depth factor must also be considered.

Table C-18 shows end distance and spacing requirements for split rings. Nomographs (reference 2) are helpful in computing spacings, end distances, and edge distances in joints where the axis (line through a row of connectors) is not parallel to the grain.

Example 9-6

What is the allowable load, F (including wind), for the connection of Fig. 9-20? The lumber is seasoned, but will be wet in service.

Load is parallel to the grain in the 2×8 member and at 53.1° to the grain in the 4×10. The members are of species group B. From Table C-19, the unadjusted allowables for one 2½-in. split ring installed on one side of the member are:

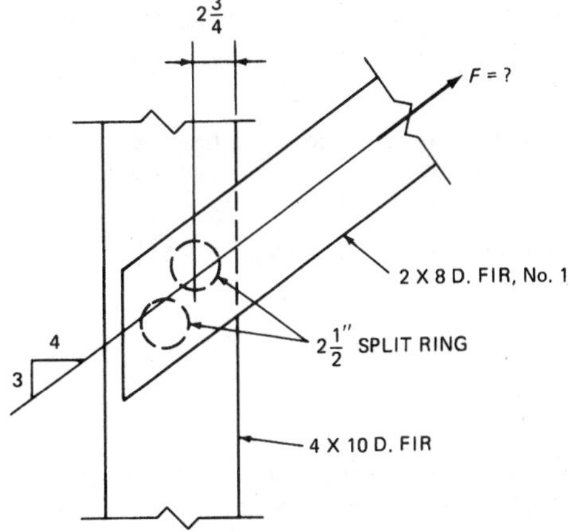

Fig. 9-20. Connection for Example 9-6.

Diagonal member, 2730 lb, parallel to the grain

Vertical member, 2730 lb parallel to grain and 1940 perpendicular to grain

By Hankinson's equation, allowable at 53.1° to grain in the vertical member = 2166 lb

The unadjusted value in the vertical member controls. Modifying for load duration and for wet service,

$$\text{Allowable } F \text{ (2 rings)} = 2(2166)(1.33)(0.67) = 3860 \text{ lb}$$

based on the connection strength. Net area of the diagonal is

$$1.5(7.25) - 0.375(2.92) - 1.125(9/16)$$
$$= 9.15 \text{ in.}^2$$

The allowable tensile stress is

$$F_t' = C_F C_M C_D F_t = 1.2(1.0)(1.33)(675)$$
$$= 1077 \text{ psi (on net section)}$$

Allowable F based on tensile strength of the diagonal is

$$F = 9.15(1077) = 9850 \text{ lb}$$

The controlling allowable F is 3860 lb.

Example 9-7

Design a double-shear splice using 2½-in. split rings to develop the maximum allowable tension in No. 1 Dense southern pine 3 × 10s, surfaced dry and used at not over 19% MC.

If one row of rings is used, the net area of the member spliced will be its gross area minus the projected area of two half-rings and of the 9/16-in. bolt hole between bottoms of ring grooves. That is, it will be the shaded area of Fig. 9-19, or

$$23.125 - 2(2.92 \times 0.375) - (2.5 - 0.75)(9/16)$$
$$= 19.95 \text{ in.}^2$$

For No. 1 Dense southern pine, Appendix Table B-4 gives $F_t' = 775$ psi on the net area. So the splice will have to be designed for 19.95(775) = 15,460 lb. The two side members, combined, must have at least as much net area as the main member, so they will have to be 1.25 in. thick or more. Use 2 × 10 side pieces.

Load is parallel to the grain in both main and side members. The allowable load per 2½-in. ring in species Group B is 2730 lb in the 2-in. side member and also in the 3-in. main member. The only adjustment needed is the group action factor, C_g. Estimating that C_g is 0.90, the approximate required N = 15,460/(0.90 × 2730) = 6.3 rings.

Try eight rings per end of the splice (one row in each face of the main member, four rings per row). Use Table C-15: A_m/A_s = 2.5/3.0 = 0.83, and A_m = 23.125 in.² Interpolating, find a factor of C_g = 0.86.

$$N \text{ required} = 15,460/(0.86 \times 2730) = 6.6 \text{ rings}$$

Use eight rings per end of splice, as shown by Fig. 9-21.

Example 9-8

What spacing and end distance are needed for the splice of Example 9-7?

Table C-18 gives spacing requirements for connectors with geometry factors of 1.0 and of 0.5. In other words, it gives spacing requirements for connectors loaded to 100% and to 50% of their full allowable values. In this case only 6.6 rings are required, but we are using 8, so the rings are loaded to only 82.5% of their allowable. Interpolating between the 3.5 in. shown for C_Δ = 0.5 (50% load) and 6.75 in. for C_Δ = 1.0 (100% load), the required spacing for this splice is

$$3.50 + (6.75 - 3.50)(82.5 - 50)/50 = 5.61 \text{ in.}$$

Use a 6-in. spacing.

Similarly, for end distance, the table shows requirements for C_Δ = 1.0 (100% full load) and C_Δ = 0.625 (62.5% full load). The interpolated requirement is 4.22 in. Use 5-in. end distance. This applies to the inner ends of the members being spliced, as well as the outer ends of the side pieces.

Example 9-9

Figure 9-22 shows the intersection of members at a bottom-chord joint of an existing roof truss. All members are species group C. The new owner of the build-

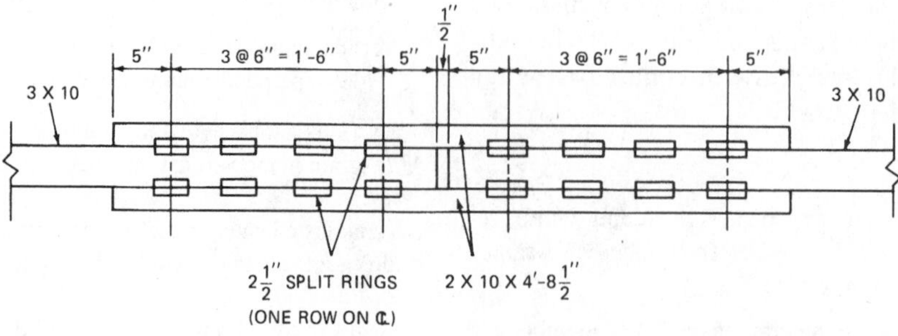

Fig. 9-21. Solution to Examples 9-7 and 9-8.

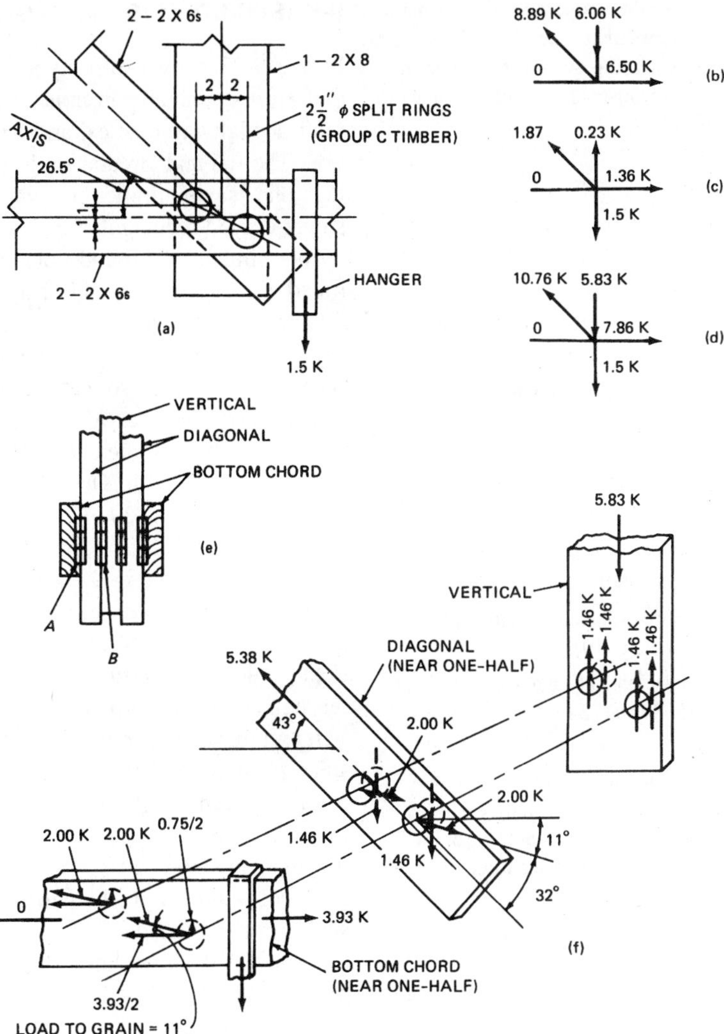

Fig. 9-22. Connection for Example 9-9.

ing wants to add a materials-handling system that will apply an additional 1.5-k load just next to the panel point, as shown by Fig. 9-22. Will the connection be satisfactory under the increased loads?

From record prints of the original construction drawings, you find that the original member forces under dead plus snow loads are as shown by Fig. 9-22b. Assume that you have computed member forces due to the new conveyor load; they are shown in Fig. 9-22c. The total member forces are as in Fig. 9-22d.

Figure 9-22e shows a cross section of the truss joint. There are a total of eight split rings, two in each plane between members. It is easiest to see what is happening in a joint of this type if we use an "exploded" view, as in Fig. 9-22f. This exploded view isolates a portion of each member, showing it as a free body with *all* forces that act on that portion—a set of forces that must be in equilibrium.

Start with the vertical. The member is being pushed downward with a 5.83-k load. That force is resisted by four rings labeled *B*. The force carried by each ring must be 5.83/4, or 1.46 k.

Next, consider a short portion of one member of the bottom chord. It is pulled to the right by 3.93 kips (one-half of 7.86 k) and downward with 0.75 k from the proposed conveyor load. To keep this free body in equilibrium requires a 4.00-k force at 11° to the horizontal, or 2.00 k in each of the two rings *A*.

This completes the solution of forces in the rings. To find the ring forces it was not necessary to consider the diagonal, but we do now know the forces applied to it, as shown by the center picture of Fig. 9-22f. (As a general rule, consider the inside and outside members first, and the intermediate ones last.) Now that we know the force in each ring, we must compare it to the allowable for each ring and in each member in which the ring is installed.

For rings *A:*

In diagonal—load is at 32° to the grain.

Allowable parallel to grain = 1760 lb

Allowable perpendicular to grain

Edge distance = 2.75 − 0.70 (scaled) = 2.05 in.

For *ED* = 1.75 in. the allowable is 0.83(1250) = 1040 lb; for 2.75 in. it is 1250 lb

Interpolating, for *ED* of 2.05 in., allowable = 1103 lb

By Hankinson's equation, at 32° to grain

Allowable = 1508 lb

In bottom chord—does not control (by inspection, Table C-19).

Adjust for snow. Controlling allowable in ring *A* = 1.15(1508) = 1734 lb.

The calculated load is 2.00 k, so rings *A* will be overloaded, and regardless of the outcome of checking rings *B*, the connection is unsatisfactory for the increased load. Just to illustrate the procedure, however, here is the check for rings *B:*

In diagonal—load-to-grain angle is 47°.

By Hankinson's equation, allowable at 47° to grain = 1335 lb

In vertical, allowable = 1760 lb (does not control).

Modified for snow, allowable per ring = 1.15(1335) = 1535 lb.

Actual load = 1460 lb. Rings *B* are OK (but rings *A* are not).

9-9. SPIKED GRIDS AND GRID PLATES

Figure 9-23 shows three types of spiked grid that are used mainly in pole construction, railroad or highway trestles, and marine construction. The flat and circular grids are for connecting one sawn timber to another, and the single-curved rectangular grid is for connecting a sawn timber to a round pole. The grids illustrated are cast in malleable iron. In one manufacturer's standard, the rectangular grids are about $4\frac{1}{8}$ in. square. Overall thickness for the flat grid is one inch, so that the spikes penetrate about $\frac{7}{16}$ in. into the wood members. To reach similar penetration, the curved grid has an overall thickness of $1\frac{3}{8}$ in. The circular one is $3\frac{1}{4}$ in. in diameter and is 1.20 in. deep.

A $\frac{3}{4}$- or 1-in. bolt passes through the members and the central hole in the grid. The bolt is in a $\frac{1}{16}$-in. oversize hole, so under light loads it merely holds the parts together; it does not resist shear unless the grid connection itself slips through bearing deformation of the wood.

It requires considerable pressure to embed the spikes fully in each member. The force needed is more than can be applied merely by tightening the bolt. Manufacturers provide special installation tools to embed the spikes. After the installation tool is removed, the permanent bolt is installed and tightened.

Table C-20 shows allowable loads for one manufacturer's spiked grids. Load duration factors for wood do not apply to the allowables shown. Instead, the allowable may be increased by 20% for load combinations that include wind or earthquake.

Minimum grid spacings allowed are 7 in. parallel, and $5\frac{1}{2}$ in. perpendicular to the grain, for all three types of grid. Minimum end distances

(a) CIRCULAR

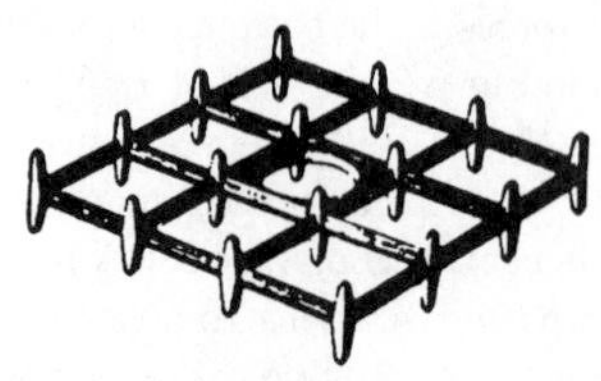

(b) FLAT

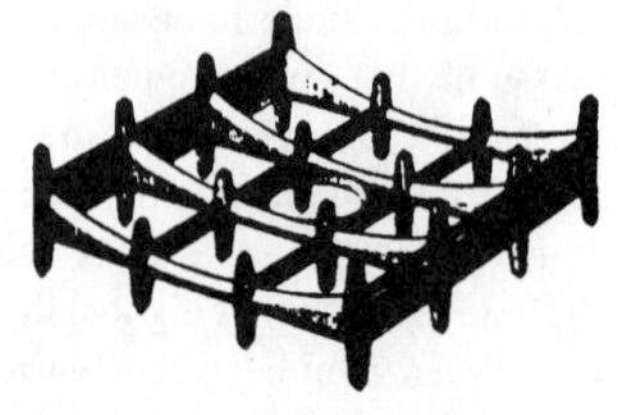

(c) SINGLE CURVE

Fig. 9-23. Spiked grids. (Courtesy of Cleveland Steel Specialty Company, Cleveland, OH.)

are 7 in. under full load or 5 in. (absolute minimum) under 85% of full load. Similarly, edge distances (for any load/grain direction) are $3^5/_8$ in. and $2^3/_4$ in. minimum.

9-10. NAILER PLATES AND TOOTHED PLATES

Nailer plates are sheet metal plates, frequently galvanized, with an array of holes through which nails may be driven. The object of the nailer plate is to make a connection more compact than it would be with only wood members. But, as load is picked up from a large number of nails, this may have the undesirable effect of overstressing the steel of the nailer plate itself. The chance of overstressing is greater when the nailer plate is used in a shear or compression connection than in a tension connection. The designer should be alert to this; it does no good to provide a connection that is excellent from the standpoint of the wood, only to have the steel parts fail.

Nailer plates are frequently used with special nails, shorter than common wire nails, and often with fluted shanks. This is usually the case when nailer plates are used with 2-in. nominal material.

Toothed plates have their own "nails" built in. The teeth are formed by bending the steel from long triangular punched holes in the sheet metal plates so that an array of closely spaced teeth project from one side of the plate. Since they are used mainly for wood truss construction, they are usually called *truss plates.* Installation is by pressing the plate firmly so that all teeth are driven into the wood. Chapter 10 covers design using truss plates.

9-11. DRIFT PINS AND DOWELS

Figure 9-24 shows fasteners that may be used to connect heavy-timber cap members to wood piles, stringers to cap members, and the like. Each type is driven into predrilled lead holes. For drift bolts and drift pins the hole diameter is $^1/_{16}$ in. less than that of the fastener. For spiral dowels (twisted steel rods), the hole diameter is about 75% of the outside diameter of the spiral. Drift pins, being merely round steel rods, can be made in a wide variety of sizes. Standard sizes of spiral dowels are from $^1/_4$ in. to $^5/_8$ in. and lengths available are by $^1/_2$-in. increments up to 8 in. and by 1-in. increments for dowels over 8 in. long.

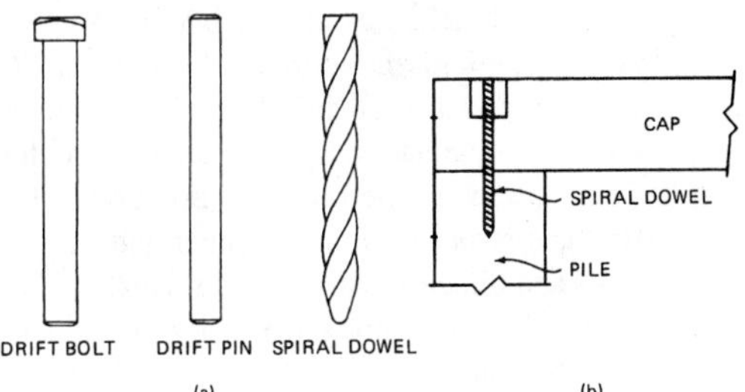

Fig. 9-24. Drift pins and dowels.

Drift pins, drift bolts, and spiral dowels have significant strength in either lateral load or withdrawal and can be used in either side grain or edge grain. Withdrawal from end grain, while not allowed by the NDS for nails or lag screws, is a common direction of loading for these fasteners, and tests show that the endgrain withdrawal value is about one-half of that for withdrawal from side grain.

Allowable loads for these fasteners are given by equations involving the same parameters as the equations for nails. Reference 4 shows the equations and gives tables of allowable loads.

REFERENCES

1. *Specification for Structural Steel Buildings,* American Institute of Steel Construction, Chicago, IL, 1989.
2. American Institute of Timber Construction, *Timber Construction Manual,* 4th ed., American Institute of Timber Construction, Wiley, NY, 1994.
3. Timoshenko, S., and S. Woinowsky-Krieger, *Theory of Plates and Shells,* 2nd ed., McGraw-Hill, New York, 1959.
4. *Western Woods Use Book,* Western Wood Products Association, Portland, OR, 1979.

PROBLEMS

All lumber sizes are nominal, S4S, unless otherwise noted. Loads are normal duration, unless otherwise noted.

9-1. Refer to the eccentrically loaded connection of Fig. 9-4. What would be the allowable load at e = 9 in. if four bolts were used in each vertical row and the spacing in the rows were still 4 in.?

9-2. Design a bracket, as shown by Fig. 9-5, to support a 9-k normal-duration load that is $2^1/_2$ in. from the face of an $8^3/_4$- × 9-in. glulam column, combination 11 Douglas fir. Use the same species in the bracket. (Solution is by trial and error. Assume a number and size of bolts and dimensions *b, c,* and *s*. Analyze the assumed joint. If not satisfactory, revise and analyze again.)

9-3. Complete the computations of Example 9-5 (Fig. 9-15) by designing the necessary welds.

9-4. Design a hanger to support rafters framing to opposite sides of a ridge beam. The rafters, 4 × 16 sawn D. fir, No. 1, slope at 4 in. in 12 in. (down away from the ridge beam), and have reactions of 3.5 k each (1.0 dead plus 2.5 snow). The ridge beam is a $6^3/_4$- × 30-in. glulam, western species combination 20F-E1.

9-5. Figure 9-25 shows roof framing and bracing members and their loads. Design a connection weldment to use with bolts (and shear plates, if necessary).

9-6. Design a split-ring connection to splice together two 2 × 12 tension members of Select Structural D. fir-larch. Use a double-shear connection. Design the splice to carry the maximum allowable tensile load for the main members. Use $2^1/_2$-in. rings.

9-7. Repeat the above problem using shear plates and metal side plates. Use $2^5/_8$-in. shear plates and $^3/_4$-in. bolts.

9-8. A 2 × 6 diagonal brace is connected to an 8 × 8 column using $2^1/_2$ in. split rings. The brace slopes 6 in./ft and has a load (from wind) of 4.5 k. The diagonal is Select Structural D. fir and the column is No. 1 D. fir. Design the connection and show a detail of the joint, including edge and end distances and spacing.

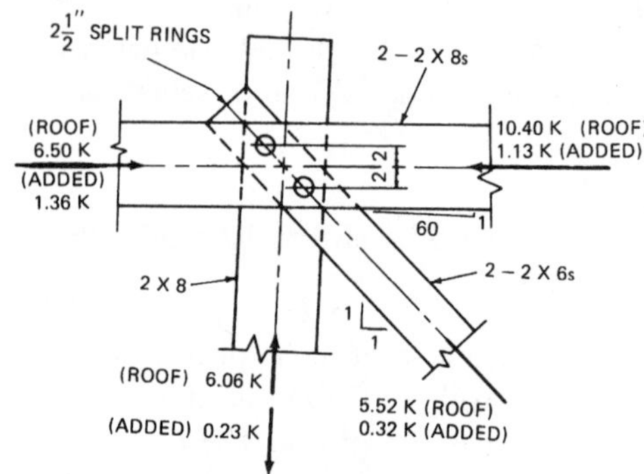

Fig. 9-26. Truss joint for Problem 9-10.

9-9. Design a split-ring connection to carry a 4.8-k load (1 k dead plus 3.8 k live) from a vertical hanger into a horizontal beam. The beam is a 6 × 12 of No. 1 D. fir, and the vertical hanger consists of two No. 1 redwood 2 × 6s, one on each side of the beam. The hanger is 4 ft from the end of the 12-ft span. Live-load duration (cumulative) is two months.

9-10. Figure 9-26 shows a top-chord joint of a truss connected with split rings. The members are all D. fir-larch, Select Structural grade. Revised use of the building causes member loads to increase above the original design loads. Is the split-ring connection satisfactory for the new increased loads?

9-11. A horizontal 8 × 14 dense D. fir timber is to be attached to the vertical piling of a trestle bent. The vertical load to be transferred to each pile is 2000 lb dead load and 5000 lb live load. The D. fir coast region wood piles are approximately 15 in. in diameter where the 8 × 14 is attached. Choose a size and number of single-curve spiked grids to connect the 8 × 14 to one pile.

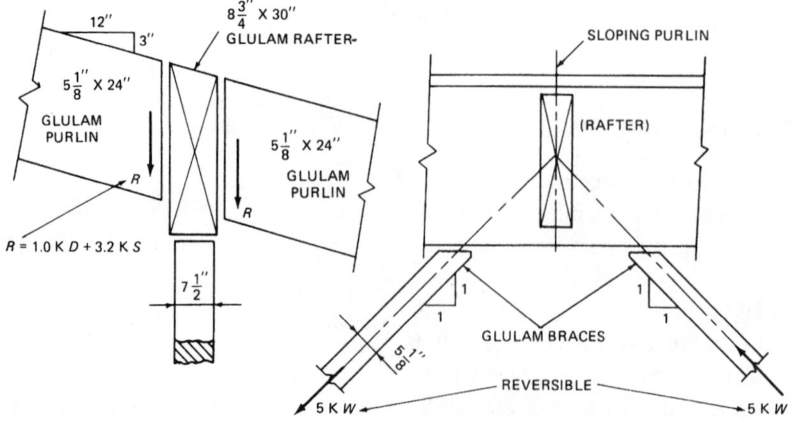

Fig. 9-25. Diagram for Problem 9-5.

10

Timber Trusses

For long or moderate spans, wood trusses are often lighter, more economical, and more practical than wood beams. In addition, a structure using trusses usually is stiffer than one using beams of the same span; that is, it has smaller deflections.

The wood truss is not a recent invention—wood trusses have been used for centuries for both buildings and bridges. For buildings, an early recorded use was during the first century B.C. (1). Recent improvements in methods of connecting members together to form a truss, however, have led to new uses for wood trusses and an increase in their popularity. This is particularly true in light-frame construction.

10-1. WOOD TRUSS TYPES

The variety of geometrical patterns that may be used is almost unlimited. Those shown by Fig. 10-1, however, have proved practical for wood trusses and are frequently used. Figure 10-1a shows two examples of the Pratt truss, a patented configuration when it was first developed. When the Pratt truss is fully loaded, its web diagonal members are in tension, while the verticals are in compression.

The sign of web member forces is the opposite for the Howe truss: Web verticals are in tension and diagonals in compression. Note that is it not the direction of member slope that distinguishes the Pratt truss from the Howe; rather, it is the sign of force in the web members.

The double-intersection Howe truss of Fig. 10-1c has diagonals in both directions. If its members were all rigid and their ends were rigidly connected, this truss would be highly statically indeterminate. The verticals, usually steel rods, are nonrigid; that is, they can resist only tension forces. The diagonals are capable of resisting compression, but their ends are connected so that they cannot resist tension. Under a particular moving-load position, one of the diagonals in a particular panel will act in compression, while the other will be inactive. When the load moves so as to change the sign of shear in the panel, the first diagonal becomes inactive and the second receives compression.

Figure 10-1d shows a Warren truss. Its web diagonals alternate in both direction and sign of stress. Whether or not there are vertical members is immaterial. Verticals are included only if needed to reduce the laterally unsupported length of chord members or to shorten the length of floor system members that bring load to the truss panel points.

In its action, the bowstring truss of Fig. 10-1e is both a tied arch and a truss. Under uniform top-chord loading, the curved top chord acts (in compression) as an arch and the bottom chord as a tie connecting the ends together, the web members carrying little or no load. Under unsymmetrical load, however, the web members carry appreciable load and the system acts more as a truss and less as an arch. The bowstring truss is useful for both wood and structural-steel construction.

The fan, scissors, and Fink trusses of Fig. 10-1f, g, and h are also well suited to wood construction. The scissors truss is often used exposed for interiors of buildings with steeply sloped roofs (see Fig. 10-2).

Wood trusses are often categorized into one of three broad groups: light-frame trusses; heavy-timber trusses; and split-ring connected trusses.

Light-frame trusses are made using members

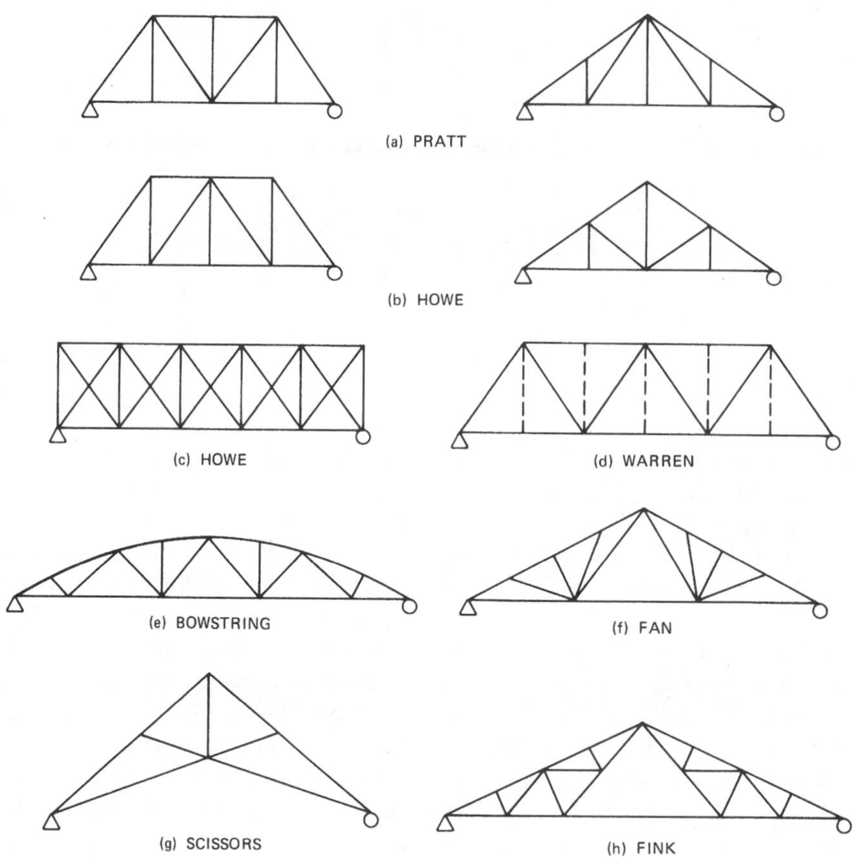

Fig. 10-1. Truss geometrical patterns.

Fig. 10-2. Interior of church building with scissors trusses formed from peeled logs. (Photograph by authors.)

of 2-in. nominal thickness, usually connected by metal nailer plates or toothed truss plates, or occasionally by nailed or glued plywood gusset plates. Whereas a few years ago sawn lumber was almost always used for sloping roof rafters, ceiling joists, and floor joists, today light-frame trusses have largely replaced the sawn-lumber members for these uses.

Light-frame trusses are especially economical for roof framing in buildings where spans are small or moderate. For floors, trusses with horizontal top and bottom chords also are often found economical. Floor trusses may provide better rigidity than sawn-lumber joists, and the openings between their members provide for passage of pipes, conduits, and ducts without cutting, as might be required if sawn-lumber joists were used.

Light-frame trusses, assembled in a fabrication shop, are complete and ready to install when shipped to the building site. Their advantages of stiffness, light weight, and low cost are partially offset by their greater potential for fire damage than heavier sawn timbers and glulams. (See Fig. 10-3.)

Heavy-timber trusses have member thicknesses up to 8 or 10 in. With truss members of such large size, connections pose a major problem. In large older structures, heavy truss connections were made using end bearing whenever possible to transfer load from one member to another and by using steel rods for selected tension members. This will be shown by a later example.

Timber connectors make possible a variety of large timber trusses that would otherwise be impractical. The larger members of these trusses are *spaced;* that is, they are built up using two or more separated elements of 2, 3, or 4-in. lumber. Lighter members are single pieces. At joints, the various elements overlap and are connected by split rings installed between adjacent elements. Such trusses are economical in their use of material and were especially popular in times of material shortage during World War II.

For best appearance (as in exposed trusses) solid members rather than spaced multiple-element members should be used. For these members glulams are often advantageous. It is attractive as well as practical to join the members of exposed trusses with bolted steel gusset plates on each face of the members (see Fig. 10-4).

Fig. 10-3. Light-frame trusses for industrial structure. (Courtesy of Lumbermate Company, St. Louis, MO.)

Fig. 10-4. Weldment at conjunction of glulam truss and transverse bracing. (Photograph by authors.)

Shear plates may be used on the inner surfaces of the gusset plates if the bearing strength of the bolt on the wood is not sufficient.

"Hybrid" trusses using mixed materials are becoming increasingly popular. Examples include floor members whose top and bottom chords are wood, while the web members are metal. Top and bottom chords of these trusses are also being fabricated with finely laminated wood (called laminated veneer lumber, or LVL) rather than sawn lumber.

10-2. LIGHT-FRAME TRUSSES, ANALYSIS

Specification ANSI/TPI 1-1995 of the Truss Plate Institute (2) defines current practice for design, fabrication, and erection of light-frame trusses. TPI 1-1995 includes design guides for three standard truss configurations (shown by Fig. 10-5), defining in detail how to determine bending moments in top- and bottom-chord members for these three patterns. After axial load and bending moment are determined, each member is designed by the criteria of the NDS (3,4). Other truss patterns are permitted, but the designer must follow a more rigorous procedure rather than the simpler empirical procedure spelled out by TPI 1-1995 for the three standard patterns.

Since the chords of light-frame trusses are continuous over two or more panels, the trusses are statically indeterminate. Each chord member has bending moments as well as axial loads. For any truss pattern other than the three standard ones shown, TPI 1-1995 requires that the analysis be made using a matrix analysis system that permits solution of both bending moments and axial loads for each member. One such system was developed by researchers at Purdue University (5), and extended by USDA's Forest Products Laboratory (6). General-purpose finite-element analysis systems such as SAP90 (7) could also be used.

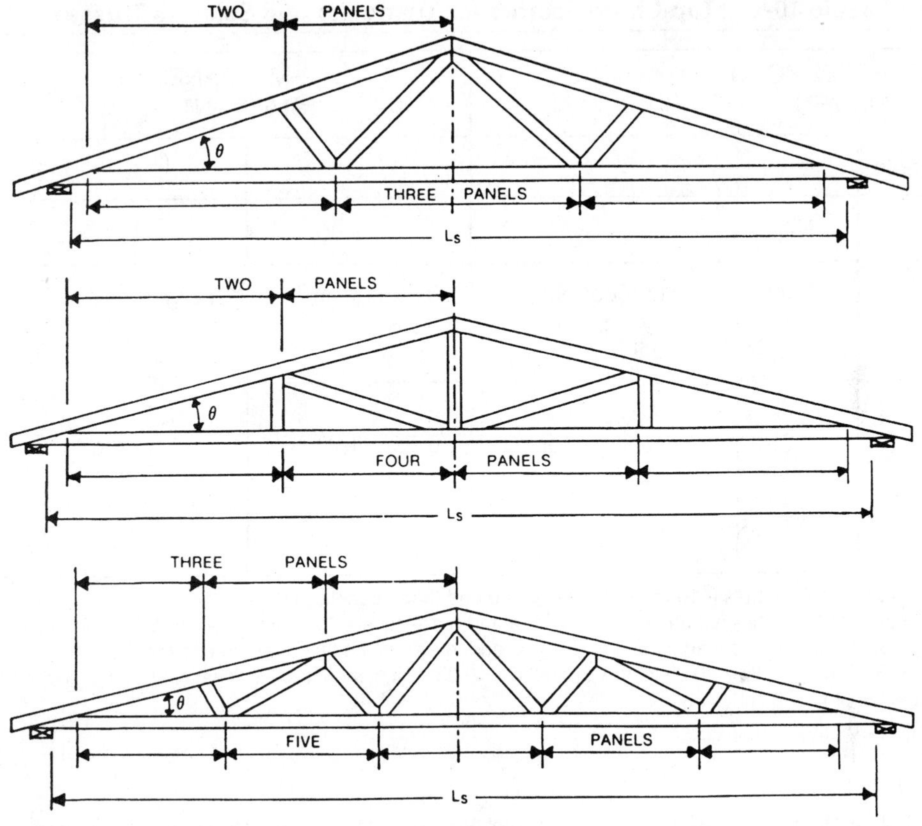

Fig. 10-5. Truss configurations covered by ANSI/TPI 1-1995 (Courtesy of Truss Plate Institute.)

In the simplified, empirical method given by TPI 1-1995, a moment factor Q is determined from a table such that QL represents the horizontal component of member length between inflection points. Tables 10-1, 10-2, and 10-3 show Q and L for the *top* chords of different types of truss. Using the appropriate Q and L, the bending moment for top-chord segments with uniformly distributed vertical load is

$$M = w\,(QL)^2/8 \qquad (10\text{-}1)$$

In these three tables, L_i is the horizontal length of the panel being investigated, and L_a is the horizontal length of the longer panel adjacent to the one being investigated.

Some of the tabulated expressions for Q seem involved and complex. Their purpose is simply to account for the lack of complete fixity of the chord members at the heel joint. Since the effective length, QL, is squared in Eq. 10-1, it is important to consider all factors that affect the location of the inflection points.

For bottom-chord segments, the procedure is the same but with factors Q and L from Table 10-4.

The above method, including Eq. 10-1 and Tables 10-1–10-4, is *only* for trusses in which the larger cross-sectional dimension of the chord is vertical. TPI 1-1995 includes a different but similar method for trusses in which the larger dimension is horizontal, that is, in which the members are in flatwise bending.

Specification TPI 1-1995 does not suggest how truss deflections should be computed, but does require that live-load deflection of roof trusses not exceed $\ell/360$ when plaster is used on the ceiling below, nor $\ell/240$ when a flexible ceiling such as drywall or a suspended ceiling is used. If there is no finished ceiling, the deflection limit is $\ell/180$ (ℓ is the truss clear span). The authors' suggestion for calculating deflection is presented later in this chapter.

Table 10-1. Top-Chord Factors for Triangular and Scissors Trusses.

PANEL POINT MOMENT			MID-PANEL MOMENT	
#Panels	Q	L	Q	L
1	NA	NA	0.90	$L_i + S_a$ ③
2	0.90	Largest of: ① $0.9\ L_i$ $\frac{(L_i+L_a)}{2}$, or $0.9\ L_a$	$0.58(\cot\Theta)^{0.23}$ ②	Largest of: ③④ $0.9(L_i+cS_a)$, $\frac{(L_i+L_a)}{2} + cS_a$, $0.9(L_a+cS_a)$
3 or more	0.85		$0.53(\cot\Theta)^{0.36}$ ②	

1. If S_t exceeds 24 inches (610 mm), add excess to end (heel) panel L_i or L_a.
2. Q = **a**$(\cos\Theta)^{\mathbf{b}}$ but shall not be less than 0.74; **a** and **b** are constants derived from PPSA analysis.
3. $S_a = S_t - B$ but not less than zero. cS_a shall be added only to the length of the end (heel) panel.
4. c = 0.5 for two panels; c = 0.33 for three panels; c = 0.25 for 4 or more panels; if neither L_i nor L_a are end (heel) panel lengths, then $cS_a = 0$.

Source: Courtesy of Truss Plate Institute.

Table 10-2. Top-Chord Factors for Monopitch and Monoscissors Trusses.

PANEL POINT MOMENT			**MID-PANEL MOMENT**	
No. of Panels	**Q**	**L**	**Q**	**L**
1	**N/A**	**N/A**	**1.00**	$L_i + S_a$ ③
2	**0.90**	**Largest of:** ① $0.9\ L_i$ or $\frac{(L_i + L_a)}{2}$ $0.9\ L_a$	**0.90** ②	**Largest of:** ③ ④ $0.9(L_i+cS_a)$ $\frac{(L_i + L_a)}{2} + cS_a$ $0.9(L_a + cS_a)$
3 or 4	**0.90**		$0.58(\cot\Theta)^{0.23}$ ②	
5 or more	**0.85**		$0.53(\cot\Theta)^{0.36}$ ②	

1. If S_t exceeds 24 inches (610 mm), add excess to end (heel) panel L_i or L_a.
2. Q = **a**$(\cot\Theta)^{\mathbf{b}}$, but shall not be less than 0.74; **a** and **b** are constants derived from PPSA analysis.
3. $S_a = S_t - B$ but not less than zero. cS_a shall be added only to the length of the end (heel) panel.
4. c = 0.5 for two panels; c = 0.33 for three panels; c = 0.25 for 4 or more panels; if neither L_i nor L_a are end (heel) panel lengths, then $cS_a = 0$.

Source: Courtesy of Truss Plate Institute.

Table 10-3. Top-Chord Factors for Chord Segments Without Heel Joints, for Top Chords with Slopes of Less than 1.5/12.

Panel Point Moment		Midpanel Moment	
Q	L	Q	L
Not Applicable[a]		**One panel** 0.95	**One panel** L_i
		Two panels 0.90	**Two or more panels** L is largest of: $(L_i + L_a)/2$, or $0.9\,L_a$, or $0.9\,L_i$
		Three or more panels 0.85	

[a]As these factors are derived from TPI-74, only mid-panel members are considered critical.
Source: Courtesy of Truss Plate Institute.

Table 10-4. Bottom-Chord Factors.

Q	L
One panel 1.0	One panel L_i
Two or more panels 1.0	Two or more panels L is largest of: $0.9\,L_i$, or $\dfrac{(L_i + L_a)}{2}$, or $0.9\,L_a$

Source: Courtesy of Truss Plate Institute.

Example 10-1

Compute design axial forces and bending moments for members of the light-frame wood roof truss of Fig. 10-6. The trusses will be installed at 24-in. centers. Tentatively, the designer is considering six equal top-chord and five equal bottom-chord panels as shown. (This is the third of the standard configurations covered by ANSI/TPI 1-1995.)

Loads are:
Applied to the top chord—

Snow	25 psf of horizontal projection
Sheathing	2.5 psf of roof surface
Roofing	6 psf of roof surface

Applied to the bottom chord—

Ceiling	8 psf
Insulation	3 psf

Member sizes are not yet known, so we must estimate the truss weight. Luckily, truss weight is a small part of the total, so a precise estimate is not needed. Assuming that all members will be 2 × 4s, a reasonably conservative weight for the truss is estimated to be about 5 lb per horizontal foot. Assume 3 lb/ft of this as top-chord load and the other 2 lb/ft as bottom-chord.

For a roof pitch of 4 in. per ft, the sloping length per foot is 12.65 in., or 1.05 times horizontal distance. Loads per *horizontal* ft applied to the top chord of one truss are:

Snow 2 × 25	= 50
Truss weight	= 3
Plywood sheathing 1.05 × 2 × 2.5	= 5
Roofing 1.05 × 2 × 6	= 13
Total (per horizontal ft)	= 71 lb

The horizontal length of each top-chord panel is 26/6 = 4.333 ft.

Distributed load along the top chord is transferred to the truss at panel points. Panel point loads are estimated treating the chord in each 4.333-ft length as a simple beam, that is, ignoring continuity. Thus, a full top panel point load is 4.333 × 71 = 308 lb. End panel points receive one-half as much from top-chord loading.

Bottom-chord loads are:

Ceiling 2 × 8	= 16 lb/ft
Insulation 2 × 3	= 6
Truss weight	= 2
Total	= 24 lb/ft

The panel length for the bottom chord is 26/5 = 5.2 ft, so a full bottom panel point load is 5.2 × 24 = 125 lb. The end bottom-chord panel point in this case is also

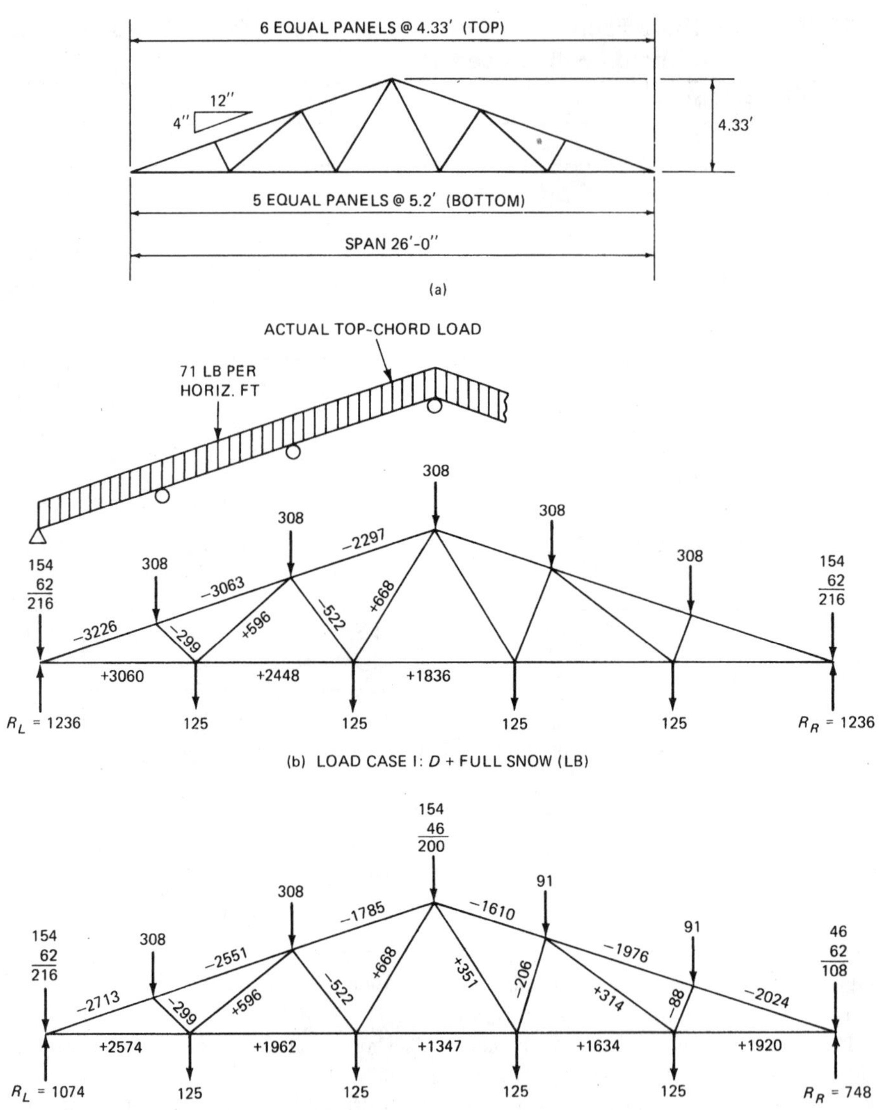

Fig. 10-6. Truss for Examples 10-1 and 10-2.

a top-chord panel point, so the end downward load is the sum of loads from a half top panel and a half bottom panel, or 308/2 + 125/2 = 216 lb. Total end reactions (not merely the net force at the end panel point) are needed to compute the end-bearing requirement, and are 1236 lb each.

Figure 10-6b summarizes the panel point loads for full, symmetrical snow load and dead load and shows the axial forces in each member due to these loads. For some members, though, full snow load *may* not cause the worst condition, so the truss is also analyzed assuming snow load on only one-half of the roof. Figure 10-6c shows the panel point loads and axial forces for this case. (Truss analysis was by the method of sections, as shown in any book on structural analysis.)

Both top and bottom chords have transverse load, so they will have bending moments as well as axial forces. Web members are all assumed to be pin-connected to the chords. Having no transverse loads, the web members have no bending moments.

For the top chord, for negative bending moment at the panel point, Table 10-1 gives factor $Q = 0.85$ for a three-panel member. For length, L, we must use the largest of the three choices shown:

1. Horizontal length of the panel being considered times 0.9
2. Average horizontal length of the panel being considered and the longer adjacent panel
3. Horizontal length of the longer adjacent panel times 0.9

This truss has equal-length top panels, so the second of the above choices is largest, and $L = 4.333$ ft. Negative bending moments at the end of top-chord panels are

$$M = 71\,(0.85 \times 4.333)^2/8 = 120 \text{ ft-lb}$$

To compute positive bending moment at the midlength of top-chord panels requires knowing dimensions S_t and S_a, dimensions of the *scarf,* that is, the diagonal cut, shown by Fig. 10-7. We do not yet know these dimensions, but can make a fairly good guess. Later, if our guess proves wrong, we must correct it and compute again. For now, knowing the pitch of the top chord and assuming 4-in. nominal depths of members, we can either sketch the heel joint to scale or calculate to see that S_t will be 10.5 in. and (if $B = 3.5$ in.) S_a will be 7 in. For the top-chord panel next to the heel joint, term cS_a must be added to its horizontal length. Using $c = 0.33$ (see table footnote) the largest of the three choices for L by Table 10-1 (for computing midpanel moment) is:

$$0.5\,(L_i + L_a) + cS_a$$

$$= 0.5\,(4.333 + 4.333) + 0.33 \times 7/12 = 4.53 \text{ ft}$$

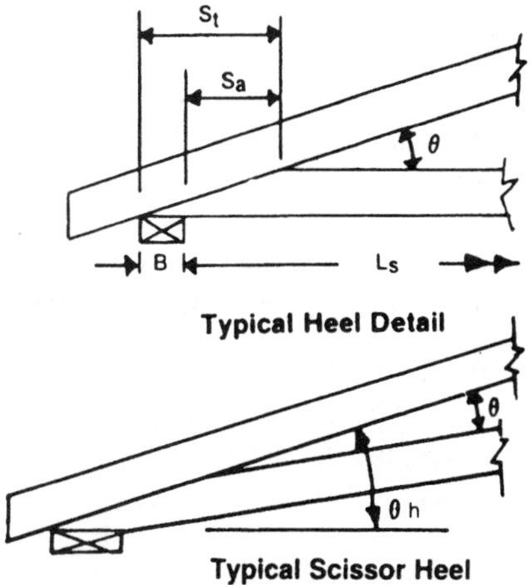

Fig. 10-7. Heel joints. (Courtesy of Truss Plate Institute.)

For the top chord, positive M (at midpanel):

$$Q = 0.53\,(\cot\theta)^{0.36} = 0.53\,(12/4)^{0.36}$$

$$= 0.787\ (> 0.74 \text{ specified minimum,} \quad OK)$$

Positive M for the top chord is

$$M = 71\,(0.787 \times 4.53)^2/8 = 113 \text{ ft-lb}$$

For the bottom chord, from Table 10-4, factor Q is 1.0 and length L is the largest of three choices shown, which is 5.2 ft. The critical bottom-chord bending moment, at the panel points, is

$$M = 24\,(5.2)^2/8 = 81 \text{ ft-lb}$$

Example 10-2 will show how top- and bottom-chord members are designed for these combined axial loads and bending moments.

10-3. LIGHT-FRAME TRUSSES, MEMBER DESIGN

Members of light-frame trusses are designed under TPI 1-1995. They may be designed using either the allowable stress method or LRFD.

Compression Members

With one exception, the TPI 1-1995 interaction and allowable stress equations are the same as those of the NDS, which were shown by Chapters 6 and 7. The one thing different for trusses is that the "buckling stiffness" factor may be considered.

For top-chord members, buckling is possible either in the plane of the truss, or perpendicular to that plane.

> For buckling in the plane of the truss, the effective buckling length, L', equals QL, as found in Tables 10-1, 10-2, and 10-3, except that for sloping members the horizontal length, L, is multiplied by sec θ (Fig. 10-8).
>
> Buckling perpendicular to the plane of the truss is restrained when roof or floor sheathing is attached along the length of the member. If sheathing is *not* attached, the effective length of the top chord is the distance between purlins (dimension L_u in

Member	Effective Dimension (d)	Effective Buckling Length (L')
Top Chord	d1 d2	See 10.1.2.3 L_u
Bottom Chord	d1 d2	L_1, L_2 L_u
Web	d1 d2	0.8 L_w L_u or 0.8 L_w

Where sheathing or ceiling is nailed to a compression chord (either top or bottom), lateral support of the supported axis shall be assumed to be continuous and $L'/d = L_u/d_2$ is neglected.

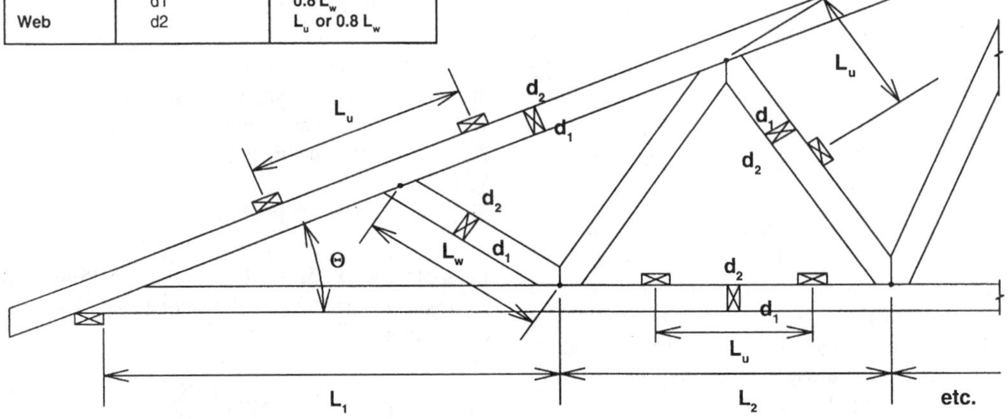

Fig. 10-8. Effective buckling lengths. (Courtesy of Truss Plate Institute.)

Fig. 10-8) or between points that are otherwise laterally braced to prevent their movement normal to the truss plane.

For web members that have compression, the effective length for buckling in the plane of the truss is 0.8 times the member length, L_w, as shown by Fig. 10-8. For buckling normal to the truss plane, the effective length is 0.8 L_w or (if a brace is attached) the length L_u shown for a web member in Fig. 10-8.

Bottom chords are seldom compression members, but if they are stressed in compression, their effective lengths are as shown in Fig. 10-8.

Factor C_T, the buckling stiffness factor, is used (along with moisture factor C_M) to adjust modulus of elasticity, E, for use in equations of Chapter 7 for allowable compressive stress in light-frame truss members. It is conservative to assume this factor to be unity, but a more economical design *may* result if the factor is solved from Eqs. 10-2 and 10-3, below. These equations apply if all four of the following are true:

1. The member is 2 in. × 4 in. or smaller.
2. Plywood sheathing, ³⁄₈ in. or thicker, is attached to the narrow face in accordance with good nailing practice.
3. The member has both bending and axial compression.
4. The truss is used under dry conditions only.

If *all four* of these conditions are met, then

$$C_T = 1 + 2300L'/kE \qquad (10\text{-}2)$$

for wood seasoned to 19% or less MC when the plywood is attached, or

$$C_T = 1 + 1200L'/kE \qquad (10\text{-}3)$$

for wood that is either partially seasoned or unseasoned when the plywood is attached. In Eqs. 10-2 and 10-3,

Effective buckling length L' used in the equations must not be more than 96 in.

Factor k is 0.82 for machine-stress-rated lumber, 0.64 for machine-evaluated lumber, and 0.59 for visually graded lumber.

Just as for compression members designed under NDS, the maximum allowable ratio L'/d is 50. Members with larger L'/d may not be used.

Allowable bending stresses, F'_b, are found by the methods of NDS, shown in Chapter 6. Top chords of trusses meeting the requirements for "repetitive members" (Section 4.3.4 of NDS) may be considered repetitive, and the allowable bending stress, F_b, may include the adjustment factor C_r. That adjustment applies *only* to the allowable bending stress, however.

For members with combined bending and axial compression, the interaction equations of Chapter 7 are used.

Tension Members

Tension members are much easier to design. The NDS interaction equations of Chapter 7 apply to truss members under combined tension and bending. Note that *both* Eqs. 7–15 and 7–16 must be used since a large bending moment (even when it acts together with axial tension) can cause compressive stresses on one extreme fiber of the member.

The maximum allowable slenderness, L'/d, for truss tension members is 80.

10-4. LIGHT-FRAME TRUSSES, CONNECTION DESIGN

Light-frame truss connections today are normally made using metal connector plates, often called *truss plates.* Truss plates are made of sheet steel (galvanized or aluminum–zinc coated) having projecting teeth punched from closely spaced rows of slots in the metal. The truss plates are used in pairs, one on each face of the lumber being connected. They are pressed (or rolled) so that the projecting teeth are forced into the wood, acting as small nails, but in a regular, closely spaced array.

Standard design values for truss plates are usually not available as they are for other connectors. Rather, individual truss manufacturers determine (for their own plate type, grade, and species of lumber) the necessary values for truss design. Specification ANSI/TPI 1-1995 spells out test procedures the manufacturer must follow in determining those design values. These include procedures for determining strength of the plates themselves, and for determining strength and stiffness of connections between the plates and the wood members. The procedures are very detailed, considering orientation of the plate axis (direction of the slots) with respect to the load direction, as well as direction of the load relative to the wood grain. Test procedures are also available in ASTM documents (8).

Critical slip for joints is considered to be 0.015 in. for movement of the plate relative to the wood.

Metal connector plate joints can reach their limit of usefulness by slipping, by the teeth breaking or pulling out of the wood, or by failure of the steel plates. The plate itself may control the limit by buckling, by tensile fracture, or by plate failure in shear. The truss designer must proportion the plates to prevent these types of failure.

Building designers ordinarily do not design the light-gauge truss connections; that responsibility is usually assigned to the *truss* designer, an employee of the truss manufacturer.

Before the wide use of truss plates, light-frame trusses were sometimes assembled using glued or nailed plywood gusset plates, or using nailer plates (plates with closely spaced arrays of holes through which nails could be driven).

Truss connections often require cutting the member ends diagonally. Such end cuts reduce the net area of the wood member. Figure 10-9 shows three such cases. Where diagonal cuts occur, the computed member stress is increased, and that increased stress on the remaining net section must be limited to the adjusted allowable axial stress for the wood.

Design Value for a Truss Plate

The design value for lateral resistance is used to determine how many teeth or how many square inches of contact area (net or gross, according to the manner of test evaluation used) are needed to transfer the member force. The design value must then be adjusted by applicable factors:

1. The usual NDS factor for load duration.
2. A 20% reduction if the moisture content at time of fabrication exceeds 19%.
3. The chemical manufacturer's recommendation for wood that is pressure-treated with a fire-retardant chemical.

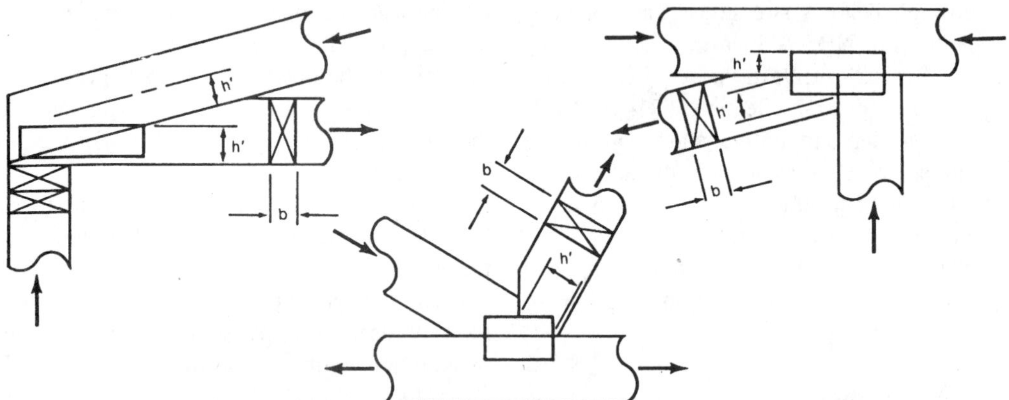

Fig. 10-9. Net widths of members with diagonal end cuts. (Courtesy of Truss Plate Institute.)

Design values are determined from the tests. The design value may be specified as either:

1. Design value per unit *gross contact area* between the plate and the wood; or
2. Design value per unit *net contact area* between the plate and the wood. (Net contact area is determined by deducting strips of area at the member end and plate edges, as shown by Fig. 10-10.)

Whichever of the above methods is used to specify the design value of the plate, that same method must be used by the truss designer to determine the contact area required between the wood and the plate.

If the design value for the plate was determined by the *gross area method,* and the teeth are pressed into the narrow face of the wood member, that design value must be reduced by 15%.

Connection of Compression Members

TPI 1-1995 requires that light-frame trusses have close-fitting members at every joint. A per-

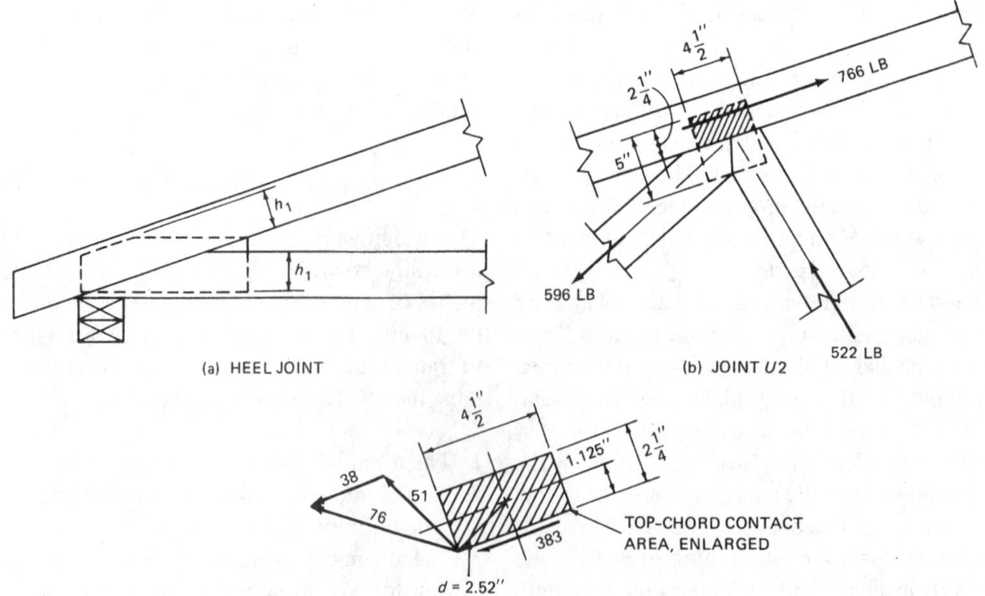

Fig. 10-10. Connections for truss of Examples 10-1 and 10-2.

fect fit would be one with actual firm contact all across the width of the member end. Fabrication is never perfect, however, so specified tolerances are given to limit the degree of imperfection.

TPI 1-1995 requires that the allowable bearing between wood members ending at truss joints be at least 100% of the force in the member that ends. Thus, a large part of the compressive force to be transferred is carried by direct contact—wood to wood. One-half (50%) of the ending-member force must be carried by the truss plate. TPI 1-1995 spells out this requirement in greater detail.

Heel-Joint Connections

Figure 10-7 shows a typical heel joint. The heel joint is the most difficult light-frame joint to design rationally. Since the two members do not intersect within the limits of the plate area, they cause moments that tend to rotate the truss plates.

TPI 1-1995 provides an empirical design method to compensate for the effects of this rotating moment. The members are assumed to have axial load only, but the plate design values (given for plates that are *not* eccentrically loaded) are reduced by the following factor:

$$H_R = 0.85 - 0.05\,(12 \tan \theta - 2.0), \text{ but not less than } 0.65 \text{ or greater than } 0.85 \qquad (10\text{-}4)$$

The reduction factor does not apply to cases with top-chord slope of 45 degrees or more.

Truss-Plate Connections—Caution

Joint design is very complex. The brief presentation above is insufficient for use by a light-frame truss designer. For actual light-frame truss-connection design, the designer should consult the complete specification TPI 1-1995.

Example 10-2

Select member sizes and design connections for the truss analyzed in Example 10-1. Use No. 1 & Btr Douglas fir-arch, (visually graded). Use the gross contact area method for connection design. Assume the design values of the truss plate (as determined by tests defined by TPI 1-1995) to be:

90 psi, based on the gross area method, for load parallel to the grain and parallel to the slots of the truss plate. (TPI 1-1995 refers to this value as V_{LRAA}.)

45 psi, for load perpendicular to the grain and to the slots of the truss plate. (TPI 1-1995 refers to this value as V_{LRAE}.)

Design values listed by NDS for this species and grade of lumber are:

Bending, 1150 psi

Compression parallel to grain, 1500 psi

Compression perpendicular to grain, 625 psi

Tension, 775 psi

Shear, 95 psi

Modulus of elasticity, 1,800,000 psi

Dead load is only about 30% of the total load per square foot, so the load combination $D + S$ controls, and all allowable stresses above (except compression perpendicular to grain) can be multiplied by $C_D = 1.15$.

Also, the trusses are so spaced and connected as to satisfy the requirements for "repetitive" members, so the adjustment factor $C_r = 1.15$ may be applied to the allowable stress for bending, F_b (only). Other adjustment factors will be computed as needed.

The top chord is laterally braced by closely spaced nails connecting the roof plywood to the truss. The chord has combined axial load and bending, so its design is by trial and error.

Top-chord selection: The maximum bending moment in the end panel of the top chord is 120 ft-lb (from Ex. 10-1), and the simultaneous axial load is 3226 lb. For bending alone, the adjusted allowable bending stress is

$$\begin{aligned} F_b' &= C_F \times C_D \times C_r \times F_b \\ &= 1.5 \times 1.15 \times 1.15 \times 1150 = 2281 \text{ psi} \end{aligned}$$

Based on bending alone, S req $= 12(120)/2281 = 0.63$ in.^{3}

Try a 2×4 (with the 4-in. dimension vertical). Next, find terms to use in the interaction equation to verify that the 2×4 is satisfactory for combined compression and bending.

Horizontal buckling is prevented by the attached roof diaphram material. For vertical buckling, the effective length of each top-chord panel is

$L' = QL \sec\theta = 0.85\,(4.333)\,(1.05) = 3.87$ ft $= 46.4$ in. (< 96, *OK*)

$L'/d = 12\,(3.87)/3.5 = 13.27$

By Eq. 10-2, the buckling stiffness factor is

$$C_T = 1.0 + 2300\,(12 \times 3.87)/(0.59 \times 1{,}800{,}000) = 1.10$$

Modulus of elasticity, E, in the allowable compressive stress computation is adjusted by this factor, giving

$$E' = 1.10\,(1{,}800{,}000) = 1{,}980{,}000 \text{ psi}$$

$$F_c^* = F_c C_D C_F = 1500\,(1.15)\,(1.15) = 1984 \text{ psi, and}$$

$$F_{cE} = 0.3\,(1{,}980{,}000)/(46.4/3.5)^2 = 3380 \text{ psi}$$

The column stability factor, C_p, is computed to be 0.838, and

$$F_c' = 0.838 \times 1984 = 1663 \text{ psi.}$$

$$f_c = 3226/(1.5 \times 3.5) = 614 \text{ psi}$$

$$f_b = 12\,(120)/3.06 = 471 \text{ psi}$$

Substituting these values in the interaction equation,

$$(614/1663)^2 + 471/[2281(1 - 614/3380)] = 0.389 < 1.00 \quad OK$$

The 2 × 4 top chord will be used provided that satisfactory connections can be designed.

Bottom-chord selection: Try a 2 × 4 for the bottom chord also. The maximum axial tension is 3060 lb, and the bending moment in this same member (from Ex. 10-1) is 81 ft-lb. The adjusted allowable tensile stress is $C_F C_D (F_t) = (1.5)\,(1.15)\,(775) = 1337$ psi. The material (drywall) attached to the bottom chord is not stiff enough to transfer load from a bottom chord that is weak in bending to adjacent stronger ones, so the bottom chord must be considered a single-use member (not repetitive). Thus, the adjusted allowable bending stress is $1.5(1.15)\,(1150) = 1984$ psi. Other adjusted allowables are the same as for the top chord. Both interaction equations shown in Chapter 7 for tension members must be used. Terms needed for these two equations are:

$$f_t = P/A = 3060/5.25 = 583 \text{ psi}$$

$$f_b = M/S = 12(81)/3.063 = 317 \text{ psi}$$

By the first equation,

$$583/1337 + 317/1984 = 0.596 < 1.0 \quad OK$$

By the second equation,

$$317 - 583 = \text{negative}, < F_b' \quad OK$$

Top and bottom chords may each be made of 2 × 4 material.

Web member selection: For simplicity, try to find a size that can be used for all members. Consider first the compression web members. In this truss, the longest compression diagonal is also the one with the largest compressive axial force (522 lb), so it will control the design for compression. After choosing a size for this member, we can then check to see if it is also satisfactory for the 668-lb maximum tension. It would probably be practical to use 2 × 4 members throughout, although a manufacturer might elect to use smaller web member sizes if, in total, it would reduce the cost, and if the narrower width did not make the truss-plate contact area too small.

For our example design, try 2 × 4. Figure 10-8 shows the effective length, L_w, of web members to be determined by the *clear* distance between the chords. For the longest compression diagonal, this net length is 3.369 ft minus approximately 3.5 in., or 36.9 in. The effective buckling length, L', is 0.8(36.9) = 29.52 in. Ratio L'/d will then be 29.52/1.5 = 19.68, which is less than the allowable limit of 50.

For the 2 × 4, $L'/d = 19.68$, and noting that C_T is not applicable,

$$F_{cE} = 0.3\,(1{,}800{,}000)/(19.68)^2 = 1394 \text{ psi}$$

$$F_C^* = C_D C_F F_c = 1.15\,(1.15)\,(1500) = 1984 \text{ psi}$$

Substitution into the equation for column stability factor gives

$$C_P = 0.560$$

$$F_c' = C_P F_c^* = (0.560)\,(1984) = 1111 \text{ psi}$$

The actual stress is only $522/(1.5 \times 3.5) = 99.4$ psi < 1111, so the 2 × 4 is OK for all web diagonals. Obviously, though, 2 × 2 members would also prove OK.

The longest tension web member is 5.053 ft minus about 3.5 in., or about 57 in. long. $L'/d = 0.8 \times 57/1.5 = 30.4$, which is well within the limit of 80 allowed for tension members. The axial stress is $668/5.25 = 127$ psi, much less than the adjusted allowable of $F_t' = F_t(C_D)\,(C_F) = 775\,(1.15)\,(1.5) = 1337$ psi.

Connections: Design of only two of the connections will be illustrated here. Other connections will be the subject of problems at the end of the chapter. (The shear and axial load values of each plate would also have to be checked in a complete design.)

Heel Joint: Figure 10-10a shows the heel joint drawn to scale. Because top and bottom chords intersect beyond the limits of the area in which truss plates can be placed, the heel-joint connection to each member is eccentrically loaded. The horizontal component of the top-chord compression and the tensile force in the bottom chord are a couple, tending to rotate the truss plate counterclockwise. To compensate for this, TPI 1-1995 allows the connection to be designed as though each member applies axial load only, but requires that the lateral resistance design value for the truss plate be multiplied by a reduction factor, H_R (Eq. 10-4).

Tan θ for this Example is 4/12, so that $H_R = 0.75$, and the reduced design value for the truss plate is 0.75 (90) (1.15) = 77.6 psi of gross contact area. This value is only for load applied parallel to the grain. Thus, it applies only to the connection of the plate to the bottom chord.

Gross contact area required for the bottom chord
$= 3060/(2 \times 77.6) \times 19.72 \text{ in.}^2$

The top-chord load is at an angle of $\tan^{-1}(4/12)$ to the truss plate axis, therefore, the design value must be computed using the Hankinson equation, as follows:

Design value

$$= V_{LRAA} V_{LRAE}/(V_{LRAA} \sin^2\theta + V_{LRAE} \cos^2\theta)$$
$$= 77.6\,(45 \times 1.15)/(77.6 \times \sin^2 18.43° + 45 \times 1.15 \times \cos^2 18.43°)$$
$$= 73.9 \text{ psi of gross contact area}$$

The scale drawing of the joint is used to determine the outline of plate that will best provide the above areas. After trying several sizes of plate, it was decided that the 4 in. × 13 in. plate shown would be satisfactory. However, there are many different satisfactory solutions. The one shown requires a diagonal cut at the corner so that the plate does not interfere with the roof plywood. This may be objectionable, but to eliminate the diagonal cut would require using a larger plate (or one with higher design values). It might also be possible to put a rectangular plate in a skew position to get the required areas without a diagonal cut. If the plate were on a skew, however, each member would require design values computed by Hankinson's equation.

Net areas of each member at this joint must be checked. The *net area* is defined as the product of member thickness and dimension h_l, as shown by Fig. 10-10a. With the plate edge ½ in. from the edge of the 2 × 4, each member has $h_l = 3$ in. Thus, the net area of either member is 1.5(3) = 4.5 in.² For the top chord, the stress on this net area is 3226/4.5 = 717 psi. This is less than the adjusted allowable compressive stress of 1.15(1500) = 1120 psi, so the top chord is satisfactory for net area. The bottom chord, also, can be shown satisfactory for stress on its net area.

Interior top-chord joint: Figure 10-10b shows, to scale, the second interior top-chord joint. This joint is highly statically indeterminate, since the members do not intersect at a common point, and since part of the force in the compression diagonal is transferred by direct bearing rather than by way of the truss plates. TPI 1-1995 does not address this condition rationally, but rather gives empirical rules for truss plate design. The authors, however, believe it can be addressed by more nearly rational means, and in the following they show how:

For the tension diagonal, the authors are assuming that each plate must carry one-half of whichever is larger: the member load, or 375 lb. (This requirement was included in previous editions of the TPI specification, but in the present TPI 1-1995 it is shown as applying only to trusses whose chord members have their larger cross-sectional dimension horizontal.)

Adjusted plate values are:

$$V'_{LRAA} = (1.15)\,(90) = 103.5, \text{ and}$$
$$V'_{LRAE} = (1.15)\,(45) = 51.8$$

Thus, for the tension diagonal shown, the required gross contact area per plate is

$$596/(2 \times 103.5) = 2.88 \text{ sq in.}$$

For the compression diagonal, TPI 1-1995 requires that the member end be capable of transferring the entire 522-lb member load. It also requires that the truss plate connection be capable of transferring at least 50% of the load, or 261 lb. The grain is at an angle of 77.4° to the plate axis (slot direction). The design value, by Hankinson's equation, is

$$(103.5)(51.8)/(103.5 \times \sin^2 77.4° + 51.8 \times \cos^2 77.4°)$$
$$= 53.1 \text{ psi of gross area}$$

The two plates must be designed to transfer at least (522)/2 = 261 lb each. The required gross contact area for each plate is 261/(2 × 53.1) = 2.46 in.²

If the top chord were spliced at this joint, and if the member ends were in firm contact with each other, the two plates (one on each face of the spliced member) would have to carry the entire chord force minus up to 50% of that force carried by end bearing.

For the top chord (not spliced at this joint) the plates must transfer the maximum *difference* between chord forces in members on opposite sides of the joint. At this joint, the maximum difference for either of the load cases shown by Fig. 10-6 is 3063 − 2297 = 766 lb, applied eccentrically to the intersection point of the two other members at this joint. In the absence of guiding rules by TPI 1-1995, the authors suggest an analysis such as shown by Fig. 10-10c.

Try two 5- × 4½-in. plates, as shown by Fig. 10-10b. This provides more than enough contact area to connect the two diagonal web members. The contact area with the top chord measures 4.5 in. by 2.25 in. Assuming that this area is subject to 766/2 = 383 lb per plate and a moment of $383e = 383\,(1.125) = 431$ in.-lb per plate, the largest shear stress *per square inch of contact area* (with the top chord) is the vector sum of P/A and Md/J, where J is the polar moment of inertia of the contact area and d is the distance from the centroid of the area to the point farthest from the centroid. (This is exactly the same as the elastic method for analyzing eccentrically loaded bolted joints in structural steel.)

$$J = I_x + I_y = (1/12)\,[(4.5(2.25)^3 + 2.25(4.5)^3] = 21.36 \text{ in.}^4$$

$$d = \sqrt{2.25^2 + 1.125^2} = 2.52 \text{ in.}$$

The maximum shear stress between one plate and the top chord, computed as suggested above, is the vector sum of

$$P/A = 383/(2.25 \times 4.5) = 38 \text{ psi}$$

and

$$Md/J = 431(2.52)/21.36 = 51 \text{ psi}$$

or 76 psi, shown graphically on Fig. 10-10c.

The resultant stress of 76 psi is at 36.9° to the plate axis, and the plate design value in that direction (by Hankinson's equation) is

$$103.5(51.8)/(103.5 \sin^2 36.9° + 51.8 \cos^2 36.9°) = 76.1 \text{ psi} \quad OK$$

10-5. BRACING AND ERECTION OF LIGHT-FRAME TRUSSES

Light-frame truss material is thin. Although properly designed light-frame trusses are strong enough to resist the *in-plane* loads for which they are designed, they are easily damaged during shipping, handling, and erection, and their member strength is often controlled by buckling. Obviously, bracing to keep the trusses straight and in their intended position is very important. Bracing is required during both truss erection and service. It is essential that *all* trusses be not merely braced to each other, but also firmly fastened to the supports at their reactions.

TPI 1-1995 spells out the bracing requirements in considerable detail. The TPI-specified bracing serves three purposes:

1. It holds the trusses in the desired position in the building.
2. It holds them upright (i.e., it prevents them from tipping).
3. It may reduce the laterally unsupported lengths of certain compression members.

Bracing is needed when the finished building is subject to design loads, but is usually even more important during construction. Whose responsibility is it to ensure that the light-frame trusses are adequately braced? Many accidents have occurred because of inadequate bracing, each party assuming that someone else was supposed to be responsible.

TPI 1-1995 now *requires* that (1) the *building* designer design and specify bracing to ensure structural safety of the completed building, and (2) the *installer* be responsible for "receipt, storage, erection, installation, field assembly, and bracing (2, p. 15)."

The truss designer assumes that each of the above parties designs and specifies the proper bracing.

The web members of adjacent trusses can buckle simultaneously, unless there is either (1) a connection to a solid end wall, or (2) diagonal bracing in the plane of the web diagonals (i.e., extending diagonally and joining the web members of three or more consecutive trusses). Top chords can all buckle simultaneously unless (1) nailed roof sheathing is present all along the truss chord (acting as a diaphram), or (2) diagonal bracing is present in the plane of the top chord and connected to the top chords.

When the first truss of any group is lifted into place, there is nothing for it to "lean against" for stability. So it must be braced temporarily by *ground bracing,* a system of *both* vertical and diagonal members that will prevent movement or rotation of the first truss in the group. The members of the ground bracing can be anchored either to the ground itself or to a previously constructed floor.

After the first truss is in place and braced temporarily to prevent it from moving or tilting, additional trusses can be connected to it by diagonal braces connected to their web members. This group of the first three or four trusses, braced to each other, becomes part of the *required permanent bracing system.* Additional trusses can then be braced by connecting them to this first group, using only members perpendicular to the plane of the trusses. Such lateral braces at the level of the top chord are connected to the web members. The initial braced group of trusses should be repeated at intervals along the building length.

After the first few trusses have been erected, fastened, and braced, permanent cross-bracing members are attached to the truss-web members. Such diagonal bracing sets should be used at each end of the building and at intervals of not over 20 ft between the building ends.

Permanent bracing is also required at the plane of the bottom chord. These members are usually parallel to the building length only, as shown in Fig. 10-11.

Bottom-chord bracing, normal to the truss plane, should extend over the full length of the building, with individual members spliced by overlapping by a length equal to one truss spacing. Since a ceiling is often applied to the bottom chord, the bracing members are usually nailed to the top surface of the bottom chord. Bottom-chord bracing members should be close to bottom-chord panel points, but not further apart than 15 ft.

Also, at the building ends, it is sometimes necessary to install horizontal braces, forming a horizontal-plane truss, to prevent the end wall of the building from being dented (pushed in or out) by wind loads.

See Chapter 5 of ANSI/TPI 1-1995 (2) for more detailed requirements for truss bracing. Be warned that trusses fail more often because of improper bracing than by any other cause.

10-6. HEAVY-TIMBER TRUSSES

Before structural steel became practical, heavy-timber trusses were commonly used for bridges and buildings. Today, building designers often select exposed, heavy-timber trusses for their attractive appearance. Such trusses are not only attractive, but because of the large size of their members, they also have good fire resistance. Glued laminated members rather than sawn timbers are used in most modern heavy-timber

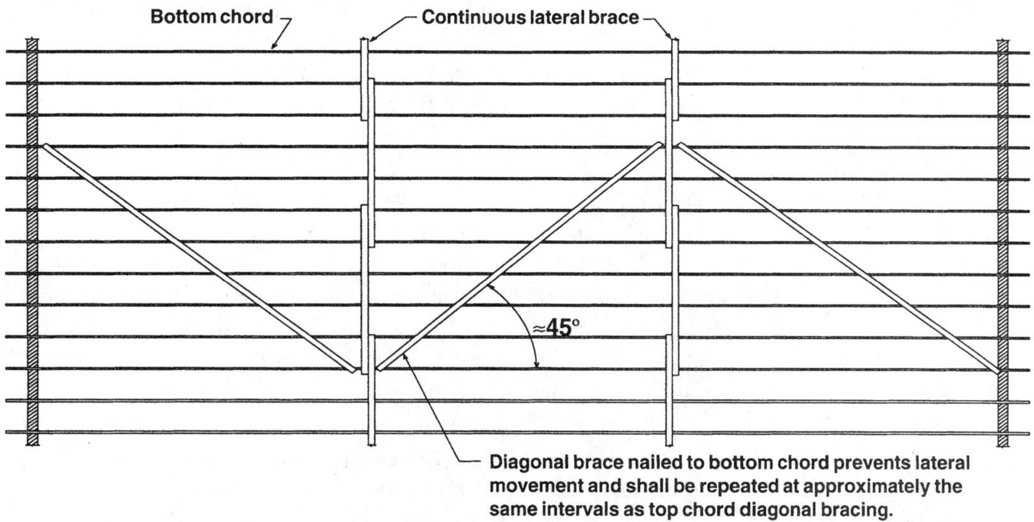

Fig. 10-11. Diagonal bracing in the plane of the bottom chord. (Courtesy of Truss Plate Institute.)

trusses. Glulams as truss members offer the same advantages over sawn timbers as they do for beams (Chapter 8).

Types of Heavy-Timber Truss

For our purpose, there are two main types of heavy-timber truss: (1) those with steel or (on very old trusses) wrought-iron vertical tension members but without gusset plates, and (2) those with steel gusset plates at every joint. The first of these we will refer to as old-style timber trusses.

10-7. OLD-STYLE HEAVY-TIMBER TRUSSES

Old-style heavy-timber trusses usually were of Howe truss configuration, shown in Fig. 10-1b. When the Howe truss is fully loaded, its web verticals are tension members and its web diagonals compression members. For this reason, the web verticals could be merely threaded metal rods.

Connections for Old-Style Heavy-Timber Trusses

Connections for the Howe truss were relatively simple. The vertical rods were connected using just large washers and nuts. Tightening the nuts compressed the ends of the diagonals firmly against the chords, holding the whole assembly together. The compression members transferred their forces to the chords by end bearing of wood against wood, or of wood against cast-iron joint blocks.

Figure 10-12 shows several examples of connections recommended by Henry Dewell (9) for old-style heavy-timber trusses. They are shown and discussed here because, in building renovation or in historic preservation work, today's structural designer may have to investigate existing old-style trusses to determine their suitability for a new use.

Figure 10-12a shows a simple heel joint, in which the entire load of the top chord was transferred to the bottom-chord member by end bearing. Note that the axes of the two members met at a point directly above the reaction, so that the reaction and two member forces were apparently concurrent. Four steps were involved in designing this joint:

1. The allowable stress in compression at 30° to the grain of the bottom chord was determined so that the 5-in. dimension could be verified as adequate. (Of course, if the top chord were of a different species or grade than the bottom chord, it is possible that bearing parallel to the grain of the top chord might control.)
2. The length of bottom chord extension to the left was determined, so that the bottom chord would not be "sliced off" by the 49,000-lb horizontal component of the top-chord force.
3. The net area below the bottom-chord notch was checked to see whether the bottom chord was adequate to resist the 49,000-lb axial load. For this purpose, Dewell considered the force to be centrally located on the net section. (The net section at the bolt hole was also considered.)
4. The exact location of the support was determined to avoid applying a moment to the heel joint. Because of the tapered ends, the top-chord force of 56,500 lb is actually applied at a point not on the centroidal axis of the top chord. (Its horizontal component is 49,000 lb acting to the left.) Also, the bottom-chord tension acts through the center of the remaining net width after the slot is cut. Thus the two 49,000-lb forces form a couple with a 4-in. arm. To resist the moment of this counterclockwise couple, the support was moved 7 in. to the left of the intersection of top- and bottom-chord centerlines.

Force transfer from chord to chord on the long diagonal cut was apparently ignored. However, if there were any force transfer there, the two members would be forced apart, and the two bolts shown would resist this tendency.

Figure 10-12b shows a similar antique connection, but in this case neither cut at the end of the top chord was perpendicular to the grain. Thus, each surface was assumed to take a component of the chord force, each surface having a different allowable compressive stress, computed by Hankinson's equation. Other features of the design are similar to those for Fig. 10-12a.

Figure 10-12c shows a heel joint using a "bolster" to serve as a bearing block, so that the top-chord member is in bearing parallel to the grain.

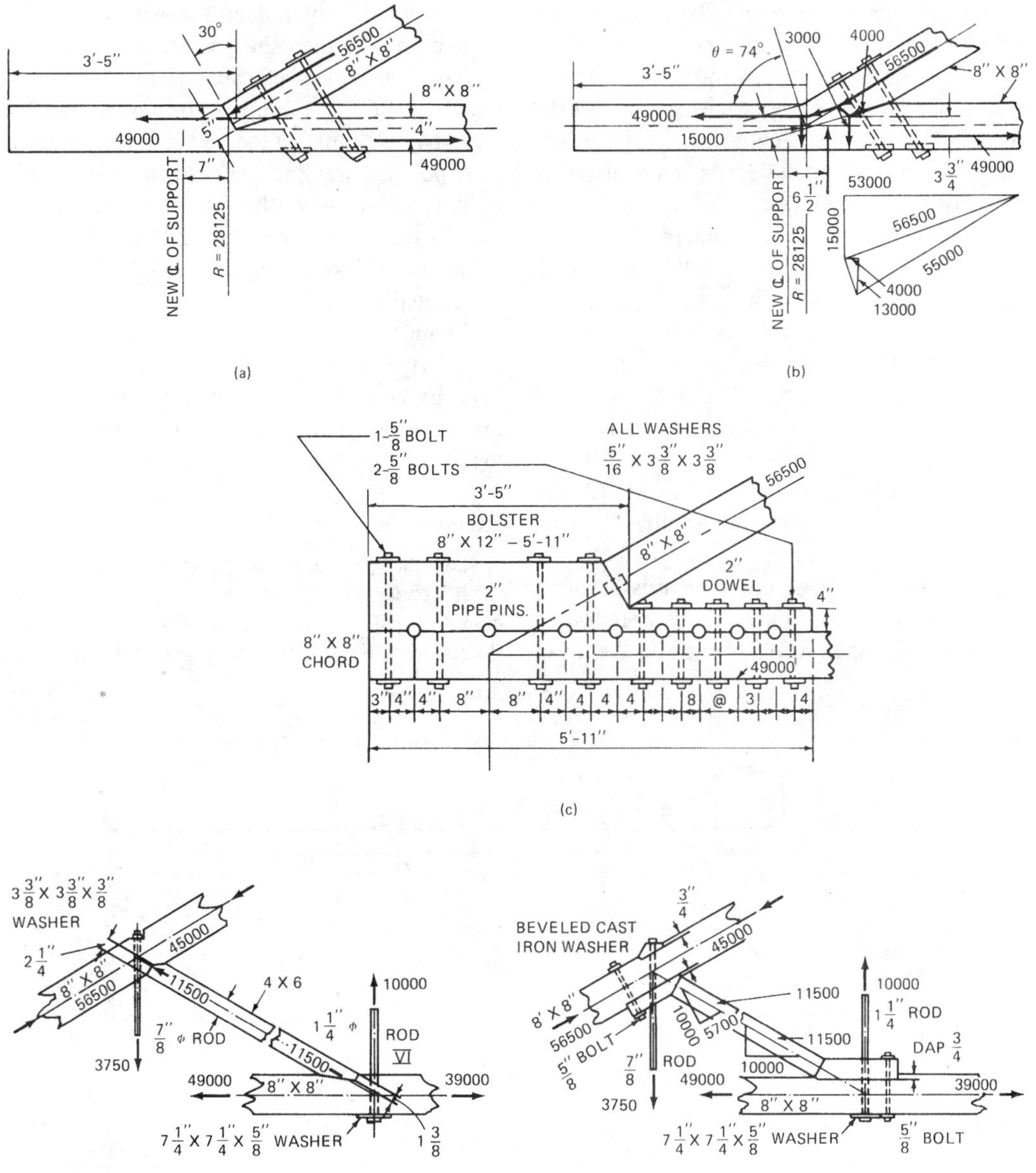

Fig. 10-12. Old-style heavy-timber truss connections. (From reference 9.)

A short steel dowel keeps the top chord and the bolster in proper alignment. Pipe pins transfer the necessary shear force from the bolster to the bottom chord, and vertical bolts tie the bolster to the bottom chord. Note that with this connection, there is no need to shift the support point from the intersection of member centerlines. Bottom-chord net area is whichever is less, net area at a bolt hole or at a groove for a pipe pin.

Figure 10-12d and e show two types of intermediate joint for top and bottom chords of Howe trusses. In the first detail the ends of the diagonal bear against the sides of notches cut into the chords. The top chord also is notched where a horizontal bearing surface is needed under the washer and nut of the threaded vertical member. The second detail uses bolster blocks to transfer the diagonal member's load to the chords. In this detail, the bolster blocks are keyed into 3/4-in. daps of the chord members. A component of the diagonal's force is transferred to the chord where the bolster bears against the end of the dap. A single bolt through each bolster block and chord holds the parts together until assembly is complete.

Of these latter two details, that of Fig. 10-12e is preferable. The joint of Fig. 10-12d, having the diagonal inserted into skew-cut chord notches, required skillful carpentry to ensure good contact between members over the entire intended bearing surfaces. The joint using the bolster was much easier to fabricate so as to get contact over the entire cross section. All that should be added to the detail of Fig. 10-12e are short dowels to key the ends of the diagonal member to the bolsters.

The vertical rods of old-style trusses often had *upset* ends, that is, the rod ends were forged to a larger diameter. Threads cut into the upset portion would not reduce the net area at the threads to less than the gross area of the unthreaded shank. Dewell cautions investigators to make certain that the upset portion was not attached by welding. At the time that these trusses were popular, welding was not nearly as reliable as it is today. Nevertheless, some fabricators, lacking the expensive equipment needed for upsetting, would often substitute welded rods of lesser quality.

Engineers or architects investigating trusses of this type would benefit by reading the 1917 book by Dewell (9) to learn from his extensive experience with these trusses and his remarkable insight into how these connections function.

Example 10-3

Assume that you are investigating an old building to see if it can carry a 55-psf live load, plus the code-required 20 psf for partitions, applied to the existing floor system. Figure 10-13, a scale drawing prepared

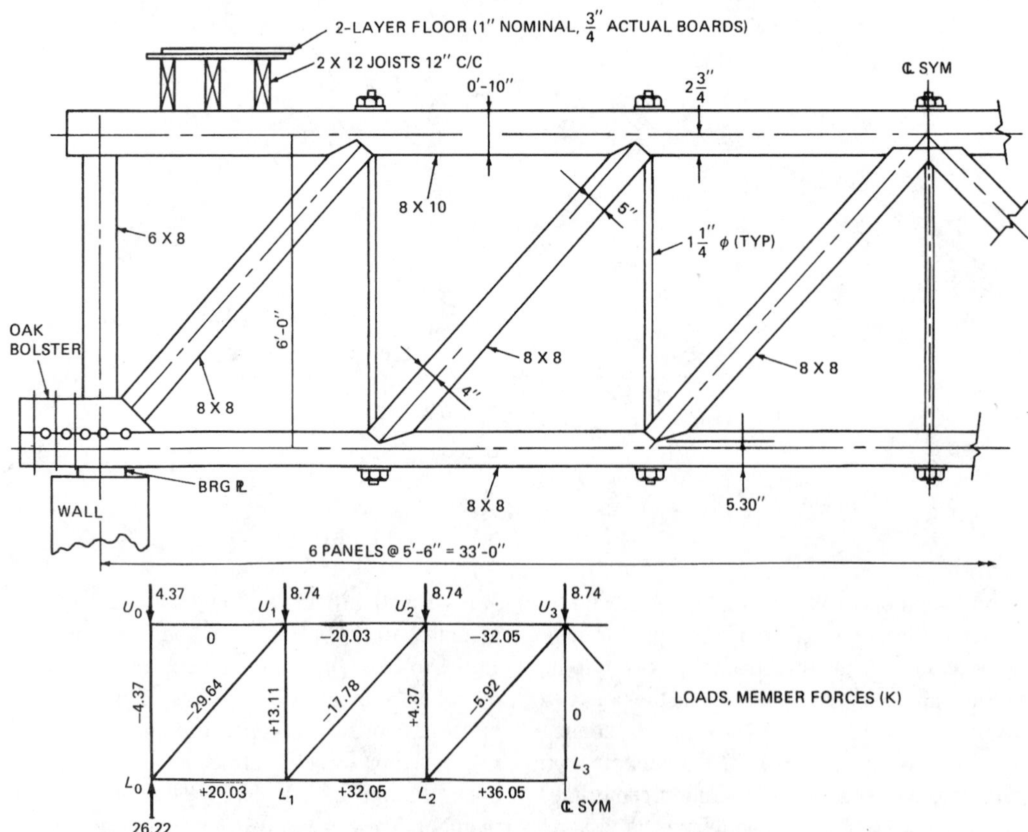

Fig. 10-13. Truss for Example 10-3.

from your field notes, shows one of several old-style trusses in the building. These trusses are 18 ft 0 in. c/c. Member sizes shown are all "full" dimensions, that is, the member is the size its name implies. The truss is some 80–90 years old, but your thorough examination of the truss shows all materials to be in good condition, with no moisture damage and no splits. The vertical rods (not upset) are wrought iron. The bottom layer of floor planks is laid diagonally, so that the floor serves as an effective diaphragm.

For this example check only the top chord, the vertical rods, the bottom chord at joint L_2, and the connection at joint L_1.

First, you must determine suitable allowable stresses. If the vertical rods were steel, allowable stresses in use for structural steel when the truss was built would probably be satisfactory. But these rods are wrought iron and today's building codes do not show allowable stresses for wrought iron. So you have shored up the truss carefully and have removed and tested one of the vertical members, intending to replace it with a modern steel rod. Your test of the removed rod shows a 30,200-psi yield strength and a 50,000-psi tensile strength. Comparing to the AISC code for structural steel (10), you note that for rods of A36 steel, the allowable tensile stress is about 0.53 times the yield strength and 0.33 times the minimum ultimate tensile strength. These same percentages, applied to the wrought iron, suggest an allowable stress of 16 ksi based on yield or 16.5 ksi based on ultimate. With these comparisons as guides, you decide, conservatively, to allow 15 ksi for tension stress on the *net* area of the threaded rod. (For today's threaded steel rods, gross area is used by the AISC Code.)

For the wood members, determining an allowable stress is not nearly so easy. (See references 1 and 11.) Species, grade, and condition must all be considered. Grade is probably unknown. Even if a grade name is known, the grading system used when this truss was built probably differs from that of today. Even worse, there may have been no grading system at all. A suitable solution would be to employ a qualified lumber grader to grade the lumber according to today's standards.

Assume that you have done this and, using your grader's answers, have decided from the NDS Supplement (4) to use the following unadjusted allowable stresses:

Bending, 1300 psi
Compression parallel to grain, 970 psi
Compression perpendicular to grain, 520 psi
Tension, 900 psi
Modulus of elasticity, 1,300,000 psi

The above allowables are for normal duration, dry use. The truss is some 80–90 years old and the building was in use for most of that time. It seems reasonable to assume that, by now, the entire ten-year period of full design loading has occurred and that continued use of the building will extend loading beyond normal duration. Therefore, you decide that it would be wise to treat any further loading as permanent, multiplying each of the above allowables for the wood by a duration factor of 0.9 (except for $F_{c\perp}$).

Your next step is to compute the loads. The floor load is

Live load	= 55 psf
Double board floor, about 35 (1.5/12)	= 4.4
Specified partition load	= 20
Total	= 79.4 psf

Weight of joist = 2 × 12 (full dimensions) × (35/144) = 5.8 lb/ft

Reaction of one joist = 9(79.4 + 5.8) = 767 lb

The load per foot of top chord is the sum of two joist reactions plus the self-weight of the chord itself. Estimated truss weight (pounds per foot of truss):

Top chord, 10(8)(34/144)	= 18.9
Bottom chord, 8(8)(34/144)	= 15.1
Wood web members, 1.38(8)(8) × (34/144)	= 20.9
Wrought-iron rods, 1.23(3.4)(7/5.5)	= 5.3
Total, per foot of truss	= 60.2 lb

(3.4 is the weight, in pounds, of a 12-in. length of wrought iron or steel with a 1.0-in.^{2} cross-sectional area.)

Dividing the last two items equally between top and bottom chords, the top chord will receive a total load of 32 + (2 × 767) = 1566 lb/ft and the bottom chord only 28 lb/ft. For practical purposes, all the load (say 1590 lb/ft) can be treated as top-chord load. Thus, a top interior panel point will receive (5.5)(1590), or 8740 lb, and end ones each one-half as much. Figure 10-13 shows the truss loads and the member forces.

Actual top-chord uniform load of about (2 × 767) + 19 = 1553 lb/ft causes vertical bending moments in the chord. (Web member and bottom-chord weights do not cause top-chord bending.) About 0.1 wl^2 is a reasonable value to use for bending moments in a continuous member of several spans, as this top chord. Thus, the estimated bending moment is

$$M = 0.1(1553)(5.5^2) = 4700 \text{ ft-lb}$$

Check the top chord in panel U_2U_3: Computed actual stresses on the net area are:

$$f_b = 12 \times (4700)/(8 \times 8.5^2/6) = 585 \text{ psi}$$

$$f_c = 32{,}050/(8.5 \times 8) = 471 \text{ psi}$$

The 1½-in. notch is on the edge of the member where the bending stress is compressive. The web member may not tightly fill the notch, however, so the net section rather than gross was used above.

Allowable stresses are computed next. The top chord is laterally braced horizontally by the floor system nailed to it. Vertically, it is free to buckle over the 5-ft 6-in. panel length.

$$\text{Ratio } L'/d = 12(5.5)/10 = 6.6$$

$$F_{cE} = 0.3(1{,}300{,}000)/(6.6)^2 = 8950 \text{ psi}$$

$$F_c^* = 0.9(970) = 873 \text{ psi}$$

Using these values, column stability factor C_p is computed to be 0.979. Terms needed in the interaction equation are f_c and f_b shown above, and

$$F'_c = 970(0.9)(0.979) = 855 \text{ psi}$$

$$F'_b = 1300(0.9) = 1170 \text{ psi}$$

Inserting these into the interaction equation,

$$(471/855)^2 + 585/[1170(1 - 471/8950)] = 0.831 < 1.0 \quad OK$$

Check wrought-iron vertical: Figure 10-13 shows the largest tensile force to be 13.11 k. For a 1¼-in.-diameter rod with U.S. Standard threads, the net (root) area is 0.969 in.² (10). Stress on this net area is

$$13{,}110/0.969 = 13{,}529 \text{ psi}$$

This does not exceed the allowable of 15,000 psi, so the rods are satisfactory.

Check bottom chord at L_2: Net area of the bottom chord at the notch is

$$5.30(8) = 42.40 \text{ in.}^2$$

The bolt hole is at about the same location as the notch, so the net area is further reduced to

$$42.40 - (1.3125 \times 5.30) = 35.44 \text{ in.}^2$$

The tensile stress on the net area is

$$36{,}050/35.44 = 1017 \text{ psi}$$

which is considerably above the allowable of 0.9(900) = 810 psi. Panel 1–2 of the bottom chord will also show overstress. If the truss is to carry the planned load, the central four panels of the bottom chord will have to be reinforced.

Check joint $L_1$1: Assuming only the square-cut end of the diagonal to be effective, the member load of 17.78 k is transferred on an area of 4 × 8 in., or 32.0 in.². The compressive stress is 17,780/32.0 = 556 psi. For the web diagonal, this is less than $F_c^* = 0.9(970)$, and obviously less than the end bearing value, F_g, so it is satisfactory. For the bottom chord, the stress is at an angle of 47.5.° to the grain. Hankinson's equation gives an allowable compressive stress of 638 psi > 556 psi. So the joint is satisfactory. (*Note:* F_g in end bearing could be used, if it were known, in place of F_c in Hankinson's equation also.)

10-8. MODERN GUSSETED HEAVY-TIMBER TRUSSES

Members of modern heavy-timber trusses are usually connected by bolts passing through the members and two steel gusset plates, one on each face of the members. For this to be practical, thickness normal to the truss plane should be alike for all members. Where the bearing strength of the bolt on the wood is not adequate, bolts may be supplemented by shear plates embedded in each face of the timber members. Advantages of the truss with double-plane gusset plates are:

1. Design flexibility. It is not necessary to conform to particular truss configurations, as was the case with the old-style trusses.
2. Less carpentry skill is needed to produce good-quality trusses than was the case with the old-style truss. Members do not have to be notched, nor do ends of web members need to be cut to precise angles to obtain end bearing against other members.
3. Double-gusseted trusses can be of very attractive appearance. For this reason, they are frequently exposed in building interiors. The truss configuration used can be selected for aesthetics, rather than just utility, although the simplest pattern is often also the most attractive.

In general, member design is for axial load only. The exception would be members having significant bending moment. Bending can result from three causes: (1) floor or roof loads applied to chords between panel points; (2) self-weight, for members with fairly large horizontal components of length; and (3) secondary stresses. Secondary bending stresses occur because the truss members are *not* pin connected, as assumed in analysis. Where members are long and slender, these secondary stresses are usually negligible. However, if the truss members have low length/depth ratios, secondary stresses may be significant. Matrix computer analysis giving end bending moments as well as axial forces is so simple and inexpensive today that a designer can easily consider all causes of bending. If significant bending moment is present, the members would be designed for combined axial load and bending.

When choosing a top-chord size, a designer must consider how the top chord is braced laterally (against movement normal to the plane of the truss). If the structural roof material has appreciable diaphram stiffness and is attached to the chord at very short intervals, the chord will be prevented from buckling normal to the truss plane. However, it still can buckle vertically over the length between panel points. In this case, its effective length can be taken as the panel length.

If the structural roof or floor material is not connected directly to the top chord, then it is usually connected to purlins, that is, to beams spanning from truss to truss. In this case, at best, the chord would be laterally supported only at points where the purlins are attached to the truss. Even then, if the purlins are not attached to a competent roof diaphragm, the top chord may not be adequately laterally supported. In this case the effective length of the top chord for buckling normal to the truss plane would be the distance between points at which an adequate bracing system frames into the truss. Usually, this distance is more than a single panel length. So the chord in this case would have two effective lengths—a single panel length for buckling vertically and a longer length (two or three panels, for example) for buckling normal to the plane of the truss. Figure 10-14 illustrates this.

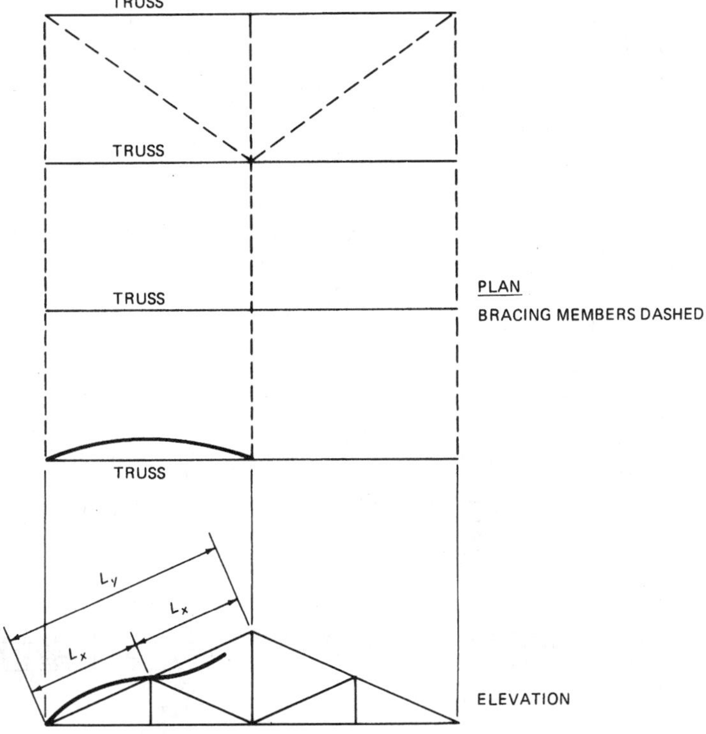

Fig. 10-14. Effective lengths for top-chord buckling.

Web compression members ordinarily are unbraced in either direction. Thus, their effective length for determining allowable stress is their actual length. The authors consider the actual length to be that measured along the web member's centerline from the center of one chord to the center of the other. Bottom-chord members and web tension members must be designed for their *net* area. Connections for gusseted trusses are designed by the methods shown in Chapters 5 and 9.

10-9. TRUSSES CONNECTED BY SPLIT RINGS

Split-ring trusses, many of which are still in use, often had fairly large spans and were used for bridges, buildings, towers, and other structures. Figure 10-15 shows an example of such a truss, and Example 9-9 illustrated the analysis of one of its split-ring joints. The same procedure could be used for analyzing the rest of the joints.

In the designing of these trusses, an important step was to determine how the various layers of adjoining members were to be "stacked." This problem is shown by Fig. 10-15. In this truss, the largest number of pieces used per member was two (for both chords and web diagonals). The vertical members had just one element each and the supporting column had three elements. All pieces overlapped each other as shown, so that no element had to be bent to a different position in the stack at one end than at the other. In the designing of the members for strength, it might be that a single element would have been good enough; however, if two were needed so that the truss could be assembled without any members being bent or being off-center in the stack, then two elements were used.

Compression members having more than one element were designed as spaced columns (see Chapter 7), or the individual elements were treated as separate solid columns. In the truss of Fig. 10-15, the top chord was designed as a spaced column, while the compression verticals were single-element solid columns.

Having to design such a truss is unlikely today, but if the need does arise, a suggested design sequence is:

1. Determine member forces for load combinations to be considered.
2. Choose a species and stress grade.
3. Plan a tentative stacking arrangement.
4. Design the top chord, considering the problems of connections, so that end- and edge-distance requirements will not be violated.
5. Design the other members.

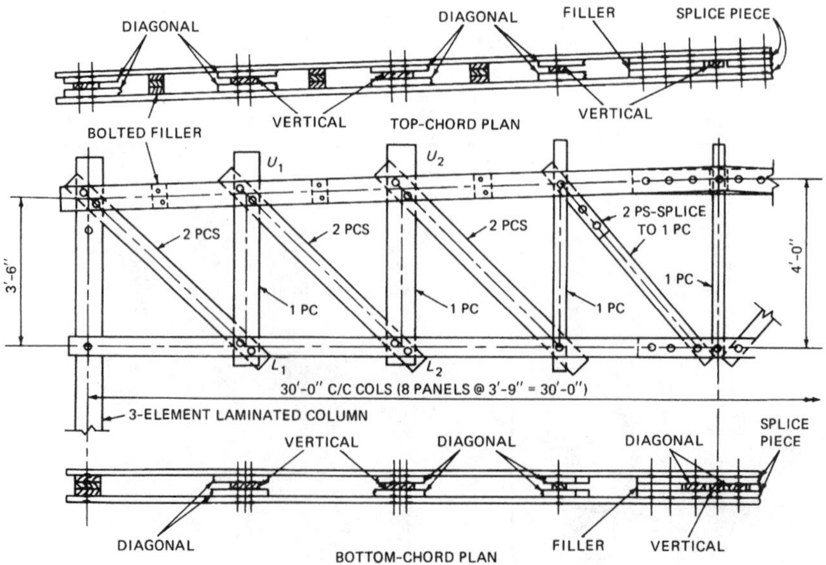

Fig. 10-15. Timber roof truss using 2½ in. split-ring connectors.

6. If the tentative stacking arrangement is still OK, design the joints.

Analysis of existing split-ring trusses is a more likely task than new design. In analyzing existing trusses, it is important to determine the condition of all members. Split-ring connections having more than one bolt often caused members to split as moisture content changed. In the rush of wartime construction, rings were sometimes inadvertently omitted. Through shrinkage, connections may have loosened. And, of course, the wood may have deteriorated. Allowable tensile stresses used in wartime were often higher than those tolerated today. And, finally, the trusses may already have been loaded for periods exceeding the ten-year normal duration, so future loading would wisely be treated as permanent when allowable stresses are selected (references 1 and 11).

10-10. TRUSS DEFLECTIONS AND CAMBER

Truss deflections should be limited for the same reasons that beam deflections are limited. However, computing deflections for trusses is slightly more trouble than for beams. As with beams, both immediate and long-term deflections must be considered. Trusses deflect because of (1) changes in member length, and (2) slip at the end connections of members. (See Fig. 10-16.)

If a computer matrix analysis method is used, the portion of deflection due to member elongation or shortening and member bending is easily determined for every joint. The computer program would probably have to be modified, however, so that it would include the effect of joint slip.

If the deflection at only a single point is needed (this is usually the case) then the virtual work method is the easiest to use. (See any book on structural analysis.) For trusses, this requires making a table showing, for each truss member, the following:

1. Force, F (axial load) due to the loading considered (lb)
2. Cross sectional area, A (in.^{2})
3. Length, L (in.)
4. Modulus of elasticity, E (psi)

Fig. 10-16. Light-frame trusses under construction load of over 100 psf. Truss deflection about one in. Note square panel at midspan for passage of duct work. (Photograph by authors.)

5. Member elongation, FL/AE (in.), either positive or negative. Modify member elongation, FL/AE, by adding or subtracting to cover slip of the end connections. The resulting total, $FL/AE + S$, is the change of distance between joints at opposite ends of the member.
6. Force, u, in each member due to a unit (virtual) load in the direction and at the location of the deflection to be computed.

The desired deflection, Δ, is then given by

$$1 \times \Delta = \Sigma\,[(FL/AE + S)\,(u)] \qquad (10\text{-}5)$$

F and S may each be either positive or negative, depending on whether the member force, F, is tension or compression. Slip, S, may involve slip at one end only or both ends, depending on where the actual member ends are. Virtual force, u, may be of either sign. The product for each member in Eq. 10-5 may also have either sign. For hybrid trusses, in which some members are metal and others are wood, appropriate values of E must be used for each member. Equation 10-5 considers only the effects of axial force and end slip in each member. If deflection at mid-length of a member is desired, bending deflection for that member must be added to the deflection of the truss joints.

The deflection solved will be the *immediate* deflection, not considering creep. Long-term deflection could be estimated by multiplying the immediate deflection by a factor, as was done in Chapter 6 for beams. But the factors given there (and by NDS) are for members in *bending,* not for members in tension or in compression, as in a truss. The authors feel, however, that using those same factors to find long-term deflections for trusses will be very conservative, especially so since the joint-slip term in Eq. 10-5 is not affected by creep.

For heavy timber trusses, reference 12 recommends using the difference between bolt diameter and the drilled hole diameter for joint slip. Dimensions S to use in calculating deflection for a truss with bolted joints would be twice this amount for members that could have such slip at each end. For split rings or shear plates, reference 12 suggests that slip be taken as

For unseasoned lumber, 0.07 in. × load ratio,
or For seasoned lumber, 0.05 in. × load ratio

The load ratio is the actual load for one connector divided by its allowable load.

Camber

Bottom-chord camber offsets the bad appearance of sagging when the truss deflects. If the top chord is horizontal, it too should be cambered. If the amount of camber is not specified by code, it should be at least equal to the long-term dead-load deflection expected. Trusses supporting a flat roof should be cambered even more to prevent ponding. To provide camber, web member lengths must be adjusted to fit the geometry of the cambered position of the panel points. Continuous top- and bottom-chord members are bent into the cambered position. Since the camber dimension is small, stresses due to this bending are usually ignored in design. In any case, these bending stresses are relieved when the truss deflects under load.

10-11. TRUSS DESIGN BY LRFD

Truss members may be chosen by load and resistance factor design (LRFD) rather than by allowable stress design (ASD). Member axial forces and bending moments are found by the procedures outlined in Section 10-2. These forces and moments are then multiplied by appropriate load factors from Table 4-1.

The members selected must satisfy interaction equations for combined bending and axial load.

The buckling stiffness factor, C_T, which was used in ASD, has a different form in LRFD. For seasoned wood the buckling stiffness factor for LRFD is

$$C_T = 1 + 2.3\,\ell_e/(K_T E'_{05}) \qquad (10\text{-}6)$$

in which K_T is 0.59 for either visually graded lumber or machine-evaluated lumber and is 0.82 for products with coefficient of variation less than or equal to 0.11. In Eq. 10-6, the effective unbraced length of the compression chord, l_e, should be expressed in inches, and the adjusted modulus of elasticity at the fifth percentile, E'_{05}, should be expressed in kips per square inch. In the equation for Euler load (Eq. 7-17), the product $C_T I_x$ replaces I_x.

The repetitive member factor from ASD has

its counterpart in LRFD, known as the load-sharing factor, C_r. For trusses (three or more) spaced not more than 24 in. on centers and fabricated using solid-sawn lumber, $C_r = 1.15$. This factor is applied *only* to the adjusted allowable bending strength for members connected by load-distributing material, such as plywood roof sheathing. It is *not* applied to the axial load strength. Ordinarily, it would apply only to the top chord of the truss.

Example 10-4

Use the LRFD method to choose the top chord of the truss of Example 10-2.

Reference strengths for the No. 1 & Btr D. fir-larch are:

For bending, 2.92 ksi
For compression parallel to grain, 3.60 ksi, and
For mean modulus of elasticity, 1.8×10^3 ksi

Member forces and bending moments must be computed separately for dead, live, and snow loads. Fortunately, however, the *loads* themselves can be factored and the worst load combination for this truss selected before starting the selection of member size. Load combinations that must be considered are (using data from Ex. 10-2, and considering the UBC minimum roof live load of 20 psf shown in Table A-3):

(I) $1.4D = 1.4(3 + 5 + 13) = 29.4$ lb per horizontal ft

(II) $1.2D + 1.6L + 0.5S = 1.2(21) + 1.6(20 \times 2) + 0.5(25 \times 2) = 114.2$ lb per horizontal ft

(III) $1.2D + 1.6S + 0.5L = 1.2(21) + 1.6(25 \times 2) + 0.5(20 \times 2) = 125.2$ lb per horizontal ft

Combinations II and III will each have the same time-effect factors, $\lambda = 0.8$, while the factor for combination I is only 0.6. Thus, it is obvious that combination III will control the design.

By direct proportion compared to Ex. 10-1 and 10-2, the factored values of axial load and bending moment for the top chord are:

$P_u = 3226(125.2/71) = 5689$ lb $= 5.69$ kips compression

$M_u = 120(125.2/71) = 211.6$ ft-lb $= 0.212$ ft-k

Try a 2 × 4, with the 4-in. dimension vertical. Check its suitability by using the LRFD interaction equation (Eq. 7-23). Terms needed in that equation are computed next.

Using mean $E' = 1800$ ksi, Eq. 7-18 gives $E'_{05} = 1090$ ksi. The buckling stiffness factor (for seasoned visually graded lumber) is

$$C_T = 1 + 2.3\,\ell_e/(0.59E'_{05})$$
$$= 1 + 2.3(46.4)/(0.59 \times 1800)$$
$$= 1.10$$

The Euler limit load, $P_{ex} = \pi^2 E'_{05}(C_T I)/(L'^2)$, with respect to the x-axis is

$$P_{ex} = \pi^2\,(1090)(1.10 \times 5.36)/(46.4^2) = 29.5 \text{ kips}$$

The resistance of a zero-length column (using $C_F = 1.15$) is

$$P'_0 = AF^*_c = (1.5 \times 3.5)\,(1.15 \times 3.60) = 21.7 \text{ kips}$$

The column stability factor is computed using the strong-axis properties since weak-axis buckling is prevented.

$$\alpha_c = \phi_s P_{ex}/\lambda\phi_c P'_0 = 0.85(29.5)/(0.8 \times 0.9 \times 21.7) = 1.60$$

Let $k = (1 + \alpha_c)/1.6 = 1.625$

$$C_p = k - \sqrt{k^2 - \alpha_c / 0.8} = 0.825$$

$$P' = AC_pF^*_c = (1.5 \times 3.5)(0.825)(1.15 \times 3.60) = 17.9 \text{ kips}$$

Items required for the bending term are

$$C_{mx} = 1.0$$

$$B_{bx} = C_{mx}/(1 - P_u/\phi_c P_{ex}) = 1.0/[1 - 5.69/(0.9 \times 29.5)] = 1.27$$

$$M_{mx} = (1.27)(0.212) = 0.269 \text{ ft-k} = 3.23 \text{ in.-k}$$

Load sharing factor $C_r = 1.15$ and size factor C_F for bending $= 1.5$.

$$F'_{bx} = 2.92C_rC_F = 2.92(1.15)(1.5) = 5.04 \text{ ksi}$$

$$M'_x = S_xF'_{bx} = 3.063(5.04) = 15.4 \text{ in.-k}$$

Substituting in the interaction equation (Eq. 7-23),

$$[5.69/(0.8 \times 0.9 \times 17.9)]^2 + 3.23/(0.8 \times 0.85 \times 15.4) = 0.195 + 0.308 = 0.503 < 1.0$$

OK

As was the case with allowable stress design, the 2 × 4 will be satisfactory for the top chord.

REFERENCES

1. *Evaluation, Maintenance and Upgrading of Wood Structures, A Guide and Commentary,* American Society of Civil Engineers, New York, 1982.
2. *National Design Standard for Metal Plate Connected Wood Truss Construction,* ANSI/TPI 1-1995, Truss Plate Institute, Madison, WI, 1995.
3. *National Design Specification for Wood Construction,* rev. 1991 Ed., American Forest & Paper Association, Washington, DC, 1993.
4. *Design Values for Wood Construction,* NDS Supplement to the 1991 edition, American Forest & Paper Association, Washington, DC, 1993.
5. *A Computerized Wood Engineering System: Purdue Plane Structures Analyzer,* USDA Forest Service Research Paper FPL 168, Forest Products Laboratory, Madison, WI, 1972.
6. *Purdue Plane Structures Analyzer-II: A Computerized Wood Engineering System.* Gen Tech. Rep. FPL40, USDA Forest Service, Forest Products Laboratory, Madison, WI, 1984.
7. Wilson, E. L., and A. Habibullah, *SAP90 Structural Analysis Programs,* Computers & Structures, Inc., Berkeley, CA, 1990.
8. *Methods of Tension Testing of Metallic Materials,* Standard E-8, American Society for Testing and Materials, West Conshohocken, PA, 1993.
9. Dewell, H. D., *Timber Framing,* Dewey Publishing Company, San Francisco, 1917.
10. *Specification for Structural Steel Buildings,* Allowable Stress Design and Plastic Design, American Institute of Steel Construction, Chicago, IL, 1989.
11. Lanius, R. M., Jr., "Evaluating Residual Strength and Repair of Structures," in Meyer, R. W. and R. M. Kellogg, *Structural Uses of Wood in Adverse Environments,* Van Nostrand Reinhold, New York, 1982.
12. Milbradt, K. P., "Timber Structures," in Gaylord, E. H., Jr., and C. N. Gaylord, *Handbook of Structural Engineering,* McGraw-Hill, New York, 1979.

PROBLEMS

10-1. Try redesigning the truss of Examples 10-1 and 10-2 (Fig. 10-6) using four equal panels rather than six for the top chord, and three rather than five for the bottom chord. Can all NDS requirements be met?

10-2. Design truss-plate connections for the other joints of the truss of Example 10-2, using the gross area method and 90 psi for allowable lateral resistance.

10-3. Design truss connections for joints of the truss of Example 10-2 (except the heel joint) using the net contact area method and an allowable lateral resistance of 95 psi.

10-4. Choose a truss configuration and design (by either ASD or LRFD, whichever your instructor specifies) a light-frame truss for a 30-ft span, without overhangs, having a roof pitch of 4 in. in 12 in. Use a 24-in. truss spacing. Assume snow load of 25 psf and roof dead loads the same as for Example 10-1. Assume that the ceiling attached to the bottom chord has an access opening (*scuttle*) through which the building owner has access to the space between the roof and the ceiling. The ceiling drywall and insulation are expected to weigh 11 lb per square ft. Would it be prudent for you as the designer to anticipate that the owner might store things in this space above the ceiling? If so, should you assume a bottom-chord load larger than 11 psf? (This is the type of decision a designer may have to make. You decide, and design the truss accordingly.)

10-5. In one area of the old building of Example 10-3 (Fig. 10-14) the truss span is only 30 ft zero in., rather than 33 ft zero in. Floor load (psf) truss member sizes, and truss spacing are the same as for Example 10-3. Check strength of members for this truss.

10-6. Design the double-gusseted scissors truss shown by Fig. 10-18. Trusses are 18 ft apart. The roof is 3-in. wood decking with roofing weighing 6 psf and insulation weighing 3 psf of roof surface. Purlins (at each top-chord panel point) weigh 340 lb each. Snow load is 35 psf of horizontal projection for a level surface. This roof has a steep slope. Determine the required member sizes, using combination 10, D. fir glulams of 5 1/8 in. width. Use either ASD or LRFD, whichever your instructor specifies.

10-7. For the truss of Problem 10-6, design joints *A* and *B,* using 7/8 in. diameter bolts and steel 1/4 in. gusset plates on each face.

10-8. For the truss of Problem 10-6, design joint *C* using 7/8 in. diameter bolts and 3/8 in. steel gusset plates on each face. (*Hint:* One of the long sloping members can be continuous through joint *C.*)

10-9. What would be the total long-term deflection for the truss you designed for Problem 10-4?

10-10. Find the total long-term deflection for the truss of Example 10-2.

11

Plywood and Similar Wood Products

One of the most important wood products for structural use is plywood. It is used for roof and floor sheathing, concrete formwork, webs of wood beams, and even the hulls of boats. It can be used to resist gravity loads or to resist lateral loads as in plywood diaphragms and shearwalls. Because of the way plywood is manufactured, it has more nearly isotropic properties and greater dimensional stability (resistance to change in dimension with moisture content) than sawn lumber. The dimensional change of plywood is only about one-tenth of that for solid timber over any given range in temperatures and relative humidities (1). Other manufactured wood products such as particleboard and insulation board have found general acceptance as building materials.

11-1. PLYWOOD PRODUCTION AND CLASSIFICATION

Plywood consists of thin wood sheets (called veneers or plies) assembled and glued together with the grain directions of adjacent layers arranged at right angles. The plies vary in thickness from $^1/_{16}$ in. to $^5/_{16}$ in. Structural plywood has an odd number of layers. In four-ply panels the two central plies have their grain in the same direction and therefore form a single layer. The standard nominal thickness of panels varies from $^1/_4$ in. to $1^1/_8$ in.

The logs from which plywood is made are moistened before the manufacturing process begins by storing them in ponds, spraying them with water, or putting them in steam vats. They are then dropped into a vat where they are kept in boiling water for six hours, so that the veneer will not break apart as it is peeled off. After the bark is removed, a saw cuts the logs typically into sections just over 8 ft long. Then a giant lathe rotates the logs, peeling off a continuous sheet of veneer, which is stored on a rack. When the veneer comes off the peel rack, it goes through a machine that clips out any bad veneer and strings it into continuous rolls. After the veneer is seasoned in a dryer to a moisture content of 5% or less, it is graded. The plies are assembled, glued, and pressed. Under heat and pressure, the glue hardens quickly and the sheets of plywood are complete. After the panel is inspected for quality of face and back, it is stamped with the grade trademark.

The standard plywood panel size is 4 × 8 ft. Nearly all plywood in the United States conforms to the U.S. Department of Commerce *Product Standard* PS 1-95 (2). A manufacturers' association, APA—The Engineered Wood Association (formerly American Plywood Association), publishes much information concerning the proper use and design properties of plywood. Most notable is the *Plywood Design Specification* (3), which contains allowable stresses and other design information.

Veneer Grades

The veneer is always seasoned to eliminate residual stresses that would occur if it were glued and hot-pressed before being allowed to dry. To classify plywood by its strength properties, the veneer itself must be graded. This grading process occurs just before the plies are assembled. The five veneer grades, in descending order of quality, are as follows:

N	No knots; may have small patches; sanded smooth; for natural finish
A	No knots; may have small patches; sanded smooth; for painting
B	Solid surface; small round knots; patches and round plugs are allowed
C-Plugged	Improved C grade
C	Small knots, knotholes, patches; lowest grade allowed for exterior use
D	Larger knots, knotholes

The veneer grades that interest the structural designer the most are grades C and D. Many interior panels have grade C veneer on the face and grade D for inner and back plies. This is why these panels are sometimes called C-D interior. They are also referred to as APA Rated Sheathing, Exposure 1 or Exposure 2.

There are two main types of plywood—interior and exterior. Interior plywood is further categorized as either Exposure 1, Exposure 2, or simply Interior. Exposure 1 is made with an exterior glue that adds shear strength to the panel. Exposure 1 plywood is recommended if the conditions in use will be dry but prolonged exposure to moisture is expected during construction. Exposure 2 panels may be used where only moderate delays in providing protection may be expected. The third type of interior plywood is not used for strictly structural purposes.

Exterior plywood is bonded with waterproof adhesives. It has superior moisture resistance and should be used for panels exposed to wet conditions or high equilibrium moisture content. Grade D veneer (the lowest grade) is not allowed in Exterior plywood. The large open knotholes in grade D veneer could allow moisture to seep in between the plies, causing them to separate.

Plywood Grades

Table 11-1 shows plywood grades. The first three rows of interior plywood and the first three rows of exterior plywood each contain four grades of plywood that are for structural use. Most of these are unsanded. These four grades as classified by APA for structural use are:

1. APA Rated Sheathing (also called C-D Interior or C-C Exterior)—a general-use structural plywood that has all C-grade veneers if exterior and C face with D inner and back veneers if interior.
2. APA Structural I Rated Sheathing (also called Structural I C-D Interior or Structural I C-C Exterior)—very strong panel using all Group I species wood and only exterior glue; veneers same as above.
3. APA Structural II Rated Sheathing (also called Structural II C-D Interior or Structural II C-C Interior or Structural II C-C Exterior)—similar to Structural I, but Group 3 species wood is allowed.
4. APA Rated Sturd-I-Floor—for combination subfloor-underlayment (see Section 11-7).

Each of these comes in Exterior type, Exposure 1 type, and Exposure 2 type, except for Structural I and II for which Exposure 2 is not available. (The Structural II grade is not very common and is difficult to obtain.) Sturd-I-Floor is the only engineered grade that has been sanded at all; it is merely touch-sanded (somewhat, but not thoroughly smoothed). Contary to what one might assume, "CDX" plywood is not an exterior plywood. It is an interior plywood assembled with exterior glue; that is, it is Exposure 1.

The other plywood grades listed in Table 11-1 are appearance grades. These grades are sanded so that they will have a pleasing appearance for cabinets, shelving, and the like. The appearance grades are generally not used in structural work, with the exception of APA B-B Plyform. This special-use exterior panel has sanded surfaces and is mill-oiled to serve as formwork for concrete. Allowable stresses are given for the appearance grades for the instances where appearance and strength are both important.

Conformance to Standards

Plywood was first manufactured under a product standard in 1966. A product standard contains voluntary requirements for producing and marketing a product. The revised version of this

Table 11-1. Guide to Use of Allowable Stress and Section Properties Tables—Interior.

	Plywood Grade	Description and Use	Typical Trademarks	Veneer Grade			Common Thicknesses	Grade Stress Level (Table 3)	Species Group	Section Property Table
				Face	Back	Inner				
INTERIOR OR PROTECTED APPLICATIONS	APA RATED SHEATHING EXP 1 or 2[3]	Unsanded sheathing grade for wall, roof, subflooring, and industrial applications such as pallets and for engineering design, with proper stresses. Manufactured with intermediate and exterior glue (1). For permanent exposure to weather or moisture only Exterior type plywood is suitable.	APA RATED SHEATHING 32/16 15/32 INCH SIZED FOR SPACING EXPOSURE 1 000 PS 1-83 C-D NER-108	C	D	D	5/16, 3/8, 15/32, 1/2, 19/32, 5/8, 23/32, 3/4	S-3 (1)	See "Key to Span Rating"	Table 4A (unsanded)
	APA STRUCTURAL I RATED SHEATHING EXP 1 or APA STRUCTURAL II[2] RATED SHEATHING EXP 1	Plywood grades to use where strength properties are of maximum importance, such as plywood-lumber components. Made with exterior glue only. STRUCTURAL I is made from all Group 1 woods. STRUCTURAL II allows Group 3 woods.	APA RATED SHEATHING STRUCTURAL I 24/0 3/8 INCH SIZED FOR SPACING EXPOSURE 1 000 PS 1-83 C-D NER-108	C	D	D	5/16, 3/8, 15/32, 1/2, 19/32, 5/8, 23/32, 3/4	S-2	Structural I use Group 1 Structural II See "Key to Span Rating"	Table 4B (unsanded) Table 4A (unsanded)
	APA RATED STURD-I-FLOOR EXP 1 or 2[3]	For combination subfloor-underlayment. Provides smooth surface for application of carpet. Possesses high concentrated and impact load resistance during construction and occupancy. Manufactured with intermediate and exterior glue. Touch-sanded (4). Available with tongue and groove.(5)	APA RATED STURD-I-FLOOR 20 oc 19/32 INCH SIZED FOR SPACING T&G NET WIDTH 47-1/2 EXPOSURE 1 000 UNDERLAYMENT PS 1-83 NER-108	C plugged	D	C & D	19/32, 5/8, 23/32, 3/4, 1-1/8 (2-4-1)	S-3 (1)	See "Key to Span Rating"	Table 4A (touch-sanded)
	APA UNDERLAYMENT EXP 1, 2 or INT	For underlayment under carpet. Available with exterior glue. Touch-sanded. Available with tongue and groove.(5)	APA UNDERLAYMENT GROUP 1 EXPOSURE 1 000 PS 1-83	C plugged	D	C & D	1/2, 19/32, 5/8, 23/32, 3/4	S-3 (1)	As Specified	Table 4A (touch-sanded)
	APA C-D PLUGGED EXP 1, 2 or INT	For built-ins, wall and ceiling tile backing. NOT for underlayment. Available with exterior glue. Touch-sanded.(5)	APA C-D PLUGGED GROUP 2 EXPOSURE 1 000 PS 1-83	C plugged	D	D	1/2, 19/32, 5/8, 23/32, 3/4	S-3 (1)	As Specified	Table 4A (touch-sanded)
	APA APPEARANCE GRADES EXP 1, 2 or INT	Generally applied where a high quality surface is required. Includes APA N-N, N-A, N-B, N-D, A-A, A-B, A-D, B-B, and B-D INT grades.(5)	APA A-D GROUP 1 EXPOSURE 1 000 PS 1-83	B or better	D or better	C & D	1/4, 11/32, 3/8, 15/32, 1/2, 19/32, 5/8, 23/32, 3/4	S-3 (1)	As Specified	Table 4A (sanded)

(Table 11-1 continued on next page.)

standard is *Product Standard* PS 1-95 (2). Until recently, most grades of plywood were manufactured under PS 1-95 or its predecessors. Now some plywood is instead manufactured under APA performance standards as set forth in *Performance Standards and Policies for Structural-Use Panels* (4). A performance standard describes how a panel must perform in end use rather than how it must be manufactured. Performance-rated panels must undergo specific tests for strength and bond integrity as detailed in reference 4 to gain trademark privilege. Besides certain plywood, structural composites (veneer faces bonded to reconstituted wood cores) and nonveneered panels (waferboard, oriented strand board, and structural particleboard) are produced under performance standards.

All Structural I and Structural II plywood sheets conform to PS 1-95 because their strength properties are of maximum importance. Sturd-I-Floor may or may not conform to PS 1-95; however, because it is a flooring panel, its performance (maximum span) is generally its significant feature. Ordinary APA Rated Sheathing (other than Structural I or II) may or may not meet all the provisions of the product standard. For engineered applications where design stresses and section properties need to be known, only panels conforming to PS 1-95 should be specified.

Grade Stamps

Performance-rated panels are acknowledged by the National Evaluation Service, a body sponsored by major building codes, under report number PRP-108 (previously NER-108). Panels stamped with PRP-108 may be used for spans

Table 11-1. (*Continued*) Guide to Use of Allowable Stress and Section Properties Tables—Exterior.

	Plywood Grade	Description and Use	Typical Trademarks	Veneer Grade: Face	Veneer Grade: Back	Veneer Grade: Inner	Common Thicknesses	Grade Stress Level (Table 3)	Species Group	Section Property Table
EXTERIOR APPLICATIONS	APA RATED SHEATHING EXT(3)	Unsanded sheathing grade with waterproof glue bond for wall, roof, subfloor and industrial applications such as pallet bins.	APA RATED SHEATHING 48/24 23/32 INCH SIZED FOR SPACING EXTERIOR 000 PS 1-83 C-C NER-108	*C*	*C*	*C*	5/16, 3/8, 15/32, 1/2, 19/32, 5/8, 23/32, 3/4	*S-1*	See "Key to Span Rating"	Table 4A (unsanded)
	APA STRUCTURAL I RATED SHEATHING EXT or APA STRUCTURAL II(2) RATED SHEATHING EXT	"Structural" is a modifier for this unsanded sheathing grade. For engineered applications in construction and industry where full exterior-type panels are required. STRUCTURAL I is made from Group 1 woods only.	APA RATED SHEATHING STRUCTURAL I 24/0 3/8 INCH SIZED FOR SPACING EXTERIOR 000 PS 1-83 C-C NER-108	*C*	*C*	*C*	5/16, 3/8, 15/32, 1/2, 19/32, 5/8, 23/32, 3/4	*S-1*	Structural I use Group 1 Structural II See "Key to Span Rating"	Table 4B (unsanded) Table 4A (unsanded)
	APA RATED STURD-I-FLOOR EXT(3)	For combination subfloor-underlayment where severe moisture conditions may be present, as in balcony decks. Possesses high concentrated and impact load resistance during construction and occupancy. Touch-sanded (4). Available with tongue and groove.(5)	APA RATED STURD-I-FLOOR 20 oc 19/32 INCH SIZED FOR SPACING EXTERIOR 000 UNDERLAYMENT PS 1-83 NER-108	*C* plugged	*C*	*C*	19/32, 5/8, 23/32, 3/4,	*S-2*	See "Key to Span Rating"	Table 4A (touch-sanded)
	APA UNDERLAYMENT EXT and APA C-C PLUGGED EXT	Underlayment for floor under where severe moisture conditions may exist. Also for controlled atmosphere rooms and many industrial applications. Touch-sanded. Available with tongue and groove.(5)	APA C-C PLUGGED GROUP 2 EXTERIOR 000 PS 1-83	*C* plugged	*C*	*C*	1/2, 19/32, 5/8, 23/32, 3/4	*S-2*	As Specified	Table 4A (touch-sanded)
	APA B-B PLYFORM CLASS I or II(2)	Concrete-form grade with high reuse factor. Sanded both sides, mill-oiled unless otherwise specified. Available in HDO. For refined design information on this special-use panel see *APA Design/Construction Guide: Concrete Forming*, Form No. V345. Design using values from this specification will result in a conservative design.(5)	APA PLYFORM B-B CLASS I EXTERIOR 000 PS 1-83	*B*	*B*	*C*	19/32, 5/8, 23/32, 3/4	*S-2*	Class I use Group 1; Class II use Group 3	Table 4A (sanded)
	APA MARINE EXT	Superior Exterior-type plywood made only with Douglas Fir or Western Larch. Special solid-core construction. Available with MDO or HDO face. Ideal for boat hull construction.	MARINE · A-A · EXT · APA · 000 · PS1-83	*A* or *B*	*A* or *B*	*B*	1/4, 3/8, 1/2, 5/8, 3/4	*A* face & back use *S-1* *B* face or back use *S-2*	Group 1	Table 4B (sanded)
	APA APPEARANCE GRADES EXT	Generally applied where a high quality surface is required. Includes APA A-A, A-B, A-C, B-B, B-C, HDO and MDO EXT.(5)	APA A-C GROUP 1 EXTERIOR 000 PS 1-83	*B* or better	*C* or better	*C*	1/4, 11/32, 3/8, 15/32, 1/2, 19/32, 5/8, 23/32, 3/4	*A* or *C* face and back use *S-1* *B* face or back use *S-2*	As Specified	Table 4A (sanded)

(1) When exterior glue is specified, i.e. Exposure 1, stress level 2 (S-2) should be used.
(2) Check local suppliers for availability of STRUCTURAL II and PLYFORM Class II grades.
(3) Properties and stresses apply only to APA RATED STURD-I-FLOOR and APA RATED SHEATHING manufactured entirely with veneers.
(4) APA RATED STURD-I-FLOOR 2-4-1 may be produced unsanded.
(5) May be modified to STRUCTURAL I. For such designation use Group 1 stresses and Table 4B section properties.

Source: Courtesy of APA—the Engineered Wood Association.

shown on the trademark stamp. APA Rated Sheathing conforming to the product standard will have PS 1-95 contained in the trademark stamp. Examples of typical grade/trademark stamps are shown in Fig. 11-1.

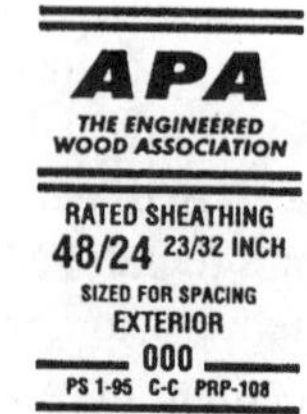

Fig. 11-1. Plywood trademark stamps. (Courtesy APA—The Engineered Wood Association.)

Species Group Number

The term *group number* in plywood refers to the wood species group from which the plywood is manufactured. The 70 species used in the United States for plywood are divided into five groups, four of which are used in the engineered grades of plywood. Species Group 1 contains the strongest varieties of wood. Mostly softwoods are used in plywood manufacture. To find the allowable stresses for a particular plywood panel, the designer must know the species group for the outer plies of the panel. Except for Structural I plywood, the inner plies do not necessarily have to belong to this same group. (Plies of Structural I must all be from species Group 1.) Allowable stresses for all panels except Structural I are based on having inner plies of the weakest species (Group 4).

Span Rating

The span-rating system indicates the performance of the plywood when used as a roof or floor sheathing, without the need for further engineered design. It gives the allowable span when the face grain is placed perpendicular to the supports (span and grain in the same direction). Usually two nominal thicknesses have the same span rating. Generally the thinner of these is easier to obtain.

The span rating consists of two numbers separated by a slash, such as 32/16. The number on the left is the maximum spacing (in inches) of supports for plywood as roof sheathing, and the one on the right is the maximum spacing for floor supports. Table 11-2 shows how to relate span rating to species group.

Building codes generally accept the span rating, so long as edge support requirements of the codes are met. For roof sheathing, edges parallel to the span usually need to be supported by either blocking, tongue-and-groove edges, or H-shaped clips; otherwise a reduced span must be used. These edge supports may be unnecessary if the actual span is smaller than the span rating. If the roof is to carry diaphragm loads (see Chapter 12) in addition to gravity loads, blocking (Fig. 6-1) may be the only permissible edge support.

Because such a large percentage of plywood is used as sheathing in light-frame construction, the span rating recommendations make it unnecessary to perform design calculations in common sheathing situations. However, the designer needs to understand the directional properties of plywood in order to be able to design for such applications as gusset plates and beam webs.

Allowable Stresses

Allowable stresses for plywood are shown by Table 11-3. These allowable stresses depend on (1) species group of the face ply, (2) condition of

Table 11-2. Key to Span Rating and Species Group.
For panels with "Span Rating" as across top, and thickness as at left, use stress for species group given in table.

Thickness (in.)	Span Rating (APA RATED SHEATHING grades)							
	12/0	16/0	20/0	24/0	32/16	40/20	48/24	
					Span Rating (STURD-I-FLOOR grades)			
					16 o.c.	20 o.c.	24 o.c.	48 o.c.
5/16	4	3	1					
3/8			4[3]	1				
15/32 & 1/2				4[3]	1[1]			
19/32 & 5/8					4[3]	1		
23/32 & 3/4						4[3]	1	
7/8							3[2]	
1-1/8								1

(1) Thicknesses not applicable to APA RATED STURD-I-FLOOR.
(2) For APA RATED STURD-I-FLOOR 24 oc, use Group 4 stresses.
(3) For STRUCTURAL II, use Group 3 stresses.

Source: Courtesy APA—the Engineered Wood Association.

Table 11-3. Allowable Stresses for Plywood (psi) conforming to U.S. Product Standard PS 1-95 for Construction and Industrial Plywood. Stresses are based on normal duration of load, and on common structural applications where panels are 24" or greater in width.

Type of Stress		Species Group of Face Ply	Grade Stress Level[1] S-1 Wet	S-1 Dry	S-2 Wet	S-2 Dry	S-3 Dry Only
EXTREME FIBER STRESS IN BENDING (F_b)		1	1430	2000	1190	1650	1650
TENSION IN PLANE OF PLIES (F_t)	F_b &	2, 3	980	1400	820	1200	1200
Face Grain Parallel or Perpendicular to Span (At 45° to Face Grain Use 1/6 F_t)	F_t	4	940	1330	780	1110	1110
COMPRESSION IN PLANE OF PLIES		1	970	1640	900	1540	1540
Parallel or Perpendicular to Face Grain	F_c	2	730	1200	680	1100	1100
		3	610	1060	580	990	990
(At 45° to Face Grain Use 1/3 F_c)		4	610	1000	580	950	950
SHEAR THROUGH THE THICKNESS[3]		1	155	190	155	190	160
Parallel or Perpendicular to Face Grain	F_v	2, 3	120	140	120	140	120
(At 45° to Face Grain Use 2 F_v)		4	110	130	110	130	115
ROLLING SHEAR (IN THE PLANE OF PLIES)		MARINE & STRUCTURAL I	63	75	63	75	—
Parallel or Perpendicular to Face Grain (At 45° to Face Grain Use 1-1/3 F_s)	F_s	ALL OTHER[2]	44	53	44	53	48
MODULUS OF RIGIDITY (OR SHEAR MODULUS) Shear in Plane Perpendicular to Plies (through the thickness) (At 45° to Face Grain Use 4G)	G	1	70,000	90,000	70,000	90,000	82,000
		2	60,000	75,000	60,000	75,000	68,000
		3	50,000	60,000	50,000	60,000	55,000
		4	45,000	50,000	45,000	50,000	45,000
BEARING (ON FACE)		1	210	340	210	340	340
Perpendicular to Plane of Plies	$F_{c\perp}$	2, 3	135	210	135	210	210
		4	105	160	105	160	160
MODULUS OF ELASTICITY IN BENDING IN PLANE OF PLIES		1	1,500,000	1,800,000	1,500,000	1,800,000	1,800,000
	E	2	1,300,000	1,500,000	1,300,000	1,500,000	1,500,000
Face Grain Parallel or Perpendicular to Span		3	1,100,000	1,200,000	1,100,000	1,200,000	1,200,000
		4	900,000	1,000,000	900,000	1,000,000	1,000,000

(1) See Table 11-1 for guide.
To qualify for stress level S-1, gluelines must be exterior and only veneer grades N, A, and C are allowed in either face or back.
For stress level S-2, gluelines must be exterior and veneer grade B, C-Plugged and D are allowed on the face or back.
Stress level S-3 includes all panels with interior or intermediate (IMG) gluelines.

(2) Reduce stresses 25% for 3-layer (4- or 5-ply) panels over 5/8" thick. Such layups are possible under PS 1-95 for APA RATED SHEATHING, APA RATED STURD-I-FLOOR, UNDERLAYMENT, C-C Plugged and C-D Plugged grades over 5/8" through 3/4" thick.

(3) Shear-through-the-thickness stresses for MARINE and SPECIAL EXTERIOR grades may be increased 33%. See Section 3.8.1 for conditions under which stresses for other grades may be increased.

Source: Courtesy of APA—the Engineered Wood Association.

use (wet or dry), and (3) grade stress level (shown by third column from the right in Table 11-1.)

Example 11-1

Find the grade stress level of ½ in. APA Rated Sheathing, Exposure 1.

See Table 11-1 for interior plywood. Footnote 1, referred to in the Grade Stress Level column, indicates that S-2 is the proper grade stress level.

11-2. CROSS-SECTIONAL PROPERTIES

Because a plywood layer has its grain at right angles to the grain in adjacent layers, the section properties are not as readily found as in sawn lumber. It is important that the designer understand how the grain direction affects section properties to be used in design.

Wood is much stronger and stiffer parallel to the grain than it is perpendicular to the grain. In a three-layer plywood panel the grain of the outer plies runs parallel to the long dimension of the panel, while the grain of the middle layer (crossband) runs perpendicular. Since the outer plies contain roughly two-thirds of the wood and these outer plies are farthest from the neutral axis, it would be advantageous for bending stresses to act in the same direction as the grain of the outer plies. This is the reason that plywood in flatwise bending should almost always be placed with the face grain perpendicular to the supports, as shown in Fig. 11-2.

Positioning the panel as in Fig. 11-2 (rather than turned 90°) so that the bending stress is acting in the same direction as face grain is called placing it in the "strong" direction. For some panels the effective moment of inertia is more than ten times greater when they are correctly placed (face grain perpendicular to the supports) than when face grain is parallel to the supports. To understand the structural behavior of plywood, it is imperative to know the *direction* of stress and its relation to *direction* of face grain.

A few terms dealing with grain direction of the layers must be defined. An inner ply (in a five- or seven-layer panel) whose grain is in the same direction as the face grain is called a *center ply.* Inner plies whose grain runs perpendicular to the face grain are called *crossbands.* The outer plies can be called the face or back interchangeably, unless they are of different grades, in which case the face is the outer ply of higher grade while the back is the outer ply of lower grade.

The effective section properties for design are shown in Table 11-4. It is important to note that *the properties listed are for a one-foot-wide strip of plywood* (see Fig. 11-2). Because these effective properties are published, the designer does not have to be concerned with the multilayer quality of the wood. These properties have been adjusted for orientation of wood fiber and for grade of veneer in the different layers and are based on minimum manufacturing requirements. There are two sets of cross-sectional properties listed—the properties for computing parallel-to-face-grain stress (in columns 4–7) and those for computing perpendicular-to-face-grain stress (in columns 8–11). In most cases, the largest stresses will act parallel to the face grain, so the parallel properties will be used. The perpendicular properties are used if the stress is acting perpendicular to the face grain, that is, if the plywood is bent in the weaker direction. Sanding can remove considerable surface wood, so the table has different properties listed for sanded, touch-sanded, and unsanded panels. Usually the structural designer will be using the data for unsanded panels.

The moments of inertia listed in the table of section properties are for stiffness (deflection) calculations only, not for calculating bending stress. They were calculated using a transformed section to account for the fact that, for a thin peeled layer, E perpendicular to the grain is only $^1/_{35}$ of its E parallel to the grain. The listed moments of inertia are for use when the panel is loaded perpendicular to the plane of the panel (flatwise bending).

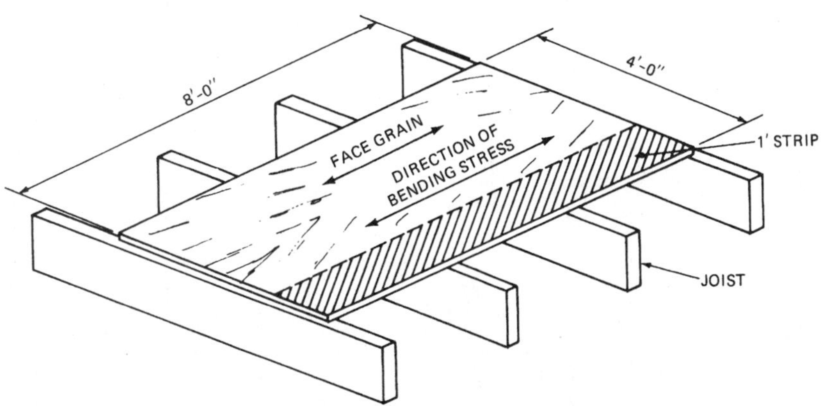

Fig. 11-2. Plywood panel with face grain perpendicular to supports.

Table 11-4. EFFECTIVE SECTION PROPERTIES FOR PLYWOOD.

Table 11-4A. Face Plies of Different Species Group from Inner Plies (Includes all Product Standard Grades except those noted in Table 11-4B.)

			Stress Applied Parallel to Face Grain				Stress Applied Perpendicular to Face Grain			
Nominal Thickness (in.)	Approximate Weight (psf)	t_s Effective Thickness For Shear (in.)	A Area (in.^{2}/ft)	I Moment of Inertia (in.^{4}/ft)	KS Effective Section Modulus (in.^{3}/ft)	Ib/Q Rolling Shear Constant (in.^{2}/ft)	A Area (in.^{2}/ft)	I Moment of Inertia (in.^{4}/ft)	KS Effective Section Modulus (in.^{3}/ft)	Ib/Q Rolling Shear Constant (in.^{2}/ft)
UNSANDED PANELS										
5/16-U	1.0	0.268	1.491	0.022	0.112	2.569	0.660	0.001	0.023	4.497
3/8 -U	1.1	0.278	1.866	0.039	0.152	3.110	0.799	0.002	0.033	5.444
15/32&1/2 -U	1.5	0.298	2.292	0.067	0.213	3.921	1.007	0.004	0.056	2.450
19/32&5/8 -U	1.8	0.319	2.330	0.121	0.379	5.004	1.285	0.010	0.091	3.106
23/32&3/4 -U	2.2	0.445	3.247	0.234	0.496	6.455	1.563	0.036	0.232	3.613
7/8 -U	2.6	0.607	3.509	0.340	0.678	7.175	1.950	0.112	0.397	4.791
1 -U	3.0	0.842	3.916	0.493	0.859	9.244	3.145	0.210	0.660	6.533
1-1/8 -U	3.3	0.859	4.725	0.676	1.047	9.960	3.079	0.288	0.768	7.931
SANDED PANELS										
1/4 -S	0.8	0.267	0.996	0.008	0.059	2.010	0.348	0.001	0.009	2.019
11/32-S	1.0	0.284	0.996	0.019	0.093	2.765	0.417	0.001	0.016	2.589
3/8 -S	1.1	0.288	1.307	0.027	0.125	3.088	0.626	0.002	0.023	3.510
15/32-S	1.4	0.421	1.947	0.066	0.214	4.113	1.204	0.006	0.067	2.434
1/2 -S	1.5	0.425	1.947	0.077	0.236	4.466	1.240	0.009	0.087	2.752
19/32-S	1.7	0.546	2.423	0.115	0.315	5.471	1.389	0.021	0.137	2.861
5/8 -S	1.8	0.550	2.475	0.129	0.339	5.824	1.528	0.027	0.164	3.119
23/32-S	2.1	0.563	2.822	0.179	0.389	6.581	1.737	0.050	0.231	3.818
3/4 -S	2.2	0.568	2.884	0.197	0.412	6.762	2.081	0.063	0.285	4.079
7/8 -S	2.6	0.586	2.942	0.278	0.515	8.050	2.651	0.104	0.394	5.078
1 -S	3.0	0.817	3.721	0.423	0.664	8.882	3.163	0.185	0.591	7.031
1-1/8 -S	3.3	0.836	3.854	0.548	0.820	9.883	3.180	0.271	0.744	8.428
TOUCH-SANDED PANELS										
1/2 -T	1.5	0.342	2.698	0.083	0.271	4.252	1.159	0.006	0.061	2.746
19/32&5/8 -T	1.8	0.408	2.354	0.123	0.327	5.346	1.555	0.016	0.135	3.220
23/32&3/4 -T	2.2	0.439	2.715	0.193	0.398	6.589	1.622	0.032	0.219	3.635
1-1/8 -T	3.3	0.839	4.548	0.633	0.977	11.258	4.067	0.272	0.743	8.535

Table 11-4B. Structural I and Marine

			Stress Applied Parallel to Face Grain				Stress Applied Perpendicular to Face Grain			
Nominal Thickness (in.)	Approximate Weight (psf)	t_s Effective Thickness For Shear (in.)	A Area (in.^{2}/ft)	I Moment of Inertia (in.^{4}/ft)	KS Effective Section Modulus (in.^{3}/ft)	Ib/Q Rolling Shear Constant (in.^{2}/ft)	A Area (in.^{2}/ft)	I Moment of Inertia (in.^{4}/ft)	KS Effective Section Modulus (in.^{3}/ft)	Ib/Q Rolling Shear Constant (in.^{2}/ft)
UNSANDED PANELS										
5/16-U	1.0	0.356	1.619	0.022	0.126	2.567	1.188	0.002	0.029	6.037
3/8 -U	1.1	0.371	2.226	0.041	0.195	3.107	1.438	0.003	0.043	7.307
15/32&1/2 -U	1.5	0.535	2.719	0.074	0.279	4.157	2.175	0.012	0.116	2.408
19/32&5/8 -U	1.8	0.707	3.464	0.154	0.437	5.685	2.742	0.045	0.240	3.072
23/32&3/4 -U	2.2	0.739	4.219	0.236	0.549	6.148	2.813	0.064	0.299	3.540
7/8 -U	2.6	0.776	4.388	0.346	0.690	6.948	3.510	0.131	0.457	4.722
1 -U	3.0	1.088	5.200	0.529	0.922	8.512	5.661	0.270	0.781	6.435
1-1/8 -U	3.3	1.118	6.654	0.751	1.164	9.061	5.542	0.408	0.999	7.833
SANDED PANELS										
1/4 -S	0.8	0.342	1.280	0.012	0.083	2.009	0.626	0.001	0.013	2.723
11/32-S	1.0	0.365	1.280	0.026	0.133	2.764	0.751	0.001	0.023	3.397
3/8 -S	1.1	0.373	1.680	0.038	0.177	3.086	1.126	0.002	0.033	4.927
15/32-S	1.4	0.537	1.947	0.067	0.246	4.107	2.168	0.009	0.093	2.405
1/2 -S	1.5	0.545	1.947	0.078	0.271	4.457	2.232	0.014	0.123	2.725
19/32-S	1.7	0.709	3.018	0.116	0.338	5.566	2.501	0.034	0.199	2.811
5/8 -S	1.8	0.717	3.112	0.131	0.361	5.934	2.751	0.045	0.238	3.073
23/32-S	2.1	0.741	3.735	0.183	0.439	6.109	3.126	0.085	0.338	3.780
3/4 -S	2.2	0.748	3.848	0.202	0.464	6.189	3.745	0.108	0.418	4.047
7/8 -S	2.6	0.778	3.952	0.288	0.569	7.539	4.772	0.179	0.579	5.046
1 -S	3.0	1.091	5.215	0.479	0.827	7.978	5.693	0.321	0.870	6.981
1-1/8 -S	3.3	1.121	5.593	0.623	0.955	8.841	5.724	0.474	1.098	8.377
TOUCH-SANDED PANELS										
1/2 -T	1.5	0.543	2.698	0.084	0.282	4.511	2.486	0.020	0.162	2.720
19/32&5/8 -T	1.8	0.707	3.127	0.124	0.349	5.500	2.799	0.050	0.259	3.183
23/32&3/4 -T	2.2	0.739	4.059	0.201	0.469	6.592	3.625	0.078	0.350	3.596

Source: Courtesy of APA—The Engineered Wood Association.

The effective section modulus, KS, is for use in bending-stress calculations for flatwise bending. This effective section modulus is not the same as I/c because KS for the perpendicular condition is calculated ignoring the outermost tension ply. (This ply adds little strength for bending that causes flexural stress normal to the face grain.) Also the constant K has been partially determined by experiment.

Sometimes loads are applied in the same plane as the plane of the panel rather than normal to the face. This is known as edge loading. The plywood web of a wood I-beam and a plywood diaphragm are examples of edge loading. If the panel is edge loaded, then the moment of inertia is found by $I = bh^3/12$, where b equals the effective thickness of either the parallel plies or the perpendicular plies (crossbands), whichever have their grain in the direction of bending stress due to the edge loading. For example, if the edge loading is at 90° to the face grain, then b is the effective thickness of the *parallel* plies; the crossbands are considered worthless in this situation because stress is perpendicular to their grain direction.

To find the effective thickness of the parallel plies, divide the area of the parallel plies (given in Table 11-4 in units of square inches per foot) by 12 in. To find the effective thickness of the perpendicular plies, divide the area of the perpendicular plies by 12. Moment of inertia calculated by the above equation may be used for either deflection or bending-stress calculations of edge-loaded panels.

The modulus of elasticity listed in the table of allowable stresses (Table 11-3) is a conservative figure that includes an allowance for average shear deflection. Therefore, if shear deflection and bending deflection are being calculated separately in a design, the modulus can be increased by 10% in calculating the bending deflection. (This 10% allowance for shear deflection is identical in intent to the 3% allowance for shear deflection that is made in determining published E values for sawn lumber.)

Example 11-2

For a $^5/_{16}$-in. Structural I unsanded panel, show how moment of inertia for stress parallel to the face grain could be determined for bending by loads acting normal to the panel face. Compare the answer to the value shown in Table 11-4B.

Figure 11-3 shows the $^5/_{16}$-in. panel makeup for a one-foot strip and the transformed section. To ensure that the moment of inertia of this transformed section is a minimum expected value, three conservative assumptions are made:

1. The overall thickness, t, is taken to be the nominal thickness of the panel minus one-half of the thickness tolerance. (PS 1–95 gives $\pm$ $^1/_{32}$ in. for unsanded panels up to $^{13}/_{16}$-in. nominal; for thicker panels the tolerance is $\pm$ 5% of nominal thickness.)
2. The thickness of the ply (or plies) having stress parallel to the grain is computed using the tabulated minimum area for that direction (Table 11-4B).
3. The thickness of plies not parallel to the bending stress is whatever must be added to the minimum for parallel-to-stress plies (assumption 2) to add up to the total, t, of assumption 1.

For this $^5/_{16}$-in. panel, minimum total thickness, t, is

$$5/16 - 0.5\,(1/32) = 0.297 \text{ in.}$$

For stress parallel to face grain, the minimum thickness of each face and back ply is one-half of the tabulated minimum area of 1.619 in.^{2}/ft, or

$$t_1 = t_3 = 0.5(1.619/12) = 0.0675 \text{ in.}$$

The inner-ply thickness is

$$t_2 = 0.297 - 2(0.0675) = 0.162 \text{ in.}$$

The 12-in. actual width of inner ply is replaced in Fig. 11-3b by $^1/_{35}$ of 12 in., the fraction $^1/_{35}$ being the ratio of E perpendicular to the grain to E parallel. For this transformed section moment of inertia, I, is

$$\begin{aligned} I &= 2[(1/12)(12)(0.0675^3) \\ &\quad + 12(0.0675)(0.1147^2)] \\ &\quad + (1/12)(12/35)(0.162^3) \\ &= 0.0220 \text{ in.}^4 \end{aligned}$$

This is identical to the value published in Table 11-4B. The procedure illustrated is for the simplest case—a panel of three plies of the same species. For all other cases, the computation is more involved. Luckily, minimum moments of inertia for the designer's use have been computed by the APA.

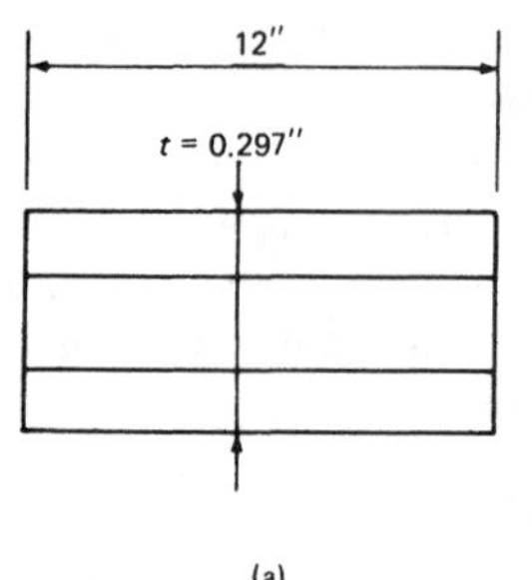

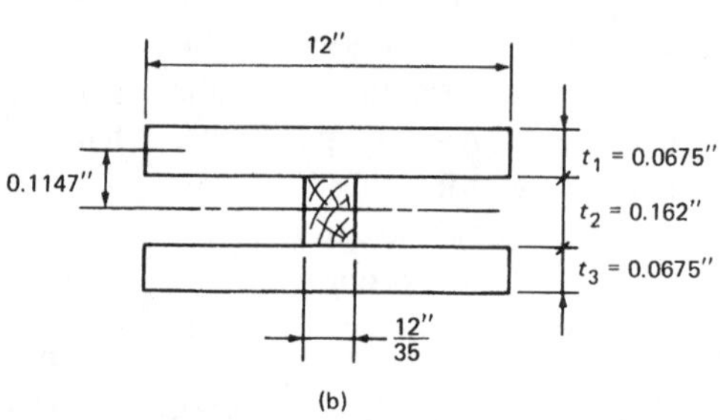

Fig. 11-3. Transformed section for Example 11-2.

$$I = 0.191(24)^3/12 = 220 \text{ in.}^4$$

$$S = I/c = 220/12 = 18.3 \text{ in.}^3$$

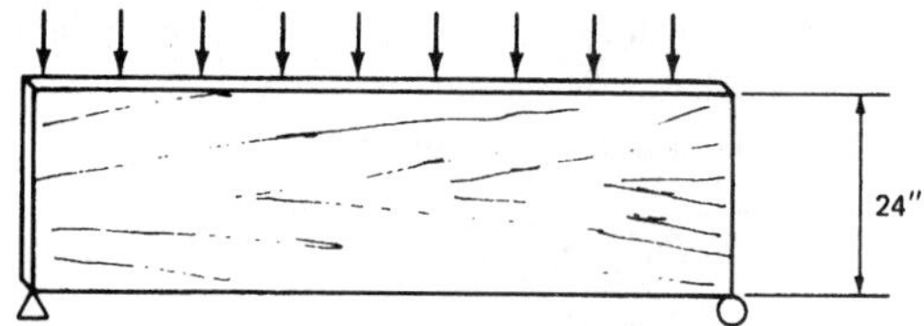

Fig. 11-4. Edge-loaded plywood for Example 11-3.

Example 11-3

An unsanded strip of ½-in. plywood whose inner and face plies are from different species groups is edge loaded with load perpendicular to the face grain. (This causes bending stress parallel to the face grain.) The strip is 24 in. wide as shown in Fig. 11-4. Find I and S.

The area of the parallel plies is 2.292 in.²/ft, so the thickness of the parallel plies is 2.292/12 = 0.191 in. Assuming the crossbands to contribute nothing to stiffness,

$$I = 0.191(24)^3/12 = 220 \text{ in.}^4$$

$$S = I/c = 220/12 = 18.3 \text{ in.}^3$$

11-3. PLYWOOD DESIGN FOR BENDING AND AXIAL LOAD

Unless the panels are span rated and used as common floor or roof sheathing, plywood will be chosen by allowable-stress design. The same NDS (5) load-duration factors apply in plywood that are used for sawn lumber. Preservative-treated panels do not have a reduction in allowable stresses. However, fire-retardant treated panels have a reduction in allowable stresses that must be obtained from the company providing the treatment, because standards have not been developed.

Flexure

For flatwise bending, one uses the effective section modulus KS to calculate bending stress. The maximum bending moment and maximum bending stress depend on the number of spans over which the panel is continuous. Appendix Table A-8 shows values of maximum bending moment and maximum shear for fully uniformly loaded beams with one, two, three, or four spans. The bending stress for plywood is calculated using bending moment and section properties each for a *one-foot wide* strip, as shown in Fig. 11-2.

Tension or Compression

To determine the allowable axial force in tension or compression, the allowable stress is multiplied by the area, using the area of only those plies whose grain is in the same direction as the tensile or compressive stress. Plies with grain at right angles to the stress are considered ineffective.

If the tensile stress is at 45° to the face grain, the allowable stress is one-sixth F_t listed in Table 11-3. The area to use in conjunction with this allowable is the full area based on nominal thickness if the panel is Structural I where plies are all from the same species group. If the panel has plies from different species groups in its makeup, then an adjustment must be made to account for the varying strengths of the plies. An adjusted area can be approximated by using 70% of the gross area. The same areas are used for compression at 45°, except that the allowable is one-third of F_c listed in Table 11-3.

The allowable tensile or compressive load at some angle other than 0° (direction of the face grain), 45°, or 90° (direction of crossbands) can be found by interpolating between the area-stress product (load) for 45° and the area-stress product for either 0° or 90°.

Section 3.3.3 of the *Plywood Design Specification* (3) recommends reducing the allowable load to 50% of the computed value for plates 8 in. or less in width, which have a greater possibility of a critical defect. (Full computed load is recommended for plates 24 in. or wider, interpolating to find the value for plates between 8 in. and 24 in. wide.) This reduction is based on the idea that a defect in a very small piece of plywood could cause collapse.

Example 11-4

Find the allowable bending stress and modulus of elasticity for the following plywood panel: ½-in. APA Rated Sheathing, Exterior, span rating of 32/16, wet use. ("Wet" allowable stresses are used if the equilibrium MC in service will be 16% or greater.)

Using the Key to Span Rating and Species Group (Table 11-2), this panel belongs to species Group 1. The first row of Table 11-1 for Exterior plywood shows that this panel is Grade Stress Level S-1. For species Group 1, Grade Stress Level S-1, and wet conditions, the allowables shown by Table 11-3 are: F_b = 1430 psi; E = 1,500,000 psi.

Example 11-5

The lower chord of an interior wood truss has a tension splice with 8 in. × 17 in. plywood gusset plates on each side as shown in Fig. 11-5. The plywood plates are oriented in the strong direction (i.e., with face grain parallel to the tensile stress). What is the allowable load for these two ½-in. Structural I, Exp 1 plates based on the tensile strength of the plywood?

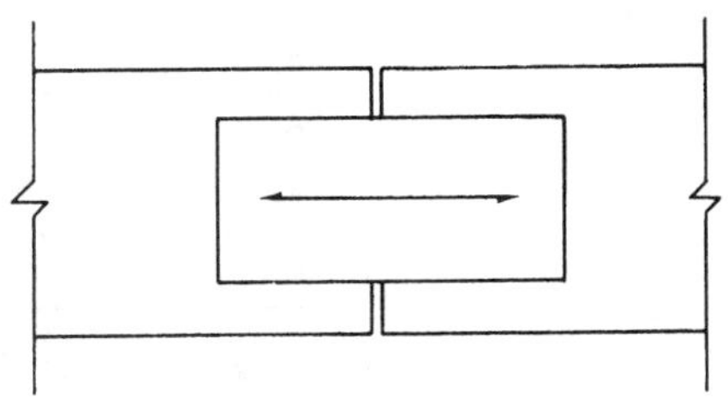

Fig. 11-5. Gusset plate for Example 11-5.

For one gusset plate,
A of parallel plies per ft = 2.719 in.^{2}
A of parallel plies per 8 in. = 2.719 (8) 12
= 1.813 in.^{2}
F_t = 1650 psi (S-2, Group 1, dry)
Using a 50% reduction for an 8-in. plate,

$$\text{Allowable } T = 0.50\,(2)\,(1650)\,(1.813) = 2990 \text{ lb for a pair of plates}$$

Example 11-6

Determine the allowable tensile load parallel to an edge on a 10 in. × 10 in. plywood plate whose face grain is at 45° to the edges of the plate. The plywood is ¾-in. Exterior APA Rated Sheathing, 48/24.

Adjusted gross area = (0.70) (3/4) (10) = 5.25 in.^{2}
Table 11-2 shows species Group 1; Table 11-1 shows grade stress level S-1, Choose F_t, for S-1 stresses, species Group 1, and wet conditions.

$$F_t/6 = 1430/6 = 238 \text{ psi}$$

$$T = (238)(5.25)$$

$$= 1250 \text{ lb, for a plate larger than 24 in.}$$

With the reduction for small plates the allowable load is

$$T = 1250 - 625(24 - 10)/(24 - 8) = 703 \text{ lb}$$

Example 11-7

Find the allowable compressive load on a 6-in.-wide plywood gusset plate if the stress is at 25° to the direction of the face grain (wet conditions of use). The plate is ½-in. Exterior APA Rated Sheathing, Species Group 4.

Table 11-1 shows that the proper Grade Stress Level is S-1.
Table 11-3 gives F_c = 610 psi.

$$\text{Allowable } C \text{ at } 45° = [(0.70)(1/2)(6)] \times [(1/3)(610)] = 427 \text{ lb}$$

Table 11-4 shows that area of parallel plies = 2.292 in.^{2}/ft.

Allowable C at 0° = (2.292 × 6/12) (610) = 699 lb. By interpolation, Allowable C (for panels 24 in. or more in width) at 25°

$$= 699 - (699 - 427)\,(25/45) = 548 \text{ lb}$$

Since the gusset plate is less than 8 in. in width, final allowable C = 548/2 = 274 lb.

Example 11-8

A construction project is under a tight time schedule. The structural grade of plywood needed by the contractor is not available in his city. For a higher price, however, appearance-grade exterior panels are available. The stamp on the panel indicates "A–C Group 1 Exterior." Will this 1/2-in. panel with supports perpendicular to face grain 24 in. c/c be sufficient for 70 psf in flatwise bending (dry use; permanent duration of load)?

Table 11-1 shows that Grade Stress Level S-1 should be used. Table 11-3 shows $F_b = 2000$ psi, so

$$F'_b = 0.9(2000) = 1800 \text{ psi}$$

Table 11-4A (sanded) shows $KS = 0.236$ in.³/ft. For a one-foot width, $w = (70 \text{ psf})(1 \text{ ft}) = 70$ lb/ft. For four spans,

$$\begin{aligned} f_b &= M/KS = (0.107\, w\ell^2)/0.236 \\ &= [(0.107)(70)(2)^2(12)]/0.236 \\ &= 1520 \text{ psi} < F'_b \qquad OK \end{aligned}$$

Example 11-9

Sheets of plywood (4 × 8 ft) are each cut into four equal pieces (2 ft dimension parallel to face grain), forming 24-in. deep web members for a lumber-flanged box beam. Plywood is 15/32-in. APA Rated Sheathing Exposure 1. Calculate the amount that one web will contribute to the total moment of inertia of the beam.

The moment of inertia of the web will be relatively small, as only the center ply is effective in bending. It is the only ply that has grain in the direction of stress. The thickness of this ply = 1.007/12 = 0.084 in.

$$\begin{aligned} 1 \text{ (one web)} &= bh^3/12 \\ &= 0.084(24)^3/12 = 97 \text{ in.}^4 \end{aligned}$$

11-4. PLYWOOD DESIGN FOR ROLLING SHEAR

Because plywood is built up in layers, two types of shear must be considered. Rolling shear is the first of these. Rolling shear occurs when the loads are normal to the surface of the plywood, as in floor sheathing. The layers of veneer tend to slide with respect to one another just like horizontal shear in a beam. A rolling-shear failure occurs only in a layer perpendicular to the face grain (a crossband). This is because shear strength is low across the grain.

Rolling shear can possibly best be visualized if one thinks of a panel whose crossband consists of a layer of toothpicks all oriented perpendicular to the face grain. The toothpicks represent the wood fibers. When the load is applied and the layers slide with respect to one another, the toothpicks are rolled and pushed past one another. Rolling shear takes its name from the rolling of these fibers, as shown in Fig. 11-6a. It should be noted that a rolling-shear failure could occur in the outer plies if the panel were placed across the supports in the weak direction (outer grain parallel to the supports).

It is conventional to use a subscript s when referring to rolling-shear stress, to prevent confusing it with shear through the thickness. The allowable rolling-shear stress. F_s, is less than half of the allowable for shear through the thickness. References 6 and 7 give further information on the modulus of rigidity in rolling shear for plywood.

Rolling-shear stress is calculated from $f_s = VQ/Ib$, with the Ib/Q being given in Table 11-4. The maximum shear V is based on a one-foot width, as in most plywood design. For certain component design (such as stressed-skin panels

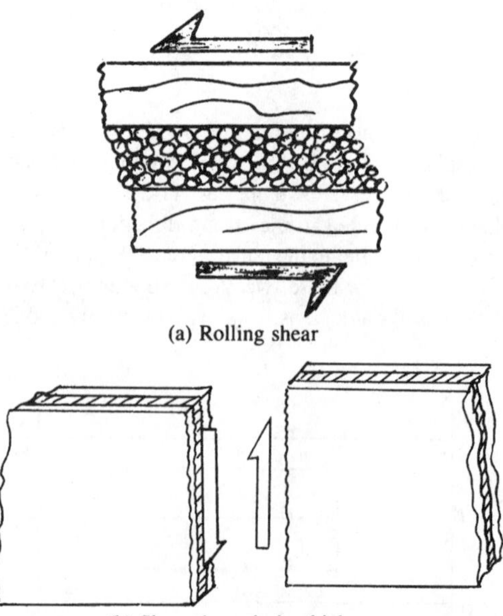

(a) Rolling shear

(b) Shear through the thickness

Fig. 11-6. Plywood in shear.

and flange-web joints in box beams) the rolling-shear allowable is reduced 50% to account for stress concentrations.

Example 11-10

A 5/8-in. Structural I, Exposure 1 standard plywood panel is loaded in flatwise bending with 70 psf. The face grain is perpendicular to the supports. For dry use with the supports at 32 in. c/c, find the actual and allowable rolling-shear stresses.

All Structural I plywood uses species Group 1 wood, so it is not necessary to use the Key to Span Rating. Table 11-1 shows that it is Grade Stress Level S-2 and is unsanded. Remember that if the face grain is perpendicular to the supports, the bending stress is parallel to the face grain. For Structural I, 5/8 in. unsanded Table 11-4 shows (for stress parallel to grain) $Ib/Q = 5.685$ in.²/ft. Also, 32 in. is one-third of the 8-ft panel length, so we have a three-span condition. Maximum shear for a three-span beam can be found in Table A-8. Assuming 2-in. nominal supports, $\ell_c =$ 30.5 in.

$$V = 0.6w\ell_c = 0.6(70 \text{ lb/ft})(30.5/12 \text{ ft})$$

$$= 107 \text{ lb (per ft of width)}$$

$$\text{Actual } f_s = V/(Ib/Q) = 107/5.685 = 18.8 \text{ psi}$$

$$\text{Allowable } F_s = 75 \text{ psi (Table 11-3)} \qquad OK$$

11-5. PLYWOOD DESIGN FOR SHEAR THROUGH THE THICKNESS

Plywood can also be stressed in shear on a plane that is perpendicular to the panel surface, that is, by loads in the plane of the panel. This is known as *shear through the thickness* (see Fig. 11-6b) because it affects the total thickness of a panel rather than just a crossband. Loading in the plane of the plywood (as in a diaphragm) can cause shear through the thickness. This same type of shear occurs in plywood-lumber built-up beams made using dimension lumber flanges and plywood webs. The plywood web of these beams is subjected to this type of shear, with the maximum shear stress at the neutral axis. Of course, this is similar to the horizontal shear that would occur in a steel wide-flange beam.

A load perpendicular to the panel surface and concentrated over a small area may punch out a core of wood; hence the name *punching shear*. Rolling of fibers past one another is not involved with this type of failure. Punching-shear calculations should be made with shear-through-the-thickness (rather than rolling-shear) allowable stresses.

When making calculations for shear through the thickness, the designer should *not* use the nominal thickness of the panel, but instead should use the effective thickness for shear as listed by Table 11-4. This effective thickness for shear is necessary to compensate for several factors. One such factor is that the shear strength of a panel increases with the number of glue lines, because the glue adds shear strength. Another factor has to do with the reduced shear strength of inner plies in mixed-species panels.

For shear at 45° to the face grain, the allowable stress, F_v, for shear through the thickness may be increased 100%. Where plywood is continuously glued (instead of mechanically fastened) to framing, other increases are allowed. If a panel is glued to continuous framing around all its edges, the increase is 33%. If the gluing is to framing at only the two edges parallel to the face grain, the increase is only 19%. Increases are cumulative with duration factor.

Example 11-11

A 5000-lb condenser is located on a plywood roof. The condenser has four legs, each with a 4 in. × 4 in. base. The plywood is 7/8-in. Exterior APA Rated Sheathing, 48/24. Find the punching shear stress and compare with the allowable.

Table 11-2 indicates that species Group 3 stresses should be used. Table 11-1 shows that it is an unsanded panel and grade stress level S-1 is to be used. The effective thickness for shear is found to be 0.607 in. from Table 11-4.

$$A = (\text{eff. } t)(\text{perim}) = 0.607(16) = 9.7 \text{ in.}^2$$

$$V = 1250 \text{ lb/leg}$$

$$f_v = 1250/9.7 = 129 \text{ psi}$$

$$F_v = 120 \text{ psi (Table 11-3, assumed wet)} \qquad NG$$

11-6. ROOF SHEATHING

Although lumber and particleboard are also used, plywood is the most common roof sheathing. Table 11-5, reproduced from the UBC (8), shows allowable spans for plywood roof sheath-

Table 11-5. Allowable Spans and Loads for Wood Structural Panel Sheathing and Single-Floor Grades Continuous over Two or More Spans with Strength Axis Perpendicular to Supports[1,2]

Sheathing Grades		Roof[3]				Floor[4]
		Maximum Span (inches)		Load[5] (pounds per square foot)		
Panel Span Rating	Panel Thickness (inches)	× 25.4 for mm		× 0.0479 for kN/m²		Maximum Span (inches)
Roof/Floor Span	× 25.4 for mm	With Edge Support[6]	Without Edge Support	Total Load	Live Load	× 25.4 for mm
12/0	5/16	12	12	40	30	0
16/0	5/16, 3/8	15	16	40	30	0
20/0	5/16, 3/8	20	20	40	30	0
24/0	3/8, 7/16, 1/2	24	20[7]	40	30	0
24/16	7/16, 1/2	24	24	50	40	16
32/16	15/32, 1/2, 5/8	32	28	40	30	16[8]
40/20	19/32, 5/8, 3/4, 7/8	40	32	40	30	20[8,9]
48/24	23/32, 3/4, 7/8	48	36	45	35	24
54/32	7/8, 1	54	40	45	35	32
60/48	7/8, 1, 1 1/8	60	48	45	35	48
Single-Floor Grades		Roof[3]				Floor[4]
		Maximum Span (inches)		Load[5] (pounds per square foot)		
Panel Span Rating (inches)	Panel Thickness (inches)	× 25.4 for mm		× 0.0479 for kN/m²		Maximum Span (inches)
× 25.4 for mm		With Edge Support[6]	Without Edge Support	Total Load	Live Load	× 25.4 for mm
16 oc	1/2, 19/32, 5/8	24	24	50	40	16[8]
20 oc	19/32, 5/8, 3/4	32	32	40	30	20[8,9]
24 oc	23/32, 3/4	48	36	35	25	24
32 oc	7/8, 1	48	40	50	40	32
48 oc	1 3/32, 1 1/8	60	48	50	50	48

[1] Applies to panels 24 inches (610 mm) or wider.
[2] Floor and roof sheathing conforming with this table shall be deemed to meet the design criteria of Section 2321.
[3] Uniform load deflection limitations 1/180 of span under live load plus dead load, 1/240 under live load only.
[4] Panel edges shall have approved tongue-and-groove joints or shall be supported with blocking unless 1/4-inch (6.4 mm) minimum thickness underlayment or 1 1/2 inches (38 mm) of approved cellular or lightweight concrete is placed over the subfloor, or finish floor is 3/4-inch (19 mm) wood strip. Allowable uniform load based on deflection of 1/360 of span is 100 pounds per square foot (psf) (4.79 kN/m²) except the span rating of 48 inches on center is based on a total load of 65 psf (3.11 kN/m).
[5] Allowable load at maximum span.
[6] Tongue-and-groove edges, panel edge clips [one midway between each support, except two equally spaced between supports 48 inches (1219 mm) on center], lumber blocking, or other. Only lumber blocking shall satisfy blocked diaphgrams requirements.
[7] For 1/2-inch (13 mm) panel, maximum span shall be 24 inches (610 mm).
[8] May be 24 inches (610 mm) on center where 3/4-inch (19 mm) wood strip flooring is installed at right angles to joist.
[9] May be 24 inches (610 mm) on center for floors where 1 1/2 inches (38 mm) of cellular or lightweight concrete is applied over the panels.
Source: Reproduced from the 1994 edition of the *Uniform Building Code*™, copyright © 1994, with the permission of the publisher, the International Conference of Building Officials.

ing oriented in the strong direction (face grain perpendicular to supports). These allowable spans could also be determined directly from the span rating of the particular panel as discussed in Section 11-1. The table also gives allowable total and live roof loads when the specified span is used. This table deals with vertical loads only and does not list requirements for diaphragm action.

To choose plywood roof sheathing for gravity loads, Table 11-5 will generally be all the designer needs. Should an unusually large snow load exist so that the table cannot be used, the roof sheathing can be chosen based on bending moment, shear, and deflection. Use deflection limits of $\ell/180$ for total load and $\ell/240$ for live load alone.

The UBC has another table (not shown here) that gives allowable loads for roof sheathing oriented in the weak direction. Preframed roof systems often use plywood in the weak direction. In the preframed system, panels have stiffeners preattached before the assembly is lifted into position. The expense of having to use thicker panels is more than offset by the speed of construction.

Unless special means are taken, panel edges parallel to the span will be unsupported. Differential movement of these free edges in adjacent panels can occur. To prevent this movement, panel free edges may be nailed to lumber blocking fitted between the roof joists. Tongue-and-groove edges and panel clips (H-shaped metal devices) are other means of providing edge support. Notice from Table 11-5 that most panels require edge support to achieve the full nominal allowable span (as given by span rating).

Example 11-12

A roof framing system consists of trusses at 24 in. c/c. The roof has a slope of 1:1. Roof $DL = 9$ psf of sloping surface and roof $SL = 30$ psf horizontal. Design plywood for sheathing loads only (not diaphragm loads).

Panel orientation: For greatest strength, plywood will be placed with face grain perpendicular to the top chords of the roof trusses.

Glue type: Although the UBC allows intermediate glue (unless the roof panel is exposed on the underside as at the edge of an overhanging roof), the authors recommend using only panels with exterior glue.

Loads:

UBC snow load reduction per degree of pitch over $20° = S/40 - 0.5 = 30\text{ psf}/40 - 0.5 = 0.25$ psf/deg over 20°

$45° - 20° = 25°$

$(25)(0.25) = 6$ psf

Snow load $= 30 - 6 = 24$ psf of horizontal projection

Design snow load $= 24/1.414 = 17.0$ psf of roof surface

Total load $= 9 + 17 = 26$ psf of roof surface

The component of this which is perpendicular to the roof surface and which causes bending is 18.4 psf.

Panel options: Using Table 11-5, any Rated Sheathing that is 15/32 in. or 1/2 in. thick with a span rating of 24/0 will be sufficient to carry 18.4 psf total load. Table 11-5 shows these thicknesses will not require blocking (unless required for diaphragm loads). *Use 15/32-in. or thicker C-C Exterior or C-D Interior Exposure 1, 24/0.*

Another panel, a 3/8-in. panel with span rating of 24/0, will also work. If this panel is used, blocking or T&G edges will be required according to the table. This panel is Structural I Rated Sheathing (see Table 11-2). *Alternatively, use 3/8-in. Structural I Exterior or Interior 24/0.*

Example 11-13

A roof framing system with beams supporting 2-in. nominal joists at 24 in. c/c is desired for a roof with a gentle slope. The structure is located in a ski area at Steamboat, Colorado, where measured snow loads have reached 175 psf. Roof dead load is 12 psf. Design plywood for vertical loads.

Loads are too great for Table 11-5 to be useful. Assume a Structural I Exterior panel and dry use, so $F'_b = 1.15(2000) = 2300$ psi, $F'_s = 1.15(75) = 86$ psi, and $E = 1{,}800{,}000$ psi. Check bending of a 12-in. width of plywood.

$$w = (175 + 12)/12 = 15.58 \text{ lb/in.}$$

$$\text{Req } KS = M/F'_b$$

$$= 0.107(15.58)(24^2)/2300$$

$$= 0.417 \text{ in.}^3/\text{ft}$$

19/32 in. req for bending (see Table 11-4)

Check shear

$$\text{Req } (Ib/Q) = 0.607\, w\ell_c/F'_s$$

$$= 0.607(15.58)(22.5)/86$$

$= 2.47 \text{ in.}^2/\text{ft}$

$5/16$ in. req for rolling shear

Check deflection

$$\ell/240 = 24/240 = 0.10 \text{ in.}$$

$$\text{Req } I = 0.0065\, w\ell^4/0.10E$$

where APA recommends ℓ

$$= \ell_c + 0.25 \text{ in.}$$

$$\text{Req } I = \frac{(175/12)(22.75)^4}{154 \times 0.10(1{,}800{,}000)}$$

$$= 0.141 \text{ in.}^4/\text{ft}$$

$19/32$ in. req for deflection under LL alone (total load won't control)

Use 19/32-in. Structural I C-C Exterior.

11-7. FLOOR SHEATHING

Many plywood floors consist of a bottom structural layer of APA Rated Sheathing called the subfloor. A layer of plywood called the underlayment may be applied over the structural subfloor to provide a smooth surface for linoleum, tile, carpet, or other finish floor. If the joints in the two plywood layers are offset, edge support (such as blocking) is not required by the *Uniform Building Code* (8). The rightmost column of Table 11-5 gives allowable spans (strong orientation) for Rated Sheathing used as subfloor. When used at their listed spans, the panels in the table are good for at least 100 psf based on an $\ell/360$ deflection limit. (See footnote 4 of Table 11-5.)

Staggered underlayment panels in the two-layer floor system are considered to have only two structural functions: preventing differential movement of free edges and resistance to dents. Typical thicknesses of underlayment for use over subfloor are 1/4, 11/32, and 3/8 in. Underlayment properties can be found in Table 11-1. For areas to be covered with vinyl tile or linoleum rather than carpet, underlayment with "sanded face" should be specified. As an alternative to underlayment, lightweight concrete or wood strip finish floor may be applied over the subfloor with no edge support required for the plywood floor sheathing.

Single-Floor grade plywood (trade name, Sturd-I-Floor) is a thicker panel that can be used in place of subfloor and underlayment to make a single-layer floor system. It comes in thicknesses up to 1 1/8 in. The panels intended for use in a single layer are shown in the bottom portion of Table 11-5 under the heading Single-Floor Grades. Span ratings for these panels range from 16 oc to 48 oc. Proper thickness can be chosen from span rating.

The thicker panels of Underlayment grade plywood are also allowed as the structural floor in single-layer floor construction. Table 11-6 gives allowable spans for Underlayment, Single-Floor grade plywood (Sturd-I-Floor), or C-C Plugged (Ext) plywood.

Example 11-14

The floor system for a junior high school consists of steel girders that support wood I-joists at 24 in. c/c. The school has movable partions between classrooms and flooring is linoleum. Floor $DL = 6$ psf, partition $DL = 20$ psf, and $LL = 40$ psf. Choose plywood floor sheathing for vertical loads.

The allowable load for subflooring listed in Table 11-5 = 100 psf. The total floor load for this case is 66 psf, so the table may be used. After finding 24 in. in the right-hand column of Table 11-5, move across the row to find a minimum 23/32-in. thickness with span rating 48/24 is necessary for the subfloor (strong orientation, of course). Interior type will be specified.

A 1/4-in. interior underlayment with joints offset from subfloor joints will be specified to be placed over the subfloor to ensure a smooth surface for the linoleum. Blocking will not be required (unless needed for lateral loads).

Use 23/32-in. or thicker Interior Rated Sheathing 48/24 for subfloor and 1/4-in. or thicker Interior Underlayment.

Alternatively, a single-layer floor system could be used. Two possible choices are Exterior Underlayment 1-in. thick of Species Group 4 wood (Table 11-6) or Sturd-I-Floor with a 24 oc rating. The single-layer floor system will require tongue-and-groove edges or blocking.

11-8. WOOD-BASED FIBER AND PARTICLE PANELS

Plywood is by far the most common wood sheet or panel for structural use. Other wood panel products are manufactured using wood particles

Table 11-6. Allowable Span for Wood Structural Panel Combination Subfloor–Underlayment (Single Floor)[1,2] (Panels continuous over two or more spans and strength axis perpendicular to supports.)

	Maximum Spacing of Joists (inches) × 25.4 for mm				
Identification	16	20	24	32	48
	Thickness (inches) × 25.4 for mm				
Species Group[3]					
1	1/2	5/8	3/4	—	—
2, 3	5/8	3/4	7/8	—	—
4	3/4	7/8	1	—	—
Span rating[4]	16 o.c.	20 o.c.	24 o.c.	32 o.c.	48 o.c.

[1] Spans limited to values shown because of possible effects of concentrated loads. Allowable uniform loads based on deflection of 1/360 of span is 100 pounds per square foot (psf) (4.79 kN/m^2), except allowable total uniform load for 1 1/8-inch (29 mm) wood structural panels over joists spaced 48 inches (1219 mm) on center is 65 psf (3.11 kN/m^2). Panel edges shall have approved tongue-and-groove joints or shall be supported with blocking, unless 1/4-inch (6.4 mm) minimum thickness underlayment or 1 1/2 inches (38 mm) of approved cellular or lightweight concrete is placed over the subfloor, or finish floor is 3/4-inch (19 mm) wood strip.

[2] Floor panels conforming with this table shall be deemed to meet the design criteria of Section 2321.

[3] Applicable to all grades of sanded exterior-type plywood. See U.B.C. Standard 23-2 for plywood species groups.

[4] Applicable to underlayment grade and C-C (plugged) plywood, and single-floor-grade wood structural panels.

Source: Reproduced from the 1994 edition of the *Uniform Building Code*™, © 1994, with the permission of the publisher, the International Conference of Building Officials.

or wood pulp. The manufacturing process causes these panels to have properties different from the original wood particles. Most of these fiber and particle panels (particleboard, insulation board, and hardboard) are used in light-frame construction, their suitability and strength having been determined over time. With the advent of performance standards, some of these panels can now be rated based on their performance in standard tests.

Particleboard

Particleboard is made of discrete wood particles and synthetic resin adhesives bonded together under heat and pressure. Sometimes particleboard is known as flakeboard, chipboard, waferboard, or shavings board, indicating the size and shape of particles used. Wood chips and particles for making particleboard may come from waste materials of sawmills and plywood and woodworking plants. Most particleboard is compressed in flat-platen presses, although a small amount is extruded through a long, wide, and thin heated die.

Wood particles may range in size from an inch or more in length to very fine. Panels may consist of a single layer or may be multilayer with higher-density top and bottom surfaces. Particles may be either uniform, graduated, or mixed.

Particleboard production began in Europe around 1941. Current consumption of particleboard is much greater in Europe (particularly West Germany and Switzerland) than it is in either the United States or Canada (1).

One of the main structural uses of particleboard is for horizontal and vertical diaphragms (see Chapter 12 for definition). The *Uniform Building Code* now allows particleboard to be used as a lateral-force-resisting diaphragm. Other structural uses are subflooring and roof sheathing. Table 11-7 shows allowable loads for particleboard when used as a roof sheathing. Allowable spans for particleboard subfloor and combined subfloor-underlayment are given in Table 11-8.

Because particleboard can be made in very large sizes (up to 8 ft by 40 ft), special structural uses have developed for large particleboard panels. Huge combination subfloor-underlayment particleboard panels for mobile homes and for factory-built homes can be handled mechanically in the factory. Separate standards (9, 10)

Table 11-7. Allowable Loads for Particleboard Roof Sheathing[1,2,3]

Grade	Thickness (inch)	Maximum on-Center Spacing of Supports (inches)	Live Load (pounds per square foot)	Total Load (pounds per square foot)
		× 25.4 for mm	× 0.0479 for kN/m²	
2-M-W	3/8[4]	16	45	65
	7/16	16	105	105
	7/16[4]	24	30	40
	1/2	16	110	150
	1/2	24	40	55

[1] Panels are continuous over two or more spans.
[2] Uniform load deflection limitations 1/180 of the span under live load plus dead load and 1/240 of the span under live load only.
[3] Roof sheathing conforming with this table shall be deemed to meet the design criteria of Section 2321.
[4] Edges shall be tongue-and-groove or supported with blocking or edge clips.
Source: Reproduced from the 1994 edition of the *Uniform Building Code*™, copyright © 1994, with the permission of the publisher, the International Conference of Building Officials.

Table 11-8. Allowable Spans for Particleboard Subfloor and Combined Subfloor–Underlayment[1,2]

Grade	Thickness (inches)	Maximum Spacing of Supports (inches)[3]	
		× 25.4 for mm	
	× 25.4 for mm	Subfloor	Combined Subfloor-Underlayment[4,5]
2-M-W	1/2	16	—
	5/8	20	16
	3/4	24	24
2-M-3	3/4	20	20

[1] All panels are continuous over two or more spans.
[2] Floor sheathing conforming with this table shall be deemed to meet the design criteria of Section 2321.
[3] Uniform deflection limitation: 1/360 of the span under 100 pounds per square foot (4.79 kN/m²) minimum load.
[4] Edges shall have tongue-and-groove joints or shall be supported with blocking. The tongue-and-groove panels are installed with the long dimension perpendicular to supports.
[5] A finish wearing surface is to be applied to the top of the panel.
Source: Reproduced from the 1994 edition of the *Uniform Building Code*™, copyright © 1994, with the permission of the publisher, the International Conference of Building Officials.

have been developed for this mobile home decking and factory-built housing decking.

Fiber-Based Panels

Fiber building boards have been in production longer than most other wood-base fiber and particle panel materials. Fiberboards are made of wood fibers interfelted together, the matting under heat and pressure producing a bond. They are manufactured from wood pulp (fibers suspended in water), and no adhesives are used in their manufacture (except in the case of tempered hardboard). Additives may be used during manufacture to make the panels more resistant to fire, moisture, or decay.

Although other manufacturing methods are possible, the pulp or slurry is usually laid out on a mesh screen during manufacture. After the water has been removed, the impression of the screen is still apparent on the back of the final board. The two broad types of fiberboards are rigid insulation board and hardboard. Insulation board is light in weight and not compressed during manufacture, whereas the denser hardboard is subject to intense pressure at high temperature. Insulation board is typically thicker than hardboard.

Sheathing-quality insulation board, the most common material for sheathing houses in the United States, is manufactured in three grades (11): regular-density (18 pcf), intermediate (22 pcf), and nail-base (25 pcf). It provides good resistance to heat flow and to racking (shearing deformation). Most building codes approve its use for sheathing wood stud walls and to serve as vertical diaphragms, provided it is properly placed and nailed and provided edge-blocking is used.

Hardboard varies in density from 33 to 90 pcf. Its strength (in all directions) is greater than the perpendicular-to-grain strength of natural wood. It has a myriad of uses—house siding, furniture backs, floor underlayment, core material for thin veneers, beam webs, facing for concrete forms for architectural concrete, door panels of autos, and so on. However, few of these applications are strictly structural. Hardboard does qualify under building codes as an approved material for wall bracing in conventional, light-frame construction; it can be used in lieu of let-in bracing (Fig. 15-1) or plywood or other sheathing.

In Europe and Australia, hardboard has been used for beam webs since the 1930s (12). However the only composite beams made commercially in the United States with wood product webs have plywood webs rather than hardboard webs. Permissible design stresses for structural use of tempered hardboard are now available in the *British Standard Code of Practice,* CP:112 (13).

Designing with fiberboard or particleboard is difficult because design properties vary with the manufacturer. For example, modulus of rupture for hardboard can vary from 1,900 to 12,500 psi. As knowledge of these products increases, there may be greater standardization. If so, the designer will be able to design without knowing in advance who the producer will be.

11-9. LRFD FOR PLYWOOD

For ASD, much of the design involving plywood or other structural wood panels such as oriented strand board (OSB) is done using span tables or span rating. For situations not covered by span tables, a design professional may use allowable stresses (Table 11-3) and section properties (Table 11-4) to determine panel bending strength, bending stiffness, shear through the thickness, and other structural properties.

For LRFD, a soft conversion procedure based on the principles of ASTM D5457 (14) has been used to develop design resistances. Tables of design resistances are available for LRFD (15, 16). One of these LRFD tables, for example, expresses moment resistance, $\lambda\phi_b M$, in units of kip-in. per ft of panel width. The designer can simply compare applied factored bending moment per ft for a particular panel with the moment resistance listed in the table. Of course, moment resistances are given for panel placement both with the face grain parallel and the face grain perpendicular to the framing.

For design of diaphragms and shear walls (see Chapter 12), tables of shear resistance, $\lambda\phi_v V$, are used. Those tables incorporate a time-effect factor of 1.0, since they are based on wind loading. For other durations of load, the table values must be adjusted.

REFERENCES

1. Desch, H. E., *Timber—Its Structure and Properties.* 5th ed., St. Martin's Press, New York, 1973.
2. *U.S. Product Standard PS 1-95 for Construction and Industrial Plywood,* National Institute for Standards and Technology, 1995.
3. *Plywood Design Specification,* APA—The Engineered Wood Association, Tacoma, WA, 1986.
4. *Performance Standards and Policies for Structural-Use Panels,* APA—The Engineered Wood Association Tacoma, WA, 1984.
5. *National Design Specification for Wood Construction,* American Forest & Paper Association. rev. 1991 Ed., Washington, DC, 1993.
6. Lee, W. C., and E. J. Biblis, "Predicting Rolling Shear Modulus of Plywood," *Forest Products Journal,* 29(4), Apr. 1979, pp. 31–34.
7. Polka, L. C., and J. Hejjes, "Effect of Moisture Content on the Mechanical Properties of Douglas Fir Plywood in Rolling Shear," *Forest Products Journal,* 27(4), Apr. 1977, pp. 49–53.
8. *Uniform Building Code,* International Conference of Building Officials, Whittier, CA, 1994.
9. *Standard for Particleboard for Mobile Home Decking.* NPA 1-70. National Particleboard Association, Oct. 1970.
10. *Standard for Particleboard for Decking for Factory-Built Housing,* NPA 2-70. National Particleboard Association, 1971.
11. *Wood Handbook,* U.S. Forest Products Laboratory, U.S. Department of Agriculture, 1974.
12. McNatt, J. D., "Hardboard-Webbed Beams: Research and Application," *Forest Products Journal,* 30(10), Oct. 1980, pp. 57–64.
13. Council for Codes of Practice, British Standards Institution, *British Standard Code of Practice,* CP:112, 1971.
14. *Standard Specification for Computing the Reference Resistance of Wood-Based Materials and Structural*

Connections for Load and Resistance Factor Design, American Society for Testing and Materials, ASTM D 5457, West Conshohocken, PA, 1993.

15. *Structural-Use Panel Design Supplement for Load and Resistance Factor Design,* APA—The Engineered Wood Association, Tacoma, WA, to be published.
16. *LRFD Manual of Wood Construction,* American Forest & Paper Association, Washington, DC, to be published.

PROBLEMS

11-1. What is the species group classification of Structural I, Exterior plywood? Of 5/16-in. APA Rated Sheathing 16/0, Exp 2? What is the Grade Stress Level of each?

11-2. What is the species group classification of 7/8-in. APA Rated Sheathing 48/24, Exterior? Of Structural I, Exp 1? What is the Grade Stress Level of each?

11-3. What is the tabular allowable bending stress (dry use) for the following plywood panels? (a) 3/8-in. APA Rated Sheathing 24/0, Exp 1; (b) 5/16-in. APA Rated Sheathing 16/0, Exp 2; (c) 1/2-in. Structural I, Exp 1.

11-4. What is the adjusted allowable compression stress (wet use) for the following plywood panels if snow loading controls? (a) 3/4-in. Structural I, Exp 1; (b) 5/16-in. APA Rated Sheathing 12/0, Exp 1; (c) 1/2-in. APA Rated Sheathing 24/0, Exterior.

11-5. What are the modulus of elasticity and modulus of rigidity for dry 5/8-in. APA Rated Sheathing 32/16, Exp 1?

11-6. If the face grain is perpendicular to the supports, what is the moment of inertia to use in figuring deflection for 1/2-in. APA Rated Sheathing 32/16, Exp 1?

11-7. Same as Problem 11-6, except face grain is parallel to the supports.

11-8. For each of the three plywood panels in Problem 11-3, find the effective section modulus, *KS,* if the face grain is perpendicular to the supports.

11-9. For each of the three plywood panels in Problem 11-4, find the moment of inertia, if the face grain is parallel to the supports.

11-10. For the plywood considered in Example 11-2, find the transformed section and compute the moment of inertia if the stress is now *perpendicular* to the face grain. Compare to the published value in Table 11-4B.

11-11. A truss tension-member splice has a nailed 6 in. × 17 in. plywood gusset plate on each side. The tensile force is parallel to the 17-in. dimension and also parallel to the face grain of the plywood. The plates are 15/32-in. C-D Interior Exp 1, 32/16. What is the allowable load (ten-year duration, dry use) for the member based on the tensile strength of the plywood? Based on the rolling-shear strength of the plywood?

11-12. The lower chord of a truss is spliced using a plywood gusset plate on each side. The plates are Structural I, Exp 1 plywood with face grain horizontal parallel to the bottom chord. The truss chord (2 × 12) carries a tension load of 2,800 lb (D = 560 lb and L = 2240). What is the minimum width of gusset plate (i.e., vertical dimension)? Conditions of use are dry.

11-13. Repeat Problem 11-11, but with the given force compression rather than tension.

11-14. Figure 11-7 shows two plywood gusset plates (dry use) joining the 2 × 4 members of a wood truss. Select a thickness of Structural I Exterior plywood to satisfy stress requirements (C_D = 1.0). Run face grain parallel to the bottom chord. Suggested sequence of steps: (a) Assuming a critical section on line *a-a,* make a free-body diagram of the portion of one gusset plate above *a-a;* (b) solve for the equilibrant shear and in-plane moment on line *a-a;* (c) solve for shear stress and flexural stress along section *a-a;* and (d) select the thickness.

11-15. A horizontal 4 × 8-ft plywood panel is loaded with 95 psf, permanent-duration load. The supports (2-in. nominal) for the panel are at 32 in. c/c with the panel having its face grain perpendicular to the supports, as shown in Fig. 11-8. The panel is 1/2-in. Structural II Exp 1, 32/16. Moisture content is estimated to be 10%. Compute the actual rolling-shear stress, taking the span as the clear span in the equation for maximum shear. Compare the actual stress to the allowable rolling-shear stress.

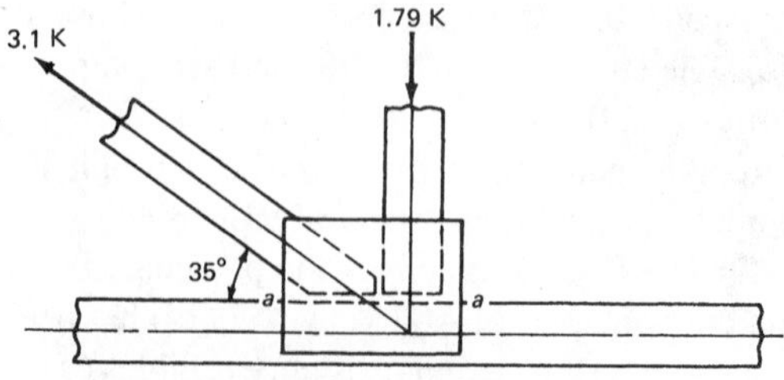

Fig. 11-7. Plywood gusset plates for Problem 11-6.

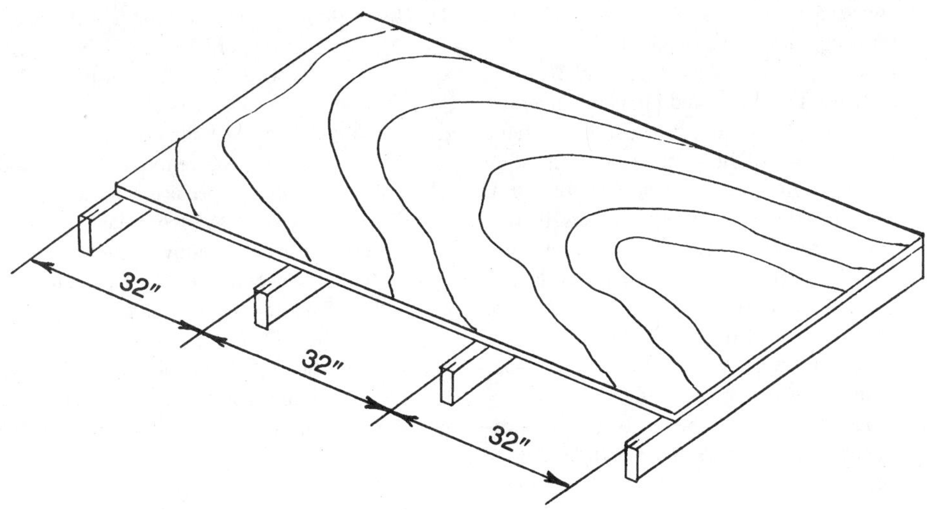

Fig. 11-8. Plywood for Problems 11-15, 11-16, 11-17, and 11-18.

11-16. For the plywood in Problem 11-15, compute the actual bending stress and compare it to the allowable bending stress. Use center to center of supports as the span.

11-17. For the plywood in Problem 11-15, compute the actual deflection, taking the span to be 0.25 in. more than the clear span. If the deflection limit under total load is 1/180 of the span, is the panel satisfactory for deflection?

11-18. A dry 4- × 8-ft plywood sheet (3/8-in. Structural I Exp 1) is supported as shown in Fig. 11-8 and loaded with $D = 3$ psf and $L = 32$ psf. What is the allowable bending stress? What is the actual bending stress? Is the panel satisfactory for bending?

11-19. Plywood is applied to 2 × 4 studs, which are at 16 in. on center. The plywood face grain is parallel to the studs. The plywood is 5/16-in. APA Rated Sheathing, Exposure 1, Species Group 3, dry. If a pressure of 10 psf (due to wind) is applied perpendicular to the wall, what will be the actual bending stress? Is this OK? In the solution consider that the plywood is continuous and is not a single, simple span. Use $C_D = 1.33$.

11-20. For the plywood in Problem 11-19, find the actual rolling-shear stress and the allowable rolling-shear stress. Is it OK for rolling shear? (In the equation for maximum shear, let span equal clear span.)

11-21. A rough plywood-lumber I-shaped beam (dry use) consists of four 2 × 4s with a plywood central web (face grain horizontal) of 1/2-in. Structural I Exp. 1 plywood. Total depth is 24 in. and the 2 × 4s are surfaced to dimensions of 1.5 in. by 3.5 in. The 2 × 4s are placed with their larger dimension vertical. Considering shear through the thickness at the neutral axis, what is the allowable total shear, V, for the beam? The load is of permanent duration.

11-22. A roof framing system consists of joists at 24 in. and plywood sheathing (strong orientation, dry use). Loads on the gently sloping roof are $D = 12$ psf and $S = 50$ psf. For these vertical loads, choose the proper thickness of Exterior APA Rated Sheathing. Table 11-5 cannot be used, because it does not show a live load as great as 50 psf for sheathing grade plywood. Therefore, base your choice on computed bending stress, rolling shear stress, and a deflection limit of $\ell/240$ for live load alone. Assume Species Group 4.

11-23. Same as Problem 11-22, except roof loads are $D = 10$ psf and $S = 30$ psf.

11-24. Rework Problem 11-23 using the UBC table, which appears here as Table 11-5. What span rating and minimum thickness are required? Will the panels require blocking?

11-25. Choose plywood for Problem 11-22 if the design snow load is increased to 80 psf.

11-26. Determine the correct thickness if the sheathing in Problem 11-23 is particleboard instead of plywood. Is blocking required?

11-27. A roof-framing system consists of rafters at 16-in. centers. The roof has a slope of 30°. Loads are $D = 9$ psf horizontal and $S = 60$ psf horizontal. Choose the proper thickness of particleboard to carry these vertical loads. Will edge support be required?

11-28. A floor system consists of joists at 16 in, c/c, a plywood subfloor, and plywood underlayment; $D = 6$ psf and $L = 40$ psf. Using Table 11-5, choose subfloor (indicate span rating and minimum thickness) and underlayment. Describe edge support requirements, if any.

11-29. The floor of Problem 11-28 is going to consist of a single layer, rather than two layers. Using the lower portion of Table 11-5, what span rating and minimum thickness is required of single-floor-grade plywood? Describe edge support requirements, if any.

11-30. If the subfloor of Problem 11-28 is of particleboard, what grade and thickness are required? Describe edge-support requirements, if any.

11-31. A floor system has ¾-in. Structural I Exp 1 plywood supported by 2 × 8 No. 1 & Btr D. fir joists at 12 in. c/c. The joists are 12 ft long and have a span of 11 ft 8 in.

(a) What is the allowable ten-year duration load per square foot if the plywood face grain is perpendicular to the joists and the plywood is merely nailed to the joists?

(b) What is the allowable ten-year duration load if the plywood face grain is *parallel* to the joists and the plywood is glued to the joists to create a composite action? (In calculating properties of the transformed composite section, neglect plies whose grain is perpendicular to the length of the joists.)

(c) Calculate the shear strength of the glue necessary to ensure composite action.

12

Diaphragm Action and Design

12-1. LATERAL LOADS ON STRUCTURES

In addition to designing for gravity loads, the designer must proportion a structure to resist lateral loads caused by wind or earthquake. After every major earthquake, there is renewed interest in improving wind and seismic provisions of building codes and in understanding better the effect of dynamic loads on structures. Designing for lateral loads is an important consideration in most structures. For residential wood structures, however, lateral-load resistance is usually ensured by conformance to code rules for structural details.

In a rectangular wood building, the most common elements resisting lateral force are horizontal and vertical diaphragms. A diaphragm is a structural member (wall, floor, or roof section) that resists lateral load in the plane of the member, thereby acting as a thin, deep beam. In a concrete building, the concrete floor slabs are horizontal diaphragms and the elevator core and/or shearwalls are the vertical diaphragms.

In wood structures, although the roof and floor sheathing are designed for gravity loads, they can also carry lateral load at minor additional cost due to nailing or blocking requirements. Plywood diaphragms can eliminate the need for diagonal bracing members.

Figure 12-1 shows the outline of a rectangular wood building under wind loading on its larger face. The plywood or other sheathing on the windward face transfers the wind load to the studs which, in turn, act as vertical beams to divide the load between the foundation the chord at roof level. The chords (also known as top plates) are boundary members perpendicular to the applied lateral load. They consist of dimension lumber around the outer boundary of the roof.

The roof (horizontal diaphragm) acts like a beam, resisting both shear and bending moments. This "beam" differs from the usual beam because its loads are not vertical, but horizontal. If we were considering a multistory building, then the floors would also be horizontal diaphragms. The chords of Fig. 12-1 act as the beam's flanges. They alone are considered to resist the wind-induced bending moment in the horizontal diaphragm. The roof sheathing and the framing members that support it act as the beam's web, resisting only the shear.

The end reactions of the horizontal diaphragm become lateral loads at the top of the two small walls in Fig. 12-1. These walls act as shearwalls. The shearwall load (called R in the figure) is applied along the length of the strut (a boundary member parallel to the applied lateral load) at the top of the shearwall. The shearwalls act as cantilever beams with their fixed ends at the foundation. Because these "cantilever beams" have small span-to-depth ratios, shear deformation is more significant than flexural deformation.

A rectangular building must be designed for lateral loads in both the transverse and the longitudinal directions. After members were chosen for wind or seismic loads normal to the large wall of Fig. 12-1, then the same design procedure would need to be repeated for load normal to the small wall. With wind in this second direction, the functions of the chords and struts would reverse. Also, the two longer walls would now be the shearwalls.

Figure 12-2 shows the shear and bending-moment diagrams for the roof diaphragm of the

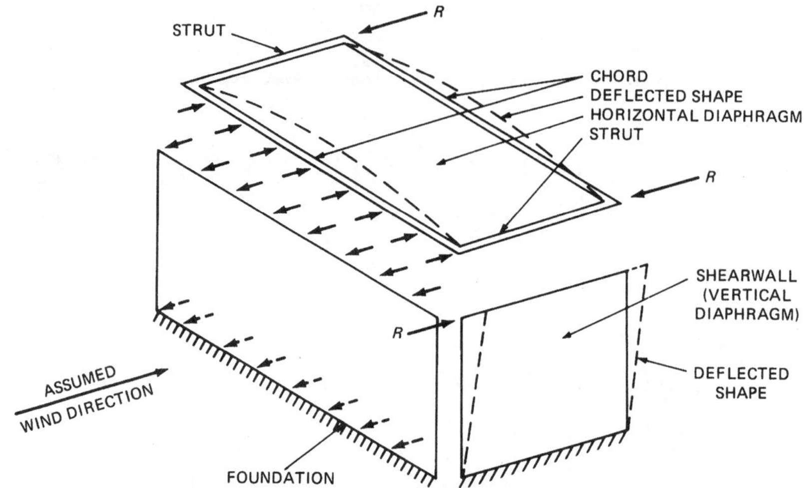

Fig. 12-1. Lateral loading of a rectangular building.

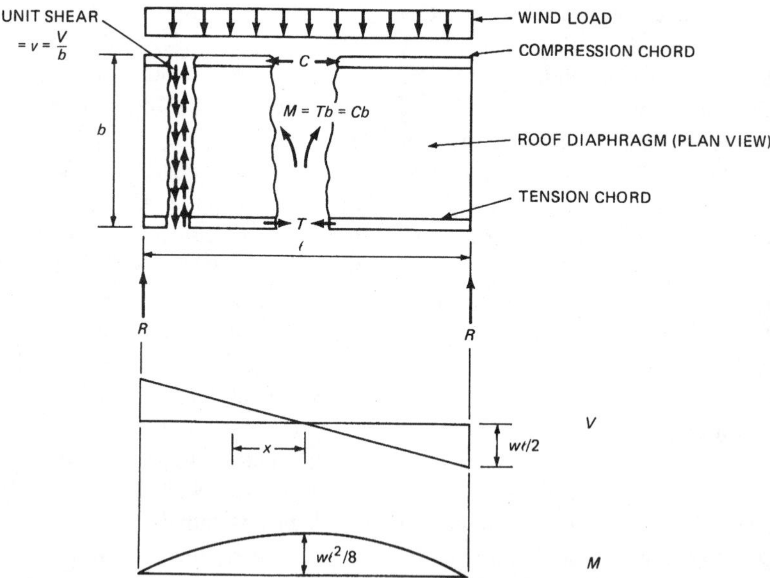

Fig. 12-2. Shear and bending-moment diagrams for roof diaphragm.

rectangular building. The variation of shear and moment at various locations along the span is no different than for a simple beam. However, for design of the chords, bending moment is resolved into a couple consisting of a tension force in one chord and an equal compression force in the other. In an ordinary beam, the variation of shear over the depth is parabolic. In a horizontal diaphragm, however, the shear has been found to be nearly uniform. Therefore, shear per unit width of diaphragm is

$$v = V/b \tag{12-1}$$

(In this chapter shear force per unit width of plywood will be called *unit shear,* expressed as pounds per foot, for example.)

Wood diaphragms, whether at service or limit states, have been the subject of considerable analytical and experimental research. References 1 and 2 contain bibliographies of literature published on wood diaphragms and are a good starting point for those interested in lumber or plywood diaphragms. Many journal articles report on diaphragm strength and stiffness as affected by such things as composite action, nail slip, adhesives, and openings.

Duration of Load Factor

The *National Design Specification* stipulates (for either member or connection design) a wind or earthquake load duration factor, C_D, of 1.6. However, the *Uniform Building Code* specifies different load duration adjustments. The UBC specifies for member design under earthquake loads $C_D = 1.33$ and for member design under wind loads $C_D = 1.6$. The UBC also requires for connection design under either E or W that the designer use $C_D = 1.33$. Because much of this diaphragm chapter depends on UBC tables, the authors have chosen to use the more conservative UBC values.

Connections

Attention to connections is important in the design of lateral-force resisting systems. Without proper connections, the various parts of the system cannot accept or transfer load. For example, columns must be properly connected to the footing, beam-type members (joists, girders, and beams) must be properly connected at their ends (for both gravity and lateral loads), and studs must be connected to wall plates and sills for lateral loads parallel and perpendicular to the wall.

Section 12.3.6 of the NDS (3) permits an adjustment factor, $C_{di} = 1.1$, as a multiplier to tabular allowable values for nails used in lateral-load resisting diaphragms, in addition to all other adjustments including the duration-of-load increase. However, this 10% increase applies only to nailing of sheathing to framing members. It cannot be applied to other diaphragm nailing such as splices and joist-to-top-plate toe nails. Therefore the adjustment is seldom directly used by the designer. In the 1986 NDS, the diaphragm factor was 1.3 and C_D for both E and W equalled 1.33. The current diaphragm factor of 1.1 was chosen so that the product $C_D \times C_{di}$ would be approximately equal under both older and more recent codes.

12-2. HORIZONTAL-DIAPHRAGM WEB DESIGN

The roof or floor diaphragm is generally sheathed in plywood, although it may be sheathed with lumber (4) or particleboard. Resistance of the sheathing to shear caused by lateral loads depends on four things: sheathing thickness and layout, nailing type and spacing, provision for blocking, and width of framing members. Resistance to shear actually is also a function of framing spacing, although this factor is ignored in design. Often plywood thickness is chosen by requirements for gravity loads, and then proper nailing and blocking are used so that lateral loads also can be carried. If lateral-load shears are large, panel thickness may not be controlled by gravity loads.

A plywood diaphragm may fail by nail heads pulling through the panel face, by nails pulling out through panel edges, by nails causing framing members to split, or by buckling of the plywood. Generally shear-through-the-thickness failure is not a factor. Sometimes staples rather than nails are used; staples are less likely than nails to cause splitting.

Table 12-1 (reproduced from 5) shows allowable shears in horizontal plywood diaphragms nailed to either D. fir-larch or southern pine framing. Notice from the table that plywood other than Structural I is assumed to have the same strength as a Structural I panel one size thinner. For example, $^{15}/_{32}$-in. C-D has the same strength as $^{3}/_{8}$-in. Structural I. The allowable shear given in Table 12-1 depends on the load case, that is, the layout of the plywood and framing members. The six load cases sketched in the table are based on whether the lateral load is parallel to or perpendicular to continuous panel joints and to unblocked edges.

A diaphragm with lateral load perpendicular to continuous panel joints (like case 1 of Table 12-1) is stronger than a diaphragm with load parallel (like cases 3–6). Case 1 is stronger than case 2 because unblocked edges are perpendicular to load in case 1. Experiments (6) indicate, too, that diaphragms with an orientation of continuous panel joints perpendicular to eccentric load have greater torsional resistance than if continuous joints are parallel to load. Orienta-

Table 12-1. Allowable Shear in Pounds per Foot for Horizontal Wood Structural Panel Diaphragms with Framing of Douglas Fir-Larch or Southern Pine.[1]

Panel Grade	Common Nail Size	Minimum Nail Penetration In Framing (inches)	Minimum Nominal Panel Thickness (inches)	Minimum Nominal Width of Framing Member (inches)	Blocked Diaphragms				Unblocked Diaphragms	
					Nail spacing (in.) at diaphragm boundaries (all cases), at continuous panel edges parallel to load (Cases 3 and 4) and at all panel edges (Cases 5 and 6) × 25.4 for mm				Nails spaced 6″ (152 mm) max. at supported edges	
					6	4	2 1/2[2]	2[2]	Case 1 (No unblocked edges or continuous joints parallel to load)	All other configurations (Cases 2, 3, 4, 5 and 6)
					Nail spacing (in.) at other panel edges × 25.4 for mm					
					6	6	4	3		
		× 25.4 for mm			× 0.0146 for N/mm					
Structural 1	*6d*	1 1/4	5/16	2	185	250	375	420	165	125
				3	210	280	420	475	185	140
	8d	1 1/2	3/8	2	270	360	530	600	240	180
				3	300	400	600	675	265	200
	10d[3]	1 5/8	15/32	2	320	425	640	730	285	215
				3	360	480	720	820	320	240
C-D, C-C Sheathing, and other grades covered in U.B.C. Standard 23-2 or 23-3	*6d*	1 1/4	5/16	2	170	225	335	380	150	110
				3	190	250	380	430	170	125
			3/8	2	185	250	375	420	165	125
				3	210	280	420	475	185	140
	8d	1 1/2	3/8	2	240	320	480	545	215	160
				3	270	360	540	610	240	180
			7/16	2	255	340	505	575	230	170
				3	285	380	570	645	255	190
			15/32	2	270	360	530	600	240	180
				3	300	400	600	675	265	200
	10d[3]	1 5/8	15/32	2	290	385	575	655	255	190
				3	325	430	650	735	290	215
			19/32	2	320	425	640	730	285	215
				3	360	480	720	820	320	240

[1] These values are for short-time loads due to wind or earthquake and must be reduced 25% for normal loading. Space nails 12 in. (305 mm) on center along intermediate framing members.
Allowable shear values for nails in framing members of other species set forth in Table 23-III-FF of Division III shall be calculated for all other grades by multiplying the shear capacities for nails in Structural I by the following factors: 0.82 for species with specific gravity greater than or equal to 0.42 but less than 0.49, and 0.65 for species with a specific gravity less than 0.42.

[2] Framing at adjoining panel edges shall be 3-inch (76 mm) nominal or wider and nails shall be staggered where nails are spaced 2 inches (51 mm) or 2 1/2 inches (64 mm) on center.

[3] Framing at adjoining panel edges shall be 3-inch (76 mm) nominal or wider and nails shall be staggered where *10d* nails having penetration into framing of more than 1 5/8 inches (41 mm) are spaced 3 inches (76 mm) or less on center.

Table 12-1. (*continued*)

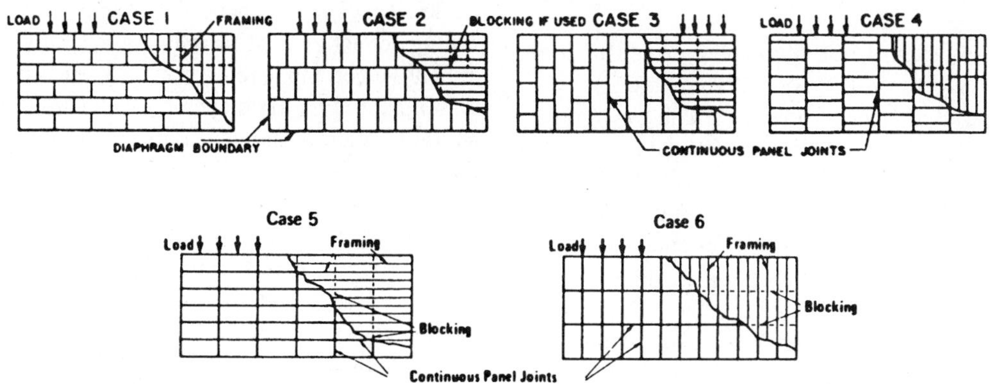

Note: Framing may be oriented in either direction for diaphragms, provided sheathing is properly designed for vertical loading.

tion of face grain is not a consideration in finding the correct load case to use. Case 2 is often the panelized system, a partially pre-framed diaphragm that can be erected quickly.

When plywood joints in the direction parallel to applied force are offset (Case 1), the allowable nail spacing along interior edges is larger than at boundary members.

For *unblocked* diaphragms, nails should be spaced at 6 in. c/c along supported plywood edges (including diaphragm boundaries; i.e., chords and struts) and at 12 in. c/c along intermediate framing members. Plywood panels with T&G edges or panel clips, although having some edge support, are considered to be unblocked because there is no member into which nails can be driven to resist shear.

Blocked diaphragms will be stronger than those with no blocking, because all four plywood edges are nailed. The nail spacing used at each edge will determine the exact allowable shear. However, minimum requirements for use as a blocked diaphragm are that all four panel edges be nailed at 6 in. c/c. Nails are spaced along intermediate framing members as for unblocked diaphragms. Blocking of a diaphragm is really a means of splicing the sheets of plywood so that they will have better continuity.

The diaphragm must be designed for both load cases—that of transverse lateral load and that of longitudinal lateral load. Not only will each cause a different shear per foot in the diaphragm, but also allowable shears will be different because plywood layout (case number) will be different for different directions of load. (For example, if wind from the south causes load parallel to continuous panel joints, wind from the west may cause load perpendicular to continuous panel joints.)

Derivation of Table Values

There may be occasions for which Table 12-1 does not cover a specific design problem. In such instances it is helpful to know how to determine the allowable shear without using the table. Based on research results (7), the allowable shear in *blocked* diaphragms is calculated as the smaller of (1) the allowable shear, V_{cp}, based on plywood shear stress and (2) the allowable shear, V_{np}, based on lateral fastener load at the boundary. The product of the following items will give V_{cp}:

1. Allowable stress for shear through the thickness of the plywood (Table 11-3 or reference 8).
2. A factor of 1.33 for load duration.
3. A factor of 12 (in./ft).
4. The effective thickness for shear for the plywood (Table 11-4).

The product of the following items will give V_{np}:

1. Lateral load design value for the particular nail (63 lb for 6*d*, 78 lb for 8*d*, and 94 lb for 10*d* or 12*d*).
2. A factor of 1.30 for diaphragm construction.
3. A factor of 1.33 for load duration.
4. The number of boundary nails per foot, except use 2.25 (experimentally derived) when there are actually 2 nails per foot.
5. A factor of 0.89 if 2-in. nominal lumber is used, *or* if two rows of fasteners are used in 3-in. nominal lumber, *or* if three rows are used in 4-in. nominal lumber.
6. A factor of 0.90 if the diaphragm consists of D. fir framing members and non-Structural-I plywood.
7. A factor of 0.85 if nails are spaced 2 in. c/c at boundary.
8. For 10*d* nails only, an additional factor of 0.85 if nails are spaced 2 in. or 2½ in. c/c at boundary when boundary members are single 2-in. nominal members.
9. For non-Douglas fir lumber, a factor of 0.82 for lumber Group C and a factor of 0.65 for lumber Group D.

For *unblocked* diaphragms, calculate the allowable shear for "basic" nailing (nailing 4 in. c/c at the boundary and 6 in. c/c at interior panel edges). Then, if the unblocked diaphragm pattern is case 1, use 67% of this value: if cases 2 through 6, use 50%. These reduction factors for unblocked diaphragms have been established by test, and no direct analytic procedure has been developed.

The ability to calculate allowable shear is very useful for highly loaded diaphragms that are not covered by Table 12-1. For instance, if a two-layer diaphragm is made by nailing the first layer to the boundary with nails at 4 in. c/c and the second layer (offset) is nailed over the first also with boundary nails at 4 in. c/c. and also penetrating into the framing member. then there are six fasteners per foot. Once this is known, the allowable shear, V_{np}, based on lateral fastener load at the boundary, can now be calculated. For these highly loaded diaphragms, an additional check for shear-through-the-thickness failure at discontinuous panel joints must be made for case 1 and 2 diaphragms. Reference 7 describes this check.

Deflection of Diaphragms

Deflection of a horizontal diaphragm (and resulting deflection of shearwalls) is generally not critical. According to reference 2, "Many offices do not determine the diaphragm deflection and building officials seldom ask for it." In order to limit deflection, the UBC limits the span-to-width ratio of plywood and particleboard horizontal diaphragms to 4:1. If the designer feels special conditions necessitate deflection calculations, references 9 and 10 give a procedure that includes the effects of nail deformation and slip at chord splices.

Continuity

When there are three or more shearwalls (rather than two), it is difficult to assess the degree of continuity of the horizontal diaphragm over the supporting shearwalls. In buildings where the horizontal and vertical diaphragms are both wood frame, the walls are probably more rigid than the floors or roof, because the walls are likely to have several layers of sheathing (siding, plywood, and a finish layer of interior gypsum board). Therefore, the shearwalls are generally considered to be rigid, unyielding supports. Assuming rigid supports, if bending deformations predominate, a horizontal diaphragm over three supports (two equal spans) will transmit five-eighths of the total lateral force to the center wall (analogous to a continuous beam action). If shear deformations predominate, then the distribution will be one-half to the center wall (as for two simple beams). The conservative approach would be to calculate both the simple-beam and continuous beam shears and use the larger value at all locations both for determining load transmitted to the vertical diaphragms and for determining the unit shear in the horizontal diaphragms at various locations. Experiments (6) seem to indicate that having continuity over a center support does not change the shear resistance of a horizontal diaphragm.

Pitched Roof Diaphragms

If a pitched roof diaphragm has a gentle slope, it can be treated as if it were horizontal. However, for more steeply pitched roof diaphragms, calculating the unit shear in the diaphragm will involve solving for vector components of force. Figure 12-3a shows one square foot of a steeply pitched roof diaphragm subject to a suction wind force, *W*, which is perpendicular to the roof surface. This is the type of wind force that would be specified by UBC Method 1 (see Section 4-5). The corresponding vector diagram is shown in part (b) of the figure, where *V* is a vertical force and *D* is the shear force in the roof diaphragm. Note that if the wind were pressure rather than suction, the vertical force would be in the opposite direction.

For ease in design, the diaphragm force, *D*, shown in Fig. 12-3b, can be broken into a horizontal component (equal to the horizontal component of *W*) and a vertical component. Because the ratio of this horizontal component to the length of the horizontal projection of the roof equals the ratio of *D* to the sloping roof length, the horizontal component of *W* can be used with the roof's horizontal projection length in the calculation of unit shear.

Code-specified seismic forces are horizontal. Figure 12-3c shows a roof diaphragm subjected to an earthquake force. The vector diagram in part (d) shows the diaphragm force, *D*. Notice that the diaphragm force is greater than the applied earthquake force.

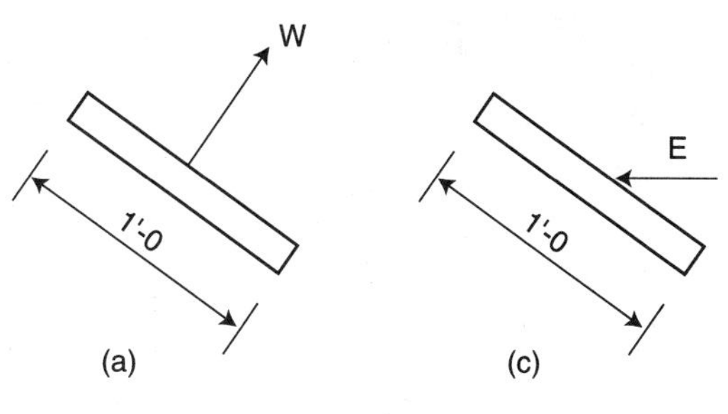

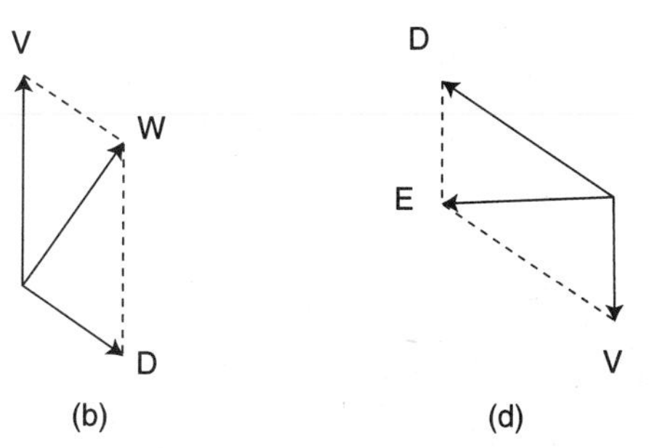

Fig. 12-3. Pitched roof diaphragms.

Nail Capacity

The allowable lateral load given in the NDS or UBC for a particular nail size is based on nailing through lumber into lumber. When nailing through plywood into lumber, a too-large nail in a thin plywood will not be able to reach its full design capacity (7). Provided the plywood and the lumber receiving the point are of similar densities, and provided the nail has full standard penetration (or more), the following plywood thicknesses have been proven by test to fully develop, for common nails, the tabular lateral nail values.

Nail Size	APA Rated Sheathing (in.)	Structural I (in.)
6*d*	3/8	5/16
8*d*	15/32	3/8
10*d*	19/32	15/32

12-3. HORIZONTAL-DIAPHRAGM CHORD DESIGN

The chords are designed for the axial tension or compression forces that make up the bending-moment couple. The axial force at any point in the chord (Fig. 12-2) can be determined by dividing the moment at that point by the lever arm of the couple (distance between chords, *b*). That is,

$$T = C = M/b \qquad (12\text{-}2)$$

The chord force varies with *M* along the chord's length. Because it is easier (and conservative), the chord and its connections are usually designed for the maximum force in the chord, even though the force may be considerably less at the location under consideration. A load duration factor of 1.33 is used in chord and connection design, because the chord force is due to either earthquake or wind.

If the shearwalls in a building with plywood horizontal diaphragms should be masonry or concrete rather than wood, then the chord may consist of continuous horizontal reinforcing bars in the wall at diaphragm level. The rebar alone is assumed to take the entire tension load, and the masonry or concrete the compression load if the "chord" happens to be in compression. (Allowable stresses for steel, concrete, and masonry are all raised by a factor of 1.33 for wind or seismic load.) For force transfer from the diaphragm to the chord, connections with anchor bolts must be well designed.

In ordinary light-frame construction, the chord consists of a double top plate (two pieces of dimension lumber) as shown in Fig. 12-4. Using two pieces allows one plate member to be continuous at the location where the other is being spliced, assuring (with adequate connections by either nails or bolts, depending on magnitude of force) that the chord force can be transferred from one piece to the other. If chord forces are large, three top plates may be necessary. It is best, of course, to locate splices as far as possible from the position of maximum moment.

Sometimes the designer may object to using double or triple top plates just to provide continuity through a splice. In this case, the designer may use a steel splice strap top and bottom to transfer the chord force across the splice.

Figure 12-4 shows that the plywood sheathing and the chord do not touch one another, emphasizing the need for good connection design. Since we do not know the direction of the lateral load, each chord must be designed for tension or compression, whichever is more severe. Usually tension is critical, because F_t generally is less than F_c. Also, a portion of compressive force can be transferred across a splice by end bearing, but a tension force cannot.

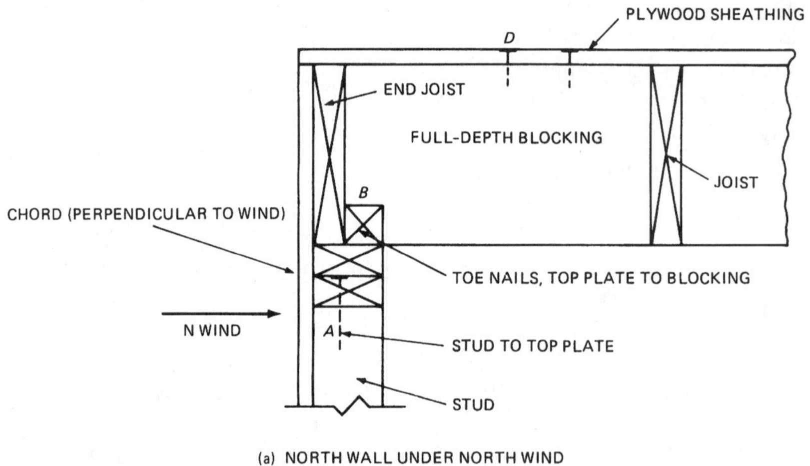

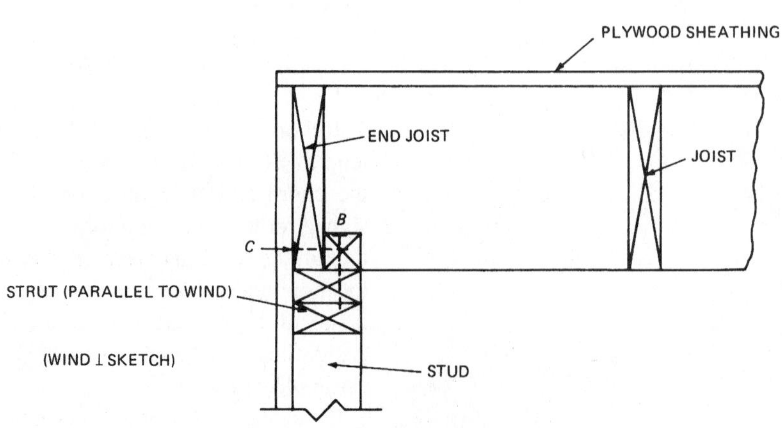

Fig. 12-4. Wall-to-roof connections.

The examples that follow are for either wind forces alone or earthquake forces alone. In actual design practice, the designer must determine specified values of both wind loads and seismic loads, and design for the more severe of these in combination with dead and other live loads. (See Section 4-6.)

Example 12-1

A one-story wood frame warehouse whose plan view is shown in Fig. 12-5a is 13 ft high. Based on vertical loads, $^{15}/_{32}$-in. C-C Exterior plywood has been chosen tentatively for sheathing of the nearly flat roof. Blocking was not necessary for vertical loads. Determine roof diaphragm requirements (using Table 12-1) for wind pressure of 35 psf on a vertical surface, assuming the plywood layout as shown by Fig. 12-5b. Design the chords.

Wind pressure on north wall: Wind pressure on the upper one-half (6.5 ft) of the wall will be transferred to the roof diaphragm, with the rest going to the foundation. Horizontal load applied to the diaphragm is

$$w = (6.5 \text{ ft})(35 \text{ psf}) = 228 \text{ lb/ft}$$

Recalling the shear diagram of Fig. 12-2,

$$\text{Max } V = (228)(88)/2 = 10{,}000 \text{ lb}$$

$$\text{Max unit } v = 10{,}000/50 = 200 \text{ lb/ft}$$

The plywood layout shown in Fig. 12-5b is like case 3. Table 12-1 shows that for $^{15}/_{32}$-in. C-C plywood with 2-in. nominal framing members, an unblocked diaphragm has an allowable unit shear (if 8*d* nails are used) of

$$180 \text{ lb/ft} < 200 \text{ lb/ft } NG$$

Blocking will be required. (Even with 10*d* nails, blocking would still be required.) Blocking can be installed over the entire roof or, alternatively, can be installed only in regions of large shear (shear > 180 lb/ft) near the east and west edges of the diaphragm. The designer must weigh the expense of blocking over the entire roof against the confusion and errors

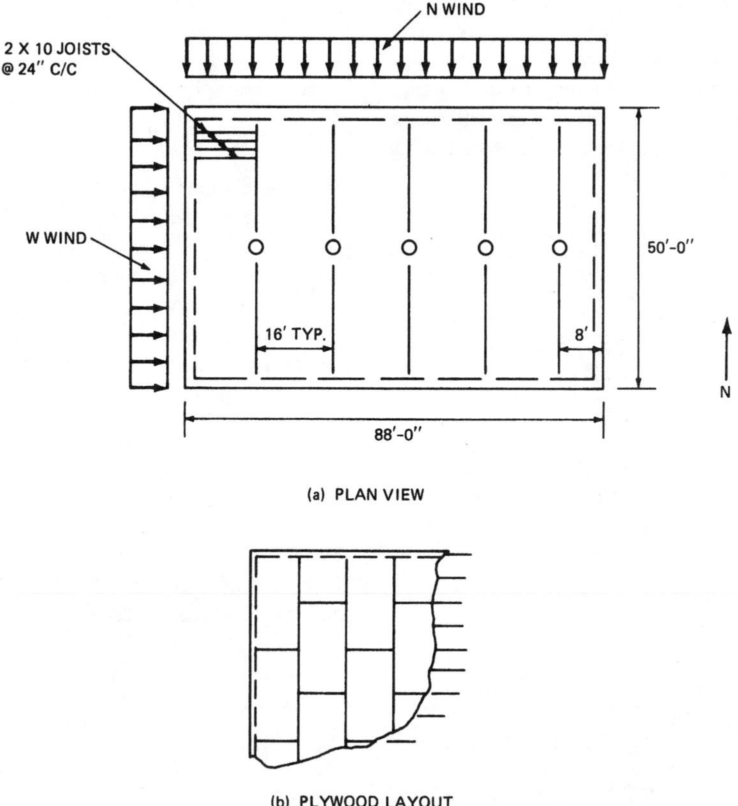

Fig. 12-5. Plan for Examples 12-1 through 12-4.

that might occur if blocking is specified only in certain regions.

By similar triangles on the shear diagram, determine distance x, measured from the building centerline to where no blocking is necessary.

$$x = 44(180/200) = 39 \text{ ft}$$

Blocking is necessary within 44 − 39 = 5 ft of the west wall and within 5 ft of the east wall. By Table 12-1, a blocked diaphragm has allowable shear of 270 lb/ft. *Use 8d nails at 6 in. c/c at framed edges and at 12 in. c/c at intermediate framing members (field nailing). Install blocking within 5 ft of the west wall and within 5 ft of the east wall.*

Wind pressure on west wall: The plywood layout is now case 1 because the load is no longer parallel to the continuous panel joints.

$$\text{Max } V = 228(50)/2 = 5700 \text{ lb}$$

$$\text{Max unit } v = 5700/88 = 65 \text{ lb/ft}$$

Obviously this will not control over the wind load from the north. It should be noted here that it would have been better to use a plywood layout with the continuous panel joints running east-west rather than north-south. Because the critical (larger) unit shear will be due to wind against the longer wall, continuous joints should be parallel to the longer wall.

Chord design (for wind from north): Recalling the moment diagram of Fig. 12-2,

$$\text{Max } M = w\ell^2/8 = (228)(88)^2/8$$
$$= 221{,}000 \text{ ft-lb}$$

By Eq. 12-2 and assuming the chords to be 50 − 6/12 = 49.5 ft apart,

$$T = C = 221{,}000/49.5 = 4460 \text{ lb}$$

(Chord force actually varies from zero at the ends to 4460 at the center of the diaphragm.) For No. 2 D. fir 2 × 6s,

$$F'_t = C_D C_F F_t = 1.33(1.3)(575) = 994 \text{ psi}$$

$$\text{Req net } A = 4460/994 = 4.49 \text{ in.}^2$$

The chord will consist of two members. The piece that, at splices, is continuous must be able to carry the entire 4460-lb force at splice locations. Try a 2 × 6.

Assuming that the connection at the splice will use 1-in.-diameter bolts,

$$\text{Net } A = 8.25 - (17/16)(1.5)$$
$$= 6.66 \text{ in.}^2 > 4.49 \text{ in.}^2 \qquad OK$$

Under compression (force C) the chord is laterally braced along its full length—vertically by the wall sheathing and horizontally by the roof sheathing. Thus $F'_c = 1.33(1.1)(1300) = 1900 > F'_t$, so tension controls.

Since the chord force is less for east or west wind, the same chord may be used on the other two locations. (Design of these top plate members is not yet final; their size may actually be controlled by the force in them when they are functioning as struts. See Section 12-6.)

Tentatively, use two No. 2 Douglas fir-larch 2 × 6s for chords.

Example 12-2

For north wind loads of Example 12-1, design the connection at splices in the south chord.

Bolt design at chord splices: Figure 12-6 shows that the bolts carry only half of the 4460-lb force through the splice when splices are large distances apart. The other half is carried directly by the continuous member.

For 1-in.-diameter bolts, main member thickness = 1.5 in., and side member thickness = 1.5 in., the single-shear design value is $Z_{\|} = 970$ lb/bolt. Guessing that two bolts will be required, no group action factor, C_g, is needed.

$$Z'_{\|} = C_D(970) = 1.33(970) = 1290 \text{ lb/bolt}$$

$$\text{Req } N = 2230/1290 = 1.7 = 2 \text{ bolts each side}$$

Even though the moment is smaller away from the center (where T is below 4460 lb), it is more practical to use two bolts each side at all splices.

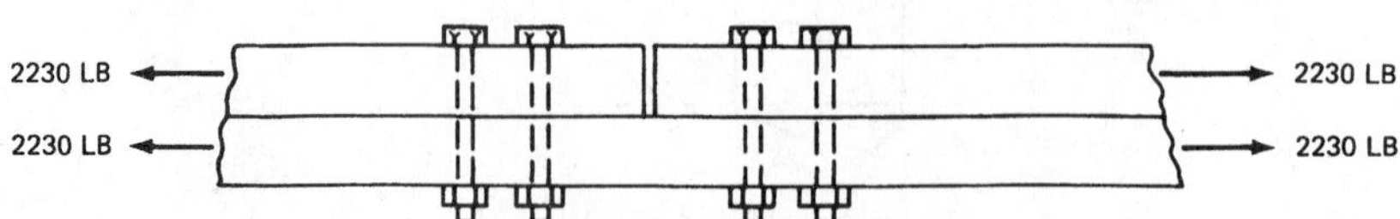

Fig. 12-6. Chord splice for Example 12-2.

Still tentatively, use chord of two 2 × 6s with two 1-in.-diameter bolts each side of each splice.

Alternative of nails at splices: Usually chord splices are only 4–6 ft apart when nails are used, rather than 10 ft or more when bolts are used. Therefore, in the nailed splice, the entire 4460-lb force would need to be transferred.

If 16*d* common nails were used to transfer load across this chord splice,

Nail value = Z = 141 lb/nail

Nail length = 3.5 in.

$12D = 12(0.162) = 1.94$ in.

Penetration into main member = 1.5 in. < 1.94

$C_d = 1.5/1.94 = 0.77$

$Z' = 0.77(1.33)(141) = 144$ lb/nail

Req $N = 4460/144 = 31$ nails per side of each splice of north or south chords.

The UBC specifies that doubled top plates be nailed with 16*d* nails at 16 in. c/c and with eight 16*d* nails at splices. Obviously, the nailing schedule given by the UBC is inadequate to cover a case such as this.

Example 12-3

Design nailed connections for the building of Example 12-1 as follows: (a) stud to top plate under north wind; (b) top plate to blocking under north wind; and (c) end joist to top plate under west wind.

a. Stud to top plate under north wind: Wind load perpendicular to the north wall at the eaves is

$$w = 228 \text{ lb/ft}$$

The studs transfer this load to the lower member of the upper plate (chord) by nails driven into end grain of the stud at location A of Fig. 12-4a.

Assuming 16*d* nails.

$$Z' = Z\, C_D\, C_{eg}$$

$$= (141)(1.33)(0.67) = 126 \text{ lb}$$

With studs at 24 in. c/c, force per stud

= (228 lb/ft)(2 ft) = 456 lb

Nails req = 456/126 = 4 nails (16*d*) per stud

Because this is a large number of nails, framing anchors might be a better solution. One manufacturer gives a design load of 390 lb for one framing anchor. Thus two such anchors (or one anchor plus nails) would be required at each end of each stud.

An equal nailed connection is required to transfer the stud reaction from the lower member to the upper member of the top plate.

b. Top plate to blocking under north wind: Directly behind point B in Fig. 12-4a are toe nails (not shown) through full-depth blocking and into the top plate. (Note that if this were an east or west wall, the toe nails would be through the *joist* and into the top plate; i.e., whether it is a joist or blocking depends on which way the roof framing runs.) These toe nails transfer the perpendicular-to-wall wind load from the top plate into the full-depth blocking.

For 16*d* nails.

$$Z' = Z\, C_D\, C_{tn}$$

$$= 141\,(1.33)(0.83) = 156 \text{ lb}$$

With blocking at 24 in. c/c, load per block = (228 lb/ft) (2 ft) = 456 lb

Nails required = 456/156 = 3 nails (16*d*) per block (or use framing anchors)

c. End joist to top plate under west wind: Nails at locations B and C of Fig. 12-4b transfer the 65-lb/ft unit shear in the roof plywood (Example 12-1) from the end joist into the top plate.

For 16*d* nails,

$$Z' = 141\,(1.33)(1.5/1.94) = 145 \text{ lb}$$

Required spacing = (145 lb)/(65 lb/ft) = 2.23 ft (at both locations B and C)

This is a large spacing. When making these same calculations for the east or west wall, the roof unit shear will be 200 lb/ft, which requires a spacing of 8.7 in. c/c. Therefore, for practicality, an 8-in. spacing will be used around the entire periphery.

Example 12-4

For the building of the previous three examples (see Fig. 12-5), calculate the maximum unit shear in the roof diaphragm for Zone 4 seismic loads in the N-S direction knowing that wall dead load = 18 psf and roof dead load = 13 psf.

For N-S seismic loads. W = dead load to diaphragm.

From roof = (88 ft) (50 ft) (13 psf) = 57,200 lb

From N & S walls = (2 walls) (6.5 ft) (88 ft) (18 psf) = 20,600 lb

$W = 57{,}200 + 20{,}600 = 77{,}800$ lb

(Weight of end walls is not included; their inertial forces affect only the end walls.)

$Z = 0.40$ (Zone 4; see Section 4–7)

$I = 1.0$

$C = 2.75$

$R_w = 8$

$V = ZICW/R_w = 0.138W$

$= 10{,}700$ lb

The uniform (seismic) load at the roof level is

$$w = V/88 = 10{,}700/88 = 122 \text{ lb/ft}$$

Maximum roof shear (near ends) is

$$w(88)\ 2 = 5370 \text{ lb}$$

Maximum unit roof shear equals

$$5370/50 = 107 \text{ lb/ft}$$

(This is smaller than that caused by wind in Example 12-1, so wind controls.)

Example 12-5

Without using Table 12-1, calculate the allowable unit shear for a blocked diaphragm that has 3/8–in. C-C plywood, 2-in. nominal Douglas fir framing, and 8*d* nails at 6 in. c/c along all panel edges.

From Table 11–3, the shear-through-the-thickness allowable stress (Grade Stress Level S-1 and species Group 4) = 130 psi and effective thickness for shear = 0.278 in.

Calculating with factors in the same order as presented previously (Section 12-2),

$$V_{cp} = (1.30)\ (1.33)\ (12)\ (0.278) = 577 \text{ lb/ft}$$

$$V_{np} = (78)\ (1.3)\ (1.33)\ (2.25)\ (0.89)\ (0.90)$$

$$= 243 \text{ lb/ft}$$

The latter, 243 lb/ft. controls.

This compares well with the value of 240 lb/ft in Table 12-1. This table was derived from allowable lateral load nail values which have changed since the table was compiled. Therefore tabular and calculated values will not agree exactly, especially for 6*d* nails (which had the largest change in allowable load).

Example 12-6

For the two-story structure in Fig. 12-7 determine loads per foot, w_1 and w_2, transferred to each horizontal diaphragm for a 30-psf wind pressure acting on the long wall. (The roof diaphragm is taken to be horizontal since it has a very small slope.)

Analyzing the wall as two simple spans rather than one continuous one, the tributary width to the second floor extends 4.5 ft below second-floor level and 4.5 ft above second-floor level; therefore,

$$w_1 = (30 \text{ psf})\ (9 \text{ ft}) = 270 \text{ lb/ft}$$

Using Method 2 (Section 4–5), the tributary width to the eaves line is the vertical projection of the roof plus 4.5 ft.

$$w_2 = (30 \text{ psf})\ (2 \text{ ft} + 4.5 \text{ ft}) = 195 \text{ lb/ft}$$

Even though the roof is sloping, chord force at the eaves due to bending moment is found by Eq. 12-2 as before.

12-4. SHEARWALL DESIGN

The load from the horizontal diaphragm is transferred through the shearwalls (vertical diaphragms) to the ground. If a wood shearwall has no openings, it is said to consist of a single shear panel. If there are window or door open-

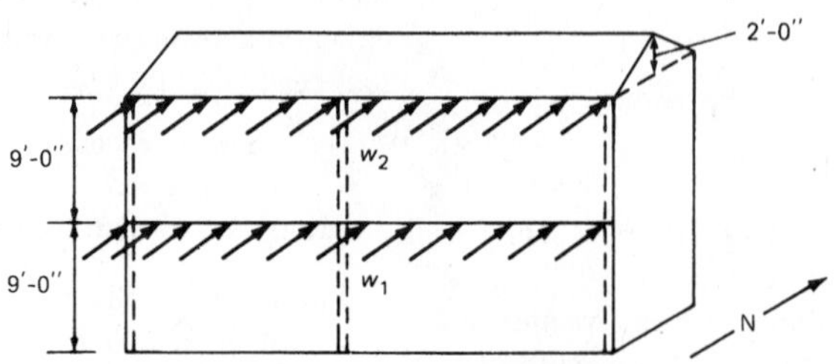

Fig. 12-7. Structure for Example 12-6.

ings, the shearwall will consist of multiple shear panels as shown in Fig. 12-8a. The height of a shearwall in a multistory building is generally accepted to be the distance between horizontal diaphragms (although a few challenge this).

For wood shearwalls, unlike concrete ones, the shear per unit width in every panel is assumed to be identical. This is because shear deformation predominates in wood walls. The total force in a panel will be in proportion to its length (horizontal dimension). The unit shear can be determined by dividing the total shear from the roof by the sum of the lengths of all shear panels.

For each panel to function as a separate element, the filler panels (Fig. 12-8a) may be nailed less heavily than the shear panels. Double full-height studs at the sides of openings separate the full-height shear panels from the filler panel. Panels are constrained by the drag struts to deflect together.

Chords of Shear Panels

The full-height studs at each side of a shear panel are the chords that resist all the bending moment for that panel. Figure 12-8b shows the cantilever bending moment in panel 1 resolved into a couple of a tension force, T_1, in the left chord and a compression force, C_1, in the right chord. From the free body of Fig. 12-8b it can be seen that

$$T_1 = C_1 = vh \tag{12-3}$$

Design of these chords is similar to design of a horizontal-diaphragm chord. Wall, floor, or roof dead load tributary to a chord can be neglected in designing the tension chord (a conservative approach), but must be considered when designing the compression chord. Because we do not know the direction of the wind, all chords are designed for both tension and compression.

Covering Materials for Shearwalls

Plywood is probably the most common sheathing material in the wood shearwall, although particleboard is becoming more popular as more efficient use of wood is being made. For plywood and particleboard shear panels, the UBC (5) specifies a maximum height-length ratio of 3.5:1 (blocked) or 2:1 (unblocked). Keeping the panels from being too tall and slender helps ensure that shear (deep beam) action will predominate and also limits deflection.

Table 12-2 shows allowable shears for shearwalls in which the plywood is applied directly to the framing or is applied over gypsum sheathing. The table also shows allowable shears for plywood panel siding (grooved panels that serve as exterior finish). Large nail heads appear unsightly on plywood siding, so Table 12-2 lists allowable shears for siding nailed with smaller-headed galvanized casing nails. Allowable shears are less when casing nails are used, because a failure due to nail head pull-through is more likely.

Table 12-2 for shearwalls was developed by

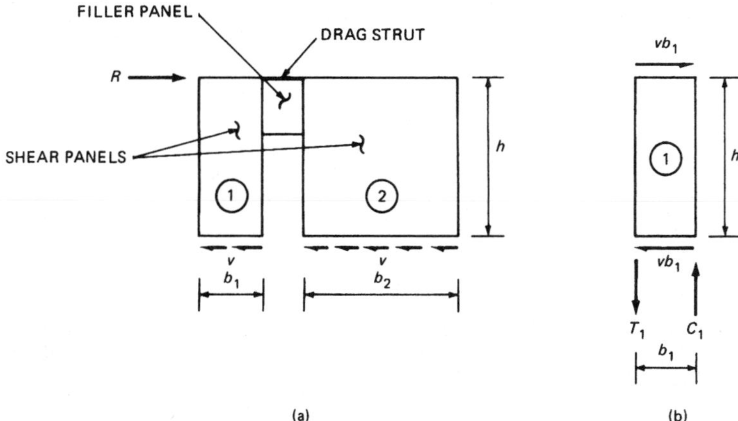

Fig. 12-8. Two-panel wood shearwall.

Table 12-2. Allowable Shear for Wind or Seismic Forces in Pounds per Foot for Wood Structural Panel Shearwalls with Framing of Douglas Fir-Larch or Southern Pine.[1,2]

Panel Grade	Minimum Nominal Panel Thickness (inches)	Minimum Nail Penetration in Framing (inches)	Panels Applied Directly to Framing: Nail Size (Common or Galvanized Box)	Nail Spacing at Panel Edges (in.) × 25.4 for mm: 6	4	3	2[3]	Panels Applied Over 1/2-inch (13 mm) or 5/8-inch (16 mm) Gypsum Sheathing: Nail Size (Common or Galvanized Box)	Nail Spacing at Panel Edges (in.) × 25.4 for mm: 6	4	3	2[3]
	× 25.4 for mm			× 0.0146 for N/mm					× 0.0146 for N/mm			
Structural I	5/16	1 1/4	6d	200	300	390	510	8d	200	300	390	510
	3/8			230[4]	360[4]	460[4]	610[4]					
	7/16	1 1/2	8d	255[4]	395[4]	505[4]	670[4]	10d[5]	280	430	550	730
	15/32			280	430	550	730					
	15/32	1 5/8	10d[5]	340	510	665	870	—	—	—	—	—
C-D, C-C Sheathing, plywood panel siding and other grades covered in U.B.C. Standard 23-2 or 23-3	5/16	1 1/4	6d	180	270	350	450	8d	180	270	350	450
	3/8			200	300	390	510		200	300	390	510
	3/8			220[4]	320[4]	410[4]	530[4]					
	7/16	1 1/2	8d	240[4]	350[4]	450[4]	585[4]	10d[5]	260	380	490	640
	15/32			260	380	490	640					
	15/32	1 5/8	10d[5]	310	460	600	770	—	—	—	—	—
	19/32			340	510	665	870					
			Nail Size (Galvanized Casing)					Nail Size (Galvanized Casing)				
Plywood panel siding in grades covered in U.B.C. Standard 23-2	5/16	1 1/4	6d	140	210	275	360	8d	140	210	275	360
	3/8	1 1/2	8d	160	240	310	410	10d[5]	160	240	310	410

[1] All panel edges backed with 2-in. (51 mm) nominal or wider framing. Panels installed either horizontally or vertically. Space nails at 6 in. (152 mm) on center along intermediate framing members for 3/8-inch (9.5 mm) and 7/16-inch (11 mm) panels installed on studs spaced 24 in. (610 mm) on center and 12 in. (305 mm) on center for other conditions and panel thicknesses. These values are for short-time loads due to wind or earthquake and must be reduced 25% for normal loading.

Allowable shear values for nails in framing members of other species set forth in Table 23-III-FF of Division III shall be calculated for all other grades by multiplying the shear capacities for nails in Structural I by the following factors: 0.82 for species with specific gravity greater than or equal to 0.42 but less than 0.49, and 0.65 for species with a specific gravity less than 0.42.

[2] Where panels are applied on both faces of a wall and nail spacing is less than 6 in. (152 mm) on center on either side, panel joints shall be offset to fall on different framing members or framing shall be 3-in. (76 mm) nominal or thicker and nails on each side shall be staggered.

[3] Framing at adjoining panel edges shall be 3-in. (76 mm) nominal or wider and nails shall be staggered where nails are spaced 2 in. (51 mm) on center.

[4] The values for 3/8-in. (9.5 mm) and 7/16-in. (11 mm) panels applied direct to framing may be increased to values shown for 15/32-in. (12 mm) panels, provided studs are spaced a maximum of 16 in. (406 mm) on center or panels are applied with long dimension across studs.

[5] Framing at adjoining panel edges shall be 3-in. (76 mm) nominal or wider and nails shall be staggered where *10d* nails having penetration into framing of more than 1 5/8 in. (41 mm) are spaced 3 in. (76 mm) or less on center.

Source: Reproduced from the 1994 edition of the *Uniform Building Code*™, copyright © 1994, with the permission of the publisher, the International Conference of Building Officials.

taking ultimate loads from test results, applying a load factor, and placing the result in the table. Therefore, the shearwall table (Table 12-2) is not identical to the one for horizontal diaphragms (Table 12-1). Eventually the two tables may be combined into one; or, possibly, a design equation may replace the tables.

For small unit shears, lath and plaster or gypsum wallboard may be used as shear-resisting materials in a wood shearwall. The maximum height-to-length ratio for such panels is 2:1 (blocked) or 1.5:1 (unblocked). Fiberboard-sheathed shear panels are covered in UBC Table 23-I-P. Reference 4 covers the use of diagonally placed lumber as a shearwall covering.

The UBC will not allow addition of the shear resistance of two different materials in a composite shearwall (a wall with plywood attached to one side of the studs and gypsum sheathing to the other). The code says gypsum must be disregarded, although tests (11) show that the contributions to strength and stiffness of the plywood and gypsum sheathings are additive.

Because light-frame residential structures have proved their performance over the years, building codes such as the UBC consider them to be "conventional construction," not requiring designed shearwalls. However, the UBC requires bracing by some method such as diagonal braces across the studs (let-in braces), 1/2-in. fiberboard, 1/2-in. gypsum board, and so on. Full details are given in UBC section 2326.11.3.

Deformations of Shearwalls

Shear deformations (racking) of shearwalls do not vary linearly with load, because load-slip nail behavior is highly nonlinear. When a wood shearwall is subjected to racking loads, the stud frame distorts as a parallelogram, the sheathing distorts to a different parallelogram, and the nails must deform. Therefore, it is nail slip that is the major factor affecting racking deformation. Stiffness of studs and sheathing plays only a secondary role. Reference 12 presents a convenient method for determining wall deformations at service loads. Gupta and Kuo (13) present an iterative model for determining racking load versus deformation (shear strain) that does not require finite-element analysis.

Example 12-7

Figure 12-9 shows the elevation of the west wall of the building considered in Example 12-1. Studs are 2 × 6 No. 2 D. fir at 24 in. c/c. Wall dead load = 18 psf and roof dead load = 13 psf. Using wind load as given in Example 12-1, choose chord sizes and sheathing for this shearwall.

The maximum height-width ratio allowed is 3.5:1. All panels meet this requirement. Noting from Fig. 12-2 that R is the reaction of the roof diaphragm which, in turn, loads the end shearwall,

$$R = w\ell/2 = 228\,(88)/2 = 10{,}000 \text{ lb}$$

Dividing this by the sum of the length of all shear panels,

$$\text{Unit } v = 10{,}000/(13.5 + 10 + 14) = 267 \text{ lb/ft}$$

From Table 12-2, *use 5/16-in. C-D or C-C plywood with 6d nails at 4 in. c/c around panel edges. Use nails at 12 in. c/c at intermediate members. Blocking is required.*

Chord design for tension: Critical tension chords will be at the small opening because these chords carry the least dead load from the roof. Due to lateral load only,

$$T = vh = (267)\,(13) = 3470 \text{ lb}$$

The width of roof tributary to the west wall is 8 ft because the joists (Fig. 12-5a) are 16 ft long.

$$\text{Roof dead load} = (13 \text{ psf})\,(8 \text{ ft}) = 104 \text{ lb/ft}$$

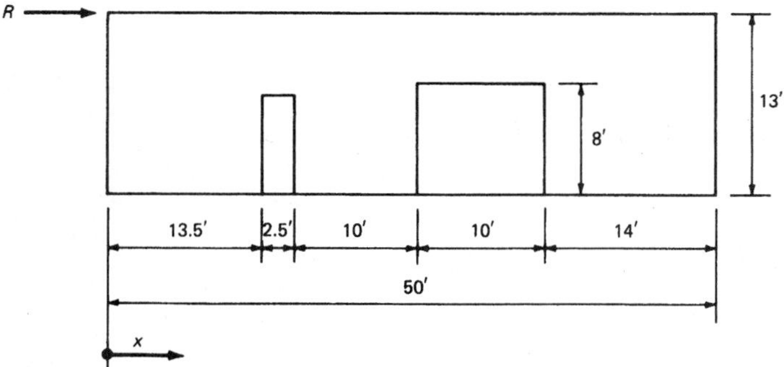

Fig. 12-9. Elevation of wall for Example 12-7.

Wall dead load when no openings = (18 psf) (13 ft) = 234 lb/ft

Wall dead load above door = (18 psf) (5 ft) = 90 lb/ft

With studs 2 ft c/c, there will be a 1-ft width of full-height wall tributary to the chord and 1.25 ft of tributary wall width above the door. The total chord tension is

$$T = 3470 - (104)\,(2.25) - (234)\,(1) - (90)\,(1.25) = 2890 \text{ lb}$$

Assuming a 7/8-in. bolt (15/16-in. hole) for connecting a metal anchor bracket to the base of the 2 × 6 vertical chord member,

$$\text{Net area} = 1.5\,(5.5 - {}^{15}\!/_{16}) = 6.84 \text{ in.}^2$$

$$f_t = 2890/6.84 = 423 \text{ psi}$$

$$F'_t = C_D C_F F_t = 1.33\,(1.3)(575) = 994 \text{ psi}$$

One 2 × 6 OK for tension

Chord design for compression: Critical compression chords will be at the large opening since these carry more roof and wall dead load than those next to the small opening. Dead loads per foot of wall width are the same as used to design a tension chord. For studs at 2 ft c/c, there will be a 1-ft width of full-height wall tributary to the chord and a 5-ft width of wall above the door tributary go the chord.

$$C = 3470 \text{ lb} + (104 \text{ lb/ft})\,(6\text{ft}) + 234(1) + 90(5) = 4780 \text{ lb}$$

Nailing of the plywood sheathing to the narrow face of the 2 × 6 chord will prevent buckling about the weak axis; therefore, the controlling ratio

$$\ell/d = 13(12)/5.5 = 28.4$$

$$F_{cE} = 0.3(1{,}600{,}000)/(28.4)^2 = 595 \text{ psi}$$

$$F_c^* = C_D C_F F_c = 1.33\,(1.1)(1300) = 1902 \text{ psi}$$

$$C_p = 0.289 \text{ (Eq. 7–5)}$$

$$F'_c = (0.289)(1902) = 550 \text{ psi}$$

For a chord of two 2 × 6s,

$$f_c = 4780/(2 \times 8.25) = 290 \text{ psi} < 550 \text{ psi}$$

OK

(Even if only one 2 × 6 were required for column strength, a chord of two members would be used to provide stiffness at the door openings.)

Checking compressive stress in the chord at the bolt-hole location near the base,

$$f_c = 4780/(2 \times 6.84) = 349 \text{ psi}$$

$$F_c^* = 1.33)(1.1)(1300) = 1902 \text{ psi} > 349 \text{ psi}$$

OK

Checking bearing perpendicular to the grain on the bottom plate,

$$f_{c\perp} = 290 \text{ psi} < F_{c\perp} = 625 \text{ psi} \qquad OK$$

Use two 2 × 6 studs for all shearwall chords.

Example 12-8

Consider again the same building as that in Fig. 12-7 and Example 12-6, recalling that the lateral load at the eaves line is 195 lb/ft and at second-floor level is 270 lb/ft. The building has two end shearwalls and a central shearwall as shown in Fig. 12-10. Calculate the lateral loads, R_1 and R_2, transferred to the central and end shearwalls. If the interior shearwall has no openings, what unit shear must the wall sheathing resist? Show the shear diagram for the floor diaphragm and for the roof diaphragm.

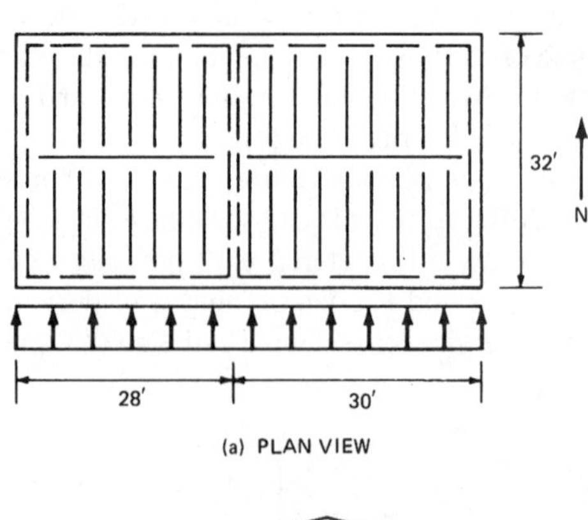

(a) PLAN VIEW

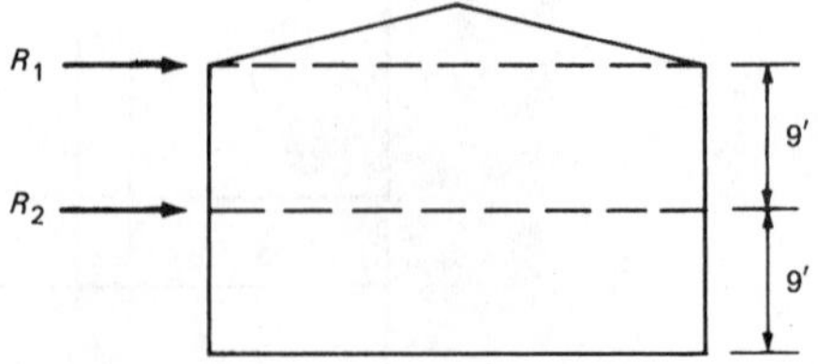

(b) ELEVATION OF ANY OF THE THREE TRANSVERSE WALLS

Fig. 12-10. Building for Example 12-8.

Calculation of lateral forces: Assuming two simple spans rather than continuous action, the tributary width to the central shearwall is 14 ft + 15 ft = 29 ft.

$$\text{Central } R_1 = (195 \text{ lb/ft})(29 \text{ ft}) = 5655 \text{ lb}$$

$$\text{Central } R_2 = (270 \text{ lb/ft})(29 \text{ ft}) = 7830 \text{ lb}$$

For the west wall, tributary width = 14 ft

$$\text{West } R_1 = 195(14) = 2730 \text{ lb}$$

$$\text{West } R_2 = 270(14) = 3780 \text{ lb}$$

For the east wall, tributary width = 15 ft

$$\text{East } R_1 = 195(15) = 2925 \text{ lb}$$

$$\text{East } R_2 = 270(15) = 4050 \text{ lb}$$

Unit shears in central wall: Central wall is 32 ft long. In upper story of central wall,

$$\text{Unit shear} = 5655/32 = 177 \text{ lb/ft}$$

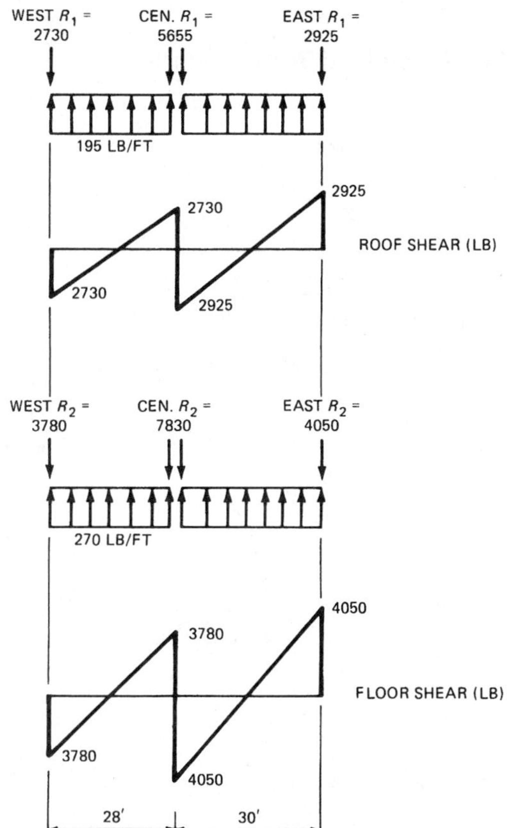

Fig. 12-11. Shear diagrams for horizontal diaphragms of Example 12-8.

In lower story of central wall.

$$\text{Unit shear} = (5655 + 7830)/32 = 421 \text{ lb/ft}$$

(Now that these unit shears are known, thickness of plywood wall sheathing and spacing of nails can be easily chosen.)

Shear diagrams for roof and floor diaphragms: The shear diagrams are shown in Fig. 12-11. Because the maximum shear in the floor diaphragm is 4050 lb, the maximum unit shear in the floor is 4050/32 = 127 lb/ft. Required plywood thickness can now be chosen for this unit shear.

12-5. ANCHORAGE OF SHEAR PANELS TO FOUNDATION

Shear panels must be anchored to the foundation so that they will not overturn because of the moment created by the lateral loads. (See also Section 4–5.) The tension chord force caused by this moment tends to lift the chord from the foundation. The typical connection used to resist this uplift is an anchor as shown in Fig. 12-12. Bolts (in single shear) connect the bracket to the chord, and an anchor bolt (in tension) connects the bracket to the concrete foundation.

The UBC specifies a factor of safety of 1.5 against overturning due to wind. This applies to the entire structure and also to any one of its individual shear panels (see Example 4–9 and 4–10). Tie-down anchors will not be necessary if the

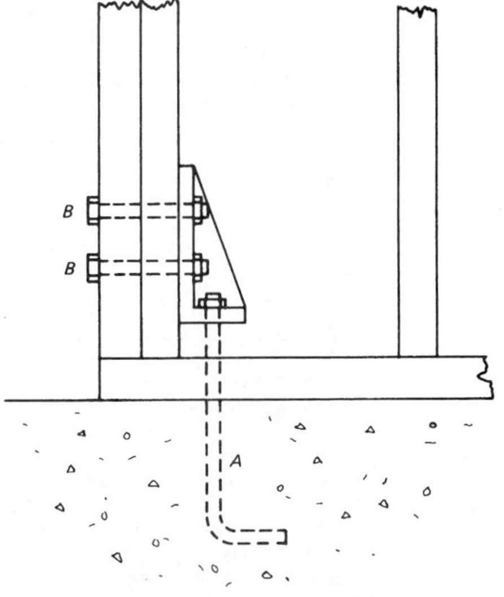

Fig. 12-12. Anchor bracket.

overturning moment is less than two-thirds the dead-load resisting moment. Anchorage for overturning will nearly always be more critical for the individual shear panel than for the entire wall.

ASD Design

Table 12-3 is used to find the allowable tension force on anchor bolts that are resisting uplift. The service (unfactored) tension load on the anchor cannot exceed the appropriate value from Table 12-3. Based on experiments with isolated single anchor bolts, the table cannot be used for multiple bolts in a closely spaced group. Oftentimes anchor brackets have allowable loads given by the manufacturer's catalog, assuming a minimum embedment and minimum bolt diameter.

Strength Design

The Strength Design method for concrete uses factored loads. This method is based on observed failure modes for anchor bolts in tension: yielding followed by fracture of the steel (long embedments); yielding and straightening of a hook; or pullout of a cone of concrete (shorter embedments). For the last cause, the failure surface is assumed to be a 45° cone (see Fig. 12-13) with uniform tensile stress distribution on the sides of the cone (uniform stress being used for simplicity). Current design procedures were developed for headed anchor studs that have a disk rather than a hook at the end. The 45° cone is assumed to branch out from the periphery of the disk. At present, the equivalent for a hooked bar is not known.

The area effective in resisting tension will be limited if the cone intersects a nearby vertical face of the concrete (too small a side-cover distance). If there are several anchor bolts in a close group, then overlapping stress cones will also limit the effective area.

When headed anchor bolts in tension are de-

Table 12-3. Allowable Service Load on Embedded Bolts (Pounds).[1,2,3]

Bolt Diameter (inches)	Minimum[4] Embedment (inches)	Edge Distance (inches)	Spacing (inches)	Minimum Concrete Strength (psi) × 0.00689 for MPa					
				$f'_c = 2{,}000$		$f'_c = 3{,}000$		$f'_c = 4{,}000$	
				Tension[5]	Shear[6]	Tension[5]	Shear[6]	Tension[5]	Shear[6]
× 25.4 for mm				× 4.5 for newtons					
1/4	2 1/2	1 1/2	3	200	500	200	500	200	500
3/8	3	2 1/4	4 1/2	500	1,100	500	1,100	500	1,100
1/2	4	3	6	950	1,250	950	1,250	950	1,250
	4	5	6	1,400	1,550	1,500	1,650	1,550	1,750
5/8	4 1/2	3 3/4	7 1/2	1,500	2,750	1,500	2,750	1,500	2,750
	4 1/2	6 1/4	7 1/2	2,050	2,900	2,200	3,000	2,400	3,050
3/4	5	4 1/2	9	2,250	2,940	2,250	3,560	2,250	3,560
	5	7 1/2	9	2,700	4,250	2,950	4,300	3,200	4,400
7/8	6	5 1/4	10 1/2	2,550	3,350	2,550	4,050	2,550	4,050
1	7	6	12	2,850	3,750	3,250	4,500	3,650	5,300
1 1/8	8	6 3/4	13 1/2	3,400	4,750	3,400	4,750	3,400	4,750
1 1/4	9	7 1/2	15	4,000	5,800	4,000	5,800	4,000	5,800

[1] Values are natural stone aggregate concrete and bolts of at least A 307 quality. Bolts shall have a standard head or an equal deformity in the embedded portion.
[2] The tabulated values are for anchors installed at the specified spacing and edge distances. Such spacing and edge distance may be reduced 50% with an equal reduction in value. Use linear interpolation for intermediate spacings and edge margins.
[3] The allowable values may be increased per Section 1603.5 for duration of loads such as wind or seismic forces.
[4] An additional 2 in. (51 mm) of embedment shall be provided for anchor bolts located in the top of columns located in Seismic Zones 2, 3, and 4.
[5] Values shown are for work without special inspection. Where special inspection is provided, values may be increased 100%.
[6] Values shown are for work with or without special inspection.
Source: Reproduced from the 1994 edition of the *Uniform Building Code*™, copyright © 1994, with the permission of the publisher, the International Conference of Building Officials.

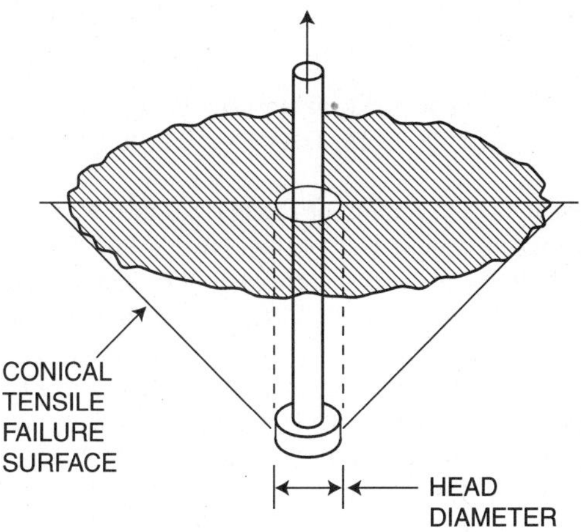

Fig. 12-13. Concrete failure cone around headed anchor bolt.

signed, both the limit state of concrete failure and the limit state of steel failure must be checked. If there is sufficient side cover and no overlapping stress cones, then the design tensile (pullout) strength of the *concrete* equals

$$\phi P_n = \phi_c P_c = \phi_c (2.8A_s)\sqrt{f'_c} \quad (12\text{-}4)$$

for normal weight concrete, where $\phi_c = 0.65$ and f'_c is the concrete compressive strength in psi. A_s is the sloping surface area of the 45° truncated cone shown in Fig. 12-13, which can be found from the formula for surface area of a full 45°-cone

$$S = \pi\sqrt{2}\, r^2 \quad (12\text{-}5)$$

The design tensile strength of the *steel* equals

$$\phi_s P_s = \phi_s A_b f'_s \quad (12\text{-}6)$$

where $\phi_s = 0.9$, A_b = area of anchor bolt, and f'_s = tensile strength (psi) of the bolt or stud. For A 307 bolts or A 108 studs, f'_s may be assumed to be 60,000 psi.

The designer may wish to make the embedded length sufficiently large so that a concrete failure is prevented altogether. To find the embedment length required to prevent concrete pullout, the upper-bound strength of the steel will be calculated without reduction by ϕ; that is,

$$P_s \text{ upper bound} = A_b f'_s \quad (12\text{-}7)$$

Then the tensile capacity of the concrete is ensured to be greater than or equal to the upper-bound tensile capacity of the steel, as follows:

$$\phi_c (2.8A_s)\sqrt{f'_c} \geq A_b f'_s$$

$$\phi_c (2.8)(\pi\sqrt{2})\,[(l_e + r_h)^2 - r_h^2]\sqrt{f'_c} \geq A_b f'_s \quad (12\text{-}8)$$

where r_h is the radius of the bolt or stud head. Equation 12-8 then can be solved for the embedment length, l_e, which ensures that a concrete failure will not occur.

All of the above concerns the *strength* of the anchorage. The strength required is equal to the calculated force under service loads multiplied by a load factor. A load factor of 1.3 to 1.7 is usually used. References 14 to 17 contain further information on design of anchor bolts for tension and shear.

Shear

In addition to transferring tension uplift to the foundation, anchor bolts must transfer the horizontal shear in the shearwall to the foundation. This is accomplished by anchor bolts (in addition to those provided for uplift) that are embedded in the concrete foundation at points along the base of the shearwall. This series of anchor bolts passes through holes in the horizontal lumber member (sill plate) at the base of the stud wall. The UBC specifies a maximum spacing of 6 ft and minimum diameter of ½ in. for these bolts. The strength of these bolts both in concrete and in wood parallel to the grain will need to be considered.

The same series of anchor bolts must also resist base reactions for perpendicular-to-wall wind load. In this case the sill bolts will bear on the wood perpendicular to the grain. Of course the bolts will be designed for whichever is the most severe condition—unit shear in the plane of the shearwall using parallel-to-grain bolt values, perpendicular-to-wall base shear using perpendicular-to-grain bolt values, or concrete failure.

For proper shear transfer, the plywood sheathing should extend down far enough that it can be securely nailed to the sill plate. The shear is transferred from wall sheathing to sill plate and then, by means of the anchor bolts, to the foundation. Experimental and analytical work (18) shows that increasing the number of nails through the plywood into the sill plate and into the adjacent header can greatly increase wall stiffness. The result is a reduction of deflection and an increase in resistance to uplift and shear loads.

Although it is common practice to choose corner bolts in a stud wall for tension alone and regularly spaced sill bolts for shear alone, there are situations when combined shear and tension on a bolt may need to be checked (at the base of columns in a post-and-beam building, for example). For ASD, the designer uses shear and tension values, V_t and P_t, from Table 12-3, and shear- and tension-applied service loads, V_s and P_s, in the following equation:

$$(P_s/P_t)^{5/3} + (V_s/V_t)^{5/3} < 1.0 \qquad (12\text{-}9)$$

The bolt is satisfactory for combined tension and shear if the left side of the expression is less than or equal to 1.0.

Example 12-9

Design tie-down anchors for the center shear panel of Example 12-7 and Fig. 12-9. Use allowable stress design and Table 12-3. $F'_c = 3000$ psi.

Design of embedded anchor bolt for uplift: Unit υ = 267 lb/ft (previously determined). Since the panel is 10 ft long, the total horizontal force at the top is

$$R = 267\ (10) = 2670 \text{ lb}$$

Total dead load equals roof dead load plus wall dead load, which equals

$$(104 \text{ lb/ft})\ (10 \text{ ft}) + (234)\ (10)$$
$$= 3380 \text{ lb (applied at panel centerline)}$$

Using a factor of safety of 1.5 against overturning,

$$1.5M = 1.5Rh = 1.5(2670)(13) = 52{,}100 \text{ ft-lb}$$

If the panel overturns, it will rotate on one lower corner. Dead load tends to resist this rotation with a moment equal to the dead load times the distance from the corner, or

Dead load resisting $M = (3380)\ (5) = 16{,}900$ ft-lb

Req M to be provided by anchors
$= 52{,}100 - 16{,}900 = 35{,}200$ ft-lb

Assuming a lever arm of slightly less than 10 ft (say 9.7 ft), the required anchor bolt force is 35,200/9.7 = 3630 lb. An embedded anchor bolt (like bolt *A* of Fig. 12-12) that has an allowable tension of 3630 lb must be provided for the connection of bracket and foundation. There will be one anchor bolt at the left chord of the panel and another 9.7 ft away at the right chord of the panel. From Table 12-3 (assuming special inspection—see Table 12-3 footnotes), *use one ⅝-in. bolt with at least 4.5 in. embedment for the bracket-to-foundation connection.*

Design of bolts connecting bracket and chord: For one ⅞-in. bolt (0.25 in.-thick bracket) in single shear due to wind in 3-in.-thick D. fir,

$$Z' = (1.33)(1670) = 2220 \text{ lb/bolt}$$

$$N \text{ req} = 3630/2220 = 2 \text{ bolts}$$

Use two ⅞-in. bolts (bolts B in Fig. 12-12) for bracket-to-stud (chord) connection.

To minimize chances for construction errors, the same bolts should be used at the ends of the other two shear panels even though the uplift force is not as great for these panels.

Example 12-10

Using ASD, design anchor bolts (sill bolts) to be distributed along the sill plate to prevent the wall of Fig. 12-9 from sliding.

These bolts must resist racking shear of 267 lb/ft (from Example 12-7) and (but not simultaneously) perpendicular-to-wall shear of 228(50/37.5) = 304 lb/ft (from Example 12-1). Both the strength of the wood and the strength of the concrete ($f'_c = 3000$ psi) must be checked. Try ⅝-in. bolts and D. fir sill (preservative treated).

Based on wood strength under racking shear of 267 lb/ft: Bolt load is parallel to the grain of the sill. For a 1.5-in. wood sill anchored to concrete, the NDS specifies that the concrete be treated as if it were a wood main member of $t_m = 2(1.5) = 3$ in. The single-shear parallel-to-grain value for D. fir with wind loading is

$$Z'_{\parallel} = 1.33(880) = 1170 \text{ lb}$$

Req pitch ≤ 1170 lb/267 lb per ft = 4.38 ft

Based on wood strength under perpendicular-to-wall shear of 304 lb/ft: Bolt load is perpendicular to the grain of the sill. Again use the tabulated value for a 3-in. main member thickness.

$$Z'_{s\perp} = 1.33\ (520) = 692 \text{ lb}$$

$$\text{Req pitch} \leq 692/304 = 2.28 \text{ ft}$$

Based on shear strength of the bolt in concrete: Anchor bolts will be at the center of the sill width (i.e., edge distance of 2.75 inches). Table 12-3 shows an allowable shear of 2750 lb if the edge distance is at least 3.75 in. This allowable is reduced (see footnotes to Table 12-3) to (2.75/3.75) (2750) = 2017 lb for our limited edge distance.

$$\text{Req pitch} \leq 2017/304 = 6.6 \text{ ft}$$

The perpendicular-to-grain strength of the wood controls and required pitch is 2.28 ft or less. With studs at 2 ft on center, a spacing of 2 ft is a practical solution.

Use 5/8-in.-diameter anchor bolts at 2 ft c/c along sill plate.

Example 12-11

A 3/4-in.-diameter headed anchor bolt (A 307) with a head of about 1 3/16 in. diameter is embedded 9 in. in concrete of compressive strength 3000 psi. Find the ultimate tensile load using Strength Design. From this value, find the approximate allowable service load if the average load factor is 1.6.

Tensile capacity as governed by concrete: The radius of the head = 0.594 in., so the projected circular area (shown shaded in Fig. 12-13) has radius 9.594 in. Surface area of the truncated cone

$$= \pi\sqrt{2}\,(9.594)^2 - \pi\sqrt{2}\,(0.594)^2 = 407 \text{ in.}^2$$

$$\phi_c P_c = 0.65\ (2.8A_s)\sqrt{f'_c}$$
$$= 0.65\ (2.8 \times 407)\sqrt{3000}$$
$$= 40{,}600 \text{ lb}$$

Tensile capacity as governed by steel:

$$\phi_s P_s = (0.9)(0.442)(60{,}000) = 23{,}900 \text{ lb}$$

The ultimate capacity, controlled by steel, is 23,900 lb. The allowable service load is 23,900/1.6 = 14,900 lb. This is much larger than any of the values on Table 12-3 for two main reasons. First, this anchor bolt has a much greater embedment than the minimum shown on the table. Second, the table values have very large factors of safety.

12-6. STRUT DESIGN

If a shearwall has no openings, the strut (top plate parallel to lateral load) can transfer the load directly to the shearwall. However, a strut above an opening ("drag" strut) must transfer the unit shear in the diaphragm near the opening to the shear panels on either side of the opening. The force in a strut will vary along the length.

Figure 12-14 shows a shearwall with a single opening. The unit shear in the roof is

$$v_R = R/(b_1 + b_2 + b_3)$$

where R is the wind load reaction of the horizontal diaphragm. The unit shear in the shearwall panels is

$$v_S = R/(b_1 + b_3)$$

Figure 12-14b shows a portion of the building isolated as a free body. For the sum of the hori-

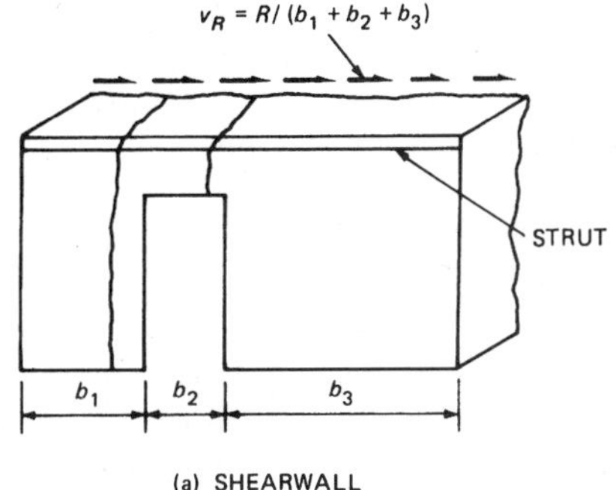

(a) SHEARWALL

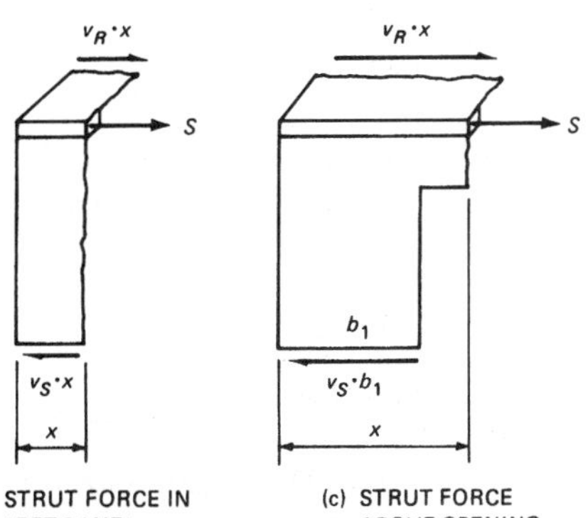

(b) STRUT FORCE IN LEFT PANEL

(c) STRUT FORCE ABOVE OPENING

Fig. 12-14. Strut force, S.

zontal forces to equal zero, a strut force, S, is needed, because unit shears v_R and v_S are not of the same magnitude. Each force shown on the free body is a function of distance x.

Figure 12-14c shows a similar free body, but cut from the building at a location through the opening. In this case, the force at the bottom of the shearwall depends on dimension b_1, but not x. Force S can be determined from these free bodies, and will vary along the length of the strut.

Splices in struts will have connections that are designed for either the strut force or the chord force (when the member serves as chord), whichever is larger.

Struts may have the combined stresses of bending due to vertical loads and tension or compression due to lateral loads. This is particularly true of a drag strut (over an opening) which may be supporting roof or floor load. Sometimes the drag strut is not continuous with the rest of the strut (top plate) because it has been made out of a larger section (or a glulam) in order to support both gravity and lateral loads. In this case, the drag force at the edge of the opening becomes important, because connections have to be designed to transfer this force from the drag strut to the other parts of the chord of the horizontal diaphragm.

Example 12-12

Recalling that $R = 10{,}000$ lb when wind is from the north, find the strut force at all points along the strut for the shearwall in Fig. 12-9. What is the maximum force in this same top plate when wind is from the west and the top plate serves as a chord?

$$v_R = 10{,}000/50 = 200 \text{ lb/ft}$$

$$v_S = 10{,}000/[13.5 + 10 + 14] = 267 \text{ lb/ft}$$

For distance x varying from zero to 13.5 ft, summing forces in the horizontal direction for a free body similar to Fig. 12-14b gives

$$S + 200x - 267x = 0$$

$$S = 67x$$

So S will vary from 0 to 904.5 lb (tension) in the first 13.5 ft of strut. Now considering the drag strut over the first opening (for $13.5 < x < 16$), summing horizontal forces gives

$$S + 200x - 267(13.5) = 0$$

$$S = 3604.5 - 200x$$

So, at $x = 16$ ft,

$$S = 404.5 \text{ lb (tension)}$$

Making similar calculations for the other segments of the strut, the variation of strut force will be that of Fig. 12-15. The sign of the forces would be reversed if the wind reversed direction. The maximum force at any point in the strut is 1075 lb. The whole strut can be conservatively designed for the larger of this 1075-lb force or the maximum force obtained when it is serving as a chord (wind perpendicular to this wall).

When serving as a chord,

$$M = w\ell^2/8 = 228(50^2)/8 = 71{,}250 \text{ ft-lb}$$

Assuming the chords to be 88 ft $-$ 6/12 $=$ 87.5 ft apart,

$$T = C = 71{,}250/87.5 = 814 \text{ lb} < 1075 \text{ lb}$$

Maximum force in the west wall top plate is due to strut force (under north wind) rather than chord force (under west wind). The procedure for choosing a top plate member for this 1075-lb force is identical to choosing the top plate member in Example 12-1 (assuming that the vertical load from the roof over the opening is carried by a separate header). The procedure for designing the connection at splices in the strut is identical to that shown in Example 12-2.

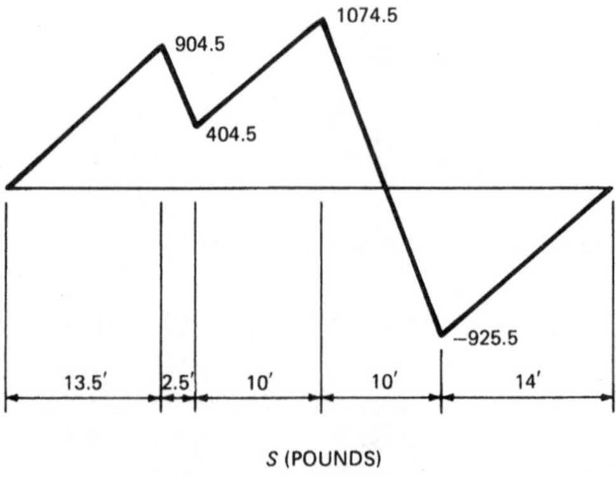

Fig. 12-15. Variation of strut force.

12-7. WOOD DIAPHRAGM WITH MASONRY WALLS

When wood floor and roof diaphragms are combined with masonry or concrete walls, additional design constraints exist. The UBC (reference 5) requires that concrete or masonry walls be designed to resist bending between diaphragm anchors where anchor spacing is greater than 4 ft. Therefore, anchors (such as steel straps) which transfer perpendicular-to-wall forces from the masonry (or concrete) wall into the body of the wood diaphragm are often spaced at 4 ft c/c. These anchors will extend from the wall some distance inward, where the force will be transferred into larger framing members of the diaphragm. The anchors must be designed for a minimum seismic force of 200 lb/ft.

The UBC requires continuous ties between opposite diaphragm chords to distribute the anchorage forces into the diaphragm. If the framing members of the horizontal diaphragm were continuous in the two perpendicular directions, the requirement would be satisfied. However, even if framing is continuous in one direction, it is highly unlikely that it will be in the other. An expensive way of providing continuity would be to splice the purlins at crossings of the girders.

Another method to provide the necessary continuity, called the subdiaphragm method (references 2 and 19), has been developed to avoid the tie splice at every beam crossing. Certain portions of the total diaphragm are chosen as *subdiaphragms* and are designed as independent diaphragms. The normal-to-wall loads are transferred from wall to subdiaphragm and then to a member that is fully continuous across the building (such as a girder).

Another requirement of the UBC is that (in seismic zones 2, 3, or 4) wood diaphragms cannot be anchored to concrete or masonry walls by using toe nails or nails in withdrawal.

A ledger (steel angle or lumber member) can be bolted with embedded anchor bolts to the concrete or masonry wall. The ledger can provide an attachment for wood roof trusses, floor joists, or other framing members to transfer diaphragm forces from the walls. Figure 12-16 shows a ledger-wall joint that has been carefully detailed to avoid cross-grain tension.

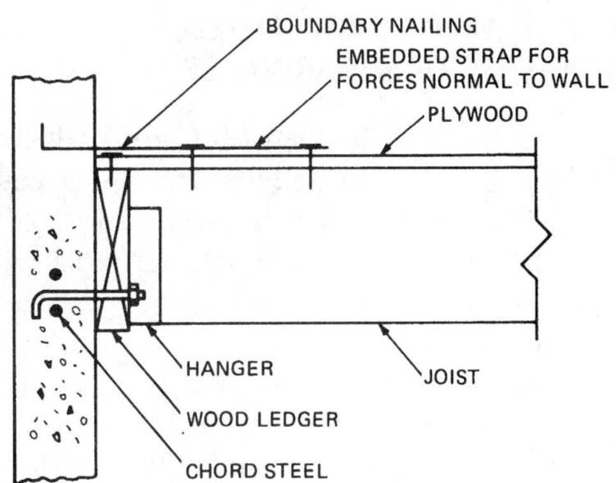

Fig. 12-16. Concrete wall–diaphragm joint.

Wood can resist very little tensile stress perpendicular to the grain. The embedded strap in Fig. 12-16 prevents this tensile bending stress from occurring. Figure 12-17 shows how the ledger would be bent across the grain if the strap were not present. This failure mode was brought to light by cross-grain tension failures of ledgers during the San Fernando earthquake (20).

Rather than the wood ledger, continuous horizontal reinforcing bars in the concrete wall will serve as the tension chord of the diaphragm. Once the flange force, T (Eq. 12-2), has been determined and multiplied by a load factor of 1.3, the required area of reinforcing steel, A_s, can be found by

$$\text{Req } A_s = 1.3\,T/(\phi f_y) \qquad (12\text{-}8)$$

where $\phi = 0.9$ for tension and f_y is the yield stress of reinforcing steel.

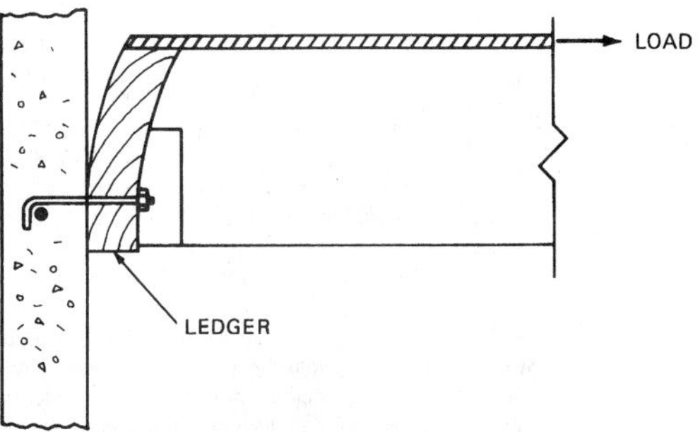

Fig. 12-17. Cross-grain tension in ledger.

12-8. PLYWOOD-SHEATHED, METAL-STUD SHEARWALLS

Since the early 1970s, the use of steel studs for wall framing has increased for both commercial and residential construction. Steel studs, cold-formed into C-shaped sections, are replacing wood studs in walls sheathed with plywood or other materials.

The structural response of both types of stud walls under combined vertical and lateral loads was tested at Oregon State University (21). Douglas fir studs of Stud grade and 20-gauge steel studs were compared, because they are used under similar conditions. The steel stud wall had its 3/8-in. plywood siding attached with screws, whereas the plywood for the wood stud wall was connected with nails. The ultimate load causing bending was 12% greater for the steel stud wall. However, all steel walls tested failed suddenly by buckling. Wood walls failed gradually and sustained additional load after the first stud failed.

Decreasing slip by applying a rigid adhesive between plywood coverings and studs increased ultimate load considerably, especially for the wood walls. The ultimate lateral (pressure) load increased 2.7 times for the wood stud wall and 1.8 times for the steel stud wall.

Steel stud walls conforming to certain specifications may be used for shearwalls as described in ICBO Report No. 2392P (22). Table 12-4 shows allowable unit shears for steel stud walls sheathed with plywood. The table is for walls with 25-gauge or thicker steel studs at 16 in. c/c maximum. Reference 22 also gives allowable loads for bearing walls and for shear walls of steel studs sheathed with gypsum wallboard.

Table 12-4. Allowable Unit Racking Shear for Wind or Seismic Forces for Steel-Stud/Plywood Vertical Diaphragms.

Wall Construction	Screw Spacing (in.)	Shear Value (lb/ft)
One side: 4-ft-wide plywood dimension vertical; joints on studs; blocking provided at horizontal joints; studs at 16 in. c/c max.	6	200[a]
	6	115[b]
	4	300[a]
	4	175[b]
One side: 8-ft-long plywood dimension vertical; joints on studs; studs at 16 in. c/c max.	6	170[a]
	6	133[b]
	4	250[a]
	4	200[b]

[a] Values when $\frac{1}{2}$-in. gypsum wallboard (screwed at 12 in. c/c) is applied to opposite side of the wall.
[b] Where plywood is applied to both sides of wall, these values may be doubled.
Note: Minimum stud to be used: 25-gauge. Minimum plywood to be used: $\frac{3}{8}$-in. Structural I. Screws shall be 1 in. long, coated, double-lead thread drywall screws, spaced as indicated with $\frac{3}{8}$-in. edge distance, in all studs at joints, and panel blocking (where blocking is indicated), bottom track and top track or wood top plate. Screw spacing to all other studs not at panel edges shall be 12 in. c/c. Values above are as reported in reference 22.

12-9. TORSION IN DIAPHRAGMS

The design examples for the previous building dealt with nearly symmetrically placed shear walls. This is the ideal situation; otherwise, torsion of the entire structure will induce increased shearing forces on certain shearwalls. Torsion is caused when the location of the resultant lateral force above any level does not coincide with the center of rigidity of the shearwalls at that level.

If there is a lack of symmetry, then torsion must be considered whether due to wind or earthquake. For seismic design, an accidental torsion due to eccentricity equal to 5% of the building dimension should be considered (reference 5). For wind design, accidental torsion need not be considered.

12-10. OTHER LOADS ON WALL SHEATHING

Discussion of wall sheathing in earlier parts of this chapter have centered on the sheathing's ability to resist unit shear due to lateral load. In addition, wall-sheathing (1) transmits wind loads to wall studs and (2) keeps studs from buckling in their weak direction. Choosing wall plywood or other sheathing for these latter structural functions is based mainly on accepted practice rather than hard design criteria. The recommendations given in the UBC for stud and nail spacing for wall sheathing are usually sufficient for residential construction. The most critical structural function of the sheathing is diaphragm (shearwall) action, so choice will be based on shearwall performance.

REFERENCES

1. Peterson, J., "Bibliography on Lumber and Wood Panel Diaphragms," *Journal of Structural Engineering, ASCE,* 109(12), Dec. 1983.
2. Applied Technology Council, *Guidelines for the Design of Horizontal Wood Diaphragms,* ATC, Berkeley, CA, 1981.
3. *National Design Specification for Wood Construction,* Rev. 1991 ed., American Forest and Paper Association, Washington, DC, 1993.
4. *Western Woods Use Book,* 2nd ed., Western Wood Products Association, Portland, OR, 1979.
5. *Uniform Building Code,* International Conference of Building Officials, Whittier, CA, 1994.
6. Tarpy, T. S., D. J. Thomas, and L. A. Soltis, "Continuous Timber Diaphragms," *Journal of Structural Engineering, ASCE,* 111(5), May 1985, pp. 992–1002.
7. Tissell, J. R., and J. R. Elliott, Research Report 138, APA—The Engineered Wood Association, Tacoma, WA, 1983.
8. *Plywood Design Specification,* APA—The Engineered Wood Association, Tacoma, WA, 1986.
9. Diaphragms, Design/Construction Guide, APA—The Engineered Wood Association, Tacoma, WA, 1995.
10. Bower, W. H., "Lateral Analysis of Plywood Diaphragms," *ASCE, Journal of the Structural Division,* 100(ST4), Apr. 1974, pp. 759–772.
11. Patton-Mallory, M., R. W. Wolfe, L. A. Soltis and R. M. Gutkowski, "Light-Frame Shear Wall Length and Opening Effects," *Journal of Structural Engineering, ASCE,* 111(10), Oct. 1985, pp. 2227–2239.
12. McCutcheon, W. J., "Racking Deformations in Wood Shear Walls," *Journal of Structural Engineering, ASCE,* 111(2), Feb. 1985, pp. 257–269.
13. Gupta, A. K., and G. P. Kuo, "Behavior of Wood-Framed Shear Walls," *Journal of Structural Engineering, ASCE,* 111(8), Aug. 1985, pp. 1722–1733.
14. Marsh, M. L., and E. G. Burdette, "Anchorage of Steel Building Components to Concrete," *Engineering Journal, AISC,* 22(1), 1st Quarter 1985.
15. Shipp, J. G., and E. R. Haninger, "Design of Headed Anchor Bolts," *Engineering Journal, AISC,* 20(2), 2nd Quarter, 1983.
16. ACI Committee 349, *Code Requirements for Nuclear Safety Related Concrete Structures* (ACI 349–80). Appendix B—Steel Embedments, American Concrete Institute, Farmington Hills MI, 1980.
17. *PCI Design Handbook,* (4th ed.), Prestressed Concrete Institute, Chicago, IL, 1992.
18. Polensek, A., and B. Schimel, "Rotational Restraint of Wood-Stud Wall Supports," *Journal of Structural Engineering, ASCE,* 112(6), June 1986, pp. 1247–1262.
19. Breyer, Donald E., *Design of Wood Structures,* McGraw-Hill, New York, 1993.
20. Diekmann, E. F., "Design Details for the Transfer of Forces in Wood Diaphragms to Vertical Elements," *Proceedings of Workshop on Design of Horizontal Wood Diaphragms,* Applied Technology Council, Berkeley, CA, 1980.
21. Polensek, A., "Strength and Stiffness of Walls with Wood and Steel Studs," *Forest Products Journal,* 27(2), 1977, pp. 45–53.
22. Report No. 2392P, International Conference of Building Officials, Whittier, CA, June, 1982.

PROBLEMS

12-1. A blocked horizontal diaphragm consists of $^{15}/_{32}$-in. Structural I plywood nailed to 2-in.-nominal D. fir-larch framing using 10*d* nails at 4 in. c/c at diaphragm boundaries. (a) Find allowable shear in the diaphragm from Table 12-1 (b) Repeat using Derivation of Table Values (Section 12-2). (Answers will be approximately equal, but not identical.)

12-2. A blocked horizontal diaphragm consists of $^{19}/_{32}$-in. C-D plywood (Species Group 1, Stress Level S-3) nailed to 3-in.-nominal D. fir framing members with 10*d* common nails. There are two lines of nails, each with spacing of 4 in. at boundaries and 3 in. at other panel edges. Calculate the allowable unit shear, without using a table, by using Derivation of Table Values (Section 12-2).

12-3. The framing members of the structure considered in Example 12-8 are 2-in. nominal D. fir. The floor diaphragm plywood panels are nailed at 6-in. on center at all panel edges. If increased wind pressures double the values shown in the shear diagrams of Figure 12-11, what thickness of Structural I plywood is required for the floor? What common nail size should be used? If C-C or C-D plywood is used, what thickness and nail size are required? Solve using a table.

12-4. Consider the central shearwall of Example 12-8 where the unit shear in the lower story of the central wall was found to be 421 lb/ft. Plywood is $^{3}/_{8}$-in. Structural I (face grain parallel to studs) and studs (D. fir) are at 24 in. c/c. Using a table, describe the required nail size and nail spacing at panel edges and at the intermediate studs. What load will the shearwall carry with this nail spacing? Will there be any changes in required nail spacing in the upper story of this wall?

12-5. For the east shearwall of Example 12-8, what is the maximum unit shear (lower story)? If the wall studs (D. fir) are at 24-in. c/c, will $^{3}/_{8}$-in. Structural I plywood be satisfactory? What nail size is required? What is the required nail spacing at panel edges and at the intermediate studs? Solve using a table.

12-6. For the north wall of Example 12-1, recall that there is a perpendicular wind load of 228 lb/ft at the eaves Framing is D. fir. For the blocking (which can be seen to be at 4 ft c/c) perpendicular to this wall, how many 16*d* common nails per blocking member (at location *D* of Fig. 12-4a) will be needed to transfer the eaves load out of the block and into the 15/32-in. plywood? Show proof for the value of C_d. Don't forget to apply the adjustment C_{di}. An approximate plywood thickness of 0.5 in. may be used for finding *Z*.

12-7. Using No. 2 D. fir-larch, what is the minimum size of top plate members for the shearwall of Example 12-12? So that one piece will always be continuous at splices, use two members for the top plate Assume holes for 5/8-in. diameter bolts at splices.

12-8. The shearwall of Example 12-12 has 2-in.-nominal top plate members of No. 2 D. fir-larch. If the top plate is to be spliced using 16*d* common nails, how many are required per side of each splice?

12-9. The warehouse of Fig. 12-18, constructed with No. 1 D. fir-larch wall and roof framing, is to be designed for a 70-mph wind load (Exposure B). The roof diaphragm and chords are at the 23-ft level and there is a 2-ft parapet. (a) Using UBC Method 2, find the maximum unit shear in the roof diaphragm (north wind). (b) Assuming 2-in.-nominal framing at 24 in. c/c E-W for the roof, what is the layout (case no.) and what thickness of Structural I plywood will be required if the roof diaphragm is to be *unblocked?* (c) What are the nail size and spacing requirements for this diaphragm?

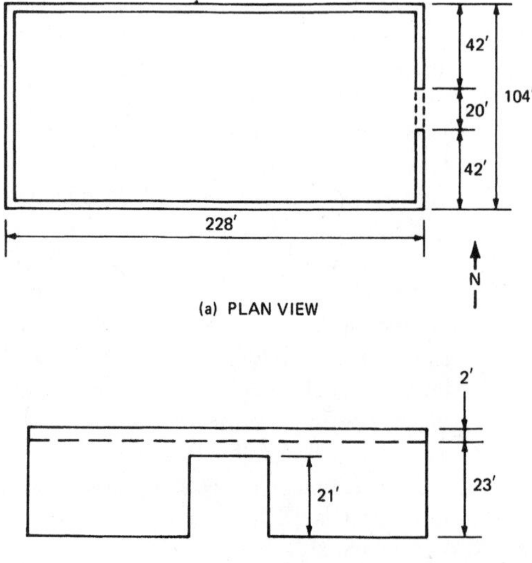

Fig. 12-18. Warehouse for Problems 12-9 to 12-13.

12-10. Assuming that the lateral north wind load to the roof diaphragm of Problem 12-9 is 260 lb/ft, select a top plate (chord) size based on maximum chord force and design a bolted connection (using 3/4-in.-diameter bolts) at splices. Should a pair of chord members be required, use three members so that two will remain effective where the third one is spliced.

12-11. Assuming that the lateral north wind load to the roof diaphragm of Problem 12-9 is 260 lb/ft, (a) determine the unit shear in the shear panels of the east wall; (b) using a table, select plywood (Structural I) and nail spacing for the wall; (c) sketch a diagram of the variation of strut force in the east wall top plate; (d) find required top-plate size based on maximum strut force.

12-12. Assuming that the lateral north wind load to the roof diaphragm of Problem 12-9 is 260 lb/ft, what is the required anchor bolt force at each end of each shear panel of the east wall? Use a factor of safety of 1.5 against overturning. Joists run E-W (Problem 12-9) and girders run N-S at 18 ft c/c. Make reasonable estimates of dead load from Appendix Table A-1. For the wall dead load include weight of 2 × 6 studs, plywood sheathing, insulation, gypsum drywall, and finish siding. For the roof dead load include weight of joists, insulation, plywood sheathing, and 5-ply felt and gravel roofing.

12-13. The warehouse of Fig. 12-18 is constructed with No. 1 D. fir-larch wall and roof framing. The roof diaphragm and chords are at the 23-ft level and there is a 2-ft parapet. Code-specified wind pressures are: for heights between zero and 15 ft—19 psf, from 15 ft to 20 ft—22 psf, and from 20 ft to 25 ft—24 psf. (a) What is the maximum unit shear in the roof diaphragm (north wind)? Assume that all lateral load above roof level goes to the roof diaphragm and that all lateral load below roof level would be distributed to the foundation and to the roof diaphragm the same as for simple beam reactions. (b) Assuming 2-in.-nominal framing at 24 in. c/c E-W for the roof, what is the layout (case no.)? (c) If the roof diaphragm is to be *blocked,* what nail spacing and size would be required if 5/16-in. Structural I plywood were used? If 3/8-in. Structural I were used?

12-14. Example 12-9 said that uplift force for chords of the wider panels is not as great as uplift for chords of the narrower panel. Prove it.

12-15. Review Example 4-10, which showed the method for calculating the necessary anchor bolt force to give a 1.5 factor of safety against overturning. The solution was based on a wall (with only small openings) acting as a single panel. For a similar single-panel wall the vertical force required in the anchor bolts at each of the two corners is 8.1 kips. For 7/8-in.-diameter bolts (A 307) with 6-in. embedment, $f'_c = 3000$ psi, and special inspection, how many anchor bolts will be needed at each corner of the wall? Use the ASD (Table) method.

12-16. Same as Problem 12-15, except use the Strength Design method. Use an average load factor of 1.6 and assume bolt-head diameter is 1.31 in.

12-17. Problem 12-15 involves a wall and the design of corner bolts for uplift. For this same wall, choose sill bolts for horizontal shear. Racking shear is 1.29 k/ft; perpendicular-to-wall shear is 0.51 k/ft. The sill plate is 2 × 6 Douglas fir. The studs are at 16 in. c/c, so a sill-bolt spacing of a multiple of 16 in. is desirable. Use 7/8 in. diameter sill bolts for the first trial.

12-18. Table 12-3 shows that the UBC requirement for minimum embedment of a 7/8-in.-diameter anchor bolt is 6 in. What additional embedment length is required to ensure that failure will be in the bolt rather than the concrete? Bolt head diameter is about 1.41 in.

12-19. Using Table 12-3, find the allowable tension service load (special inspection) on a 5/8-in.-diameter anchor bolt (1.0-in.-diameter head) with 5-in. embedment, and concrete $f'_c =$ 3,000. Next, rework using Strength Design and an average load factor of 1.7. Compare the solutions and give possible reasons for the way they compare.

12-20. A 5/8-in.-diameter anchor bolt (1.0-in.-diameter head) embedded in concrete with compressive strength of 3000 psi is subjected to a shear load of 1700 lb and a tension load of 1450 lb. Special inspection will not be employed. Will the bolt be satisfactory for combined tension and shear?

12-21. In a two-story building, the second-story shearwall chord is not continuous with the first-story chord. A threaded rod has been chosen for the continuity tie. The second-story (vertical) chord force of 6100 lb must be transferred to the first-story chord. Check the adequacy of the detail shown in Fig. 12-19 for maintaining continuity. What diameter threaded A36 rod is needed? (Use an allowable tensile stress for A36 steel of 1.33 × 19 = 25.3 ksi, based on gross area of the rod.) Check the tension capacity of the wood chord and the capacity of the pair of 1-in.-diameter bolts. If either is insufficient, suggest alternatives.

12-22. The chord of a roof diaphragm is to be spliced using sheet metal straps. Design steel straps (one on each of the two wide faces) to transfer an 1800 lb tension load across a butt splice of a No. 1 D. fir 2 × 4. Use 16 gauge (t = 0.0598 in.) and an allowable tensile stress of 18 ksi. Assume nails are adequate.

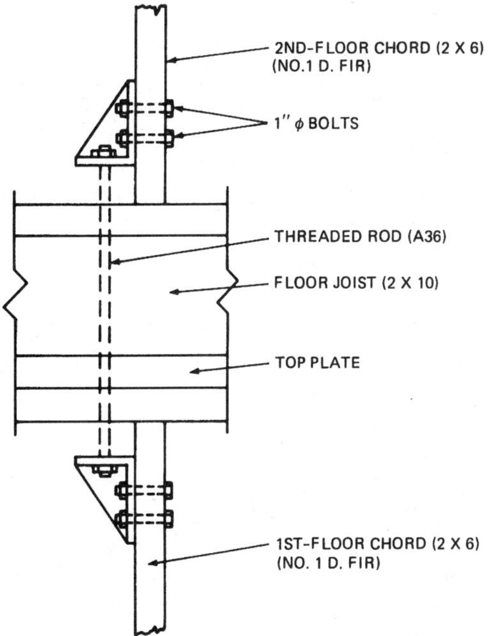

Fig. 12-19. Continuity tie for Problem 12-21.

13

Built-Up and Composite Members

To make efficient use of materials, built-up and composite members of wood are common. Plywood and lumber can be used in an all-wood assembly to form a plywood-lumber box beam or a stressed-skin panel. Wood can also be used in combination with other materials, such as steel or honeycombed paper, to form structural components.

Many of these built-up or composite members, such as joists with I-shaped cross sections, are proprietary in nature. These members are frequently used in larger residences, apartments, and commercial buildings.

13-1. PLYWOOD-LUMBER BEAMS

Plywood-lumber beams have plywood webs glued to dimension lumber flanges. As shown in Fig. 13-1, these beams may be box- or I-shaped. The cross section may be varied along the length of the beam by adding extra webs in areas of high shear near the supports. In a similar manner to a steel wide flange, the flanges of the plywood-lumber beam resist most of the bending load, while the web resists the shear.

Allowable Stresses for Design

Because each flange has nearly uniform compression or nearly uniform tension, the allowable stress in bending for the flange lumber will be the smaller of the table values for F_c or F_t, rather than F_b. *The Plywood Design Specification* (1) allows for an increase in the allowable horizontal (shear-through-the-thickness) shear stress if the plywood is rigidly glued to continuous framing members. If the framing is glued to only two panel edges parallel to the face grain, the increase is 19%. For gluing to framing around all panel edges, the increase is 33%. Because of stress concentrations, the allowable rolling-shear stress at the flange-web joint is reduced by 50% (2).

The moment of inertia used to check bending stress is the net *I*, consisting of (in a two-web beam) contributions from the flanges and one web. Because of staggered web splices, only one of the two webs is effective in bending at any one section.

Deflection

Deflection of a plywood-lumber beam can be approximated by multiplying the bending deflection by a factor to account for shear deflection. As a function of span/depth ratio, the factors (2) are

Span/Depth	Factor
10	1.5
15	1.2
20	1.0

Intermediate factors may be obtained by interpolation. Use the elastic modulus of the flanges in the calculations. Deflection limits for floor beams are $\ell/360$ for live load and $\ell/240$ for total load. For roof beams the limits are $\ell/240$ for live load and $\ell/180$ for total load (2).

Stiffeners

Bearing stiffeners are required at the beam ends and also at points of other concentrated loads or reactions. The cross-sectional dimension of the bearing stiffener parallel to the span will be de-

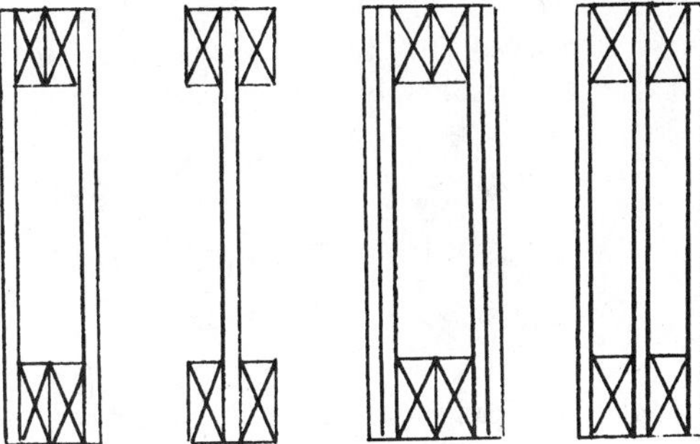

Fig. 13-1. Plywood-lumber beam sections.

termined based on the rolling-shear strength of the plywood web and the bearing strength perpendicular to the grain for the flange that the stiffener bears on. Intermediate stiffeners of nominal 2-in. lumber are spaced at a maximum of 48-in. c/c to provide stability.

Example 13-1

Design plywood-lumber box beams to span between two masonry walls. The span is 30 ft and beams will be at 8 ft c/c. Use hem-fir Select Structural flanges and Structural I Interior plywood webs. The beams support a roof system (mild slope) having the following loads:

Suspended ceiling	8 psf
Roofing	7 psf
Insulation	2 psf
Decking and beams	9 psf
Snow	30 psf
	56 psf

Total w = 56 (8 ft) = 448 lb/ft; maximum M = (448)(30²)/8 = 50,400 ft-lb. Choose a trial section:

Use depth approx = span/10 = 30(12)/10 = 36 in.

Base F_t for flanges is 900 psi

Assuming 2 × 8 flanges, the allowable bending stress is

$$F'_b = C_D C_F F_t = (1.15)(1.2)(900)$$
$$= 1240 \text{ psi}$$

For 2 × 8s and total depth of 36 in., the moment arm between the centroid of the flanges is 36 − 7.25 = 28.75 in. Equating internal and external moments,

$$1240\, A_f(28.75) = 50{,}400(12)$$

$$\text{Req } A_f = 17.0 \text{ in.}^2$$

Area provided by two 2 × 8s = 2(10.875) = 21.8 in.²

Try two 2 × 8s for each flange.

Choose plywood based on the thickness effective for shear, t_s. The following empirical equation can be used to estimate the required thickness effective for shear.

$$\text{Req } t_s = 5V/(4hF'_v)$$

where h is overall depth. For species Group 1 and Grade Stress Level S-2,

$$F'_v = 1.15(190) = 218 \text{ psi}$$

$$V = 448(30)/2 = 6720 \text{ lb}$$

$$\text{Req } t_s = 5(6720)/[4(36)(218)] = 1.07 \text{ in.}$$

t_s provided by two ¾-in. panels = 2(0.739) = 1.5 in.

Try ¾-in. Structural I plywood for each outside web. Figure 13-2 shows the trial section. The flanges have been surfaced so that a pair of 2 × 8s is 7 in. deep by

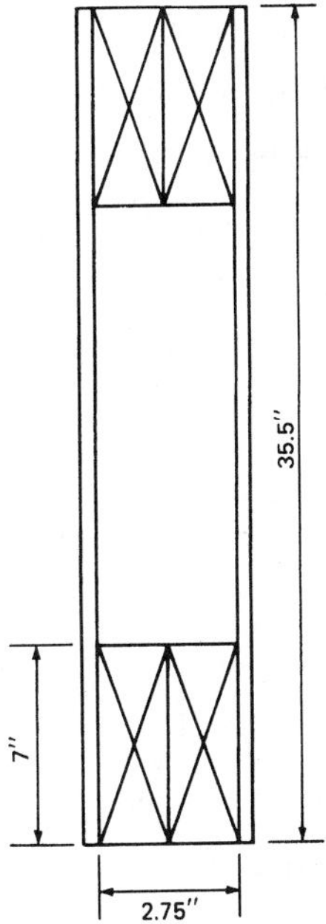

Fig. 13-2. Plywood-lumber beam for Example 13-1.

2.75 in. wide. Surfacing has decreased the net overall depth to 35.5 in.

Bending check:

$$I_{flanges} = 2[(1/12)(2.75)(7)^3 + 2.75(7)(14.25^2)]$$
$$= 7975 \text{ in.}^4$$

Specifying that the plywood face grain is horizontal, the thickness of plywood effective in resisting bending is $A_{\parallel}/12 = 4.219/12 = 0.352$ in.

$$I_{\text{one web}} = (1/12)(0.352)(35.5)^3 = 1312 \text{ in.}^4$$
$$I_{\text{total}} = 7975 + 2(1312) = 10{,}600 \text{ in.}^4$$
$$I_{\text{net}} = 7975 + 1312 = 9290 \text{ in.}^4$$

For the plywood, allowable bending stress, F'_c is 1.15(1540) = 1770 psi. (The tabular value F_b is not used, because it is for flatwise bending only.)

The allowable bending stress for the flange lumber is 1240 psi. (See above.) The 1240-psi value controls.

$$I \text{ req} = Mc/F_c{}' = 50{,}400(12)(17.75)/1240$$
$$= 8657 \text{ in.}^4$$
$$I \text{ provided} = I_{\text{net}} = 9290 \text{ in.}^4 \qquad OK$$

Shear check:

$$Q_{flanges} = bd(h/2 - d/2)$$
$$= (2.75)(7)(35.5/2 - 7/2) = 274.3 \text{ in.}^3$$
$$Q_{\text{webs}} = 2t(h/2)(h/4)$$
$$= 2(4.219/12)(17.75)(8.875)$$
$$= 110.8 \text{ in.}^3$$
$$Q_{\text{total}} = 274.3 + 110.8 = 385.1 \text{ in.}^3$$

Horizontal shear: For continuous glue along framing parallel to face grain, allowable horizontal shear stress may be increased 19%. The allowable shear force is

$$V_h = F'_v I_t t_s / Q$$
$$= (1.19 \times 1.15 \times 190)(10{,}600)$$
$$\times (2 \times 0.739)/385.1$$
$$= 10{,}580 \text{ lb}$$

Actual $V = 448(15) = 6720$ lb *OK*

Flange-web (rolling) shear: The basic rolling-shear stress is reduced 50% for flange-web shear (sect. 3.8.2, ref 1). The allowable shear force is

$$V_s = 2F'_s \, dI_t/Q_{\text{flanges}}$$
$$= 2(0.50 \times 1.15 \times 75)(7)(10{,}600)/274.3$$
$$= 23{,}300 \text{ lb} > 6720 \text{ lb} \qquad OK$$

Deflection: Span/depth ratio = 10:1, so factor = 1.5

$$\text{Total-load } \Delta = 1.5(5)\, w\ell^4/(384\, EI_t)$$
$$= \frac{1.5(5)(448)(30^4)(1728)}{384(1{,}600{,}000)(10{,}600)}$$
$$= 0.72 \text{ in.}$$
$$\text{Total-load } \Delta \text{ limit} = \ell/180 = 30(12)/180$$
$$= 2.0 \text{ in.} \qquad OK$$
$$\text{Live-load } \Delta = (30/56)(0.72) = 0.39 \text{ in.}$$
$$< \ell/240 = 1.5 \text{ in. } OK$$

Bearing stiffeners: Stiffener width required for bearing of stiffener perpendicular to grain of flange:

$$\text{End reaction} = R = 448(15) = 6720 \text{ lb}$$
$$F_{c\perp} \text{ (of flange lumber)} = 405 \text{ psi}$$
$$\text{Req bearing area} = R/F'_{c\perp} = 6720/405$$
$$= 16.6 \text{ in.}^2$$

Using a pair of 2× stiffeners (surfaced to 1³/₈ in. each), the required stiffener width is

$$16.6/(2 \times 1.375) = 6.04 \text{ in.}^2$$

Double 2 × 8s are OK for bearing perpendicular to grain.

Stiffener width required for rolling shear at each of the two stiffener-web interfaces:

$$\text{Req surface area} = R/F'_c = 6720/(75 \times 1.15 \times 0.50)$$
$$= 156 \text{ in.}^2$$
$$\text{Req stiffener width} = 156/(2 \times 35.5)$$
$$= 2.20 \text{ in.}$$

Use double 2 × 8s at each end for bearing stiffeners.

13-2. STRESSED-SKIN PANELS

Stressed-skin panels are wood assemblies consisting of a plywood skin securely glued to 2 × stringers (webs) so that the unit acts compositely. With composite action, no slippage occurs between the skin and stringers. Horizontal shears are developed along the glue line as the parts are constrained to deflect together. Stressed-skin panels are structurally efficient, resulting in less overall weight for the roof or floor system.

Typical panel sections are shown in Fig. 13-3. Panels may have two skins, a single skin, or a single skin with 1- or 2-in. lumber for the opposite flange (T-flange panel). One-skin and T-flange panels are useful where access to space for piping and ductwork is necessary. Skins are usually a minimum of $^5/_{16}$ in. thick. Thinner plywood may have adequate strength, but may also have an undesirable bowed appearance (3).

In addition to skins and stringers, panels have headers at their ends and usually have blocking to give edge support to the plywood and lateral support to the stringers. The blocking also backs up splice plates at butt joints in the plywood and helps distribute concentrated loads.

Because the stringer lumber can be obtained in relatively long lengths, a single stressed-skin panel may be made of two or more sheets of plywood (4 ft wide by a multiple of 8 ft long), which cuts on handling and erection costs. Often stressed-skin panels are continuous over three or more supports.

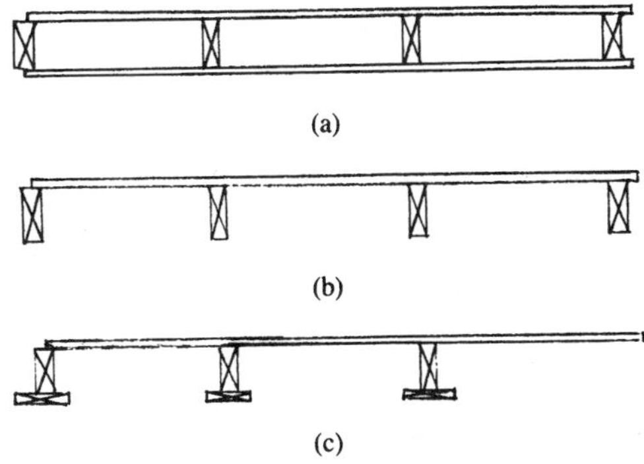

Fig. 13-3. Stressed-skin panels: (a) two-skin, (b) one-skin, (c) T-flange.

Bending

Because the plywood and stringers usually have different moduli of elasticity, the net effective moment of inertia for bending is calculated by the transformed-section method. The properties of the whole 4-ft-wide assembly are used. Only those plywood plies whose grain is parallel to the span of the assembly are used in calculating section properties. The larger cross-sectional dimension of the stringers will be $^1/_8$ in. less than the sawn lumber width because of smoothing of the gluing surfaces.

If the spacing between stringers is large, not all of the skin is effective in resisting bending. Basic spacings, b, (of plywood effective in bending) are listed in Table 13-1 for three-layer plywood. The effective width of the flange (skin) cannot exceed $t + b$, where t is stringer thickness; that is, the effective part of the skin extends no further than $b/2$ out from either face of the stringer. Only the effective portions of the skins are used in determining the net moment of inertia for bending-stress calculations.

Because the skins have nearly uniform tension or compression, allowable extreme fiber stress in bending is either F_c' or F_b'. Repetitive-member allowable bending stresses may be used for stringers (if closer than 24 in. apart). Allowable bending stress in the plywood is subject to a modification, C, to provide a proper factor of safety against buckling of the skin (3). Where c is the clear spacing between stringers,

For $c/b \leq 0.5$, $C = 1.0$
For $0.5 < c/b < 1.0$, $C = 0.67\,(2 - c/b)$
For $c/b \geq 1.0$, $C = 0.67$

Table 13-1. Basic Spacing, *b*, of Unsanded Plywood Effective in Bending for Stressed-Skin Panels.

Thickness (in.)	Basic Spacing, b (in.)	
	Face Grain Parallel to Stringers	Face Grain ⊥ to Stringers
$\frac{5}{16}$	12	13
$\frac{3}{8}$	14	17
$\frac{15}{32}$, $\frac{1}{2}$	18	21
$\frac{19}{32}$, $\frac{5}{8}$	23	22
$\frac{23}{32}$, $\frac{3}{4}$	31	29

Deflection

Deflection (of either the whole assembly or of the skin) often controls the design of stressed-skin panels. In calculations for deflection, using the published values for E of plywood and stringers (which have an increase to account for shear deflection) results in overly conservative assembly deflection values. To obtain a bending-only modulus when designing stressed-skin panels, the published value of E for plywood should be increased 10%, and for lumber increased 3%. The gross moment of inertia for deflection calculations is also found by the transformed-section method.

For checking horizontal shear, rolling shear, and deflection, the section properties should be based on the gross section of all material having its grain parallel with the direction of principal stress.

Example 13-2

A trial section has been chosen for a stressed-skin roof panel (two-skin) for a 12-ft simple span with loads of $D = 11$ psf and $S = 30$ psf. The deflection limit under live load is $\ell/240$ for both the assembly as a whole and for the top skin alone. The assembly will have overall dimensions of 4 ft by 12 ft. The trial section is composed of the following:

Top skin: 15/32-in. Rated Sheathing (C-D), Exp 1, 32/16

Stringers: Four 2 × 6 stringers running parallel to face grain (No. 1 & Btr D. fir-larch)

Bottom skin: 5/16-in. Rated Sheathing (C-D), Exp 1, 20/0

Because the assembly has a length of 12 ft, the plywood will have to be spliced once. Clear distance between stringers calculates as 13.9 in. Check the trial section for bending and deflection.

Properties of a 4-ft width of top skin are

$$A_{\parallel} = 2.292(4) = 9.17 \text{ in.}^2$$

$$I_{\parallel} = 0.067(4) = 0.268 \text{ in.}^4$$

$$I_{\perp} = 0.004(4) = 0.016 \text{ in.}^4$$

$$E = 1{,}800{,}000(1.10) = 1{,}980{,}000 \text{ psi}$$

Properties of four stringers after resurfacing are

$$A = 8.06(4) = 32.2 \text{ in.}^2$$

$$I = 19.4(4) = 77.6 \text{ in.}^4$$

$$E = 1{,}800{,}000(1.03) = 1{,}850{,}000 \text{ psi}$$

Properties of a 4-ft width of bottom skin are

$$A_{\parallel} = 1.491(4) = 5.96 \text{ in.}^2$$

$$I_{\parallel} = 0.022(4) = 0.088 \text{ in.}^4$$

$$E = 1{,}980{,}000 \text{ psi}$$

$$n = E \text{ stringer}/E \text{ plywood} = 1.85/1.98 = 0.934$$

Find N.A. of the gross transformed section (see Fig. 13-4a): Because the bending stress in the assembly runs parallel to the plywood face grain, only the areas of those plies whose grain is parallel to the face grain are used. The stringer material is transformed into material having the stiffness of plywood. Therefore the transformed area of the stringers is 32.2(0.934) = 30.07 in.²

	A	y	Ay
Top skin	9.17	5.922	54.30
Stringers	30.07	3.000	90.21
Bottom skin	5.96	0.156	0.93
	45.20		145.44

To the N.A., $\bar{y} = 145.44/45.20 = 3.22$ in.

Using this neutral axis, the moment of inertia of the gross transformed section, I_g, can be found.

I_0	A	d	d^2	Ad^2	$I_0 + Ad^2$
0.268	9.17	2.702	7.30	66.9	67.2
77.6(0.934)	30.07	0.22	0.05	1.5	74.0
0.088	5.96	3.064	9.39	56.0	56.1
					I_g = 197.3

$$EI_g = (1{,}980{,}000)(197.3)$$
$$= 3.91 \times 10^8 \text{ lb-in.}^2 \text{ per 4-ft width}$$

Check deflection of the assembly: Bending deflection (using a 4′ width) is

$$\Delta_b = 5(4w)\ell^4/(384EI_g)$$

where w = live load in psf

$$= \frac{5(4 \times 30)(12^4)(1728)}{384(391{,}000{,}000)}$$

$$= 0.14 \text{ in.}$$

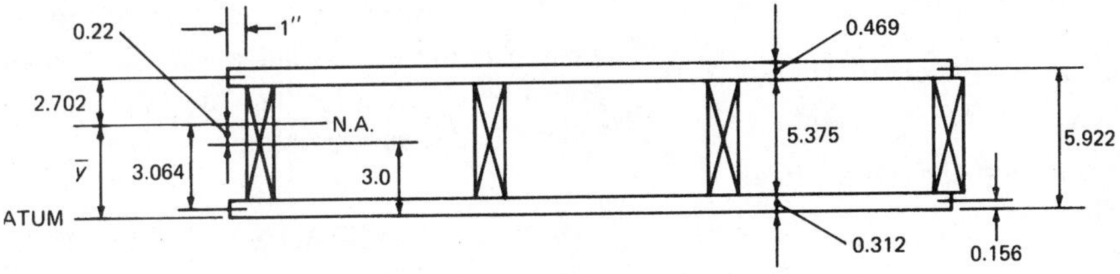

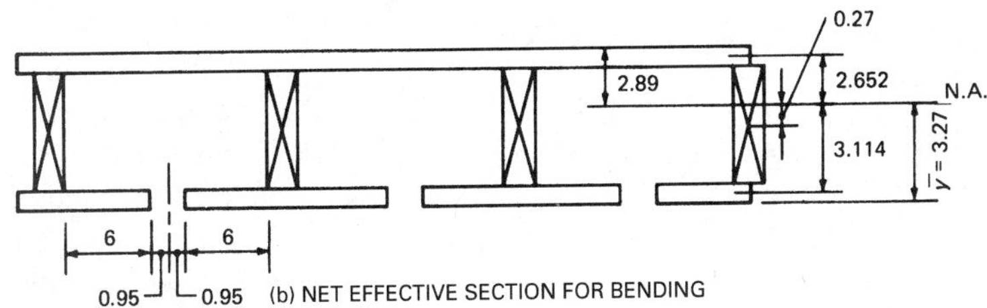

Fig. 13-4. Stressed-skin panel for Example 13-2.

For shear deflection, the properties of the stringers are used assuming $G = 0.06E$. The shear deflection is given by

$$\Delta_s = 0.15(4w)\ell^2 / AG$$

$$= \frac{0.15(4 \times 30)(12^2)(12)}{(32.2)(0.06 \times 1{,}850{,}000)}$$

$$= 0.01 \text{ in.}$$

Total deflection = 0.15 in.

$$\ell/240 = 0.60 \text{ in.} \qquad OK$$

Check deflection of the top skin: Considering bending between stringers of the top skin acting alone, stress is perpendicular to the face grain. Therefore the correct moment of inertia to calculate deflection of the skin is $I_\perp = 0.016$. For a 4-ft width, assuming the plywood acts like a beam with fixed ends, the live load deflection is

$$\Delta = (4w)(\ell^4) / (384EI)$$

$$= \frac{(4 \times 30)(13.9 / 12)^4(1728)}{384(1{,}800{,}000)(0.016)}$$

$$= 0.03\text{in.} < 13.9 / 240 = 0.06 \text{ in.} \qquad OK$$

Determine skin effective in bending: For unsanded plywood with face grain parallel to stringers,

$b = 18$ in. for top skin

$b = 12$ in. for bottom skin (Table 13-1)

For the top skin, the effective width beyond the face of the stringer on either side = 18/2 = 9 in. However this effective distance cannot be more than half the clear span = 13.9/2 = 6.95 in. So the entire top skin is effective in bending. For the bottom skin, only 12/2 = 6 in. beyond each stringer edge is effective, so there will be three ineffective strips of width 2(0.95) = 1.9 in. (see Fig. 13-4b). This leaves 48 − 1.9(3) = 42.3 in. effective; that is, 42.3/48 of the bottom skin is effective.

Find N.A. of the section effective in bending: The effective area of the bottom flange is 5.96(42.3/48) = 5.25.

A	y	Ay
9.17	5.922	54.30
30.07	3.000	90.21
5.25	0.156	0.82
44.49		145.33

$$\bar{y} = 145.33/44.49 = 3.27 \text{ in.}$$

Calculate net effective (in bending) moment of inertia:

I_0	A	d	d^2	Ad^2	$I_0 + Ad^2$
0.268	9.17	2.652	7.03	64.5	64.8
77.6(0.934)	30.07	0.27	0.07	2.1	74.6
0.088(42.3/48)	5.25	3.114	9.70	50.9	51.0
					$I_n = 190.4$

Find allowable bending stress:

For top skin, $c/b = 13.9/18 = 0.77$, so reduction $C = 0.82$

For bottom skin $c/b = 13.9/12 = 1.16$, so $C = 0.67$

Top skin (in compression): $F'_c = (0.82)(1540) = 1260$ psi (plywood species 1; Grade Stress Level S-2)

Bottom skin (in tension): $F'_t = (0.67)(1650) = 1100$ psi

Find actual bending stress: For a 4-ft width,

$$M = (4w)(\ell^2)/8$$

$$= (4 \times 41)(12)^2/8 = 2950 \text{ ft-lb}$$

At top,

$$f_b = Mc/I_n = (2950)(12)(2.89)/190.4$$

$$= 537 \text{ psi} < 1260 \text{ psi} \qquad OK$$

At bottom,

$$f_b = (2950)(12)(3.27)/190.4$$

$$= 608 \text{ psi} < 1100 \qquad OK$$

Trial assembly is OK for deflection and bending. (For a complete design, splice plates would need to be designed and checks on rolling shear and horizontal shear would be required.)

13-3. SANDWICH PANELS

A sandwich panel is made up of a plywood face and back bonded to a thicker central core. Often the core is paper honeycomb, but may be a variety of other materials such as plastic foam or formed metal. The facing is not necessarily plywood, but may be particleboard, hardboard, or sheet aluminum.

The face and back provide the bending strength and rigidity, the core providing shear resistance. It is an economical assembly because the core material is cheaper than the facing material and the faces are advantageously positioned apart from one another. Sandwich construction results in a lightweight member which, by proper selection of materials, can be moisture resistant, heat resistant, and fire resistant.

The sandwich panel must be designed to resist bending loads and shear. The panel also must be checked for buckling, wrinkling of the faces, and shear crimping.

13-4. WOOD BEAMS REINFORCED WITH METAL

Wood beams occasionally have to be reinforced with metal plates, either to improve strength or to reduce deflection. Some of the ways in which this may be done are shown by Fig. 13-5. The vertically laminated (flitch) beam may be used with original construction to make possible a larger span than would be possible with an all-wood beam of equal depth. The wood beam with reinforcing plates top and bottom is not so common, although its strength and stiffness can be equal to that of a steel beam of equal depth. Ordinarily, it would be simpler and more effective to use a steel beam rather than reinforcing a wood beam in this manner. The wood beam with a steel plate on the bottom only, however, could be used for adding to the strength or stiffness of a wood beam already installed in a structure.

To design any of these types, the transformed-section method must be used. The transformed section is an imaginary section of stiffness equal to that of the actual reinforced beam. Its dimensions and properties are found by imagining the steel area to be replaced by an elastically equivalent area of wood. The elastically equivalent area of wood is the area of steel multiplied by the ratio of the modulus of elasticity of steel to that of wood.

Designers should be cautioned that the variation in modulus of elasticity of wood affects the assumptions made. The coefficient of variation

Fig. 13-5. Steel-reinforced beams.

of E in wood is around 25–30%. For example, extreme values in a batch of wood with a nominal E of 1,800,000 psi may range from 1,000,000 psi to 3,000,000 psi.

The action of the reinforced beams is based on the fact that the deflected shape of the wood and steel portions are compatible. For the flitch beam, for example, both the wood and the steel plate bend to the same deflected shape. If the loads include long-term loads, a lesser E-value should be used for the wood.

Example 13-3

Two 3 × 12 No. 1 Douglas fir beams are connected to a 1/4-in. × 11-in. steel plate, as shown by Fig. 13-6, to form a flitch beam. Assuming full lateral support, what is the allowable normal-duration bending moment?

Modulus of elasticity, E, for the steel is 29,000,000 psi, and E for the wood is 1,700,000 psi. The modular ratio is 29/1.7 = 17.1. The transformed section is found by replacing the 1/4-in. thickness of steel with wood 17.1 times this thick, or 4.275-in. thick, as shown by Fig. 13-6. (Note that the imaginary wood that is substituted for the steel must be at the same location relative to the neutral axis as the actual steel; thus its total height normal to that axis is still 11 in.) The transformed section has a moment of inertia of

$$I_t = 2(2.5)(11.25^3)/12 + 4.275(11^3)/12 = 1067 \text{ in.}^4$$

Bending stress on the transformed section is $f_b = My/I_t$. The allowable bending stress in the wood is 1000 psi ($C_F = 1.0$). In the steel it is 22,000 psi for A36 steel. The allowable bending moment, M, must be calculated two ways: first assuming that the extreme fiber of the wood will be stressed to 1000 psi; and second, assuming that the extreme fiber of the steel will be stressed to 22,000 psi. Based on the wood,

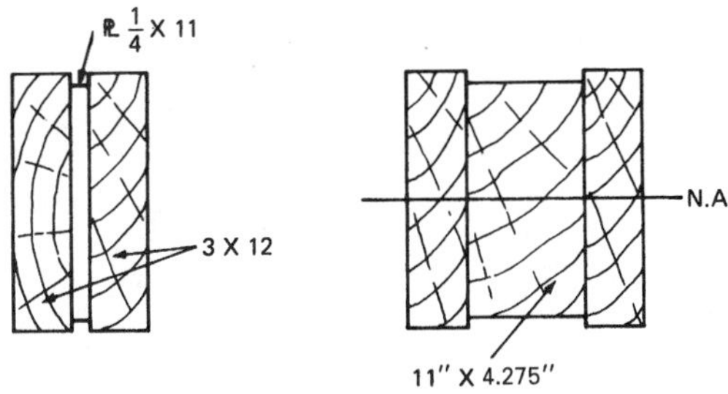

Fig. 13-6. Flitch beam for Example 13-3.

$$M = F'_b(I/c) = 1000(1067/5.625)$$
$$= 190{,}000 \text{ in.-lb, or } 15.8 \text{ ft-k}$$

Based on the steel, assume a stress at the top of the transformed steel to be 22,000/17.1, or 1287 psi at 5.5 in. from the neutral axis. Thus

$$M = 1287(1067)/5.5$$
$$= 250{,}000 \text{ in.-lb, or } 20.8 \text{ ft-k}$$

The lower value, 15.8 ft-k, controls. The wood will be fully stressed when M = 15.8 ft-k, but the steel stress will be less than 22,000 psi.

What to do for connections between the wood and the steel depends on how the vertical loads are applied to the beam and whether or not the steel plate bears on the supports at the end. The need for connections is minimized if both the steel and the wood bear on the end supports. If they do, and if the load also bears on all three pieces, there should be no shearing load on the bolts holding the parts together.

Example 13-4

A 6 × 12 beam, Dense No. 2 (WWPA) D. fir-larch, in place in a building, needs to be strengthened to carry increased load. Decking for the floor above is attached to the top surface, but the bottom surface is accessible for adding a reinforcing plate. The beam spans 15 ft, is laterally supported, and carries a controlling load of normal duration. How much increased capacity can be had by shoring up the beam to remove all load and then attaching a steel plate 3/16 in. × 4 in. to the bottom only?

In this case, the transformed section is unsymmetrical. The modular ratio n = 29/1.4 = 20.7. The plate is replaced by imaginary wood of total width 4(20.7) = 82.8 in., as shown by Fig. 13-7. The centroid of the transformed section is at distance y below the top of the wood beam, and is computed as follows:

Item	Area	y	Ay
Wood 6 × 12	63.25 in.^{2}	5.75	363.7
Imaginary wood 3/16 × 82.8 in.	15.52	11.59	179.9
Totals	78.77 in^2		543.6

Distance of centroid below top = 543.6/78.77 = 6.90 in.

This centroid of the transformed area is the neutral axis for bending of the combined section. The moment of inertia, I_t, for the transformed section is the sum of

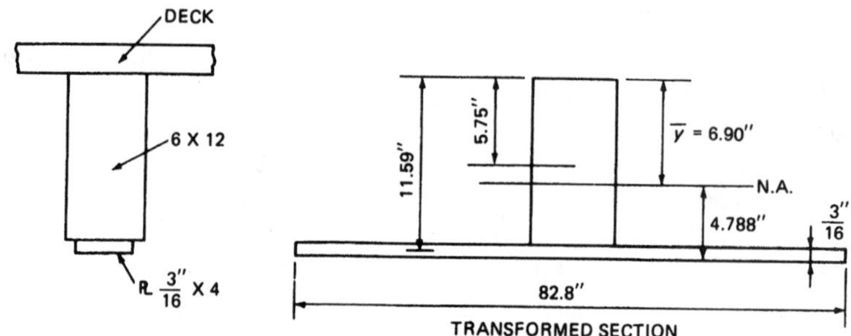

Fig. 13-7. Reinforced beam for Example 13-4.

$$
\begin{aligned}
(1/12)(5.5)(11.5)^3 &= 697.1\\
5.5(11.5)(5.75 - 6.90)^2 &= 83.6\\
(1/12)(82.8)(3/16)^3 &= 0.0\\
15.52(11.59 - 6.90)^2 &= \underline{341.4}\\
I_t &= 1122 \text{ in.}^4
\end{aligned}
$$

Flexural stress in the wood will be highest at the top, where distance $y = 6.90$ in. Flexural stress in the steel will be largest at $y = 4.788$ in. and will be n times the theoretical stress in the transformed section at the same location. Or, the stress in the imaginary material will be 1/20.7 times the actual steel stress.

Based on wood (with $F_b' = 1000$ psi),

$$
\begin{aligned}
\text{Allowable } M &= 1000(1122)/6.90\\
&= 162{,}600 \text{ in.-lb, or } 13.6 \text{ ft-k}
\end{aligned}
$$

Based on steel,

$$
\begin{aligned}
\text{Allowable } M &= (22{,}000/20.7)(1122)/4.788\\
&= 249{,}100 \text{ in.-lb, or } 20.8 \text{ ft-k}
\end{aligned}
$$

Stress in the wood controls, and the allowable M for the reinforced section is 13.6 ft-k.

Without the added steel plate, the section modulus of the beam would be 121.2 in.^{3}, and

$$\text{Allowable } M = 1000(121.2)/12{,}000 = 10.1 \text{ ft-k}$$

Adding the plate caused only a 35% increase in bending-moment capacity. A single plate is seldom very effective, but is often the only way available to strengthen a beam.

Connecting the plate to the wood beam is another part of the total design problem. To determine the required spacing of lag screws, bolts, or nails, the designer must compute the longitudinal shear stress *for the transformed section* at the junction of the steel and the wood, and then choose the proper size and spacing of connectors to resist this shear stress. The maximum shear stress in the wood member is at the centroid of the transformed section, and this too must be checked.

REFERENCES

1. *Plywood Design Specification,* APA—The Engineered Wood Association, Tacoma, WA, 1986.
2. *Design and Fabrication of Plywood-Lumber Beams,* APA—The Engineered Wood Association, Tacoma, WA, 1992.
3. *Design and Fabrication of Plywood Stressed-Skin Panels,* APA—The Engineered Wood Association, Tacoma, WA, 1990.

PROBLEMS

13-1. Design plywood-lumber roof beams of 30-in. depth (surfaced to 29.5 in. actual) to span 22 ft. Use flanges of No. 1 hem-fir and webs of Structural I plywood. These box beams are spaced at 7 ft c/c. The applied dead loads total 22 psf, and snow load is 40 psf.

13-2. Design plywood-lumber I-shaped beams for floor loading of 40 psf live load and 11 psf dead load. Beams span 14 ft and are placed at 4-ft centers. Use No. 2 D. fir-larch flanges and Structural I plywood webs. Try 16-in.-deep beams, which (because beams under 24-in. depth have 3/8 in. removed during surfacing) will have actual depth of 15.625 in. after surfacing.

13-3. A stressed-skin panel is 4 ft wide and spans 16 ft. The top skin is 1/2-in. Structural I Exterior plywood; the bottom skin is similar, except 3/8 in. thick. The four stringers (which run parallel to the plywood face grain) are 2 × 6 D. fir-larch No. 2. Clear distance between stringers is 13.9 in. Dead load is 15 psf, and snow live load is 35 psf. Using a live-load deflection limit of $\ell/240$, check the assembly for bending and deflection.

13-4. What is the maximum allowable total load ($D + S$) for the stressed-skin panel of Problem 13-3, based on bending? Based on live-load deflection?

13-5. Check the panel of Problem 13-3 for horizontal shear at the neutral axis of the gross section and for rolling shear at the junction of the inner-face ply and the adjoining cross-band.

13-6. Will the stressed-skin panel of Problem 13-3 be OK for bending and deflection if the bottom skin is replaced by 1 × 6s as in Fig. 13-3c? Thickness of each 1 × 6 bottom flange is reduced from 0.75 in. to 0.6875 in. because of surfacing. Clear distance between stringers will now be $48/3 - 1.5 = 14.5$ in. Assume that design values of the 1 × 6s are the same as for the 2 × 6 stringers.

13-7. Determine a spacing of 6-in. × 5/8-in. lag screws to connect the 3/16 × 4-in. plate to the 6 × 12 in Example 13-4. *(Hint:* Since allowable moment is 13,600 ft-lb $= w\ell^2/8$, the maximum uniform load will be 484 lb/ft.)

13-8. Assume that it is found possible to add another similar plate to the top of the beam of Example 13-4. By what percentage will this increase the bending capacity of the 6 × 12? For a span of 15 ft, what is the allowable load per foot, based on bending? Will shear now control, and if so, what is the allowable load?

13-9. Two 2 × 12s of Select Structural D. fir-larch (full lateral support) are used on a 15-ft span to carry a normal-duration load of 350 lb/ft. Design a flitch beam to do the job.

13-10. A beam with 5-ft tributary width has full lateral support and a 17-ft span. The dead plus snow load totals 39 psf, not including beam self-wt. Will two No. 1 hem-fir 2 × 10s be sufficient to carry the load? If not, will a flitch beam using the 2 × 10s and a 1/4-in. × 9-in. steel plate work?

13-11. A plywood-lumber beam cross section and its loading are shown in Fig. 13-8. Plywood is APA Rated Sheathing, 1/2-in., 24/0, exterior, dry. Webs are glued continuously to flanges. Webs are oriented with face grain vertical. At any location along the length of the beam, only one of the webs contains a splice. The flanges are composed of Select Structural hem-fir 2 × 8s that have been surfaced to a thickness of 1.375-in. The load case of $D + L$ controls. Find the actual maximum bending stress and compare it to the allowable bending stress. Is the beam OK in bending?

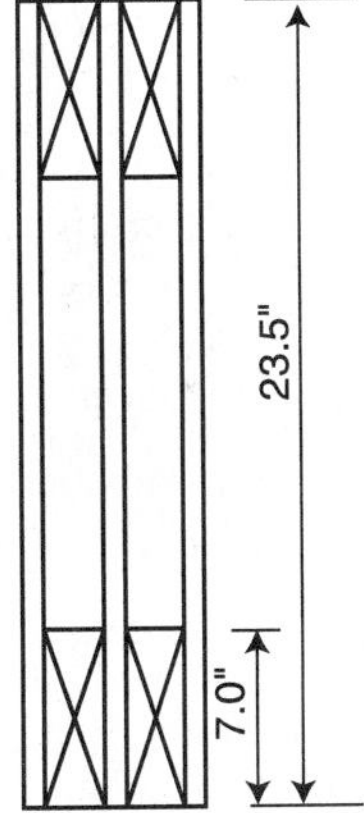

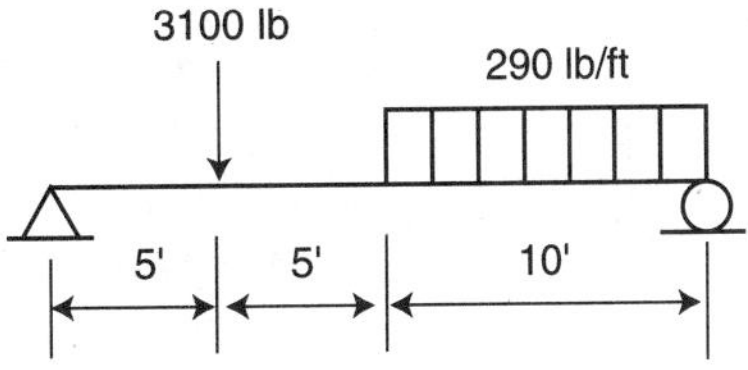

Fig. 13-8. Beam cross section for Problem 13-11.

14

Formwork for Concrete

14-1. INTRODUCTION

One of the important uses of lumber and structural plywood is one in which the wood does not even remain as a part of the completed structure. Plywood and lumber used as concrete formwork are, of course, stripped (removed) after the concrete hardens. Wood formwork can be used to make a simple concrete wall or column or a complex shell roof.

Economy is a major consideration in the design of concrete formwork, since the formwork may represent about half the total cost of a concrete structure. Formwork economy depends to a large degree on the concrete designer. Keeping beam and column sizes the same for several floors will permit reuse of the forms from floor to floor. Columns can be kept the same size over several floors by merely changing the amount of steel reinforcing and concrete compressive strength. Planning for reuse of the forms is an important money-saving measure. With careful removal and proper cleaning and oiling after each use, a plywood form can be reused 30 times or more. Before forms are used they are coated with a form-release agent (or oil) that keeps the concrete from sticking to the forms and reduces the absorption of water into the plywood.

Years ago formwork consisted of many lumber planks, the evidence of this still remaining in the form of horizontal stripes visible on old retaining and foundation walls. (See Fig. 14-1.) Today most of the forms use special types of plywood that are reusable and labor saving. The plywood forms also result in large joint-free, smooth concrete surfaces and can be curved for forming curved grade beams and walls. Steel, plastic, and paper-fiber forms are also used in building concrete structures. Wood forms, however, are the most common.

14-2. PLYWOOD FOR FORMWORK

The most common type of plywood for formwork is B-B Plyform, which has both face and back of grade B veneers. Grade B is a better grade than the C or D grade veneer in most structural plywood. It has the solid surface necessary for wet concrete. Plyform comes in Class I or Class II, both of which are exterior type. The more commonly available Plyform Class I has Group 1 species faces, Group 1 or 2 crossbands, and Group 1, 2, 3, or 4 centers. Class II is somewhat lower quality. Plyform is sanded on both sides, mill-oiled, and edge-sealed to provide protection against moisture. Because it is sanded, it will have different section properties than other structural plywoods of the same nominal thickness.

The most common thicknesses are 5/8 in. and 3/4 in. Table 11-1 shows that Plyform is Grade Stress Level S-2. For determining allowable stress, Plyform Class I is designated as Group 1 species and Class II as Group 3 species. Other exterior types of plywood may also be used for formwork, especially when surface evenness and form reuse are not of prime concern.

A coated plywood with a very smooth surface, called high-density overlay (HDO), is also used. The overlay consists of a cellulose fiber sheet, having no less than 45% resin solids, which is fused under heat and pressure. Overlaid plywood is sometimes referred to as plastic-coated, but should be distinguished from plywood that is user-coated with a plastic com-

Fig. 14-1. Imprint left on concrete retaining wall by use of board forms instead of plywood. (Photograph by authors.)

pound. HDO is used where a smooth, grainless concrete surface is desired, because the resin blocks out the grain and knot pattern. HDO is often reused 200 times.

Allowable stresses for plywood used in formwork are generally modified with a duration of load factor of 1.25. The designer may need to adjust this for forms that will be reused. APA—The Engineered Wood Association also recommends modifying plywood allowable stresses with an "experience factor" of 1.30 (1). Stresses for "wet" plywood should be used.

For dimension lumber members that support the plywood sheathing, allowable bending stresses for single-member use should be used (2). The repetitive-member stresses should be used only for assemblies such as crane-handled panels carefully constructed for reuse.

14-3. DESIGN FORCES

Formwork must be designed to support both vertical and lateral loads. Forms for slabs or beams must support the dead load of the concrete and reinforcing steel in addition to the vertical live loads of workers, materials, and equipment on freshly hardened slabs. Live loads during construction can often be more severe than those imposed after occupancy of the building. The form designer must be alert to unusual loads that might be applied while the structure is under construction. A *minimum* live load of 50 psf is recommended by ACI Committee 347, Formwork for Concrete.

For walls and columns it is the lateral force (lateral pressure) that constitutes the design load. The amount of this pressure due to the fresh concrete depends on pour rate, job temperature, slump, cement type, concrete density, method of vibration, and height of form. A fast pour rate will result in a higher pressure. Concrete that is placed in cold weather will remain unhardened for a longer time and to a greater depth; thus, forms for cold-weather concrete should be designed for higher pressures.

The ACI publishes recommended equations (experimentally determined) for calculating lateral concrete pressures. In the equations that follow, p is the maximum equivalent liquid pressure, in pounds per square foot, at any elevation in the form; R is the rate of concrete placement, in feet per hour; T is the temperature, in degrees Fahrenheit, of the concrete in the forms; and h is the maximum height, in feet, of fresh concrete above the point being considered. The equations are for normal weight, internally vibrated concrete, having slump no greater than 4 in., of Type I cement, with no admixtures.

For columns:

$$p = 150 + 9000R/T \qquad (14\text{-}1)$$

but not over 3000 psf or $150h$

For walls with $R < 7$ ft/hr:

$$p = 150 + 9000R/T \qquad (14\text{-}2)$$

but not over 2000 psf or $150h$

For walls with R of 7-10 ft/hr:

$$p = 150 + 43{,}400/T + 2800\,R/T \qquad (14\text{-}3)$$

but not over 2000 psf or $150h$

For walls with $R > 10$ ft/hr:

$$p = 150h \qquad (14\text{-}4)$$

The maximum pressure computed from any of these four equations cannot be taken as less than 600 psf. (Adjustments to these equations for external vibration when consolidating, superplasticizing admixtures, and others are too varied to discuss here in this basic treatment.)

14-4. TERMINOLOGY

Figure 14-2 shows the location of the various formwork members in a wall form, and Fig. 14-3 the parts of a slab form. The designer needs to be familiar with names of the parts of the typical formwork assembly. These terms are defined below.

Sheathing: The plywood that forms the surface against which the wet concrete is placed.

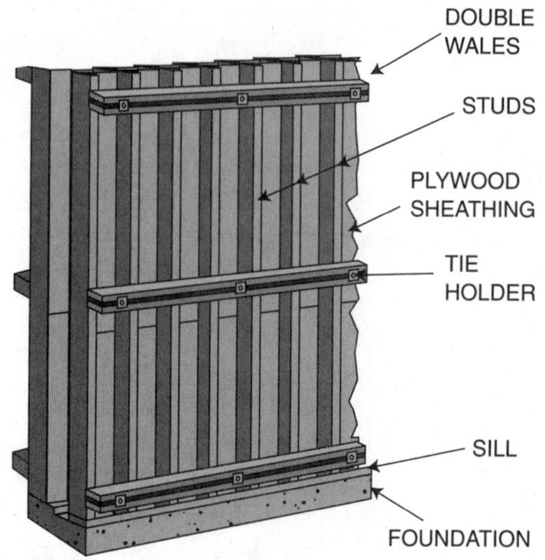

Fig. 14-2. Typical wall form.

Studs: The vertical members (usually 2 ×'s) that support the plywood sheathing in wall forms.

Wales: The horizontal members that support the studs. Each wale usually consists of two timbers, so that holes do not have to be drilled in a wale member to accept the ties (*see ties* below).

Ties: Metal hardware that ties together the opposite sides of the form, preventing them from separating under lateral pressure of the concrete. The form ties pass through the entire assembly and may (or may not) also spread the sheathing the correct distance apart. A particular type of tie called a *snap tie* is commonly used. A snap tie is twisted off just below the concrete surface after the concrete has hardened, the resulting hole then being filled with mortar.

Tie wedge or tie holder: The part of the tie assembly (often wedge-shaped) that is external to the wall form and presses against the wales. It is driven to put tension into the ties.

Joists: The horizontal beams that support the plywood sheathing for floor or roof slab formwork.

Stringers: Beams running perpendicular to and supporting the joists of floor or roof slab formwork.

Shores: The vertical posts or struts that support the roof or floor formwork assembly.

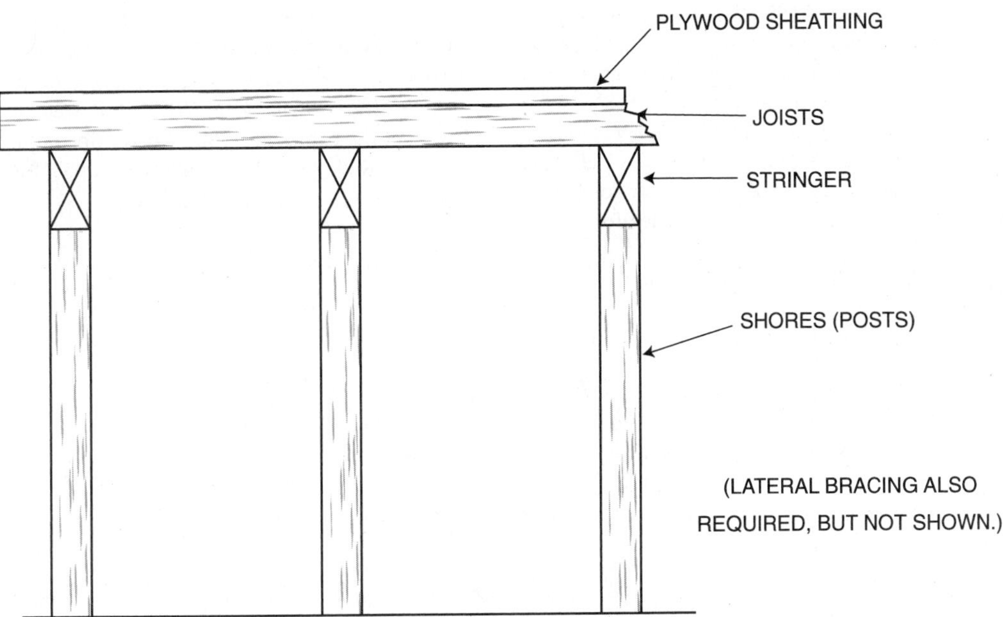

Fig. 14-3. Formwork for concrete slab.

14-5. WALL FORMWORK

The maximum lateral pressures on wall and column forms are given by Eqs. 14-1 to 14-4. These pressures occur in the lower part of the forms, the pressure near the top tapering to zero. However, it is usually more practical to assume this maximum pressure to be uniform over the entire height of the form. The wale and tie spacings could be increased near the top of the wall forms if the designer desired.

Steps in the design procedure for wall forms are as follows:

1. Calculate the design lateral pressure.
2. Check plywood sheathing for bending, deflection, and shear. Either the sheathing thickness or the spacing of its supports (i.e., the stud spacing) will be predetermined based on economy and what materials are available. If the plywood thickness is fixed, the bending check will consist of finding the maximum allowable span of the sheathing. This, of course, will be the stud spacing. But, if the stud spacing is fixed, the bending check will consist of finding the required section modulus of plywood based on bending stresses. The deflection limit is $\ell/360$ for the plywood and also for the lumber members.
3. Design studs, considering bending, shear, and deflection.
4. Design wales, considering bending, shear, and deflection.
5. Choose ties and spacing.
6. Check bearing stresses (studs on wales and tie holders on wales).
7. Design lateral bracing.

Example 14-1

Design wall forms for a 12-ft-high wall, concreted at 3 ft/hr. The concrete has a temperature of 60°F and is internally vibrated; that is, a concrete vibrator is immersed in the wet concrete but not applied to the form itself. B-B, Class I, $^3/_4$-in. Plyform is available and will not be reused. Ties with an allowable tensile force of 3000 lb are on hand. Use construction grade Douglas fir for the studs and wales.

From Eq. 14-2,

$$\text{Max pressure} = p = 150 + 9000(3)/60$$
$$= 600 \text{ psf}$$

The lateral pressure varies (as a straight line) with height from the top of the concrete surface. Pressure is γh, *where γ is the specific weight of concrete* = 150 pcf. At 1 ft from the top of the form the pressure will be 150 psf, at 2 ft it will be 300 psf, and so on. The maximum pressure of 600 psf is reached at only 4 ft from the top of the form. At greater depth, the maximum is used, so it is logical to design the entire height of form for a uniform pressure of 600 psf.

Check sheathing for bending: For species Group 1, Grade Stress Level S-2, the base (wet) $F_b = 1190$ psi. This is adjusted by an "experience factor" of 1.3 and by a duration of load factor of 1.25. Therefore,

$$F'_b = 1190(1.3)(1.25) = 1930 \text{ psi}$$

For sanded 3/4-in. plywood, Table 11-4A shows

$$KS = 0.412 \text{ in.}^3/\text{ft}$$

(for a panel with 8-ft dimension horizontal; i.e. with bending stress parallel to the face grain).

Assuming continuous over three spans (Table A-8),

$$M = F'_b(KS) = w\ell^2/10$$

$$w = (600 \text{ psf})(1 \text{ ft}) = 600 \text{ lb/ft} = 50 \text{ lb/in.}$$

$$\ell = \sqrt{10(1930)(0.412)/50} = 12.6 \text{ in.}$$

(maximum allowable span of plywood based on bending)

Check sheathing for deflection: Deflection should be limited to $\ell/360$; I (from Table 11-4A) is 0.197 in.^{4}/ft.

$$\Delta\text{max} = w\ell^4/145EI = \ell/360$$

$$\ell = \sqrt[3]{145EI/360w}$$

$$= \sqrt[3]{145(1.5 \times 10^6)(0.197)/[360(50)]}$$

$$= 13.4 \text{ in.}$$

(maximum allowable span of plywood based on a deflection limit of $\ell/360$)

Check sheathing for rolling shear: From Table 11-4A, $Ib/Q = 6.762$

$$V = 0.6w\ell$$

$$f_s = VQ/Ib$$

$$F'_s = 44(1.3)(1.25) = 71.5 \text{ psi}$$

Set $F'_s = f_s$

$$71.5 = 0.6(50)(\ell)/6.762$$

$$\ell = 16.1 \text{ in. (max span based on shear)}$$

So bending controls and the maximum allowable span of the plywood (maximum spacing of the studs) is 12.6 in. The stud spacing should be an even fractional part of the 96-in. horizontal dimension of the plywood. *Use studs at 12 in. c/c.*

Stud design for bending: With studs spaced at 12 in. c/c, the tributary width to a single stud is 1 ft, so $w = 600$ psf $\times$ 1 ft $= 600$ lb/ft $= 50$ lb/in. Assuming 2 $\times$ 4s and single-member use,

$$F'_b = 1000(1.25) = 1250 \text{ psi and } S_x = 3.06 \text{ in.}^3$$

$$\ell = \sqrt{10(12.50)(3.06)/50} = 27.7 \text{ in.}$$

(max span based on bending)

Stud design for deflection:

$$I = 5.36 \text{ in.}^4 \quad \text{and} \quad E = 1{,}500{,}000 \text{ psi}$$

Using a deflection limit of $\ell/360$,

$$\ell = 40.1 \text{ in.} \quad \text{(max span based on deflection)}$$

Stud design for shear:
Hurd (2) recommends multiplying the allowable shear stress by a load-duration factor of 1.25 and by a "two-beam" factor of 1.5. Therefore,

$$F'_v = 95(1.25)(1.5) = 178 \text{ psi}$$

Shear at a distance d from the support is

$$V = 0.6w\ell_c - wd$$

$$f_v = 1.5V/A$$

$$= \frac{1.5(0.6w\ell - wd)}{bd}$$

With $d = 3.5$ in., $b = 1.5$ in., and $w = 50$ lb/in., and setting the allowable stress equal to the actual stress, the maximum stud clear span based on shear is

$$\ell_c = 26.6 \text{ in.}$$

With doubled wales of 2 $\times$'s, the approximate maximum span is

$$\ell = 26.6 + 2(1.5)$$

$$= 29.6 \text{ in.} \quad \text{(max span based on shear)}$$

Bending controls and the maximum allowable span of the studs (spacing of the wales) is 27.7 in. *Place top and bottom wales one foot from the top and bottom of the form (a common construction practice) and space other wales at 24 in. c/c.*

Design wales for bending: The tributary width to a single wale is 2 ft, so

$$w = (600 \text{ psf})(2 \text{ ft}) = 1200 \text{ lb/ft} = 100 \text{ lb/in.}$$

Assuming ties are spaced at 24 in. c/c,

$$\text{Req } S = 0.1w\ell^2/F'_b$$
$$= 0.1(100)(24^2)/1250 = 4.61 \text{ in.}^3$$

$$S \text{ provided by doubled } 2 \times 4\text{s} = 6.12 \text{ in.}^3$$

Design wales for shear:

$$F'_v = 178 \text{ psi}$$

$$f_v = \frac{1.5(0.6(100)(24) - (100)(3.5))}{(2 \times 1.5)(3.5)}$$
$$= 156 \text{ psi}$$

Wales of doubled 2×4s are OK if ties are spaced at 24 in. c/c.

Check ties:

Req spacing of ties = tie capacity/wale load

= 3000 lb/1200 lb per ft

= 2.5 ft > 2 ft *OK*

Check bearing stresses: Two locations need to be investigated for bearing (compression perpendicular to grain): (1) where the tie wedges bear on the wales and (2) where the studs bear on the wales. This check is left as a student problem.

Lateral bracing: The forms will require lateral bracing for the local code-specified wind loads. These braces could be wood struts on a single side, in which case they must resist either compression or tension. Guy wire bracing on both sides of the wall could also be used.

14-6. FORMWORK FOR ROOF AND FLOOR SLABS

There is no accepted sequence for design of formwork for slabs. The design sequence may depend upon the fact that a particular sheathing is available, that a particular set of shores is available, or that a certain construction module is used. In the design, however, the plywood, joists, and stringers must each be checked for bending, deflection, and shear. Also, the shores must be designed (including lateral braces) and bearing stresses checked.

The shores (posts) that support the stringers are designed exactly as any other wood column (Chapter 7). Each shore supports the load on the tributary formwork area extending halfway to the adjacent shore on all sides. Because shores are usually reused repeatedly, no increase in allowable stress for load duration should be used. Bearing stress in the member the shore supports can often be the governing design criterion. The bearing area may be less than the full area of the shore if a narrow member rests on the shore. Metal or hardwood plates attached to the top of the shores may be used to extend the bearing area and reduce the bearing stress.

Example 14-2

A 9-in.-thick floor slab of normal weight concrete is to be formed using 3/4-in. B-B Class II Plyform. Bays are 15 ft × 15 ft. Joists are construction grade D. fir-larch. Determine the spacing of the joists and the stringers. Forms will be used only once.

Determine loads:

Concrete 150(9/12)	= 113 psf
Recommended live load	= 50 psf
Est. form wt.	= 4 psf
Total	= 167 psf

Plywood for bending: For species Group 3, F_b = 820 psi (wet)

$$w = 167 \text{ lb/ft} = 13.9 \text{ lb/in. per ft width}$$
$$F'_b = 820(1.3)(1.25) = 1332 \text{ psi}$$
$$F'_b(KS) = w\ell^2/10$$

Effective section modulus is $KS = 0.412$ in.^{3}/ft width

$$\ell = \sqrt{10(1332)(0.412)/13.9}$$
$$= 19.9 \text{ in.}$$

Plywood for deflection: For stress parallel to face grain,

$$I = 0.197 \text{ in.}^4\text{/ft}$$
$$w\ell^4/145EI = \ell/360$$
$$\ell = \sqrt[3]{145(1,100,000(0.197)/[360(13.9)]}$$
$$= 18.4 \text{ in.}$$

Plywood for rolling shear:

$$V = 0.6w\ell$$

$$f_s = VQ/Ib = 0.6w\ell/6.762$$

$$F'_s = 44(1.3)(1.25) = 71.5 \text{ psi}$$

$$71.5 = 0.6(13.9)\ell/6.762$$

$$\ell = 58.0 \text{ in.}$$

Deflection governs and joist spacing can be no more than 18.4 in. *Use joists at 16 in. c/c* in order to make joist spacing an even fractional part of the 96-in. plywood dimension.

Joist for bending: Assume 2 × 4 joists. The tributary width to a joist is 16 in.

$$w = 167(16/12) = 223 \text{ lb/ft}$$

$$= 18.6 \text{ lb/in. on a 1-ft width}$$

$$F'_b = (1000)(1.25) = 1250 \text{ psi} \quad \text{and}$$

$$S = 3.06 \text{ in.}^3$$

$$\ell = \sqrt{10(1250)(3.06) / 18.6}$$

$$= 45.3 \text{ in.}$$

Joist for deflection:

$$w\ell^4/145EI = \ell/360$$

$$\ell = \sqrt[3]{145(1{,}500{,}000)(5.36) / [360(18.6)]}$$

$$= 55.8 \text{ in.}$$

Joist for shear:

$$F'_v = 178 \text{ psi (see Example 14-1)}$$

$$V = 0.6\, w\ell_c - wd$$

$$f_v = 1.5V/bd = 178$$

$$\ell_c = 61.7 \text{ in.}$$

$$\ell = 61.7 + 1.5 = 63.2 \text{ in.}$$

Beading controls and maximum span of joists (spacing of stringers) is 45.3 in.

For uniform spacing across the bay, *use stringer spacing of approximately 45 in.* for four equal spaces. Based on strength and stiffness of stringers, the spacing between shores (maximum stringer span) can be determined in a similar manner.

REFERENCES

1. *Concrete Forming,* APA—The Engineered Wood Association, Tacoma, WA, 1994.
2. Hurd, M. K., *Formwork for Concrete,* 6th ed., American Concrete Institute, Farmington Hills, MI, 1995.

PROBLEMS

14-1. Design the forms for a 11-ft-high concrete wall for which the concrete is to be placed at a rate of 3 ft/hr at a temperature of 70°F. On hand is $^5/_8$-in. B-B, Class I Plyform. Use No. 2 D. fir-larch for studs and wales. Ties can carry 4000 lb. Contact area of each tie wedge is 4 in.^{2} For the purpose of finding bearing area factor, C_b, assume that bearing length of tie wedges is 2 in.

14-2. Design the wall forms for Problem 14-1, except that the wall is concreted at the rate of 4 ft/hr and Plyform is $^3/_4$-in.

14-3. If the contact area between tie wedge and doubled wale of Example 14-1 is as shaded in Fig. 14-4, compare actual bearing stress with the adjusted allowable stress. Be sure to include the bearing area factor, C_b.

14-4. Is the junction of stud and wale for Example 14-1 satisfactory in bearing? Include the bearing area factor, C_b, in your solution. Make a sketch of the contact area.

14-5. Design forms to support an 8-in. concrete floor slab using $^{23}/_{32}$-in. B-B, Class I Plyform (placed in its strong orientation). Forms will be used only once. Use No. 2 D. fir framing members and shoring. Ceiling height is 9 ft, and bays are 22 ft × 22 ft. To find the stringer span, assume that the shore spacing is two times the joist spacing. Design the stringer as if three-span continuous. When

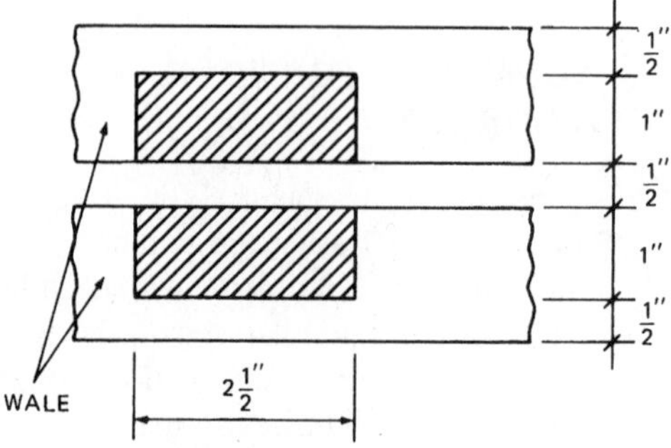

Fig. 14-4. Bearing area of tie wedge for Problem 14-3.

choosing shores, use $C_D = 1.0$ since shores are repeatedly reused.

14-6. Find the maximum practical shore spacing for Example 14-2 assuming stringers are 2 × 6s of No. 2 D. fir.

14-7. Design shores for Example 14-2 using No. 2 D. fir-larch. Clear height between floors is 9.5 ft. Assume shores are spaced every 10 ft along the length of the stringer. If stringers are also No. 2 D. fir-larch, what bearing area is required at an interior support for the stringer?

14-8. Forms are being designed for a concrete wall. The maximum pressure due to wet concrete is 365 psf. Sheathing is B-B, Class I Plyform, supported by 2 × 4 studs at 16-in. centers. If the plywood is placed with the face grain horizontal, what is the required plywood thickness (based on bending)?

14-9. If ¾-in. sheathing is used in Problem 14-8, will it be OK for rolling shear and for a deflection limit of $\ell/360$ under total load?

15

Miscellaneous Structure Types

Light-frame construction, pole buildings, and timber bridges were introduced in Chapter 1 but essentially ignored thereafter. Each of these deserves further attention, so this chapter covers them in greater detail. It also introduces the subject of wood foundations. Design features of the other principal building types (heavy timber, and post and beam) have already been discussed.

Timber bridges are not covered here in great detail. To do so would double or triple the length of this book. However, some attention should be given to this important use of wood, so the authors have elected to discuss the common timber bridge types and to give a few design examples to illustrate design features that differ somewhat from those for wood buildings.

15-1. LIGHT-FRAME CONSTRUCTION

Figures 15-1 and 15-2 show two types of light-frame construction. The platform frame of Fig. 15-1 is by far the more common. It is the simpler to construct, therefore usually less expensive. It can be built one story at a time. As one story is completed, wall framing for the story above can be assembled using the completed story below as a work platform.

A disadvantage of platform framing is that its total vertical shrinkage is the sum of the longitudinal shrinkage of the studs plus the transverse shrinkage of the horizontal members. The latter, shrinkage (or swelling) in width of the horizontal members, can be much larger than the change of length of the studs. Thus, the total vertical dimension change is fairly concentrated at the horizontal members. This can be troublesome, leading to cracking in outside finish materials, such as brick veneer or stucco. Also, in buildings of more than two stories, it may cause difficulties with rigid plumbing and mechanical systems. Problems of vertical shrinkage in platform framing can be lessened by use of I-shaped joists having plywood webs.

Vertical shrinkage is better controlled by the balloon system of Fig. 15-2. It includes only the longitudinal shrinkage of the vertical members, which extend over the entire two stories. Transverse shrinkage of the horizontal members does not add to the total. The balloon frame is the preferred type where the exterior finish surface is to be a brittle material without expansion or contraction joints at story levels (such as masonry veneer or stucco). A disadvantage is that it requires longer studs, and a two-story or three-story preassembled wall is much harder to lift into position than a one-story wall assembly of the platform type.

Frequently, for small buildings of either type of light-frame construction, little engineering analysis is done. However, with increasing material scarcity and material costs, builders are encouraged to find ways of using material more efficiently. As a result, engineering analysis is becoming more important, and the future may see more sophisticated designs in light-frame construction. Examples already evident are the substitution of parallel-chord trusses or of shop-fabricated wood beams for sawn-lumber joists, and substitution of prefabricated wall panels for the conventional studs and sheathing.

If traditional light-frame structures are examined using modern structural-design criteria, they may be discovered in some respects to be overdesigned; that is, they are stronger than necessary for the expected loads. In other respects,

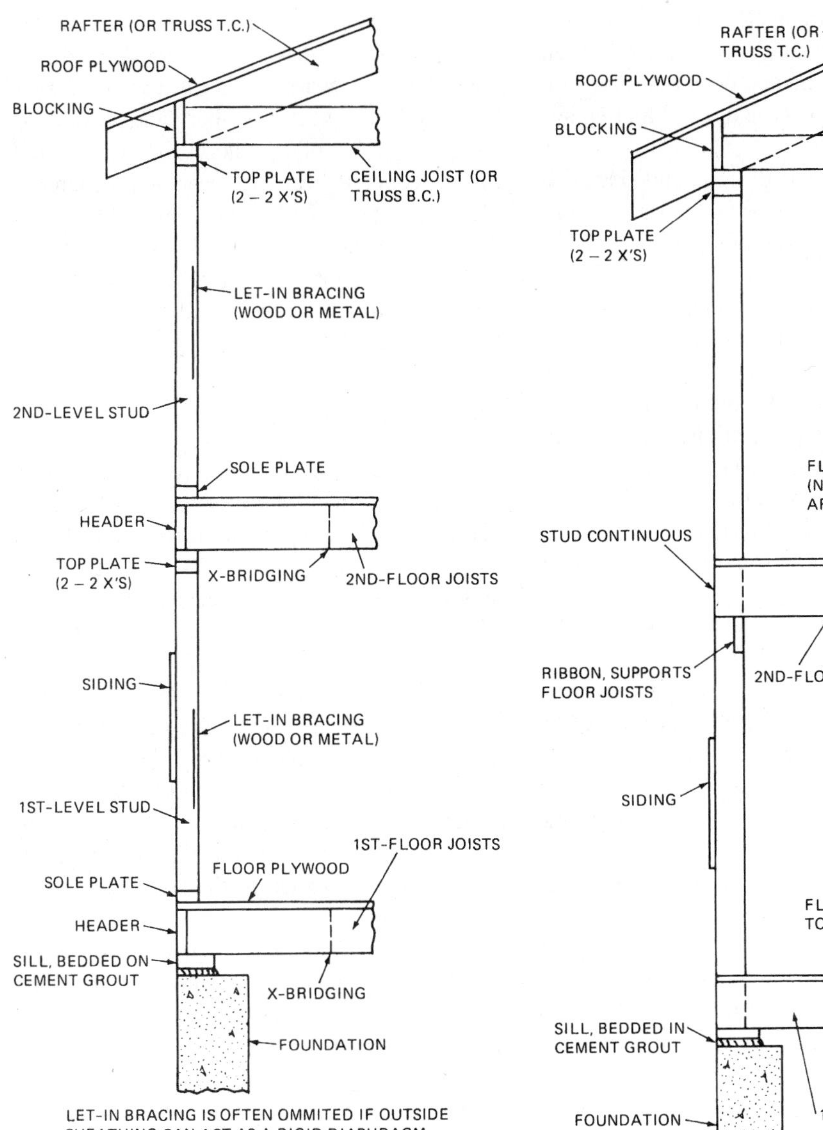

Fig. 15-1. Platform framing: cross section through exterior wall.

Fig. 15-2. Balloon framing: cross section through exterior wall.

however, they may be found to be not strong enough. One example of the latter is the plate or sill, the 2-in. nominal horizontal member on which rafters, joists, or ends of the wall studs bear. Bearing stresses normal to the grain of the plate frequently exceed the allowables for compression perpendicular to the grain. Yet we use these 2-in. nominal plates without problems. Why? The answer is that a 0.04-in. deformation (bearing failure, as it is defined by the code) probably occurs, but collapse does not. A deformation this small is normally undetectable and is considered harmless in the finished structure.

Most design decisions for light-frame structures can be made using tables. Some engineering design knowledge went into preparing these tables, and so did experience. Thus, the designer may usually use the tables with confidence, but should be alert to any limitations placed on their use. The UBC (1) allows design using tables alone for light-frame structures up to two stories high. For higher buildings it requires engineered design—not for structural purposes, but to ensure that vertical shrinkage does not raise trouble for mechanical, electrical, or plumbing systems.

Much of light-frame design (and design of other types, too) is limited by module sizes—length and width of available plywood sheets, for example. Thus, the builder is encouraged to choose stud, joist, and truss spacings that are even fractions of those dimensions.

Nailing schedules tell how to connect the members of light-frame buildings together. These schedules give connections that are satisfactory in most cases. However, the designer should give special attention to connections wherever unusual conditions occur. Under these conditions, better-than-normal connections may be required.

As one example of the latter, consider buildings with roof overhangs. Standard nailing schedules are based on the assumption that roof uplift forces are relatively small and are transferred by nails from one member to the next: from roof sheathing to rafter, to top plate, to wall sheathing, to bottom plate, to anchor bolt, and finally to the foundation. In localities with high winds, this system of nailing may prove inadequate for roof members with large overhangs. For such cases, metal timber connectors may be necessary. For extreme cases, a better solution is to provide full-height steel anchor rods, running all the way from the roof member to the foundation anchor bolts. If this is done, uplift forces are transferred directly from the rafters to the foundation, bypassing the sequence of nailed members occurring between the roof and the foundation.

Substructures

Light-frame buildings are usually built on foundations of either masonry or concrete. If the foundation is a wall, the first wood member is a sill plate (or sill), not engineered, placed on a leveling bed of grout (cement mortar). The sill plate is usually of 2-in. nominal thickness and 4-in. (or more) nominal width. The UBC (1) requires these plates to be either redwood or preservative-treated lumber. Anchor bolts connect the plate to the foundation.

If the-light-frame building is supported on isolated footings, then stub columns or pedestals will support a beam—either a glulam, a wood beam built by nailing together (side by side) the required number of 2-in. nominal members, or a steel or concrete beam. If the beam is concrete, a properly anchored sill plate is placed on top of the beam. If the beam is steel, a wood sill plate is bolted to the top flange of the steel beam. The first-floor structure is mounted on the sill plate or on the beam as shown by Fig. 15-3.

Wall Structure

Wall studs are connected to a sole plate, usually the same width as the studs. A double plate, made of two 2-in. nominal pieces is attached to the top end of the studs. The wall assembly must have bracing to resist shearing loads parallel to the length of the wall. This bracing may be *let-in* bracing (as shown by Fig. 15-1), which may be either metal or wood. If the bracing is wood, it is of 1×4 nominal size. It should be set into recesses cut into the studs and preferably also into both the top and bottom plates. (Figure 15-1 shows it let into the studs only.) As an alternate to let-in bracing, certain types of rigid sheathing, nailed to the studs and plates, may be used.

In the United States, plywood and other sheathing materials are used almost to the exclusion of let-in bracing. Also, in the United States, the wall studs, plates, and at least the end panels of plywood sheathing are assembled on a flat working surface, and then the entire assembly is tilted to the desired position on the structure. If let-in bracing is used, however, installation of sheathing is frequently left until the wall framing is raised into its vertical position.

Solid blocking, horizontal pieces of 2×4 nailed between studs (usually at mid-height) was formerly considered to have structural value, reducing the weak-axis ℓ/d ratio of the studs. Attached sheathing materials are now realized to be much more effective for this purpose than blocking. Thus, blocking between studs is often omitted, unless it is needed in a particular location for edge nailing of sheathing panels, as fire blocking, or as support for nonstructural elements such as electrical boxes.

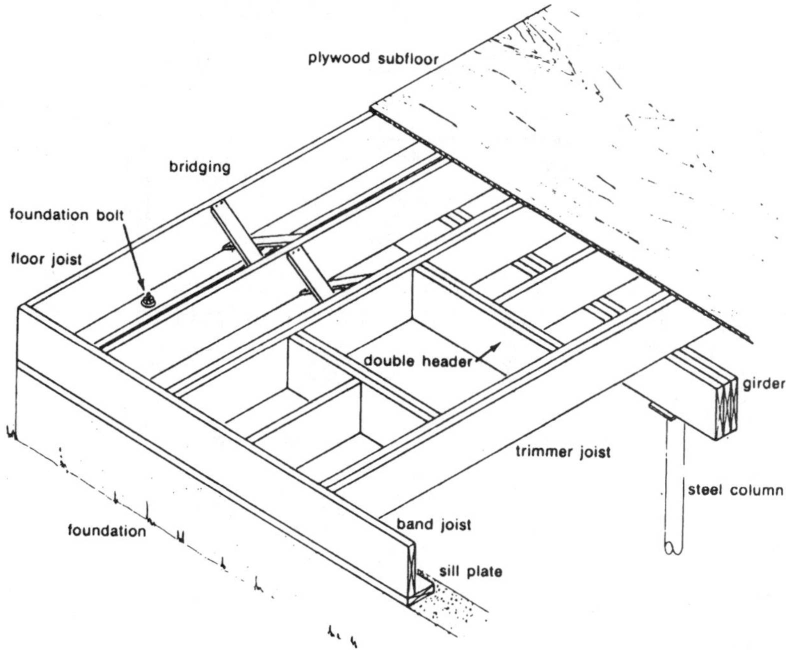

Fig. 15-3. Connection of light-frame building to substructure. (Courtesy of American Society of Civil Engineers, D. H. Percival and S. K. Suddarth, "Light-Frame Construction," *Wood Structures, a Design Guide and Commentary,* 1975.)

15-2. POLE BUILDINGS

Pole buildings have been used since ancient times. Early American Indians in the Missouri valley used living tree trunks as poles for buildings of up to 40-ft diameter, with low side walls and a center pole about 14 ft high. In the Fiji Islands, deeply buried poles were used to help buildings resist the forces of frequent hurricanes, the structures acting as rigid frames but with considerable flexibility. In Japan, pole buildings with relatively light walls were found to be a good answer to the lateral forces of earthquakes.

Modern pole construction, dating from the 1950s, became feasible when preservative treatment materials and methods were developed to prevent rotting of wood in contact with the ground. Recent pole construction was originally for agricultural buildings but now often finds use for residences, commercial and industrial buildings, churches, and schools. It is claimed that there are some 600,000 pole buildings in the United States and Canada (2).

The typical pole building has vertical poles, embedded in the ground, serving both as columns and as members to resist lateral forces. The poles are usually round members, but might instead be rectangular sawn timbers or glulams. The floor and roof structures attached to and supported by the poles may be strictly of light-frame construction or may include glulam members or trusses.

The poles, being set into the ground, *must* be treated with preservative. The American Wood Preservers' Association standards (3) specify appropriate preservatives and methods of treatment.

There are two main types of pole building. In the first type, poles extend only to the bottom of the first-floor framing, as in Fig. 15-4, and serve merely as foundations for the platform-type frame structure above. In the second type, the poles extend all the way to the roof, as in Fig. 15-5, in which case they serve as building columns also. The connection of the top of the poles to the roof structure may be either pinned (nonrigid) or made rigid by means of bracing. In either type of pole building, bents as shown are normally in the range from 10 to 20 ft apart (i.e., the bays are 10–20 ft long). Spans of up to 100

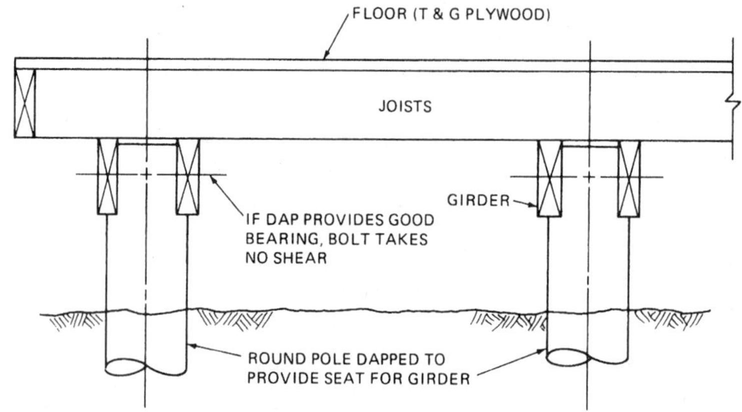

Fig. 15-4. Platform framing for pole building.

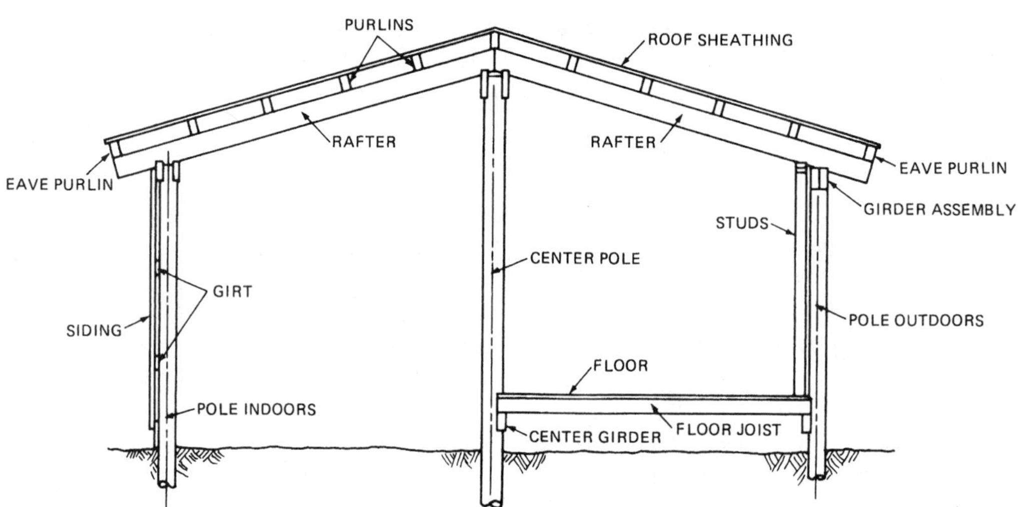

Fig. 15-5. Two methods of framing a pole building with poles extending to roof.

ft have been built where roof trusses are used, but 60 ft is the usual practical limit (4).

Advantages of Pole Construction

1. Less costly excavation may be needed than for buildings with concrete foundations or foundation walls. Normally the only needs are round drilled holes to receive the poles, plus the usual digging for water lines and sewers.

2. After the poles are placed, the roof can be installed immediately, providing shelter under which construction may continue.

3. If the poles go all the way to the roof, the walls need to carry only their own weight. Thus, walls might be of lighter construction. For example, there would be no need to provide the usual 16-in. stud spacing (local building code permitting, of course).

4. Pole construction is good for "difficult" locations, such as hillsides, swampy ground, permafrost areas, and seaside areas where high wind and waves may be present.

Arrangements of Poles and Walls

Figure 15-6 shows three ways of locating walls relative to the poles. In the first, the pole actually becomes a part of the wall structure. This system appears simplest, but it has disadvantages. Round timbers used for poles are seldom straight, and sealing the joint between a crooked pole and the wall material is difficult. With the poles inside, as in Fig. 15-6b, otherwise useful

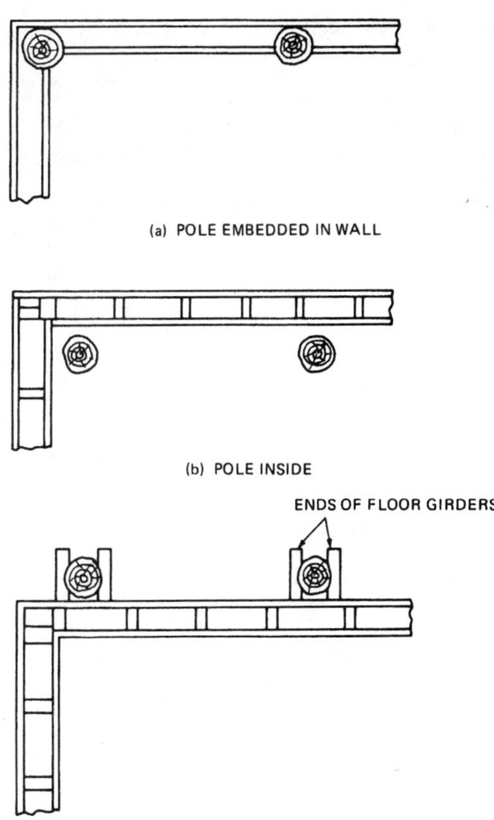

Fig. 15-6. Three positions of pole relative to wall.

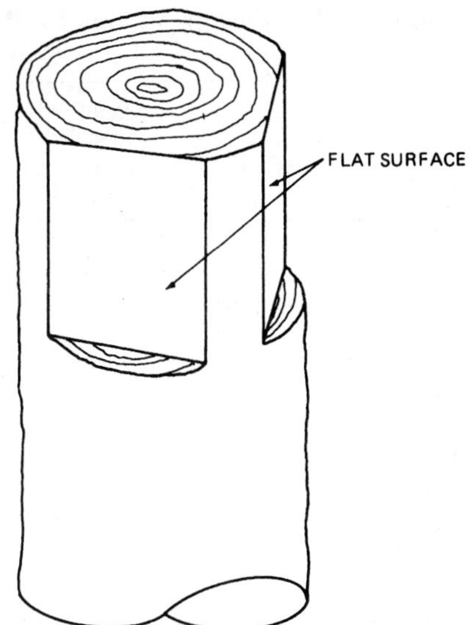

Fig. 15-7. Pole dapped to provide seat and flat faces for connecting girders (corner pole shown).

space inside the building is consumed. If the appearance of the round pole inside the building is objectionable, inside wall finish material (drywall, for example) can be placed around the pole to enclose it. Sealing against weather is simplest of all with the arrangement in Fig. 15-6c, in which the poles are outside the building. This scheme also keeps the possibly odorous and toxic preservative chemicals outside the building. Exposed ends of floor girders may require treatment.

Connections

Horizontal structural members may be connected to the poles using bolts or timber connectors. If possible, the poles should be *dapped,* that is, notched as shown by Fig. 15-7. If the dap is carefully prepared and the horizontal cut deep enough, the vertical reaction of the beam is resisted by direct bearing of the beam on the pole, so that the bolts merely ensure that the beam does not shift sidewise and slide off its seat. In this case the bolts carry no load. If large enough daps are not possible, then the bolts (or timber connectors) would be in vertical shear, carrying the entire reaction of the beam into the pole.

Spikes might be used instead of bolts. If spikes are used, however, it will be preferable to transfer the vertical reaction by beam bearing on the dapped pole, using the spikes only to hold the parts in contact.

Allowable Design Stresses for Poles

Table 15-1 shows allowable stresses for poles that are treated with preservative. The stresses are for single-member use. Increase of F_b for repetitive-member use is not allowed. Also, the form factor for round members has already been considered in formulating the table; it should *not* be used again to modify the tabulated value of F_b. The allowables are to be adjusted for load duration, however.

Distribution of Horizontal Load Among Poles

This part of the design is merely structural analysis. For many configurations of pole building, the analysis may be aided by simplifying as-

Table 15-1. Design Values for Preservatively Treated Poles,[a,b] Wet and Dry Conditions of Use.

Species	Bending F_b (psi)	Compression Parallel to Grain F_c (psi)	Modulus of Elasticity E (psi)	Compression Perpendicular to Grain $F_{c\perp}$ (psi)	Horizontal Shear F_v (psi)	End Grain in Bearing[c] F_g (psi)
Douglas fir, coast	1,850	1,000[d]	1,500,000	375	115	1,200
Jack pine	1,500	800	1,100,000	280	95	870
Lodgepole pine	1,350	700	1,100,000	240	85	870
Northern white cedar	1,050	525	600,000	225	80	670
Ponderosa pine	1,300	650	1,000,000	320	90	820
Red or Norway pine	1,450	725	1,300,000	265	85	790
Southern pine	1,700	900[d]	1,500,000	320	105	1,120
Western red cedar	1,350	750	900,000	255	95	940
Western hemlock	1,650	900	1,300,000	245	115	1,120
Western larch	2,050	1,075	1,500,000	375	120	1,210

[a]The design values are based on ASTM D 2899-86, *Method for Establishing Design Stresses for Round Timber Piles*. The values are for single-member uses of poles and assume that the conditioning prior to teatment was in accordance with ASTM D 1760-86a. If the poles are conditioned by air or kiln drying only prior to treatment, the design values, with the exception of modulus of elasticity, may be increased by application of the modifying factor for seasoning conditioning, C_{co}, which is 1.18 for southern pine and 1.11 for other species.
[b]The compression perpendicular to grain design values $F_{c\perp}$ are based on the design values for sawn lumber under wet conditions of use, which have been reduced for conditioning. For dry conditions of use, multiply by 1.5.
[c]End grain in bearing design values, F_g, are based on the design values for sawn lumber used in a wet location and reduced for conditioning. When used in a dry location, they may be increased by 10% (C_M = 1.10).
[d]Design values for compression parallel to grain, F_c, in Douglas fir and southern pine may be increased 0.2% for each foot of length from the tip of the pole to the critical section. This increase shall not exceed 10%. The modification factor for location of critical section, C_{cs} = 1 + 0.002L, where L = distance (ft) from the tip to the critical section; C_{cs} shall not exceed 1.10.
Source: Copyright ASTM. Reprinted with permission.

sumptions, such as those shown by Patterson (5). The shear in each pole depends on pole stiffness and the location of horizontal loads on the structure. Poles resist horizontal loads, acting as embedded cantilevers. The roof system is usually sufficiently rigid that the tops of the poles all deflect by the same amount. The same is true for floor systems and for the pole deflections where the floors are attached. Stating these conditions mathematically permits solving for the amount of horizontal load resisted by each pole and for the horizontal forces transmitted by the floor and roof systems.

Poles may act as simple cantilevers, having the superstructure above pin-connected to the poles (nonrigid connection), or they may have some degree of top fixity by being connected more rigidly to the superstructure. Means of providing a degree of top fixity would include connecting the pole to fairly deep trusses, glulams, or deep wall panels, or using bracing.

If the pole tops are free to rotate with respect to the roof structure, the analysis of wind load distribution to poles that support only a roof and walls is fairly simple, as shown below.

Figure 15-8a shows wind loads on a simple two-pole bent, with the poles pin-connected to the roof structure above. Girts (horizontal beams spanning from pole to pole) support the wall and transfer its wind loads to the poles. These loads on the poles may be treated as uniformly distributed loads. A redundant horizontal force, F, in the roof system causes both poles to deflect equally at the top. Figure 15-8b shows free bodies of the left and right poles, with the loads acting directly on each pole, and the unknown redundant force F. In this case, with uniform lateral load over the full pole height, the lateral deflections of each pole are easily computed using superposition of expressions for deflection of a cantilever. Thus, *for this case,* the lateral deflections of the left and right poles are:

$$\Delta_1 = w_1h^4/8EI - Fh^3/3EI$$

$$\Delta_2 = w_2h^4/8EI + Fh^3/3EI$$

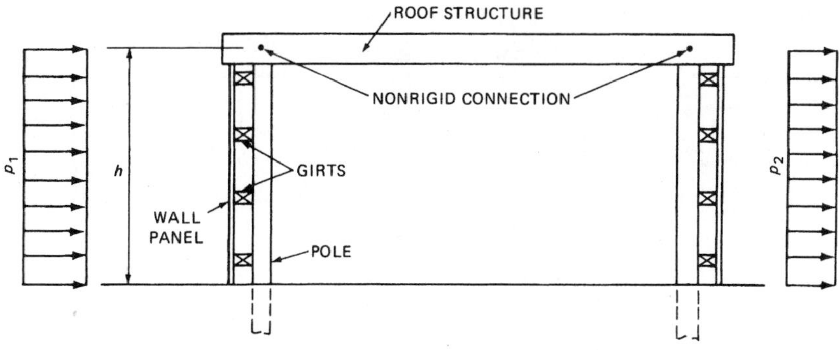

(a) CROSS SECTION (SCHEMATIC) AND WIND PRESSURES ON WALLS (WIND UPLIFT ON ROOF NOT SHOWN)

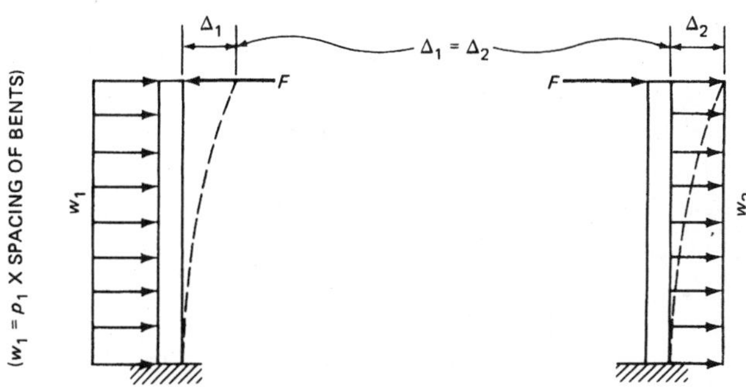

(b) LOADS ON POLES AND POLE DEFLECTIONS

Fig. 15-8. Distribution of wind load to poles.

Equating the two deflections and solving for F gives

$$F = (w_1 - w_2)(3h/16) \qquad (15\text{-}1)$$

Knowing the value of the redundant force, F, we can now solve for the base shear and bending moment for each pole.

Figure 15-9 shows another simple case, a building with outside poles and supporting one floor as well as a roof. Assuming that wind forces on the exposed poles are negligible compared to those on the broad expanse of the light-frame walls, wind loads are transferred to the poles at roof and floor levels only. Since both poles deflect equally at each level and there is negligible distributed load applied along the pole length, *each pole is bent to the same shape.* Thus, the base shears are the same for each pole, and analysis is even simpler than in the first example above.

For most other conditions (such as poles of unequal stiffness or length to the ground line, or more than two poles, or a frame in which the poles are braced to the roof system) the analysis is based on similar principles, but is more complex (5).

Foundation

Space around the embedded pole must be filled with well-compacted soil, gravel, or sand. This backfill is important for three reasons. Skin friction (friction between the pole cylindrical surface and the soil) increases the ability of the pole to transfer vertical force to the soil. The backfill also raises the resistance to lateral loads, and (by reducing the lever arm of lateral loads) reduces the bending moment for which the pole must be designed. Finally, the skin friction also resists uplift forces on the pole.

Greatest economy usually results if the pole itself can transfer the entire vertical load to the soil through end bearing and side friction. How-

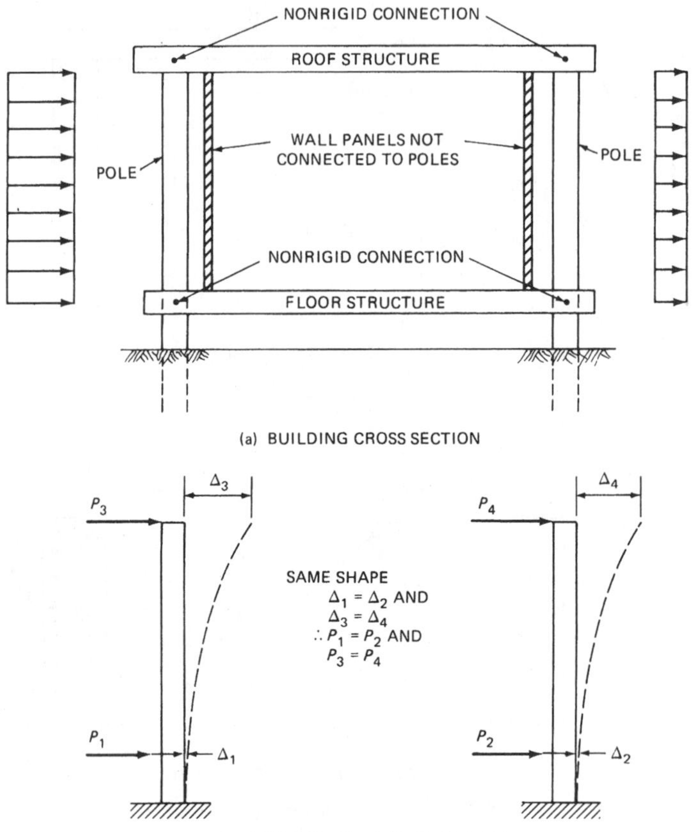

Fig. 15-9. Distribution of shear to poles with concentrated loads only.

ever, if this requires a pole that is either too long or too large in diameter, then other means may be used. The simplest solution in this case is to drill a hole larger than required for the pole itself, casting concrete in the bottom of the hole to serve as a footing, or *punch pad* as shown by Fig. 15-10. The footing must be large enough in diameter to transfer the load to the soil by bearing. It must be thick enough that the end of the pole will not punch through the footing (punching shear failure).

Another alternative is to drill an oversize hole, filling the space around the pole with concrete, all the way to the ground surface, or to the original ground surface, whichever is less. When this is done, it provides an end bearing area larger than that of the pole, and also a larger perimeter for transferring load to the soil by skin friction.

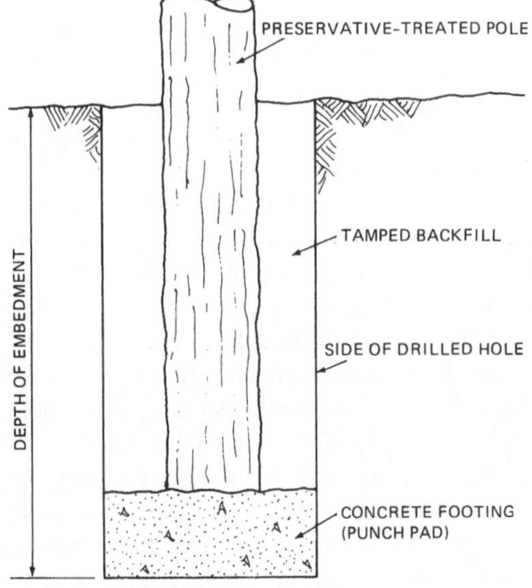

Fig. 15-10. Pole with concrete footing.

Table 15-2. Allowable Foundation and Lateral Pressure.

Class of Materials[1]	Allowable Foundation Pressure (psf)[2] × 0.0479 for kPa	Lateral Bearing lbs./sq.ft./ft. of Depth Below Natural Grade[3] × 0.157 for kPa per meter	Lateral Sliding[4] Coefficient[5]	Lateral Sliding[4] Resistance (psf)[6] × 0.0479 for kPa
1. Massive crystalline bedrock	4,000	1,200	.70	
2. Sedimentary and foliated rock	2,000	400	.35	
3. Sandy gravel and/or gravel (GW and GP)	2,000	200	.35	
4. Sand, silty sand, clayey sand, silty gravel and clayey gravel (SW, SP, SM, SC, GM and GC)	1,500	150	.25	
5. Clay, sandy clay, silty clay and clayey silt (CL, ML, MH and CH)	1,000[7]	100		130

[1] For soil classifications OL, OH, and PT (i.e., organic clays and peat), a foundation investigation shall be required.
[2] All values of allowable foundation pressure are for footings having a minimum width of 12 in. (305 mm) and a minimum depth of 12 in. (305 mm) into natural grade. Except as in Footnote 7 below, increase of 20% allowed for each additional foot (305 mm) of width or depth to a maximum value of three times the designated value.
[3] May be increased the amount of the designated value for each additional foot (305 mm) of depth to a maximum of 15 times the designated value. Isolated poles for uses such as flagpoles or signs and poles used to support buildings which are not adversely affected by a 1/2-in. (13 mm) motion at ground surface due to short-term lateral loads may be designed using lateral bearing values equal to two times the tabulated values.
[4] Lateral bearing and lateral sliding resistance may be combined.
[5] Coefficient to be multiplied by the dead load.
[6] Lateral sliding resistance value to be multiplied by the contact area. In no case shall the lateral sliding resistance exceed one-half the dead load.
[7] No increase for width is allowed.
Source: Reproduced from the 1994 edition of the *Uniform Building Code*™, copyright © 1994, with the permission of the publisher, the International Conference of Building Officials.

Pole Embedment

This is largely a soil mechanics problem rather than one of timber design. To determine necessary embedment requires knowing allowable soil properties. In the absence of field tests, codes such as the UBC (1) often give allowable pressure for various types of soil. Table 15-2, taken from the UBC, is one of these.

Derivations of equations for required embedment depth may be found in texts on soil mechanics. Examples of such equations (from reference 6) are shown below: Eq. 15-2 for cases in which there is lateral restraint at the ground line (such as would be provided by a concrete slab surrounding the pole), and Eq. 15-3 for cases in which there is no such restraint. Figure 15-11 shows dimensions referred to in these equations.

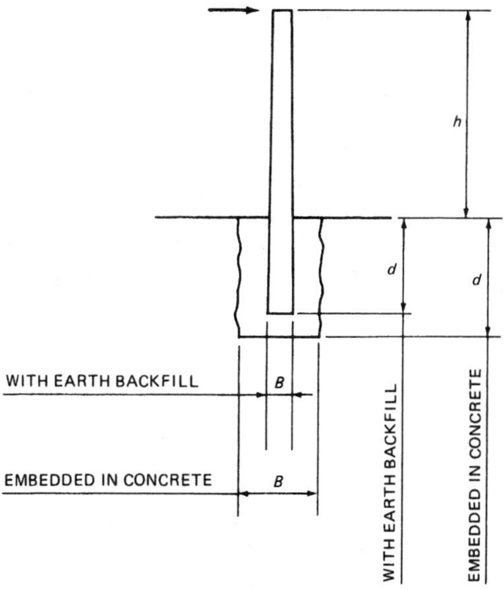

Fig. 15-11. Depth of embedment.

$$d = 1.63 \sqrt[3]{\frac{Ph}{BS_4}} \qquad (15\text{-}2)$$

$$d = 1.97 \sqrt[3]{\frac{P(h + 0.93d)}{BS_4}} \qquad (15\text{-}3)$$

Definitions of other terms in the two equations are:

P = lateral force applied to pole (lb)
S_4 = modified allowable lateral soil pressure as adjusted for load duration and allowable movement (footnote 3 on Table 15-2) (psf per ft of depth)
B = bottom-end diameter of round pole, or diameter of concrete casing (if used), or diagonal dimension of a square pole (ft).

In Eq. 15-3, each side of the equation includes the unknown whose value is sought. Thus, it is necessary to solve by successive approximation (trial and error), assuming a value for the unknown d before each trial. This will be shown by Example 15-1.

Pole Standards

Table 15-3 shows published standard dimensions of poles of several classes. Notice that pole circumferences, rather than diameters, are given by the table. This is because it is easy to measure the circumference of a somewhat round pole, but harder to measure its diameter.

Table 15-3. Dimensions of Douglas Fir (Both Types) and Southern Pine Poles (Based on a Fiber Stress of 8000 psi).

Class		1	2	3	4	5	6	7	9	10
Minimum Circumference at Top (in.)		27	25	23	21	19	17	15	15	12
Length of Pole (ft)	Groundline Distance from Butt[a] (ft)	Minimum Circumferences at 6 Feet from Butt (in.)								
20	4	31.0	29.0	27.0	25.0	23.0	**21.0**	**19.5**	**17.5**	**14.0**
25	5	33.5	31.5	29.5	27.5	25.5	**23.0**	**21.5**	**19.5**	**15.0**
30	5½	36.5	**34.0**	**32.0**	**29.5**	**27.5**	**25.0**	**23.5**	**20.5**	
35	6	39.0	**36.5**	**34.0**	**31.5**	**29.0**	**27.0**	**25.0**		
40	6	**41.0**	**38.5**	**36.0**	**33.5**	**31.0**	**28.5**	26.5		
45	6½	**43.0**	**40.5**	**37.5**	**35.0**	**32.5**	30.0	28.0		
50	7	**45.0**	**42.0**	**39.0**	36.5	34.0	31.5	29.0		
55	7½	**46.5**	**43.5**	**40.5**	38.0	35.0	32.5			
60	8	**48.0**	**45.0**	**42.0**	39.0	36.0	33.5			
65	8½	**49.5**	**46.5**	**43.5**	40.5	37.5				
70	9	**51.0**	**48.0**	**45.0**	41.5	38.5				
75	9½	**52.5**	**49.0**	**46.0**	43.0					
80	10	**54.0**	**50.5**	47.0	44.0					
85	10½	**55.0**	**51.5**	48.0						
90	11	**56.0**	**53.0**	49.0						
95	11	**57.0**	**54.0**	50.0						
100	11	**58.5**	**55.0**	51.0						
105	12	**59.5**	**56.0**	52.0						
110	12	**60.5**	**57.0**	53.0						
115	12	**61.5**	**58.0**							
120	12	**62.5**	**59.0**							
125	12	**63.5**	**59.5**							

[a]The figures in this column are intended for use only when a definition of groundline is necessary in order to apply requirements relating to scars, straightness, etc.

Note: Classes and lengths for which circumferences at 6 ft from the butt are listed in boldface type are the preferred standard sizes. Those shown in light type are included for engineering purposes only.

Source: Courtesy American Wood Preservers Institute.

Suggested Design Procedure

Reference 4 laments with respect to pole buildings, "Unfortunately, because of their simplicity, they are often built without adequate design." Often standard plans are followed, without special attention to details that raise unique problems at a particular site. It is preferable to have all parts of the structural design, especially the foundation, checked by a qualified designer.

Design procedure may vary according to number of poles in a bent, whether the poles are all of the same size, whether the terrain is level or sloping, the relative stiffness of poles and superstructure, and the rigidity of the connection of floor and roof structures to the poles. The procedure suggested here should be satisfactory for most cases, though.

1. Compute design loads on the structure, separately for dead, live, snow, and wind load. (Usually, a combination that includes wind will control the pole design.) Consider both wind pressure on the windward side of the building and suction on the leeward side. Also compute wind forces (if they are significant) on exposed parts of the poles. This requires estimating the pole diameters.

2. Compute distribution of the total wind forces among the poles. Each pole will have shear due to a portion of the total wind force.

3. Estimate the pole circumference and determine the required pole embedment. This requires knowing the soil conditions and allowable values for both direct vertical bearing and lateral passive pressure. These values should preferably be determined by a soils engineer, but often the local building code suggests suitable values.

4. Design the poles for whichever load combination is critical. For poles near the outside wall lines, this is usually a combination that includes wind, in which case the load-duration factor applies to the allowables for both bending and compression (or tension) parallel to grain. (In buildings having an interior row of poles as well as rows along each wall line, the controlling load combination for the interior poles is often the one that causes the greatest vertical load on the pole, such as full snow load but no wind.) In any case, *interaction of bending and axial load must be considered.* Maximum bending moment in the pole occurs at the point of zero horizontal shear. For a pole embedded in earth, this will be at a point below the ground surface; often it is assumed to be at one-fourth of the embedment depth below the surface. If the pole is embedded in concrete or is firmly supported by a concrete slab cast around the pole at surface level, the maximum bending moment may be assumed to occur at the top of the concrete.

5. If a diameter different from the assumed diameter is required, recycle to step 3, above.

6. Design the concrete footing, if one is needed.

Example 15-1

Choose poles for the building shown by Fig. 15-12. Pole bents are 12 ft apart; dead load plus roof live load applies 2000 lb to each pole. No bracing is used at the tops of the poles, so the roof structure can be assumed pin-connected to each pole; that is, roof members can transfer only vertical and horizontal loads (not moment) to the poles. Treated southern pine poles are available. Soil is classified as sandy gravel.

Using the exposure and wind speed of the locale, the designer used the building code to compute wind pressures for these conditions. They were 16.3 psf inward for the windward wall and 11.0 psf outward

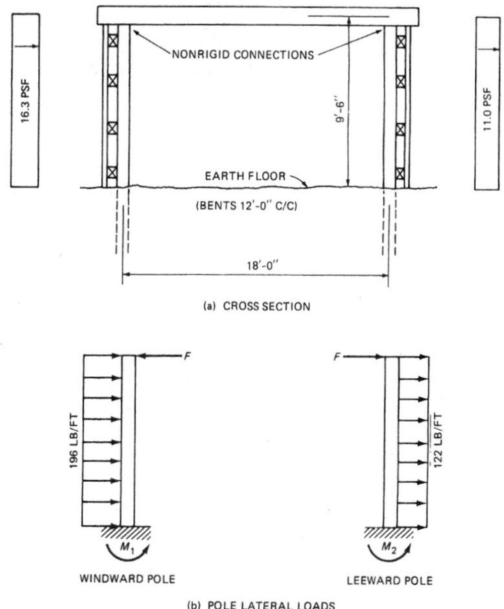

Fig. 15-12. Pole building for Example 15-1.

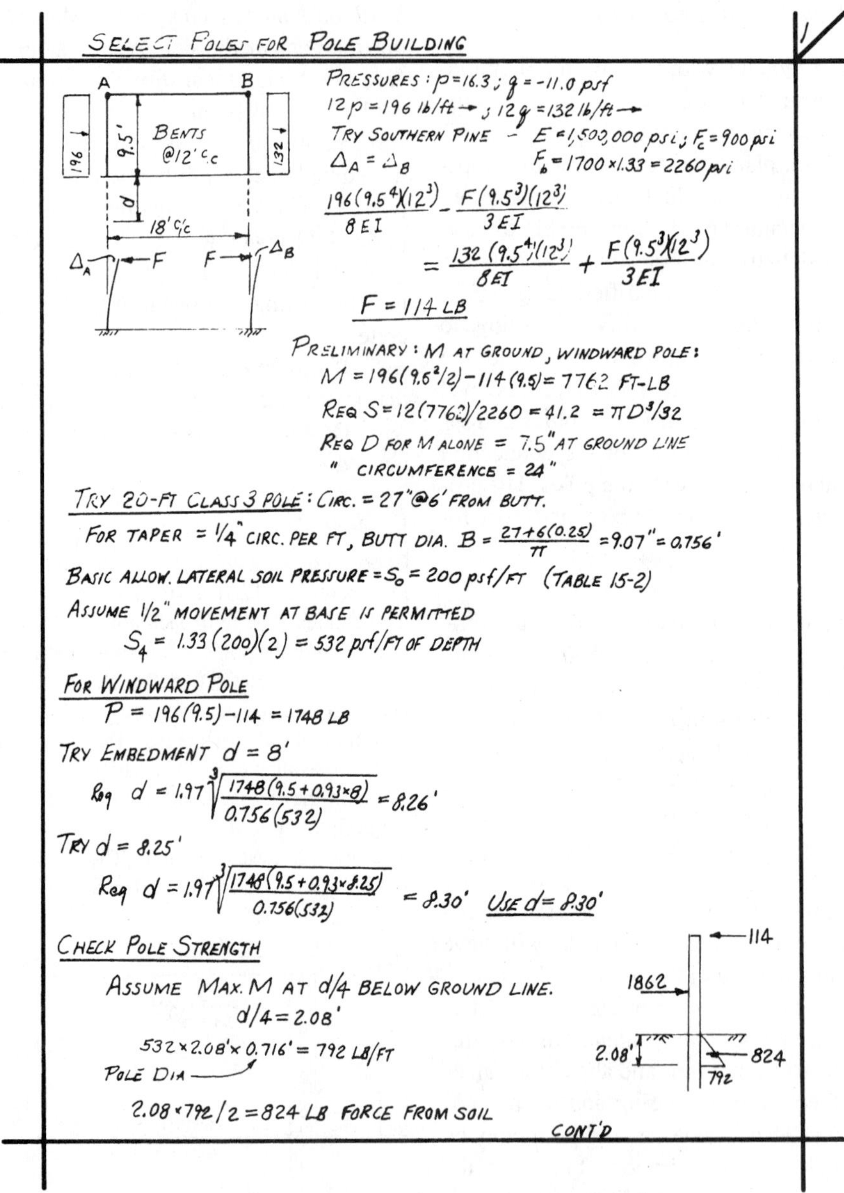

Fig. 15-13. Designer's computations for Example 15-1.

(suction) for the lee wall. Multiplying pressures by the tributary wall width (12 ft) gives the distributed wind load per foot for each pole. Starting with this information, the solution is shown in Fig. 15-13 as a designer's computation sheet.

15-3. WOOD FOUNDATIONS

Light-frame buildings are usually supported by either concrete or masonry walls which, in turn, rest on concrete wall-type footings (Figs. 15-1 and 15-2). An alternative, now feasible using preservative treatment, is to use wood for both foundation wall and footing. This has been done since the early 1960s and the practice is recognized by many building codes. Wood foundations are of several varieties: (1) poles of pole-type buildings, (2) box-beam foundation, (3) wall foundation system, and (4) bent foundation system.

POLE SELECTION - CONT'D 2

$M = 1862(9.5/2 + 2.08) - 114(9.5 + 2.08) - 824(2.08/3) = 10{,}830$ FT-LB

CIRCUMF = 27" @ 2.08' BELOW GROUND

$\therefore S = 62.3 \text{ in}^3$

$f_b = M/S = 12 \times 10{,}830/63.2 = 2086 \text{ psi} < F_b = 2260$

AXIAL LOAD = 2000 − 822 WIND UPLIFT = 1178 LB

AT 2.08' BELOW, POLE CANNOT BUCKLE

$\therefore F_c = 900(1.33) = 1200 \text{ psi}$

$f_c = 1178/58.1 = 20 \text{ psi}$ (58.1 = A)

$\frac{20}{1200} + \frac{2086}{2260} = 0.94 < 1.00$ OK

CHECK COLUMN BUCKLING

AT 1/3 FROM TOP (AFTER CUTTING TO 17.8' LONG) CIRCUMF = 25.53"
AND DIA = 8.13" USE 7.2" EQUIV. SQUARE

$\frac{K_e L}{D} = \frac{2.1 \times 12(11.58)}{7.2} = 40.5$ LONG

$F_c' = 0.3(1{,}500{,}000)/40.5^2 = 274 \text{ psi}$

$f_c = \frac{2000 + 822}{51.8} = 54.5 \text{ psi} < F_c'$

LEEWARD COL. OK FOR AXIAL LD ALONE
" " HAS SMALLER M @ 2.08'

$\therefore$ USE 20' CLASS 3 POLE - SOUTHERN PINE

Fig. 15-13. Designer's computations for Example 15-1, continued.

All four of these types are discussed in reference 4, but only the third type, also known as the *permanent wood foundation system,* is covered here. Over 50,000 buildings have been built in the United States using this type of foundation.

Untreated wood may be used in foundations, but only if it will remain well below the groundwater table at all times. Perhaps the best known of many examples is New York's famous Brooklyn Bridge. The massive timber caissons of this bridge form permanent components of the supporting piers. Being always covered with water, they are protected against all agents of deterioration. Another, but much smaller, example familiar to the authors is a truss bridge in Denver. Its piers are of sandstone blocks, extending to about 2 ft below the water line. The stone piers rest on a grillage consisting of two layers of 12 × 12 timbers (in alternating directions) capped and tied together by a diagonal layer of 3 × 12s. In England, timber piles that had been below the water line for over 2000 years were pulled up, and the oak wood was found to be well preserved and reusable for other purposes.

The modern building foundation using treated lumber differs from these in that it is built entirely above the water line and must be protected so that the wood will remain as dry as possible. Most advantages claimed for using wood foundations for buildings relate to economy: Foundations as well as superstructure are installed by carpenters, so that there is no need to schedule a masonry crew; work on the foundation can continue in freezing weather, and construction is faster.

The permanent wood foundation system uses pressure-treated standard 2-in. nominal lumber and pressure-treated exterior plywood. Figure 15-14 shows a typical example of a basement wall foundation. Reference 7 shows configurations for

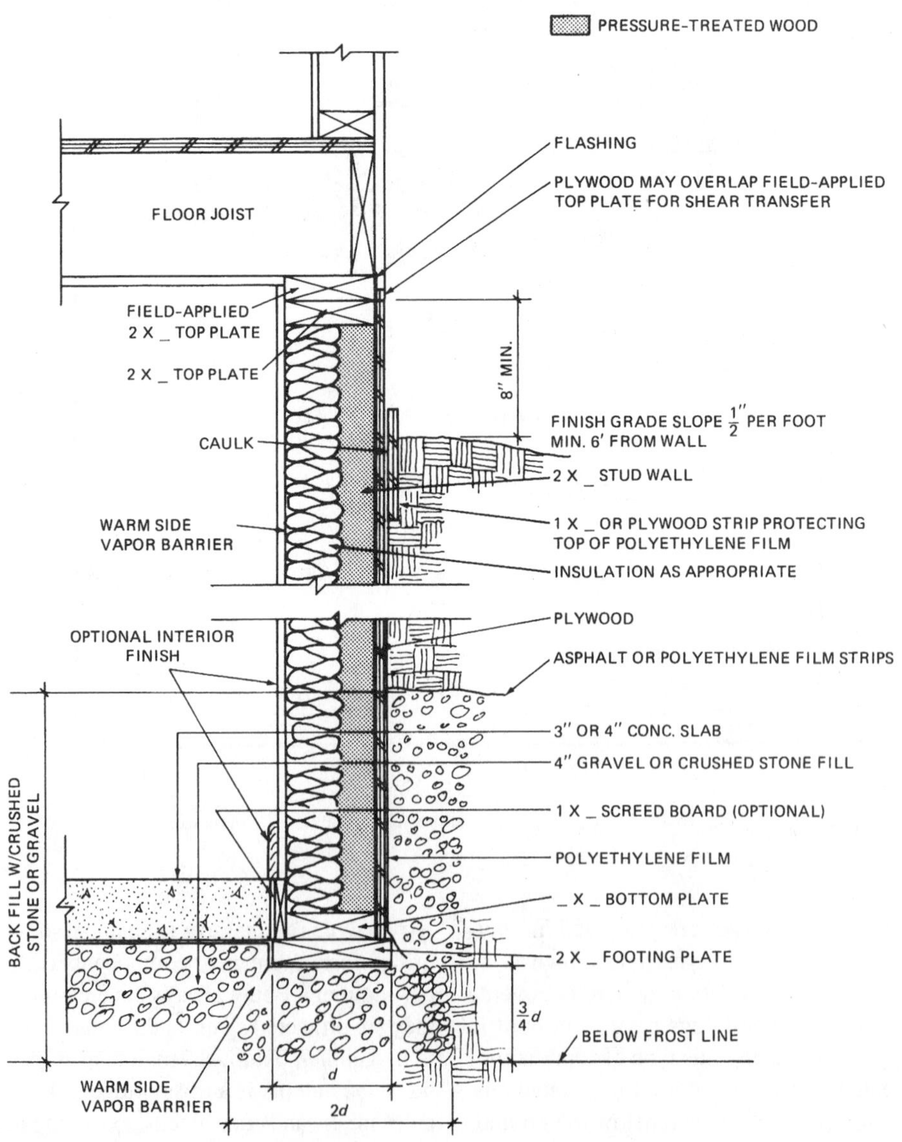

Fig. 15-14. Wood foundation wall for basement. (Courtesy of American Forest & Paper Association, Washington, D.C.)

several other applications and gives specifications for the materials and their installation.

The plywood sheathing on the outside of the foundation wall must be pressure treated with preservative. Similarly, the studs, footing plate, bottom plate, and top plate must be pressure treated. The plywood sheathing, studs, top plate, and bottom plate are preassembled, moved into place, and connected to the wood footing plate. A field-applied top plate and building elements above it need not be pressure treated.

With this type of foundation, it is essential to drain water away from the area and to prevent moisture from being absorbed by the footing plate or the wall foundation assembly. Nevertheless, the footing and the sheathing are designed using allowable stresses for wet use. Nails and other fasteners used in the pressure-treated wood parts must be of silicon bronze, copper, or stainless steel, but galvanized nails are permitted in certain locations. Reference 7 should be used to determine the proper types of fasteners for each application and the proper measures to protect against moisture.

Outside backfill material applies a horizontal force to the wall. Thus, the wall must act as a vertical beam as well as carrying the vertical load from the structure above. The base of the wall is free to rotate, so the wall cannot act effectively as a cantilever. Therefore, it acts as a simple beam, subject to horizontal distributed load from the backfill and having horizontal reactions at the base and at the first-floor structure. The reaction at the first floor causes a compressive force in the floor joists, and the connection of these joists to the top plate must be able to transfer this force. At the bottom, a concrete basement floor slab (or a wood basement floor system) provides the necessary reaction, preventing the bottom of the wall from sliding inward under the backfill pressure.

If the horizontal loads acting on opposite sides of the building are not equal and opposite, then the end walls of the building receive racking loads; that is, they must act as vertical diaphragms to carry the unbalanced load to the foundation. An unbalance of lateral loads will always be present when wind is considered or when the depth of earth fill on opposite walls is not the same.

Suggested Design Sequence

1. Choose a size and spacing of wall studs. This is a trial-and-error procedure. In the United States, the horizontal length of a panel will usually be 8 ft, the length of a sheet of plywood, so 12 in., 16 in., or 24 in., would be suitable stud spacings. The interaction of axial load and bending moment must be considered. Stud bending is caused by the horizontal pressure of the earth backfill against the plywood sheathing. The lateral soil pressure should preferably be determined by a geotechnical engineer, but again, local codes may suggest suitable values. The equivalent fluid pressure method is usually used to compute the lateral design pressures.

Figure 15-15 shows the pressure variation in pounds per square foot. Since the stud spacing may not yet be known, it is best at this stage to determine all wall loads, shears, and bending moments *per foot of wall length.* Top and bottom reactions to the wall pressures must be found next. Maximum bending moment occurs where the shear in the stud is zero. (Finding the location of zero shear and then the maximum bending moment is illustrated by Example 15-2.) Reference 7 uses dimension H for the span of the vertical beam, but the authors believe that to use the dimension shown as HH in Fig. 15-15 is more realistic. This is the distance between supporting members, from the center of the basement slab to the center of the first-floor structure.

2. Check the adequacy of the stud size and spacing using the interaction equation for members with combined axial compression and bending moment (Chapter 7). It is not easy to know in advance which load combination will control, so all reasonable ones should be checked. This may sound like a lot of work, but it is really quite easy if the designer will tabulate the terms going into the interaction equation. Again, Example 15-2 will show how. Notice that the lateral earth pressure is assumed to be a permanent load. Notice also that the stud design is based on dry use, even though the footing and sheathing designs are based on wet use.

Having found a size and spacing that are OK for bending, check to see whether the studs are satisfactory in shear. The critical section for

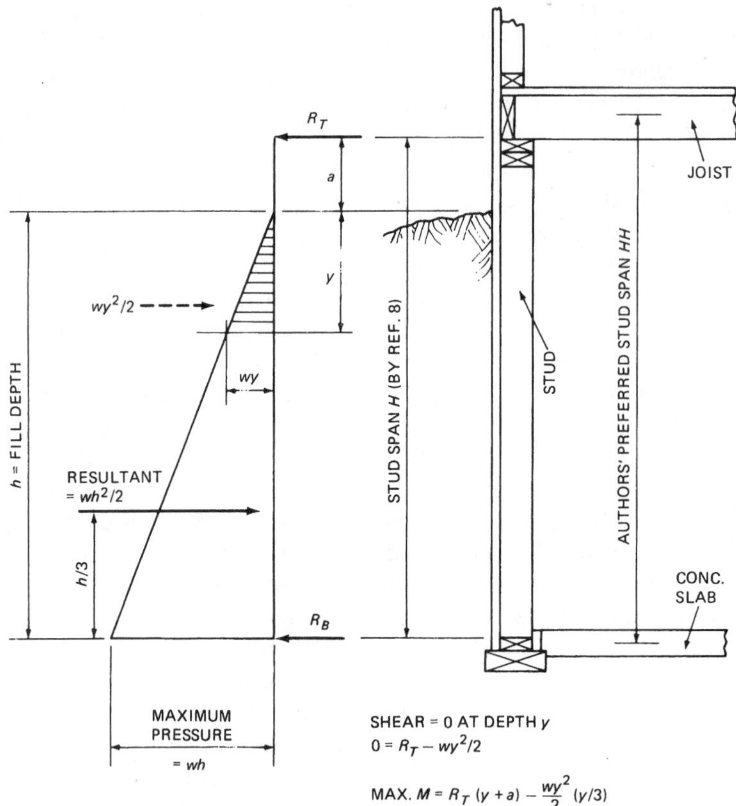

Fig. 15-15. Lateral pressure on foundation wall.

shear will be at a distance equal to stud depth from the bottom (same as for a beam under the NDS).

3. Check the selected studs to be sure that they are not overloaded in compression perpendicular to the grain where they push horizontally against the concrete basement slab (or wood floor system) to transfer the necessary bottom reaction. The allowable stress *is* adjusted for permanent load duration. (This differs from NDS requirements.)

4. Check to determine whether or not the selected studs will overstress the bottom plate in compression perpendicular to the grain. Again, the allowable for the bottom plate *is* adjusted for load duration and for wet use.

5. Knowing the stud size, and knowing that the bottom plate will probably be of the same width, determine an allowable soil bearing pressure and find a required width of footing plate. The allowable pressure should preferably be specified by a geotechnical engineer, but in many cases local building codes will suggest values that are adequate for the area where the building will be located. Allowable soil bearing pressures are *not* adjusted for load duration. The footing width required (inches)

$$= \frac{12(\text{load per ft of wall})}{\text{allowable soil pressure}}$$

For the footing plate, select a standard width of lumber that is equal to or larger than the required width. In further calculations use the *actual* soil pressure rather than the allowable.

If the footing is wider than the bottom plate, it will have transverse bending moment due to the short cantilever that projects beyond the edge of the plate. This bending moment will cause flexural stresses perpendicular to the grain, and the stress on the bottom of the projecting cantilever will be tension. The NDS gives no allowable stress for tension due to bending in this direction (perpendicular to the grain), but reference 8 suggests allowing an amount equal to one-third of the allowable shear stress.

If the footing selected is found to be overstressed, it can be made thicker, or it can be reinforced with plywood. (*Reinforced* is a poor word, however, since the recommended design procedure is to assume that the plywood alone resists the entire bending moment.)

6. Choose plywood for the foundation sheathing, considering bending due to the active soil pressure, and racking. The sheathing acts as a beam, spanning from stud to stud, and is subject to a uniform load (from soil pressure) in that distance. The horizontal bending moment is largest near the bottom (where soil pressure is the highest) and reduces to zero at the level of the top of the soil. The lower edge of the sheathing overlaps the bottom plate, so the plywood there is in two-way bending. Horizontal bending is assumed to be maximum at about ½ ft above the bottom edge, so bending moment (per foot of width) at that depth is used to compute the required plywood thickness. The horizontal bending moment per foot width of plywood at ½ ft above the bottom edge is

$$M = w(h - 0.5)s^2/8 \qquad (15\text{-}4)$$

for single spans or for two-span panels, and

$$M = w(h - 0.5)s^2/10 \qquad (15\text{-}5)$$

for panels that are continuous over three or more spans.

The required effective section modulus per foot width (vertical) of plywood is

$$KS = M/F'_b \qquad (15\text{-}6)$$

In the above equations

F'_b = allowable bending stress adjusted for wet conditions and permanent duration
w = equivalent-fluid weight of soil
s = stud spacing
h = height of backfill (ft)
All with compatible units

The racking force is caused by unbalanced earth loads and by wind (or seismic) loads. Unless the building has diaphragm-type walls crossing the building between the ends, the entire racking force must be resisted by the two end walls alone. If the end walls have openings, that fact must be considered in determining the amount of racking force (shear through the thickness) per linear foot of wall (Chap. 12). The thickness of plywood selected for lateral soil pressure should now be checked using Table 15-4 to determine whether it is satisfactory for racking, and if so, what size and spacing of nails are required at panel edges (7).

7. Provide adequate connections, as follows:

a. Wall studs to top plate. Nails or framing anchors must be sufficient to transmit the inward horizontal reaction at the top of the studs. (Consider load-duration factor.)
b. Field-applied plate to top plate. Nails must transmit force equal to the top reaction of the studs.
c. Floor framing to field-applied plate. Nails or framing anchors connecting the floor system to the field-applied top plate must be able to transfer load at least equal to the top reaction of the studs. (As an alternative, the top end of the studs may transfer their load directly to the floor system by bearing against the ends of the joists. In this case, the studs must be checked for compression perpendicular to the grain.)

Many other details and connection requirements are shown by reference 7.

Example 15-2

Design a wood foundation and basement wall structure for the building shown by Fig. 15-16. Roof trusses (not shown) are parallel to the short walls; thus, the long walls are bearing walls and support all of the roof load.

Loads in pounds per foot of each long wall are:

From roof, 220 dead (includes ceiling weight) and 270 snow
From first floor, 120 dead and 360 live
Wall, first floor to roof, 160 dead

Wind load from building above = 200 lb/ft, with resultant at 15 ft above footing.

Table 15-4. Recommended Allowable Design Shear for Lumber-Framed Foundation Walls Sheathed With $\frac{1}{2}$, $\frac{5}{8}$, and $\frac{3}{4}$ Inch Plywood (Pounds per Foot of Wall).[a, b, c]

	Plywood[a]			Nail Spacing at Panel Edges (in.)[e]				
				All Panel Edges Supported				One or More Panel Edges Not Supported
Loading	Minimum Thickness (in.)	Grade	Nail Size[d]	6	4	$2\frac{1}{2}$	2	6
Wind or wind plus unequal backfill	$\frac{1}{2}$	All grades	8*d*	280	430	640	730	215
	$\frac{1}{2}$	All grades	10*d*	310	460	640	730	230
	$\frac{1}{2}$	Structural I grades	10*d*	340	510	690	780	255
	$\frac{5}{8}$, $\frac{3}{4}$	All grades	10*d*	340	510	690	780	255
Unequal backfill	$\frac{1}{2}$	All grades	8*d*	190	290	430	490	145
	$\frac{1}{2}$	All grades	10*d*	210	310	430	490	155
	$\frac{1}{2}$	Structural I grades	10*d*	230	345	465	530	175
	$\frac{5}{8}$, $\frac{3}{4}$	All grades	10*d*	230	345	465	530	175

[a]Values apply to all plywood conforming to Article 2.2.1 except plywood made with Group 5 species (basswood, poplar, and balsam).
[b]Values shown are for studs of species/grade combinations B-1, B-2, B-3 from Table 3 (Group II per National Design Specification 8.1A). Use 81% of these design shears for combinations C-1, C-2, C-3, D-1, D-2, D-3 in Table 3 (NDS Group III). For NDS Group IV studs with 10*d* nails use 65% of these values, or 62% for Group IV with 8*d* nails.
[c]Design shears for $\frac{3}{8}$-in. plywood used in crawl space construction shall be 90% of the values for 8*d* nails. This adjustment is cumulative with the adjustment per footnote b.
[d]Nails to be hot-dipped zinc-coated steel; or copper, silicon bronze, or stainless steel Types 304 or 316 having equivalent lateral load capacity. Values apply to standard common wire nails or box nails of the following minimum sizes: 6*d*—2.5" × 0.113" (0.281" head); 10*d*—3" × 0.128" (0.312" head).
[e]All panel edges backed with 2-in. nominal or wider framing, except where noted. Space nails at 12 in. c/c along intermediate framing members.
Source: Courtesy of American Forest and Paper Association, Washington, DC.

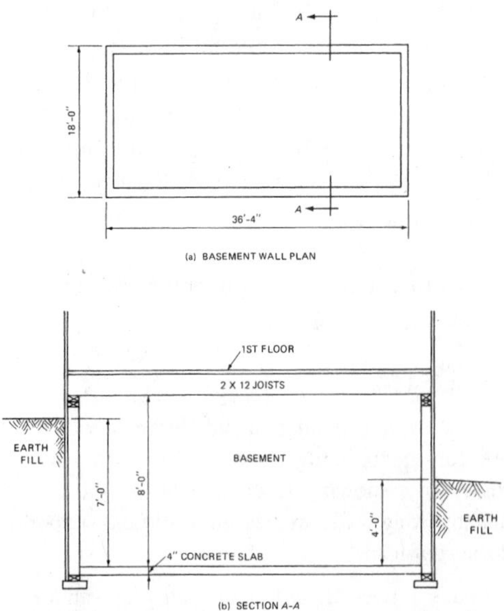

Fig. 15-16. Building for Example 15-2.

There is active lateral soil pressure; assume it is equal to pressure from a 30-pcf equivalent fluid. Use NDS allowable stresses for No. 2 D. fir-larch. The geotechnical engineer has specified an allowable soil bearing pressure of 2000 psf.

Figure 15-17 shows the designer's solution. Footing loads are calculated first. Load combinations considered by the designer were only: D alone, $D + L$; $D + L + S$; and $D + L + W$. The designer considered simultaneous full snow load and full wind load to be improbable. (However, some local building codes might require considering a combination including wind and a reduced snow load, or an unsymmetrical snow load.)

After finding the vertical loads per foot of wall, the designer next considered the lateral active pressure of the earth backfill against the stud wall. For this, she considered only the wall with the greater backfill depth. (Bending moments for the other wall will be smaller.)

Up to this point the design calculations are all in terms of force per unit length (ft) of wall. For studs at 16 in. c/c, all of these values are 16/12 times as large. Composite action of the studs and the plywood sheathing is not considered, so the studs alone must resist the entire vertical load and bending moment. The studs can buckle with respect to their major axis, but the plywood sheathing prevents them from buckling in their weak direction. After finding a stud size and spacing that were satisfactory for combined axial load and bending, the designer verified that they were also OK in longitudi-

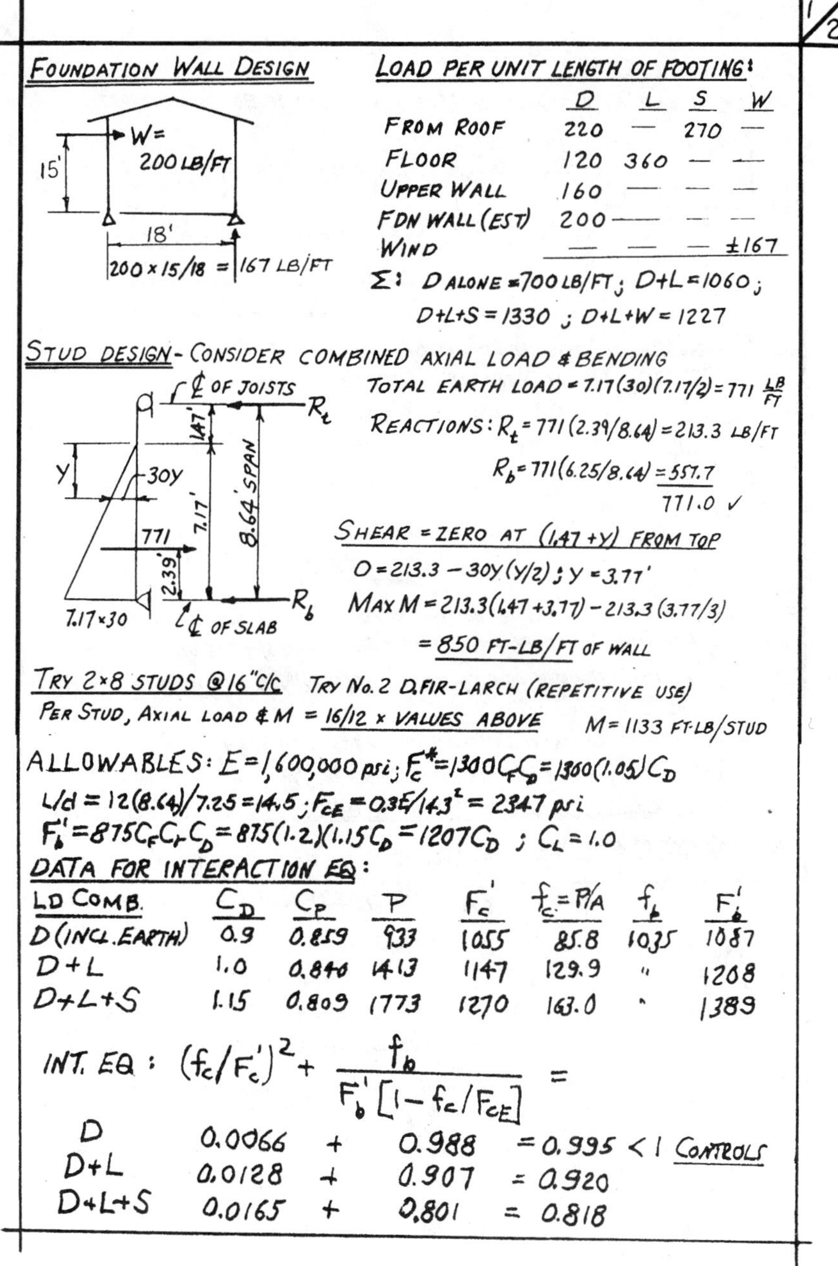

Fig. 15-17. Designer's calculations for Example 15-2.

nal shear. (The designer should also have considered end bearing and lateral bearing, but decided instead to inflict this task on the reader, by means of Problem 15-5.)

Finally, the designer selected a footing size and a thickness and grade of plywood sheathing. The table referred to for checking racking strength of the end walls shows that *all* edges of the sheathing must be nailed either to studs or to blocking. The sheathing will have its face grain horizontal. The plywood sheet is only 4 ft. 0 in. wide, so this will require that solid 2 × 4 (or larger) blocking be nailed between the studs, 48 in. from the bottom, so that the horizontal edges of the plywood sheets can be supported.

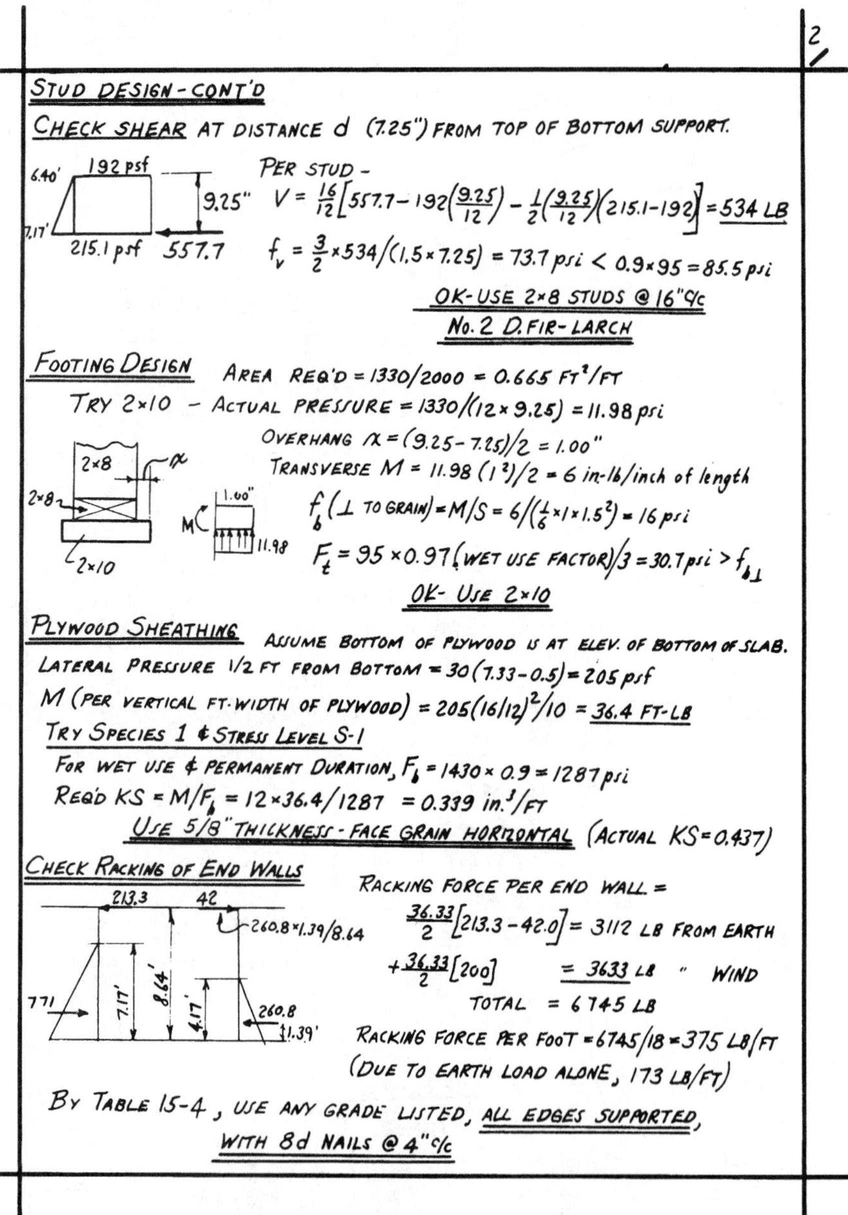

2

STUD DESIGN - CONT'D

CHECK SHEAR AT DISTANCE d (7.25") FROM TOP OF BOTTOM SUPPORT.

6.40' 192 psf 9.25" 7.17' 215.1 psf 557.7

PER STUD -

$V = \frac{16}{12}\left[557.7 - 192\left(\frac{9.25}{12}\right) - \frac{1}{2}\left(\frac{9.25}{12}\right)(215.1 - 192)\right] = 534$ LB

$f_v = \frac{3}{2} \times 534/(1.5 \times 7.25) = 73.7$ psi $< 0.9 \times 95 = 85.5$ psi

OK - USE 2×8 STUDS @ 16" c/c
No. 2 D.FIR-LARCH

FOOTING DESIGN AREA REQ'D = 1330/2000 = 0.665 FT²/FT

TRY 2×10 - ACTUAL PRESSURE = 1330/(12 × 9.25) = 11.98 psi

OVERHANG $x = (9.25 - 7.25)/2 = 1.00"$

TRANSVERSE $M = 11.98\,(1^2)/2 = 6$ in-lb/inch of length

$f_b(\perp$ TO GRAIN$) = M/S = 6/(\frac{1}{6} \times 1 \times 1.5^2) = 16$ psi

2×8 2×8 2×10 M 1.00" 11.98

$F_t = 95 \times 0.97$ (WET USE FACTOR)$/3 = 30.7$ psi $> f_{b\perp}$

OK - USE 2×10

PLYWOOD SHEATHING ASSUME BOTTOM OF PLYWOOD IS AT ELEV. OF BOTTOM OF SLAB.

LATERAL PRESSURE 1/2 FT FROM BOTTOM = 30(7.33 - 0.5) = 205 psf

M (PER VERTICAL FT-WIDTH OF PLYWOOD) = $205(16/12)^2/10 = 36.4$ FT-LB

TRY SPECIES 1 & STRESS LEVEL S-1

FOR WET USE & PERMANENT DURATION, $F_b = 1430 \times 0.9 = 1287$ psi

REQ'D $KS = M/F_b = 12 \times 36.4/1287 = 0.339$ in.³/FT

USE 5/8" THICKNESS - FACE GRAIN HORIZONTAL (ACTUAL KS = 0.437)

CHECK RACKING OF END WALLS

213.3 42 260.8×1.39/8.64 771 7.17' 8.64' 4.17' 260.8 1.39'

RACKING FORCE PER END WALL =

$\frac{36.33}{2}[213.3 - 42.0] = 3112$ LB FROM EARTH

$+\frac{36.33}{2}[200] = 3633$ LB " WIND

TOTAL = 6745 LB

RACKING FORCE PER FOOT = 6745/18 = 375 LB/FT

(DUE TO EARTH LOAD ALONE, 173 LB/FT)

BY TABLE 15-4, USE ANY GRADE LISTED, ALL EDGES SUPPORTED, WITH 8d NAILS @ 4" c/c

Fig. 15-17. Designer's calculations for Example 15-2, continued.

15-4. TIMBER BRIDGES

Timber bridges have been used for centuries, yet wood's potential for modern bridge construction is not generally appreciated. Concern about deterioration is probably the main reason for selecting steel or reinforced concrete rather than wood for bridge construction. But, if the wood is properly treated with preservative and the bridge is properly designed, a timber bridge will give trouble-free service for many years. Given the rate at which steel bridges corrode and concrete bridges crack and spall, there is every reason to consider wood for new bridges of short or moderate spans.

Advantages of wood for bridges are many. Wood is not adversely affected by deicing agents such as salt. It often more economical than other materials. Construction is easier and can be done in weather that would not be suitable for concrete work, for steel painting, or for welding. Fairly large prefabricated wood assemblies can be installed easily without long delays to traffic. That wood bridges are durable when properly designed and properly protected from the elements is proved by experience with covered bridges. There are still several hundred in use in the United States, most of them over 100 years old (8), and all built before the days of preservative treatment.

Timber bridge design is not covered in a comprehensive manner here; to do that would require much more space, perhaps several volumes. Rather, the subject of timber bridge design is merely introduced, with examples to illustrate a few of the main ways in which wood bridge design differs from that for buildings. Most of the principles for wood building design given earlier in this book apply also to bridge design.

Bridges can be designed by either the allowable stress design (ASD) or the load and resistance factor design (LRFD) method. The American Association of State Highway and Transportation Officials (AASHTO) in its 1994 edition of *Standard Specifications for Highway Bridges* shows both methods (9). Examples shown in this chapter using the ASD method could also be done by LRFD, following (with slight changes) the procedures illustrated in Chapters 5 through 9 of this book for LRFD design of beams, columns, and connections.

The AASHTO procedure for design by LRFD differs from that for ASD, however, as follows:

1. Under AASHTO, four different limit states must be considered: strength, service, extreme event, and fatigue and fracture.
2. Impact loads are computed in a different manner for LRFD design of wood bridges.
3. A deflection limit of $\ell/425$ due to live load is required.
4. Base resistance values given by AASHTO are for wet use and a two-month load duration. Thus, for dry use the base value is *increased;* that is, the value of C_M is > 1.0. For load duration greater than two months, the base value is decreased; thus the numerical values of C_D are different.

Only those parts of the AASHTO specifications (9) needed to accompany the design examples are given here. A designer intending to do wood bridge design should consult the complete AASHTO specification and the many references on timber bridges given at the end of this chapter.

Types of Wood Bridges

Timber bridges serve many purposes; they carry pedestrian, roadway, or railway traffic. Special-purpose bridges include those carrying pipelines, flumes, or conveyors. Similarly, timber bridges are of many structural types. Even suspension bridges (except for the main cables, of course) are built of wood.

Trestles. Wood trestles are used for both highways and railways. They are the most common type of wood bridge for railway use. Trestles consist of longitudinal beams (stringers) and cross ties, supported by closely spaced timber bents. (See Fig. 15-18.) Each bent has four or more vertical members, either wood posts resting on concrete footings or wood piling driven into the ground to serve as both foundation and vertical member of the bent. Usually, the outer post (or pile) on each side is battered (slightly sloped) so that the bent will be able to resist lateral load. The posts (or piles) are capped with a heavy transverse member, 12 × 12 or larger, and the stringers rest on this cap. Solid blocking placed between the stringers at their ends holds them in line, prevents them from rotating, and serves as a fire stop. Similar blocking at the one-third points of the stringer spans restrains the stringers against lateral buckling.

Girder Bridges. The type of timber bridge most widely used for highways is the bridge with a transverse deck supported on several longitudinal stringers, which may reach across the entire span of the bridge. Girder bridges may also have the deck spanning between transverse floor beams, which in turn are supported by longitudinal girders. The girders may be either

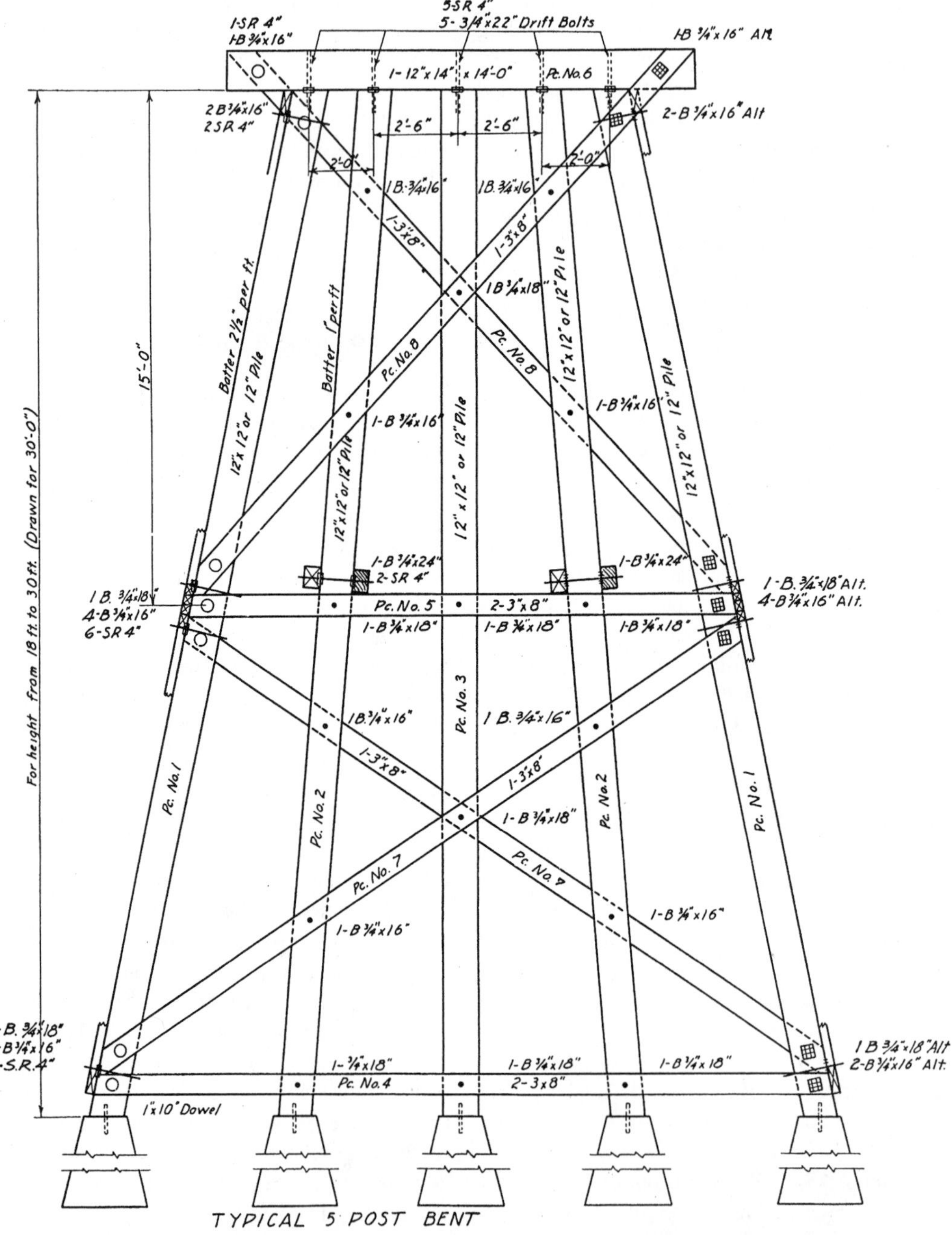

Fig. 15-18. Typical bent for railway trestle. (Courtesy of Cleveland Steel Specialty Company, Cleveland, OH.)

sawn-timber members or glulams. The choice will depend on the span. Thirty feet is probably the maximum practical span for sawn timbers, but with glulam girders, clear spans of up to 130 ft have been possible.

Truss Bridges. Wood truss bridges are practical for spans as high as 250 ft (reference 4) and may be either through-truss (see Fig. 15-19) or deck-truss bridges. Deck-truss bridges are usually the more economical.

Fig. 15-19. Pedestrian and pipeline bridge of 170-ft clear span. Bowstring trusses are made entirely of glulams. Top chord is braced by outrigged diagonals attached to ends of floor beams. (Courtesy of Western Wood Structures, Inc.)

Truss design involves no principles not already covered in Chapter 10. Perhaps the main difference between trusses for bridges and those for buildings is that sturdier cross frames are required between bridge trusses to prevent top-chord buckling. Also, horizontal-plane bracing to resist lateral loads is necessary. This lateral bracing system usually consists of a truss in a horizontal plane at the level of the top chord and another at bottom-chord level. The lateral-system trusses usually share chord members with the main, vertical trusses. (See Fig. 15-20.) The transverse floor beams act as members of one of the lateral-system trusses.

Top-chord bracing for a deck-truss bridge is easy to provide, since it will lie just below the bridge deck. If the deck is rigid enough to act as a horizontal diaphragm, it could possibly serve to brace the top chord against lateral buckling and also to resist lateral loads from traffic, wind, and other causes. This does not justify omitting a top-chord lateral system, however, since one will be needed to hold the assembly of trusses, floor beams, and stringers in line during construction.

For a through-truss bridge, however, the top-chord bracing must be high enough above the roadway to provide necessary clearance for vehicles. A compromise is the half-through or "pony truss" bridge, in which the top chord is laterally braced by outrigged diagonal members, located on the outer side of the truss and connecting the top chord to projecting ends of the transverse floor beams.

Panel lengths for bridge trusses are determined by what is a practical and economical length for the longitudinal stringers, that is, the spacing of the transverse floorbeams.

Wood trusses have declined in popularity, largely because of the difficulty of protecting the members and connections from the weather. However, using glulams makes possible longer chord members, with fewer splices. When the glulam chords are used with web members of the same thickness and with steel gusset plates on each side (as discussed in Section 10-8 for buildings), problems of weathering should be reduced. This type of construction and perhaps the increasing use of exotic, durable wood species may bring about a return in the popularity of truss bridges for new construction.

Other Structural Types. Arch bridges are possible and practical using glued laminated members. Such bridges may use either two-hinged or three-hinged arches, depending on the spans.

For short spans, bridges may consist of

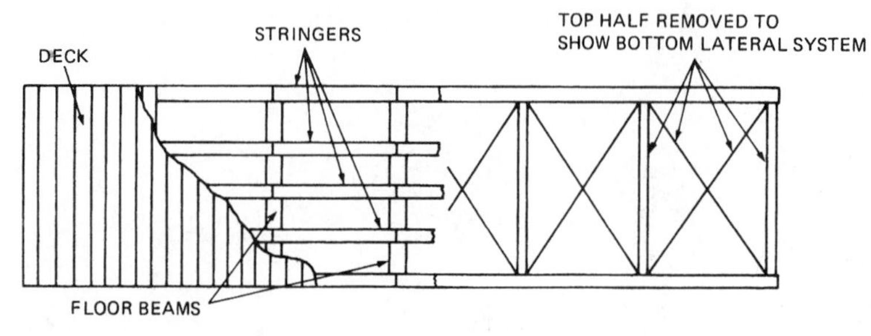

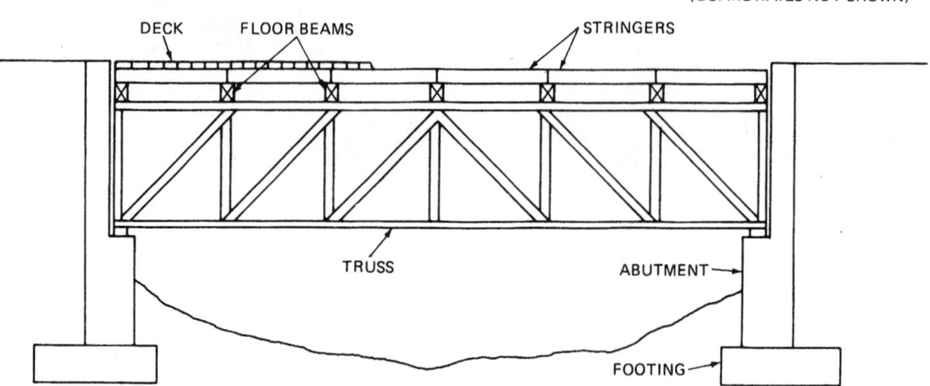

Fig. 15-20. Deck-truss bridge—schematic only, showing "anatomy."

merely a deck with no other supporting members. The longitudinal deck bridge would be most effective if adjacent elements could be connected to each other well enough to ensure that the deck acts as a structural plate. Though this ideal situation probably cannot be reached, progress toward it is being made. Devices being considered to make the performance of such decks more effective are transverse distributor beams, doweling of adjacent panels, and transverse post-tensioning.

Wood may be used for bridges in combination with other materials; for example, a wood deck may be used with a system of steel beams and girders or with steel trusses.

15-5. BRIDGE DECK TYPES

Wood bridge decks include plank decks, nailed laminated decks, glued laminated decks, and composite decks. Which type to select depends on traffic density, expected life of the bridge, and economics.

Plank Decks

The simplest deck is the plank deck, shown by 15-21a. Unfortunately, the plank deck is suitable only for light traffic or for temporary structures. It consists of transverse planks, set flatwise and spiked to the stringers. When unseasoned planks are used, they are placed in side-by-side contact. As the planks shrink, gaps form between adjacent planks. Where seasoned planks are used, they should be placed with a gap between planks, so as to allow for swelling that will occur in wet weather.

To protect the planks from abrasion, longitudinal running planks are usually placed above the transverse ones, along the line the vehicle wheels will follow. Occasionally, the longitudinal planks are placed all across the width of the bridge to create a two-layer deck.

The problem with the plank deck is that its open spaces allow water to accumulate on the stringers below. If the stringers are not treated with preservative, rotting is a problem. An as-

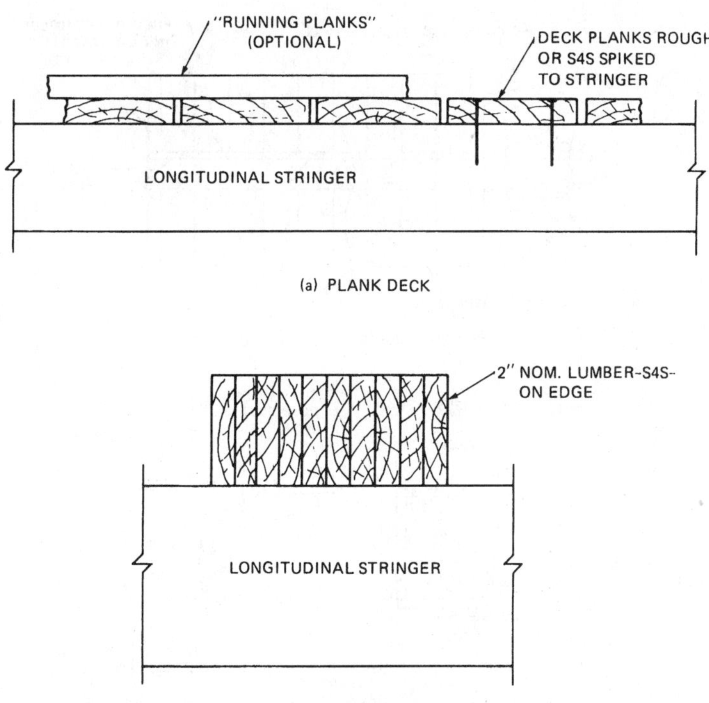

Fig. 15-21. Two types of wood deck.

phaltic wearing surface may be added to the plank deck but does little to protect the wood from moisture. Since the adjacent planks deflect independently as the load rolls across, the asphalt surface cracks and moisture penetrates easily.

Nailed Laminated Decks

Nailed laminated decks are made using 2-in. nominal planks placed with their larger cross-sectional dimension vertical, as shown by Fig. 15-21b. The planks are nailed to each other with 30*d* nails, long enough to penetrate three planks. Each plank is also toe nailed to alternate stringers. Compared to the plank deck, the nailed laminated deck gives a little better protection to the stringers below and can be both stronger and stiffer.

Nailed laminated decks have the disadvantage of the nails loosening with repeated loading by traffic and with alternate wetting and drying.

When the laminations run parallel to the bridge span, they are sometimes "stress-laminated"; that is, pressed together firmly by transverse post-tensioning rods that pass through all the laminations and are anchored at the sides of the bridge deck. Nails that have loosened because of wood shrinkage and swelling cannot be reached to drive again. However, transverse post-tensioning rods can be retightened to counter the effect of prestress loss due to shrinkage or to creep (10).

Figure 15-22 shows two methods of post-tensioning: one for new work, and one for repairing old nailed laminated decks. In the latter, the upper steel rods are embedded in the pavement material, and as long as the pavement is intact, they are protected from damage by bridge traffic.

Glued Laminated Decks

The glued laminated deck is also made of 2-in. nominal lumber placed with its larger cross-sectional dimension vertical; that is, it is vertically laminated. The depth of the deck varies from 4- to 16-in. nominal according to the width of the

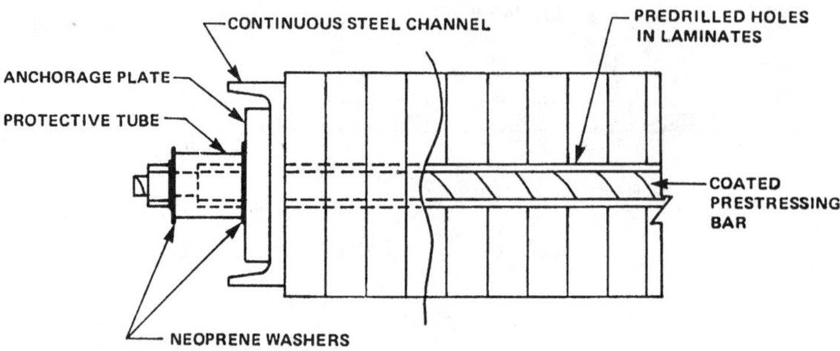

(a) Prestress Detail for New Designs

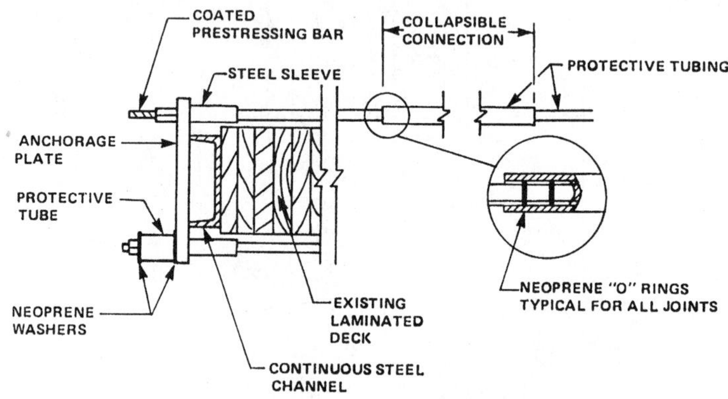

(b)Prestress Detail for Rehabilitation

Fig. 15-22. Post-tensioning laminated decks. (Courtesy of Ontario Ministry of Transportation and Communications.)

lumber used. As for glulam beams, material is planed from both surfaces after gluing, so as to remove irregularities. Thus, the deck depths available are the same as the standard widths available for glulam beams (Chapter 8).

The deck is preassembled into panels 3–4 ft wide and of length equal to the width of the bridge. Panels can be fully shop fabricated, with all necessary holes drilled, and the panel can then be treated with preservative before delivery to the bridge site.

The glued laminated deck is stiffer than either of the other deck types discussed above. Being stiffer, it may allow the use of a wider spacing between stringers or beams. Since it provides an almost solid surface, without gaps, it offers the best protection to the supporting wood members below. Tests show (11) that glued laminated decks are as much as 45% stronger than nailed laminated decks of the same thickness.

Tests show that the glued laminated decks are more effective than nailed laminated ones for distributing concentrated wheel loads, deflection directly under the load being much less than it is with an equivalent thickness of nailed laminated deck. When the laminations are merely nailed to each other, there is slip between adjacent laminations. There is no slip when they are glued together.

Glued laminated panels may be doweled to each other in the bridge deck, as shown in Fig. 15-23, or they may be undoweled. The purpose of the steel dowels is to force each panel to act together with the adjacent ones rather than deflecting independently. Design procedures for doweled glued laminated decks were developed at the Forest Products Laboratory (11) and have become the basis of the AASHTO design method (9).

Glued laminated decks find application not only for new bridges, but also for repair and upgrading of existing bridges (12). Nondoweled construction is often used. The laminated panels in this type construction are connected by a clip

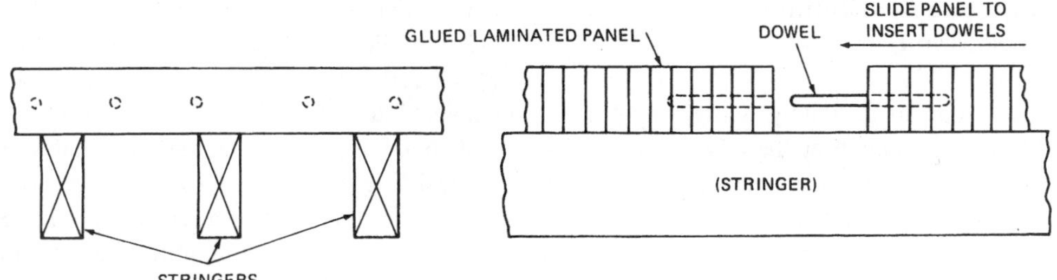

Fig. 15-23. Glued laminated panel deck.

Fig. 15-24. Connection detail. Attachment of nailed laminated deck to glulam crossbeams of pedestrian bridge. (Photograph by authors.)

system (5). Clips show on the underside of a pedestrian bridge in Fig. 15-24.

15-6. BRIDGE DESIGN CRITERIA

The AASHTO specifications (9) give design criteria for highway bridges.

Highway Bridge Loads

Truck wheel loads, as defined by AASHTO, are a specified percentage of the gross weight of the truck and its contents. For example, each front wheel carries 10% of the gross weight and each rear wheel carries 40%. The width of traffic lane required is 12 ft. If a bridge has more than two lanes, the AASHTO specification allows reducing the design load used so as to consider the lesser probability that several lanes will be fully loaded at the same time. If there are three lanes, only 90% of full load need be considered; if there are four or more lanes, only 75% need be used in design.

Wheel Load Distribution to Deck Members

The first step in designing a highway bridge deck is to determine how the wheel load is distributed; that is, over what area is the wheel load actually applied, and what width of deck participates in resisting this load?

Figure 15-25a shows two views of a wheel load and gives the length and width of its area of contact with the bridge deck. Assuming that the tire air pressure is 100 psi, the area required (square inches) to carry the wheel load is $0.01P$ (P is the wheel load in pounds). AASHTO specifies that the width of the tire "footprint" is to be 2.5 times its length. Figure 15-25b shows this wheel footprint as the loaded area on a deck whose members run transverse to the traffic and rest on longitudinal stringers.

The wheel load deforms the deck, causing a dishlike depression around the loaded area. Contour lines drawn on the deck surface in Fig. 15-25b indicate the dished shape. The beam strip directly under the load deflects and, being connected to adjacent strips, causes them to bend also, but to lesser degree. The total necessary resisting moment is provided by combined action of the strip immediately beneath the load and strips alongside. Computing the resistance is facilitated by assuming the wheel load to be distributed over an effective width of strip that is defined by the AASHTO specifications (9).

Just what width may be assumed to be effective depends on the manner of construction. The effective width is largest for decks whose adjacent elements are most firmly connected to each other. It is the largest for glued laminated decks and least for decks that consist of merely adjacent planks that are not connected to each other.

AASHTO gives two different sets of requirements for distribution of wheel loads to an effective width of deck; one is for a deck structure

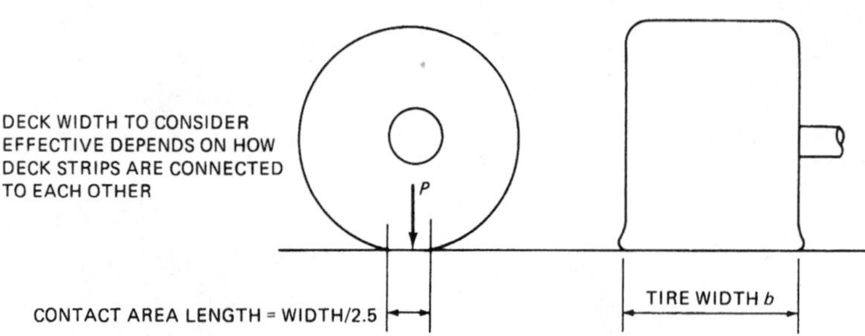

(a) LENGTH AND WIDTH OF TIRE IN CONTACT WITH DECK

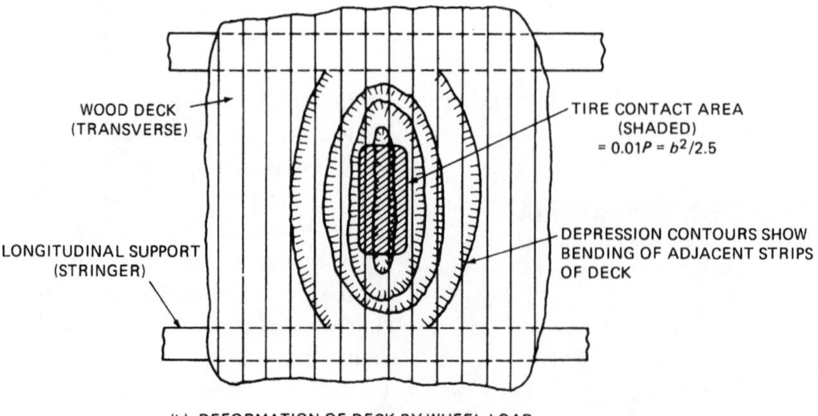

(b) DEFORMATION OF DECK BY WHEEL LOAD

Fig. 15-25. Distribution of wheel load on transverse wood deck.

that is transverse to the direction of traffic movement; the other is for a deck structure parallel to traffic movement. The rules are summarized here.

For transverse flooring: In the direction of the flooring span, the wheel load is distributed over the width of the tire. Normal to the span (i.e., in the direction of traffic) the load shall be distributed over a length as shown below.

For plank floors, distribution is over the width of plank only. For nailed laminated floors with laminated panels not interconnected, distribution is over a 15-in. width, but not more than the width of a laminated panel. For noninterconnected glued laminated panels, distribution is over a width equal to 15 in. plus the floor thickness, but not more than the panel width. For interconnected glued laminated panel floors, the wheel load may be distributed over 15 in. plus twice the floor thickness (but not more than the panel width) provided the floor panel is not less than 6 in. thick and has adequate shear transfer between panels.

For longitudinal flooring: In the flooring span direction, no distribution is allowed; the wheel load is considered as a point load.

Normal to the flooring span (i.e., transverse to the traffic direction) the load is distributed over the width of a plank for a plank floor. For noninterconnected nailed laminated or glued laminated panels, use the width of tire plus floor thickness, but not more than the panel width. For adequately interconnected nailed laminated or glued laminated panels, use the wheel width plus twice the floor thickness.

Deck Spans

In either direction, the span of the deck should be taken as the smaller of:

Clear span plus one-half the width of the supporting beam

Clear span plus the floor thickness

Wheel Load Distribution to Longitudinal Stringers

Figure 15-26 shows several bridge stringers and one wheel load near the center of the stringer span. The stringer immediately beneath the wheel load deflects, and the stiffness of the deck members (not shown) causes adjacent stringers to deflect also. Thus, the stringer beneath the

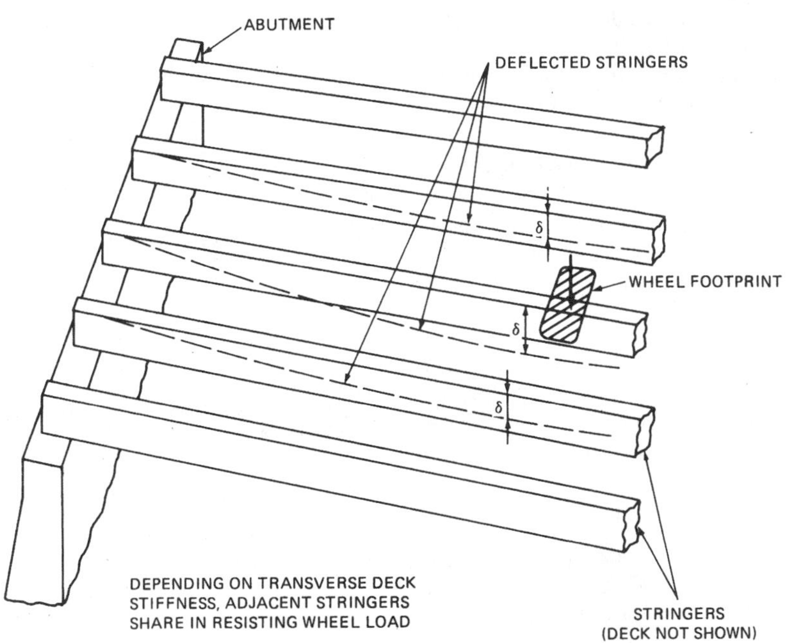

Fig. 15-26. Distribution of wheel load among longitudinal stringers.

load is getting help from the adjacent stringers; they are participating in resisting the bending moment due to the load shown.

AASHTO defines the effect of this participation by a table showing what fraction of the wheel load is to be assumed as resisted by a single beam. Table 15-5 shows those portions of the AASHTO table that apply to the design of timber bridges.

Table 15-6 summarizes the AASHTO load distributions for transverse glued laminated timber decks. Distributions for design of longitudinal glued laminated decks are recommended by reference 13.

Example 15-3

A rural bridge consisting of a simple plank deck on wood stringers is to carry occasional H-15-44 trucks. A one-lane bridge will be sufficient. Not over 20 truck passages per day are expected, and it is intended that the bridge will have only an eight-year useful life. Design the bridge using ASD. Use concrete abutments (end supports for the stringers). Provide a running strip of longitudinal planks and an 8 × 8 rough wood curb. The clear span between abutments will be 18 ft 0 in.

For allowable stress design, load-duration factors are the same under AASHTO as under NDS. AASHTO defines the load duration for live load as two months. Thus, the AASHTO live-load duration factor is 1.15, the same as for snow load under NDS.

Deck plank design: The bridge cross section will be as shown by Fig. 15-27a. As a first step in designing the deck, the designer must decide how many stringers to use and must estimate the stringer width. If either guess proves wrong, the designer must start over with the correct assumptions. For this bridge, the designer has decided to try eight stringers (seven spaces between stringers) and estimates a stringer width of 9½ in.

Assuming 1.79 ft c/c of stringers, the clear distance between stringers is $1.79 - 9.5/12 = 1.00$ ft

Assuming that 3-in. rough planks will be used, the deck span will be the smaller of

$$1.00 + 0.5(9.5/12) = 1.40 \text{ ft} \quad \text{or}$$

$$1.00 + 3/12 = 1.25 \text{ ft} \quad \textit{Controls}$$

For an H-15 truck (15 tons gross wt), one rear wheel load is 0.4 (30,000) = 12,000 lb. The 12,000-lb wheel load is to be distributed over an area of 0.01 ×

Table 15-5. Distribution of Wheel Loads in Longitudinal Beams.

Kind of Floor	Bridge Designed for One Traffic Lane	Bridge Designed for Two or More Traffic Lanes
Timber:[a]		
Plank[b]	$S/4.0$	$S/3.75$
Nail laminated[c] 4″ thick or multiple layer[d] floors over 5″ thick	$S/4.5$	$S/4.0$
Nail laminated[c] 6″ or more thick	$S/5.0$ If S exceeds 5′ use footnote f	$S/4.25$ If S exceeds 6.5′ use footnote f
Glued laminated[e] panels on glued laminated stringers		
4″ thick	$S/4.5$	$S/4.0$
6″ or more thick	$S/6.0$	$S/5.0$
	If S exceeds 6′ use footnote f	If S exceeds 7.5′ use footnote f
On steel stringers		
4″ thick	$S/4.5$	$S/4.0$
6″ or more thick	$S/5.25$	$S/4.5$
	If S exceeds 5.5′ use footnote f	If S exceeds 7′ use footnote f
Concrete:		
On timber stringers	$S/6.0$	$S/5.0$
	If S exceeds 6′ use footnote f	If S exceeds 10′ use footnote f

S = average stringer spacing in feet.

[a]Timber dimensions shown are for nominal thickness.

[b]Plank floors consist of pieces of lumber laid edge to edge with the wide faces bearing on the supports (see Article 20.17—Division II).

[c]Nail laminated floors consist of pieces of lumber laid face to face with the narrow edges bearing on the supports, each piece being nailed to the preceding piece (see Article 20.18—Division II).

[d]Muliple-layer floors consist of two or more layers of planks, each layer being laid at an angle to the other (see Article 20.17—Division II).

[e]Glued laminated panel floors consist of vertically glued laminated members with the narrow edges of the laminations bearing on the supports (see Article 20.1.1—Division II).

[f]In this case the load on each stringer shall be the reaction of the wheel loads, assuming the flooring between the stringers to act as a simple beam.

Source: Standard Specifications for Highway Bridges, Washington, DC: The American Association of State Highway and Transportation Officials, copyright 1994. Used by permission.

Table 15-6. AASHTO Method for Interconnected Glued Laminated Decks.

3.25.1.3 One design method for interconnected glued laminated panel floors is as follows: For glued laminated panel decks using vertically laminated lumber with the panel placed in a transverse direction to the stringers and with panels interconnected using steel dowels, the determination of the deck thickness shall be based on the following equations for maximum unit primary moment and shear.[a] The maximum shear is for a wheel position assumed to be 15 in. or less from the centerline of the support. The maximum moment is for a wheel position assumed to be centered between the supports.

$$M_x = P(0.51 \log_{10} s - K) \qquad (3\text{-}23)$$

$$R_x = .034P \qquad (3\text{-}24)$$

Thus

$$t = \sqrt{\frac{6M_x}{F_b}} \qquad (3\text{-}25)$$

or

$$t = \frac{3R_x}{2F_v} \text{ whichever is greater} \qquad (3\text{-}26)$$

where

M_x = primary bending moment in inch-pounds per inch
R_x = primary shear in pounds per inch
x = denotes direction perpendicular to longitudinal stringers
P = design wheel load in pounds
s = effective deck span in inches
t = deck thickness, in inches, based on moment or shear, whichever controls
K = design constant depending on design load as follows:

H 15 $K = 0.47$
H 20 $K = 0.51$

F_b = allowable bending stress, in pounds per square inch, based on load applied parallel to the wide face of the laminations
F_v = allowable shear stress, in pounds per square inch, based on load applied parallel to the wide face of the laminations

3.25.1.4 The determination of the minimum size and spacing required of the steel dowels required to transfer the load between panels shall be based on the following equation:

$$n = \frac{1000}{\sigma_{PL}} \times \left[\frac{\overline{R_y}}{R_D} + \frac{\overline{M_y}}{M_D}\right] \qquad (3\text{-}27)$$

where

n = number of steel dowels required for the given spans
σ_{PL} = proportional limit stress perpendicular to grain (for Douglas fir or southern pine, use 1000 psi)
$\overline{R_y}$ = total secondary shear transferred, in pounds, determined by the relationship

$$\overline{R_y} = 6Ps/1{,}000 \text{ for } s \le 50 \text{ in.} \qquad (3\text{-}28)$$

or

$$\overline{R_y} = \frac{P}{2s}(s - 20) \text{ for } s > 50 \text{ in.} \qquad (3\text{-}29)$$

$\overline{M_y}$ = total secondary moment transferred, in inch-pound, determined by the relationship

$$\overline{M_y} = \frac{Ps}{1600}(s - 10) \text{ for } s \le 50 \text{ in.} \qquad (3\text{-}30)$$

$$\overline{M_y} = \frac{Ps}{20}\frac{(s - 30)}{(s - 10)} \text{ for } s > 50 \text{ in.} \qquad (3\text{-}31)$$

R_D and M_D = shear and moment capacities, respectively, as given in the following table:

Diameter of Dowel (in.)	Shear Capacity R_D (lb.)	Moment Capacity M_D (in.-lb.)	Steel Stress Coefficients C_R (1/in.[2])	C_M (1/in.[3])	Total Dowel Length Required (in.)
0.5	600	850	36.9	81.5	8.50
0.625	800	1340	22.3	41.7	10.00
0.75	1020	1960	14.8	24.1	11.50
0.875	1260	2720	10.5	15.2	13.00
1.0	1520	3630	7.75	10.2	14.50
1.125	1790	4680	5.94	7.15	15.50
1.25	2100	5950	4.69	5.22	17.00
1.375	2420	7360	3.78	3.92	18.00
1.5	2770	8990	3.11	3.02	19.50

3.25.1.5 In addition, the dowels shall be checked to ensure that the allowable stress of the steel is not exceeded using the following equation:

$$\sigma = \frac{1}{n}(C_R\overline{R_y} + C_M\overline{M_y}) \qquad (3\text{-}32)$$

where

σ = minimum yield point of steel pins in pounds per square inch (see Table 10.32.1A);
$n, \overline{R_y}, \overline{M_y}$ = as previously defined;
C_R, C_M = steel stress coefficients as given in preceding table.

[a]The equations are developed for deck panel spans equal to or greater than the width of the tire (as specified in Article 3.30), but not greater than 200 in.

Source: Standard Specifications for Highway Bridges, Washington, DC: The American Association of State Highway and Transportation Officials, copyright 1994. Used by permission.

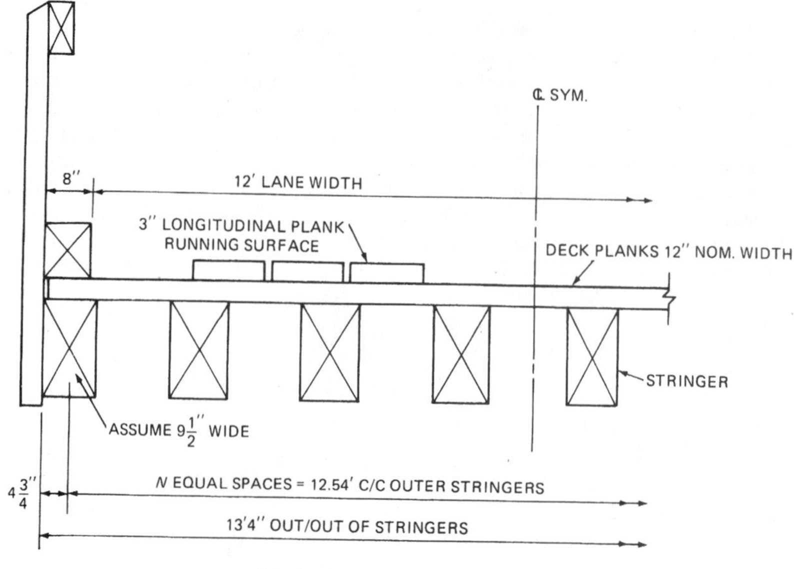

(a) CROSS SECTION OF BRIDGE

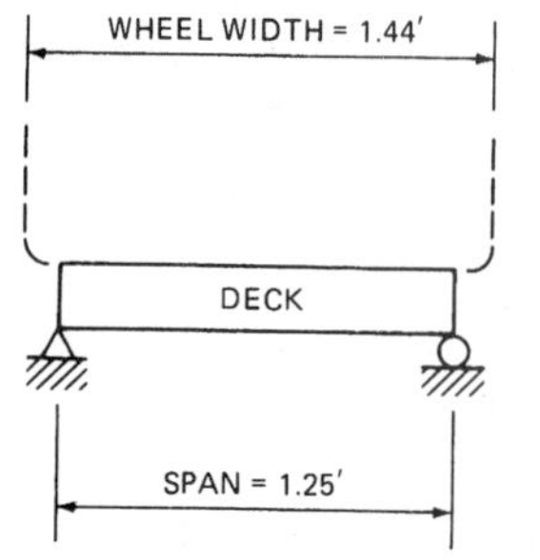

(b) WHEEL IN POSITION FOR MAXIMUM POSITIVE BENDING MOMENT IN SIMPLE-SPAN DECK

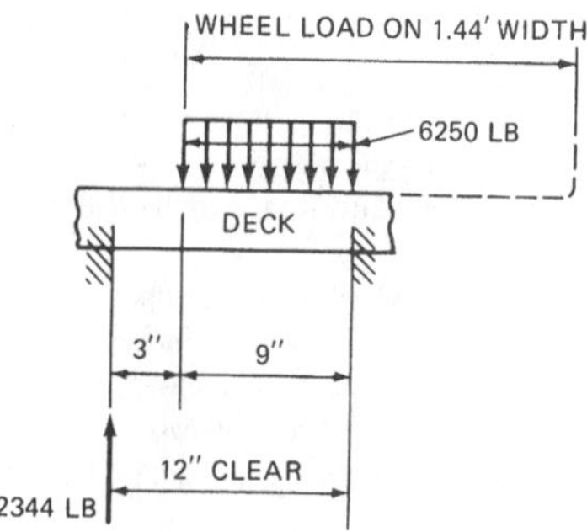

(c) WHEEL POSITIONED TO CAUSE MAXIMUM SHEAR IN DECK

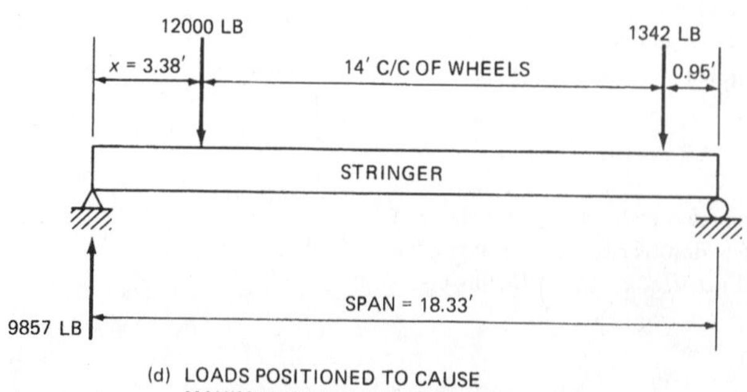

(d) LOADS POSITIONED TO CAUSE MAXIMUM SHEAR IN A STRINGER

Fig. 15-27. Plank-deck bridge for Example 15-3.

12,000, or 120 in.^{2} Setting this equal to tire width, b, times footprint length, $b/2.5$, width $b = 17.3$ in. or 1.44 ft. As shown by Fig. 15-27b, the 12,000-lb wheel load is distributed over a distance larger than the deck span. The live load per unit length is 12,000/1.44 = 8333 lb/ft. The deck is continuous over several supports, so the maximum bending moment is not the same as for a simple beam. To relieve the designer of having to consider each such case individually, AASHTO gives the rules shown in Table 15-7. These permit modifying the simple bending moment to obtain reasonably safe values for the positive and negative bending moments in continuous deck spans. (The values shown are intended for use with composite wood/concrete spans, but should be reasonably satisfactory for plank spans also.)

$$\text{Simple moment} = w\ell^2/8 = 8333\,(1.25^2)/8$$
$$= 1628 \text{ ft-lb due to live load alone}$$

AASHTO specifies that timber (either treated or untreated) be assumed to weigh 50 pcf. Thus, the design weight for the planks is about 50 × 3/12 = 13 lb/ft. Simple dead-load bending moment is 13(1.25^2)/8 = 3 ft-lb.

Correcting for continuity, positive M for the end span equals

$$(0.7 \times 3) + (0.85 \times 1628) = 1386 \text{ ft-lb}$$

Try No. 2 southern pine for the planks. The base allowable must be adjusted for load duration, flatwise use, and use under moist conditions. For 12-in. nominal plank widths, AASHTO tables show $F_b = 975$ psi, $C_{fu} = 1.2$, and C_M for bending = 1.0.

Table 15-7. Maximum Bending Moments—Percent of Simple-Span Moment.[a]

	Maximum Uniform Dead Load Moments				Maximum Live Load Moments			
	Wood Subdeck		Composite Slab		Concentrated Load		Uniform Load	
Span	Pos.	Neg.	Pos.	Neg.	Pos.	Neg.	Pos.	Neg.
Interior	50	50	55	45	75	25	75	55
End	70	60	70	60	85	30	85	65
2-span[b]	65	70	60	75	85	30	80	75

[a]Shown by AASHTO for composite wood-concrete members.
[b]Continuous beam of two equal spans.
Source: Standard Specifications for Highway Bridges, Washington, DC: The American Association of State Highway and Transportation Officials, copyright 1994. Used by permission.

$$F'_b = 975(1.15)(1.2)(1.0) = 1346 \text{ psi}$$

$$\text{Req } S = M/F'_b = 12 \times 1386/1346$$
$$= 12.37 \text{ in.}^3 \text{ per plank}$$

Actual S for a 3 × 12 rough is 18.0 in.^{3} *OK for bending.* (Narrower planks could be used, but would not save lumber. Use 12 in.)

Check 3 × 12 planks for shear: AASHTO specifies that the critical section and starting position for the edge of the wheel load shall be at whichever distance from the face of the support is smaller: three times the plank thickness, or one-fourth of the span. For this case these distances are 9 in. and 3 in., respectively. The latter controls.

Clear distance between stringer faces is 1.0 ft. Figure 15-27c shows the wheel located to cause the largest possible shear at that distance. With the wheel load in that position, the load actually on the span is 8333(9/12) = 6250 lb, and the shear (equal to the left-end reaction) is 6250 (4.5/12) = 2344 lb. Dead load shear adds about 3 lb, for a total $V = 2347$ lb. The unit shear stress is

$$f_v = 1.5(2347)/(3 \times 12) = 98 \text{ psi}$$

The adjusted allowable shear stress (using AASHTO's 1.0 for C_M) is

$$F_v = 90(1.15)(1.0) = 104 \text{ psi} > f_v$$

Use 3 × 12 rough-sawn planks.

Stringer design: Assume that that the steel bearing plates at each end will be set back 2 in. from the face of the concrete abutments, and that the required bearing length will be 3 in. at each end. If so, the span for bending of the stringer will be

$$18 + 2(2/12) + 2(3/2)/12 = 18.58 \text{ ft}$$

and the clear span for calculating shear, 18.33 ft.

Table 15-5 shows $S/4.0$ to be the fraction of a wheel load to be resisted (in bending) by one stringer. Spacing S is 1.79 ft, so the fraction to be used is 1.79/4 = 0.45. The bridge span is so short that the largest bending moment will occur when only the rear wheels are on the bridge and located at midspan. The live-load bending moment for one stringer will be

$$M = 0.45(12,000)(18.58)/4 = 25,100 \text{ ft-lb}$$

Dead load carried by one stringer is:

$$\begin{aligned} \text{Deck } 50(3/12)(1.79) &= 22.4 \\ \text{Running strip (same)} &= 22.4 \\ \text{Stringer (est)} &= \underline{50.0} \\ \text{Total } w_d &= 95 \text{ lb/ft} \end{aligned}$$

Dead-load $M = 95(18.58^2)/8 = 4100$ ft-lb. Total bending moment is

$$M = 29{,}200 \text{ ft-lb}$$

Try No. 1 SR mixed southern pine, for which AASHTO gives $F_b = 1350$ psi and $F_v = 110$ psi for this grade in sizes 5 in. × 5 in. and larger for either dry or wet use. This base allowable must be adjusted for load duration and size.

$$F'_b = 1350(1.15)\ C_F = 1550\ C_F$$

Assuming $C_F = 1.0$,

$$\text{Req } S = 12(29{,}200)/1550 = 226 \text{ in.}^3$$

Try 10 × 14 (S4S)

$$C_F = (12/13.5)^{1/9} = 0.987$$

$$F'_b = 0.987(1550) = 1530 \text{ psi}$$

$$\text{Actual } f_b = 12(29{,}200)/288.6 = 1214 \text{ psi} \qquad OK$$

Check for shear: Place the entire 12,000-lb wheel load above a stringer, as shown by Fig. 15-27d. To cause maximum shear in the stringer, this load must be placed at distance x from one end, where x is the smaller of three times the stringer depth or one-fourth of the span. Assuming the the stringer depth will be 13.5 in., the two values are 18.58/4 = 4.64 ft and 3 × 13.5/12 = 3.38 ft. The latter value controls.

For shear, no distribution to other stringers is allowed for the heavy rear wheel close to the end of the span, but the smaller wheel load (from the front axle) may be distributed among stringers. The load on this stringer, coming from the smaller wheel load of (0.1)(30,000) = 3000 lb, will be (Table 15-5)

$$3000(1.79)/4 = 1342 \text{ lb}$$

and the computed live-load shear will be 9857 lb as shown in Fig. 15-27d.

Shear at distance x from the left end due to dead load will be

$$95(18.33/2 - 3.38) = 550 \text{ lb}$$

and the total shear is the sum, or 10,410 lb.

$$F'_v = 110(1.15) = 126 \text{ psi}$$

$$\text{Actual } f_v = 1.5(10{,}410)/(9.5 \times 13.5)$$

$$= 122 \text{ psi} < F'_v \qquad OK$$

Example 15-4

Consider the same bridge as in Example 15-3, but design a transverse nail-laminated deck with a 2-in. asphaltic wearing surface. (Assume 150 pcf for the weight of this material.)

Figure 15-28a shows a cross section for this bridge. Again, to design the deck it is necessary to estimate the stringer width and the number of stringers. In this case, the designer assumed that there would be five stringers, equally spaced. The distance between stringers is 3.14 ft c/c or 2.35 ft clear. The thickness of the deck is not yet known, but it must be used to determine the deck span. Assuming that the deck will consist of 6-in. nominal lumber (5.5 in. thick) and that the stringer width will be 9.5 in., the span will be the smaller of

$$2.35 + \tfrac{1}{2}(9.5/12) = 2.75 \text{ ft} \quad \text{or} \qquad \textit{Controls}$$

$$2.35 + 5.5/12 = 2.81 \text{ ft}$$

For noninterconnected transverse nail-laminated decks, the wheel load may be assumed distributed over a 15-in. width of deck. Figure 15-28b shows one span of the deck, with the entire 12,000-lb wheel load centered on the deck span. Dead load on the 15-in. width is estimated to be

$$\begin{aligned} \text{Deck } (15/12)(50)(5.5/12) &= 29 \text{ lb/ft} \\ \text{Wearing surface } (15/12)(150)(2/12) &= \underline{31} \text{ lb/ft} \\ \text{Total dead load per 15-in. width} &= 60 \text{ lb/ft} \end{aligned}$$

The simple bending moments are

$$6000(2.75/2 - 1.44/4) = 6090 \text{ ft-lb from live load}$$

$$60(2.75^2)/8 = 57 \text{ ft-lb from dead load}$$

Correcting for continuity, the maximum positive bending moment is

$$0.85(6090) + 0.7(57) = 5220 \text{ ft-lb per 15-in. width}$$

Assuming that the asphalt surface will keep the deck dry, the allowable stresses need be adjusted only for

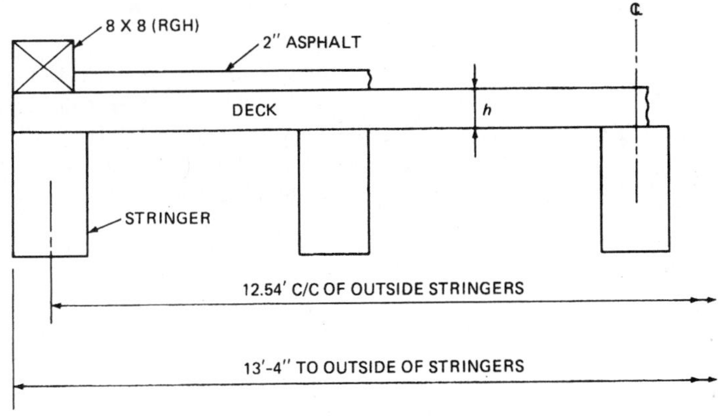

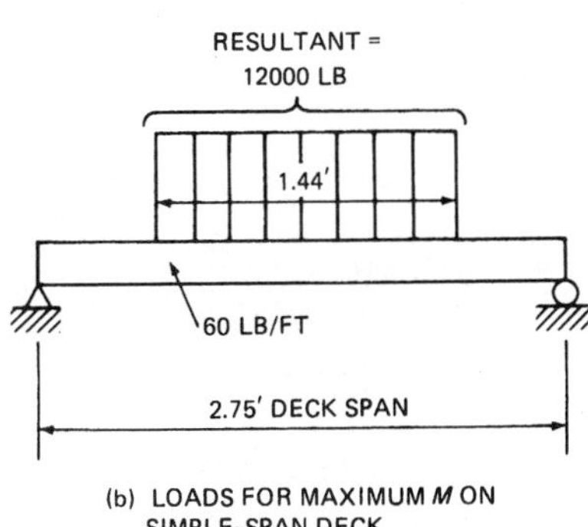

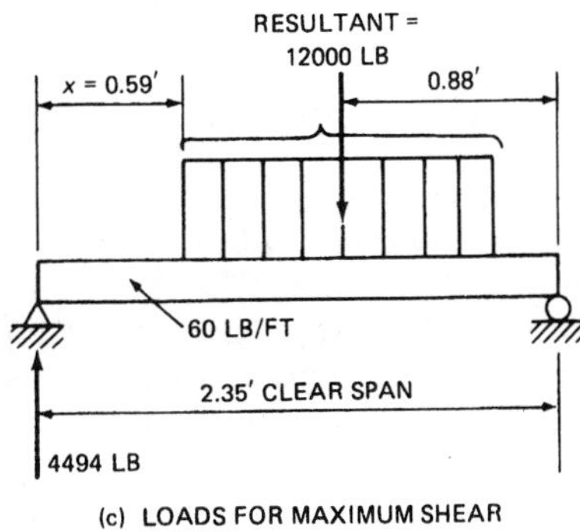

Fig. 15-28. Nailed laminated deck for Example 15-4.

load duration. For No. 2 southern pine, assuming 2 × 6 laminations, the adjusted allowables will be

$$F_b' = 1250(1.15) = 1438 \text{ psi}$$

$$F_v' = 90(1.15) = 104 \text{ psi}$$

Required section modulus is

$$S = 12(5220)/1438 = 43.6 \text{ in.}^3 \text{ per 15-in. width}$$

The required thickness is

$$\sqrt{6(43.6)/15} = 4.18 \text{ in.}$$

Try 2 × 6 laminations

Check shear: The critical section for shear is at whichever distance from the support is less,

3 × deck thickness of 5.5 in. = 16.5 in., or

Clear span/4 = 12(2.35)/4 = 7.05 in. = 0.59 ft *Controls*

Figure 15-28c shows the 12,000-lb wheel load in this position. The shear is

$$V = 12{,}000(0.88)/2.35 + 60(2.35/2 - 0.59)$$
$$= 4530 \text{ lb}$$

Shear stress is

$$f_v = 1.5(4530)/(15 \times 5.5) = 82 \text{ psi} < F_v' \qquad OK$$

Example 15-5

A two-lane bridge to carry H-15 truck loads spans 30 ft and has a 26-ft clear roadway width. The designer is considering using a 3-in. asphalt wearing surface on a transverse deck of interconnected glued laminated panels, supported by glulam stringers. One of the designer's trial designs assumed the bridge cross section shown by Fig. 15-29. Design the glued laminated

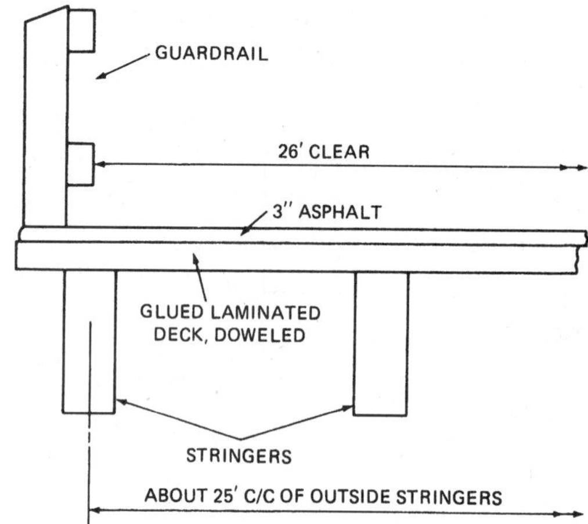

Fig. 15-29. Bridge for Example 15-5.

deck, assuming interconnected panels. Use $C_D = 1.15$ for live load.

If four stringers are used, they will be 8 ft 4 in. c/c. Assume that the deck will be 6.75 in. thick and the stringers 10.75 in. wide. (These are both standard widths for glulams; see Chapter 8.) With these assumptions, the clear distance between stringers will be 7.44 ft. The deck span will be the lesser of

$$7.44 + \tfrac{1}{2}(10.75/12) = 7.89 \text{ ft or} \qquad \textit{Controls}$$

$$7.44 + 6.75/12 = 8.00 \text{ ft}$$

From this point on, bending moments and shears must be computed using the empirical equations of the AASHTO specification. These are shown by Table 15-6. Using the equations shown, the live-load bending moment is

$$M_x = P(0.51 \log_{10} s - K)$$

in which $s = 12(7.89) = 94.7$ in. and K (defined by AASHTO) is 0.47 for H-15 loading.

$$M_x = 12{,}000(0.51 \times 1.976 - 0.47)$$
$$= 6453 \text{ ft-lb/ft}$$

Deck dead load is

Estimated deck weight 50(6.75)/12	= 28 psf
Asphalt 150(3/12)	= <u>38</u> psf
Total	= 66 psf

Simple dead-load bending moment is

$$66(7.89^2)/8 = 514 \text{ ft-lb/ft}$$

(*Note:* AASHTO is written to imply that dead-load bending moment may be ignored. The authors think it should be considered.) Using the same fraction of simple dead-load moment as for composite slabs, the total bending moment is

$$(0.7)(514) + 6453 = 6810 \text{ ft-lb/ft}$$

Allowable stresses are given by Appendix Table B-8 for glulams loaded parallel to the wide face of the laminations.

Assuming southern pine combination 20F-V1, the tabulated design values are

$$F_b = 1450 \text{ psi and } F_v = 175 \text{ psi}$$

$$F_b' = 1.15\,F_b - 1668 \text{ psi}$$

$$F_v' = 1.15\,F_v = 201 \text{ psi}$$

The required section modulus is

$$S \text{ req} = 12(6810)/1668 = 49.0 \text{ in.}^3\text{/ft width}$$

and the deck thickness required is

$$t = [6(49.0)/12]^{0.5} = 4.95 \text{ in.}$$

6¾-in. deck OK for bending

Check shear: The AASHTO empirical method uses shear equal to $R_x = 0.034P$, which figures in this case to $0.034(12{,}000) = 408$ lb *per inch* of width. (For this check, AASHTO permits ignoring dead-load shear. The authors agree, since including it would make but a trivial difference.) Setting the computed actual shear stress per inch of width equal to the allowable, the thickness required for shear is

$$t \text{ req} = 1.5\,R_x/F_v' = 1.5(408)/201 = 3.04 \text{ in.}$$

Use 6¾-in. deck

Determine required size and spacing of dowels: The method of determining these, shown by Table 15-6, is based on analysis by the method of beam on an elastic foundation (14) and has been confirmed by experiment.

Secondary shear occurs between adjacent panels due to their interaction with one another. For a span of 94.7 in., this secondary shear is

$$R_y = P(s - 20)/2s$$
$$= 12{,}000(74.7)/(2 \times 94.7) = 4730 \text{ lb}$$

The secondary moment transferred is

$$M_y = \frac{Ps(s - 30)}{20(s - 10)} = \frac{12,000(94.7)(64.7)}{20(84.7)}$$

$$= 43{,}400 \text{ in.-lb}$$

Try 1.0-in.-diameter dowels. Table 15-6 shows moment and shear capacities for these dowels as M_D = 3630 in.-lb and R_D = 1520 lb, respectively. All values needed to determine the required number of dowels are now known. Substituting in the equation for n,

$$\text{Req } n = [1000/1000] \times [(4730/1520) + (43{,}400/3630)]$$

$$= 15.0$$

With this deck thickness and dowel diameter, the dowel spacing will be 12(8.33 ft)/15 = 6½ in., which is not practical. If the dowel size is increased to 1½-in. diameter, the dowels can be spaced 15 in. c/c, which is better. The problem of dowel spacing could be alleviated also by using one more stringer, reducing the spacing of stringers to 6 ft 3 in. c/c.

15-7. COMPOSITE DECKS

In a composite deck, a layer of concrete is placed above the wood layer of the deck. The concrete layer resists the flexural compression and the wood the tension. An advantage of the composite deck is that the structural concrete may serve also as the wearing surface.

There are two types of composite deck, the composite T-beam deck and the composite slab deck. The two are shown by Fig. 15-30.

T-Beam Deck

In the T-beam type, shown by Fig. 15-30a, wood stringers form the stems of the T-beams and the concrete slab forms the flanges. Horizontal shear is transferred from stringer to slab through notches cut into the top of the stringers and through mechanical fasteners driven into the wood stringers. The slab, in addition to participating as the main compression element of the longitudinal composite T-beam, acts as a continuous beam spanning transversely from stringer to stringer. To construct this type of deck requires building concrete forms between the stringers.

Composite Slab Deck

Figure 15-30b shows the composite slab bridge deck. (This deck is useful also for floors of buildings carrying heavy live loads.) It consists of 2-in. nominal lumber in the vertical position and nail laminated, with a concrete slab cast above. The top surface of the wood portion has 2-in.-deep grooves, formed either by using two different widths of lumber for the laminations or by staggering the position of laminations to create a 2-in. offset. Longitudinal shear is transferred from timber to concrete by slots cut into the projecting laminations or by metal shear developers driven into the top of the wood, or by both.

An advantage of the composite slab deck is that the wood portion serves as the form on which to cast the concrete; no additional support is needed. If a long (parallel to the laminations) deck is needed, individual laminations can be cut to differing lengths, so that the pattern of staggered overlapping members is like a large-scale finger joint.

Since the concrete slab serves merely as the compression part of a composite member, the concrete requires reinforcement only for (a) resisting negative bending (tension on the top side) of the deck, as at locations over intermediate supports; and (b) resisting temperature and shrinkage stresses.

Design of Composite Slab Decks

In the composite slab deck, as shown by Fig. 15-30b, the timber portion alone must support its own weight plus the weight of the concrete. It may also have to support some additional load caused by construction equipment. Only after the concrete hardens does composite action occur. Composite action is effective only for carrying live load or such additional dead load that is applied to the bridge after the concrete hardens.

The effective sections and the stresses at various stages are shown graphically by Fig. 15-31. Under positive bending, the concrete is in compression and is fully effective. The concrete and

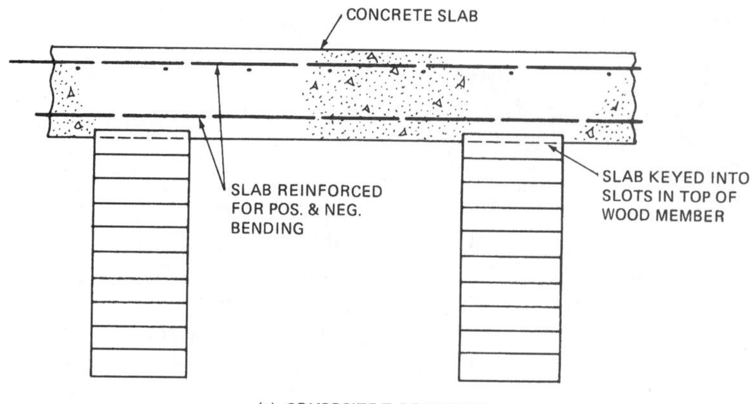

(a) COMPOSITE T-BEAM DECK

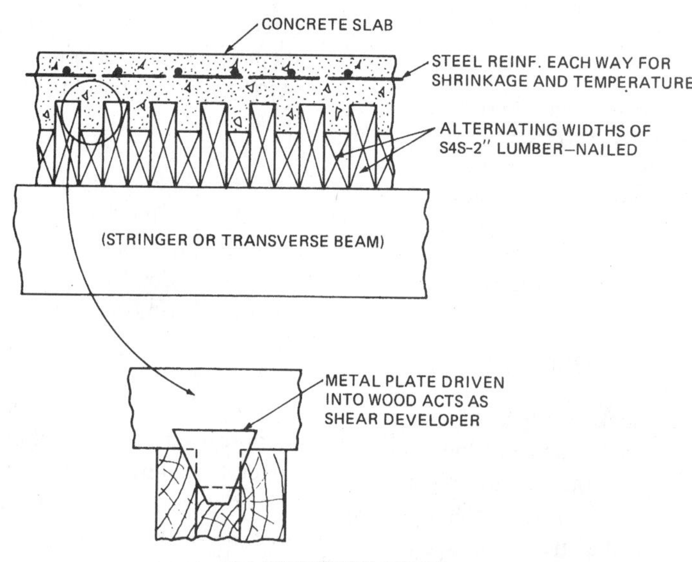

(b) COMPOSITE SLAB DECK

Fig. 15-30. Types of composite deck.

the timber will have different values for modulus of elasticity, so one of the materials is imagined to be replaced with an elastically equivalent area of the other to create the transformed section, as shown by Fig. 15-31b. In this section, the concrete has been replaced by imaginary timber, having area equal to that of the concrete times the modular ratio, n (ratio $n = E_c/E_w$).

Stresses caused by the permanent loads present while the concrete is curing are shown by Fig. 15-31c. These are merely the flexural stresses, My/I, on the timber deck alone. The concrete at this stage is unstressed. When live load is added as traffic crosses the bridge, the entire composite section resists additional bending moment, and the stresses added are as shown by Fig. 15-31d. The position of the neutral axis for these added stresses may be within the wood, at the wood-concrete junction, or in the concrete, depending on the thicknesses of the two portions and upon the modular ratio, $n = E_c/E_w$. The total stresses, shown by Fig. 15-31e, are merely the sum of the stresses shown for dead load alone and for live load alone.

In regions where bending causes tension on the top, reinforcing must be placed in the slab and the concrete is ineffective. The transformed section for this case is shown by Fig. 15-31f.

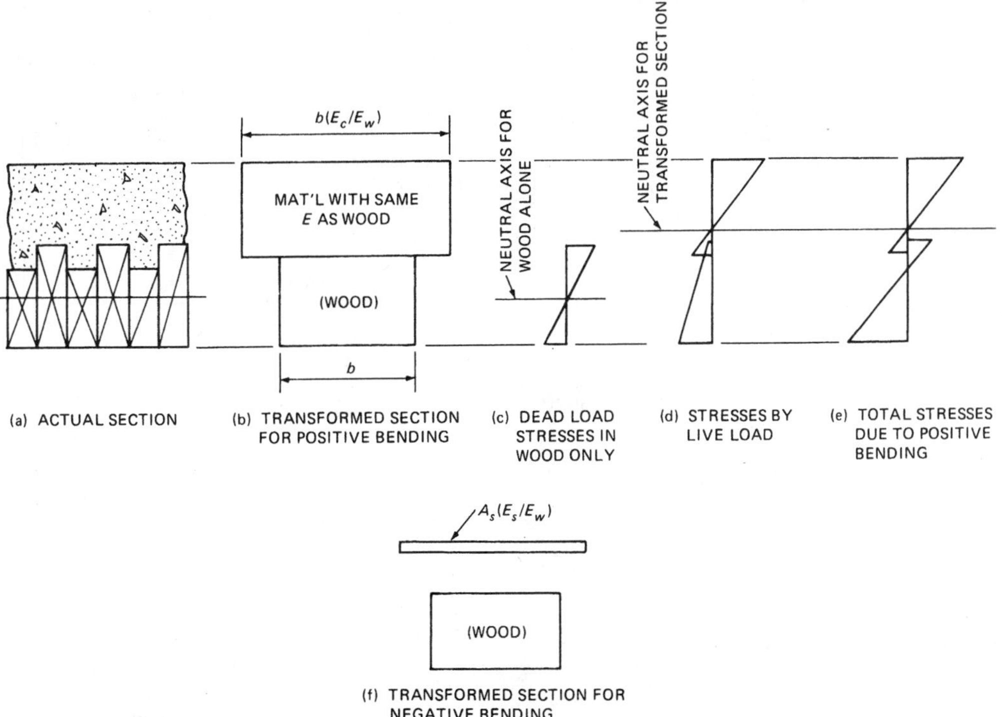

Fig. 15-31. Transformed sections and stresses in composite slab deck.

Here, the steel is imagined to be replaced by equivalent wood having n times as much area and using a modular ratio $n = E_s/E_w$.

For a much more comprehensive coverage of composite deck design, see reference 15.

15-8. CONCLUSION

The authors have attempted to cover adequately many types of wood structure and to consider various wood-based construction materials. We realize, however, that many more subjects could have been addressed were it not for limited space and limited time to complete this manuscript. For example, topics intentionally omitted include lamella roof structures, shells, domes, and folded plates.

As long as structural designers are innovative, the list of possible structural forms will continue to grow. And as long as researchers remain interested in conserving resources and using them to best advantage, the varieties of new structural materials—many of them wood-based—will also grow. Future editions of this book may include information on those structural forms and new wood-based construction materials that prove important.

REFERENCES

1. *Uniform Building Code,* International Conference of Building Officials, Whittier, CA, 1994.
2. Wolfe, R., *Low Cost Pole Construction,* Garden Way Publishing Co., VT, 1980.
3. *Book of Standards,* American Wood Preservers Association, Stevensville, MD, 1991.
4. *Wood Structures, A Design Guide and Commentary,* American Society of Civil Engineers, New York, 1975.
5. Patterson D. *Pole Building Design,* American Wood Preservers Institute, McLean, VA, 1981.
6. American Institute of Timber Construction, *Timber Construction Manual,* 4th ed., Wiley, NY, 1994.
7. *Permanent Wood Foundation System,* American Forest and Paper Association, Washington, DC, 1988.
8. *Evaluation, Maintenance, and Upgrading of Wood Structures, A Guide and Commentary,* American Society of Civil Engineers, New York, 1982.
9. *Standard Specifications for Highway Bridges,* American Association of State Highway and Transportation Officials, Washington, DC, 1994.
10. Taylor, R. J., B. Batchelor, and K. VanDalen, *Prestressed Wood Bridges,* Ontario Ministry of Trans-

portation and Communications, Downsview, Ontario, Canada, 1983.

11. McCutcheon, W. J., and R. L. Tuomi, "Design Procedure for Glued-Laminated Bridge Decks," *Forest Products Journal,* 23(6), June 1973, pp. 36–42.
12. Barnhardt, J., "Ohio's Experiences with Treated Timber for Bridge Construction," Transportation Research Record 1053, Transportation Research Board, National Research Council, Washington, DC, 1986.
13. Wipe, C., F. W. Klaiber, and W. W. Sanders, "Load Distribution Criteria for Glued Laminated Longitudinal Timber Deck Highway Bridges," Transportation Research Record 1053, Transportation Research Board, National Research Council, Washington, DC, 1986.
14. Hetenyi, M., *Beams on Elastic Foundation,* University of Michigan Press, Ann Arbor, 1946.
15. Gurfinkel, G., *Wood Engineering,* Southern Forest Products Association, New Orleans, 1973.

PROBLEMS

15-1. Choose pole sizes and depths of embedment for the pole building shown by Fig. 15-32. Assume loads as follows: roof dead load applied to pole = 1200 lb; snow load = 25 psf; wind load = 15 psf pressure on windward wall and 7 psf suction on leeward wall. Soil is classed as silty gravel (see Table 15-2). Use D. fir poles.

15-2. Choose pole size and embedment for the building shown by Fig. 15-33. Roof dead load is 12 psf and snow load is 20 psf. Wind pressure is as shown for Problem 15-1. Soil is silty gravel. Consider the combination of $W + S/2$. Note that bending may control for the outside poles, but design of the inside ones *may* be controlled by the maximum vertical load. Use southern pine.

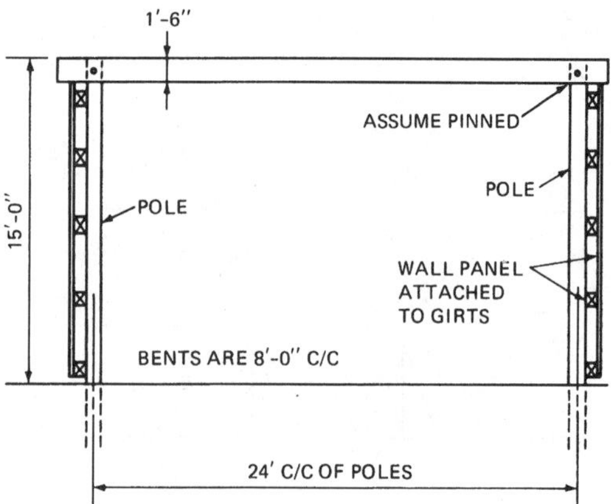

Fig. 15-32. Pole building for Problem 15-1.

15-3. Check the pole of Example 15-1 for end bearing. The allowable pressure is given in Table 15-2. Design concrete footing if required.

15-4. For the wood foundation of Example 15-2 (and Figs. 15-16 and 15-17), determine the size and spacing of nails required to connect: (a) upper end of the studs to the top plate; (b) top plate to field-installed plate; and (c) field-installed plate to floor system.

15-5. For the wood foundation of Example 15-2, determine whether or not the studs chosen are satisfactory: (a) in bearing (compression perpendicular to the grain) where they press against the edge of the concrete slab; and (b) in end bearing against the bottom plate.

15-6. Design a wood foundation system for the building shown by Figure 15-16, but with the clear height of the basement increased to 9 ft 0 in.

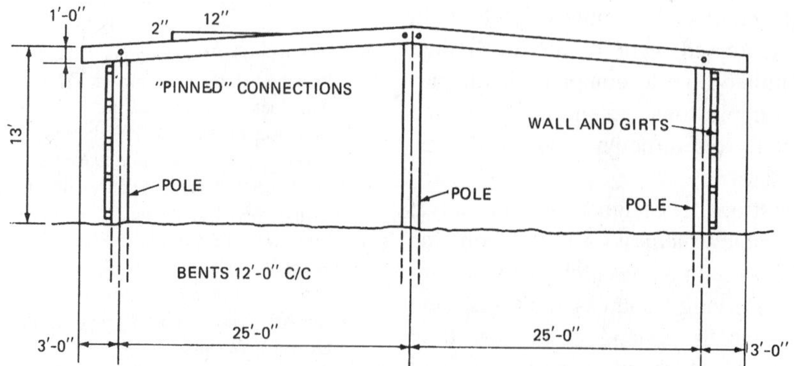

Fig. 15-33. Interior bent. Pole building for Problem 15-2.

15-7. Example 15-2 did not consider the load combination involving full wind along with live load and snow load on only one side of the peaked roof. Check the solution shown by Fig. 15-17 to determine whether or not the foundation designed will be satisfactory. Assume that the unsymmetrical snow load causes a downward force of 210 lb per foot of wall on the lee side.

15.8. Check the stringers of Example 15-3 for required bearing length and for deflection. Note that bearing length must be checked using the maximum *reaction,* not maximum shear. Note also that load-duration factors do not apply to allowable compression stress perpendicular to the grain.

15-9. Redesign the plank bridge of Example 15-3, but using S4S Douglas fir planks and using glulams for the stringers. Change the number of stringers if so directed by your instructor.

15-10. Repeat Example 15-5 using five equally spaced stringers to see if a practical spacing of 1-in. dowels results.

15-11. Design stringers for the bridge of Problem 15-10. After selecting for flexure and shear, determine the required bearing length at each end, and check the deflection.

16

Wood Durability, Protection, and Preservation

Wood has three shortcomings that tend to limit its structural use: (1) it is subject to deterioration by weathering and by attack of various organisms, (2) it is combustible, and (3) it is subject to swelling and shrinking.

Yet, in spite of these shortcomings, properly designed wood structures can be as long-lived as either steel-frame or reinforced-concrete structures, can resist fire better than unprotected steel structures, and can accommodate the swelling and shrinkage without distress. As is true for each of the materials (wood, structural steel, or reinforced concrete), to ensure successful performance and long life of the structure requires that the designer understand the behavior of the material and plan so as to avoid problems that its properties may raise. The builder also has an important role in that success, and so does the final user.

The first requirement in learning to design for best permanence and durability in any material is to identify the problems: In the case of wood, this means to identify the various agents of destruction.

16-1. CAUSES OF DETERIORATION

Borrowing terms from the science of physical geology, wood deterioration might be divided into two classes: decomposition, in which the chemical makeup is affected; and disintegration, in which the wood is merely broken apart physically but not changed chemically. Of course, both types can occur together.

Decomposition of wood is brought about by the action of bacteria, fungi, molds, chemicals, or fire. These may affect the wood while still in the tree, during lumber production and storage, or after it is installed in the structure. Decomposition may result in changes of appearance only or may result in structural weakening of the wood. The common causes of disintegration are insects, marine borers, interaction of treated wood with metals, and weathering. Disintegration almost always weakens the wood. Mechanical damage by abrasion during normal use might be considered to be a human-caused type of disintegration.

Microorganisms Affecting Wood

Four classes of microorganism affecting wood are wood-destroying fungi, wood-staining fungi, molds, and bacteria. The first class is the most important from a structural-strength standpoint. The wood-destroying fungi feed on materials in the cell walls. Thus, they destroy the cell walls and drastically reduce all strength properties of the wood. The resulting damage, called *decay,* can occur at any time: in the live tree, in the forest after the tree is cut, during processing and storage of the lumber, or in the completed structure.

It is reasonable to assume that all wood at some time or another has been infected with the spores of these fungi. All that is necessary to cause the wood to decay is to provide the conditions necessary for these spores to grow. Conversely, all that is necessary to prevent these spores from growing into fungi that will destroy the wood is to eliminate one or more of the things they need in order to grow—a food supply, air, moisture, and suitable temperature.

Decay can also occur as wood-destroying fungi spread directly from the soil or from infected wood into undamaged wood. This occurs where the wood parts of a structure are too close to the ground. It occurs also in lumber storage where the lumber is separated by infected stickers or where logs are stored on the ground.

Decay is serious even though it may affect only a small portion of the total volume of wood in a structural member. For example, a long timber beam may be rendered completely useless when only a small volume of wood has decayed near the location of maximum shearing or maximum bearing stress. A wood column that is decayed near its lower end is completely useless, even though most of its length consists of high-strength, undeteriorated wood. The bolted joint in a wood truss cannot transfer load effectively from one member to another if the wood bearing against one of the bolts is rotten; thus, the entire truss may be worthless even though the volume of defective wood is quite small.

After wood is infected by airborne spores, microscopic, threadlike structures grow in the wood. These structures, known as hyphae, can form on the wood surface or in openings such as surface checks. They can also penetrate the cell walls, passing through the wood from one cell to another. In the early stages of decay, the presence of the hyphae may not be obvious. At more advanced stages, however, the wood may become soft and will have lost almost entirely all of its structural value.

Three recognized general types of decay are white rot, brown rot, and soft rot (1). Fungi causing white rot consume both carbohydrates and lignin from the cell walls, leaving the wood a lighter color. Fungi causing brown rot remove the cellulose and hemicellulose, and the crumbly brown residue consists largely of the lignin from the cell walls. The bending strength is reduced markedly even in the early stages of decay by brown rot fungi. In soft rot, the surface of the wood becomes softened, but the depth of damaged wood is not so great. The rotted surface can be scraped away from the undamaged wood beneath.

The above classification relates only to the general nature of the rotted wood. Each type of decay can be caused by many species of fungus. Some can destroy only particular species of wood, while others can affect a wide variety of wood species.

A term frequently heard is *dry rot*. This term is normally used to describe the final stages of brown rot. It is also a misnomer, as wood will not decay if it is kept dry, regardless of whether it is infected or not. However, wood in which the growth of fungi has been arrested because of dry conditions may return to active growth again once the level of moisture increases. Further, these fungi have the ability to transport moisture from one location to another in the wood. Thus, fungi that started to develop in wood in damp locations (such as in basements or where the wood is in contact with soil) may transport moisture as far as 20 ft into wood in other parts of a structure. They may also form a coating of porous, rootlike strands over masonry to reach wood that is in drier locations.

Effect of Decay on Properties

Table 16-1 (from reference 2) shows expected strength losses in wood that has been partially destroyed by decay. The measure used to define the degree of decay is the percent loss of weight. The table is for softwoods only and for just one type of decay. Much additional research would be needed to allow reliable determination of the strength remaining in actual structural members. The striking feature of the table is that it shows that bending strength, and compression and ten-

Table 16-1. Probable Strength of Wood in Early Stage of Decay (5 to 10% weight loss).

Strength Property	Probable Remaining Strength (% of Original Strength)
Static bending	30
Impact bending	20
Modulus of elasticity	30
Compression parallel to grain	55
Tension parallel to grain	40
Compression perpendicular to grain	40
Shear	80

Based on information in Reference 16-2, courtesy Van Nostrand Reinhold Company.

sion strengths parallel to the grain—the identical strength properties that most often control our design of structural members—are the ones most severely affected by decay.

Resistance to Decay

The natural resistance to decay varies among woods of different species. As with most other wood properties, decay resistance also varies between trees of the same species, and even among pieces from the same tree. For example, lumber from the sapwood of most trees is easily decayed, while the resistance of lumber made from the heartwood is much higher. This is due to toxic substances called *extractives* which are deposited in the wood as it changes from sapwood to heartwood. These substances might be regarded as natural preservatives, protecting the wood by poisoning the food supply of the attacking microorganism. In the living tree, however, the heartwood is more susceptible to decay than the sapwood. This is probably because the sapwood contains so much water that the amount of air present is insufficient for growth of the fungi.

Some species of wood commonly used for lumber, notably redwood, junipers, and cedars, are generally considered to be very durable. Thus redwood is often specified as the species to use for sill plates, the first piece of wood above the foundation in light-frame construction, and cedar is often preferred for exterior siding. Table 16-2 shows a listing of some of the common species according to approximate average resistance to decay. Certain dense tropical hardwoods (bongassi, basrolocus, and bilinga) are believed to have superior resistance to deterioration when used in climates such as those of North America.

Necessary Elements for Decay

For decay-causing fungi to grow four things are necessary: a food supply, suitable temperature, sufficient moisture, and oxygen.

Table 16-2. Grouping of Some Domestic Woods According to Heartwood Decay.

Resistant or Very Resistant	Moderately Resistant	Slightly or Nonresistant
Baldcypress	Baldcypress	Alder
(old growth)[a]	(young growth)[a]	Ashes
Catalpa	Douglas fir	Aspens
Cedars	Honey locust	Basswood
Cherry, black	Larch, western	Beech
Chestnut	Oak, swamp chestnut	Birches
Cypress, Arizona	Pine, eastern white[a]	Buckeye
Junipers	Southern pine:	Butternut
Locust, black[b]	Longleaf[a]	Cottonwood
Mesquite	Slash[a]	Elms
Mulberry, red[b]	Tamarack	Hackberry
Oaks:		Hemlocks
Bur		Hickories
Chestnut		Magnolia
Gambel		Maples
Oregon white		Oak (red and black species)
Post		Pines (other than longleaf, slash, and eastern white)
White		Poplars
Osage orange[b]		Spruces
Redwood		Sweetgum
Sassafras		True firs (western and eastern)
Walnut, black		Willows
Yew, Pacific[b]		Yellow poplar

[a]The southern and eastern pines and baldcypress are now largely second growth, with a large proportion of sapwood. Consequently, substantial quantities of heartwood lumber of these species are not available.
[b]These woods have exceptionally high decay resistance.
Source: Courtesy Van Nostrand Reinhold Co.

Food Supply. The food supply is the wood itself. Different species of microorganism show preferences for different species of wood and for different compounds or structural parts of the wood.

Suitable Temperature. Temperatures under which wood-destroying fungi may develop vary over a fairly wide range. Normally, the fungus grows most vigorously during the hotter and more moist parts of the year. Optimum growth temperatures for most species of wood-destroying fungi are between 68°F and 97°F (reference 1). As the temperatures drop to low levels, the fungus becomes dormant but is not killed. Growth resumes when the temperature rises to an acceptable level. Temperatures higher than those suitable for fungal growth will retard growth significantly and will even kill the fungus. This is fortunate, as the temperatures reached in normal kiln drying of lumber are sufficient to kill the fungi or spores already present.

However, using high temperatures to kill fungi is of little use in infected completed structures. To raise the temperature of an entire structure to a suitable degree and to maintain that temperature long enough to kill the fungi is impractical. Furthermore, even if the infected wood parts were to be sterilized by heat, there would be no assurance that reinfection would not occur if the wood were exposed again to fungal spores.

Sufficient Moisture. For most wood-destroying fungi to grow, the moisture content of the wood must be at or above the fiber saturation point, but a few species will cause decay with less moisture. Water in excess of that causing fiber saturation is present as free water in the cell cavities. This free water carries enzymes of the fungi and promotes the deterioration of the wood. Without this free water, the enzymes cannot enter the cell walls and growth of the fungus is arrested. Consequently, wood that has been properly dried or air seasoned will not decay provided its moisture content is kept below the fiber saturation point (FSP) at every location within the piece. If any one location has water in excess of the FSP, however, decay may occur at that particular location and the entire piece may become useless.

Notice, however, that the dry-rot fungi are an exception to the above. These fungi can *carry* water to drier parts of the wood, moistening it and causing it to decay.

Oxygen. A small amount of air is all that is needed to promote the growth of wood-destroying fungi. Conditions in normal stockpiles or in a completed wood structure are sufficient. Even the air present in the cell cavities of wood is sometimes sufficient for growth of the fungi.

It is probably because air is excluded by the large amount of water present in sapwood of a living tree that its sapwood is more resistant to decay than its heartwood. Wood piling that is driven into the ground but remains below the water table may last for hundreds of years, even though thoroughly saturated. It is the exclusion of air that prevents these piles from decaying. Should the top of the pile be above the water table or should the water table become lowered, decay will probably commence.

Insects

Insects that damage wood include termites, powder-post beetles, and carpenter ants. Of these, the termites are by far the most important.

Termites. Termites are frequently called "white ants," but are not ants and generally are not white. Worldwide, there are more than 2100 species of termite. The 41 species found in North America can be divided according to their living habits into three groups: subterranean termites, dry-wood termites, and damp-wood termites.

Subterranean termites are the most troublesome in wood structures. They live in moist ground, entering wood that is either near the ground or in contact with it. They will enter either sound or decaying wood. They construct tunnels of mud and partially digested wood fibers upward from the ground to reach the wood. Once they enter the wood, they can eat their way upwards to as high as the second or third story of a building. Using the cellulose of the cell walls as their food, and generally eating parallel to the grain, they create interconnected galleries of burrows in the wood, hollowing out the interior but rarely penetrating the surface of the lumber. Thus, the damage they do is often not detected until serious weakening has occurred. Subterranean termites are found in almost every part of the United States and cause considerable damage to embedded poles, bridge timbers, and foundation members of buildings.

Dry-wood termites seldom enter the ground, but live almost entirely in the wood, from which they obtain sufficient moisture. The winged form of reproductive adult can attack sound wood directly from the outside. Once they have entered, they plug the entrance with chewed wood and proceed inside to develop a new colony of termites. Since they occasionally open the wood again to the outside to discharge pellets of waste material, it is sometimes possible to detect their presence before the wood member in question has been severely damaged. Since dry-wood termites can live in wood with as low as 10–12% MC, they may inhabit well-seasoned or dried lumber. Dry-wood termites are found in the United States only along a narrow strip along the southern border and along the Atlantic coast as far north as Virginia.

Damp-wood termites also enter wood from the air. They require considerable moisture to thrive. Consequently, they usually are active in only decaying wood or very moist wood. They can penetrate sound wood from adjoining decayed wood, however.

Powder-post and Other Beetles. There are many species of powder-post beetle, some of which attack wood in dying trees or freshly cut green wood. Others can attack wood with moisture content as low as 10% (3). Beetles cause most damage in hardwoods, although occasional population explosions of the pine-bark beetle cause considerable damage in western softwoods.

Beetles deposit eggs in open pores of cut lumber or through the bark of standing trees. The larvae that hatch burrow through the wood, using certain wood components as their food. After metamorphosis, the emerging adult beetles proceed to reproduce, infecting new wood at some distance away or perhaps even laying their eggs in the same timber from which they emerged. The damage by beetle attack is a series of irregular burrows, sometimes just following the surface under the bark, but in most cases penetrating the solid wood in random directions. The pine-bark beetle not only leaves burrows, but also infects the tree with a wood-staining fungus. Even though this does not seriously affect the structural properties of the wood from infected trees, it does severely reduce its value and salability.

Ants. Ants use wood particles for shelter rather than food. Normally, they confine their activities to wood that has already been softened by decay.

Marine Borers

Marine borers are of two classes—wood-boring mollusks (related to clams and oysters) and crustaceans (related to crabs and lobsters). Damage by marine borers is common throughout salt or brackish water regions of the world, especially so in warm regions where the water temperature is 50°F or above (4). In North America, it is common along the Pacific, Gulf, and Atlantic coasts. Marine borers can completely destroy untreated timber piles in less than a year.

Mollusks. Three important genera of mollusks that damage wood on coasts of the United States are *Teredo, Bankia,* and *Martesia.* There are a number of species of each genus. Species of the first two are wormlike in form and are commonly called shipworms. Members of the third genus are more like clams in general appearance.

Free-swimming larvae of shipworms have small bivalve shells that may be used for protection. After the larva attaches itself to a piece of wood, it uses these to eat its way inside. Once inside the piece of wood, the shipworm remains there for life, growing rapidly in both diameter and length. Using the wood as part of its food supply, it extends its burrows, often completely riddling the interior of the wood. The wood pile, for example, may become practically a hollow shell with only the small entrance hole visible on the outside.

Wood-boring crustaceans do not remain in the wood for life, so the outside of a timber attacked by them will show numerous holes. The damage is sometimes superficial but may be so thorough that the entire surface can be removed by mechanical action of the water. This exposes new wood, and action of the crustacean borers continues. In this manner, the structural value of the wood member can be completely destroyed. Of the three species of wood-boring crustacean found in United States coastal water, *Limnoria* is the most destructive.

Weathering

Several types of defect that can be caused by exposure to the weather are raised grain, checks, splits, large cracks, warping, twisting, loosening of nails, roughened surface, and change of color. The wood surface often becomes friable, with splinters and other fragments falling away. These defects can occur if the wood is not protected from wetting and drying and from exposure to wind or to sunlight. (See Fig. 16-1.)

Weathering is not the same as decay, although the defects caused by weathering may provide conditions conducive to decay. The splits and raised grain, for example, provide easy entrance into the wood by fungal spores or insects.

Changes due to weathering may be classified as those due to chemical, mechanical, and light energies (5). The principal cause of weathering changes is repeated wetting and drying. Wood is hygroscopic and wood fibers near the surface respond quickly to changes in atmospheric humidity, as well as to wetting by rain, dew, or snow, whereas those away from the surface respond slowly, if at all. The result is that wood fibers near the exposed surface have varying moisture content (MC), swelling and shrinking, while those below the surface remain at more nearly constant size. This causes a steep gradient in MC and thus a steep gradient in stresses. Swelling and shrinkage are greatest perpendicular to the grain, and strength is the least. Thus, the alternating tensile and compressive stresses that result normal to the grain cause checks and splitting parallel to the grain. With prolonged exposure, the small surface checks become larger, even culminating in splits that go through the entire thickness of the piece. (See Fig. 16-2.)

Raising of the grain, often the first sign of weathering, is due to differential swelling and shrinking of springwood and summerwood.

Erosion due to blowing grit or sand (sometimes referred to as natural sandblasting) often occurs in drier localities having high winds, such as the mountain areas of the western United States. The moving sand abrades grooves where the softer springwood is at the surface, leaving

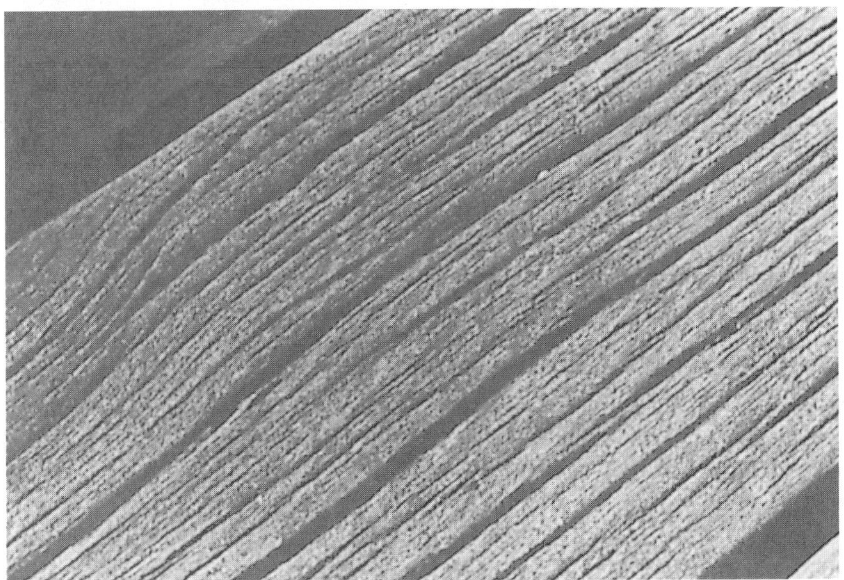

Fig. 16-1. Badly weathered redwood. Note splits and eroded surface. (Photograph by authors.)

Fig. 16-2. Splits and checks in end of glulam exposed to weather. (Photograph by authors.)

ridges where the harder summerwood is exposed. (Where this appearance is considered desirable, wood is sometimes intentionally sandblasted to create the impression of weathered wood.

Sunlight. Both ultraviolet and visible parts of sunlight cause darkening of wood and chemical degradation of the lignin. Some of the products of lignin degradation are soluble, and when these products are washed away by rainwater, the wood becomes more gray in color (5). Grayed wood is sometimes considered attractive, and chemicals are sometimes used intentionally to accelerate the color change. For example, redwood can be given a surface treatment that quickly gives it the silvery grey appearance of old, weathered redwood.

Effect of Weathering on Strength. The few studies that have been done indicate that weathering has little effect on modulus of rupture, strength in compression parallel to the grain, and modulus of elasticity. Only toughness, abrasion resistance, and (for very old wood) splitting and impact strengths decrease.

Weathering of wood products such as plywood and particleboard involves weathering of the adhesives as well as the wood veneers or particles, and as this is written, little information is available regarding the effect of weathering on strength.

Preventing Deterioration

Damage from the various agents of deterioration described above can be prevented by either (1) preservative treatment or (2) preventative construction and maintenance.

16-2. PRESERVATIVE TREATMENT

The science and practice of wood preservation is so involved and complex that the most that can be done in this short chapter is to present a few of the fundamentals and describe a few of the many methods used to treat wood.

As pointed out earlier, four essential conditions are required for growth of wood-destroying fungus. Most wood preservatives poison the food supply, thus eliminating one of these essentials. This type of treatment can also be effective as protection against attack by certain insects.

To a degree, some preservative treatments also protect by waterproofing the wood. If the interior of the piece of wood has less moisture than is needed for a damaging fungus to grow, then waterproofing will prevent the moisture content from rising to the danger level. Stains and paints can also provide this kind of protection, but often they are of short-term value and must be replaced at frequent intervals.

To be an effective wood preservative a chemical must be:

1. Toxic to the organisms that can destroy wood.
2. Soluble to some degree in the body fluids of the organism it is intended to repel or kill.
3. Capable of penetrating the wood to considerable depth, so that wood at the bottom of checks and splits will be protected.
4. Permanent, not leaching out of the wood during the years it is expected to remain in service.

There is no perfect solution, however, since different organisms have different body fluids and their reactions to various chemicals differ.

Preservative Chemicals

Generally, water-soluble preservatives soak into the cell walls, but creosote and similar ones only fill (or partially fill) the cell cavities. Reference 1 lists the principal chemicals used today in preservative treatment and gives criteria for selecting which chemical to use for various applications. The structural designer should probably not make the selection without seeking the advice of experts in wood preservation. However, the designer should be aware that wood preservatives are poisonous, because special precautions may be needed where persons, animals, or foodstuffs may come in contact with the treated wood. Some have an odor that makes them unsuited for use inside of most buildings. They may also ooze out of the timber, so that personal or animal contact with treated exterior timbers may be undesirable. Some preservatives can be used for wood that will be stained or painted, while others cannot.

Treatment Methods

Treatments include pressure methods and nonpressure methods, and there are several varieties of each. The nonpressure methods generally are both less expensive and less effective than pressure methods.

16-3. NONPRESSURE TREATMENT METHODS

There are dozens of these methods, some of them proprietary.

1. Surface treatments include brushing or spraying. The preservative is drawn into the wood only by capillary attraction, and the degree of penetration is slight. Coal tar creosote or oil-borne solutions of toxic chemicals are generally used. The advantage of brushing or spraying is that it is cheap and requires no extensive physical plant or equipment. The disadvantage is that the degree of protection is minimal. Brushing, however, is a necessary additional treatment to be used where wood treated by better means is cut during fabrication or installation. This occurs particularly at cut ends, where, fortunately, capillary attraction may cause slightly better penetration.

2. Dipping is merely immersing in a bath of the preservative for a few minutes, hoping that the chemical will have sufficient penetration. It is more costly than brushing or spraying since it requires a large vessel and considerably more preservative. However, it gives better assurance that all parts of the surface are treated. Dipping is useful with finished products such as window frames, sash, and doors. Penetration is about the same as for brushing or spraying, poor in side grain but better in end grain.

3. Steeping is merely dipping for longer periods of time, for several days to a few weeks. If steeping is used for seasoned lumber, both water and dissolved preservative are absorbed by the wood. When steeping is used for green wood, the liquid in which the preservative is dissolved cannot enter the already saturated wood. In this case the preservative enters the wood by diffusion, the preservative salt leaving the liquid in the vat and entering the water already present in the green wood. Again, penetration is not very good. To counter this, the solution used is more concentrated than is used in the pressure processes. One would think that penetration would be better for seasoned wood than for green wood, but this is not the case. The chemical entering the green wood by diffusion continues to spread within the wood after the wood is removed from the vat, as long as the moisture content of the wood remains fairly high.

Reference 1 lists 24 other nonpressure processes.

16-4. PRESSURE TREATMENT METHODS

There are numerous pressure processes, many of them proprietary. Broadly, they may be classified as empty-cell processes and full-cell processes. In each type of process, the wood is treated in a large, cylindrical tank into which the preservative may be entered and placed under pressure to force it into the wood.

Empty-Cell Processes

In empty-cell processes the preservative tends to coat the walls of the wood cells and to be absorbed by them, but after the wood is removed from the treatment tank the cells themselves are essentially empty, little preservative remaining in the cell cavities. Usually creosote or preservative oils are used, although the process may also be used with aqueous solutions of preservatives. The processes are used where good penetration is needed but maximum absorption is not.

A typical empty-cell process has the following steps:

1. After placing the wood in the tank and closing the end covers, the tank is filled with preservative at atmospheric pressure.
2. Pressure is applied to the fluid in the tank, forcing the preservative into the cell cavities. Pressures as high as 250 psi, but normally not over 150 psi, are used. These pressures are held for a few hours.
3. The pressure is released and the tank drained.
4. A partial vacuum is applied and held for about 30 minutes to remove free preservative.

In one of the empty-cell processes, air pressure of 25–75 psi is entered into the tank first and then the preservative is entered under pressure. Air in the cell cavities is initially compressed and it helps to increase the percent of preservative recovered during the vacuum step.

The steps described above are applied to seasoned (air-dried) wood. If green wood is to be treated, it must be conditioned first. One way of doing this is by steaming followed by partial vacuum; another is by boiling under a partial vacuum.

Full-Cell Processes

Full-cell treatment is used to treat wood with creosote or either oil or aqueous solutions of various salts. The amount of chemical retained by the wood is greater than with empty-cell processes. Thus, one would expect the full-cell process to be both more expensive and more effective than the empty-cell process, provided the depth of penetration is the same for each.

In full-cell processes, the preservative coats and is absorbed by the cell walls, just as in the empty-cell process, but also the cell cavities themselves contain a significant amount of free preservative. The difference between empty-cell and full-cell processes is that the full-cell process begins (after the wood is enclosed in the tank) with application of a partial vacuum to remove as much as possible of the air present in the wood cell cavities. The partial vacuum created in the cell cavities helps to draw the preservative into the wood. Also, less preservative is pushed out of the cell cavities during the final vacuum step.

Reference 1 describes numerous variations to the full-cell process.

16-5. EFFECTIVENESS OF TREATMENT

Penetration and retention are the two measures by which the effectiveness of treatment with a particular preservative may be measured. Penetration is the more important of the two measures, but it is hard to verify without damaging the wood.

The average retention (preservative weight per unit volume of wood) is easily measured by weighing the timber before and after treatment. Though this shows how much preservative is retained, it does not show how it is distributed. For an entire batch being treated, retention can be measured by the difference between the volume of preservative introduced into the vessel and the volume remaining after treatment.

For a given treatment, both penetration and retention vary with species, between sapwood and heartwood, and with density, condition, and moisture content. Retention also varies with size of cross section of the wood, so, when possible, different sizes of timber are treated separately.

Depth of penetration cannot be observed on the outside surface. It can be measured only by boring, crosscutting, or splitting the treated piece of lumber. Whichever method is used, the wood is damaged.

A small-diameter increment borer (a tool, not an insect) can be used to remove a small cylindrical portion for observation. Since the wood is variable, however, it would be necessary to make such borings at several locations to determine the least penetration. As an alternative, 6 holes may be made using ordinary wood bits, measuring the penetration by observing the color of the chips removed at various depths. For wood treated with creosote or similar materials, however, this is not a good method; the preservative material may be worked deeper into the wood by the bit, giving a false impression of penetration.

In either method, the hole opens up the interior of the wood for infection. Consequently, the holes must be filled with tight-fitting, treated wood plugs.

In treated lumber that is longer than necessary, penetration depth is shown at sections where the piece is cut to proper length before installation. This shows the depth only at the location of the cut. The depth shown at a cut near the end of the treated length piece is normally greater than at locations farther from the end. Again, such a cut exposes untreated wood, and infection can occur in spite of the preservative treatment. If the preservative treatment is to be effective, it is essential to brush or spray all such areas with preservative before completing the installation. This applies to end cuts, bolt holes, grooves for timber connectors, and cuts or notches made for end bearing or other reasons. It is especially important in wood for foundations.

The most accurate method of determining penetration is to split the treated member along its entire length. This gives a better view of penetration, but unfortunately destroys the piece of wood. Performed on a sufficient number of pieces from a treated batch, however, it could serve as an effective means of quality control.

Amount of Retention Needed

The amounts of retention specified vary with species, size of timber, and use. Higher retention (in pounds per cubic foot) is usually specified for small timbers than for large ones. The amount needed depends on what the wood is to be used for. For piling above the water table and for piling in salt water, the retention must be fairly high. For wood for most buildings, it can be less.

The ideal situation would be to have all cell walls completely saturated with the preservative. This is difficult (or impossible) to do, so we settle for lesser practical retentions. Reference 6 gives a more nearly complete list of recommendations for various applications, wood species, and chemicals.

Penetration Requirements

In glued laminated members, it is unsafe to assume that the preservative penetrates (normal to the lamination) beyond the first glue line, although it might penetrate laterally somewhat further.

For plywood and particleboard, treatment of the *edges* (including the edges that are cut during installation) is important in preventing deterioration by weathering.

For sawn timbers, specifications usually require that all (or a high percentage) of the sapwood be impregnated, as well as some of the heartwood. A designer preparing specifications should consult either reference 6 or 7 for guidance and should be careful that the specifications are feasible for the species and sizes involved.

16-6. USING PROPER DESIGN DETAILS TO PREVENT DECAY

The presence and continued use of so many early U.S. colonial buildings and old timber bridges, all constructed without preservative treatment, suggest that properly detailed wood structures can have long life, even in moist environments that one would expect to encourage decay. The secret? Merely design so as to shed water away from the wood, to prevent moisture-laden vapors from condensing in the wood, and to drain or ventilate spaces where water may accumulate.

"There is, of course, a right way and a wrong way to do everything with wood, and pursuing the wrong way will eventually lead to trouble" (2). The main agents of wood deterioration are decay and termite damage, and good construction details can go a long way toward protecting from either agent.

It is reasonable to assume that all wood as it is received for construction has already been infected, so decay can proceed if conditions are right. The wood substance itself is the food supply. The three additional things needed are air, proper temperature, and water. To eliminate any one of these is to prevent decay. In most situations where wood is used, air is available and the temperatures are suitable for decay to occur. The one necessity that we can control is water; eliminate water and decay cannot occur.

What Are Some of the Improper Details?

The list of improper details is probably not complete, and never will be. As new types of construction are attempted, new errors will be made. What is important is to learn from those errors, as has been done for the ones listed here.

1. Untreated wood in contact with the ground. In most cases, the MC of wood touching the ground rises sufficiently to allow fungal spores to grow. It also allows easy access for termites. Most building codes specify the minimum distance to be provided between the ground and untreated wood.
2. Details that can trap water from plumbing leaks, seepage, or condensation.
3. Contact between wood and a porous material that can draw water from the ground up to the wood. (A too-low concrete foundation wall, for example.)
4. Inadequate ventilation of enclosed spaces that may accumulate moisture. Examples would be enclosed soffits (overhanging roofs with the underside covered), crawl spaces beneath wood floors next to the ground, and heavily insulated but enclosed attics.
5. Inadequate flashing. Flashing is a membrane, usually metal, placed at edges and at changes in direction of roofing material so as to direct the flow of water from one surface to the next and prevent it from getting under the roofing. A common error is to have the flashing too narrow.
6. Using no vapor barrier where one is needed, or having the vapor barrier in the wrong location. A vapor barrier can be explained using Fig. 16-3. Assume that the interior is heated and the air outside is cold. The interior air has a higher relative humidity than the colder outside air. Vapor tends to move, through the wall, from inside toward outside. When it contacts material whose temperature is below the dew point, the vapor condenses. The water that forms is then absorbed by whatever permeable material is present: insulation, wood studs, or plywood siding. The vapor barrier should stop the movement of the vapor at a point where the temperature of the wall material is above the dew point. This means that the vapor barrier should be located at the warm side of the wall, as shown by Fig. 16-3. A frequent error is to place it near the outside, in which case it does more harm than good.
7. Broken membranes, such as building paper (a pervious layer used on the outside of the structural sheathing of a wall), flashing, or vapor barrier. These are frequently damaged during installation or by workers doing subsequent work.

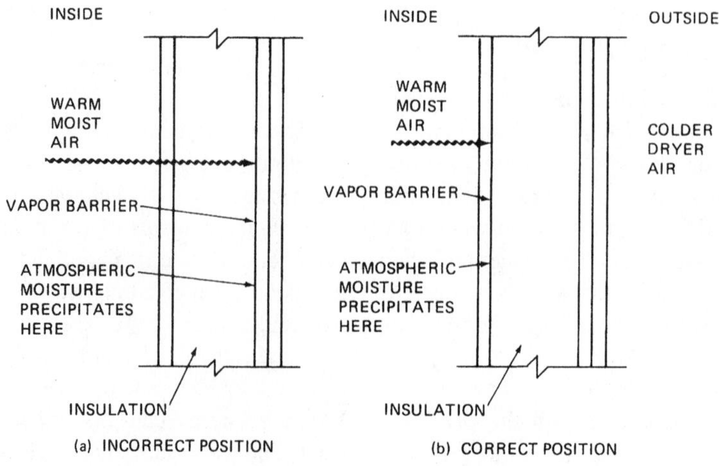

Fig. 16-3. Vapor barrier location.

8. Use of an impervious membrane where a pervious one should be used. An example is in some of the modern combination insulation and sheathing boards used instead of plywood for exterior wall sheathing. Often these have a layer of aluminum foil on the outside face. This is fine for reflecting heat, but the foil may form an impervious barrier to the outward passage of water vapor. If there is a break in the inside vapor barrier, or if there is no such barrier, then the exterior surface acts as an incorrectly placed vapor barrier. Water accumulates and decay follows. Realizing this potential difficulty, some builders deliberately knock holes in the combination sheathing to provide ventilation to the wall (2).

9. Unprotected timbers projecting beyond a building roof so that they are exposed to weathering. (See Fig. 16-4.) The situation is even worse if the timbers are horizontal, in which case rain falling on the top surface of the timber can be carried by capillary attraction into the contact between the timber and the roof decking, encouraging decay of both members.

What Details Would Be Better?

1. The sequence of wood pieces should shed water completely at the first opportunity, rather than passing it along from one piece to another. One way of doing this is to make sure that the lower edge of an exposed piece is below that of the adjacent part, as in Fig. 16-5a. Alternatives would be to provide a groove beyond which water cannot move (Fig. 16-5b) or flashing to shed the water directly.

2. Use a gutter system to prevent the water from touching any wood at the roof eaves.

3. Wherever the average January temperature is 35°F or less, use a vapor barrier *on the warm side* of exterior walls of heated buildings (3).

4. Ventilate closed spaces so that water that does enter can evaporate rather than accumulate. This rule applies especially to crawl spaces where wood floor systems are close to the ground beneath the building. It seems logical to enclose the area all around to conserve heat, yet to do so is just inviting trouble. Adequate venti-

Fig. 16-4. Improper detail caused all of these projecting beams to split badly. (Photograph by authors.)

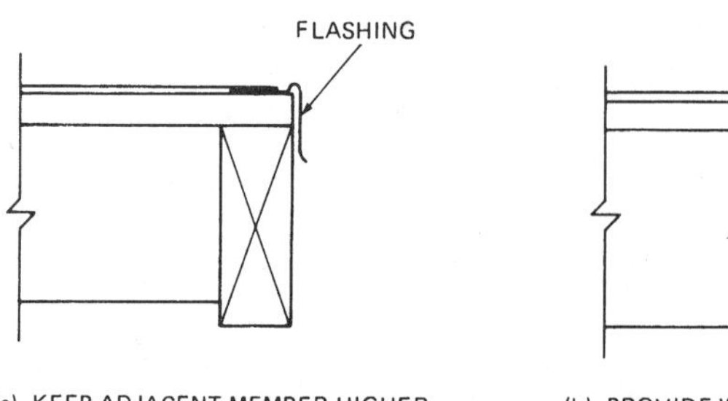

Fig. 16-5. Details to avoid wetting of adjacent member by rainwater.

lation must be provided and the area must be kept open. The ground provides water vapor, which condenses on the underside of wood members near the perimeter of the area. In addition, an impervious plastic cover may be placed over the soil to retard movement of water from the soil into the crawl space air.

5. Protect wood beams that extend beyond the cover of the roof. Flashing placed over the projecting timbers should allow air space between the wood and the protecting metal. The flashing should be so formed that water will be shed and will not pass by capillary attraction along the length or across the top of the beam.

6. Bases of unprotected wood posts or columns should be elevated above surfaces on which water may collect. Columns bearing directly on concrete floors, for example, are very prone to rotting at the base. Even though the column is an interior one that is protected from rain water, the concrete below will transmit soil moisture into the end grain of the wood. Figure 9-10 shows a suitable column base detail.

References 2 and 3 give additional recommendations.

16-7. TERMITE PROTECTION

Measures that may be used to prevent termite damage are:

1. Remove all tree stumps, roots, wood scraps, wood stakes, and concrete formwork from the building site. Do not bury any of these in the backfill. (The reason for this precaution appears at first to be merely cosmetic, but the real purpose is to prevent these scraps from decaying. Wood with brown rot is an attractive food for termites.)

2. Use only treated lumber within 18 in. of the top of the soil or highest backfill (8). In some localities, more extensive use of pressure-treated lumber is advised. Such an area is Hawaii, where all framing materials in light-frame stud walls should be treated to protect against the Formosan termite. In Puerto Rico pressure treatment is advised for all structural members to prevent attack by dry-wood termites (3).

3. Use solid concrete foundation walls, or masonry walls with portland cement mortar (to avoid cracks along which the termites can travel). Extend these foundation walls to at least 18 in. above the soil or highest backfill.

4. Poison the soil along each face of the foundation to kill termites before they can build tunnels to the wood above.

5. Use poison bait blocks around the foundation, both inside and outside. The poison used with this method has been banned for environmental reasons, but poison may return as a viable method when a satisfactory alternate chemical has been found.

6. Use termite shields. These are barriers of sheet metal to prevent the termites from reaching the wood by building tunnels on masonry or concrete foundations, or along sewer and water pipes or electrical conduits that extend from the ground to the wood above. A 2-in. width of the sheet metal barrier should slope downward at 45° on all sides of the wood being protected and should be completely sealed at all corners by soldering or crimping, so that there is no way the termite can pass the barrier except by crawling

down the bottom of the 2-in. sloping width and back up on the top surface.

Termite shields are now considered to be not as effective as using treated lumber (3). They are accepted by some building codes but not by the U.S. Minimum Property Standard.

16-8. FIRE DAMAGE

Untreated wood is flammable, yet wood structures are not necessarily more easily destroyed by fire than buildings of other materials. Further, the maximum fire hazard is usually from burning of building contents rather than the building itself. Even though a fire begins in the building contents, a prolonged fire will certainly damage the building also.

Two distinct problems in materials used for buildings are fire resistance and surface flammability. Fire resistance is the ability of the material to restrict being penetrated by the fire and to retain structural strength. Surface flammability measures how fast the flame will spread along the material surface. Ignition speed for the wood depends on temperature, whether the heat source is a flame or merely hot air, and the length of time of exposure to the elevated temperature.

The spread of fire in a wood structure depends on flammability, on size and arrangement of the wood members, and on details that allow or restrict air movement through and around the wood members. Generally, the larger the cross section of a wood member, the longer it will take to destroy the member. (Thus, five 2 × 10s would be destroyed more quickly than a single 8 × 10 of the same strength.) The burning wood develops a layer of "char" (black charcoal) on the burning surfaces, and this insulates the wood beneath to retard it from burning. The char layer proceeds through the burning wood at an average rate of about $1\frac{1}{2}$ in. per hour (4).

For fire considerations, wood structures are classified as (1) heavy-timber construction, (2) ordinary construction, and (3) wood frame construction. Heavy-timber construction was defined in Chapter 1. Ordinary construction has masonry exterior walls and wood framing members smaller than those meeting the limitations for heavy-timber construction. Wood frame construction is the same, but has wood-framed exterior walls.

In heavy-timber construction buildings, the char protects the main structural members so that, after the fire is extinguished, sufficient strength is usually present to carry at least all of the dead load. Thus the building does not collapse. Depending on actual design loads and member sizes, it may have sufficient remaining strength to allow continued use. Or it may be possible to reinforce, replace, or strengthen members selectively, as needed. In calculating the remaining strength in large timbers, however, the designer should realize the char depth is greater at the corners of the member than on the flat surfaces.

Subjected to a similar fire, however, a building with an unprotected steel frame would probably collapse completely, the yield strength and modulus of elasticity of its members being reduced by the high temperatures. The steel could be protected, of course, by fire-protective coatings or by enclosure in masonry, concrete, or other fire-resistant material. Similarly, wood parts can be treated to make them harder to ignite and to retard their burning.

16-9. DESIGN TO MINIMIZE FIRE DANGER

The most important step in minimizing fire danger is, of course, to protect the building *contents* against becoming ignited. Beyond this, there are two main categories of protection for the building itself: (1) appropriate design details to prevent ignition and the spread of fire from one part of the building to another and (2) protection or treatment of the wood to make ignition harder and retard burning.

Details

Reference 4 gives a good summary list of the detail design deficiencies responsible for spread of fire, heat, and smoke in buildings. Most of these deficiencies are addressed by building codes such as the UBC (9). The deficiencies in general consist of openings that allow heat, flame, or smoke to travel from one part of the building to another. Examples include unprotected wall openings for dumbwaiters, stairways without doors at floor levels, framed walls without firestops at concealed spaces below floors or ceilings, and improper anchorage of structural framing members in masonry walls. Figure 6-14 shows ways of framing floor or roof joists into masonry walls. The first and second of these details allow the floor of a burning building to collapse without bringing down the masonry wall at the same time.

Other design deficiencies include the lack of fire-resistant walls between adjoining buildings, the lack of protective doors in passageways between adjacent buildings, and the lack of separation between living areas and areas such as shops or boiler rooms. These requirements apply to buildings of any material—steel, concrete, masonry, or wood.

For some building uses, codes require fire doors that close automatically in case of fire. Automatic sprinklers may be required, but their function is mainly to extinguish fire in the building contents; thus their need is more dictated by building use than building material.

The designer should observe carefully all requirements of building codes to avoid most of these deficiencies.

Protection or Treatment

Using formulas published by the Council of American Building Officials, the designer may calculate the required size of members necessary for fire resistance. The formulas give the fire-resistance rating in minutes for timber beams and columns with smaller dimensions not less than 6 in. nominal. The formulas are:

For beams—

Exposed to fire on all four surfaces

$$2.54Zb(4 - 2b/d) \qquad (16\text{-}1)$$

Exposed to fire on only three surfaces

$$2.54Zb(4 - b/d) \qquad (16\text{-}2)$$

For columns—

Exposed to fire on all four surfaces

$$2.54Zd(3 - d/b) \qquad (16\text{-}3)$$

Exposed to fire on only three surfaces

$$2.54Zd(3 - d/2b) \qquad (16\text{-}4)$$

In each of the above

- b = the horizontal dimension of a beam cross section, or the *larger* side dimension of a column
- d = the vertical depth of a beam, or the *small* side dimension of a column
- Z = a "load" factor that depends on the percent of allowable load actually carried by the member; Figure 16-6 shows how to determine Z.

Using the above expressions, the designer may estimate the member dimensions needed for unprotected heavy-timber construction.

For ordinary and wood frame construction, the steps to protect against fire include, first, following all building code requirements and avoiding construction detail deficiencies. The next step, if it is judged necessary, is to reduce flammability of the wood. This can be done in three ways:

1. Encase the wood members in protective material such as gypsum plaster or plasterboard.
2. Apply an intumescent coating, painted onto the exposed surfaces, which expands when heated to produce an insulating layer. This treatment is better reserved for upgrading existing buildings. (For new buildings, treatment before the wood is installed is better.)

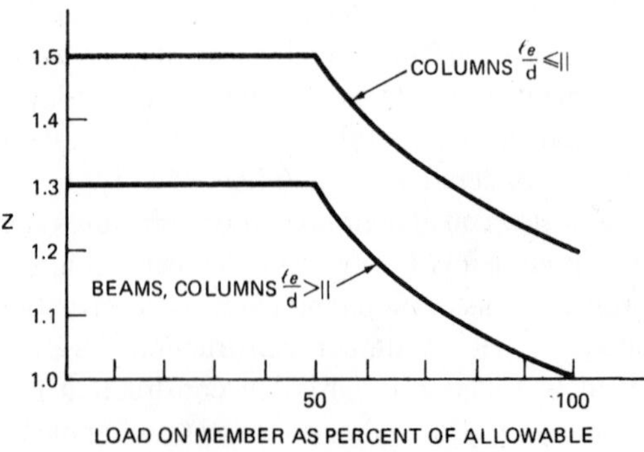

Fig. 16-6. Factor Z for fire-resistance equations. (Courtesy of the American Institute of Timber Construction (AITC).)

3. Use fire-retardant treatment, either for all wood parts or for those parts whose size or exposure makes them the most likely to ignite and fuel a fire.

Fire-retardant treatment has been in use since the fourth century B.C. (1). Modern fire-retardant treatment (it should not be called fireproofing) consists of pressure treatment with aqueous solutions of various salts, followed by drying to remove the water and to lower the MC to the desired level (19% or less for lumber not over 2 in. thick, and 15% or less for plywood). Chemicals used most widely are combinations including zinc chloride, ammonium sulfate, borax, or boric acid, with lesser amounts of sodium dichromate or diammonium phosphate (6). The effectiveness of fire-retardant treatment depends on the chemicals used and on the amount of chemical retained by the wood.

Some of the salts used in treatment are hygroscopic, so retardant-treated wood should not be used where the relative humidity will be 80% or more for long periods. To do so would cause the MC of the wood to rise to an undesirable level. (For localities where the average relative humidity is 70% or more, the designer should specify an exterior nonhygroscopic chemical.) Also, if water contacts the treated members, there can be leaching of the salt. This can be retarded by surface painting. However, if treated wood is to be stained, painted, or varnished, the designer should investigate carefully before specifying treated wood to be certain that the intended finish can be applied successfully.

Fire-retardant treatment does two things that are important to the structural designer. It increases the weight of the member, and it may decrease the allowable stresses. Weight of the member depends on the weight of chemical retained per unit volume of wood. The NDS (10) requires that the allowable stresses be reduced to values specified by the company providing the treatment.

16-10. EVALUATION AND REPAIR OF WOOD STRUCTURES

Structural designers are often called on to evaluate existing structures and either to design repair measures to correct structural deficiencies detected or to determine the capability of the structure to serve some new purpose, perhaps with increased loading. In the case of wood structures, this occurs frequently as architects and developers attempt to preserve existing structures, adapting them to new uses.

What things should the designer look for in evaluating an existing wood structure? The following lists many of them:

1. Evidence of past changes in use. Revised framing might indicate this. Revised use might also be detected by comparing the present structure to that illustrated by the drawings, both the originals and those prepared during past remodeling.
2. Evidence of past repairs. These might be detected by the presence of obviously newer lumber in some members or might be evidenced by "beefed-up" connections. Any such repair (unless it was a step made to increase the load capacity beyond that intended by the original designer) probably indicates some past structural distress.
3. Evidence of foundation problems. Unequal settlement of foundations can cause cracks in walls, poor fit of doors and windows, and sloping floors.
4. Roof surface irregularities. These might be the result of foundation problems, but more often indicate a weakness in certain roof members.
5. Abnormal deflections of beam members. This could be the result of creep. If the deflections are continuing to increase without increase in load, the possibility of collapse due to tertiary creep should be considered. (See Chapter 2.)
6. Fire damage. Should the wood show evidence of fire damage, the metal parts should also be examined.
7. Evidence of chemical deterioration or rust of metal fittings.
8. Loosened connectors. Loosening often occurs because of wood shrinkage. (If shrinkage is expected, an obviously wise maintenance step is to tighten the connectors after wood shrinkage has occurred.)
9. Splits in the wood at bolted connections.

10. Water stains on the wood, or evidence of chemical deterioration.
11. Ends of beams that rest on masonry or concrete walls and are totally enclosed to prevent adequate ventilation. Rotting may be detected at such locations.
12. Moist wood at boxed-in connection details. Fig. 9-10e, for example, especially if the weep holes are missing or blocked.)
13. Decay at the ends of posts or columns that rest directly upon concrete floors.
14. Longitudinal separations at the glue line of glued laminated members. Separation in the glue itself, rather than the wood, could be serious and might be a warning of more serious failure.

The above and several others are discussed in greater detail in reference 11.

Repairs to correct structural distress range from simple maintenance to prevent further deterioration to complete replacement of the distressed component. Repair techniques are not discussed here, but are covered in some detail by reference 11.

REFERENCES

1. Hunt, G. M., and G. A. Garratt, *Wood Preservation,* 3rd ed., McGraw-Hill, New York, 1967.
2. Wilcox, W. W., and A. F. Rosenberg, "Architectural and Construction Deficiencies Contributing to Decay of Wood in Buildings," in Meyer, R. W., and R. M. Kellogg, *Structural Uses of Wood in Adverse Environments,* Van Nostrand Reinhold, New York, 1982.
3. DeGroot, R. C., and G. R. Esenther, "Microbiological and Entomological Effects," in Meyer, R. W., and R. M. Kellogg, *Structural Uses of Wood in Adverse Environments,* Van Nostrand Reinhold, New York, 1982.
4. *Wood Structures, A Design Guide and Commentary,* American Society of Civil Engineers, New York, 1975.
5. Feist, W. C., "Weathering of Wood in Structural Uses," in Meyer, R. W., and R. M. Kellogg, *Structural Uses of Wood in Adverse Environments,* Van Nostrand Reinhold, New York, 1982.
6. *AWPA Standards,* American Wood Preservers' Association, Stevensville, MD, 1991.
7. *Federal Specification—Wood Preservation Treating Practices,* Federal Specification TT-W-571, General Services Administration, Washington, DC.
8. American Institute of Timber Construction, *Timber Construction Manual,* 4th ed., Wiley, New York, 1994.
9. *Uniform Building Code,* International Conference of Building Officials, Whittier, CA, 1994.
10. *National Design Specification for Wood Construction,* rev. 1991 Ed., American Forest & Paper Association, Washington, DC, 1993.
11. *Evaluation, Maintenance and Upgrading of Wood Structures, A Guide and Commentary,* American Society of Civil Engineers, New York, 1982.

PROBLEMS

16-1. Number 2, 8 × 8 D. fir-larch timbers are specified to be pressure treated with creosote, with 10-pcf specified retention. What will be the probable weight per linear foot of the treated timbers?

16-2. Re-solve Problem 6-15, assuming that all members will have fire-retardant treatment that reduces E and all allowable stresses by 15%.

16-3. From the standpoint or subjects of this chapter, tell what is wrong with each of the design details shown by Fig. 16-7. For each, show a better detail.

16-4. A residential building used a nailed laminated floor system (such as shown by Fig. 15-21b for a bridge) of 20-ft span. In one location, the planks forming this nailed laminated floor were extended 5 ft to the outside to form a cantilever balcony. A membrane placed to protect the balcony structure developed cracks, the wood rotted, and the balcony collapsed. Explain what went wrong and suggest a better detail. (Could alternate laminations have been eliminated from the balcony without jeopardizing strength?)

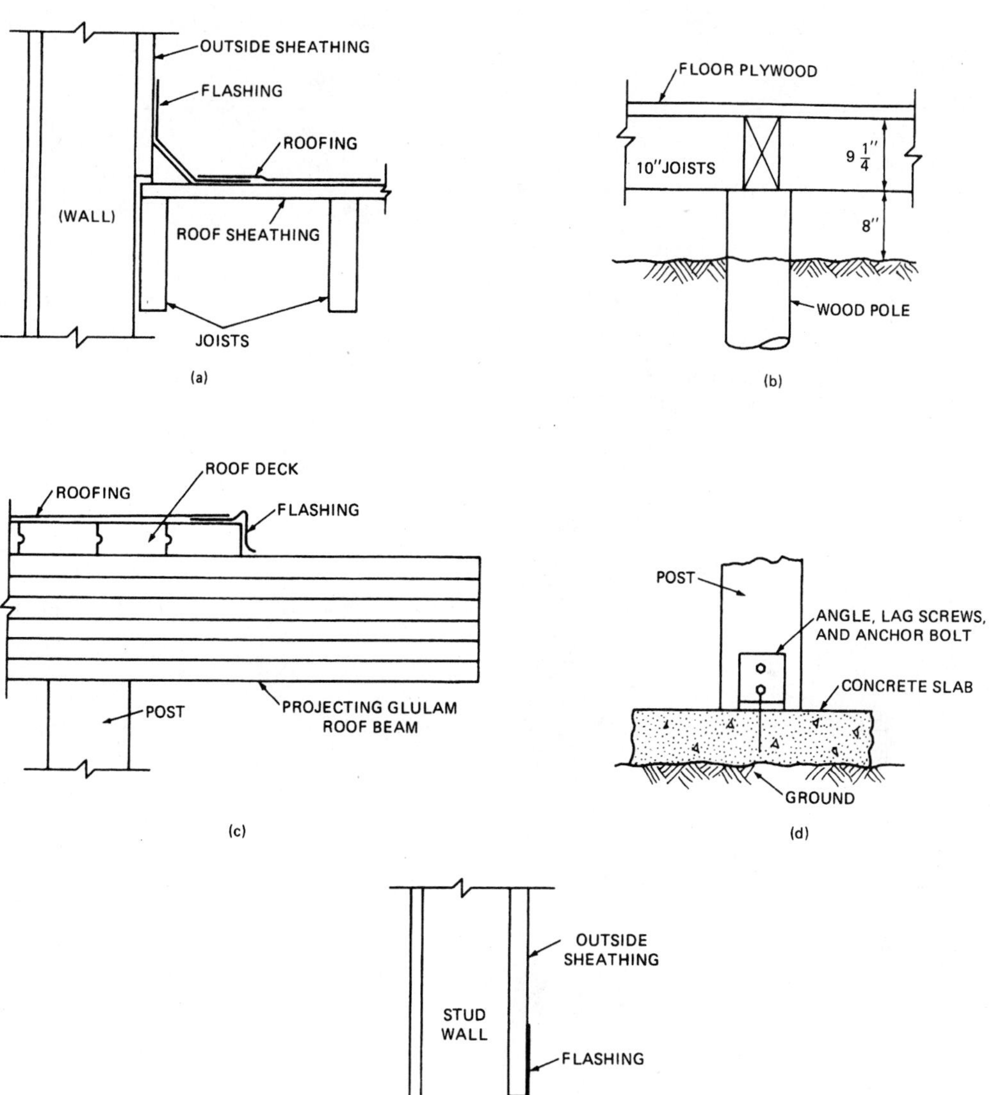

Fig. 16-7. Defective details for Problem 16-3.

Appendix A

Table A-1. Weights of Building Materials.

Materials	Weight (psf)	Materials	Weight (psf)
Ceilings		Partitions	
Channel suspended system	1	Clay Tile	
Lathing and plastering	See Partitions	3 in.	17
Acoustical fiber tile	1	4 in.	18
		6 in.	28
Floors		8 in.	34
Steel Deck	See Manufacturer	10 in.	40
Concrete-reinforced 1 in.		Gypsum block	
Stone	$12\frac{1}{2}$	2 in.	$9\frac{1}{2}$
Slag	$11\frac{1}{2}$	3 in.	$10\frac{1}{2}$
Lightweight	6 to 10	4 in.	$12\frac{1}{2}$
		5 in.	14
Concrete—plain 1 in.		6 in.	$18\frac{1}{2}$
Stone	12	Wood studs 2 × 4	
Slag	11	12–16 in. o.c.	2
Lightweight	3 to 9	Steel partitions	4
		Plaster 1 in.	
Fills 1 in.		Cement	10
Gypsum	6	Gypsum	5
Sand	8	Lathing	
Cinders	4	Metal	$\frac{1}{2}$
		Gypsum board $\frac{1}{2}$ in.	2
Finishes			
Terrazzo 1 in.	13	Walls	
Ceramic or quarry tile $\frac{3}{4}$ in.	10	Brick	
		4 in.	40
Linoleum $\frac{1}{4}$ in.	1	8 in.	80
Mastic $\frac{3}{4}$ in.	9	12 in.	120
Hardwood $\frac{7}{8}$ in.	4	Hollow concrete block (heavy aggregate)	
Softwood $\frac{3}{4}$ in.	$2\frac{1}{2}$		
		4 in.	30
Roofs		6 in.	43
Copper or tin	1	8 in.	55
3-ply ready roofing	1	$12\frac{1}{2}$ in.	80
3-ply felt and gravel	$5\frac{1}{2}$	Hollow concrete block (light aggregate)	
5-ply felt and gravel	6		
		4 in.	21
Shingles		6 in.	30
Wood	2	8 in.	38
Asphalt	3	12 in.	55
Clay tile	9 to 14	Clay tile (load bearing)	
Slate $\frac{1}{4}$	10	4 in.	25
		6 in.	30

Table A-1. (continued)

Materials	Weight (psf)	Materials	Weight (psf)
Sheathing		8 in.	33
Wood $\frac{3}{4}$ in.	3	12 in.	45
Gypsum 1 in.	4	Stone 4 in.	55
		Glass block 4 in.	18
Insulation 1 in.		Windows, glass, frame & sash	8
Loose	$\frac{1}{2}$	Curtain walls	See Manufacturer
Poured in place	2	Structural glass 1 in.	15
Rigid	$1\frac{1}{2}$	Corrugated cement asbestos $\frac{1}{4}$ in.	3

Source: Courtesy of American Institute of Steel Construction.

Table A-2. Uniform and Concentrated Loads. (from UBC Table 16-A)

Use or Occupancy		Uniform Load[1] (pounds per square foot)	Concentrated Load (pounds)
Category	Description	× 0.0479 for kN/m²	× 0.004 48 for kN
1. Access floor systems	Office use	50	2,000[2]
	Computer use	100	2,000[2]
2. Armories		150	0
3. Assembly areas[3] and auditoriums and balconies therewith	Fixed seating areas	50	0
	Movable seating and other areas	100	0
	Stage areas and enclosed platforms	125	0
4. Cornices and marquees		60[4]	0
5. Exit facilities[5]		100	0[6]
6. Garages	General storage and/or repair	100	[7]
	Private or pleasure-type motor vehicle storage	50	[7]
7. Hospitals	Wards and rooms	40	1,000[2]
8. Libraries	Reading rooms	60	1,000[2]
	Stack rooms	125	1,500[2]
9. Manufacturing	Light	75	2,000[2]
	Heavy	125	3,000[2]
10. Offices		50	2,000[2]
11. Printing plants	Press rooms	150	2,500[2]
	Composing and linotype rooms	100	2,000[2]
12. Residential[8]	Basic floor area	40	0[6]
	Exterior balconies	60[4]	0
	Decks	40[4]	0
13. Restrooms[9]			
14. Reviewing stands, grandstands, bleachers, and folding and telescoping seating		100	0
15. Roof decks	Same as area served or for the type of occupancy accommodated		
16. Schools	Classrooms	40	1,000[2]
17. Sidewalks and driveways	Public access	250	[7]
18. Storage	Light	125	
	Heavy	250	
19. Stores		100	3,000[2]
20. Pedestrian bridges and walkways		100	

[1] See Section 1606 for live-load reductions.
[2] See Section 1604.3, first paragraph, for area of load application.
[3] Assembly areas include such occupancies as dance halls, drill rooms, gymnasiums, playgrounds, plazas, terraces, and similar occupancies which are generally accessible to the public.
[4] When snow loads occur that are in excess of the design conditions, the structure shall be designed to support the loads due to the increased loads caused by drift buildup or a greater snow design as determined by the building official. See Section 1605.4. For special-purpose roofs, see Section 1605.5.
[5] Exit facilities shall include such uses as corridors serving an occupant load of 10 or more persons, exterior exit balconies, stairways, fire escapes, and similar uses.
[6] Individual stair treads shall be designed to support a 300-pound (1.33 kN) concentrated load placed in a position which would cause maximum stress. Stair stringers may be designed for the uniform load set forth in the table.
[7] See Section 1604.3, second paragraph, for concentrated loads. See Table 16-B for vehicle barriers.
[8] Residential occupancies include private dwellings, apartments, and hotel guest rooms.
[9] Rest room loads shall not be less than the load for the occupancy with which they are associated, but need not exceed 50 pounds per square foot (2.4 kN/m²).

Source: Reproduced from the 1994 edition of the *Uniform Building Code*™, copyright © 1994, with the permission of the publisher, the International Conference of Building Officials.

Table A-3. Minimum Roof Live Loads[1] (psf) (From UBC Table 16-C)

Roof Slope	Method 1			Method 2		
	Tributary Loaded Area in Square Feet for Any Structural Member × 0.0929 for m²			Uniform Load[2] (pounds per square foot)	Rate of Reduction r (percentage)	Maximum Reduction R (percentage)
	0 to 200	201 to 600	Over 600			
	Uniform Load (pounds per square foot)					
	× 0.0479 for kN/m²					
1. Flat[3] or rise less than 4 units vertical in 12 units horizontal (33.3% slope). Arch or dome with rise less than one-eighth of span	20	16	12	20	.08	40
2. Rise 4 units vertical to less than 12 units vertical in 12 units horizontal (33% to less than 100% slope). Arch or dome with rise one-eighth of span to less than three-eighths of span	16	14	12	16	.06	25
3. Rise 12 units vertical in 12 units horizontal (100% slope) and greater. Arch or dome with rise three-eighths of span or greater	12	12	12	12	No reductions permitted	
4. Awnings except cloth covered[4]	5	5	5	5	No reductions permitted	
5. Greenhouses, lath houses, and agricultural buildings[5]	10	10	10	10	No reductions permitted	

[1] Where snow loads occur, the roof structure shall be designed for such loads as determined by the building official. See Section 1605.4. For special-purpose roofs, see Section 1605.5.

[2] See Section 1606 for live-load reductions. The rate of reduction r in Section 1606 Formula (6-1) shall be as indicated in the table. The maximum reduction R shall not exceed the value indicated in the table.

[3] A flat roof is any roof with a slope of less than 1/4 unit vertical in 12 units horizontal (2% slope). The live load for flat roofs is in addition to the ponding load required by Section 1605.6.

[4] As defined in Section 3206.

[5] See Section 1605.5 for concentrated load requirements for greenhouse roof members.

Source: Reproduced from the 1994 edition of the *Uniform Building Code*™, copyright © 1994, with the permission of the publisher, the International Conference of Building Officials.

Table A-4. Combined Height, Exposure, and Gust Factor Coefficient (C_e).[1] (From UBC Table 16-G)

Height Above Average Level of Adjoining Ground (feet) × 304.8 for mm	Exposure D	Exposure C	Exposure B
0–15	1.39	1.06	0.62
20	1.45	1.13	0.67
25	1.50	1.19	0.72
30	1.54	1.23	0.76
40	1.62	1.31	0.84
60	1.73	1.43	0.95
80	1.81	1.53	1.04
100	1.88	1.61	1.13
120	1.93	1.67	1.20
160	2.02	1.79	1.31
200	2.10	1.87	1.42
300	2.23	2.05	1.63
400	2.34	2.19	1.80

[1] Values for intermediate heights above 15 feet (4572 mm) may be interpolated.
Source: Reproduced from the 1994 edition of the *Uniform Building Code*™, copyright © 1994, with the permission of the publisher, the International Conference of Building Officials.

Table A-5. Pressure Coefficients, C_q. (Excerpts from UBC Table 16-H)

Structure or Part Thereof	Description	C_q Factor
1. Primary frames and systems	**Method 1** (Normal force method)	
	Walls:	
	Windward wall	0.8 inward
	Leeward wall	0.5 outward
	Roofs[1]:	
	Wind perpendicular to ridge	
	Leeward roof or flat roof	0.7 outward
	Windward roof	
	less than 2:12 (16.7%)	0.7 outward
	Slope 2:12 (16.7%) to less than 9:12 (75%)	0.9 outward or 0.3 inward
	Slope 9:12 (75%) to 12:12 (100%)	0.4 inward
	Slope >12:12 (100%)	0.7 inward
	Wind parallel to ridge and flat roofs	0.7 outward
	Method 2 (Projected area method)	1.3 horizontal any direction
	On vertical projected area	
	Structures 40 feet (12 192 mm) or less in height	1.4 horizontal any direction
	Structures over 40 feet (12 192 mm) in height	
	On horizontal projected area[1]	0.7 upward
2. Elements and components not in areas of discontinuity[2]	Wall elements	
	All structures	1.2 inward
	Enclosed and unenclosed structures	1.2 outward
	Partially enclosed structures	1.6 outward
	Parapets walls	1.3 inward or outward
	Roof elements[3]	
	Enclosed and unenclosed structures	
	Slope <7:12 (58.3%)	
	Slope 7:12 (58.3%) to 12:12 (100%)	1.3 outward
		1.3 outward or inward
	Partially enclosed structures	
	Slope <2:12 (16.7%)	1.7 outward
	Slope 2:12 (16.7%) to 7:12 (58.3%)	1.6 outward or 0.8 inward
	Slope >7:12 (58.3%) to 12:12 (100%)	1.7 outward or inward
3. Elements and components in areas of discontinuities[2,4,5]	Wall corners[6]	1.5 outward or 1.2 inward
	Roof eaves, rakes or ridges without overhangs[6]	
	Slope <2:12 (16.7%)	2.3 upward
	Slope 2:12 (16.7%) to 7:12 (58.3%)	2.6 outward
	Slope >7:12 (58.3%) to 12:12 (100%)	1.6 outward
	For slopes less than 2:12 (16.7%)	
	Overhangs at roof eaves, rakes or ridges, and canopies	0.5 added to values above

[1] For one story or the top story of multistory partially enclosed structures, an additional value of 0.5 shall be added to the outward C_q. The most critical combination shall be used for design. For definition of open structures, see Section 1613.

[2] C_q values listed are for 10-square-foot (0.93 m^2) tributary areas. For tributary areas of 100 square feet (9.29 m^2), the value of 0.3 may be subtracted from C_q, except for areas at discontinuities with slopes less than 7 units vertical in 12 units horizontal (58.3% slope) where the value of 0.8 may be subtracted from C_q. Interpolation may be used for tributary areas between 10 and 100 square feet (0.93 m^2 and 9.29 m^2). For tributary areas greater than 1,000 square feet (92.9 m^2), use primary frame values.

[3] For slopes greater than 12 units vertical in 12 units horizontal (100% slope), use wall element values.

[4] Local pressures shall apply over a distance from the discontinuity of 10 feet (3048 mm) or 0.1 times the least width of the structure, whichever is smaller.

[5] Discontinuities at wall corners or roof ridges are defined as discontinuous breaks in the surface where the included interior angle measures 170 degrees or less.

[6] Load is to be applied on either side of discontinuity but not simultaneously on both sides.

Source: Reproduced from the 1994 edition of the *Uniform Building Code*™, copyright © 1994, with the permission of the publisher, the International Conference of Building Officials.

Table A-6. Wind Stagnation Pressure (q_s) at Standard Height of 33 ft. (UBC Table 16-F)

Basic wind speed (mph)[1] (× 1.61 for km/h)	70	80	90	100	110	120	130
Pressure q_s (psf) (× 0.0479 for kN/m²)	12.6	16.4	20.8	25.6	31.0	36.9	43.3

[1] Wind speed from Section 1615.
Source: Reproduced from the 1994 edition of the *Uniform Building Code*™, copyright © 1994, with the permission of the publisher, the International Conference of Building Officials.

Table A-7. Site Coefficients.[1] (UBC Table 16-J)

Type	Description	*S* Factor
S_1	A soil profile with either: (a) A rock-like material characterized by a shear-wave velocity greater than 2,500 feet per second (762 m/s) or by other suitable means of classification, or (b) Medium-dense to dense or medium-stiff to stiff soil conditions, where soil depth is less than 200 feet (60 960 mm).	1.0
S_2	A soil profile with predominantly medium-dense to dense or medium-stiff to stiff soil conditions, where the soil depth exceeds 200 feet (60 960 mm).	1.2
S_3	A soil profile containing more than 20 feet (6096 mm) of soft to medium-stiff clay but not more than 40 feet (12 192 mm) of soft clay.	1.5
S_4	A soil profile containing more than 40 feet (12 192 mm) of soft clay characterized by a shear wave velocity less than 500 feet per second (152.4 m/s).	2.0

[1] The site factor shall be established from properly substantiated geotechnical data. In locations where the soil properties are not known in sufficient detail to determine the soil profile type, soil profile S_3 shall be used. Soil profile S_4 need not be assumed unless the building official determines that soil profile S_4 may be present at the site, or in the event that soil profile S_4 is established by geotechnical data.
Source: Reproduced from the 1994 edition of the *Uniform Building Code*™, copyright © 1994, with the permission of the publisher, the International Conference of Building Officials.

Table A-8. Maximum Beam Bending Moments, Shears, and Deflections.

	Maximum Moment	Maximum Shear	Maximum Deflection
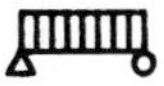	$0.125w\ell^2$	$0.5w\ell$	$5w\ell^4/(384EI)$
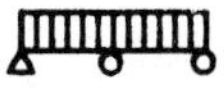	$0.125w\ell^2$	$0.625w\ell$	$w\ell^4/(185EI)$
	$0.1w\ell^2$	$0.6w\ell$	$w\ell^4/(145EI)$
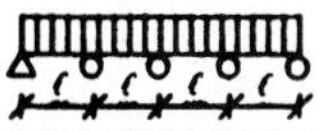	$0.107w\ell^2$	$0.607w\ell$	$w\ell^4/(154EI)$

Appendix B

Table B-1. Properties of Sawn-Timber Sections.

Nominal Size $b \times d$ (in.)	Standard Dressed Size (S4S) $b \times d$ (in.)	Area of Section A (in.^{2})	X-X Axis: Moment of Inertia I (in.^{4})	X-X Axis: S_x (in.^{3})	Y-Y Axis: Moment of Inertia I (in.^{4})	Y-Y Axis: S_y (in.^{3})	Board Measure per Lineal Foot	Weight (lb/ft): 30 pcf	Weight (lb/ft): 40 pcf	Weight (lb/ft): 50 pcf
1 × 3	$\frac{3}{4} \times 2\frac{1}{2}$	1.875	0.977	0.78	0.088	0.234	$\frac{1}{4}$	0.391	0.521	0.651
1 × 4	$\frac{3}{4} \times 3\frac{1}{2}$	2.625	2.680	1.53	0.123	0.328	$\frac{1}{3}$	0.547	0.729	0.911
1 × 6	$\frac{3}{4} \times 5\frac{1}{2}$	4.125	10.40	3.78	0.193	0.516	$\frac{1}{2}$	0.859	1.146	1.432
1 × 8	$\frac{3}{4} \times 7\frac{1}{4}$	5.438	23.82	6.57	0.255	0.680	$\frac{2}{3}$	1.133	1.510	1.888
1 × 10	$\frac{3}{4} \times 9\frac{1}{4}$	6.938	49.47	10.70	0.325	0.867	$\frac{5}{6}$	1.445	1.927	2.409
1 × 12	$\frac{3}{4} \times 11\frac{1}{4}$	8.438	88.99	15.82	0.396	1.055	1	1.758	2.344	2.930
2 × 3	$1\frac{1}{2} \times 2\frac{1}{2}$	3.750	1.953	1.56	0.703	0.938	$\frac{1}{2}$	0.781	1.042	1.302
2 × 4	$1\frac{1}{2} \times 3\frac{1}{2}$	5.250	5.359	3.06	0.984	1.313	$\frac{2}{3}$	1.094	1.458	1.823
2 × 6	$1\frac{1}{2} \times 5\frac{1}{2}$	8.250	20.80	7.56	1.547	2.063	1	1.719	2.292	2.865
2 × 8	$1\frac{1}{2} \times 7\frac{1}{4}$	10.88	47.64	13.14	2.039	2.719	$1\frac{1}{3}$	2.266	3.021	3.776
2 × 10	$1\frac{1}{2} \times 9\frac{1}{4}$	13.88	98.93	21.39	2.602	3.469	$1\frac{2}{3}$	2.891	3.854	4.818
2 × 12	$1\frac{1}{2} \times 11\frac{1}{4}$	16.88	178.0	31.64	3.164	4.219	2	3.516	4.688	5.859
2 × 14	$1\frac{1}{2} \times 13\frac{1}{4}$	19.88	290.8	43.89	3.727	4.969	$2\frac{1}{3}$	4.141	5.521	6.901
3 × 4	$2\frac{1}{2} \times 3\frac{1}{2}$	8.750	8.932	5.10	4.557	3.646	1	1.823	2.431	3.038
3 × 6	$2\frac{1}{2} \times 5\frac{1}{2}$	13.75	34.66	12.60	7.161	5.729	$1\frac{1}{2}$	2.865	3.819	4.774
3 × 8	$2\frac{1}{2} \times 7\frac{1}{4}$	18.12	79.39	21.90	9.440	7.552	2	3.776	5.035	6.293
3 × 10	$2\frac{1}{2} \times 9\frac{1}{4}$	23.12	164.9	35.65	12.04	9.635	$2\frac{1}{2}$	4.818	6.424	8.030
3 × 12	$2\frac{1}{2} \times 11\frac{1}{4}$	28.12	296.6	52.73	14.65	11.72	3	5.859	7.813	9.766
3 × 14	$2\frac{1}{2} \times 13\frac{1}{4}$	33.12	484.6	73.15	17.25	13.80	$3\frac{1}{2}$	6.901	9.201	11.50
3 × 16	$2\frac{1}{2} \times 15\frac{1}{4}$	38.12	738.9	96.90	19.86	15.88	4	7.943	10.590	13.24
4 × 4	$3\frac{1}{2} \times 3\frac{1}{2}$	12.25	12.50	7.15	12.50	7.15	$1\frac{1}{3}$	2.552	3.403	4.253
4 × 6	$3\frac{1}{2} \times 5\frac{1}{2}$	19.25	48.53	17.65	19.65	11.23	2	4.010	5.347	6.684
4 × 8	$3\frac{1}{2} \times 7\frac{1}{4}$	25.38	111.1	30.66	25.90	14.80	$2\frac{2}{3}$	5.286	7.049	8.811
4 × 10	$3\frac{1}{2} \times 9\frac{1}{4}$	32.38	230.8	49.91	33.05	18.88	$3\frac{1}{3}$	6.745	8.933	11.24
4 × 12	$3\frac{1}{2} \times 11\frac{1}{4}$	39.38	415.3	73.83	40.20	22.97	4	8.203	10.94	13.67
4 × 14	$3\frac{1}{2} \times 13\frac{1}{4}$	46.38	678.5	102.4	47.34	27.05	$4\frac{2}{3}$	9.657	12.88	16.09
4 × 16	$3\frac{1}{2} \times 15\frac{1}{4}$	53.38	1,034	135.7	54.49	31.14	$5\frac{1}{3}$	11.12	14.83	18.54
6 × 6	$5\frac{1}{2} \times 5\frac{1}{2}$	30.25	76.26	27.73	76.26	27.73	3	6.302	8.403	10.50
6 × 8	$5\frac{1}{2} \times 7\frac{1}{2}$	41.25	193.4	51.56	104.0	37.81	4	8.594	11.46	14.32
6 × 10	$5\frac{1}{2} \times 9\frac{1}{2}$	52.25	393.0	82.73	131.7	47.90	5	10.88	14.51	18.14
6 × 12	$5\frac{1}{2} \times 11\frac{1}{2}$	63.25	697.1	121.3	159.4	57.98	6	13.18	17.57	21.96
6 × 14	$5\frac{1}{2} \times 13\frac{1}{2}$	74.25	1,128	167.1	187.2	68.06	7	15.47	20.62	25.78
6 × 16	$5\frac{1}{2} \times 15\frac{1}{2}$	85.25	1,707	220.2	214.9	78.15	8	17.76	23.68	29.60
6 × 18	$5\frac{1}{2} \times 17\frac{1}{2}$	96.25	2,456	280.7	242.6	88.23	9	20.05	26.74	33.42
6 × 20	$5\frac{1}{2} \times 19\frac{1}{2}$	107.2	3,398	348.6	270.4	98.31	10	22.34	29.79	37.24
6 × 22	$5\frac{1}{2} \times 21\frac{1}{2}$	118.2	4,555	423.7	298.1	103.4	11	24.64	32.85	41.06
6 × 24	$5\frac{1}{2} \times 23\frac{1}{2}$	129.2	5,948	506.2	325.8	118.5	12	26.93	35.90	44.88
8 × 8	$7\frac{1}{2} \times 7\frac{1}{2}$	56.25	263.7	70.31	263.7	70.31	$5\frac{1}{3}$	11.72	15.62	19.53
8 × 10	$7\frac{1}{2} \times 9\frac{1}{2}$	71.25	535.9	112.8	334.0	89.06	$6\frac{2}{3}$	14.84	19.79	24.74

Table B-1. (continued)

Nominal Size $b \times d$ (in.)	Standard Dressed Size (S4S) $b \times d$ (in.)	Area of Section A (in.^{2})	X-X Axis Moment of Inertia I (in.^{4})	X-X Axis S_x (in.^{3})	Y-Y Axis Moment of Inertia I (in.^{4})	Y-Y Axis S_y (in.^{3})	Board Measure per Lineal Foot	Weight (lb/ft) 30 pcf	Weight (lb/ft) 40 pcf	Weight (lb/ft) 50 pcf
8 × 12	$7\frac{1}{2} \times 11\frac{1}{2}$	86.25	950.5	165.3	404.3	107.8	8	17.97	23.96	29.95
8 × 14	$7\frac{1}{2} \times 13\frac{1}{2}$	101.2	1,538	227.8	474.6	126.6	$9\frac{1}{3}$	21.09	28.12	35.16
8 × 16	$7\frac{1}{2} \times 15\frac{1}{2}$	116.2	2,327	300.3	544.9	145.3	$10\frac{2}{3}$	24.22	32.29	40.36
8 × 18	$7\frac{1}{2} \times 17\frac{1}{2}$	131.2	3,350	382.8	615.2	164.1	12	27.34	36.46	45.57
8 × 20	$7\frac{1}{2} \times 19\frac{1}{2}$	146.2	4,634	475.3	685.5	182.8	$13\frac{1}{3}$	30.47	40.62	50.78
8 × 22	$7\frac{1}{2} \times 21\frac{1}{2}$	161.2	6,211	577.8	755.9	201.6	$14\frac{2}{3}$	33.59	44.79	55.99
8 × 24	$7\frac{1}{2} \times 23\frac{1}{2}$	176.2	8,111	690.3	826.2	220.3	16	36.72	48.96	61.20
10 × 10	$9\frac{1}{2} \times 9\frac{1}{2}$	90.25	678.8	142.9	678.8	142.9	$8\frac{1}{3}$	18.80	25.07	31.34
10 × 12	$9\frac{1}{2} \times 11\frac{1}{2}$	109.2	1,204	209.4	821.7	173.0	10	22.76	30.35	37.93
10 × 14	$9\frac{1}{2} \times 13\frac{1}{2}$	128.2	1,948	288.6	964.6	203.1	$11\frac{2}{3}$	26.72	35.62	44.53
10 × 16	$9\frac{1}{2} \times 15\frac{1}{2}$	147.2	2,948	380.4	1,107	233.1	$13\frac{1}{3}$	30.68	40.90	51.13
10 × 18	$9\frac{1}{2} \times 17\frac{1}{2}$	166.2	4,243	484.9	1,250	263.2	15	34.64	46.18	57.73
10 × 20	$9\frac{1}{2} \times 19\frac{1}{2}$	185.2	5,870	602.1	1,393	293.3	$16\frac{2}{3}$	38.59	51.46	64.32
10 × 22	$9\frac{1}{2} \times 21\frac{1}{2}$	204.2	7,868	731.9	1,536	323.4	$18\frac{1}{3}$	42.55	56.74	70.92
10 × 24	$9\frac{1}{2} \times 23\frac{1}{2}$	223.2	10,274	874.4	1,679	353.5	20	46.51	62.01	77.52
12 × 12	$11\frac{1}{2} \times 11\frac{1}{2}$	132.2	1,458	253.5	1,458	253.5	12	27.55	36.74	45.92
12 × 14	$11\frac{1}{2} \times 13\frac{1}{2}$	155.2	2,358	349.3	1,711	297.6	14	32.34	43.12	53.91
12 × 16	$11\frac{1}{2} \times 15\frac{1}{2}$	178.2	3,569	460.5	1,964	341.6	16	37.14	49.51	61.89
12 × 18	$11\frac{1}{2} \times 17\frac{1}{2}$	201.2	5,136	587.0	2,218	385.7	18	41.93	55.90	69.88
12 × 20	$11\frac{1}{2} \times 19\frac{1}{2}$	224.2	7,106	728.8	2,471	429.8	20	46.72	62.29	77.86
12 × 22	$11\frac{1}{2} \times 21\frac{1}{2}$	247.2	9,524	886.0	2,725	473.9	22	51.51	68.68	85.85
12 × 24	$11\frac{1}{2} \times 23\frac{1}{2}$	270.2	12,437	1,058	2,978	518.0	24	56.30	75.07	93.84
14 × 16	$13\frac{1}{2} \times 15\frac{1}{2}$	209.2	4,189	540.6	3,178	470.8	$18\frac{2}{3}$	43.59	58.12	72.66
14 × 18	$13\frac{1}{2} \times 17\frac{1}{2}$	236.2	6,029	689.1	3,588	531.6	21	49.22	65.62	82.03
14 × 20	$13\frac{1}{2} \times 19\frac{1}{2}$	263.2	8,342	855.6	3,998	592.3	$23\frac{1}{3}$	58.84	73.12	91.41
14 × 22	$13\frac{1}{2} \times 21\frac{1}{2}$	290.2	11,181	1,040	4,408	653.1	$25\frac{2}{3}$	60.47	80.62	100.8
14 × 24	$13\frac{1}{2} \times 23\frac{1}{2}$	317.2	14,600	1,243	4,818	713.8	28	66.09	88.12	110.2

Table B-2. Properties of Glued Laminated Timber Sections. For glulams with $1^1/_2''$ laminations only. Units are: section modulus, *S*, in.^{3}; moment of inertia, *I*, in.^{4}.

Glulam Width		2 1/2″		3 1/8″		5 1/8″	
Depth (in.)	No. of Lams	*S*	*I*	*S*	*I*	*S*	*I*
3	2	3.8	5.6	4.7	7.0		
4 1/2	3	8.4	19.0	10.5	23.7	17.3	38.9
6	4	15.0	45.0	18.8	56.2	30.8	92.2
7 1/2	5	23.4	87.9	29.3	110	48.0	180
9	6	33.8	152	42.2	190	69.2	311
10 1/2	7	45.9	241	57.4	301	94.2	494
12	8	60.0	360	75.0	450	123	738
13 1/2	9	75.9	513	94.9	641	156	1,051
15	10	93.8	703	117	879	192	1,441
16 1/2	11			142	1,170	232	1,919
18	12			169	1,519	277	2,491
19 1/2	13			198	1,931	325	3,167
21	14			230	2,412	377	3,955
22 1/2	15			264	2,966	432	4,865
24	16			300	3,600	492	5,904
25 1/2	17			339	4,318	555	7,082
27	18			380	5,126	623	8,406
28 1/2	19			423	6,028	694	9,887
30	20			469	7,031	769	11,530
31 1/2	21					848	13,350
33	22					930	15,350
34 1/2	23					1,017	17,540
36	24					1,107	19,930

Table B-2. (*continued*)

Glulam Width		6 3/4″		8 3/4″		10 3/4″	
Depth (in.)	No. of Lams	*S*	*I*	*S*	*I*	*S*	*I*
6	4	40.5	122				
7 1/2	5	63.3	237				
9	6	91.1	410	118	532		
10 1/2	7	124	651	161	844	198	1,037
12	8	162	972	210	1,260	258	1,548
13 1/2	9	205	1,384	266	1,794	327	2,204
15	10	253	1,898	328	2,461	403	3,023
16 1/2	11	306	2,527	397	3,276	488	4,024
18	12	364	3,280	472	4,252	580	5,224
19 1/2	13	428	4,171	554	5,407	681	6,642
21	14	496	5,209	643	6,753	790	8,296
22 1/2	15	570	6,407	738	8,306	907	10,200
24	16	648	7,776	840	10,080	1,032	12,380
25 1/2	17	732	9,327	948	12,090	1,165	14,850
27	18	820	11,070	1,063	14,350	1,306	17,630
28 1/2	19	914	13,020	1,185	16,880	1,455	20,740
30	20	1,012	15,190	1,312	19,690	1,612	24,190
31 1/2	21	1,116	17,580	1,447	22,790	1,778	28,000
33	22	1,225	20,210	1,588	26,200	1,951	32,190
34 1/2	23	1,339	23,100	1,736	29,940	2,132	36,790
36	24	1,458	26,240	1,890	34,020	2,322	41,800
37 1/2	25	1,582	29,660	2,051	38,450	2,520	47,240
39	26	1,711	33,370	2,218	43,250	2,725	53,140
40 1/2	27	1,845	37,370	2,392	48,440	2,939	59,510
42	28	1,984	41,670	2,572	54,020	3,160	66,370
43 1/2	29	2,129	46,300	2,760	60,020	3,390	73,740
45	30	2,278	51,260	2,953	66,450	3,628	81,630
46 1/2	31	2,433	56,560	3,153	73,310	3,874	90,070
48	32	2,592	62,200	3,360	80,640	4,128	99,070
49 1/2	33			3,573	88,440	4,390	108,650
51	34			3,793	96,720	4,660	118,800
52 1/2	35			4,020	105,500	4,938	129,600
54	36			4,252	114,800	5,224	141,100
55 1/2	37			4,492	124,700	5,519	153,100
57	38			4,738	135,000	5,821	165,900
58 1/2	39			4,991	146,000	6,132	179,300
60	40			5,250	157,500	6,450	193,500
61 1/2	41			5,516	169,600	6,777	208,400
63	42			5,788	182,300	7,111	224,000
64 1/2	43					7,454	240,400
66	44					7,804	257,500
67 1/2	45					8,163	275,500
69	46					8,530	294,300
70 1/2	47					8,905	313,900
72	48					9,288	334,400
73 1/2	49					9,679	355,700
75	50					10,078	377,900

Table B-2. (*continued*)

Glulam Width		12 1/4″		14 1/4″	
Depth (in.)	No. of Lams	*S*	*I*	*S*	*I*
12	8	294	1,764		
13 1/2	9	372	2,512	433	2,922
15	10	459	3,445	534	4,008
16 1/2	11	556	4,586	647	5,334
18	12	662	5,954	770	6,926
19 1/2	13	776	7,569	903	8,805
21	14	900	9,454	1,047	11,000
22 1/2	15	1,034	11,630	1,202	13,530
24	16	1,176	14,110	1,368	16,420
25 1/2	17	1,328	16,930	1,544	19,690
27	18	1,488	20,090	1,731	23,370
28 1/2	19	1,658	23,630	1,929	27,490
30	20	1,838	27,560	2,138	32,060
31 1/2	21	2,026	31,910	2,357	37,120
33	22	2,223	36,690	2,586	42,680
34 1/2	23	2,430	41,920	2,827	48,760
36	24	2,646	47,630	3,078	55,400
37 1/2	25	2,871	53,830	3,340	62,620
39	26	3,105	60,550	3,612	70,440
40 1/2	27	3,349	67,810	3,896	78,890
42	28	3,601	75,630	4,190	87,980
43 1/2	29	3,863	84,030	4,494	97,750
45	30	4,134	93,020	4,809	108,200
46 1/2	31	4,415	102,600	5,135	119,400
48	32	4,704	112,900	5,472	131,300
49 1/2	33	5,003	123,800	5,819	144,000
51	34	5,310	135,400	6,177	157,500
52 1/2	35	5,627	147,700	6,546	171,800
54	36	5,954	160,700	6,926	187,000
55 1/2	37	6,289	174,500	7,316	203,000
57	38	6,633	189,100	7,716	219,900
58 1/2	39	6,987	204,400	8,128	237,700
60	40	7,350	220,500	8,550	256,500
61 1/2	41	7,722	237,500	8,983	276,200
63	42	8,103	255,300	9,426	296,900
64 1/2	43	8,494	273,900	9,881	318,600
66	44	8,893	293,500	10,350	341,400
67 1/2	45	9,302	314,000	10,820	365,200
69	46	9,720	335,400	11,310	390,100
70 1/2	47	10,150	357,700	11,800	416,100
72	48	10,580	381,000	12,310	443,200
73 1/2	49	11,030	405,300	12,830	471,500
75	50	11,480	430,700	13,360	501,000
76 1/2	51	11,950	457,000	13,900	531,600
78	52	12,420	484,400	14,450	563,500
79 1/2	53	12,900	512,900	15,010	596,700
81	54	13,400	542,500	15,580	631,100
82 1/2	55	13,900	573,200	16,160	666,800
84	56	14,410	605,100	16,760	703,800
85 1/2	57			17,360	742,200
87	58			17,980	782,000
88 1/2	59			18,600	823,100
90	60			19,240	865,700
91 1/2	61			19,880	909,700
93	62			20,540	955,200
94 1/2	63			21,210	1,002,000
96	64			21,890	1,051,000

Table B-3. Design Values for Visually Graded Dimension Lumber and Decking.

Part A: Dimension Lumber

(Tabulated design values are for normal load duration and dry service conditions.)

Species and Commercial Grade	Size Classification	Design Values in Pounds per Square Inch (psi)						Grading Rules Agency
		Bending F_b	Tension Parallel to Grain F_t	Shear Parallel to Grain F_v	Compression Perpendicular to Grain $F_{c\perp}$	Compression Parallel to Grain F_c	Modulus of Elasticity E	
DOUGLAS FIR-LARCH								
Select structural		1,450	1,000	95	625	1,700	1,900,000	
No. 1 and better	2″ to 4″ thick	1,150	775	95	625	1,500	1,800,000	
No. 1		1,000	675	95	625	1,450	1,700,000	
No. 2	2″ and wider	875	575	95	625	1,300	1,600,000	WCLIB
No. 3		500	325	95	625	750	1,400,000	WWPA
Stud		675	450	95	625	825	1,400,000	
Construction	2″ to 4″ thick	1,000	650	95	625	1,600	1,500,000	
Standard		550	375	95	625	1,350	1,400,000	
Utility	2″ to 4″ wide	275	175	95	625	875	1,300,000	
HEM-FIR								
Select structural		1,400	900	75	405	1,500	1,600,000	
No. 1 and better	2″ to 4″ thick	1,050	700	75	405	1,350	1,500,000	
No. 1		950	600	75	405	1,300	1,500,000	
No. 2	2″ and wider	850	500	75	405	1,250	1,300,000	WCLIB
No. 3		500	300	75	405	725	1,200,000	WWPA
Stud		675	400	75	405	800	1,200,000	
Construction	2″ to 4″ thick	975	575	75	405	1,500	1,300,000	
Standard		550	325	75	405	1,300	1,200,000	
Utility	2″ to 4″ wide	250	150	75	405	850	1,100,000	
REDWOOD								
Clear structural		1,750	1,000	145	650	1,850	1,400,000	
Select structural		1,350	800	80	650	1,500	1,400,000	
Select structural, open grain		1,100	625	80	425	1,100	1,100,000	
No. 1	2″ to 4″ thick	975	575	80	650	1,200	1,300,000	
No. 1, open grain		775	450	80	425	900	1,100,000	
No. 2	2″ and wider	925	525	80	650	950	1,200,000	RIS
No. 2, open grain		725	425	80	425	700	1,000,000	
No. 3		525	300	80	650	550	1,100,000	
No. 3, open grain		425	250	80	425	400	900,000	
Stud		575	325	80	425	450	900,000	
Construction	2″ to 4″ thick	825	475	80	425	925	900,000	
Standard		450	275	80	425	725	900,000	
Utility	2″ to 4″ wide	225	125	80	425	475	800,000	

Table B-3 (*continued*)

1. **Lumber Dimensions.** Tabulated design values are applicable to lumber that will be used under dry conditions such as in most covered structures. For 2-inch- (51 mm) to 4-inch-thick (102 mm) lumber, the DRY dressed sizes shall be used regardless of the moisture content at the time of manufacture or use. In calculating design values, the natural gain in strength and stiffness that occurs as lumber dries has been taken into consideration as well as the reduction in size that occurs when unseasoned lumber shrinks. The gain in load-carrying capacity due to increased strength and stiffness resulting from drying more than offsets the design effect of size reductions due to shrinkage.
2. **Stress-rated Boards.** Stress-rated boards of nominal 1-inch (25 mm), 1 1/4-inch (32 mm) and 1 1/2-inch (38 mm) thickness, 2 inches (51 mm) and wider, of most species, are permitted the design values shown for select structural, No. 1 and better, No. 1, No. 2, No. 3, stud, construction, standard, utility, clear heart structural and clear structural grades as shown in the 2-inch- (51 mm) to 4-inch-thick (102 mm) categories herein, when graded in accordance with the stress-rated board provisions in the applicable grading rules. Information on stress-rated board grades applicable to the various species is available from the respective grading rules agencies. Information on additional design values may also be available from the respective grading agencies.
3. **Size Factor, C_F.** Tabulated bending, tension and compression-parallel-to-grain design values for dimension lumber 2 inches to 4 inches (51 mm to 102 mm) thick shall be multiplied by the following size factors:

Size Factors, C_F

Grades	Width (inches) × 25.4 for mm	F_b Thickness 2″ and 3″ (51 mm and 76 mm)	F_b Thickness 4″ (102 mm)	F_t	F_c
Select structural, No. 1 and better No. 1, No. 2, No. 3	2, 3 and 4	1.5	1.5	1.5	1.15
	5	1.4	1.4	1.4	1.1
	6	1.3	1.3	1.3	1.1
	8	1.2	1.3	1.2	1.05
	10	1.1	1.2	1.1	1.0
	12	1.0	1.1	1.0	1.0
	14 and wider	0.9	1.0	0.9	0.9
Stud	2, 3 and 4	1.1	1.1	1.1	1.05
	5 and 6	1.0	1.0	1.0	1.0
Construction and Standard	2, 3, and 4	1.0	1.0	1.0	1.0
Utility	4	1.0	1.0	1.0	1.0
	2 and 3	0.4	—	0.4	0.6

4. **Repetitive Member Factor, C_r.** Bending design values, F_b, for dimension lumber 2 inches (51 mm) to 4 inches (102 mm) thick shall be multiplied by the repetitive member factor, $C_r = 1.15$, when such members are used as joists, truss chords, rafters, studs, planks, decking or similar members which are in contact or spaced not more than 24 inches (610 mm) on center, are not less than three in number and are joined by floor, roof or other load-distributing elements adequate to support the design load.
5. **Flat-use Factor, C_{fu}.** Bending design values adjusted by size factors are based on edgewise use (load applied to narrow face). When dimension lumber is used flatwise (load applied to wide face), the bending design value, F_b, shall also be multiplied by the following flat-use factors:

Flat-Use Factors, C_{fu}

Width (inches) × 25.4 for mm	Thickness 2″ and 3″ (51 mm and 76 mm)	Thickness 4″ (102 mm)
2 and 3	1.0	—
4	1.1	1.0
5	1.1	1.05
6	1.15	1.05
8	1.15	1.05
10 and wider	1.2	1.1

6. **Wet Service Factor, C_M.** When dimension lumber is used where moisture content will exceed 19% for an extended time period, design values shall be multiplied by the appropriate wet service factors from the following table:

Wet-Service Factors, C_M

F_b	F_t	F_v	$F_{c\perp}$	F_c	E
0.85*	1.0	0.97	0.67	0.8**	0.9

*when $(F_b)(C_F) \leq 1{,}150$ psi (7.92 N/mm²), $C_M = 1.0$
**when $(F_c)(C_F) \leq 750$ psi (5.17 N/mm²), $C_M = 1.0$

7. **Shear Stress Factor, C_H.** Tabulated shear design values parallel to grain have been reduced to allow for the occurrence of splits, checks and shakes. Tabulated shear design values parallel to grain, F_v, shall be permitted to be multiplied by the shear-stress factors specified in the following table when length of split, or size of check or shake is known and no increase in them is anticipated. When shear-stress factors are used for redwood, a tabulated design value of $F_V = 80$ psi (0.55 N/mm²) shall be assigned for all grades of redwood dimension lumber. Shear stress factors shall be permitted to be linearly interpolated.

Shear Stress Factors, C_H

Length of split on wide face of 2-inch (51 mm) (nominal) lumber	C_H	Length of split on wide face of 3-inch (76 mm) (nominal) and thicker lumber	C_H	Size of shake* in 2-inch (51 mm) (nominal) and thicker lumber	C_H
no split	2.00	no split	2.00	no shake	2.00
1/2 by wide face	1.67	1/2 by narrow face	1.67	1/6 by narrow face	1.67
3/4 by wide face	1.50	3/4 by narrow face	1.50	1/4 by narrow face	1.50
1 by wide face	1.33	1 by narrow face	1.33	1/3 by narrow face	1.33
1 1/2 by wide face or more	1.00	1 1/2 by narrow face or more	1.00	1/2 by narrow face or more	1.00
				*Shake is measured at the end between lines enclosing the shake and perpendicular to the loaded face.	

Table B-3 (*continued*)

Part B: Decking

(Tabulated design values are for normal load duration and dry service conditions, unless specified otherwise.)

Species and Commercial Grade	Size Classification × 25.4 for mm	Design Values in Pounds per Square Inch (psi) × 0.00689 for N/mm² — Bending: Single Member F_b	Bending: Repetitive Member $(F_c)(C_r)$	Compression Perpendicular to Grain $F_{c\perp}$	Modulus of Elasticity E	Grading Rules Agency
DOUGLAS FIR-LARCH						
Select dex	2″ to 4″ thick	1,750	2,000	625	1,800,000	WCLIB
Commercial dex	6″ to 8″ wide	1,450	1,650	625	1,700,000	
Select	2″ to 4″ thick	—	2,000	—	1,800,000	WWPA
Commercial	4″ to 12″ wide	—	1,650	—	1,700,000	
HEM-FIR						
Select dex	2″ to 4″ thick	1,400	1,600	405	1,500,000	WCLIB
Commercial dex	6″ to 8″ wide	1,150	1,350	405	1,400,000	
Select	2″ to 4″ thick	—	1,600	—	1,500,000	WWPA
Commercial	4″ to 12″ wide	—	1,350	—	1,400,000	
REDWOOD						
Select, close grain	2″ thick	1,850	2,150	—	1,400,000	RIS
Select		1,450	1,700	—	1,100,000	
Commercial	6″ and wider	1,200	1,350	—	1,000,000	
Deck heart and Deck common	2″ thick 4″ wide	400	450	420	900,000	
	2″ thick 6″ wide	700	800	420	900,000	

1. **Lumber Dimensions.** Tabulated design values are applicable to lumber that will be used under dry conditions such as in most covered structures. For 2-inch- (51 mm) to 4-inch-thick (102 mm) lumber, the DRY dressed sizes shall be used regardless of the moisture content at the time of manufacture or use. In calculating design values, the natural gain in strength and stiffness that occurs as lumber dries has been taken into consideration as well as the reduction in size that occurs when unseasoned lumber shrinks. The gains in load-carrying capacity due to increased strength and stiffness resulting from drying more than offsets the design effect of size reductions due to shrinkage.
2. **Size Factor, C_F.** Bending design values for all species of decking except redwood are based on 4-inch-thick decking. When 2-inch-thick (51 mm) or 3-inch-thick (76 mm) decking is used, the bending design values, F_b, for all species except redwood shall be multiplied by the following size factors:

Size Factors, C_F

Thickness (inches)	C_F
2	1.10
3	1.04

3. **Flat-use Factor, C_{fu}.** Tabulated bending design values, F_b, for decking have already been adjusted for flatwise usage (load applied to wide face).
4. **Repetitive Member Factor, C_r.** Tabulated bending design values for repetitive member uses (F_b C_r) for decking have already been multiplied by the repetitive member factor, C_r.
5. **Wet-Service Factor, C_M.** When decking is used where moisture content will exceed 19% for an extended time period, design values shall be multiplied by the appropriate wet-service factors from the following table (for southern pine use tabulated design values for wet-service conditions without further adjustment):

Wet Service Factors, C_M

F_b	$F_{c\perp}$	E
0.85*	0.67	0.9

*when $(F_b)(C_F) \leq 1{,}150$ psi (7.92 N/mm²), $C_M = 1.0$

Source: Reproduced from the 1994 edition of the *Uniform Building Code*™, copyright © 1994, with the permission of the publisher, the International Conference of Building Officials.

Table B-4. Design Values for Southern Pine Dimension Lumber.

Species and Commercial Grade	Size Classification	Design Values in Pounds per Square Inch (psi)						Grading Rules Agency
		Bending F_b	Tension Parallel to Grain F_t	Shear Parallel to Grain F_v	Compression Perpendicular to Grain $F_{c\perp}$	Compression Parallel to Grain F_c	Modulus of Elasticity E	
SOUTHERN PINE								
Dense select structural		3,050	1,650	100	660	2,250	1,900,000	
Select structural		2,850	1,600	100	565	2,100	1,800,000	
Non-dense select-structural		2,650	1,350	100	480	1,950	1,700,000	
No. 1 dense	2″ to 4″ thick	2,000	1,100	100	660	2,000	1,800,000	
No. 1		1,850	1,050	100	565	1,850	1,700,000	
No. 1 non-dense	2″ to 4″ wide	1,700	900	100	480	1,700	1,600,000	
No. 2 dense		1,700	875	90	660	1,850	1,700,000	
No. 2		1,500	825	90	565	1,650	1,600,000	
No. 2 non-dense		1,350	775	90	480	1,600	1,400,000	
No. 3		850	475	90	565	975	1,400,000	
Stud		875	500	90	565	975	1,400,000	
Construction	2″ to 4″ thick	1,100	625	100	565	1,800	1,500,000	
Standard		625	350	90	565	1,500	1,300,000	SPIB
Utility	4″ wide	300	175	90	565	975	1,300,000	
Dense select structural		2,700	1,500	90	660	2,150	1,900,000	
Select structural		2,550	1,400	90	565	2,000	1,800,000	
Non-dense select structural		2,350	1,200	90	480	1,850	1,700,000	
No. 1 dense	2″ to 4″ thick	1,750	950	90	660	1,900	1,800,000	
No. 1		1,650	900	90	565	1,750	1,700,000	
No. 1 non-dense	5″ to 6″ wide	1,500	800	90	480	1,600	1,600,000	
No. 2 dense		1,450	775	90	660	1,750	1,700,000	
No. 2		1,250	725	90	565	1,600	1,600,000	
No. 2 non-dense		1,150	675	90	480	1,500	1,400,000	
No. 3		750	425	90	565	925	1,400,000	
Stud		775	425	90	565	925	1,400,000	
Dense select structural		2,450	1,350	90	660	2,050	1,900,000	
Select structural		2,300	1,300	90	565	1,900	1,800,000	
Non-dense select structural		2,100	1,100	90	480	1,750	1,700,000	
No. 1 dense	2″ to 4″ thick	1,650	875	90	660	1,800	1,800,000	
No. 1	8″ wide	1,500	825	90	565	1,650	1,700,000	
No. 1 non-dense		1,350	725	90	480	1,550	1,600,000	
No. 2 dense		1,400	675	90	660	1,700	1,700,000	
No. 2		1,200	650	90	565	1,550	1,600,000	
No. 2 non-dense		1,100	600	90	480	1,450	1,400,000	
No. 3		700	400	90	565	875	1,400,000	
Dense select structural		2,150	1,200	90	660	2,000	1,900,000	
Select structural		2,050	1,100	90	565	1,850	1,800,000	
Non-dense select structural		1,850	950	90	480	1,750	1,700,000	
No. 1 dense	2″ to 4″ thick	1,450	775	90	660	1,750	1,800,000	
No. 1	10″ wide	1,300	725	90	565	1,600	1,700,000	SPIB
No. 1 non-dense		1,200	650	90	480	1,500	1,600,000	
No. 2 dense		1,200	625	90	660	1,650	1,700,000	
No. 2		1,050	575	90	565	1,500	1,600,000	
No. 2 non-dense		950	550	90	480	1,400	1,400,000	
No. 3		650	325	90	565	850	1,400,000	
Dense select structural		2,050	1,100	90	660	1,950	1,900,000	
Select structural		1,900	1,050	90	565	1,800	1,800,000	
Non-dense select structural		1,750	900	90	480	1,700	1,700,000	
No. 1 dense	2″ to 4″ thick	1,350	725	90	660	1,700	1,800,000	
No. 1		1,250	675	90	565	1,600	1,700,000	
No. 1 non-dense	12″ wide	1,150	600	90	480	1,500	1,600,000	
No. 2 dense		1,150	575	90	660	1,600	1,700,000	
No. 2		975	550	90	565	1,450	1,600,000	
No. 2 non-dense		900	525	90	480	1,350	1,400,000	
No. 3		575	325	90	565	825	1,400,000	

Table B-4 (continued)

Species and Commercial Grade	Size Classification	Bending F_b	Tension Parallel to Grain F_t	Shear Parallel to Grain F_v	Compression Perpendicular to Grain $F_{c\perp}$	Compression Parallel to Grain F_c	Modulus of Elasticity E	Grading Rules Agency
		Design Values in Pounds per Square Inch (psi)						
SOUTHERN PINE		(Dry service conditions–19 percent or less moisture content)						
Dense structural 86	2″ to 4″ thick	2,600	1,750	155	660	2,000	1,800,000	
Dense structural 72		2,200	1,450	130	660	1,650	1,800,000	SPIB
Dense structural 65	2″ and wider	2,000	1,300	115	660	1,500	1,800,000	
SOUTHERN PINE		(Wet service conditions)						
Dense structural 86		2,100	1,400	145	440	1,300	1,600,000	
Dense structural 72	2½″ to 4″ thick	1,750	1,200	120	440	1,100	1,600,000	SPIB
Dense structural 65	2½″ and wider	1,600	1,050	110	440	1,000	1,600,000	

Notes:

1. **Lumber Dimensions.** Tabulated design values are applicable to lumber that will be used under dry conditions such as in most covered structures. For 2-inch- (51 mm) to 4-inch-thick (102 mm) lumber, the DRY dressed sizes shall be used regardless of the moisture content at the time of manufacture or use. In calculating design values, the natural gain in strength and stiffness that occurs as lumber dries has been taken into consideration as well as the reduction in size that occurs when unseasoned lumber shrinks. The gain in load-carrying capacity due to increased strength and stiffness resulting from drying more than offsets the design effect of size reductions due to shrinkage.
2. **Size Factor, C_F.** Appropriate size adjustment factors have already been incorporated in the tabulated design values for most thicknesses of southern pine and mixed southern pine dimension lumber. For dimension lumber 4 inches (102 mm) thick, 8 inches (203 mm) and wider (all grades except dense structural 86, dense structural 72 and dense structural 65), tabulated bending design values, F_b, shall be permitted to be multiplied by the size factor, $C_F = 1.1$. For dimension lumber wider than 12 inches (305 mm) (all grades except dense structural 86, dense structural 72 and dense structural 65), tabulated bending, tension and compression-parallel-to-grain design values for 12-inch-wide (305 mm) lumber shall be multiplied by the size factor, $C_F = 0.9$. When the depth, d, of dense structural 86, dense structural 72 or dense structural 65 dimension lumber exceeds 12 inches (305 mm), the tabulated bending design value, F_b, shall be multiplied by the following size factor:

$$C_F = (12/d)^{1/9} \qquad \text{For SI: } C_F = (304.8/d)^{1/9}$$

3. **Repetitive Member Factor, C_r.** Bending design values, F_b, for dimension lumber 2 inches (51 mm) to 4 inches (102 mm) thick shall be multiplied by the repetitive member factor, $C_r = 1.15$, when such members are used as joists, truss chords, rafters, studs, planks, decking or similar members which are in contact or spaced not more than 24 inches (610 mm) on center, are not less than three in number and are joined by floor, roof or other load-distributing elements adequate to support the design load.
4. **Flat-use Factor, C_{fu}.** Bending design values adjusted by size factors are based on edgewise use (load applied to narrow face). When dimension lumber is used flatwise (load applied to wide face), the bending design value, F_b, shall also be multiplied by the following flat-use factors:

FLAT-USE FACTORS, C_{fu}

Width	Thickness	
× 25.4 for mm	2″ and 3″ (51 mm and 76 mm)	4″ (102 mm)
2″ and 3″	1.0	—
4″	1.1	1.0
5″	1.1	1.05
6″	1.15	1.05
8″	1.15	1.05
10″ and wider	1.2	1.1

5. **Wet Service Factor, C_M.** When dimension lumber is used where moisture content will exceed 19 percent for an extended time period, design values shall be multiplied by the appropriate wet service factors from the following table (for dense structural 86, dense structural 72 and dense structural 65 use tabulated design values for wet service conditions without further adjustment):

WET SERVICE FACTORS, C_M

F_b	F_t	F_v	$F_{c\perp}$	F_c	E
0.85*	1.0	0.97	0.67	0.8**	0.9

*when $(F_b)(C_F) \leq 1{,}150$ psi (7.92 N/mm²), $C_M = 1.0$
**when $F_c \leq 750$ psi (5.17 N/mm²), $C_M = 1.0$

6. **Shear Stress Factor, C_H.** Tabulated shear design values parallel to grain have been reduced to allow for the occurrence of splits, checks and shakes. Tabulated shear design values parallel to grain, F_v, shall be permitted to be multiplied by the shear stress factors specified in the following table when length of split, or size of check or shake is known and no increase in them is anticipated. When shear stress factors are used for southern pine and mixed southern pine, a tabulated design value of $F_V = 90$ psi (0.62 N/mm²) shall be assigned for all grades of southern pine and mixed southern pine dimension lumber. Shear stress factors shall be permitted to be linearly interpolated.

Table B-4 *(continued)*

SHEAR STRESS FACTORS, C_H

Length of split on wide face of 2-inch (51 mm) (nominal) lumber	C_H	Length of split on wide face of 3-inch (76 mm) (nominal) and thicker lumber	C_H	Size of shake* in 2-inch (51 mm) (nominal) and thicker lumber	C_H
no split	2.00	no split	2.00	no shake	2.00
1/2 by wide face	1.67	1/2 by narrow face	1.67	1/6 by narrow face	1.67
3/4 by wide face	1.50	3/4 by narrow face	1.50	1/4 by narrow face	1.50
1 by wide face	1.33	1 by narrow face	1.33	1/3 by narrow face	1.33
1 1/2 by wide face or more	1.00	1 1/2 by narrow face or more	1.00	1/2 by narrow face or more	1.00
				*Shake is measured at the end between lines enclosing the shake and perpendicular to the loaded face.	

Source: Reproduced from the 1994 edition of the *Uniform Building Code*™, copyright © 1994, with the permission of the publisher, the International Conference of Building Officials.

Table B-5. Design Values for Timbers (5 in. × 5 in. or larger).

(Tabulated design values are for normal load duration and dry service conditions, unless specified otherwise.)

Species and Commercial Grade	Size Classification	Design Values in Pounds Per Square Inch (psi)						Grading Rules Agency
		Bending F_b	Tension Parallel to Grain F_t	Shear Parallel to Grain F_v	Compression Perpendicular to Grain $F_{c\perp}$	Compression Parallel to Grain F_c	Modulus of Elasticity E	
DOUGLAS FIR-LARCH								
Dense select structural	Beams and stringers	1,900	1,100	85	730	1,300	1,700,000	WCLIB
Select structural		1,600	950	85	625	1,100	1,600,000	
Dense No. 1		1,550	775	85	730	1,100	1,700,000	
No. 1		1,350	675	85	625	925	1,600,000	
No. 2		875	425	85	625	600	1,300,000	
Dense select structural	Posts and timbers	1,750	1,150	85	730	1,350	1,700,000	WCLIB
Select structural		1,500	1,000	85	625	1,150	1,600,000	
Dense No. 1		1,400	950	85	730	1,200	1,700,000	
No. 1		1,200	825	85	625	1,000	1,600,000	
No. 2		750	475	85	625	700	1,300,000	
Dense select structural	Beams and stringers	1,850	1,110	85	730	1,300	1,700,000	WWPA
Select structural		1,600	950	85	625	1,100	1,600,000	
Dense No. 1		1,550	775	85	730	1,100	1,700,000	
No. 1		1,350	675	85	625	925	1,600,000	
Dense No. 2		1,000	500	85	730	700	1,400,000	
No. 2		875	425	85	625	600	1,300,000	
Dense select structural	Posts and timbers	1,750	1,150	85	730	1,350	1,700,000	
Select structural		1,500	1,000	85	625	1,150	1,600,000	
Dense No. 1		1,400	950	85	730	1,200	1,700,000	
No. 1		1,200	825	85	625	1,000	1,600,000	
Dense No. 2		800	550	85	730	550	1,400,000	
No. 2		700	475	85	625	475	1,300,000	

Table B-5. *(continued)*

(Tabulated design values are for normal load duration and dry service conditions, unless specified otherwise.)

Species and Commercial Grade	Size Classification	Design Values in Pounds Per Square Inch (psi)						Grading Rules Agency
		Bending F_b	Tension Parallel to Grain F_t	Shear Parallel to Grain F_v	Compression Perpendicular to Grain $F_{c\perp}$	Compression Parallel to Grain F_c	Modulus of Elasticity E	
HEM-FIR								
Select structural	Beams and stringers	1,300	750	70	405	925	1,300,000	WCLIB
No. 1		1,050	525	70	405	750	1,300,000	
No. 2		675	350	70	405	500	1,100,000	
Select structural	Posts and timbers	1,200	800	70	405	975	1,300,000	
No. 1		975	650	70	405	850	1,300,000	
No. 2		575	375	70	405	575	1,100,000	
Select structural	Beams and stringers	1,250	725	70	405	925	1,300,000	WWPA
No. 1		1,050	525	70	405	775	1,300,000	
No. 2		675	325	70	405	475	1,100,000	
Select structural	Posts and timbers	1,200	800	70	405	975	1,300,000	
No. 1		950	650	70	405	850	1,300,000	
No. 2		525	350	70	405	375	1,100,000	
REDWOOD								
Clear heart structural or clear structural	5″ × 5″ and larger (127 mm × 127 mm)	1,850	1,250	135	650	1,650	1,300,000	RIS
Select structural No. 1		1,400	950	95	650	1,200	1,300,000	
No. 2		1,200	800	95	650	1,050	1,300,000	
		975	650	95	650	900	1,100,000	

Notes:

1. **Lumber Dimensions.** Tabulated design values are applicable to lumber that will be used under dry conditions such as in most covered structures. For 5-inch (127 mm) and thicker lumber, the GREEN dressed sizes shall be permitted to be used (see Table 23-I-A-1) because design values have been adjusted to compensate for any loss in size by shrinkage which may occur.
2. **Size Factor, C_F.** When the depth, d, of a beam, stringer, post or timber exceeds 12 inches (305 mm), the tabulated bending design value, F_b, shall be multiplied by the following size factor:

$$C_F = (12/d)^{1/9} \qquad \text{For SI:} \quad C_F = (304.8/d)^{1/9}$$

Table B-5. *(continued)*

3. **Wet Service Factor, C_M.** When timbers are used where moisture content will exceed 19 percent for an extended time period, design values shall be multiplied by the appropriate wet service factors from the following table (for southern pine and mixed southern pine use tabulated design values without further adjustment):

WET SERVICE FACTORS, C_M

F_b	F_t	F_v	$F_{c\perp}$	F_c	E
1.00	1.00	1.00	0.67	0.91	1.00

4. **Shear Stress Factor, C_H.** Tabulated shear design values parallel to grain have been reduced to allow for the occurrence of splits, checks and shakes. Tabulated shear design values parallel to grain, F_v, shall be permitted to be multiplied by the shear stress factors specified in the following table when length of split, or size of check or shake is known and no increase in them is anticipated. When shear stress factors are used for redwood, southern pine and mixed southern pine, a tabulated design value of F_v = 80 psi (0.55 N/mm^2) shall be assigned for all grades of redwood and a tabulated design value of F_v = 90 psi (0.62 N/mm^2) shall be assigned for all grades of southern pine and mixed southern pine. Shear stress factors shall be permitted to be linearly interpolated.

SHEAR STRESS FACTORS, C_H

Length of split on wide face of 5-inch (127 mm) (nominal) and thicker lumber	C_H	Size of shake* in 5-inch (127 mm) (nominal) and thicker lumber	C_H
no split	2.00	no shake	2.00
1/2 by narrow face	1.67	1/6 by narrow face	1.67
3/4 by narrow face	1.50	1/4 by narrow face	1.50
1 by narrow face	1.33	1/3 by narrow face	1.33
1 1/2 by narrow face or more	1.00	1/2 by narrow face or more	1.00
		*Shake is measured at the end between lines enclosing the shake and perpendicular to the loaded face.	

Source: Reproduced from the 1994 edition of the *Uniform Building Code*™, copyright © 1994, with the permission of the publisher, the International Conference of Building Officials.

Table B-6. Design Allowable Stresses and Modulus of Elasticity for Mechanically Graded Dimension Lumber.

Species and Commercial Grade	Size Classification × 25.4 for mm	Design Values in Pounds Per Square Inch (psi) × 0.00689 for mm			
		Bending F_b	Tension Parallel to Grain F_t	Compression Parallel to Grain F_c	Modulus of Elasticity E
MACHINE STRESS-RATED (MSR) LUMBER					
900f-1.0E	2″ and less in thickness 2″ and wider	900	350	1,050	1,000,000
1200f-1.2E		1,200	600	1,400	1,200,000
1350f-1.3E		1,350	750	1,600	1,300,000
1450f-1.3E		1,450	800	1,625	1,300,000
1500f-1.3E		1,500	900	1,650	1,300,000
1500f-1.4E		1,500	900	1,650	1,400,000
1650f-1.4E		1,650	1,020	1,700	1,400,000
1650f-1.5E		1,650	1,020	1,700	1,500,000
1800f-1.6E		1,800	1,175	1,750	1,600,000
1950f-1.5E		1,950	1,375	1,800	1,500,000
1950f-1.7E		1,950	1,375	1,800	1,700,000
2100f-1.8E		2,100	1,575	1,875	1,800,000
2250f-1.6E		2,250	1,750	1,925	1,600,000
2250f-1.9E		2,250	1,750	1,925	1,900,000
2400f-1.7E		2,400	1,925	1,975	1,700,000
2400f-2.0E		2,400	1,925	1,975	2,000,000
2550f-2.1E		2,550	2,050	2,025	2,100,000
2700f-2.2E		2,700	2,150	2,100	2,200,000
2850f-2.3E		2,850	2,300	2,150	2,300,000
3000f-2.4E		3,000	2,400	2,200	2,400,000
3150f-2.5E		3,150	2,500	2,250	2,500,000
3300f-2.6E		3,300	2,650	2,325	2,600,000
900f-1.2E	2″ and less in thickness 6″ and wider	900	350	1,050	1,200,000
1200f-1.5E		1,200	600	1,400	1,500,000
1350f-1.8E		1,350	750	1,600	1,800,000
1500f-1.8E		1,500	900	1,650	1,800,000
1800f-2.1E		1,800	1,175	1,750	2,100,000
MACHINE EVALUATED LUMBER (MEL)					
M-10	2″ and less in thickness 2″ and wider	1,400	800	1,600	1,200,000
M-11		1,550	850	1,650	1,500,000
M-12		1,600	850	1,700	1,600,000
M-13		1,600	950	1,700	1,400,000
M-14		1,800	1,000	1,750	1,700,000
M-15		1,800	1,100	1,750	1,500,000
M-16		1,800	1,300	1,750	1,500,000
M-17		1,950	1,300	2,050	1,700,000
M-18		2,000	1,200	1,850	1,800,000
M-19		2,000	1,300	1,850	1,600,000
M-20		2,000	1,600	2,100	1,900,000
M-21		2,300	1,400	1,950	1,900,000
M-22		2,350	1,500	1,950	1,700,000
M-23		2,400	1,900	2,000	1,800,000
M-24		2,700	1,800	2,100	1,900,000
M-25		2,750	2,000	2,100	2,200,000
M-26		2,800	1,800	2,150	2,000,000
M-27		3,000	2,000	2,400	2,100,000

Table B-6. (*continued*)

Notes:

1. **Lumber Dimensions.** Tabulated design values are applicable to lumber that will be used under dry conditions such as in most covered structures. For 2-inch (51 mm) to 4-inch-thick (102 mm) lumber, the DRY dressed sizes shall be used regardless of the moisture content at the time of manufacture or use. In calculating design values, the natural gain in strength and stiffness that occurs as lumber dries has been taken into consideration as well as the reduction in size that occurs when unseasoned lumber shrinks. The gain in load-carrying capacity due to increased strength and stiffness resulting from drying more than offsets the design effect of size reductions due to shrinkage.
2. **Shear Parallel to Grain, F_v, and Compression Perpendicular to Grain, $F_{c\perp}$.** Design values for shear parallel to grain, F_v, and compression perpendicular to grain, $F_{c\perp}$, are identical to the design values given in Tables 23-I-A-1 and 23-I-A-2 for No. 2 visually graded lumber of the appropriate species.
3. **Modulus of Elasticity, *E*.** For any given bending design value, F_b, the average modulus of elasticity, E, may vary depending on species, timber source or other variables. The E value included in the F_b-E grade designations in Table 23-I-A-3 are those usually associated with each F_b level. Grade stamps may show higher or lower E values [in increments of 100,000 psi (689 N/mm^2)] if machine rating indicates the assignment is appropriate. When the E value shown on a grade stamp differs from the E value in Table 23-I-A-3, the E value shown on the grade stamp shall be used for design. The tabulated F_b, F_t and F_c values associated with the designated F_b value shall be used for design.
4. Refer to Table B-3 (B-4 for S. pine) for the adjustments to the allowable stresses: repetitive member factor, C_r, flat-use factor, C_{fu}, wet service factor, C_M, and shear stress factor, C_H.

Source: Reproduced from the 1994 edition of the *Uniform Building Code*™, copyright © 1994, with the permission of the publisher, the International Conference of Building Officials.

Table B-7. Allowable Unit Stresses for Structural Glued Laminated Timber for Normal Load Duration.[1,2,3]

Members stressed principally in bending with load applied perpendicular to the wide faces of the laminations

		Bending About X-X Axis						Bending About Y-Y Axis					Axially Loaded		
		Loaded Perpendicular to Wide Faces of Laminations						Loaded Parallel to Wide Faces of Laminations							
		Extreme Fiber in Bending F_{bxx}		Compression Perpendicular to Grain $F_{c\perp xx}$ [6]											
Combination Symbol[4]	Species Outer Laminations/ Core Laminations[5]	Tension Zone Stressed in Tension[7,8] psi	Compression Zone Stressed in Tension[9] psi	Tension Face psi	Compression Face psi	Horizontal Shear F_{vxx} psi	Modulus of Elasticity E_{xx} [6] × 10^6 psi	Extreme Fiber in Bending[7,10] F_{byy} psi	Compression Perpendicular to Grain $F_{c\perp yy}$ [6] psi	Horizontal Shear F_{vyy} psi	Horizontal Shear F_{vyy} psi (For members with multiple-piece laminations which are not edge glued)[11]	Modulus of Elasticity E_{yy} [6] × 10^6 psi	Tension Parallel to Grain F_t psi	Compression Parallel to Grain F_c psi	Modulus of Elasticity E^6 × 10^6 psi
		× 0.00689 for N/mm²													
1	2	3	4	5	6	7	8	9	10	11	12	13	14	15	16
Visually Graded Western Species															
The following four combinations are not balanced and are for either dry or wet use.															
16F-V1 16F-V2 16F-V3 16F-V8	DF/WW HF/HF DF/DF DFS/DFS	1,600	800	560[12,13] 500[16] 560[12,13] 650	560[12,13] 375[16] 560 500	140[14,15] 155 165 165	1.3 1.4 1.5 1.2	950 1,250 1,450 1,200	255 375 560 500	130 135 145 145	65 70 75 75	1.1 1.3 1.5 1.1	675 875 950 825	975 1,300 1,550 1,350	1.1 1.3 1.5 1.1
The following two combinations are intended for straight or slightly cambered members for dry use and industrial appearance.[17]															
16F-V4 16F-V5	DF/N3WW DF/M3DF	1,600	800	650 650	560[12] 560[12]	90[13,16] 90[19]	1.5[18] 1.6	900 1,000	255 470	130 135	65 70	1.3[18] 1.5	650 750	600 875	1.3 1.5
The following two combinations are balanced and are intended for members continuous or cantilevered over supports and provide equal capacity in both positive and negative bending.															
16F-V6 16F-V7	DF/DF HF/HF	1,600	1,600	560[12,13] 375[16]	560[12] 375[16]	165 155	1.5 1.4	1,450 1,200	560 375	145 135	75 70	1.4 1.3	950 850	1,550 1,350	1.5 1.3

(continued)

Table B-7. *(continued)*

Members stressed principally in bending with load applied perpendicular to the wide faces of the laminations

Combination Symbol[4]	Species Outer Laminations/ Core Laminations[5]	Bending About X-X Axis: Loaded Perpendicular to Wide Faces of Laminations — Extreme Fiber in Bending F_{bxx}: Tension Zone Stressed in Tension[7,8] psi	Extreme Fiber in Bending F_{bxx}: Compression Zone Stressed in Tension[9] psi	Compression Perpendicular to Grain $F_{c\perp xx}$[6]: Tension Face psi	Compression Perpendicular to Grain $F_{c\perp xx}$[6]: Compression Face psi	Horizontal Shear F_{vxx} psi	Modulus of Elasticity E_{xx}[6] × 10^6 psi	Bending About Y-Y Axis: Loaded Parallel to Wide Faces of Laminations — Extreme Fiber in Bending[7,10] F_{byy} psi	Compression Perpendicular to Grain $F_{c\perp yy}$[6] psi	Horizontal Shear F_{vyy} psi	Horizontal Shear F_{vyy} psi (For members with multiple-piece laminations which are not edge glued)[11]	Modulus of Elasticity E_{yy}[6] × 10^6 psi	Axially Loaded: Tension Parallel to Grain F_t psi	Compression Parallel to Grain F_c psi	Modulus of Elasticity E[6] × 10^6 psi
		× 0.00689 for N/mm²													
1	2	3	4	5	6	7	8	9	10	11	12	13	14	15	16
Visually Graded Western Species															
The following seven combinations are not balanced and are for either dry or wet use.															
20F-V1	DF/WW			650	560[12]	140[14,15]	1.4[18]	1,000	255	130	65	1.2[18]	750	1,000	1.2
20F-V2	HF/HF			500[16]	375[16]	155	1.5	1,200	375	135	70	1.4	950	1,350	1.4
20F-V3	DF/DF			650	560[12]	165	1.6	1,450	560	145	75	1.5	1,000	1,550	1.5
20F-V4	DF/DF	2,000	1,000	590[12,13]	560[12]	165	1.6	1,450	560	145	75	1.6	1,000	1,550	1.6
20F-V10	DF/HF			650	560	155	1.5	1,300	375	135	70	1.4	950	1,500	1.4
20F-V11	DFS/DFS			650	500	165	1.3	1,400	500	145	75	1.1	900	1,400	1.1
20F-V12	AC/AC			560	560	190	1.5	1,200	470	165	80	1.4	900	1,500	1.4
The following two combinations are intended for straight or slightly cambered members for dry use and industrial appearance.[17]															
20F-V5	DF/N3WW	2,000	1,000	650	560[12]	90[14,20]	1.6	1,000	255	135	70	1.3	750	725	1.3
20F-V6	DF/M3DF			650	560[12]	90[19]	1.6	1,000	470	135	70	1.5	775	900	1.5

(continued)

Table B-7. (*continued*)

Members stressed principally in bending with load applied perpendicular to the wide faces of the laminations

Combination Symbol[4]	Species Outer Laminations/ Core Laminations[5]	Bending About X-X Axis: Loaded Perpendicular to Wide Faces of Laminations — Extreme Fiber in Bending F_{bxx}: Tension Zone Stressed in Tension[7,8] psi	Extreme Fiber in Bending F_{bxx}: Compression Zone Stressed in Tension[9] psi	Compression Perpendicular to Grain $F_{c\perp xx}$[6]: Tension Face psi	Compression Perpendicular to Grain $F_{c\perp xx}$[6]: Compression Face psi	Horizontal Shear F_{vxx} psi	Modulus of Elasticity E_{xx}[6] × 10^6 psi	Bending About Y-Y Axis: Loaded Parallel to Wide Faces of Laminations — Extreme Fiber in Bending[7,10] F_{byy} psi	Compression Perpendicular to Grain $F_{c\perp yy}$[6] psi	Horizontal Shear F_{vyy} psi	Horizontal Shear F_{vyy} psi (For members with multiple-piece laminations which are not edge glued)[11]	Modulus of Elasticity E_{yy}[6] × 10^6 psi	Axially Loaded: Tension Parallel to Grain F_t psi	Compression Parallel to Grain F_c psi	Modulus of Elasticity E[6] × 10^6 psi
		× 0.00689 for N/mm²													
1	2	3	4	5	6	7	8	9	10	11	12	13	14	15	16
Visually Graded Western Species															
The following three combinations are balanced and are intended for members continuous or cantilevered over supports and provide equal capacity in both positive and negative bending.															
20F-V7	DF/DF	2,000	2,000	650	650	165	1.6	1,450	560	145	75	1.6	1,000	1,600	1.6
20F-V8	DF/DF			590[12,13]	590[12,13]	165	1.7	1,450	560	145	75	1.6	1,000	1,600	1.6
20F-V9	HF/HF			500[16]	500[16]	155	1.5	1,400	375	135	70	1.4	975	1,400	1.4
The following five combinations are not balanced and are for either dry or wet use.															
22F-V1	DF/WW	2,200	1,100	650	560[12]	140[14]	1.6[18]	1,050	255	130	65	1.3[18]	850	1,100	1.3
22F-V2	HF/HF			500[16]	500[16]	155	1.5	1,250	375	135	70	1.4	950	1,350	1.4
22F-V3	DF/DF			650	560[16]	165	1.7	1,450	560	145	75	1.6	1,050	1,500	1.6
22F-V4	DF/DF			590[12,13]	560[12]	165	1.7	1,450	560	145	75	1.6	1,000	1,550	1.6
22F-V10	DF/DFS			650	560[12]	165	1.6	1,600	500	145	75	1.3	1,000	1,400	1.3
The following two combinations are intended for straight or slightly cambered members for dry use and industrial appearance.[17]															
22F-V5	DF/3WW	2,200	1,100	650	560[12]	90[14,20]	1.6[18]	1,100	255	135	75	1.4[18]	800	725	1.4
22F-V6	DF/M3DF			650	560[12]	90[19]	1.7	1,250	470	135	75	1.6	900	925	1.6

(*continued*)

Table B-7. *(continued)*

Members stressed principally in bending with load applied perpendicular to the wide faces of the laminations

		Bending About X-X Axis						Bending About Y-Y Axis					Axially Loaded		
		Loaded Perpendicular to Wide Faces of Laminations						Loaded Parallel to Wide Faces of Laminations							
		Extreme Fiber in Bending F_{bxx}		Compression Perpendicular to Grain $F_{c\perp xx}$[6]											
Combination Symbol[4]	Species Outer Laminations/ Core Laminations[5]	Tension Zone Stressed in Tension[7,8] psi	Compression Zone Stressed in Tension[9] psi	Tension Face psi	Compression Face psi	Horizontal Shear F_{vxx} psi	Modulus of Elasticity E_{xx}[6] × 10^6 psi	Extreme Fiber in Bending[7,10] F_{byy} psi	Compression Perpendicular to Grain $F_{c\perp yy}$[6] psi	Horizontal Shear F_{vyy} psi	Horizontal Shear F_{vyy} psi (For members with multiple-piece laminations which are not edge glued)[11]	Modulus of Elasticity E_{yy}[6] × 10^6 psi	Tension Parallel to Grain F_t psi	Compression Parallel to Grain F_c psi	Modulus of Elasticity E[6] × 10^6 psi
		× 0.00689 for N/mm²													
1	2	3	4	5	6	7	8	9	10	11	12	13	14	15	16
Visually Graded Western Species															
The following three combinations are balanced and are intended for members continuous or cantilevered over supports and provide equal capacity in both positive and negative bending.															
22F-V7	DF/DF	2,200	2,200	650	650	165	1.8	1,450	560	145	75	1.6	1,100	1,650	1.6
22F-V8	DF/DF			590[12,13]	590[12,13]	165	1.7	1,450	560	145	75	1.6	1,050	1,650	1.6
22F-V9	HF/HF			500[16]	500[16]	155	1.5	1,250	375	135	70	1.4	975	1,400	1.4
The following six combinations are not balanced and are for either dry or wet use.															
24F-V1	DF/WW	2,400	1,200	650	650	140[15]	1.7[18]	1,250	255	130	70	1.4[18]	1,000	1,300	1.4
24F-V2	HF/HF			500[16]	500[16]	155	1.5	1,250	375	135	70	1.4	950	1,300	1.4
24F-V3	DF/DF			650	560[12]	165	1.8	1,500	560	145	75	1.6	1,100	1,600	1.6
24F-V4	DF/DF			650	650	165	1.8	1,500	560	145	75	1.6	1,150	1,650	1.6
24F-V5	DF/HF			650	650	155	1.7	1,350	375	140	70	1.5	1,100	1,450	1.5
24F-V11	DF/DFS			650	560[12]	165	1.7	1,600	500	145	75	1.4	1,150	1,700	1.4
The following two combinations are intended for straight or slightly cambered members for dry use and industrial appearance.[17]															
24F-V6	DF/N3WW	2,400	1,200	650	560[12]	90[2,3,14,20]	1.7[18]	1,200	255	140	70	1.5[18]	950	800	1.5
24F-V7	DF/M3DF			650	560[12]	90[19]	1.7	1,250	470	135	70	1.6	900	950	1.6

(continued)

Table B-7. *(continued)*

Members stressed principally in bending with load applied perpendicular to the wide faces of the laminations

		Bending About X-X Axis						Bending About Y-Y Axis					Axially Loaded		
		Loaded Perpendicular to Wide Faces of Laminations						Loaded Parallel to Wide Faces of Laminations							
		Extreme Fiber in Bending F_{bxx}		Compression Perpendicular to Grain $F_{c\perp xx}$ [6]											
Combination Symbol[4]	Species Outer Laminations/ Core Laminations[5]	Tension Zone Stressed in Tension[7,8] psi	Compression Zone Stressed in Tension[9] psi	Tension Face psi	Compression Face psi	Horizontal Shear F_{vxx} psi	Modulus of Elasticity E_{xx} [6] × 10^6 psi	Extreme Fiber in Bending[7,10] F_{byy} psi	Compression Perpendicular to Grain $F_{c\perp yy}$ [6] psi	Horizontal Shear F_{vyy} psi	Horizontal Shear F_{vyy} psi (For members with multiple-piece laminations which are not edge glued)[11]	Modulus of Elasticity E_{yy} [6] × 10^6 psi	Tension Parallel to Grain F_t psi	Compression Parallel to Grain F_c psi	Modulus of Elasticity E[6] × 10^6 psi
		× 0.00689 for N/mm²													
1	2	3	4	5	6	7	8	9	10	11	12	13	14	15	16
Visually Graded Western Species															
The following three combinations are balanced and are intended for members continuous or cantilevered over supports and provide equal capacity in both positive and negative bending.															
24F-V8	DF/DF	2,400	2,400	650	650	165	1.8	1,450	560	145	75	1.6	1,100	1,650	1.6
24F-V9	HF/HF			500[16]	500[16]	155	1.5	1,500	375	135	70	1.4	1,000	1,450	1.4
24F-V10	DF/HF			650	650	155	1.8	1,400	375	140	70	1.6	1,150	1,600	1.6
Wet-use factors[2]		0.8	0.8	0.667	0.667	0.875	0.833	0.8	0.667	0.875	0.875	0.833	0.8	0.73	0.833
E-Rated Western Species															
The following three combinations are not balanced and are for either dry or wet use.															
16F-E1	WW/WW	1,600	800	255[21]	255[21]	140[14,15]	1.3[18]	1,050	255	125	65	1.2[18]	725	925	1.2
16F-E2[22]	HF/HF			500[23]	500[23]	155	1.4	1,250	375	135	70	1.3	825	1,200	1.3
16F-E3	DF/DF			650	650	165	1.6	1,450	560	145	75	1.5	975	1,600	1.5

(continued)

Table B-7. *(continued)*

Members stressed principally in bending with load applied perpendicular to the wide faces of the laminations

		Bending About X-X Axis						Bending About Y-Y Axis					Axially Loaded		
		Loaded Perpendicular to Wide Faces of Laminations						Loaded Parallel to Wide Faces of Laminations							
		Extreme Fiber in Bending F_{bxx}		Compression Perpendicular to Grain $F_{c\perp xx}$ [6]											
Combination Symbol[4]	Species Outer Laminations/ Core Laminations[5]	Tension Zone Stressed in Tension[7,8] psi	Compression Zone Stressed in Tension[9] psi	Tension Face psi	Compression Face psi	Horizontal Shear F_{vxx} psi	Modulus of Elasticity E_{xx} [6] × 10[6] psi	Extreme Fiber in Bending[7,10] F_{byy} psi	Compression Perpendicular to Grain $F_{c\perp yy}$ [6] psi	Horizontal Shear F_{vyy} psi	Horizontal Shear F_{vyy} psi (For members with multiple-piece laminations which are not edge glued)[11]	Modulus of Elasticity E_{yy} [6] × 10[6] psi	Tension Parallel to Grain F_t psi	Compression Parallel to Grain F_c psi	Modulus of Elasticity E[6] × 10[6] psi
		× 0.00689 for N/mm²													
1	2	3	4	5	6	7	8	9	10	11	12	13	14	15	16
E-Rated Western Species															
The following two combinations are intended for straight or slightly cambered members for dry use and industrial appearance.[17]															
16F-E4	DF/N3WW	1,600	800	650	650	90[2,3,14,20]	1.6[18]	900	255	130	65	1.3[18]	675	675	1.3
16F-E5	DF/M3DF			650	650	90[19]	1.6	1,050	470	135	70	1.5	700	900	1.5
The following two combinations are balanced and are intended for members continuous or cantilevered over supports and provide equal capacity in both positive and negative bending.															
16F-E6	DF/DF	1,600	1,600	650	650	165	1.6	1,500	560	145	75	1.5	1,000	1,600	1.5
16F-E7[22]	HF/HF			500[23]	500[23]	155	1.4	1,250	375	135	70	1.3	850	1,150	1.3
The following four combinations are not balanced and are for either dry or wet use.															
20F-E1	WW/WW	2,000	1,000	255[21]	255[22]	140[14,15]	1.6[18]	1,100	255	125	65	1.3[18]	800	1,050	1.3
20F-E2[22]	HF/HF			500[23]	500[23]	155	1.6	1,400	375	135	70	1.4	925	1,550	1.4
20F-E3	DF/DF			650	650	165	1.7	1,550	560	145	75	1.6	1,050	1,650	1.6
20F-E8	ES/ES			450	450	145	1.5	1,400	300	125	65	1.4	800	1,000	1.4

(continued)

Table B-7. *(continued)*

Members stressed principally in bending with load applied perpendicular to the wide faces of the laminations

Combination Symbol[4]	Species Outer Laminations/ Core Laminations[5]	Bending About X-X Axis — Loaded Perpendicular to Wide Faces of Laminations — Extreme Fiber in Bending F_{bxx} — Tension Zone Stressed in Tension[7,8] psi	Extreme Fiber in Bending F_{bxx} — Compression Zone Stressed in Tension[9] psi	Compression Perpendicular to Grain $F_{c\perp xx}$[6] — Tension Face psi	Compression Perpendicular to Grain $F_{c\perp xx}$[6] — Compression Face psi	Horizontal Shear F_{vxx} psi	Modulus of Elasticity E_{xx}[6] × 10^6 psi	Bending About Y-Y Axis — Loaded Parallel to Wide Faces of Laminations — Extreme Fiber in Bending[7,10] F_{byy} psi	Compression Perpendicular to Grain $F_{c\perp yy}$[6] psi	Horizontal Shear F_{vyy} psi	Horizontal Shear F_{vyy} psi (For members with multiple-piece laminations which are not edge glued)[11]	Modulus of Elasticity E_{yy}[6] × 10^6 psi	Axially Loaded — Tension Parallel to Grain F_t psi	Compression Parallel to Grain F_c psi	Modulus of Elasticity E[6] × 10^6 psi
		× 0.00689 for N/mm²													
1	2	3	4	5	6	7	8	9	10	11	12	13	14	15	16
E-Rated Western Species															
The following two combinations are balanced and are intended for members continuous or cantilevered over supports and provide equal capacity in both positive and negative bending.															
20F-E6 20F-E7[22]	DF/DF HF/HF	2,000	2,000	650 500[23]	650 500[23]	165 155	1.7 1.6	1,600 1,500	560 375	145 135	75 70	1.6 1.4	1,150 1,050	1,650 1,550	1.6 1.4
The following two combinations are not balanced and are for either dry or wet use.															
22F-E1 22F-E2[22]	DF/DF HF/HF	2,200	1,100	650 500[23]	650 500[23]	165 155	1.7 1.6	1,550 1,400	560 375	145 135	75 70	1.6 1.4	1,050 950	1,600 1,400	1.6 1.4
The following two combinations are intended for straight or slightly cambered members for dry use and industrial appearance.[20]															
22F-E3 22F-E4	DF/N3WW DF/N3DF	2,200	1,100	650 650	650 650	90[2,3,14,20] 90[19]	1.7[18] 1.8	1,250 1,350	255 470	135 135	70 70	1.4[18] 1.6	825 950	750 950	1.4 1.6
The following two combinations are balanced and are intended for members continuous or cantilevered over supports and provide equal capacity in both positive and negative bending.															
22F-E5 22F-E6[22]	DF/DF HF/HF	2,200	2,200	650 500[23]	650 500[23]	165 155	1.7 1.7	1,650 1,550	560 375	145 135	75 70	1.6 1.5	1,100 1,050	1,650 1,500	1.6 1.5

(continued)

Table B-7. *(continued)*

Members stressed principally in bending with load applied perpendicular to the wide faces of the laminations

Combination Symbol[4]	Species Outer Laminations/ Core Laminations[5]	Bending About X-X Axis — Loaded Perpendicular to Wide Faces of Laminations: Extreme Fiber in Bending F_{bxx}: Tension Zone Stressed in Tension[7,8] psi	Extreme Fiber in Bending F_{bxx}: Compression Zone Stressed in Tension[9] psi	Compression Perpendicular to Grain $F_{c\perp xx}$[6]: Tension Face psi	Compression Perpendicular to Grain $F_{c\perp xx}$[6]: Compression Face psi	Horizontal Shear F_{vxx} psi	Modulus of Elasticity E_{xx}[6] × 10^6 psi	Bending About Y-Y Axis — Loaded Parallel to Wide Faces of Laminations: Extreme Fiber in Bending[7,10] F_{byy} psi	Compression Perpendicular to Grain $F_{c\perp yy}$[6] psi	Horizontal Shear F_{vyy} psi	Horizontal Shear F_{vyy} psi (For members with multiple-piece laminations which are not edge glued)[11]	Modulus of Elasticity E_{yy}[6] × 10^6 psi	Axially Loaded: Tension Parallel to Grain F_t psi	Compression Parallel to Grain F_c psi	Modulus of Elasticity E[6] × 10^6 psi
		× 0.00689 for N/mm²													
1	2	3	4	5	6	7	8	9	10	11	12	13	14	15	16
E-Rated Western Species															
The following six combinations are not balanced and are for either dry or wet use.															
24F-E1	DF/DF			650	650	165	1.8	1,550	560	145	75	1.6	1,100	1,600	1.6
24F-E2[22]	HF/HF			500[23]	500[23]	155	1.7	1,300	375	135	70	1.5	850	1,400	1.5
24F-E3	DF/HF	2,400	1,200	650	500[23]	155	1.8	1,500	375	135	70	1.5	1,050	1,550	1.5
24F-E4	DF/DF			650	650	165	1.8	1,650	560	145	75	1.7	1,100	1.700	1.7
24F-E5	DF/DF			650	650	165	1.8	1,650	560	145	75	1.6	1,100	1,550	1.6
24F-E6[22]	HF/WW			500[23]	500[23]	140[14,15]	1.8[18]	1,100	255	130	65	1.4[18]	750	1,250	1.4
The following five combinations are intended for straight or slightly cambered members for dry use and industrial appearance.[17]															
24F-E14	DF/DF			650	650	165	1.8	1,450	560	145	75	1.6	950	1,600	1.6
24F-E15	HF/HF			500[23]	500[23]	155	1.8	1,300	375	135	70	1.5	950	1,200	1.5
24F-E7	DF/N3WW	2,400	1,200	650	650	90[2,3,14,20]	1.9[18]	1,400	255	135	70	1.6[18]	975	875	1.6
24F-E8	DF/N3DF			650	650	90[19]	1.9	1,400	470	135	70	1.7	1,000	1,050	1.7
24F-E9[22]	HF/N3HF			500[23]	500[23]	90[20]	1.8	1,350	375	135	70	1.6	950	825	1.6

(continued)

Table B-7. *(continued)*

Members stressed principally in bending with load applied perpendicular to the wide faces of the laminations

		Bending About X-X Axis						Bending About Y-Y Axis					Axially Loaded		
		Loaded Perpendicular to Wide Faces of Laminations						Loaded Parallel to Wide Faces of Laminations							
		Extreme Fiber in Bending F_{bxx}		Compression Perpendicular to Grain $F_{c\perp xx}{}^{6}$											
Combination Symbol[4]	Species Outer Laminations/ Core Laminations[5]	Tension Zone Stressed in Tension[7,8] psi	Compression Zone Stressed in Tension[9] psi	Tension Face psi	Compression Face psi	Horizontal Shear F_{vxx} psi	Modulus of Elasticity $E_{xx}{}^{6}$ × 10⁶ psi	Extreme Fiber in Bending[7,10] F_{byy} psi	Compression Perpendicular to Grain $F_{c\perp yy}{}^{6}$ psi	Horizontal Shear F_{vyy} psi	Horizontal Shear F_{vyy} psi (For members with multiple-piece laminations which are not edge glued)[11]	Modulus of Elasticity $E_{yy}{}^{6}$ × 10⁶ psi	Tension Parallel to Grain F_t psi	Compression Parallel to Grain F_c psi	Modulus of Elasticity E^6 × 10⁶ psi
		× 0.00689 for N/mm²													
1	2	3	4	5	6	7	8	9	10	11	12	13	14	15	16
E-Rated Western Species															
The following eight combinations are balanced and are intended for members continuous or cantilevered over supports and provide equal capacity in both positive and negative bending.															
24F-E10	DF/DF			650	650	165	1.9	1,850	560	145	75	1.7	1,300	1,750	1.7
24F-E11[22]	HF/HF			500[23]	500[23]	155	1.8	1,600	375	135	70	1.5	1,150	1,550	1.5
24F-E12	DF/HF			650	650	155	1.9	1,750	375	135	70	1.6	1,200	1,600	1.6
24F-E13	DF/DF	2,400	2,400	650	650	165	1.8	1,950	560	145	75	1.7	1,250	1,700	1.7
24F-E16	HF/HF			500[23]	500[23]	155	1.7	1,300	375	135	70	1.5	850	1,400	1.5
24F-E17	HF/WW			500[23]	500[23]	140[14,15]	1.8[18]	1,100	255	130[14]	65[14]	1.4[18]	750	1,250	1.4
24F-E18	DF/DF			650	650	165	1.8	1,450	560	145	75	1.6	950	1,600	1.6
24-E19	HF/HF			500[23]	500[23]	155	1.8	1,300	375	135	70	1.5	950	1,200	1.5
Wet-use factors[2]		0.8	0.8	0.667	0.667	0.875	0.833	0.8	0.667	0.875	0.875	0.833	0.8	0.73	0.833

[1] The combinations in this table are applicable to members consisting of four or more laminations and are intended primarily for members stressed in bending due to loads applied perpendicular to the wide faces of the laminations. Design values are tabulated, however, for loading both perpendicular and parallel to the wide faces of the laminations. For combinations and design values applicable to members loaded primarily axially or parallel to the wide faces of the laminations, see Table 23-I-C-1, Part B. For members of two or three laminations, see Table 23-I-C-1, Part B.

[2] The tabulated design values are for dry conditions of use. To obtain wet-use design values, multiply the tabulated values by the factors shown at the end of the table.

[3] The tabulated design values are for normal duration of loading. For other durations of loading, see Section 2304.3.4.

[4] The combination symbols relate to a specific combination of grades and species that will provide the design values shown for the combinations. The first two numbers in the combination symbol correspond to the design value in bending shown in Column 3. The letter in the combination symbol (either a "V" or an "E") indicates whether the combination is made from visually graded (V) or E-rated (E) lumber in the outer zones.

Table B-7. *(continued)*

[5] The symbols used for species are AC = Alaska cedar, DF = Douglas fir-larch, DFS = Douglas fir south, HF = hem-fir, WW = softwood species, SP = southern pine and ES = eastern spruce. (N3 refers to No. 3 structural joists and planks or structural light-framing grade.) Softwood species (WW) and eastern spruce are included in the general category of western species although eastern spruce and some softwood species are produced in other areas.

[6] The duration of load modification factors given in Section 2304.3.4 shall not apply.

[7] The tabulated design values in bending are applicable to members 12 inches (305 mm) or less in depth. For members greater than 12 inches (305 mm) in depth, the requirements of Section 2312.4.5 apply.

[8] The design values in bending about the x-x axis (F_{bx}) in this column for bending members shall be multiplied by 0.75 when the member is manufactured without the required special tension laminations.

[9] Design values in this column are for extreme fiber stress in bending when the member is loaded such that the compression zone laminations are subjected to tensile stresses. The values in this column may be increased 200 pounds per square inch (psi) (1.38 N/mm^2) where end-joint spacing restrictions are applied to the compression zone when stressed in tension.

[10] Footnote 7 to Table 23-I-C-1, Part B, also applies.

[11] These values for horizontal shear, F_{vyy}, apply to members manufactured using multiple-piece laminations with unbonded edge joints. For members manufactured using single-piece laminations or using multiple-piece laminations with bonded edge joints, the horizontal shear values in Column 11 apply.

[12] Where specified, this value may be increased to 650 psi (4.48 N/mm^2) by providing in the bearing area at least one dense 2-inch (51 mm) nominal thickness lamination of Douglas fir-larch for western species combinations, or southern pine for southern pine combinations. These dense laminations must be backed by a medium-grain lamination of the same species.

[13] For bending members greater than 15 inches (381 mm) in depth, the design value for compression stress perpendicular to grain is 650 psi (4.48 N/mm^2) on the tension face.

[14] When Douglas fir south is used in place of all of the western wood laminations required in western species combinations 16F-V1, 16F-V4, 20F-V1, 20F-V5, 22F-V1, 22F-V5, 24F-V1, 24F-V6, 16F-E1, 16F-E4, 20F-E1, 20F-E4, 22F-E3, 24F-E6 and 24F-E7, the design value for horizontal shear is the same as for combinations using all Douglas fir-larch [F_{vxx} = 165 psi (1.14 N/mm^2) and F_{vyy} = 145 psi (1.00 N/mm^2) for L3; and F_{vxx} = 90 psi (0.62 N/mm^2) and F_{vyy} = 135 psi (0.93 N/mm^2) for N3].

[15] The following species may be used for softwood species (WW), provided the design values in horizontal shear in Column 7 (F_{vx}) and in Column 11 (F_{vy}) are reduced by 10 psi (0.07 N/mm^2), and the design values in horizontal shear in Column 12 (F_{vy}) are reduced by 5 psi (0.03 N/mm^2): coast sitka spruce, coast species, eastern white pine (north) and western white pine.

[16] Where specified, this value may be increased by providing at least two 2-inch (51 mm) nominal thickness Douglas fir-larch laminations in the bearing area. The compression-perpendicular-to-grain design values for Douglas fir-larch are 560 psi (3.86 N/mm^2) for medium grain and 650 psi (4.48 N/mm^2) for dense.

[17] These combinations are for dry conditions of use only because they may contain wane. They are recommended for industrial appearance grade and for straight or slightly cambered members only. If wane is omitted these restrictions do not apply.

[18] The following species may be used for softwood species (WW), provided the design values in modulus of elasticity (E_x and F_y) in Columns 8 and 13 are reduced by 100,000 psi (689 N/mm^2): western cedars, western cedars (north), white woods (western woods) and California redwood–open grain.

[19] This value may be increased to 110 psi (0.76 N/mm^2) when the member does not contain coarse-grain material; to 140 psi (0.96 N/mm^2) when the member does not contain wane on both sides or the member does not contain coarse-grain material and wane on one side; to 165 psi (1.14 N/mm^2) when the member does not contain coarse-grain material and wane on both sides.

[20] This value may be increased to 140 psi (0.96 N/mm^2) for softwood species (WW) and to 155 psi (1.07 N/mm^2) for hem-fir when the member does not contain wane on both sides; to 115 psi (0.79 N/mm^2) for softwood species (WW) and to 130 psi (0.90 N/mm^2) for hem-fir when the member does not contain wane on one side.

[21] The compression-perpendicular-to-grain design value of 255 psi (1.76 N/mm^2) is based on the lowest strength species of the western woods group. If at least one 2-inch (51 mm) nominal thickness lamination of E-rated hem-fir with the same E value, or E-rated Douglas fir-larch 200,000 psi (1378 N/mm^2) higher in modulus of elasticity (E) than that specified is used in the bearing area on the face of the member subjected to the compression-perpendicular-to-grain stress, $F_{c\perp}$ may be increased to 375 psi (2.58 N/mm^2). If at least two 2-inch (51 mm) nominal thickness laminations of E-rated hem-fir with the same E value, or E-rated Douglas fir-larch 200,000 psi (1378 N/mm^2) higher in modulus of elasticity than that specified are used in the bearing area on the face of the member subjected to the compression-perpendicular-to-grain stress, $F_{c\perp}$ may be increased to 500 psi (3.45 N/mm^2).

[22] E-rated Douglas fir-larch 200,000 psi (1378 N/mm^2) higher in modulus of elasticity may be substituted for the specified E-rated hem-fir.

[23] Where specified, this value may be increased to 650 psi (4.48 N/mm^2) by providing in the bearing area at least one 2-inch (51 mm) nominal thickness lamination of Douglas fir-larch for western species combinations, or one 2-inch (51 mm) nominal thickness lamination for southern pine of southern pine combinations having a modulus of elasticity (E) value 200,000 psi (1378 N/mm^2) higher than the E value specified.

[24] This value may be increased to 140 psi (0.96 N/mm^2) when the member does not contain coarse-grain material or when the member does not contain wane on both sides; to 165 psi (1.14 N/mm^2) when members do not contain coarse-grain material or wane on one side; or to 200 psi (1.38 N/mm^2) when the member does not contain both coarse-grain material and wane on both sides of the member.

Source: Reproduced from the 1994 edition of the *Uniform Building Code*™, copyright © 1994, with the permission of the publisher, the International Conference of Building Officials.

Table B-8. Allowable Unit Stresses for Structural Glued Laminated Timber for Normal Load Duration.[1,2,3]

Members stressed principally in axial tension or compression or in bending with load applied parallel to the wide faces of the laminations

					Axially Loaded			Bending About Y-Y Axis							Bending About X-X Axis		
								Loaded Parallel to Wide Faces of Laminations							Loaded Perpendicular to Wide Faces of Laminations		
					Tension Parallel to Grain F_t	Compression Parallel to Grain F_c		Extreme Fiber in Bending[7] F_{byy}			Horizontal Shear[8] F_{vyy}				Extreme Fiber in Bending[9] F_{bxx}		
Combination Symbol	Species[4]	Grade[5]	Modulus of Elasticity E[6] $\times 10^6$ psi	Compression Perpendicular to Grain $F_{c\perp}$ psi	2 or More Lams psi	4 or More Lams psi	2 or 3 Lams psi	4 or More Lams psi	3 Lams psi	2 Lams psi	4 or More Lams (For members with multi-ple-piece laminations.)[10]	4 or More Lams psi	3 Lams psi	2 Lams psi	2 Lams to 15 in. Deep[11] psi	4 or More Lams[12,13] psi	Horizontal Shear F_{vxx} 2 or More Lams psi
					$\times$ 0.00689 for N/mm^2												
1	2	3	4	5	6	7	8	9	10	11	12	13	14	15	16	17	18
Visually Graded Western Species																	
1		L3	1.5	560[14]	900	1,550	1,200	1,450	1,250	1,000	75	145	135	125	1,250	1,500	165
2		L2	1.7	560[14]	1,250	1,900	1,600	1,800	1,600	1,300	75	145	135	125	1,700	2,000	165
3		L2D	1.8	650	1,450	2,300	1,850	2,100	1,850	1,550	75	145	135	125	2,000	2,300	165
4		L1CL	1.9	590[14]	1,400	2,100	1,900	2,200	2,000	1,650	75	145	135	125	1,900	2,200	165
5		L1	2.0	650	1,600	2,400	2,100	2,400	2,100	1,800	75	145	135	125	2,200	2,400	165
6		N3C	1.4	470	350	875	550	550	550	550	60	120	115	105	450	—	140
7	DF	N3M	1.5	560	900	1,550	700	1,450	1,250	1,000	75	145	135	125	1,000	—	165
8		N2	1.6	560[14]	1,000	1,550	1,150	1,600	1,550	1,300	75	145	135	125	1,350	1,600	165
9		N2D	1.8	650	1,150	1,800	1,350	1,850	1,800	1,500	75	145	135	125	1,600	1,850	165
10		N1	1.8	560[14]	1,300	1,950	1,450	1,950	1,750	1,500	75	145	135	125	1,750	2,100	165
11		N1D	2.0	650	1,500	2,300	1,700	2,300	2,100	1,750	75	145	135	125	2,100	2,400	165
12		SS	1.8	560[14]	1,400	1,950	1,650	2,100	1,950	1,650	75	145	135	125	1,900	2,200	165
13		SSD	2.0	650	1,600	2,300	1,950	2,400	2,300	1,950	75	145	135	125	2,200	2,400	165

(continued)

Table B-8. (continued)

Members stressed principally in axial tension or compression or in bending with load applied parallel to the wide faces of the laminations

					Axially Loaded			Bending About Y-Y Axis — Loaded Parallel to Wide Faces of Laminations							Bending About X-X Axis — Loaded Perpendicular to Wide Faces of Laminations		
					Tension Parallel to Grain F_t	Compression Parallel to Grain F_c		Extreme Fiber in Bending[7] F_{byy}			Horizontal Shear[8] F_{vyy}				Extreme Fiber in Bending[9] F_{bxx}		
Combination Symbol	Species[4]	Grade[5]	Modulus of Elasticity E[6] $\times 10^6$ psi	Compression Perpendicular to Grain $F_{c\perp}$ psi	2 or More Lams psi	4 or More Lams psi	2 or 3 Lams psi	4 or More Lams psi	3 Lams psi	2 Lams psi	4 or More Lams (For members with multi-ple-piece laminations.)[10]	4 or More Lams psi	3 Lams psi	2 Lams psi	2 Lams to 15 in. Deep[11] psi	4 or More Lams[12,13] psi	Horizontal Shear F_{vxx} 2 or More Lams psi
			× 0.00689 for N/mm²														
1	2	3	4	5	6	7	8	9	10	11	12	13	14	15	16	17	18
Visually Graded Western Species																	
14	HF	L3	1.3	375[14]	800	1,100	975	1,200	1,050	850	70	135	130	115	1,100	1,300	155
15		L2	1.4	375[14]	1,050	1,350	1,300	1,500	1,350	1,100	70	135	130	115	1,450	1,700	155
16		L1	1.6	375[14]	1,200	1,500	1,450	1,750	1,550	1,300	70	135	130	115	1,600	1,900	155
17		L1D	1.7	500	1,400	1,750	1,700	2,000	1,850	1,550	70	135	130	115	1,900	2,200	155
18		N3	1.3	375	425	900	575	700	700	700	70	135	130	115	575	—	155
19		N2	1.4	375[14]	850	1,300	975	1,350	1,300	1,100	70	135	130	115	1,150	1,350	155
20		N1	1.6	375[14]	975	1,450	1,250	1,550	1,500	1,250	70	135	130	115	1,350	1,550	155
21		SS	1.6	375[14]	1,100	1,450	1,350	1,750	1,650	1,400	70	135	130	115	1,500	1,750	155
22	WW	L3	1.0[15]	255	525	850	675	800	700	550	60[16]	120[16]	115[16]	105[16]	725	850	140[16]
23		N3	1.0[15]	255	275	625	450	450	450	450	60[16]	120[16]	115[16]	105[16]	400	—	140[16]
24		N2	1.1[15]	255	550	900	700	900	875	725	60[16]	120[16]	115[16]	105[16]	775	900	140[16]
25		N1	1.2[15]	255	650	1,000	875	1,050	1,000	850	60[16]	120[16]	115[16]	105[16]	875	1,050	140[16]
26		SS	1.2[15]	255	750	1,000	1,000	1,150	1,100	925	60[16]	120[16]	115[16]	105[16]	1,000	1,150	140[16]
59	DFS	L3	1.1	500	800	1,400	1,050	1,200	1,050	850	75	145	135	125	1,050	1,250	165
60		L2	1.3	500	1,050	1,750	1,400	1,750	1,550	1,150	75	145	135	125	1,450	1,700	165
61		L1	1.5	650	1,350	2,200	1,850	2,000	1,800	1,500	75	145	135	125	1,850	2,200	165

(continued)

Table B-8. *(continued)*

Members stressed principally in axial tension or compression or in bending with load applied parallel to the wide faces of the laminations

Combination Symbol	Species[4]	Grade[5]	Modulus of Elasticity E[6] $\times 10^6$ psi	Compression Perpendicular to Grain $F_{c\perp}$ psi	Axially Loaded: Tension Parallel to Grain F_t, 2 or More Lams psi	Axially Loaded: Compression Parallel to Grain F_c, 4 or More Lams psi	Compression Parallel to Grain F_c, 2 or 3 Lams psi	Bending About Y-Y Axis (Loaded Parallel to Wide Faces of Laminations): Extreme Fiber in Bending[7] F_{byy}, 4 or More Lams psi	F_{byy}, 3 Lams psi	F_{byy}, 2 Lams psi	Horizontal Shear[8] F_{vyy}, 4 or More Lams (For members with multi-ple-piece laminations.)[10]	F_{vyy}, 4 or More Lams psi	F_{vyy}, 3 Lams psi	F_{vyy}, 2 Lams psi	Bending About X-X Axis (Loaded Perpendicular to Wide Faces of Laminations): Extreme Fiber in Bending[9] F_{bxx}, 2 Lams to 15 in. Deep[11] psi	F_{bxx}, 4 or More Lams[12,13] psi	Horizontal Shear F_{vxx} 2 or More Lams psi
			× 0.00689 for N/mm²														
1	2	3	4	5	6	7	8	9	10	11	12	13	14	15	16	17	18
Visually Graded Western Species																	
64	ES	D	1.3	300	575	925	850	950	925	875	65	125	120	110	800	950	145
65	ES	D4	1.5	450	825	1,050	1,050	1,350	1,350	1,350	65	125	120	110	1,150	1,350	145
66	ES	C4	1.5	450	975	1,550	1,200	1,750	1,750	1,700	65	125	120	110	1,400	1,650	145
67	ES	C6	1.7	450	1,100	1,850	1,450	1,950	1,950	1,900	65	125	120	110	1,550	1,800	145
68	ES	B	1.7	450	1,350	1,850	1,800	2,250	2,200	2,150	65	125	120	110	1,900	2,250	145
69 AC	ES	L3	1.3	470	700	1,150	1,150	1,000	875	700	80	165	160	140	1,000	1,150	190
70 AC	ES	L2	1.4	470	1,000	1,450	1,550	1,250	1,100	925	80	165	160	140	1,350	1,550	190
71 AC	ES	L1D	1.7	560	1,250	1,900	2,050	1,650	1,500	1,250	80	165	160	140	1,700	2,000	190
72 AC	ES	L1S	1.7	560	1,250	1,900	2,050	1,650	1,500	1,250	80	165	160	140	1,700	2,000	190

(continued)

Table B-8. *(continued)*

Members stressed principally in axial tension or compression or in bending with load applied parallel to the wide faces of the laminations

					Axially Loaded			Bending About Y-Y Axis: Loaded Parallel to Wide Faces of Laminations							Bending About X-X Axis: Loaded Perpendicular to Wide Faces of Laminations		
					Tension Parallel to Grain F_t	Compression Parallel to Grain F_c		Extreme Fiber in Bending[7] F_{byy}			Horizontal Shear[8] F_{vyy}				Extreme Fiber in Bending[9] F_{bxx}		
Combination Symbol	Species[4]	Grade[5]	Modulus of Elasticity E[6] $\times 10^6$ psi	Compression Perpendicular to Grain $F_{c\perp}$ psi	2 or More Lams psi	4 or More Lams psi	2 or 3 Lams psi	4 or More Lams psi	3 Lams psi	2 Lams psi	4 or More Lams (For members with multi-ple-piece laminations.)[10]	4 or More Lams psi	3 Lams psi	2 Lams psi	2 Lams to 15 in. Deep[11] psi	4 or More Lams[12,13] psi	Horizontal Shear F_{vxx} 2 or More Lams psi
				× 0.00689 for N/mm²													
1	2	3	4	5	6	7	8	9	10	11	12	13	14	15	16	17	18
E-Rated Western Species																	
27	DF	1/2-1.8E	1.8	650	900	1,750	1,200	1,450	1,250	1,000	75	145	135	125	1,250	1,500	165
28	DF	1/2-2.0E	2.0	650	1,100	2,000	1,400	1,450	1,250	1,000	75	145	135	125	1,500	1,750	165
29	DF	1/2-2.2E	2.2	650	1,250	2,300	1,550	1,650	1,400	1,150	75	145	135	125	1,700	2,000	165
30	DF	1/6-1.8E	1.8	650	1,550	2,100	1,700	2,400	2,400	2,100	75	145	135	125	1,800	2,100	165
31	DF	1/6-2.0E	2.0	650	1,800	2,400	1,900	2,400	2,400	2,400	75	145	135	125	2,100	2,400	165
32	DF	1/6-2.2E	2.2	650	1,800	2,400	2,100	2,400	2,400	2,400	75	145	135	125	2,300	2,400	165
62	DF	1/2-2.1E	2.1	650	1,150	2,200	1,500	1,550	1,350	1,100	75	145	135	125	1,600	1,900	165
63	DF	1/6-2.1E	2.1	650	1,800	2,400	2,000	2,400	2,400	2,400	75	145	135	125	2,200	2,400	165
33	HF	1/2-1.5E	1.5	500	800	1,050	950	1,200	1,050	850	70	135	130	115	1,100	1,300	155
34	HF	1/2-1.8E	1.8	500	900	1,300	1,200	1,450	1,250	1,000	70	135	130	115	1,250	1,500	155
35	HF	1/2-2.0E	2.0	500	1,100	1,550	1,400	1,450	1,250	1,000	70	135	130	115	1,500	1,750	155
36	HF	1/4-1.5E	1.5	500	1,200	1,450	1,300	2,100	1,900	1,700	70	135	130	115	1,400	1,650	155
37	HF	1/6-1.8E	1.8	500	1,550	1,950	1,700	2,400	2,400	2,100	70	135	130	115	1,800	2,100	155
38	HF	1/6-2.0E	2.0	500	1,800	2,400	1,900	2,400	2,400	2,400	70	135	130	115	2,100	2,400	155

(continued)

Table B-8. *(continued)*

Members stressed principally in axial tension or compression or in bending with load applied parallel to the wide faces of the laminations

					Axially Loaded			Bending About Y-Y Axis — Loaded Parallel to Wide Faces of Laminations							Bending About X-X Axis — Loaded Perpendicular to Wide Faces of Laminations		
					Tension Parallel to Grain F_t	Compression Parallel to Grain F_c		Extreme Fiber in Bending[7] F_{byy}			Horizontal Shear[8] F_{vyy}				Extreme Fiber in Bending[9] F_{bxx}		
Combination Symbol	Species[4]	Grade[5]	Modulus of Elasticity E[6] $\times 10^6$ psi	Compression Perpendicular to Grain $F_{c\perp}$ psi	2 or More Lams psi	4 or More Lams psi	2 or 3 Lams psi	4 or More Lams psi	3 Lams psi	2 Lams psi	4 or More Lams (For members with multi-piece laminations.)[10]	4 or More Lams psi	3 Lams psi	2 Lams psi	2 Lams to 15 in. Deep[11] psi	4 or More Lams[12,13] psi	Horizontal Shear F_{vxx} 2 or More Lams psi
			$\times$ 0.00689 for N/mm²														
1	2	3	4	5	6	7	8	9	10	11	12	13	14	15	16	17	18
E-Rated Western Species																	
39		1/2-1.5E	1.5	255	800	1,200	950	1,200	1,050	850	60[16]	120[16]	115[16]	105[16]	1,100	1,300	140[16]
40	WW	1/2-1.8E	1.8	255	900	1,500	1,200	1,450	1,250	1,000	60[16]	120[16]	115[16]	105[16]	1,250	1,500	140[16]
41		1/2-2.0E	2.0	255	1,100	1,750	1,400	1,450	1,250	1,000	60[16]	120[16]	115[16]	105[16]	1,500	1,750	140[16]
42		1/4-1.5E	1.5	255	1,200	1,550	1,300	2,100	1,900	1,700	60[16]	120[16]	115[16]	105[16]	1,400	1,650	140[16]
43	WW	1/6-1.8E	1.8	255	1,550	1,950	1,700	2,400	2,400	2,100	60[16]	120[16]	115[16]	105[16]	1,800	2,100	140[16]
44		1/6-2.0E	2.0	255	1,800	2,200	1,900	2,400	2,400	2,400	60[16]	120[16]	115[16]	105[16]	2,100	2,400	140[16]

(continued)

Table B-8. *(continued)*

Members stressed principally in axial tension or compression or in bending with load applied parallel to the wide faces of the laminations

					Axially Loaded			Bending About Y-Y Axis: Loaded Parallel to Wide Faces of Laminations							Bending About X-X Axis: Loaded Perpendicular to Wide Faces of Laminations		
					Tension Parallel to Grain F_t	Compression Parallel to Grain F_c		Extreme Fiber in Bending[7] F_{byy}			Horizontal Shear[8] F_{vyy}				Extreme Fiber in Bending[9] F_{bxx}		
Combination Symbol	Species[4]	Grade[5]	Modulus of Elasticity E[6] $\times 10^6$ psi	Compression Perpendicular to Grain $F_{c\perp}$ psi	2 or More Lams psi	4 or More Lams psi	2 or 3 Lams psi	4 or More Lams psi	3 Lams psi	2 Lams psi	4 or More Lams (For members with multiple-piece laminations.)[10]	4 or More Lams psi	3 Lams psi	2 Lams psi	2 Lams to 15 in. Deep[11] psi	4 or More Lams[12,13] psi	Horizontal Shear F_{vxx} 2 or More Lams psi
					× 0.00689 for N/mm²												
1	2	3	4	5	6	7	8	9	10	11	12	13	14	15	16	17	18
								Visually Graded Southern Pine									
45		N3C	1.1	470	325	850	550	550	550	550	60	120	115	105	450	—	140
46		N3M	1.3	560	900	1,500	675	1,450	1,250	1,000	90	175	165	150	1,000	—	200
47		N2M[17]	1.4	560[14]	1,200	1,900	1,150	1,750	1,550	1,300	90	175	165	150	1,400	1,600	200
48	SP	N2D[17]	1.7	650	1,400	2,200	1,350	2,000	1,800	1,500	90	175	165	150	1,600	1,900	200
49		N1M[17]	1.7	560[14]	1,350	2,100	1,450	1,950	1,750	1,500	90	175	165	150	1,800	2,100	200
50		N1D[17]	1.9	650	1,550	2,300	1,700	2,300	2,100	1,750	90	175	165	150	2,100	2,400	200
51		SSM	1.7	560[14]	1,300	1,900	1,600	2,100	1,950	1,650	90	175	165	150	1,750	2,100	200
52		SSD	1.9	650	1,500	2,200	1,850	2,400	2,300	1,950	90	175	165	150	2,100	2,400	200
								E-Rated Southern Pine									
53		1/2-1.8E	1.8	650	900	1,900	1,200	1,450	1,250	1,000	90	175	165	150	1,250	1,500	200
54		1/2-2.0E	2.0	650	1,100	2,300	1,400	1,450	1,250	1,000	90	175	165	150	1,600	1,750	200
55	SP	1/2-2.2E	2.2	650	1,250	2,400	1,550	1,650	1,400	1,150	90	175	165	150	1,700	2,000	200
56		1/6-1.8E	1.8	650	1,550	1,850	1,700	2,400	2,400	2,100	90	175	165	150	1,800	2,100	200
57		1/6-2.0E	2.0	650	1,800	2,400	1,900	2,400	2,400	2,400	90	175	165	150	2,100	2,400	200
58		1/6-2.2E	2.2	650	1,800	2,400	2,100	2,400	2,400	2,400	90	175	165	150	2,300	2,400	200
Wet-use factors[2]		—	0.833	0.53	0.8	0.73	0.73	0.8	0.8	0.8	0.875	0.875	0.875	0.875	0.8	0.875	0.875

(continued)

Table B-8. *(continued)*

[1] The combinations in this table are intended primarily for members loaded either axially or in bending with the loads acting parallel to the wide faces of the laminations. Design values for bending due to loading applied perpendicular to the wide faces of the laminations are also included; however, the combinations in Table 23-I-C-1, Part A, are usually better suited for this condition of loading. The design values for bending about the x-x axis (F_{bxx}) shown in Column 16 are for members from two laminations to 15 inches (381 mm) deep without tension laminations. Design values approximately 15 percent higher for members with four or more laminations are shown in Column 17. These higher design values, however, require special tension laminations which may not be readily available.

[2] The tabulated design values are for dry conditions of use. To obtain wet-use design values, multiply the tabulated values by the factors shown at the end of the table.

[3] The tabulated design values are for normal duration of loading. For other durations of loading, see Section 2304.3.4.

[4] The symbols used for species are AC = Alaska cedar, DF = Douglas fir-larch, DFS = Douglas fir south, ES = eastern spruce, HF = hem-fir, WW = softwood species, and SP = southern pine.

[5] Grade designations are as follows: softwood species (WW) and eastern spruce are included in general category of western species although eastern spruce and some softwood species are produced in other areas.

Visually Graded Western Species
L1 is L1 laminating grade (dense for Douglas-fir-larch and Douglas fir south).
L1D is L1 dense laminating grade for hem-fir.
L1S is a special grade of Alaska cedar.
L1CL is L1 close grain laminating grade.
L2D is L2 laminating grade (dense).
L2 is L2 laminating grade (medium grain).
L3 is L3 laminating grade (medium grain for Douglas fir-larch, Douglas fir south and hem-fir).
SSD is dense select structural, structural joists and planks, or structural light-framing grade (dense).
SS is select structural, structural joists and planks, or structural light-framing grade (medium grain for Douglas fir-larch).
N1D is dense No. 1 structural joists and planks, or structural light-framing grade (dense).
N1 is No. 1 structural joists and planks, or structural light-framing grade (medium grain for Douglas fir-larch).
N2D is dense No. 2 structural joists and planks or structural light-framing grade (medium grain for Douglas fir-larch dense).
N2 is No. 2 structural joists and planks or structural light-framing grade (medium grain for Douglas fir-larch).
N3M is No. 3 structural joists and planks, or structural light-framing grade (medium grain).
N3C is No. 3 structural joists and planks, or structural light-framing grade (coarse grain).
N3 is No. 3 structural joists and planks, or structural light-framing grade.

Visually Graded Southern Pine
SSD is dense select structural, structural joists and planks, or structural light-framing grade (dense).
SSM is select structural, structural joists and planks, or structural light-framing grade (medium grain).
N1D is No. 1 dense structural joists and planks or structural light-framing grade or No. 1 boards graded as dense.
N1M is No. 1 structural joists and planks, or structural light-framing grade or No. 1 boards all with a medium grain rate of growth.
N2D is No. 2 dense structural joists and planks, or structural light-framing grade or No. 2 boards graded as dense.
N2M is No. 2 structural joists and planks, or structural light-framing grade or No. 2 boards all with medium grain rate of growth.
N3M is No. 3 structural joists and planks, or structural light-framing grade or No. 3 boards all with a medium grain rate of growth.
N3C is No. 3 structural joists and planks, or structural light-framing grade or No. 3 boards all with coarse grain rate of growth.

(continued)

Table B-8. (*continued*)

E-Rated Grades—All Species

1/6-2.2E has 1/6 edge characteristic with 2.2E.
1/6-2.1E has 1/6 edge characteristic with 2.1E.
1/6-2.0E has 1/6 edge characteristic with 2.0E.
1/6-1.8E has 1/6 edge characteristic with 1.8E.
1/4-1.5E has 1/4 edge characteristic with 1.5E.
1/2-2.2E, 1/2-2.1E, 1/2-2.0E, 1/2-1.8E are E-rated grades with edge characteristics occupying up to one half of cross section.

[6] The duration of load modification factors given in Section 2304.3.4 shall not apply.

[7] The values of F_{byy} were calculated based on members 12 inches (305 mm) in depth (bending about *Y-Y* axis). When the depth is less than 12 inches (305 mm), the values of F_{byy} can be increased by multiplying by the following factors.

DEPTH, (inches) × 25.4 for mm	MULTIPLYING FACTOR
10.75	1.01
8.75	1.04
6.75	1.07
5.125	1.10
3.125	1.16

[8] The design values in horizontal shear contained in this table are based on members without wane.

[9] The tabulated design values in bending are applicable to members 12 inches (305 mm) or less in depth. For members greater than 12 inches (305 mm) in depth, the requirements of Section 2312.4.5 apply.

[10] These values for horizontal shear, F_{vyy}, apply to members manufactured using multiple-piece laminations with unbonded edge joints. For members using single-piece laminations or using multiple-piece laminations with bonded edge joints, the horizontal shear values tabulated in Columns 13, 14 and 15 apply.

[11] The design values in Column 16 are for members of from two laminations to 15 inches (381 mm) in depth without tension laminations.

[12] The design values in Column 17 are for members of four or more laminations in depth and require special tension laminations. When these values are used in design and the member is specified by combination symbol, the designer should also specify the required design value in bending.

[13] When special tension laminations are not used, the design values in bending about the x-x axis (F_{bx}) shall be multiplied by 0.75 for bending members over 15 inches (381 mm) deep. For bending members 15 inches (381 mm) and less in depth, use the design values in Column 16.

[14] When tension laminations are used to obtain the design value for F_{bxx} shown in Column 16, the compression perpendicular to grain value, $F_{c\perp}$, for the tension face may be increased to 650 psi (4.48 N/mm²) for Douglas fir-larch and southern pine, and to 500 psi (3.45 N/mm²) for hem-fir because the tension laminations are required to be dense.

[15] The following species may be used for softwood species (WW), provided the modulus of elasticity, *E*, is reduced by 100,000 psi (689 N/mm²): western cedars, western cedars (north), white woods (western woods) and California redwood–open grain.

[16] The following species may be used for softwood species (WW) provided the design values in horizontal shear in Column 12 (F_{vy}) are reduced by 5 psi (0.03 N/mm²) and the design values in horizontal shear in Columns 13, 14 and 15 (F_{vy}) and in Column 18 (F_{vx}) are reduced by 10 psi (0.06 N/mm²): coast sitka spruce, coast species, western white pine and eastern white pine.

Source: Reproduced from the 1994 edition of the *Uniform Building Code*™, copyright © 1994, with the permission of the publisher, the International Conference of Building Officials.

Table B-9. Allowable Parallel-to-Grain Compressive Stress for End Grain Bearing, F_g (psi).

Species	Wet service	Dry service		
		Sawn lumber @ MC ≤ 19%		
		≤4″ thick	>4″ thick	Glulams
Douglas fir-larch	1340	1480	2020	2350
Douglas fir-south	1220	1340	1820	2130
Eastern hemlock	1140	1260	1710	2000
Eastern hemlock-tamarack	1150	1270	1730	2020
Eastern softwoods	890	980	1340	1560
Engelmann spruce-alpine fir	810	890	1220	1420
Hem-fir	1110	1220	1670	1940
Lodgepole pine	970	1060	1450	1690
Red oak	1060	1160	1590	1850
Redwood, close grain	1560	1720	2270	2620
Redwood, open grain	1150	1270	1670	1940
Sitka spruce	990	1090	1480	1730
Southern pine	1320	1450	1970	2300
Spruce-pine-fir	940	1040	1410	1650
Western cedars	1040	1140	1820	1750

Source: Reproduced from the 1994 edition of the *Uniform Building Code*™, copyright © 1994, with the permission of the publisher, the International Conference of Building Officials.

Table B-10. Reference Strengths for Douglas Fir-Larch Timbers 5 in. × 5 in. or Larger (in kips per square inch (ksi)).

(Tabulated design values are for normal load duration and dry service conditions, unless specified otherwise.)

Species and Commercial Grade	Size Classification	Bending F_b	Tension Parallel to grain F_t	Shear Parallel to Grain F_v	Compression Perpendicular to Grain $F_{c\perp}$	Compression Parallel to Grain F_c	Grading Rules Agency
		$\phi = 0.85$	$\phi = 0.80$	$\phi = 0.75$	$\phi = 0.90$	$\phi = 0.90$	
Dense select structural	Beams and stringers	4.70	3.00	0.24	1.40	3.12	WWPA
Select structural		4.07	2.57	0.24	1.20	2.64	
Dense No. 1		3.94	2.09	0.24	1.40	2.64	
No. 1		3.43	1.82	0.24	1.20	2.22	
Dense No. 2		2.54	1.35	0.24	1.40	1.68	
No. 2		2.22	1.15	0.24	1.20	1.44	
Dense select structural	Posts and timbers	4.45	3.11	0.24	1.40	3.24	WWPA
Select structural		3.81	2.70	0.24	1.20	2.76	
Dense No. 1		3.56	2.57	0.24	1.40	2.88	
No. 1		3.05	2.23	0.24	1.20	2.40	
Dense No. 2		2.03	1.49	0.24	1.40	1.32	
No. 2		1.78	1.28	0.24	1.20	1.14	

Notes:

1. The reference strengths above are for use with the LRFD method as outlined in *Standard for LRFD for Engineered Wood Construction.*
2. The reference strengths were obtained by multiplying the ASD design values from Table B-5 by $2.16/\phi$ where ϕ is the resistance factor for the stress property being obtained. To obtain the $F_{c\perp}$ reference strengths an additional multiplier of 0.80 was used to convert to pre-1982 values. (See *Standard for LRFD for Engineered Wood Construction*, p. 73.)
3. These values must be multiplied by the product of all applicable adjustment factors.

Appendix C

Table C-1. Specific Gravity of Species.

Species	Specific Gravity
Douglas fir-larch	0.50
Douglas fir-larch (north)	0.49
Douglas fir-south	0.46
Eastern hemlock	0.41
Eastern hemlock-tamarack	0.41
Eastern softwoods	0.36
Engelmann spruce-lodgepole pine (MSR only, 1650f and higher grades)	0.46
Engelmann spruce-lodgepole pine (MSR only, 1500f and lower grades)	0.38
Hem-fir	0.43
Mixed southern pine	0.51
Red oak	0.67
Redwood, close grain	0.44
Redwood, open grain	0.37
Sitka spruce	0.43
Southern pine	0.55
Spruce-pine-fir	0.42
Western cedars	0.36

Specific Gravity based on weight and volume when oven dry.
Source: Reproduced from the 1994 edition of the *Uniform Building Code*™, copyright © 1994, with the permission of the publisher, the International Conference of Building Officials.

Table C-2. Wet Service Factors, C_M, for Connections.

Fastener Type	Condition of Wood[1] At Time of Fabrication	Condition of Wood[1] In Service	Wet Service Factor C_M
Split ring or shear plate[2] connectors	Dry	Dry	1.0
	Partially seasoned	Dry	see Footnote 3
	Wet	Dry	0.8
	Dry or wet	Partially seasoned or wet	0.67
Bolts or lag screws	Dry	Dry	1.0
	Partially seasoned or wet	Dry	see Footnote 4
	Dry or wet	Exposed to weather	0.75
	Dry or wet	Wet	0.67
Laterally loaded drift bolts or drift pins	Dry or wet	Dry	1.0
	Dry or wet	Partially seasoned or wet or subject to wetting and drying	0.7
Wood screws	Dry or wet	Dry	1.0
	Dry or wet	Exposed to weather	0.75
	Dry or wet	Wet	0.67
Common wire nails, box nails or common wire spikes			
—Withdrawal loads	Dry	Dry	1.0
	Partially seasoned or wet	Wet	1.0
	Partially seasoned or wet	Dry	0.25
	Dry	Subject to wetting and drying	0.25
—Lateral loads	Dry	Dry	1.0
	Partially seasoned or wet	Dry or wet	0.75
	Dry	Partially seasoned or wet	0.75
Threaded hardened steel nails	Dry or wet	Dry or wet	1.0
Metal connector plates	Dry	Dry	1.0
	Partially seasoned or wet	Dry or wet	0.8

[1] Conditions of wood are defined as follows for determining wet service factors for connections:

Dry wood has a moisture content ≤ 19 percent.
Wet wood has a moisture content ≥ 30 percent (approximate fiber saturation point).
Partially seasoned wood has 19 percent < moisture content < 30 percent.
"Exposed to weather" means that the wood will vary in moisture content from dry to partially seasoned, but is not expected to reach the fiber saturation point at times when the connection is supporting full design load.
"Subject to wetting and drying" means that the wood will vary in moisture content from dry to partially seasoned or wet, or vice versa, with consequent effects on the tightness of the connection.

[2] For split ring or shear plate connectors, moisture content limitations apply to a depth of 3/4 inch (19 mm) below the surface of the wood.

[3] When split ring or shear plate connectors are installed in wood that is partially seasoned at the time of fabrication, but that will be dry before full design load is applied, proportional intermediate wet service factors shall be permitted to be used.

[4] When bolts or lag screws are installed in wood that is wet at the time of fabrication, but that will be dry before full design load is applied, the following wet service factors, C_M, shall apply:

Arrangement of bolts or lag screws	C_M
One fastener only, or Two or more fasteners placed in a single row parallel to grain, or Fasteners placed in two or more rows parallel to grain with separate splice plates for each row	1.0
All other arrangements	0.4

When bolts or lag screws are installed in wood that is partially seasoned at the time of fabrication, but that will be dry before full design load is applied, proportional intermediate wet service factors shall be permitted to be used.

Source: Reproduced from the 1994 edition of the *Uniform Building Code*™, copyright © 1994, with the permission of the publisher, the International Conference of Building Officials.

Table C-3. Nail Design Values (Z) for Single-Shear Connections of Two Wood Members of Identical Species.[1,2]

Part A—Common Wire Nails

Side Member Thickness t_s inches	Nail Length L inches	Nail Diameter D inches	Pennyweight	$G = 0.50$ Douglas Fir-Larch Z lbs.
1/2	2	0.113	6d	59
	2 1/2	0.131	8d	76
	3	0.148	10d	90
	3 1/4	0.148	12d	90
	3 1/2	0.162	16d	105
	4	0.192	20d	124
	4 1/2	0.207	30d	134
	5	0.225	40d	147
	5 1/2	0.244	50d	151
	6	0.263	60d	171
5/8	2	0.113	6d	66
	2 1/2	0.131	8d	82
	3	0.148	10d	97
	3 1/4	0.148	12d	97
	3 1/2	0.162	16d	112
	4	0.192	20d	130
	4 1/2	0.207	30d	140
	5	0.225	40d	151
	5 1/2	0.244	50d	155
	6	0.263	60d	175
3/4	2 1/2	0.131	8d	90
	3	0.148	10d	105
	3 1/4	0.148	12d	105
	3 1/2	0.162	16d	121
	4	0.192	20d	138
	4 1/2	0.207	30d	147
	5	0.225	40d	158
	5 1/2	0.244	50d	162
	6	0.263	60d	181
1	3	0.148	10d	118
	3 1/4	0.148	12d	118
	3 1/2	0.162	16d	141
	4	0.192	20d	159
	4 1/2	0.207	30d	167
	5	0.225	40d	177
	5 1/2	0.244	50d	181
	6	0.263	60d	199
1 1/4	3 1/4	0.148	12d	118
	3 1/2	0.162	16d	141
	4	0.192	20d	170
	4 1/2	0.207	30d	186
	5	0.225	40d	200
	5 1/2	0.244	50d	204
	6	0.263	60d	222

(continued)

Table C-3. (*continued*)

Part A—Common Wire Nails

Side Member Thickness t_s inches	Nail Length L inches	Nail Diameter D inches	Pennyweight	$G = 0.50$ Douglas Fir-Larch Z lbs.
1½	3½	0.162	16d	141
	4	0.192	20d	170
	4½	0.207	30d	186
	5	0.225	40d	205
	5 ½	0.244	50d	211
	6	0.263	60d	240

[1] Tabulated lateral design values (Z) for nailed connections shall be multiplied by all applicable adjustment factors (see UBC Table 23-III-A).

[2] Tabulated lateral design values (Z) are for common wire nails inserted in side grain with nail axis perpendicular to wood fibers, and with the following nail bending yield strength (F_{yb}):

F_{yb} = 100,000 psi (690 N/mm²) for 0.113-inch- (2.9 mm) and 0.131-inch-diameter (3.3 mm) common wire nails.

F_{yb} = 90,000 psi (621 N/mm²) for 0.148-inch- (3.8 mm) and 0.162-inch-diameter (4.1 mm) common wire nails.

F_{yb} = 80,000 psi (552 N/mm²) for 0.192-inch- (4.9 mm), 0.207-inch- (5.3 mm) and 0.225-inch-diameter (5.7 mm) common wire nails.

F_{yb} = 70,000 psi (483 N/mm²) for 0.244-inch- (6.2 mm) and 0.263-inch-diameter (6.7 mm) common wire nails.

Part B—Box Nails

Side Member Thickness t_s inches	Nail Length L inches	Nail Diameter D inches	Pennyweight	$G = 0.50$ Douglas Fir-Larch Z lbs.	$G = 0.43$ Hem-Fir Z lbs.
½	2	0.099	6d	48	39
	2½	0.113	8d	59	49
	3	0.128	10d	73	61
	3¼	0.128	12d	73	61
	3½	0.135	16d	79	66
	4	0.148	20d	90	75
	4½	0.148	30d	90	75
	5	0.162	40d	105	89
⅝	2	0.099	6d	55	44
	2½	0.113	8d	66	53
	3	0.128	10d	79	65
	3¼	0.128	12d	79	65
	3½	0.135	16d	86	71
	4	0.148	20d	97	80
	4½	0.148	30d	97	80
	5	0.162	40d	112	93
¾	2	0.099	6d	55	48
	2½	0.113	8d	72	58
	3	0.128	10d	87	70
	3¼	0.128	12d	87	70
	3½	0.135	16d	94	76
	4	0.148	20d	105	85
	4½	0.148	30d	105	85
	5	0.162	40d	121	99

(continued)

Table C-3. (*continued*)

Part B—Box Nails

Side Member Thickness t_s inches	Nail Length L inches	Nail Diameter D inches	Pennyweight	$G = 0.50$ Douglas Fir-Larch Z lbs.	$G = 0.43$ Hem-Fir Z lbs.
1	2½	0.113	8d	72	63
	3	0.128	10d	93	80
	3¼	0.128	12d	93	80
	3½	0.135	16d	103	89
	4	0.148	20d	118	99
	4½	0.148	30d	118	99
	5	0.162	40d	141	113
1¼	3	0.128	10d	93	80
	3¼	0.128	12d	93	80
	3½	0.135	16d	103	89
	4	0.148	20d	118	102
	4½	0.148	30d	118	102
	5	0.162	40d	141	122
1½	3¼	0.128	12d	93	80
	3½	0.135	16d	103	89
	4	0.148	20d	118	102
	4½	0.148	30d	118	102
	5	0.162	40d	141	122

[1] Tabulated lateral design values (Z) for nailed connections shall be multiplied by all applicable adjustment factors (see UBC Table 23-III-A).

[2] Tabulated lateral design values (Z) are for box nails inserted in side grain with nail axis perpendicular to wood fibers, and with the following nail bending yield strength (F_{yb}):

F_{yb} = 100,000 psi (690 N/mm^2) for 0.099-inch- (2.5 mm), 0.113-inch- (2.9 mm), 0.128-inch- (3.3 mm) and 0.135-inch-diameter (3.4 mm) box nails.

F_{yb} = 90,000 psi (621 N/mm^2) for 0.148-inch- (3.8 mm) and 0.162-inch-diameter (4.1 mm) box nails.

Source: Reproduced from the 1994 edition of the *Uniform Building Code*™, copyright © 1994, with the permission of the publisher, the International Conference of Building Officials.

Table C-4. Common Wire Nail Design Values (Z) for Single-Shear Connections with One Steel Side Plate.[1,2,3]

Steel Side Plate	Nail Length L inches	Nail Diameter D inches	Pennyweight	$G = 0.50$ Douglas Fir-Larch Z lbs.	$G = 0.43$ Hem-Fir Z lbs.
3 gage $t_s = 0.239''$	5½	0.244	50d	245	218
	6	0.263	60d	271	241
7 gage $t_s = 0.179''$	4	0.192	20d	189	168
	4½	0.207	30d	202	179
	5	0.225	40d	217	193
	5½	0.244	50d	223	198
	6	0.263	60d	249	221
10 gage $t_s = 0.134''$	2	0.113	6d	84	75
	2½	0.131	8d	107	95
	3	0.148	10d	127	113
	3¼	0.148	12d	127	113
	3½	0.162	16d	149	132
	4	0.192	20d	174	155
	4½	0.207	30d	188	167
	5	0.225	40d	205	181
	5½	0.244	50d	210	186
	6	0.263	60d	237	210
12 gage $t_s = 0.105''$	2	0.113	6d	78	69
	2½	0.131	8d	100	89
	3	0.148	10d	120	106
	3¼	0.148	12d	120	106
	3½	0.162	16d	141	125
	4	0.192	20d	168	148
	4½	0.207	30d	182	161
	5	0.225	40d	199	176
	5½	0.244	50d	205	181
	6	0.263	60d	232	205
14 gage $t_s = 0.075''$	2	0.113	6d	72	64
	2½	0.131	8d	95	84
	3	0.148	10d	115	101
	3¼	0.148	12d	115	101
	3½	0.162	16d	136	120
	4	0.192	20d	163	144
	4½	0.207	30d	178	157
	5	0.225	40d	196	172
	5½	0.244	50d	202	177
	6	0.263	60d	229	201

(continued)

Table C-4. (*continued*)

Steel Side Plate	Nail Length L inches	Nail Diameter D inches	Pennyweight	$G = 0.50$ Douglas Fir-Larch Z lbs.	$G = 0.43$ Hem-Fir Z lbs.
16 gage $t_s = 0.06''$	2	0.113	6d	70	62
	$2^1/_2$	0.131	8d	94	83
	3	0.148	10d	113	100
	$3^1/_4$	0.148	12d	113	100
	$3^1/_2$	0.162	16d	135	119
	4	0.192	20d	162	142
	$4^1/_2$	0.207	30d	177	156
18 gage $t_s = 0.048''$	2	0.113	6d	69	61
	$2^1/_2$	0.131	8d	93	82
	3	0.148	10d	112	99
	$3^1/_4$	0.148	12d	112	99
	$3^1/_2$	0.162	16d	134	118

[1] Tabulated lateral design values (Z) for nailed connections shall be multiplied by all applicable adjustment factors (see UBC Table 23-III-A).

[2] Tabulated lateral design values (Z) are for common wire nails inserted in side grain with nail axis perpendicular to wood fibers, and with the following nail bending yield strength (F_{yb})

F_{yb} = 100,000 psi (690 N/mm²) for 0.113-inch- (2.9 mm) and 0.131-inch-diameter (3.3 mm) common wire nails.
F_{yb} = 90,000 psi (621 N/mm²) for 0.148-inch- (3.8 mm) and 0.162-inch-diameter (4.1 mm) common wire nails.
F_{yb} = 80,000 psi (552 N/mm²) for 0.192-inch- (4.9 mm), 0.131-inch- (3.3 mm) and 0.225-inch-diameter (5.7 mm) common wire nails.
F_{yb} = 70,000 psi (483 N/mm²) for 0.224-inch- (5.7 mm) and 0.263-inch-diameter (6.7 mm) common wire nails.

[3] Tabulated lateral design values (Z) are based on a dowel-bearing strength (F_e) of 45,000 psi (310 N/mm²) for ASTM A 446, Grade A steel.

Source: Reproduced from the 1994 edition of the *Uniform Building Code*™, copyright © 1994, with the permission of the publisher, the International Conference of Building Officials.

Table C-5. Dimensions of Standard Lag Screws.

D—Unthreaded shank diameter.
D_r—Root diameter of threaded portion.
W—Width of head across flats.
H—Height of head.

S—Unthreaded shank length.
T—Thread length[1].
E—Length of tapered tip.
N—Number of threads per inch (25 mm).

Nominal Length, *L* (inches)		Unthreaded Shank Diameter, *D* (inches)										
× 25.4 for mm		1/4″	5/16″	3/8″	7/16″	1/2″	5/8″	3/4″	7/8″	1″	1 1/8″	1 1/4″
	D_r	0.173	0.227	0.265	0.328	0.371	0.471	0.579	0.683	0.780	0.887	1.012
	E	5/32	3/16	7/32	9/32	5/16	13/32	1/2	19/32	11/16	25/32	7/8
	H	11/64	7/32	1/4	19/64	11/32	27/64	1/2	37/64	43/64	3/4	27/32
	W	7/16	1/2	9/16	5/8	3/4	15/16	1 1/8	1 5/16	1 1/2	1 11/16	1 7/8
	N	10	9	7	7	6	5	4 1/2	4	3 1/2	3 1/4	3 1/4
1	*S*	1/4	1/4	1/4	1/4	1/4						
	T	3/4	3/4	3/4	3/4	3/4						
	T-E	19/32	9/16	17/32	15/32	7/16						
1 1/2	*S*	1/4	1/4	1/4	1/4	1/4						
	T	1 1/4	1 1/4	1 1/4	1 1/4	1 1/4						
	T-E	1 3/32	1 1/16	1 1/32	31/32	15/16						
2	*S*	1/2	1/2	1/2	1/2	1/2	1/2					
	T	1 1/2	1 1/2	1 1/2	1 1/2	1 1/2	1 1/2					
	T-E	1 11/32	1 5/16	1 9/32	1 7/32	1 3/16	1 3/32					
2 1/2	*S*	3/4	3/4	3/4	3/4	3/4	3/4					
	T	1 3/4	1 3/4	1 3/4	1 3/4	1 3/4	1 3/4					
	T-E	1 19/32	1 9/16	1 17/32	1 15/32	1 7/16	1 11/32					
3	*S*	1	1	1	1	1	1	1	1	1		
	T	2	2	2	2	2	2	2	2	2		
	T-E	1 27/32	1 13/16	1 25/32	1 23/32	1 11/16	1 19/32	1 1/2	1 13/32	1 5/16		
4	*S*	1 1/2	1 1/2	1 1/2	1 1/2	1 1/2	1 1/2	1 1/2	1 1/2	1 1/2	1 1/2	1 1/2
	T	2 1/2	2 1/2	2 1/2	2 1/2	2 1/2	2 1/2	2 1/2	2 1/2	2 1/2	2 1/2	2 1/2
	T-E	2 11/32	2 5/16	2 9/32	2 7/32	2 3/16	2 3/32	2	1 29/32	1 13/16	1 23/32	1 5/8

(continued)

Table C-5. *(continued)*

D—Unthreaded shank diameter.
D_r—Root diameter of threaded portion.
W—Width of head across flats.
H—Height of head.

S—Unthreaded shank length.
T—Thread length[1].
E—Length of tapered tip.
N—Number of threads per inch (25 mm).

Nominal Length, *L* (inches)		Unthreaded Shank Diameter, *D* (inches)										
× 25.4 for mm		1/4″	5/16″	3/8″	7/16″	1/2″	5/8″	3/4″	7/8″	1″	1 1/8″	1 1/4″
5	*S*	2	2	2	2	2	2	2	2	2	2	2
	T	3	3	3	3	3	3	3	3	3	3	3
	T-E	2 27/32	2 13/16	2 25/32	2 23/32	2 11/16	2 19/32	2 1/2	2 13/32	2 5/16	2 7/32	2 1/8
6	*S*	2 1/2	2 1/2	2 1/2	2 1/2	2 1/2	2 1/2	2 1/2	2 1/2	2 1/2	2 1/2	2 1/2
	T	3 1/2	3 1/2	3 1/2	3 1/2	3 1/2	3 1/2	3 1/2	3 1/2	3 1/2	3 1/2	3 1/2
	T-E	3 11/32	3 5/16	3 9/32	3 7/32	3 3/16	3 3/32	3	2 29/32	2 13/16	2 23/32	2 5/8
7	*S*	3	3	3	3	3	3	3	3	3	3	3
	T	4	4	4	4	4	4	4	4	4	4	4
	T-E	3 27/32	3 13/16	3 25/32	3 23/32	3 11/16	3 19/32	3 1/2	3 13/32	3 5/16	3 7/32	3 1/8
8	*S*	3 1/2	3 1/2	3 1/2	3 1/2	3 1/2	3 1/2	3 1/2	3 1/2	3 1/2	3 1/2	3 1/2
	T	4 1/2	4 1/2	4 1/2	4 1/2	4 1/2	4 1/2	4 1/2	4 1/2	4 1/2	4 1/2	4 1/2
	T-E	4 11/32	4 5/16	4 9/32	4 7/32	4 3/16	4 3/32	4	3 29/32	3 13/16	3 23/32	3 5/8
9	*S*	4	4	4	4	4	4	4	4	4	4	4
	T	5	5	5	5	5	5	5	5	5	5	5
	T-E	4 27/32	4 13/16	4 25/32	4 23/32	4 11/16	4 19/32	4 1/2	4 13/32	4 5/16	4 7/32	4 1/8
10	*S*	4 1/2	4 1/2	4 1/2	4 1/2	4 1/2	4 1/2	4 1/2	4 1/2	4 1/2	4 1/2	4 1/2
	T	5 1/2	5 1/2	5 1/2	5 1/2	5 1/2	5 1/2	5 1/2	5 1/2	5 1/2	5 1/2	5 1/2
	T-E	5 11/32	5 5/16	5 9/32	5 7/32	5 3/16	5 3/32	5	4 29/32	4 13/16	4 23/32	4 5/8
11	*S*	5	5	5	5	5	5	5	5	5	5	5
	T	6	6	6	6	6	6	6	6	6	6	6
	T-E	5 27/32	5 13/16	5 25/32	5 23/32	5 11/16	5 19/32	5 1/2	5 13/32	5 5/16	5 7/32	5 1/8
12	*S*	6	6	6	6	6	6	6	6	6	6	6
	T	6	6	6	6	6	6	6	6	6	6	6
	T-E	5 27/32	5 13/16	5 25/32	5 23/32	5 11/16	5 19/32	5 1/2	5 13/32	5 5/16	5 7/32	5 1/8

[1] Thread length (*T*) for intermediate nominal lag screw lengths (*L*) is 6 inches (152 mm), or one half the nominal lag screw length plus 0.5 inch (13 mm), whichever is less.

Source: Reproduced from the 1994 edition of the *Uniform Building Code*™, copyright © 1994, with the permission of the publisher, the International Conference of Building Officials.

Table C-6. Lag Screw Design Values (Z) for Single-Shear Connections with Both Members of Identical Species.[1,2]

Side Member Thickness t_s inches	Lag Screw Diameter D inches	$G = 0.50$ Douglas Fir-Larch			$G = 0.43$ Hem-Fir		
		$Z_{\parallel}$ lbs.	$Z_{s\perp}$ lbs.	$Z_{m\perp}$ lbs.	$Z_{\parallel}$ lbs.	$Z_{s\perp}$ lbs.	$Z_{m\perp}$ lbs.
3/4	1/4	180	130	140	160	110	120
	5/16	250	170	190	230	150	170
	3/8	310	210	220	280	170	200
1 1/2	1/4	220	170	170	210	150	150
	5/16	320	230	240	300	200	210
	3/8	400	270	290	370	230	260
	7/16	540	320	380	480	270	340
	1/2	660	380	460	580	330	400
	5/8	920	530	620	820	420	550
	3/4	1,240	590	820	1,120	460	720
	7/8	1,620	630	1,040	1,470	500	920
	1	2,060	680	1,290	1,800	540	1,150
	1 1/8	2,360	710	1,560	2,030	570	1,400
	1 1/4	2,630	750	1,860	2,250	600	1,670
2 1/2	1/4	220	170	170	210	150	150
	5/16	320	240	240	300	210	210
	3/8	400	290	290	370	260	260
	7/16	550	380	380	510	340	340
	1/2	710	480	480	660	420	440
	5/8	1,120	630	730	1,030	540	660
	3/4	1,570	800	1,020	1,390	690	910
	7/8	1,950	1,000	1,280	1,740	830	1,110
	1	2,390	1,130	1,530	2,140	900	1,340
	1 1/8	2,880	1,180	1,800	2,600	960	1,590
	1 1/4	3,430	1,250	2,100	3,110	1,000	1,850

[1] Tabulated lateral design values (Z) for lag screw connections shall be multiplied by all applicable adjustment factors (see UBC Table 23-III-A).
[2] Tabulated lateral design values (Z) are for "full diameter" lag screws inserted in side grain with lag screw axis perpendicular to wood fibers, and with the following lag screw bending yield strengths (F_{yb}):
F_{yb} = 70,000 psi (482 N/mm^2) for D = 1/4 inch (6.4 mm)
F_{yb} = 60,000 psi (413 N/mm^2) for D = 5/16 inch (7.9 mm)
F_{yb} = 45,000 psi (310 N/mm^2) for $D \geq$ 3/8 inch (9.5 mm)

Source: Reproduced from the 1994 edition of the *Uniform Building Code*™, copyright © 1994, with the permission of the publisher, the International Conference of Building Officials.

Table C-7. Lag Screw Design Values (Z) for Single-Shear Connections with 1/4-in. A36 Steel Side Plate.[1,2,3]

Steel Side Plate t_s inches	Lag Screw Diameter D inches	$G = 0.50$ Douglas Fir-Larch $Z_{\parallel}$ lbs.	$Z_{\perp}$ lbs.	$G = 0.43$ Hem-Fir $Z_{\parallel}$ lbs.	$Z_{\perp}$ lbs.
	1/4	300	220	280	200
	5/16	400	280	370	250
	3/8	490	330	460	300
	7/16	620	480	580	370
	1/2	780	560	730	440
1/4	5/8	1,140	790	1,070	610
	3/4	1,600	1,060	1,490	820
	7/8	2,130	1,640	1,990	1,050
	1	2,750	1,990	2,570	1,310
	1 1/8	3,460	2,380	3,230	1,600
	1 1/4	4,260	2,800	3,970	1,910

[1] Tabulated lateral design values (Z) of lag screw connections shall be multiplied by all applicable adjustment factors (see UBC Table 23-III-A).

[2] Tabulated lateral design values (Z) are for "full diameter" lag screws inserted in side grain with lag screw axis perpendicular to wood fibers, and with the following lag screw bending yield strengths (F_{yb}).
$F_{yb} = 70{,}000$ psi (482 N/mm²) for D = 1/4 inch (6.4 mm)
$F_{yb} = 60{,}000$ psi (413 N/mm²) for D = 5/16 inch (7.9 mm)
$F_{yb} = 45{,}000$ psi (310 N/mm²) for D ≥ 3/8 inch (9.5 mm)

[3] Tabulated lateral design values (Z) are based on dowel-bearing strength (F_e) of 58,000 psi (400 N/mm²) for ASTM A 36 steel and 45,000 psi (310 N/mm²) for ASTM A 446 Grade A steel.
Source: Reproduced from the 1994 edition of the *Uniform Building Code*™, copyright © 1994, with the permission of the publisher, the International Conference of Building Officials.

Table C-8. Group-Action Factors, C_g, for Bolt or Lag Screw Connections with Wood Side Members.[2]

A_s/A_m^1	A_s^1 in.²	Number of Fasteners in a Row: 2	3	4	5	6	7	8	9	10	11	12
	5	0.98	0.92	0.84	0.75	0.68	0.61	0.55	0.50	0.45	0.41	0.38
	12	0.99	0.96	0.92	0.87	0.81	0.76	0.70	0.65	0.61	0.57	0.53
0.5	20	0.99	0.98	0.95	0.91	0.87	0.83	0.78	0.74	0.70	0.66	0.62
	28	1.00	0.98	0.96	0.93	0.90	0.87	0.83	0.79	0.76	0.72	0.69
	40	1.00	0.99	0.97	0.95	0.93	0.90	0.87	0.84	0.81	0.78	0.75
	64	1.00	0.99	0.98	0.97	0.95	0.93	0.91	0.89	0.87	0.84	0.82
	5	1.00	0.97	0.91	0.85	0.78	0.71	0.64	0.59	0.54	0.49	0.45
	12	1.00	0.99	0.96	0.93	0.88	0.84	0.79	0.74	0.70	0.65	0.61
1	20	1.00	0.99	0.98	0.95	0.92	0.89	0.86	0.82	0.78	0.75	0.71
	28	1.00	0.99	0.98	0.97	0.94	0.92	0.89	0.86	0.83	0.80	0.77
	40	1.00	1.00	0.99	0.98	0.96	0.94	0.92	0.90	0.87	0.85	0.82
	64	1.00	1.00	0.99	0.98	0.97	0.96	0.95	0.93	0.91	0.90	0.88

[1] When $A_s/A_m > 1.0$, use A_m/A_s and use A_m instead of A_s.

[2] Tabulated group action factors (C_g) are conservative for $D < 1$ inch (25 mm), $s < 4$ inches (102 mm) or $E > 1{,}400{,}000$ psi (9646 N/mm²).
Source: Reproduced from the 1994 edition of the *Uniform Building Code*™, copyright © 1994, with the permission of the publisher, the International Conference of Building Officials.

Table C-9. Group-Action Factors, C_g, for Bolt or Lag Screw Connections with Steel Side Plates.[1]

		Number of Fasteners in a Row										
A_m/A_s	A_m in.^{2}	2	3	4	5	6	7	8	9	10	11	12
12	5	0.97	0.89	0.80	0.70	0.62	0.55	0.49	0.44	0.40	0.37	0.34
	8	0.98	0.93	0.85	0.77	0.70	0.63	0.57	0.52	0.47	0.43	0.40
	16	0.99	0.96	0.92	0.86	0.80	0.75	0.69	0.64	0.60	0.55	0.52
	24	0.99	0.97	0.94	0.90	0.85	0.81	0.76	0.71	0.67	0.63	0.59
	40	1.00	0.98	0.96	0.94	0.90	0.87	0.83	0.79	0.76	0.72	0.69
	64	1.00	0.99	0.98	0.96	0.94	0.91	0.88	0.86	0.83	0.80	0.77
	120	1.00	0.99	0.99	0.98	0.96	0.95	0.93	0.91	0.90	0.87	0.85
	200	1.00	1.00	0.99	0.99	0.98	0.97	0.96	0.95	0.93	0.92	0.90
18	5	0.99	0.93	0.85	0.76	0.68	0.61	0.54	0.49	0.44	0.41	0.37
	8	0.99	0.95	0.90	0.83	0.75	0.69	0.62	0.57	0.52	0.48	0.44
	16	1.00	0.98	0.94	0.90	0.85	0.79	0.74	0.69	0.65	0.60	0.56
	24	1.00	0.98	0.96	0.93	0.89	0.85	0.80	0.76	0.72	0.68	0.64
	40	1.00	0.99	0.97	0.95	0.93	0.90	0.87	0.83	0.80	0.77	0.73
	64	1.00	0.99	0.98	0.97	0.95	0.93	0.91	0.89	0.86	0.83	0.81
	120	1.00	1.00	0.99	0.98	0.97	0.96	0.95	0.93	0.92	0.90	0.88
	200	1.00	1.00	0.99	0.99	0.98	0.98	0.97	0.96	0.95	0.94	0.92
24	40	1.00	0.99	0.97	0.95	0.93	0.89	0.86	0.83	0.79	0.76	0.72
	64	1.00	0.99	0.98	0.97	0.95	0.93	0.91	0.88	0.85	0.83	0.80
	120	1.00	1.00	0.99	0.98	0.97	0.96	0.95	0.93	0.91	0.90	0.88
	200	1.00	1.00	0.99	0.99	0.98	0.98	0.97	0.96	0.95	0.93	0.92
30	40	1.00	0.98	0.96	0.93	0.89	0.85	0.81	0.77	0.73	0.69	0.65
	64	1.00	0.99	0.97	0.95	0.93	0.90	0.87	0.83	0.80	0.77	0.73
	120	1.00	0.99	0.99	0.97	0.96	0.94	0.92	0.90	0.88	0.85	0.83
	200	1.00	1.00	0.99	0.98	0.97	0.96	0.95	0.94	0.92	0.90	0.89
35	40	0.99	0.97	0.94	0.91	0.86	0.82	0.77	0.73	0.68	0.64	0.60
	64	1.00	0.98	0.96	0.94	0.91	0.87	0.84	0.80	0.76	0.73	0.69
	120	1.00	0.99	0.98	0.97	0.95	0.92	0.90	0.88	0.85	0.82	0.79
	200	1.00	0.99	0.99	0.98	0.97	0.95	0.94	0.92	0.90	0.88	0.86

[1] Tabulated group action factors (C_g) are conservative for $D < 1$ inch (25 mm) or $s < 4$ inches (102 mm).

Source: Reproduced from the 1994 edition of the *Uniform Building Code*™, copyright © 1994, with the permission of the publisher, the International Conference of Building Officials.

Table C-10. Bolt Design Values (Z) for Single-Shear Connections of Two Wood Members of Identical Species.[1,2]

Thickness			$G = 0.50$ Douglas Fir-Larch			$G = 0.43$ Hem-Fir		
Main Member t_m inches	Side Member t_s inches	Bolt Diameter D inches	$Z_{\parallel}$ lbs.	$Z_{s\perp}$ lbs.	$Z_{m\perp}$ lbs.	$Z_{\parallel}$ lbs.	$Z_{s\perp}$ lbs.	$Z_{m\perp}$ lbs.
1 1/2	1 1/2	1/2	480	300	300	410	250	250
		5/8	600	360	360	520	300	300
		3/4	720	420	420	620	350	350
		7/8	850	470	470	720	390	390
		1	970	530	530	830	440	440
2 1/2	1 1/2	1/2	610	370	370	550	320	310
		5/8	850	520	430	730	420	360
		3/4	1,020	590	500	870	460	410
		7/8	1,190	630	550	1,020	500	450
		1	1,360	680	610	1,160	540	500
3	1 1/2	1/2	610	370	420	550	320	350
		5/8	880	520	480	790	420	400
		3/4	1,190	590	550	1,020	460	450
		7/8	1,390	630	610	1,190	500	500
		1	1,590	680	670	1,360	540	550
3 1/2	1 1/2	1/2	610	370	430	550	320	380
		5/8	880	520	540	790	420	440
		3/4	1,200	590	610	1,100	460	500
		7/8	1,590	630	680	1,370	500	550
		1	1,830	680	740	1,570	540	600
	3 1/2	1/2	720	490	490	660	440	440
		5/8	1,120	700	700	1,040	600	600
		3/4	1,610	870	870	1,450	740	740
		7/8	1,970	1,060	1,060	1,690	910	910
		1	2,260	1,230	1,230	1,930	1,030	1,030
5 1/2	1 1/2	5/8	880	520	590	790	420	530
		3/4	1,200	590	790	1,100	460	700
		7/8	1,590	630	980	1,460	500	780
		1	2,050	680	1,060	1,800	540	860
	3 1/2	5/8	1,120	700	730	1,040	600	660
		3/4	1,610	870	1,030	1,490	740	920
		7/8	2,190	1,060	1,260	1,950	920	1,030
		1	2,660	1,290	1,390	2,370	1,140	1,150
7 1/2	1 1/2	5/8	880	520	590	790	420	530
		3/4	1,200	590	790	1,100	460	700
		7/8	1,590	630	1,010	1,460	500	900
		1	2,050	680	1,270	1,800	540	1,130
	3 1/2	5/8	1,120	700	730	1,040	600	660
		3/4	1,610	870	1,030	1,490	740	920
		7/8	2,190	1,060	1,360	1,950	920	1,210
		1	2,660	1,290	1,630	2,370	1,140	1,340

[1] Tabulated lateral design values (Z) for bolted connections shall be multiplied by all applicable adjustment factors (see UBC Table 23-III-A).

[2] Tabulated lateral design values (Z) are for "full diameter" bolts with a bending yield strength (F_{yb}) of 45,000 psi (310 N/mm^2).

Source: Reproduced from the 1994 edition of the *Uniform Building Code*™, copyright © 1994, with the permission of the publisher, the International Conference of Building Officials.

Table C-11. Bolt Design Values (*Z*) for Single-Shear Connections of One Wood Member to One 1/4″ Steel Side Plate.[1,2,3]

Thickness			G = 0.50 Douglas Fir-Larch		G = 0.43 Hem-Fir	
Main Member t_m inches	Steel Side Plate t_s inches	Bolt Diameter *D* inches	$Z_{\parallel}$ lbs.	$Z_{\perp}$ lbs.	$Z_{\parallel}$ lbs.	$Z_{\perp}$ lbs.
		1/2	530	270	470	240
		5/8	660	320	590	270
1 1/2	1/4	3/4	800	360	700	310
		7/8	930	400	820	340
		1	1,060	440	940	380
		1/2	750	390	700	320
		5/8	1,010	440	880	370
2 1/2	1/4	3/4	1,210	490	1,050	410
		7/8	1,410	540	1,230	450
		1	1,620	590	1,410	490
		1/2	750	450	710	370
		5/8	1,130	510	1,040	420
3	1/4	3/4	1,430	570	1,240	470
		7/8	1,670	620	1,450	510
		1	1,910	670	1,660	560
		1/2	750	470	710	430
		5/8	1,130	580	1,050	480
3 1/2	1/4	3/4	1,580	650	1,440	530
		7/8	1,940	710	1,680	570
		1	2,210	760	1,910	630
		5/8	1,130	660	1,050	600
4 1/2	1/4	3/4	1,580	820	1,480	660
		7/8	2,130	880	1,990	710
		1	2,760	950	2,440	770
		5/8	1,130	660	1,050	600
5 1/2	1/4	3/4	1,580	900	1,480	790
		7/8	2,130	1,070	1,990	860
		1	2,760	1,150	2,580	930
		5/8	1,130	660	1,050	600
7 1/2	1/4	3/4	1,580	900	1,480	810
		7/8	2,130	1,160	1,990	1,040
		1	2,760	1,460	2,580	1,250
		3/4	1,580	900	1,480	810
9 1/2	1/4	7/8	2,130	1,160	1,990	1,040
		1	2,760	1,460	2,580	1,310
11 1/2	1/4	7/8	2,130	1,160	1,990	1,040
		1	2,760	1,460	2,580	1,310
13 1/2	1/4	1	2,760	1,460	2,580	1,310

[1] Tabulated lateral design values (*Z*) for bolted connections shall be multiplied by all applicable adjustment factors (see UBC Table 23-III-A).

[2] Tabulated lateral design values (*Z*) are for "full diameter" bolts with a bending yield strength (F_{yb}) of 45,000 psi (310 N/mm²).

[3] Tabulated lateral design values (*Z*) are based on a dowel-bearing strength (F_e) of 58,000 psi (400 N/mm²) for ASTM A 36 steel.

Source: Reproduced from the 1994 edition of the *Uniform Building Code*™, copyright © 1994, with the permission of the publisher, the International Conference of Building Officials.

Table C-12. Bolt Design Values (Z) for Double-Shear Connections of Three Wood Members of Identical Species.[1,2]

Thickness			$G = 0.50$ Douglas Fir-Larch			$G = 0.43$ Hem-Fir		
Main Member t_m inches	Side Member t_s inches	Bolt Diameter D inches	$Z_{\parallel}$ lbs.	$Z_{s\perp}$ lbs.	$Z_{m\perp}$ lbs.	$Z_{\parallel}$ lbs.	$Z_{s\perp}$ lbs.	$Z_{m\perp}$ lbs.
1½	1½	½	1,050	730	470	900	650	380
		⅝	1,310	1,040	530	1,130	840	420
		¾	1,580	1,170	590	1,350	920	460
		⅞	1,840	1,260	630	1,580	1,000	500
		1	2,100	1,350	680	1,800	1,080	540
2½	1½	½	1,230	730	790	1,100	650	640
		⅝	1,760	1,040	880	1,590	840	700
		¾	2,400	1,170	980	2,190	920	770
		⅞	3,060	1,260	1,050	2,630	1,000	830
		1	3,500	1,350	1,130	3,000	1,080	900
3½	1½	½	1,230	730	860	1,100	650	760
		⅝	1,760	1,040	1,190	1,590	840	980
		¾	2,400	1,170	1,370	2,190	920	1,080
		⅞	3,180	1,260	1,470	2,920	1,000	1,160
		1	4,090	1,350	1,580	3,600	1,080	1,260
	3½	½	1,430	970	970	1,330	880	880
		⅝	2,240	1,410	1,230	2,070	1,190	980
		¾	3,220	1,750	1,370	2,980	1,490	1,080
		⅞	4,290	2,130	1,470	3,680	1,840	1,160
		1	4,900	2,580	1,580	4,200	2,280	1,260

[1] Tabulated lateral design values (Z) for bolted connections shall be multiplied by all applicable adjustment factors (see UBC Table 23-III-A).

[2] Tabulated lateral design values (Z) are for "full diameter" bolts with a bending yield strength (F_{yb}) of 45,000 psi (310 N/mm²).

Source: Reproduced from the 1994 edition of the *Uniform Building Code*™, copyright © 1994, with the permission of the publisher, the International Conference of Building Officials.

Table C-13. Bolt Design Values (Z) for Double-Shear Connections of One Wood Member to Two 1/4″ (A36) Steel Side Plates.[1,2,3]

Thickness			$G = 0.50$ Douglas Fir-Larch		$G = 0.43$ Hem-Fir	
Main Member t_m inches	Steel Side Plate t_s inches	Bolt Diameter D inches	$Z_{\parallel}$ lbs.	$Z_{\perp}$ lbs.	$Z_{\parallel}$ lbs.	$Z_{\perp}$ lbs.
		1/2	1,050	470	900	380
		5/8	1,310	530	1,130	420
1 1/2	1/4	3/4	1,580	590	1,350	460
		7/8	1,840	630	1,580	500
		1	2,100	680	1,800	540
		1/2	1,510	790	1,410	640
		5/8	2,190	880	1,880	700
2 1/2	1/4	3/4	2,630	980	2,250	770
		7/8	3,060	1,050	2,630	830
		1	3,500	1,130	3,000	900
		1/2	1,510	940	1,410	860
		5/8	2,250	1,230	2,110	980
3 1/2	1/4	3/4	3,170	1,370	2,960	1,080
		7/8	4,260	1,470	3,680	1,160
		1	4,900	1,580	4,200	1,260

[1] Tabulated lateral design values (Z) for bolted connections shall be multiplied by all applicable adjustment factors (see UBC Table 23-III-A).

[2] Tabulated lateral design values (Z) are for "full diameter" bolts with a bending yield strength (F_{yb}) of 45,000 psi (310 N/mm^2).

[3] Tabulated lateral design values (Z) are based on a dowel-bearing strength (F_e) of 58,000 psi (400 N/mm^2) for ASTM A 36 steel.

Source: Reproduced from the 1994 edition of the *Uniform Building Code*™, copyright © 1994, with the permission of the publisher, the International Conference of Building Officials.

Table C-14. Wood Screw Design Values (Z) for Single-Shear Connections of Two Wood Members of Identical Species.[1,2]

Side Member Thickness t_s inches	Wood Screw Diameter D inches	Wood Screw Gage	$G = 0.50$ Douglas Fir-Larch Z lbs.	$G = 0.43$ Hem-Fir Z lbs.
	0.138	6d	92	73
	0.151	7d	102	82
	0.164	8d	114	92
	0.177	9d	124	101
	0.190	10d	127	103
3/4	0.216	12d	140	115
	0242	14d	146	121
	0.268	16d	169	141
	0.294	18d	187	156
	0.320	20d	216	182
	0.372	24d	252	212
	0.138	6d	101	87
	0.151	7d	115	99
	0.164	8d	135	117
	0.177	9d	152	132
	0.190	10d	157	136
1 1/2	0.216	12d	183	159
	0.242	14d	195	170
	0.268	16d	233	192
	0.294	18d	260	212
	0.320	20d	300	238
	0.372	24d	349	277

[1] Tabulated lateral design values (Z) for wood screw connections shall be multiplied by all applicable adjustment factors (see UBC Table 23-III-A).

[2] Tabulated lateral design values (Z) are for cut thread wood screws inserted in side grain with wood screw axis perpendicular to wood fibers, and with the following wood screw bending yield strength (F_{yb}):

F_{yb} = 100,000 psi (690 N/mm²) for 6d wood screws.
F_{yb} = 90,000 psi (621 N/mm²) for 7d, 8d, and 9d wood screws.
F_{yb} = 80,000 psi (552 N/mm²) for 10d and 12d wood screws.
F_{yb} = 70,000 psi (483 N/mm²) for 14d and 16d wood screws.
F_{yb} = 60,000 psi (414 N/mm²) for 18d and 20d wood screws.
F_{yb} = 45,000 psi (310 N/mm²) for 24d wood screws.

Source: Reproduced from the 1994 edition of the *Uniform Building Code*™, copyright © 1994, with the permission of the publisher, the International Conference of Building Officials.

Table C-15. Group-Action Factors, C_g, for 4-in. Split-Ring or Shear-Plate Connectors with Wood Side Members.[1,2]

		Number of Fasteners in a Row										
A_s/A_m^2	A_s^1 in.^{2}	2	3	4	5	6	7	8	9	10	11	12
0.5	5	0.90	0.73	0.59	0.48	0.41	0.35	0.31	0.27	0.25	0.22	0.20
	12	0.95	0.83	0.71	0.60	0.52	0.45	0.40	0.36	0.32	0.29	0.27
	20	0.97	0.88	0.78	0.69	0.60	0.53	0.47	0.43	0.39	0.35	0.32
	28	0.97	0.91	0.82	0.74	0.66	0.59	0.53	0.48	0.44	0.40	0.37
	40	0.98	0.93	0.86	0.79	0.72	0.65	0.59	0.54	0.49	0.45	0.42
	64	0.99	0.95	0.91	0.85	0.79	0.73	0.67	0.62	0.58	0.54	0.50
1	5	1.00	0.87	0.72	0.59	0.50	0.43	0.38	0.34	0.30	0.28	0.25
	12	1.00	0.93	0.83	0.72	0.63	0.55	0.48	0.43	0.39	0.36	0.33
	20	1.00	0.95	0.88	0.79	0.71	0.63	0.57	0.51	0.46	0.42	0.39
	28	1.00	0.97	0.91	0.83	0.76	0.69	0.62	0.57	0.52	0.47	0.44
	40	1.00	0.98	0.93	0.87	0.81	0.75	0.69	0.63	0.58	0.54	0.50
	64	1.00	0.98	0.95	0.91	0.87	0.82	0.77	0.72	0.67	0.62	0.58

[1] Tabulated group action factors (C_g) are conservative for 2½-inch (64 mm) split ring connectors, 2⅝-inch (67 mm) shear plate connectors, $s < 9$ inches (229 mm) or $E > 1{,}400{,}000$ psi (9646 N/mm²).

[2] When $A_s/A_m > 1.0$, use A_m/A_s and use A_m instead of A_s.

Source: Reproduced from the 1994 edition of the *Uniform Building Code*™, copyright © 1994, with the permission of the publisher, the International Conference of Building Officials.

Table C-16. Group-Action Factors for 4-in. Shear-Plate Connectors with Steel Side Members.[1]

		Number of Fasteners in a Row						
A_m/A_s	A_m in.^{2}	2	3	4	5	6	7	8
	5	0.91	0.75	0.60	0.50	0.42	0.36	0.31
	8	0.94	0.80	0.67	0.56	0.47	0.41	0.36
	16	0.96	0.87	0.76	0.66	0.58	0.51	0.45
12	24	0.97	0.90	0.82	0.73	0.64	0.57	0.51
	40	0.98	0.94	0.87	0.80	0.73	0.66	0.60
	64	0.99	0.96	0.91	0.86	0.80	0.74	0.69
	120	0.99	0.98	0.95	0.91	0.87	0.83	0.79
	200	1.00	0.99	0.97	0.95	0.92	0.89	0.85
	5	0.97	0.83	0.68	0.56	0.47	0.41	0.36
	8	0.98	0.87	0.74	0.62	0.53	0.46	0.40
	16	0.99	0.92	0.82	0.73	0.64	0.56	0.50
18	24	0.99	0.94	0.87	0.78	0.70	0.63	0.57
	40	0.99	0.96	0.91	0.85	0.78	0.72	0.66
	64	1.00	0.97	0.94	0.89	0.84	0.79	0.74
	120	1.00	0.99	0.97	0.94	0.90	0.87	0.83
	200	1.00	0.99	0.98	0.96	0.94	0.91	0.89
	40	1.00	0.96	0.91	0.84	0.77	0.71	0.65
	64	1.00	0.98	0.94	0.89	0.84	0.78	0.73
24	120	1.00	0.99	0.96	0.94	0.90	0.86	0.82
	200	1.00	0.99	0.98	0.96	0.94	0.91	0.88
	40	0.99	0.93	0.86	0.78	0.70	0.63	0.57
30	64	0.99	0.96	0.90	0.84	0.78	0.71	0.66
	120	0.99	0.98	0.94	0.90	0.86	0.81	0.76
	200	1.00	0.98	0.96	0.94	0.91	0.87	0.83
	40	0.98	0.91	0.83	0.74	0.66	0.59	0.53
35	64	0.99	0.94	0.88	0.81	0.73	0.67	0.61
	120	0.99	0.97	0.93	0.88	0.82	0.77	0.72
	200	1.00	0.98	0.95	0.92	0.88	0.84	0.80

[1] Tabulated group action factors (C_g) are conservative for 2⅝-inch (67 mm) shear plate connectors for $s < 9$ inches (229 mm).

Source: Reproduced from the 1994 edition of the *Uniform Building Code*™, copyright © 1994, with the permission of the publisher, the International Conference of Building Officials.

Table C-17. Design Values for Shear Plates (For One Shear Plate and Bolt in Single Shear).[1,2,3]

Shear Plate Diameter, inches	Bolt Diameter, inches	Number of Faces of Member with Connectors on Same Bolt	Net Thickness of Member, inches	Loaded Parallel to Grain (0 degrees): Design value, P, per connector unit and bolt, pounds				Loaded Perpendicular to Grain (90 degrees): Design value, Q, per connector unit and bolt, pounds			
				× 4.45 for N							
× 25.4 for mm			× 25.4 for mm	Group A species	Group B species	Group C species	Group D species	Group A species	Group B species	Group C species	Group D species
2⁵⁄₈	³⁄₄	1	1¹⁄₂ minimum	3,110*	2,670	2,220	2,010	2,170	1,860	1,550	1,330
		2	1¹⁄₂ minimum	2,420	2,080	1,730	1,500	1,690	1,450	1,210	1,040
			2	3,190*	2,730	2,270	1,960	2,220	1,910	1,580	1,370
			2¹⁄₂ or thicker	3,330*	2,860	2,380	2,060	2,320	1,990	1,650	1,440
4	³⁄₄ or ⁷⁄₈	1	1¹⁄₂ minimum	4,370	3,750	3,130	2,700	3,040	2,620	2,170	1,860
			1³⁄₄ or thicker	5,090*	4,360	3,640	3,140	3,540	3,040	2,530	2,200
		2	1³⁄₄ minimum	3,390	2,910	2,420	2,090	2,360	2,020	1,680	1,410
			2	3,790	3,240	2,700	2,330	2,640	2,260	1,880	1,630
			2¹⁄₂	4,310	3,690	3,080	2,660	3,000	2,550	2,140	1,850
			3	4,830*	4,140	3,450	2,980	3,360	2,880	2,400	2,060
			3¹⁄₂ or thicker	5,030*	4,320	3,600	3,110	3,500	3,000	2,510	2,160

[1] Tabulated lateral design values (P, Q) for shear plate connector units shall be multiplied to all applicable adjustment factors (see UBC Table 23-III-A).

[2] Allowable design values for shear plate connector units shall not exceed the following:

(a) 2⁵⁄₈-inch (67 mm) shear plate — 2,900 pounds (12 905 N)
(b) 4-inch (102 mm) shear plate with ³⁄₄-inch (19 mm) bolt — 4,400 pounds (19 580 N)
(c) 4-inch (102 mm) shear plate with ⁷⁄₈-inch (22 mm) bolt — 6,000 pounds (26 700 N)

The design values in Footnote 2 shall be permitted to be increased in accordance with the American Institute of Steel Construction (AISC) *Manual of Steel Construction,* 9th edition, Section A5.2 "Wind and Seismic Stresses," except when design loads have already been reduced by load combination factors.

* These loads exceed those permitted by Footnote 2, but are needed for determination of design values for other angles of load to grain. Footnote 2 limitations apply in all cases.

Source: Reproduced from the 1994 edition of the *Uniform Building Code*™, copyright © 1994, with the permission of the publisher, the International Conference of Building Officials.

Table C-18. Required Connector Spacing and End Distance.

Required Spacing —		
For $2\frac{1}{2}$" split ring or $2\frac{5}{8}$" shear plate		
Load Parallel to Grain	$C_\Delta = 1.0$	$C_\Delta = 0.5$
Spacing parallel to grain	$6\frac{3}{4}$"	$3\frac{1}{2}$"(min.)
Spacing perpendicular to grain	$3\frac{1}{2}$"	$3\frac{1}{2}$"(min.)
Load Perpendicular to Grain		
Spacing parallel to grain	$3\frac{1}{2}$"	$3\frac{1}{2}$"(min.)
Spacing perpendicular to grain	$4\frac{1}{4}$"	$3\frac{1}{2}$"(min.)
Required Spacing —		
For 4" split ring or 4" shear plate		
Load Parallel to Grain	$C_\Delta = 1.0$	$C_\Delta = 0.5$
Spacing parallel to grain	9"	5"(min.)
Spacing perpendicular to grain	5" (min.)	5"(min.)
Load Perpendicular to Grain		
Spacing parallel to grain	5"	5"(min.)
Spacing perpendicular to grain	6"	5"(min.)
Required End Distance—		
For $2\frac{1}{2}$" split ring or $2\frac{5}{8}$" shear plate		
Load Parallel to Grain	$C_\Delta = 1.0$	$C_\Delta = 0.625$
Tension member	$5\frac{1}{2}$"	$2\frac{3}{4}$"(min.)
Compression member	4"	$2\frac{1}{2}$"(min.)
Load Perpendicular to Grain		
Tension or compr. member	$5\frac{1}{2}$"	$2\frac{3}{4}$"(min.)
Required End Distance —		
For 4" split ring or 4" shear plate		
Load Parallel to Grain		
Tension member	7"	$3\frac{1}{2}$"(min.)
Compression member	$5\frac{1}{2}$"	$3\frac{1}{4}$"(min.)
Load Perpendicular to Grain		
Tension or compr. member	7"	$3\frac{1}{2}$"(min.)

Reprinted with permission of the American Forest & Paper Association, Washington, D.C.

Table C-19. Design Allowable Loads for Split Rings.

Minimum Wood Thickness	Rings in		Allowable load, lb (unmodified) — Species grouping A	B	C	D	Edge distance (min.) — Load direction: Parallel to Grain	Normal to Grain (Loaded edge)
			Per $2\frac{1}{2}$" diameter split ring with $\frac{1}{2}$" dia bolt					
1"	One	P	2630	2270	1900	1640	$1\frac{3}{4}$"	
	Face	Q	1580	1350	1130	970		$1\frac{3}{4}$" min.
			to	to	to	to		
		Q	1900	1620	1350	1160		$2\frac{3}{4}$" or more
$1\frac{1}{2}$"	One	P	3160	2730	2290	1960	$1\frac{3}{4}$"	
	Face	Q	1900	1620	1350	1160		$1\frac{3}{4}$" min.
			to	to	to	to		
		Q	2280	1940	1620	1390		$2\frac{3}{4}$" or more
$1\frac{1}{2}$"	Both	P	2430	2100	1760	1510	$1\frac{3}{4}$"	
	faces	Q	1460	1250	1040	890		$1\frac{3}{4}$" min.
			to	to	to	to		
		Q	1750	1500	1250	1070		$2\frac{3}{4}$" or more
2"	Both	P	3160	2730	2290	1960	$1\frac{3}{4}$"	
	faces	Q	1900	1620	1350	1160		$1\frac{3}{4}$" min.
			to	to	to	to		
		Q	2280	1940	1620	1390		$2\frac{3}{4}$" or more
			Per 4" diameter split ring with $\frac{3}{4}$" dia bolt					
1"	One	P	4090	3510	2920	2520	$2\frac{3}{4}$"	
	Face	Q	2370	2030	1700	1470		$2\frac{3}{4}$" min.
			to	to	to	to		
		Q	2840	2440	2040	1760		$3\frac{3}{4}$" or more
$1\frac{1}{2}$"	One	P	6020	5160	4280	3710	$2\frac{3}{4}$"	
	Face	Q	3490	2990	2490	2150		$2\frac{3}{4}$" min.
			to	to	to	to		
		Q	4180	3590	2990	2580		$3\frac{3}{4}$" or more
$1\frac{1}{2}$"	Both	P	4110	3520	2940	2540	$2\frac{3}{4}$"	
	faces	Q	2480	2040	1700	1470		$2\frac{3}{4}$" min.
			to	to	to	to		
		Q	2980	2450	2040	1760		$3\frac{3}{4}$" or more
$1\frac{5}{8}$"	One	P	6140	5260	4380	3790	$2\frac{3}{4}$"	
	face	Q	3560	3050	2540	2190		$2\frac{3}{4}$" min.
			to	to	to	to		
		Q	4270	3660	3050	2630		$3\frac{3}{4}$" or more
2"	Both	P	4950	4250	3540	3050	$2\frac{3}{4}$"	
	faces	Q	2870	2470	2050	1770		$2\frac{3}{4}$" min.
			to	to	to	to		
		Q	3440	2960	2460	2120		$3\frac{3}{4}$" or more
$2\frac{1}{2}$"	Both	P	5830	5000	4160	3600	$2\frac{3}{4}$"	
	faces	Q	3380	2900	2410	2080		$2\frac{3}{4}$" min.
			to	to	to	to		
		Q	4050	3480	2890	2500		$3\frac{3}{4}$" or more
3"	Both	P	6140	5260	4380	3790	$2\frac{3}{4}$"	
	faces	Q	3560	3050	2540	2190		$2\frac{3}{4}$" min.
			to	to	to	to		
		Q	4270	3660	3050	2630		$3\frac{3}{4}$" or more

P is the allowable load parallel to the grain.
Q is the allowable load normal to the grain.
For intermediate loaded edge distances, interpolate to obtain allowable Q.
For load-to-grain angles between 0 and 90 degrees, use Hankinson's formula.
All values above are for one ring, installed in seasoned wood that will remain dry in service. Allowables are for normal load duration.

Table C-20. Design and Load Data for Spike-Grid Connectors.

SPIKE-GRID			
Type	Flat	Single curve	Circular
Size, square	$4\frac{1}{8}''$	$4\frac{1}{8}''$	$3\frac{1}{4}''$
Total depth of grids, maximum	1″	1.38″	1.20″
Diameter of bolt hole	1.06″	1.06″	1.33″
Weight, per 100 grids, lbs	50	75	26
LUMBER DIMENSIONS, minimum recommended for installation of flat grids			
Face width	$5\frac{1}{2}''$	$5\frac{1}{2}''$	$5\frac{1}{2}''$
Thickness			
Grids one face only	$1\frac{1}{2}''$	$1\frac{1}{2}''$	$1\frac{1}{2}''$
Grids opposite in both faces	$2\frac{1}{2}''$	—	$2\frac{1}{2}''$
Minimum diameter of pile for curved grids	—	10″	—
BOLT, diameter	$\frac{3}{4}''$ or 1″	$\frac{3}{4}''$ or 1″	$\frac{3}{4}''$ or 1″
BOLT HOLE, diameter in timber	$\frac{13}{16}''$ or $1\frac{1}{16}''$	$\frac{13}{16}''$ or $1\frac{1}{16}''$	$\frac{13}{16}''$ or $1\frac{1}{16}''$
WASHERS			
Round, cast or malleable iron		Standard Size for Bolt Diameter Used.	
Square plate		$3'' \times 3'' \times \frac{3}{8}''$ Punched for Bolt Diameter Used.	
SPACING OF GRIDS, minimum, center to center			
0°–30° angle of load to grain			
Spacing parallel to grain	7″	7″	7″
Spacing perpendicular to grain	$5\frac{1}{2}''$	$5\frac{1}{2}''$	$5\frac{1}{2}''$
30°–90° angle of load to grain			
Spacing parallel or perpendicular to grain	$5\frac{1}{2}''$	$5\frac{1}{2}''$	$5\frac{1}{2}''$
END DISTANCES, center of grid to end of piece (tension or compression members)			
Standard	7″	7″	7″
Minimum, reduce loads 15%	5″	5″	5″
EDGE DISTANCES, center of grid to edge of piece			
Load applied at any angle to grain			
Standard	$3\frac{5}{8}''$	$3\frac{5}{8}''$	$3\frac{5}{8}''$
Minimum, reduce loads 15%	$2\frac{3}{4}''$	$2\frac{3}{4}''$	$2\frac{3}{4}''$
PROJECTED AREA for portion of one grid within member, square inches	2.06	2.06	1.95

Connector Load Grouping of Species When Structurally Graded

Connector Load Grouping

Group A Species	Group B Species	Group C Species
Douglas fir (dense)	Douglas fir (coast region)	Cypress, southern and tidewater red
Oak, red and white	Larch, western	Hemlock, west coast
Pine, southern (dense)	Pine, southern	Pine, Norway
		Redwood

Design Loads[1] for One Spike-Grid and Bolt in Single Shear

Group A			Group B			Group C		
Type of Grid	Bolt Diameter	Allowable Load	Type of Grid	Bolt Diameter	Allowable Load	Type of Grid	Bolt Diameter	Allowable Load
Flat	$\frac{3}{4}''$	3900#	Flat	$\frac{3}{4}''$	3500#	Flat	$\frac{3}{4}''$	3000#
	1″	4200#		1″	3800#		1″	3300#
Single Curve	$\frac{3}{4}''$	4200#	Single Curve	$\frac{3}{4}''$	3800#	Single Curve	$\frac{3}{4}''$	3200#
	1″	4500#		1″	4100#		1″	3500#
Circular	$\frac{3}{4}''$	3500#	Circular	$\frac{3}{4}''$	3100#	Circular	$\frac{3}{4}''$	2600#
	1″	3800#		1″	3400#		1″	2900#

[1]Allowable loads on spike-grids same for all angles of load to grain.
Source: Courtesy of Cleveland Steel Specialty Company, Cleveland, OH.

Index